THE TESTS OF TIME

//

THE TESTS OF TIME

//

Readings in the Development of Physical Theory

//

EDITED BY

Lisa M. Dolling

Arthur F. Gianelli

Glenn N. Statile

PRINCETON UNIVERSITY PRESS

PRINCETON AND OXFORD

**Information on permissions for the use of specific texts in this anthology appear in the
section immediately following the index**

ISBN: 0-691-09084-X
ISBN (pbk.): 0-691-09085-8

Library of Congress Cataloging-in-Publication Data
The tests of time: readings in the development of physical theory/edited by Lisa M. Dolling,
Arthur F. Gianelli, Glenn N. Statile.
p.   cm.
Includes bibliographical references and index.
ISBN 0-691-09084-X (alk. paper)—ISBN 0-691-09085-8 (pbk.: alk. paper)
1. Physical sciences—History. 2. Science—Philosophy. I. Dolling, Lisa M., 1962–
II. Gianelli, Arthur F. III, 1939–   . Statile, Glenn N., 1956–
QC125 .T47   2003
500.2'90—dc21   2002074976

British Library Cataloging-in-Publication Data is available

This book has been composed in Berkeley Book with Stone Sans Display

Printed on acid-free paper.∞

www.pupress.princeton.edu

Printed in the United States of America

1 3 5 7 9 10 8 6 4 2

///

We dedicate this book to those extraordinary philosopher-scientists
who have contributed to the development of physical theory.

///

Contents

//

Preface   xv

Acknowledgments   xix

Philosophical Introduction: *Philosophy of Science and Physical Theory*   xxi

PART I   The Heliocentric Theory   1

Introduction   3

1. ARISTOTLE
   The Physical Foundation for the Geocentric Universe   13

2. ARISTARCHUS
   An Early Version of Heliocentrism   26

3. CLAUDIUS PTOLEMY
   The Case for Geocentrism   29

4. NICHOLAUS COPERNICUS
   First Thoughts on Heliocentrism   38

5. NICHOLAUS COPERNICUS
   The Heliocentric Theory   42

6. TYCHO BRAHE
   The Supernova of 1572   66

7. TYCHO BRAHE
   Observational Evidence against the Aristotelian Cosmology   69

8. JOHANNES KEPLER
   The Sun as the Source of Planetary Motions   70

9. GALILEO GALILEI
   Telescopic Observations in Support of Copernicus   81

10. JOHANNES KEPLER
    The Superiority of the Copernican System   99

11. GALILEO GALILEI
The Coherence of the Copernican Theory 109

12. ISAAC NEWTON
The Physical Foundations of Heliocentrism 114

13. JOHN HERSCHEL
The Discovery of Stellar Parallax 128

Selected Bibliography 129

PART II Electromagnetic Field Theory 131

Introduction 133

1. WILLIAM GILBERT
The Properties of Magnets 145

2. CHARLES COULOMB
The Law of Electric Force 151

3. HANS CHRISTIAN OERSTED
The Effect of a Current of Electricity on a Magnetic Needle 153

4. ANDRÉ MARIE AMPÈRE
A Positivist Approach to Electromagnetism 157

5. ISAAC NEWTON
The Particle Theory of Light 162

6. CHRISTIAAN HUYGENS
The Wave Theory of Light 167

7. THOMAS YOUNG
The Vindication of the Wave Theory of Light 179

8. AUGUSTIN FRESNEL AND DOMINIQUE ARAGO
The Transverse Nature of Light Waves 201

9. MICHAEL FARADAY
Electromagnetic Induction 209

10. MICHAEL FARADAY
The Concept of an Electromagnetic Field 214

11. JAMES CLERK MAXWELL
The Theory of the Electromagnetic Field 224

12. JAMES CLERK MAXWELL
The Electromagnetic Theory of Light 232

13. JAMES CLERK MAXWELL
The Medium for Electromagnetic Action 235

14. HEINRICH HERTZ
The Production of Electromagnetic Waves 245

Selected Bibliography 251

PART III   The Theory of Relativity   253

Introduction   255

  1. JAMES CLERK MAXWELL
    The Ether   265

  2. ALBERT MICHELSON
    The Ether and Optical Experiments   273

  3. GEORGE F. FITZGERALD
    The Contraction Hypothesis   285

  4. HENDRIK A. LORENTZ
    The Contraction Hypothesis   286

  5. HENRI POINCARÉ
    A Prelude to Relativity   289

  6. ALBERT EINSTEIN
    The Postulates of the Special Theory of Relativity   299

  7. HERMAN MINKOWSKI
    The Space-Time Continuum   304

  8. ALBERT EINSTEIN
    The Foundation of the General Theory of Relativity   308

  9. ALBERT EINSTEIN
    The Ramifications of the Special and General Theories of Relativity   313

 10. ARTHUR EDDINGTON
    The Bending of Light Rays   330

 11. ALBERT EINSTEIN
    Ether and Relativity   340

 12. ALBERT EINSTEIN
    Later Comments on General Relativity   347

 13. ALBERT EINSTEIN
    $E = MC^2$ 350

Selected Bibliography   355

PART IV   Quantum Theory   357

Introduction   359

Historical and Conceptual Development   371

  1. MAX PLANCK
    The Quantum Hypothesis   373

  2. ALBERT EINSTEIN
    The Photon   380

3. NIELS BOHR
   The Quantum Character of the Atom 383

4. LOUIS DE BROGLIE
   The Wave Nature of the Electron 387

5. NIELS BOHR
   Complementarity and the New Quantum Theory 392

6. NIELS BOHR
   The Debate with Einstein 406

7. ALBERT EINSTEIN
   Response to Bohr 432

8. WERNER HEISENBERG
   A Brief History of Quantum Theory 437

9. WERNER HEISENBERG
   The Copenhagen Interpretation 446

10. ERWIN SCHRÖDINGER
    The Fundamental Idea of Wave Mechanics 454

11. ERWIN SCHRÖDINGER
    Are There Quantum Jumps? 465

12. P.A.M. DIRAC
    The Conceptual Difficulties of Quantum Theory 472

13. JOHN A. WHEELER
    Observer-Created Reality 484

The Completeness Debate 493

14. ALBERT EINSTEIN, BORIS PODOLSKY, AND NATHAN ROSEN
    The EPR Paradox 495

15. ALBERT EINSTEIN
    The Argument for Incompleteness 501

16. NIELS BOHR
    Response to EPR 507

17. DAVID BOHM
    The Hidden Variables Hypothesis 515

18. J. STEWART BELL
    Comment on the Hidden Variables Hypothesis 526

19. J. STEWART BELL
    A Conceptual Analysis of the EPR Thought Experiment of David Bohm 532

20. ABNER SHIMONY
   Philosophical Reflections on the Completeness Debate 540

Selected Bibliography 544

PART V Big Bang Cosmological Theory 545

Introduction 547

1. HENRIETTA LEAVITT
   Variables in the Magellanic Clouds 556

2. HENRIETTA LEAVITT
   The Variability-Luminosity Relationship 559

3. VESTO SLIPHER
   The Radial Velocity of the Andromeda Nebula 561

4. VESTO SLIPHER
   The Discovery of the Red Shift of Nebulae 564

5. HARLOW SHAPLEY
   The Measurement of Great Distances 567

6. WILLEM DE SITTER
   Relativity and Cosmology 573

7. EDWIN HUBBLE
   The Structure of the Universe 592

8. EDWIN HUBBLE
   The Velocity-Distance Relation 602

9. ARTHUR EDDINGTON
   The Expanding Universe 611

10. GEORGES LEMAÎTRE
   The Beginning of Big Bang Cosmology 625

11. ARNO PENZIAS AND ROBERT WILSON
   The Discovery of Background Radiation 640

12. R. H. DICKE, P.J.E. PEEBLES, P. G. ROLL, D. T. WILKINSON
   An Explanation of the Penzias and Wilson Discovery 642

13. STEVEN WEINBERG
   The Cosmic Microwave Radiation Background 646

14. ALAN GUTH AND PAUL STEINHARDT
   The Inflationary Universe 652

15. GEORGE SMOOT AND KEAY DAVIDSON
   Wrinkles in Time 666

16. STEPHEN HAWKING
   The Edge of Spacetime   677

Selected Bibliography   684

Epilogue   685

HELGE KRAGH
   Physical Theory: Present and Future   685

Sources of the Readings   693

Index of Names   699

Index of Concepts   707

Permissions Acknowledgments   709

Parts of science—theoretical physics for example—have an austerely splendid intrinsic beauty and power, they represent magnificently impressive intellectual accomplishment, they resemble great poetry and great music; perhaps the most impressive intellectual accomplishment of humankind is . . . theoretical physics from Newton to the present.
   —ALVIN PLANTINGA

The mainspring of science is the conviction that by honest, imaginative inquiry we can build up a system of ideas about Nature which has some legitimate claim to "reality." That being so, we can never make less than a three-fold demand of science: its explanatory techniques must be not only (in Copernicus' words) "consistent with the numerical records"; they must also be acceptable—for the time being, at any rate—as "absolute" and "pleasing to the mind."
   —STEPHEN TOULMIN

One can imagine a category of experiments that *refute* well-accepted theories, theories that have become part of the standard consensus of physics. *Under this category I can find no examples whatever in the past one hundred years.* . . . In this century no theory that has been generally accepted as valid by the world of physics has turned out simply to be a *mistake.*
   —STEVEN WEINBERG

Serious scholars claim that there is no such thing as progress and assert that science is but a collection of opinions, as socially conditioned as the weathervane world of Paris couture. Far too many students accept the easy belief that they need not bother learning much science, since a revolution will soon disprove all that is currently accepted anyway. In such a climate it may be worth affirming that science really is progressive and cumulative, and that well-established theories, though they may turn out to be subsets of larger and farther-reaching ones—as happened when Newtonian mechanics was incorporated by Einstein into general relativity—are seldom proved wrong.
   —TIMOTHY FERRIS

Preface

//

One of the greatest intellectual achievements of humankind has indeed been the development of physical theory. The products of this development, the theories of physics, astronomy, and cosmology, have been shown to possess an overwhelming power, both explanatory and predictive. As we begin the twenty-first century, these theories give us a rich and profound picture of the physical world, representing the deepest levels of human understanding. This book chronicles the development of five such theories: the Heliocentric Theory, Electromagnetic Field Theory, the Special and General Theories of Relativity, Quantum Theory, and Big Bang Theory.

The development and acceptance of each of these theories is a fascinating story of the interplay of empirical evidence and theoretical insight, and of the dialogue and debate of great minds. Although historical accounts of their development abound, these accounts never seem to capture the richness and the substance of their subjects. This book lets the intellectual giants responsible for these theories tell the story in their own words. To that end, appropriate readings have been chosen from the works of the scientists who made significant contributions to the development of each theory. In many cases the original paper or the chapter of the ground-breaking book is included so that the story can be told as it happened.

Choosing to use the works of the scientists themselves has a number of valuable benefits. First, and perhaps most important, it tends to humanize the enterprise of natural science. Nearly all the authors included in this volume evince more than a practitioner's interest in their respective sciences. They often show concern for the philosophical implications of their endeavors, even exhibiting a penchant for the metaphysical. Second, the works of these scientists turn out to be fluid, often poetic and artfully descriptive, resulting in a unique type of great literature. Finally, since the readings represent moments of great discovery, the reader becomes privy to the creative process, witnessing the expression of great insights as only their authors could express.

Although the development of each of these theories is unique, there are several common themes running through them that contribute to their fascinating character and define their development. The establishment of each of these theories was accompanied by a revolution in which a previously prominent theory was rejected once and for all. Heliocentrism, which is at the heart of the classic scientific revolution, came to prominence at the expense of the Geocentric Theory, which was the accepted worldview for almost 2,000 years. The development of Electromagnetic Field Theory, along with the concept of electromagnetic waves, involved the falsification of the material particle, mechanical wave, and action-at-a-distance theories of light and electromagnetic phenomena. Relativity Theory rejected the Newtonian notions of absolute space and absolute time. Quantum Theory took its leave from the strict determinism of Newtonian mechanics, and indeed from classical science in general, and Big Bang Cosmology could only be realized after the "static universe" idea had been discarded.

In addition, the development of each of these theories was aided at critical moments by extraordinary experiments and observations. Although anticlimactic because everyone had come to expect it, the observation of stellar parallax by Bessel in 1838 firmly established Heliocentrism. The production of electromagnetic waves other than light by Heinrich Hertz guaranteed Electromagnetic Field Theory and its author, James Clerk Maxwell, a prominent place in the history of science. The observation of the bending of light rays as they passed by the sun in 1919 did the same for Relativity and Einstein. Several experiments were crucial to Quantum Theory. The Compton Effect, which showed that X-rays interacted with matter as if they were particulate, and the experiments of Davisson and Germer, which showed that electrons behaved like waves, were certainly significant. But perhaps most crucial were the experiments of Alain Aspect and his colleagues in the 1980s that corroborated the predictions of quantum mechanics and ended the Completeness Debate. The acceptance of Big Bang Cosmology was guaranteed by Penzias and Wilson's discovery of the radiation echo in 1964, and by the COBE satellite's detection of ripples in that radiation echo in 1992.

But perhaps the most intriguing common thread is the counterintuitive element contained in each of the theories. Each had to overcome an earlier theory whose concepts were, in general, quite intuitive and in conformity with common sense. We have no direct experience of the earth's motion, yet Heliocentrism requires that we view the earth as spinning like a top and hurtling through space at great speed. Electromagnetic Field Theory asks us to accept the existence of a nonmechanical medium that is unobservable in principle. Relativity Theory tells us that the structure of space and the passage of time are relative to the motions of observers and to the strengths of gravitational fields. Quantum Theory asks us to believe that all entities behave like particles *and* waves, albeit at different times, that there is a significant measure of indeterminism in nature, and that certain quantum entities can communicate instantaneously across great distances. Big Bang Cosmology tells us that the stuff out of which all the forty to fifty billion known galaxies were made was, at a moment in the past, compressed to the size of the head of a pin, that space, time, and matter literally came into existence at that moment, and that the known laws of nature broke down at that moment. In a very real sense the development of these theories can be seen as a movement from the intuitive to the counterintuitive.

The common themes reveal to us that the development of these five physical theories has been a unique form of human drama. The brilliant cast of characters in this drama, the scientists whose writings appear in this book, have put these theories, as you will see, to the most severe tests imaginable—theoretical, experimental, and certainly philosophical. Because the theories have successfully met these tests over long periods of time, they have become the foundation of our scientific picture of the physical world. They carry with them a sense of permanence and will continue to be a significant part of our scientific worldview for the indefinite future.

Organization

Although the readings were chosen on the basis of their importance to the development of each theory, decisions had to be made about when to end the story. For Heliocentrism this was simple because its story was essentially complete in the nineteenth century. But the other four theories are still works in progress. We decided to end Relativity Theory with the work of Einstein because he so dominated the story of its development. Quantum Theory is tracked through the end of the Completeness Debate because that debate exposed the philosophical essence of the theory. Big Bang Cosmological Theory secured its prominent position when the confirming data from the COBE satellite was received. Only Electromagnetic Field Theory presented a problem. In the 125 years since Maxwell, the sciences of electricity and magnetism have gone through considerable development and change. Contemporary classical field theory is quite different from that of Maxwell. The decision to end the section with Faraday, Maxwell, and Hertz was thus made for practical as well as philosophical reasons. On the practical side, proceeding further could not have been done properly without doubling the size of an already long section. But more important, it can be argued that the work of Faraday, Maxwell, and Hertz established the existence of a nonmechanical, unobservable-in-principle reality that represented a decisive blow to the Newtonian worldview. As such, ending with their work gives the section a valuable coherence and intelligibility.

Audiences

We prepared the book with three audiences in mind. We tried to make it sufficiently substantive that working scientists would find it informative, and sufficiently accessible that a lay person with an interest in science would find it enjoyable and enlightening. But it is our hope that the book will also be of use in philosophy of science courses at both the graduate and undergraduate levels. Students in these courses often do not read the works of the scientists themselves, which is unfortunate. Because we have targeted philosophy of science students, we have included an introduction that briefly discusses the philosophical issues that have arisen in the development of these physical theories.

Structure

Each section is introduced by an essay that contains a precise description of the theory in question and that recounts the development of the theory in broad strokes. Each reading is preceded by a brief scientific biography of the author and a description of what follows. The readings themselves have been edited in order to make them clear and accessible without compromising their integrity, and mathematical analysis has been included only when absolutely essential to the intelligibility of the story.

Contributors

We have included selections from obviously important contributors but also from those who are often not represented in books of this type. For example, included in the section on Heliocentrism are readings by the ancient heliocentrist Aristarchus and the extraordinary sixteenth-century astronomer Tycho Brahe; André Ampère and Thomas Young are found in the section on Electromagnetism; as are Albert Michelson, George Fitzgerald, and Herman Minkowski in the section on Relativity; David Bohm and John Stewart Bell in the section on Quantum Theory; and Henrietta Leavitt, the discoverer of Cepheid variables, and Vesto Slipher, the discoverer of the red shift, in the section on Big Bang Theory. The inclusion of these authors allows for a more complete and interesting story of the theories' development.

Accessibility

As indicated, there has been a conscious effort to make the selections accessible to as large a readership as possible. However, there is a gradation in the difficulty of the selections as the book proceeds. Rather than viewing this as a problem, we see it as both natural and positive. The simple fact is that as physical theory has developed, it has become more complex both conceptually and mathematically. The gradation in the difficulty of the selections mirrors the gradation in the sophistication and complexity of physical theory as it has developed.

Acknowledgments

//

We owe a considerable debt of gratitude to so many people who helped us in the preparation of this book. Among them are three of the most important philosophers of science of the twentieth century. Robert Cohen read the entire manuscript and made invaluable comments and suggestions for each chapter. Stephen Toulmin also read the manuscript and was very generous with his encouragement and advice. Ernan McMullin made extremely helpful suggestions concerning Part 1 and would have done more had we asked. We are humbled by their assistance and tried to meet their standards as best we could. The imperfections that remain in the book come strictly from us.

Our thanks go also to astronomer and historian of science Steven Dick and philosopher of science Martin Tamny for their kind words of encouragement.

We owe a special debt of gratitude to Princeton University Press, particularly to physical science editors Trevor Lipscombe and Joe Wisnovsky, production editor Deborah Tegarden, and others on the staff who worked on the book. We are honored by their recognition and grateful for their assistance with an often unwieldly manuscript.

Our secretaries, Mary Twomey and Lucille Hartmann, worked tirelessly and enthusiastically in the preparation of the manuscript and, as one might imagine, are quite happy that the book is finally finished. We cannot thank them enough for their skills and their help.

Our thanks also go to Jamie Manson, Jennifer Scuro, and Jim Lynch, whose unique talents were quite helpful in the preparation of the manuscript.

To Scott Gianelli we owe particular thanks for suggesting what we feel is a very appropriate title. Scott is himself a budding physicist and admits to being inspired by the authors herein.

We wish to make special mention of Lisa Dolling's mentor, Marx Wartofsky, who passed away while the book was in preparation. His kind words were always an inspiration to us.

Finally, we wish to thank those family members whose support and encouragement saw us through this project: Lisa Dolling's husband, Peter Muccio, her daughter, Kaitlin Muccio, and her parents, Mari and George Dolling; Art Gianelli's wife, Barbara Gianelli, and family members Flo, Art, and Rosemarie Gianelli and Ethel and Jim Scott; and Glenn Statile's wife, Agnes Statile, and his parents Nicholas and Rosemarie Statile. We know that without them the book would not have been possible.

Lisa Dolling

Arthur Gianelli

Glenn Statile

Philosophical Introduction

///

PHILOSOPHY OF SCIENCE AND PHYSICAL THEORY

The scientific literature presented in this book possesses a philosophical richness unparalleled in the history of science. Of the many philosophical issues touched upon in this literature, the following three are among the most significant:

1. The Existence of Theoretical Entities
2. The Problem of Scientific Discovery
3. The Evaluation of Scientific Theories

Readers will assuredly profit by employing these issues as interpretive categories to guide them. Whether the things referred to by theories are real or not is relevant to both their truth and value. Further, familiarity with the creative approaches great thinkers have taken to overcome existing scientific problems, as well as with the evaluation of theories as they progress toward their final formulation, can be of immense value to both the beginning and professional student of the history and philosophy of science.

In this introduction the issue of the existence of the entities actually referred to by the five physical theories is considered first. This is followed by a discussion of the problematics and pragmatics of scientific discovery as a creative enterprise. Finally the focus shifts to the explanatory and predictive worth of these theories, as well as to the issue of their progressive character in relation to those theories that they supersede.

The Existence of Theoretical Entities

What is the ontological status of the micro-objects, processes, states, fields, and other entities referred to by physical theories? Can they be viewed as in some sense real, as truly describing the world, or are they merely convenient fictions of the theory-making process? These are the kindred questions that concern us here. Furthermore, these questions are perennial. From the Forms of Plato, to the *adequatio ad rem* vocabulary of medieval scholasticism, to the Enlightenment metaphor of the extent to which scientific theories mirror the underlying nature of physical reality, the issue of the correspondence between theory and fact has long been at the heart of epistemological, metaphysical, and scientific concerns. At stake is the attitude we adopt toward the objects of science. This constitutes the well-known *Realism vs. AntiRealism* debate.

The Heliocentric Theory

The developmental history of the Heliocentric Theory, from the time of the establishment of geocentrism in Greek antiquity to its heliocentric transformation by Copernicus, modification by Kepler, corroboration by Galileo, and ultimate vindication by Newton, provides an ample opportunity for highlighting the issue of realism as regards theoretical entities. Our goal here is not to resolve this difficult issue, but only to examine it in the setting of specific physical theories.

In his unsigned preface to the *De Revolutionibus*, Osiander reminds the reader of a lesson usually reserved for logical debates. Both a (True → True) and a (False → True) implication are to be logically construed in the same way, as true statements. That is, even if the Heliocentric Theory of Copernicus is only hypothetical, and possibly quite literally false, it can nevertheless, in principle, still accurately and truly describe astronomical appearances. The instrumental lesson here is that the truth of a theory, and the theoretical entities that compose it, are not necessary conditions of its theoretical utility. While the tools of discovery can perform a catalytic role in the development of a descriptively valid theory of reality, they themselves are not required to be rooted in the bedrock of actual fact.

On the other hand, a number of twentieth-century philosophers have argued in favor of realism as regards theoretical entities. Hilary Putnam has stated that only a realist attitude toward theoretical entities prevents the success of science from being other than miraculous, while J.J.C. Smart has suggested that an antirealist stance toward theoretical entities would amount to an implicit acknowledgment that scientific laws are the result of cosmic coincidence. And yet, despite mountains of evidence in favor of heliocentrism, and its universal acceptance, no person can actually claim to have genuinely observed the entire heliocentric arrangement of the solar system in the same way that one can when it is completely reduced to pictorial form.

Mathematical devices such as the epicycle, utilized by both Ptolemy and Copernicus, were inventions devised and used with the intent of salvaging circular celestial motions. Until proven false by Kepler and Newton, such celestial motions were deemed to represent the real motions of planets. The crystalline spheres of Aristotle, on the other hand, represented a physical attribute of geocentric theory whose purported existence could not survive precise astronomical observation. The careful scrutiny of stellar and cometary phenomena by Tycho Brahe undermined the status of the spheres as real objects, and acted as a stimulus of discovery for a new post-Aristotelian revolution in physics.

The main problem for heliocentrism was of course not so much the centrality of the sun but the mobility of the earth. But these articles of heliocentric faith have been found to be in accord with both reason and fact, thus converging upon the reality and seeming irrefutability of the Heliocentric Theory. It should be briefly noted here that within the philosophy of science, it is not correct to speak of an unproblematic demarcation between theoretically derived entities and those that we factually observe. We will not, however, belabor this point. Let us just note that factual and theoretical statements and beliefs are interdependent and intertwined. Establishing the empirical basis of our scientific commitments, especially given the mediation of technologically assisted instruments of observation, such as the telescope and microscope, is a difficult epistemological task.

The Copernican commitment to a moving earth theory, for example, not only allowed for a better overall account of retrograde planetary motions than did the Ptolemaic theory, but also enabled an estimation of the still problematic distances to the planets. Copernican sore spots, such as those of stellar parallax, and the purported inconsistencies with both Aristotelian physics and Scripture, have all since been eliminated. From the perspective of solar centrality, Kepler's attempts to delineate a centripetal force emanating from the sun can also be understood as corroborative of his own realistic interpretation of the Heliocentric Theory.

Johannes Kepler pulled no punches when it came to affirming his commitment to the reality of heliocentrism. In *Astronomia Nova* (1609), he declared that he had founded astronomy not upon fictive hypotheses, but upon physical causes. It was Kepler who revealed the identity of Osiander, thus indicating that it was not Copernicus who, in a moment of cowardice, had failed to realistically endorse his own heliocentric theory. Ironically, the empirically based epistemology stemming from Newton's *Principia* (1687), which secured a kind of secular imprimatur for the views of Kepler and Galileo, would also serve as the cornerstone for the Humean revival of skepticism concerning the reality of causes.

Electromagnetic Theory

Any philosophical assessment of the theoretical entities posited by Maxwell's Electromagnetic Theory will need to focus upon the following two theoretical concerns.

1. The existence of electromagnetic fields and waves, which encompass electrical, magnetic, and optical phenomena.
2. The existence of an ethereal medium capable of providing a physically viable means of electromagnetic transmission.

Post-Maxwellian theoretical and experimental work, by Poynting and Hertz among others, would lead to strong support for (1). The negative result of the Michelson-Morley interferometer experiment in 1887 undermined confidence in (2), as did the later Einsteinian redeployment of classical electrodynamics around the postulate that the speed of light (c) remains the same for all inertial systems despite any discrepancies owing to relative motion between a source of illumination and an observer.

It is not unfair to say that from the time of "On Physical Lines of Force" (1861)—in which the ethereal model of electromagnetic transmission is described in great detail—until "A Dynamical Model of the Electromagnetic Field" (1865)—in which the equations of the electromagnetic field are presented in a way that do not rely upon the existence of an ethereal medium of transmission—Maxwell came to recognize that the existential claim of (1) need not depend upon his own individualistic description of an ethereal model (2). But while subsequent scientific work has provided strong support for the view that a realistic commitment to (2) is not a necessary condition for the theoretical efficacy of field theory (1), this does not mean that Maxwell himself took an overall skeptical position regarding the existence of some ethereal medium as necessary for the transmission of electromagnetic waves. One need only note Maxwell's affirmation of the reality of some ethereal substance in his contribution on this subject for the ninth edition of the *Encyclopedia Britannica*. Neither was he alone. Albert Michelson would continue to search for the ether for the remainder of his life. And not only did Einstein not banish the ether from physics, despite its superfluity from the perspective of Special Relativity, he would later argue in favor of its existence in 1920 at the University of Leyden. In the *Treatise on Electricity and Magnetism* (1873), Maxwell writes: "The problem of determining the mechanism required to establish a given species of connexion between the motions of the parts of a system always admits of an *infinite number of solutions*."

In his 1856 memoir regarding Faraday's so-called lines of force, prior to the working out of his mature position, Maxwell would recommend a flexible methodological attitude as regards electromagnetic commitments. If we adhere only to the mathematical formulation of a theory and bracket its physical implications, then "we entirely lose sight of the phenomena to be explained." Conversely, blind commitment to a physical hypothesis can pervert truth. For between any physical analogy and the underlying physical reality it represents, there will always be a degree of conceptual remainder. But although individual analogies were meant by Maxwell to be viewed as "illustrative, not as explanatory," he persisted in viewing all energy, "literally," and as mechanical in nature.

André Ampère, whose steady-state law describing the mutual forces of current-carrying wires upon each other was eventually augmented by Maxwell to allow for the moving charges needed for the generation of electromagnetic waves, adopted an instrumentalist attitude toward the existence of underlying electromagnetic causes. Ampère thus was willing to settle for an action-at-a-distance account of electromagnetic phenomena. The problem of theoretically determining whether there exists a medium or mechanical means of continuous transmission for various phenomena had been of concern to electrical and magnetic theory since the time of the Greeks. For Coulomb and others, the inverse square relation governing the gravitational force served as a fruitful analogy for the

forces governing electrical phenomena. Additionally, the accompanying Newtonian attitude, concerning the avoidance of unwarranted hypotheses in ascribing physical causes to phenomena such as gravity, provided a sort of theoretical
sanction for action-at-a-distance theories. This also would have its impact on
succeeding generations of scientists, including Ampère.

Hans Christian Oersted's 1819 discovery, however, that a current-carrying
wire will cause a magnetic reaction perpendicular to the length of the wire,
required a retreat from the action-at-a-distance mentality that had been on the
rise since the time of Coulomb's 1785 demonstration that an inverse square
relation described the electrostatic force between separated point charges. In
retrospect, with the subsequent theoretical advancement of electromagnetic theory by both Faraday and Maxwell, Ampère's instrumental confidence in this case
turns out to have been premature.

Unlike the case of the gravitational force, Michael Faraday would argue for
the actual physical existence of magnetic lines of force. These forces, when varied, induce electric currents. According to Maxwell, Faraday's conception of
physical lines of force provided the key that would unlock the secrets to the
science of electricity. Maxwell proceeded to mathematize Faraday's conception
of lines of force into a fully articulated field theory. This theory, as exemplified in
the four famous field equations of Maxwell, which encompassed all the previous
advances of electrical and magnetic theory, described all electrical and magnetic
phenomena in relation to a field-theoretic framework.

Ampère's antirealist attitude concerning the possible causal underpinnings of
electromagnetic phenomena thus was no longer tenable. For Maxwell, the electromagnetic field could be identified with the dynamic or energetic conditioning
of space. All electrical and magnetic phenomena are transmitted as the result of
contiguous modifications of the field. In Maxwell's mature theory, electrical,
magnetic, and optical phenomena are unified under the mathematical formalism
and interpretation of the field equations. In the ethereal model the field was
designed as an elaborate swirling or vortical apparatus that was sensitive to
specific engineering considerations needed for the mechanical propagation of
electromagnetic waves such as compressibility and heat loss.

At the beginning of the nineteenth century, Thomas Young, Augustin Fresnel,
and others had already provided classic experimental evidence in favor of the
wave nature of light, as previously staked out by Huygens in his *Treatise on Light*
(1690). When it was shown that the wave theory of light could explain polarization if light waves were transverse, meaning that they vibrated at right angles to
their direction of motion, the Newtonian paradigm as established in the *Opticks*
(1704), which was commonly credited as holding that light was particulate in
nature, had to be abandoned. Maxwell's theory predicted that the speed of electromagnetic waves is equal to the square root of the ratio of the constants of
proportionality appearing in Coulomb's law and the Biot-Savart law for the magnetic field, respectively. That the magnitude of this speed was numerically equal
to the speed of light in a vacuum was added evidence that the speculative view
of the German idealist philosopher F. W. Schelling, that all physical forces were

emanations of a single underlying force, was not far-fetched. That these speeds were independently obtained lent further credence to the unification of the sciences of optics and electromagnetism.

Of significance in the evolution of the conviction that electromagnetic phenomena required something more than an action-at-a-distance approach was the discovery of the curvature exhibited by magnetic lines of force by Faraday. It was difficult to conceive of such curvilinear or nonlinear motion without a notion of force that was tied to some physically causal framework that was essentially mechanical in nature. But such a consensus was by no means unanimous. Wilhelm Weber's electromagnetic theory might still be classified in terms of action-at-a-distance, despite the fact that for Weber electromagnetic transmission was not instantaneous. Oersted's discovery of the perpendicularity of the force exerted by a current-carrying wire upon a magnetic pole had the same implication as Faraday's later discovery of circular lines of force, for such perpendicularity also requires an explanation of how such a twisting motion relative to the direction of the current can occur.

Maxwell's field equations implied the existence of electric and magnetic waves that are orthogonally oriented to each other as well as to their direction of motion, and which propagate through space at the incredibly fast but finite speed of light. Such waves are the net result of interactions between changing electric and magnetic fields. Hence both the oscillation and the acceleration of electric charges are crucial to the generation of electromagnetic waves.

The year 1887 was a significant year for electromagnetic theory. At approximately the same time that Michelson and Morley were failing to detect the luminiferous ether, Heinrich Hertz was beginning to provide experimental confirmation of the existence of Maxwell's electromagnetic waves. He determined that certain oscillating currents generate electromagnetic waves. Such waves, furthermore, possessed all the characteristics of light: reflection, refraction, interference, polarization, finite speed, near equivalent speed, and so on, except for visibility. Moreover, these waves could be distinguished from their electromotive source or cause, for "such forces persist in space even after the causes which have given rise to them have disappeared." Indisputably, the experimental results of Hertz inspired a professional confidence toward a realistic interpretation of the electromagnetic waves implied by the field equations of Maxwell. It was only a short time until these Hertzean waves would become better and more popularly known as radio waves.

The Theory of Relativity

Maxwell's theory of electromagnetic phenomena provided a kind of classical closure to the longstanding debate over whether such phenomena were to be viewed as substantial in themselves, as properties of some substance, or as constituents of some kind of energetic process. Einstein's Special and General Theories of Relativity revisited a metaphysically related problem, that of the well-known Newton-Leibniz debate concerning the substantival reality (Newton) vs.

the relationality (Leibniz, Berkeley, Mach) of space and time. It is upon this relativistic theme that we will focus our attention.

In the theory of Special Relativity (SR), as articulated by Einstein in 1905, time functions as an independent parameter, to be coordinated with space, and not as an integrated feature of space-time. Such is also the case in Newtonian mechanics. Minkowski would later show how SR can be validly reformulated so that space-time is construed as a continuum. Hence the meaning of Minkowski's declaration that "space and time disappear as separate concepts." In Minkowski's postulation of the *Absolute World*, space and time can only be conceptually distinguished; in reality, for Minkowski, they are interdependent. Einstein echoes this integrated spatio-temporal perspective in the context of General Relativity (GR) when he states that the objectivity of space and time as independent concepts is stripped of all physical meaning by the principle of general covariance. And yet in extensions of GR, which universalizes the more limited inertial perspective of SR, time and space are sometimes coordinated and sometimes not. Time is linked to spatial hypersurfaces (hyper = more than 3 dimensions) in the standard Friedmann models of GR, whereas time is distinguished from space in the description of the space-time metric given by Robertson-Walker. From an epistemological perspective, in which observation, intuition, evidence, measurement, and proof techniques all play an important role in the analysis and evaluation of space and time, both SR and GR can legitimately be said to have radically altered the Newtonian worldview. From a metaphysical perspective, however, there is still no definitive relativistic verdict as to the reality and constitution of space and time.

Special Relativity resolves conflicts between the principle of relativity, which imposes a physical equivalence upon all inertial frames of reference, and the Maxwell-Lorentz theory of electrodynamics, by drawing out the implications of the postulate that decrees the constancy of the speed of light. While this leads to contractions of length and dilations of time in accordance with the set of Lorentz transformations, this does not mean that space and time do not and cannot really and physically exist. On the other hand, the relativity of simultaneity which arises in SR does at least point to a definite interdependence between space and time, even though, as previously noted, time still functions as an independent parameter in SR.

Special Relativity did not necessarily issue from a metaphysical renunciation of spatial and temporal substantivalism (realism) per se, but did at least in part stem from a critique of the implications of the notions of absolute space and time upon which Newtonian mechanics in particular was based. While SR posits that the concept of a luminiferous ether is superfluous, a 1920 address given at the University of Leyden entitled "Ether and Relativity" clearly articulates Einstein's realist post-GR inclinations as to the existence, or at least the possibility, of some nonmechanical kind of ethereal substance. In addition, in this address Einstein actually equates such a substance with the Newtonian conception of space. This standing confusion as to Einstein's stance in reference to the ether and the null result of the Michelson-Morley experiment (1887) brings up an interesting question. Did Einstein's views on the issues of substantivalism (realism) vs. relationism (antirealism) as concerns space and time in particular, and as concerns realism vs. antirealism as a metaphysical attitude in general, change in any way during the course of his own development as a scientific thinker?

Various comments of Einstein throughout his life lead one to believe that he had already abandoned the concept of a substantive ethereal medium either before or while working out the details of SR. Einstein's intellectual debt to Humean skepticism and Machian positivism is well documented. As A. Fine has noted, this debt literally leaps "out of the pages of the 1905 paper on special relativity." In SR simultaneity is operationally defined in good positivist fashion. In the aforementioned GR paper, dating from 1916, Einstein's comments concerning the integration of space and time, which required the elimination of space and time as independent concepts, are laden with verificationist language. And yet we know that Einstein will later defend the realist thesis against Bohr and the entire quantum community in what will become known as the hidden variables debate. Was this the result of some inexplicable conversion to realism, or was it rather the result of Einstein's owning up to the realist implications of GR, in much the same way that he would eventually be forced to own up to his mistake about having theoretically tampered with GR in order to avoid the implications of a cosmological expansion? And concerning GR, did not Einstein indeed adopt a realist attitude toward the existence of a four-dimensional space-time manifold?

Let us now reflect upon SR. In the historical introduction to Part III the real is identified with the relative. For SR it is really the case that such properties as position, speed, length, time, duration, and simultaneity are relative in that they quantitatively vary when measured from different inertial perspectives. It is equally true, however, to claim that in SR the real is to be identified with what is invariant.

Given a spatio-temporal location for $Event_1$ of (x,y,z,t) and a spatio-temporal location for $Event_2$ of (x',y',z',t'), with $c =$ the speed of light; then according to Minkowski, the space-time separation between these two events, given by $\{(x-x')^2 + (y-y')^2 + (z-z')^2 - c^2(t-t')^2\}$, is invariant. While such an invariance in no way supports a realist interpretation of space, time, or space-time; nevertheless, it does tend to undercut the claim that SR supports antisubstantivalism or classic Leibnizian relationalism. Substantivalists can claim that Newton's disentanglement of space from time had indeed been erroneous, and, if a substantivalist commitment to space and time were indeed dependent upon this separation, then it too would be suspect. But Einstein and Minkowski have shown that there is another basis upon which to support the substantivalist thesis that there is something substantial underlying our observations of and interactions with the world. As W. H. Newton-Smith points out, SR is neutral on the issue of realism because substantivalists of the Newtonian variety have been shown by Einstein to have been wrong only on the details of their central thesis. They are not necessarily wrong about the thesis itself.

Let us now consider GR. Does GR conform to the relationalist or anti-Newtonian principle of Ernst Mach that attributes the inertial forces of bodies to accelerations that are relative to the combined masses of all the stars in the universe? Informed speculation would seem to answer this question in the negative.

Einstein's Principle of Equivalence equates acceleration with gravity. The field equations of GR, Einstein's theory of gravity, still admit of solutions that ascribe an actual structure to space-time in the absence of all existing matter and energy. Massive objects thus modify space-time; they do not define it. The curvature of space-time is the result of mass/energy deformations upon a flat space-time metric. But according to Mach's principle, which depends upon an existing distribution of matter and energy, such zero matter/energy solutions should be impossible. In 1917 the Dutch scientist Willem de Sitter advanced just such a vacuum solution. In 1949 the logician Kurt Gödel demonstrated a solution of GR in which the universe as a whole undergoes a rotation that can be detected. Such solutions, not all dealing with boundary conditions of zero matter and energy, make it difficult to view GR as in accord with the relationalist principle of Mach. A final verdict on GR in relation to realism has not yet been achieved.

Quantum Theory

Quantum considerations have been a part of the theoretical debate over the fundamental character of matter and energy ever since the introduction of the quantum concept by Max Planck in 1900. At the dawn of the Scientific Revolution, in the seventeenth century, the introduction of discrete algebraic techniques to complement the continuous character of geometrical representation, and the introduction of new mathematical insights into the nature of continuity, infinity, and infinitesimals, led to unprecedented advances in physical theory. In the twentieth century the rise of Quantum Theory (QT) shed light upon as well as stimulated skepticism about the physical implications of both discontinuous phenomena and the realm of the infinitesimally small. Whether Bohr's Principle of Complementarity (1927) signifies something deep about such mutually exclusive but jointly necessary attributes of fundamental reality as particle/wave duality, or whether such exclusivity is primarily a precipitate of notational and conceptual incompatibilities derived from insufficient modes of discourse and description, remains a crucial philosophical problem for QT.

Quantum Theory represents a greater shock to our classically trained intuitions than even Relativity Theory. It is a common oversimplification in reconstructing the history of QT to pit the realist minority, spearheaded first by Einstein and later by David Bohm, against the instrumentalist majority view represented by the Copenhagen Interpretation espoused by the Bohr contingent. Neither Heisenberg nor Schrödinger, for example, were born to the manor of such uncompromising instrumentalism as QT has since become. Heisenberg argued for something like a Craig-theoretic reduction of the theoretical terms of QT. In the abstract to his seminal 1925 paper, Heisenberg boldly states that what he is seeking is a "quantum-theoretical mechanics based exclusively on relations between quantities observable in principle." Schrödinger's route to wave mechanics was fraught with metaphysical ambivalence. Also of note is the instrumentalist apostasy of de Broglie. His slumbering realist commitments, subsequent to his postulation of a pilot wave (1927) that was plainly at odds with Bohr's conception of complementarity, were reawakened in 1952 by Bohm's formulation of a hidden variables approach to QT.

Such wavering of opinion within QT, moreover, was not limited solely to issues at the metaphysical fringe. Not only are there various rival and semantically nonequivalent interpretations of QT; there is also some doubt regarding the syntactic equivalence of its different formalizations. In 1926 Schrödinger offered a proof of the formal equivalence between his own wave mechanics and the recently invented matrix mechanics of Heisenberg (1925). In 1964, however, P.A.M. Dirac would contend that there is no such equivalence. Such formalistic discrepancies are relevant for those who view a connection between theoretical formalization and interpretation.

It is also a bit ironic that a strong paradigmatic commitment to instrumentalism within QT should be thought to be consistent with the prediction of unknown particles. Among the most successful of such predictions to date have been Dirac's prediction of the positron (1928), and the Zweig/Gell-Mann prediction of quarks (1963–64). Although the inequalities of Bell and the experimental confirmations of Aspect et al. seem to have finally overcome the persistent attempts of Einstein to salvage quantum realism, QT still has battles to face on the ontological horizon. Advocates of Super String Theory, for example, are cautiously optimistic about the possibility that the theoretical hegemony of the Standard Model of QT will someday be completely assimilated by a Final Theory of Everything.

The Copenhagen Interpretation of QT purports to pass a sentence of death on classical ontology. It asserts the thesis that QT is both consistent and complete as formulated. The wave function (Ψ), which is calculated by means of the wave equation of Schrödinger, contains all the information that can in principle be known about the state of a system. We thus cannot hope, even in principle, to penetrate more deeply into the fabric of reality. Essentially, there are no hidden variables that would enable us to determinately specify and/or further unravel the secrets underlying such quantum features as probabilistic predictions (Ψ^2); superpositions of quantum states; uncertainty relations between certain conjugate pairs (position/momentum; time/energy); complementarity (particle/wave duality); and non-local interactions (Bell's theorem). Advocates of the Copenhagen Interpretation of QT regard Humean metaphysical skepticism as fully vindicated in that our intuitions of causality are not substantiated by quantum reality. Moreover, even axioms of logic and mathematics, such as the Law of the Excluded Middle and the principle of commutativity, are put on notice by the counterintuitive results of the quantum world.

Dirac, in *The Principles of Quantum Mechanics*, homes in on the quantum problem of size. What is large and what is small? Dirac suggests a definition of size that can be classified as a kind of quantum operationalism. Thus smallness, for Dirac, is an absolute concept that characterizes contexts and conditions in which the Uncertainty relations are not measurably negligible. Our classical intuitions, however, based upon macroscopic perceptions, are negligibly impacted by quantum Uncertainty. This has led to a top-down reductionism that explains the perceptually accessible large or macroscopic in terms of the perceptually inaccessible small or microscopic. But unlike the subalternation relation of classical logic, the truth about physical reality does not flow downward. It is incorrect to assume that microscopic determinism follows from macroscopic determinism. Quantum Theory asserts that the reverse is true. Uncertainties at the quantum level are transmitted upward to the realm of our everyday perceptions, where they are about as noticeable to our common sense intuition as the gaps between our Newtonian expectations and the corrections of SR. This is why it is true to say that the Uncertainty relations are metaphysically intrinsic to physical reality. We cannot get beyond them based upon a notion of epistemological limitation that can in principle be overcome with extended technological proficiency to probe into the depths of the quantum realm.

John von Neumann showed how the Heisenberg Uncertainty relations alone were sufficient for deriving all of QT. Uncertainty stipulates that the act of observing, or interacting with, or measuring some physical process prevents us from ever achieving an arbitrarily high degree of precision about certain phenomena, properties, or parameters of reality. Philosophically, however, we can say that Einstein-Podolsky-Rosen (EPR) were correct in 1935 in pointing out the unlicensed leap of faith embraced by advocates of the Copenhagen Interpretation. Why are Heisenberg's Uncertainty relations a universal limitation on all our knowledge? Why must EPR have been generally wrong, and the Copenhagen Interpretation generally correct, when their battle over the issue of locality and Uncertainty was fought over a thought experiment dealing with electron spin in particular? Philosophically, it is not unfair to say that the Copenhagen Interpretation gained in legitimacy as a result of the efforts of Bell and Aspect, but it had not yet earned its status as a quantum dogma to be defended from the very infancy of QT. A reverse parallel can be drawn here with the case of the philosophical debate over the Heliocentric Theory between Galileo and Bellarmine. Galilean realism was not as entitled to certitude in 1615 as it would later be after Newton. Likewise, Copenhagen instrumentalism became much more philosophically defensible with the outcome of Bell's work, but not so completely as in the case of the defensibility of the post-Newtonian realist attitude toward the Heliocentric Theory. In terms of a possible historical reconstruction, as articulated by physicist J. Cushing, Bohm's realist construal of QT need not have been a reaction to the Copenhagen Interpretation. Bohm's hidden variables approach could easily have been formulated during the late 1920s. Its main drawback, that of violating the limit on the speed of light imposed by SR, presented a different kind of problem.

Let us now reflect upon the classic *Gedanken* challenge to the Copenhagen Interpretation put forward by Schrödinger in 1935: the famous cat paradox. The crux of the problem to instrumentalism and the completeness of QT revolves around the absurdity that a mutually exclusive superposition of possible outcomes, such as opposite spin orientations or the status of a cat as alive or dead, can be simultaneously true to varying degrees until some measurement or observation collapses the relevant wave function in favor of one alternative or another. Realists want to know how this discontinuous transition is brought about by the act of measurement.

Schrödinger's experiment draws upon the analogy between our intuition that a cat must be either dead or alive, and that an electron, for example, cannot be spinning in two different directions simultaneously. Prior to measurement, then, how can we falsify the contention of Copenhagen QT that a cat sequestered in a sealed metal box has a probability of being both dead and alive until we check upon its status? Likewise, how is it that an as-yet undetected electron is spinning both upward and downward with equal likelihood?

This quantum theoretical paradox illustrates the difficulty in achieving an understanding of the proper relationship between truth and knowledge. Truth is customarily held to be a necessary but not sufficient condition for knowledge claims in epistemological theory. Hence if we know something, then it must be true. But if something is true we still might not know it. This primacy of the epistemological perspective seems to parallel the primacy of the role accorded to measurement and observation in the assignment of a distinct value, whose metaphysical reality we come to know as true only after the collapse of a particular wave function. It has always been the case in philosophy to note the gap between what we know and how things are. Quantum Theory introduces a new twist in touting an (in principle) unbridgeable gap between our knowledge of a system and its actual physical state and degree of internal interconnectedness.

The stress placed by the Schrödinger paradox of a live/dead cat upon the instrumentalism and the completeness of Copenhagen QT has to do with its focus upon the discrete or binary character of existence, which must have either one of two distinct options. This presents a problem in mapping such a discontinuous feature of reality upon a continuous probability function. Richard Feynman's famous sum-over-paths approach to QT, which allows for the infinite number of trajectories that an electron might take during a double-slit experiment, also violates what we would normally construe as the physical impossibility of an electron being in more than one place at the same time.

What this signifies is that science, in the case of QT, has learned to value prediction even in the absence of further explanation. In Aristotelian science the exact opposite was true. In the Newtonian paradigm both explanation and prediction were crucial to theoretical success. The logic of modern explanation theory is complicated by its need to be sensitive to probabilities. With the rise of Copenhagen instrumentalism we witness, not so much a sanction for sustained metaphysical skepticism, but the discovery of a new respect for inexplicable predictions.

Big Bang Theory

Our Standard Model of Big Bang Theory (BBT) traces our knowledge of the origin and development of the universe back to the Planck Time, 10^{-43} seconds after the primeval explosion which lends its name to the theory. The precise moment of the Big Bang itself is not known. It could be determined if we could calculate the exact magnitudes of such parameters as Hubble's constant (H), which gives a measure of the present expansion rate of the universe, and the cosmological constant (Λ), which represents the remaining vacuum energy from the inflationary period. A value of ($\Lambda > 0$), for instance, would allow for an increase in the expansion rate of the universe, since (Λ) exerts a repulsion that counteracts gravity. But this gap in our knowledge of *when* the Big Bang occurred does not undermine the mountain of evidence in favor of the claim that it did occur. With our other physical theories, the core ontological features brought us into contact with such issues as configuration, size, observability, process, substantiality, relationality, and so on. But with BBT the core existential issue is that of a primal event.

Theoretical Support for BBT
1. BBT is implied by GR
2. Discovery of the expanding universe
3. Discovery of Cosmic Background Radiation (CBR)
4. Abundance of light elements in the early universe. Current Mass: (Helium/Hydrogen) = ¼
5. Guth's Inflationary hypothesis, which explains
   a. Horizon problem
   b. Flatness problem
   c. Absence of magnetic monopoles
6. COBE confirmation of CBR anisotropy

Our confidence in BBT clashes with Steven Weinberg's candid remark as to whether "we really know what we are talking about" in regard to the physics of the early universe. Nevertheless, the discovery of the CBR by Penzias and Wilson in 1965 is the critical factor that elevates BBT over the Steady State Theory (SST) of Hoyle, Bondi, and Gold. Why does this discovery, and not that of the expanding universe or GR, throw the weight of evidence in favor of BBT? This redounds to the question of why the Hubble confirmation *of an expanding* universe in 1929, as implied by GR, was deemed by some as insufficient for deciding between BBT and SST, whereas the discovery of the existence of a *uniform* microwave background radiation of 2.74K bombarding the earth from all directions was viewed as sufficient. Why is CBR often poetically referred to as the *echo of creation*, whereas some consider an expanding universe as equally consistent with an eternalist model of the universe?

By extrapolation of the Hubble equation linking velocity of stellar recession with observed distance from the earth (V = HD), we arrive at a first moment of time in which all matter is concentrated into a very small volume. A thermodynamic extrapolation in relation to the CBR also can be said to converge upon a common point of origin. Thus we can say that CBR lent support to a major inference of an expanding universe.

Steady State Theory posits that the engine of cosmic expansion is not the energy unleashed at the beginning of time at the moment of the Big Bang, but an eternal process whereby a negative energy field of some sort is converted into new matter. Such a continuous process of material creation is intended to avert the thermodynamic heat death that would be expected in view of the second law of thermodynamics. In this manner, the universe retains a constant density and appearance over time. There is no logical contradiction involved in SST. Its initial loss of favor had more to do with the corroboration that CBR and the current mass of helium and hydrogen provided GR and the expanding universe than with any falsification of SST per se. The current mass ratio of (helium/hydrogen = ¼) is the very result we would expect from the type of hot early universe environment predicted by BBT. Therefore, with no independent corroboration of its own, SST fell out of favor, and BBT became established as the lone, but still insecure, paradigm of cosmological theory.

The elevation of BBT over SST did not mean that BBT had overcome all explanatory problems. It had not. To view BBT and SST as the only two cosmological possibilities would be to impose a false dichotomy, although even prescientific speculation was divided between creationist and eternalist models of the universe. And even if we grant that BBT must be correct, there are still a host of different Big Bang models to choose from.

As our outline indicates, the Inflationary hypothesis, as articulated by Guth, Steinhardt, and Linde et al. from the early 1980s onward, was successful in explaining a number of important problems, all of which were troublesome for BBT. But none of these Inflation based explanations would be worth anything if the legitimacy of the Inflationary hypothesis itself was undermined. According to Inflation, the CBR must possess the signature of radiation irregularities that would be reflective of the possibility of future structure-forming in the universe. This would only be the case if CBR could be shown to possess sufficient density variation or anisotropy in its signal. According to the COBE findings of Smoot et al. in 1992, the CBR exhibits a sufficient degree of variation in its thermodynamic profile to satisfy this Inflationary requirement.

What is the connection between the Inflationary hypothesis and what F. Zwicky once referred to as the missing *dunkle Materie*, or dark matter? Let us first define these terms.

Inflation a period of exponential expansion lasting from about 10^{-36} to 10^{-34} seconds after the Big Bang in which the universe increases in size by a factor of about 10^{50}

Dark Matter mass that must be postulated to account for gravitational instabilities. It possesses an extremely high mass-to-luminosity ratio.

Inflation was introduced as a conjecture that could explain certain still-unanswered puzzles of BBT. Three such problems are enumerated in the outline above. Inflation not only explains the so-called flatness problem of BBT; by virtue of its exponential stretching out of any initial spatial curvature, it also predicts it. So while the preceding supports for BBT in the outline point toward the actuality of a Big Bang moment, there was a problem as to how known physics (GR and QT) and unificationist physics (Grand Unified Theories [GUT]) could explain the current state of the universe (horizon problem, absence of magnetic monopoles, large scale structure, etc.). Thus, in general terms, an expansionist period was invoked to salvage BBT. But we still do not know the precise mechanism whereby such an expansion is accomplished. While explaining the horizon problem, the seeming coincidence of widely separated homogeneous regions of space not linked by light signals, we find that inflation creates a problem for itself. How can expansion occur at such a rate so as to disconnect regions of space that were once previously connected by light signals?

One suggestion for the Inflationary hypothesis compares the onset of Inflation to a GUT-scale symmetry-breaking phase transition. In such a scenario there might be a disruption of the opposing tension between an entropy field featuring a kinetic energy density, which enables expansion and cooling, and a vacuum energy density , which opposes it. Likewise, termination of the Inflationary period might occur when the symmetry that holds between the weak and electromagnetic forces is broken. Simply put, when the universe no longer behaves as if the stress-energy tensor is controlled by the cosmological constant (Λ), it then reverts to normal pre-inflationary expansion.

All this is obviously hypothetical. Further complications arise when it is realized that a BBT that includes Inflation yields a Robertson-Walker curvature term equal to zero. Such a state for the space-time metric coincides with a value of Omega (Ω) = 1. Since $\Omega = p/p_c$, where p equals the actual mass density of the universe and p_c equals the critical density between expansion and contraction, this leads to the Inflationary prediction that the universe is dominated by some form of dark matter. Otherwise $\Omega \neq 1$. Thus Inflation is consistent with the findings of Ostriker, Peebles, Rubin et al., which showed that the explanation of galactic stability requires an invisible gravitational source, perhaps comprising more than 90 percent of the mass of the universe.

Our evidence for the existence of dark matter is indirect. Given that its existence is predicted by BBT + Inflation, its detection is vital for both, as far as our outline-based line of argumentation, which stresses the dependency of BBT upon Inflation, is concerned. Our current evidence seems to support a value of $\Omega \leq 1$. If $\Omega < 1$, then the Inflationary requirement of a flat space will require that ($\Lambda > o$). Furthermore, there is conflicting evidence as to the age of the universe: sometimes (Age$_{Hubble}$ < Age$_{stars}$). In terms of the parameters (H, Ω, Λ) and the estimated ages of globular clusters, there are five major BBT models that are consistent with current data, at least two of which are inconsistent with Inflation. Add to this that there are various Inflationary alternatives for which space-time is curved, and we must conclude that although (BBT) is currently paradigmatic for cosmological theorizing, its details are by no means theoretically secured.

Solving the problem of detecting the missing mass of the universe constitutes possibly the biggest challenge facing BBT today. There are many candidates for the mysterious dark matter, but only two major categories: baryonic (which includes protons and neutrons) and nonbaryonic. On the basis of the CBR the mass of the universe was believed to have been predominantly baryonic. With

Inflation this inference has been seriously weakened. If baryonic, then dark matter is at least in principle accessible to telescopic discovery. If nonbaryonic, the composition of dark matter will need to be ascertained by experimentation at extremely high energies here on earth.

The Problem of Scientific Discovery

Ever since Plato's *Meno*, epistemological theory has come to differentiate between discovery and invention. Are scientific hypotheses and theories discovered or invented? Additionally, is there such a thing as a logic of discovery? And if it is true, as Karl Popper vigorously argued, that there is no such thing as a method or general algorithm of discovery, then surely there exist various pragmatic tools that serve as catalysts of discovery. It is in this latter sense that we refer to the existence not of a logic of discovery, but of a logistics of discovery.

Bacon, Descartes, Galileo, and Newton were pioneers of the scientific method who would be among the first to admit that we cannot lay down sufficient and necessary rules that would enable us to automatically solve a scientific problem. Twentieth-century phenomenology has tried to delve beneath the surface of psychologistic cognition to a logical core of concepts that govern rational thinking, but has done so at the expense of a proliferation of the logical distinctions upon which such thinking is based. If we accept the inevitable problematics involved in the possibility of devising a general rule of procedure for theory construction, then we can focus our attention upon those pragmatic measures that have assisted scientists in the creative development of successful scientific theories.

Neither inductive logics, such as those formulated by Carnap and Reichenbach, nor the so-called Hypothetico-Deductive method, in which hypotheses both serve as the premises for previously obtained results and as a source of new experimental predictions, describe actual scientific procedure. Even the famous *experimentum crucis* of Baconian invention does not serve a purely discovery-oriented function, if one considers the goals of experimental corroboration and falsification as not being included within the purview of discovery as an originating process. When existing hypotheses or theories are put to the test, then the critical experiments that are designed to test them can be characterized more in terms of a contribution to the completeness of a hypothesis or theory than as a creative component of theory construction itself. On the other hand, experimental results, sometimes serendipitous, are crucial for the creative work of theory construction. Nevertheless, it is a philosophical difficulty as to where to draw the line of demarcation between the theoretical phases of discovery and attempted corroboration. It is equally difficult to establish the mutual relationship between theory and experiment.

Certainly the experimental researches of Faraday and Rutherford were crucial for the formation of Maxwell's Electromagnetic Theory (1865) and Bohr's composite Classical/Quantum theory of the atom (1913), respectively. The failure of Michelson/Morley (1887) to discover the luminiferous ether was significant but not crucial to the development of SR. Eddington's celebrated expedition to photograph a solar eclipse in 1919 led to the spectacular corroboration of a core prediction of GR. The experimental results of Compton and Davisson/Germer in the 1920's were pivotal in corroborating the experimentally irreducible complementarity of electromagnetic phenomena. Bell's thought-experimental approach, later experimentally confirmed by Aspect et al., was devised with the original intent of vindicating Einstein's realist intuitions as to the proper interpretation of QT. Ironically, his work led to a corroboration of the instrumentalist Copenhagen Interpretation of QT. The double-slit experiment has performed double service, as a critical experiment in the development of both Electromagnetic and Quantum Theory. Interestingly, Feynman went so far as to describe the double-slit experiment as the defining experiment for at least attempting to understand the complexities, implications, and absurdities of QT. Simply, then, experiment has played and continues to play not only an important role in the overall work of scientific discovery, but one that is multidimensional.

Let us now focus upon the use of analogy as a tool of discovery. In logical discourse analogy is described as a species of inductive thinking, involving the formulation of a general correspondence between two conceptual structures in which there is always a degree of conceptual remainder. In requiring an inductivist leap of faith concerning all physical instantiations of a particular concept, or complex conceptual model, analogical thinking can often be fallacious. In the history of scientific discovery the use of analogy has also been extremely fruitful.

Our references to analogical thinking in the history of science here will not aspire to any rigorous definitional standard, for it is not in the nature of analogical thinking to be held up to any such standard of rigor. In current QT, hypotheses are said to be renormalizable if they can be described using Feynman diagrams. If they are not so renormalizable, they will often be rejected, or looked upon with a lesser degree of confidence, for this reason alone. Such a diagrammatic approach to scientific thinking qualifies as analogical. If we view mathematical relationships and structures as analogs of those that occur in nature, then Schrödinger's attempt to eliminate the negative energy solutions in Dirac's electron theory for failing to conform to Lorentzian invariance is cut from the same cloth. N. R. Campbell, in *Physics, the Elements* (1920), argued that there exists a necessary connection between scientific theories and models, which for our purposes, unlike the distinction made between models and analogies by someone like Pierre Duhem, will be viewed as being continuous with analogical comparison. Einstein's Principle of Equivalence, in which acceleration and gravity are treated as mathematically interchangeable, represents a stunning insight into the analogous character of the forces exhibited in nature. Newton's Law of Universal Gravitation provided the analog for the Electrostatic Law discovered by Coulomb, while Newtonian/Hamiltonian mechanics performed the same function in regard to the formulation of both matrix and wave mechanics in the 1920s. Huygens developed a wave theory of light, as did both Grimaldi and Hooke, from the prevailing view that sound is a wave phenomenon. Many problems in applied acoustics have been solved as a result of the analogous connection between acoustics and preexisting electric circuit theory. Both Einstein and Perrin extended the analogy between ideal gases and solutions to colloidal solutions and suspensions. Among the most celebrated examples of analogical induction in the history of modern science was Kepler's generalization that an elliptical orbit for Mars also applied to the other planets in his *Astronomia Nova* (1609).

The mathematization of physics in the seventeenth century made it possible for mathematical considerations to prove fruitful for physical theory. Planck's fortunate employment of a curve-fitting technique while working out the details of black-body radiation, by which he found a formula that fit all his experimental observations, serves as a case in point—although, as Edward Witten points out, a current disadvantage of Super String Theory is that it possesses no guiding physical intuition analogous to that of Einstein's Principle of Equivalence. And while it is doubtful that Maxwell was led by mainly mathematical considerations in his reflections regarding displacement current, one may nevertheless reconstruct such a possible path. In the case of Heisenberg in the mid 1920s, for example, his publications betray no hint of the scientific models and analogies that were instrumental in guiding his thinking. Unlike Kepler, he presented his results in a form that bore no signs of the travails of its birth. Newton did much the same thing in the *Principia* in covering up the intellectual path he trod toward his principle of inertia. To recover this path we must consult *De Motu*. In any case, analogy has played a crucial role in the forging of physical theory.

The Evaluation of Scientific Theories

The efficacy of a theory is related to its explanatory scope and to its predictive accuracy. As Leibniz conveys in his correspondence, nothing commends a hypothesis more to working scientists than its predictive success. We have seen, however, that classical deterministic and reductionistic explanation may not be possible, even in principle, for QT. Nevertheless, when a theoretical prediction is corroborated, experimentally or otherwise, this can only bolster our confidence in such a theory. When phenomena within the purview of a particular theory resist explanation, or do not conform to theoretical expectations, then our confidence in such a theory begins to wane. How many disconfirming instances or explanatory anomalies suffice to warrant a theoretical overhaul or replacement is not something known with any degree of certainty, but differs from theory to theory.

If a single counterexample sufficed in common scientific praxis to undo a successful theory, then Newtonian physical theory would have been more seriously jeopardized by the aberration in the orbit of the planet Mercury at perihelion. General Relativity recorded its first success by explaining this orbital anomaly. but we still refer to Newtonian theory as having been assimilated by Relativity Theory. It lives on as an approximation. We do not speak of Newtonian theory as having been refuted. That a seemingly fruitful theoretical framework will dig in its heels despite immediate and serious shortcomings is well exemplified by the persisting influence of Bohr's original theory of the atom after 1913. What is important, as Popper points out, is that a theory be falsifiable in principle. It must be able to be proven wrong, unlike a psychological theory such as psychoanalysis, which many view as interpretively closed, meaning that it can explanatorily counter any external criticism to its theoretical claims. But what happens to a falsifiable theory when confronted with a counterexample to its claims will depend upon a number of different things.

The problematic of the transition from one theory to another has become an important area of inquiry within the philosophy of science over the last generation or so. Are such transitions better described as continuous or discontinuous in character? Are they more revolutionary than reformatory? It is interesting in this regard to note that at the dawn of modernity we refer to a revolution in science, but a reformation in religion. The main tenets of the Christian religion remained essentially the same, while the traditional claims of Ptolemaic astronomy and Aristotelian science were superseded entirely by different and undeniably better theories.

Why is it that successor theories (T_2) are superior to the theories they replace (T_1)? A continuing goal of the philosophy of science has been to flesh out just what such superiority entails. Let us consider two examples borrowed from L. Laudan.

1. Newtonian celestial mechanics (T_2) did not explain why the planets all revolved in the same direction around the sun. Cartesian cosmological theory (T_1) explained this orbital similarity as the result of the planets being carried by a swirling vortex extending outward from the sun. Thus T_2, although a better theory than T_1, did not completely prove itself as explanatorily complete.
2. Benjamin Franklin's one-fluid theory could not account for the repulsion exhibited between bodies possessing a negative charge. Earlier in the 1740s the Abbe Nollet had explained this repulsion in terms of an electrical vortex. Despite its explanatory omission, the Franklin theory succeeded in establishing itself over the theory of Nollet.

Both examples illustrate that the theory that wins out over its rival need not be explanatorily complete relative to its rival in every respect. The decision to transfer scientific allegiance from some T_1 to some T_2 is thus not altogether based upon an itemized notion of superiority, where T_2 must be superior to T_1 on an item-by-item basis. Hence, if we were ever to devise a method for choosing between two or more competing theories, we would have to be able to precisely calibrate the importance and relevance of numerous theoretical elements. In the history of the competition between the Heliocentric and Geocentric Theories, for example, it seems that the relevance of the failure to detect stellar parallax became less important as time went by. At the time of Aristarchus such a failure was a highly relevant piece of falsifying ammunition for proponents of geocentrism. By the time of Bessel's detection of stellar parallax in 1838, no advocate of the well-entrenched Heliocentric Theory saw any particular reason to be jubilant over this long-awaited corroborative piece of information.

It is interesting to note a particular asymmetry as well as a possible point of symmetry between the first and the last of our five physical theories. Heliocentric Theory acted as a catalyst for overthrowing the hegemony of Aristotelian science, thus leading to the new science created by Newton and the other giants upon whose shoulders he stood. BBT, on the other hand, seemed to usher forth from GR, like Athena from the head of Zeus. Nowadays, both GR and QT are involved in the process of deepening our knowledge about the Big Bang event and the cosmological process that ensued from it. This differing relationship with reference to physical theory represents a point of asymmetry between these two theories. From the point of view of a possible symmetry between these two cosmological theories we can point to the fact that contemporary cosmology provides us with a laboratory for exploring and coming to understand the physics of unification. Super String theorists, for example, evaluate cosmological conditions in their attempt to forge an adequate theoretical framework that can overcome the failures of GR and QT to cope with all the requirements of a comprehensive physical theory. Hence, like Heliocentric Theory, BBT also can be said to function as both a stimulus and sponsor for new insights into physical theory.

Just over half a century ago, Carl Hempel, Paul Oppenheim, and Hans Reichenbach et al. commented upon the structural similarity between explanation and prediction. While explanation focuses upon past occurrences, prediction is directed toward the future. A fully adequate explanation is one that might also have led to a successful prediction of some phenomenon if it were not already known. Recall that the Inflationary hypothesis of BBT is said to both explain and predict the flatness of the curvature of space. This structural similarity between explanation and prediction is predicated upon similar inferential connections between what is to be explained (*explanandum*), or predicted, and the set of premises and laws (*explanans*) that carry the burden of explanation, or prediction. For Hempelian explanation, logical inference is analogous to physical causation. Hempel's so-called Cumulative model of theory change, in which scientific progress is extended in a linear way as the result of the continuous accumulation of new knowledge, is rooted in his understanding of the structure of explanation. The progressive character of scientific knowledge is the result of the ongoing concatenation of causally related explanans and explananda.

There are various other models of theory change that oppose the legitimacy of the cumulative model of theoretical development. The focal point of such opposition is the logical structure of the transition that occurs when one theory is superseded by another. A further problem is that of the possible incommensurability between a predecessor and a successor theory. Such a problem arises, for example, when individuals are speaking different languages. Communication becomes impossible without a means of translation. For example, someone reading a history of modern physics would be both perplexed and misinformed if the term electron is not precisely defined. The reader might reasonably wonder whether such a reference is to the electron of any number of different individuals: Thomson, Lorentz, Bohr, Millikan, Dirac, and so on.

Thomas Kuhn's theory of the structure of scientific revolutions lends its name to the so-called revolutionary model of theoretical development. Kuhn's model stresses the discontinuity or incommensurability between T_1 and T_2. Popper's evolutionary model, in which inherited theories (T_1) must somehow attempt to absorb shocks due to theoretically intractable or anomalous phenomena, is another well known model of theory change. Laudan's gradualist model of theoretical development suggests that it is wrong to limit ourselves to shifts in the theoretical content of scientific theories alone when assessing the problem of theory change. For it is possible that theory changes might not be accompanied by corresponding changes in the methodology and/or the goals of science. We know that in the seventeenth century, for example, new methods led both to a transformation of physical theory and to an undermining of the teleological basis of Aristotelian physical science. But do paradigmatic shifts in twentieth-century physical theory automatically inspire corresponding shifts in methodology or a renewal of interest in teleology?

Our Descartes/Newton and Nollet/Franklin examples exhibit the sheer untenability of any view that espouses a straightforward and unnuanced cumulative theory of how scientific knowledge progresses. But such a legitimate criticism of naïve or oversimplistic cumulativity, however, stems from a too-zealous concentration upon all the individual details of a particular theory. When we take such a *microscopic* approach to the inspection of theoretical components, it is tough to see how any successor theory could prove itself superior in every particular area of theoretical comparison. When Mandelbrot probed the coastline of Britain from a fractal perspective, it turned out to be even more jagged than anyone had ever imagined. So it is with attempts to find a perfectly smooth cumulative fit between related theories. From a macroscopic perspective, on the other hand, all rational practitioners of science would concur that Newtonian heliocentrism was superior to and represented undeniable progress over the Cartesian version of the Heliocentric Theory. It is, however, interesting to note that the deterministic structure of cumulative progression is in general accord with classical, or prequantum, intuitions, whereas Kuhn's revolutionary model of theoretical development seems to resonate with the discontinuities and indeterminacies that pervade QT.

Thus the type of rationality associated with our major physical theories seems to be connected to the current alternatives for interpreting scientific progress. Variations on the evolutionary model, such as that of the more dialectically and historically sensitive model developed by Imre Lakatos, attempt to mediate between the opposing poles of an overly simplistic cumulativity on the one hand, and an overly exaggerated commitment to the notion of revolutionary changes to existing paradigms on the other. The big picture presented by each of the five physical theories in this book seems to square with the view of what we might characterize as a kind of intermediate cumulativity, an overall and long-term continuous theoretical development that negotiates between the extremes, and that allows for revolutionary change and progress at the same time.

Given this intermediate cumulativity, we feel that the five physical theories whose developments are portrayed in this book have achieved a kind of security that guarantees them a role in the scientific worldview for the indefinite future. Their worst fate, which has already happened to Electromagnetic Field Theory, is that they might be subsumed into a higher theory. They have survived the tests of time.

PART I

THE HELIOCENTRIC THEORY

//

INTRODUCTION

////////////////////////////////////

Although Heliocentric Theory is well known, describing it without the use of unwarranted or unjustified assumptions is not easy. Simply put, the theory suggests that the earth has two motions, a rotation on an axis and an orbital motion about the sun. Further, it maintains that the sun is central to, although not exactly in the center of, the orbits of all those heavenly bodies known as the planets, of which the earth is one. The physical reference frame used to determine the motions of this "solar" system is the frame of the fixed stars, bodies that do not appear to change their positions relative to one another. In this theory the dual motions attributed to the earth are considered to be in some sense real.

The Geocentric View of Eudoxus

Although not really interested in astronomy, the philosopher Plato had a great influence on the course of its early history. Because he perceived the heavens to be more perfect than the earth, Plato urged astronomers to describe celestial motions in terms of the most perfect of geometrical shapes, the circle. In fact, for Plato, the most perfect motion would be uniform circular motion, motion in a circle at a constant rate of speed.

One of Plato's pupils, Eudoxus of Cnidus (409 B.C.–356 B.C.), was the first astronomer to follow Plato's recommendation. Blending careful observation with sophisticated mathematical constructs, Eudoxus sought to describe the motions of the heavens in terms of a series of concentric spherical shells, with the earth geometrically at the center of those shells. His model consisted of twenty-seven spheres, three each for the sun and the moon and four for each of the five known planets; Mercury, Venus, Mars, Jupiter, and Saturn. The final sphere carried all the "fixed" stars and presumably contained the whole universe. Each sphere turned on an axis at a uniform rate of speed and was attached to adjacent spheres at its axis. Since the axes of the spheres were in different planes and since the spheres could transmit their motions to one another through the axes, Eudoxus was able to "account" for the rather complicated motions that had been observed—for example, the retrograde motions of Mars, Jupiter, and Saturn. In this process the earth was not only considered to be in the center of these shells but also totally immobile and the reference point of all perceived motions.

The Aristotelian Cosmology

As mentioned, Eudoxus was somewhat successful in explaining observations of the heavens. Another of Plato's students, the philosopher Aristotle, found that the theory of Eudoxus fit nicely with his overall philosophy of nature, and he transformed it into a physical theory. Aristotle saw nature as dynamic, powerful, and teleological. Each object in nature tended toward a certain set of ends or goals, and nature as a whole tended toward a goal that was the result of the motion of the individual objects. What this meant was that each of the four inanimate elements—earth, water, air, and fire—had a natural place or equilibrium position in the universe, a place where it belonged. If nature were left alone Aristotle felt each element would move "naturally" to its proper place and nature as a whole would achieve an ideal structure.

Using this understanding of natural place and natural motion, Aristotle built a picture of the universe as a whole. Since for him there was no such thing as empty space, every object had to be surrounded by some medium. By observation it could be seen that heavy objects move naturally toward the surface of the earth and light objects move away from the earth's surface; for example, a rock falls in air, and air bubbles rise to the surface of water. More precisely, these motions appeared to be along lines perpendicular to the surface of the earth.

Now the ancient Greeks had become convinced through observations of lunar eclipses and measurements of shadows cast by sticks in the ground at various locations that the earth as a body was a sphere. Thus, the downward motions of heavy objects were directed toward the center of the earth, while the upward motions of light objects were directed away from the center of the earth. Further, since the earth as a body was made of the heaviest element—earth—its natural place would be in the center of the natural world. The center of the earth was literally, then, the center of the universe and the reference point for all motions.

The other three elements, water, air, and fire, arranged themselves accordingly in the vicinity of the spherical earth and below the moon. In this sublunar region of the universe, natural objects could undergo radical (substantial) changes. But such changes did not appear to take place in the heavens. In fact, the only change that did take place in the heavens was change of place—the heavens seemed to move in circles about the earth. Because substantial changes did not take place in the heavens and because their motion was circular, Aristotle decided that the heavens were more perfect than the sublunar region and were made of a fifth element called the ether. Since the heavens were more perfect it was fitting that they would have the most perfect motion, namely, uniform circular motion.

The marriage between Aristotle's philosophy of nature and the model of the universe of Eudoxus can now be seen. The shells of Eudoxus became hollow transparent spheres that carried the sun, moon, and five planets around the earth. Aristotle used fifty-five crystalline shells in all. Rather, Aristotle needed these additions because his model was, in effect, an interconnected quasi-mechanical system. He used a set of counteracting spheres in order to offset the one-way effects of the outer spherical shell motions upon the movements of inner shells and the planetary bodies they carried. The outermost sphere in this system, called the celestial sphere, carried all the stars and enclosed a finite universe.

In the eternal universe of Aristotle, the earth was perfectly at rest. If it ever had any translational motion, it would have eventually come to rest in the center, its natural place. Further, it could not be spinning on an axis because circular motion was only appropriate for the more perfect heavens, not for the imperfect entities of the sublunar region. The sun, the moon, and the planets all maintained the same distance from the earth at all times. Because the stars were all attached to the celestial sphere they maintained the same distance from the earth and the same relationship to one another at all times.

Early Theories of a Moving Earth

Despite the fact that a stationary earth theory is consistent with ordinary experience, there were some contemporaries of Aristotle who were willing to speculate that the earth might in reality be moving. Heraclides of Pontus (388 B.C.–315 B.C.) suggested that the daily motion of the stars could be accounted for equally well by the rotation of the earth on an axis. He maintained that the rotation of the earth would be far less *violent*, an Aristotelian technical term for unnatural motion, than any rotation of the much larger celestial sphere carrying the stars. Heraclides also thought that the motions of the so-called lesser planets, Mercury and Venus, could be better explained if they revolved around the sun rather than the earth.

The first complete heliocentric theory came from Aristarchus of Samos (310 B.C.–250 B.C.), who is known as the Copernicus of antiquity. In addition to its rotational motion, Aristarchus suggested that the earth itself was also a planet that moved in a circular orbit about the sun. In fact, he placed the sun at the center of the celestial sphere, which in turn he thought to be at rest.

The heliocentric theory devised by Aristarchus attempted to deal with two continuing problems whose solution had still not been adequately resolved by the existing geocentric theory. These problems were

1. *Retrograde motion of the planets.* Some planets in their eastward movement through the stars can be perceived to slow down, come to a standstill, reverse direction, and eventually begin to move again in their original direction.
2. *Variations in planetary brightness during the year.*

Critical problems, however, also beset the heliocentric theory of Aristarchus. One such problem was the failure to observe variation in stellar brightness. If the earth actually orbited the sun, then the stars should appear to vary in brightness, since they would no longer be equidistant from the earth. Further, if the earth moved among the fixed stars, the angular separation of stars should vary. However, this phenomenon, called stellar parallax, was not observed. To explain the absence of stellar parallax within the framework of heliocentrism would require the hypothesis that the earth's orbit be insignificantly small relative to the distances of the stars. This was a move the ancients were unwilling to make, especially because the heliocentric theory was at odds with our common experience of motion. If the heliocentric theory is correct, the earth and its inhabitants would have to be spinning like a top and hurtling through space at incredible rates of speed. Yet there is no experience of these motions. A reasonable person in ancient Greece, looking at all the evidence, would have to side with geocentrism.

The Geocentrism of Claudius Ptolemy

The geocentric theory achieved its greatest expression in the publication of Claudius Ptolemy's *Almagest*. This book, which was written in Alexandria, Egypt, in about A.D. 150, is the greatest surviving astronomical work bequeathed to us from antiquity. Ptolemy, however, did not simply inherit Aristotle's version of the geocentric theory. In the interim, other important contributions were made with regard to geocentric theory and method. Two individuals worthy of note in preparing the way for Ptolemaic astronomy were Apollonius of Perga, who lived in the latter half of the third century B.C., and Hipparchus of Nicea (second century B.C.), whom many scholars of astronomy rank as the greatest astronomer of antiquity.

Apollonius is credited with being the Greek mathematician who made the greatest contribution to the study of the conic sections; namely, the figures of the ellipse, parabola, and hyperbola. He is also credited with having invented the mathematical constructs called epicycles and deferents that became important parts of the Ptolemaic and post-Ptolemaic descriptive framework of geocentric, as well as Copernican heliocentric, cosmology. Without getting into the philosophical debate about actual ontological commitments, one can safely say that Apollonius not only contributed important aspects of the mathematical apparatus of ensuing geocentric theory, but did so in such a way as to bracket the issue of the physical existence of spherical shells.

Hipparchus made extraordinary contributions to astronomical methodology as well as to theoretical astronomy. He is credited with having invented or at least developed trigonometry into an important tool for numerically calculating the relationships that exist between and among the geometrical figures used to represent celestial motions. He also made extensive use of the Apollonian epicycles and deferents, and of eccentric circles, setting the stage for Ptolemy. His great contributions to lunar and solar astronomy enabled him to contribute to our understanding of the precession of the equinoxes, a phenomenon in which the sun, in its journey around the celestial plane known as the ecliptic, returns to a particular point on the ecliptic in advance of where it is expected to be relative to the background of the stars.

The completion of the mathematization of astronomy fell to Ptolemy in the second century A.D. For Ptolemy, astronomy was a mathematical exercise designed to "save the phenomena," to account for the observations of the activities of the heavenly bodies by use of mathematical hypotheses concerning their motions. This he accomplished with great success in the *Almagest*, in which Ptolemy rejected the heliocentrism of Aristarchus because stellar parallax was not observable and because the stars did not vary in brightness. He affirmed the Aristotelian arguments that the earth is completely at rest and in the center of the universe. Computationally it was possible for the earth to occupy other positions in the universe, but Ptolemy concluded that if the earth was not in the center of the universe, then the "order of things" would be "fundamentally upset." Although he agreed with Aristotle and Plato that the heavens move spherically, he eliminated all the hollow transparent spheres, but not the celestial sphere. He considered the sun, moon, and five planets to be independent bodies, while he kept the stars attached to the celestial sphere that enclosed the universe.

To account for solar, lunar, and planetary motions Ptolemy used the following mathematical devices:

1. Eccentric Circles.
   No celestial bodies move with simple uniform circular motions. Both the sun and the moon appear to speed up and slow down, while the planets at times appear to move in opposing directions. Eccentric circles are circular paths of motion that are meant to be observed from some internal point displaced from the circle's center. This allows for better approximations of celestial motions, as celestial objects, such as the sun, appear to move faster and slower, toward perigee and apogee, while supposedly orbiting uniformly around the circumference of the circle.
2. Epicycle/Deferent System.
   This system is far more powerful than that of eccentrics, although Apolloinus had already shown them to be geometrically equivalent, with the latter being a special case of the former. In this system a celestial body revolves uniformly around a smaller circle called an epicycle, the center of which itself revolves uniformly around a larger circle called a deferent. The observer is then situated at the center of the deferent circle. Ptolemy considered the epicycle/deferent system to be more powerful than eccentric circles since they possess a greater degree of freedom for representing observed motions. Epicycles were found to explain variations in brightness much better than eccentrics. Epicyclic movement consisted of a kind of looping motion that enabled Ptolemy to explain the retrograde motions of the outer planets.

3. Equant Point.

Ptolemy introduced the device known as the equant, a point displaced from the center of a deferent circle. The earth is positioned equidistant from this center, but on the opposite side from the equant. With the help of the equant, Ptolemy was able to work out the motions of the planets. Copernicus would later eliminate the equant from his astronomical repertoire of mathematical techniques because he thought it to be in violation of the Platonic ideal to preserve uniform circular motion. Relative to the equant Ptolemy could conserve uniform angular motion, in that a revolving body would sweep out equal areas around the equant in equal times. As a result, however, observation from the perspective of the earth would no longer be uniform. It can accurately be said that the equant point, in allowing for greater mathematical manipulation of epicyclic movement, contributed to Ptolemy's view that the epicycle/deferent system was superior to the eccentric.

Using these mathematical devices, Ptolemy was able to produce the first comprehensive and systematic quantitative account of celestial motions. Ptolemy's astronomical ambitions included not only accounting for past and present celestial motions, but also predicting future celestial and planetary motions as well. As new and more precise data became available, Ptolemy's eccentrics, equants, and epicycles would be adjusted to fit that data if necessary. All in all, the theory of Ptolemy became quite accurate and extremely useful (e.g., it could be used as a basis for keeping track of time). If Ptolemaic geocentric theory was to be dethroned and replaced by another theory, it would have to be for powerful and practical reasons. For in addition to its entrenchment and pedigree within the scientific community, the Ptolemaic-Aristotelian cosmological model had been embraced as a theory of the heavens by Christianity.

The Copernican Heliocentrism:
A Revolution in Astronomy and Physics

Between the time of Ptolemy and Copernicus, astronomy underwent more than a millennium of normal geocentric activity in which the principles of the theory went relatively unquestioned, and in which most of the research and investigative work was aimed at applying the theory and not undermining it. There were some minor rumblings about geocentrism during this time, most notably from Nicholas Oresme in the fourteenth century and Nicholas of Cusa in the fifteenth. However, neither man rejected geocentrism. But Oresme's views at least can be characterized as a definite stepping stone on the conceptual path to heliocentrism. He argued in his *Le Livre du ciel et du monde* that observed astronomical phenomena can be explained in a computationally equivalent manner by the assumption of a rotating earth or by the rotation of the heavens about the earth. He further argued that it would not be possible by reason or experience to confirm either hypothesis. Concerning the argument that only a motionless earth squared with Aristotelian physics, Oresme pointed out that such physical reasoning was based upon a theory of motion that had never been confirmed.

Despite his concerns, Oresme never really considered committing himself to the reality of a rotating earth. By the turn of the sixteenth century, geocentrism was as entrenched as a scientific paradigm could possibly be. However, it was also at this time that the first difficulties with Ptolemy's astronomy began to appear. For example, predictions of planetary locations were becoming noticeably inaccurate.

Early in the sixteenth century a Polish clergyman name Nicholaus Copernicus set out to revise the existing astronomical tables and remove the inaccuracies therein. He was a realist who believed that if he could hit upon "the one true form of the heavens," he could bring perfect accuracy to astronomy. But Copernicus quickly realized that the problems with Ptolemy's astronomy involved much more than the sizes and speeds of epicycles. To him, Ptolemy's astronomy seemed to be internally inconsistent and incoherent. In his efforts to "save the phenomena," Ptolemy had sometimes used incompatible hypotheses for the same heavenly body to account for different sets of data. In addition, his system showed a strange kind of incoherence in the orbital times of the heavenly bodies. The moon, which was closest to the earth, completed its orbit in four weeks, while the sun, which was farther away, took only one day. Jupiter, one of the outer planets, spent 300 years completing its trips around the earth, while the celestial sphere, the farthest body from the earth, completed its revolution in just one day.

As a realist, Copernicus felt that the elimination of these internal problems would lead to a greater precision and accuracy in astronomy. Placing the sun in the center of the universe and giving the earth two motions, a rotation on an axis and an orbit about the sun, he found that he could account for all the data associated with each heavenly body without resorting to inconsistent hypotheses. Further, he found that, with the sun as the central reference point, the orbital speeds of the planets varied inversely with their distance from the sun, and that the apparent retrograde motion of the outer planets was due to these variations in orbital speeds. He was even able to eliminate the awkward equant points and return astronomy to the "strict" use of uniform circular motion. Certainly Copernicus's system had an internal coherence that Ptolemy's system did not. But he still had to use a complex array of cycles and epicycles, and so his theory retained an air of artificiality.

Despite his success in bringing coherence and consistency to astronomy, Copernicus was not able to shed new light on the problems that plagued a moving earth theory. To explain why we do not experience any dynamical effects due to the motion of the earth, he suggested that the atmosphere and everything in it participates in these motions, but he did not tell us why this happened. The earth does not disintegrate as it rotates for the same reason that the celestial sphere did not disintegrate in the old view, whatever that reason was. Stellar parallax was not observed, he said, because of the immense size of the universe, ignoring the attendant problems with this view. But the biggest problem for Copernicus and his moving earth theory was this: in the end his astronomical system was not significantly simpler nor more accurate than the revised Ptolemaic view of his time. Since he had set the accuracy of prediction as the crucial test for his theory, Copernicus realized that he did not have a strong enough argument to get the approval of the scientific community.

Copernicus developed his heliocentric theory between 1510 and 1514. During that time he prepared his *Commentariolus*, or First Commentary, which contained his basic assumptions and arguments. Although Copernicus circulated the *Commentariolus* among some of his friends and students, he refrained from publishing it primarily because its basic hypothesis was at odds with Christian theology. During the next twenty years, Copernicus prepared his definitive work, *De Revolutionibus Orbium Caelestium*. In 1540, an enthusiastic disciple Georg Joachim Rheticus, published a brief technical description of Copernicus's heliocentric system called *Narratio Prima*, which set the stage for the publication of *De Revolutionibus* in 1543. When it was finally published, it was accompanied by an unsigned preface written by a friend of Copernicus named Andreas Osiander. In the preface Osiander sought to deflect any potential criticisms from Aristotelians and theologians by suggesting that the heliocentric theory was not necessarily true but was of considerable instrumental value; it provided, for example, a correct basis for calculation and was "more convenient" to use. Once published, the *De Revolutionibus* did not cause much of a stir in either religious or scientific circles.

Brahe and Kepler

Copernicus died believing that the moving earth hypothesis was correct. But he had not convinced the scientific community. It took the next hundred years and the work of four exceptional scientists to tip the argument in his favor.

Tycho Brahe was a first-rate astronomer of the late sixteenth century who continually made systematic and sophisticated observations and measurements of the heavens over long periods of time. From his observations he was able to conclude that if the earth moved in orbit about the sun, as Copernicus suggested, it would travel approximately 200 million miles in a year. Brahe could not understand why, if the earth traveled so far, that stellar parallax would not be observed. In the end he offered a hybrid theory in which the other planets (Mercury, Venus, Mars, Jupiter, and Saturn) all orbited the sun while the sun orbited the stationary earth. The celestial sphere was retained to carry the stars around the earth.

Although he could not accept the Copernican hypothesis, Brahe did make two extraordinary observations that began the process that led to the ultimate rejection of the Aristotelian-Ptolemaic universe. In 1572 he observed a bright new star that remained visible even during the day for a brief period of time. We now know that star to have been a supernova. What these phenomena showed clearly was that changes did occur in the heavens, even radical changes. The heavens were not the immutable region that Aristotle had envisioned.

In 1577 Brahe began careful calculations of the motions of comets. After three years of study he was able to prove that the comets moved well into the heavens beyond the orbits of the moon and several of the planets. This was further evidence that the hollow transparent spheres of Aristotle did not exist.

But perhaps the greatest contribution of Brahe to astronomy was the great fund of information about the heavens that he passed on to his research assistant Johannes Kepler. Unlike his teacher, Kepler was a convinced Copernican. Upon studying the planetary records within the perspective of a moving earth theory, Kepler confirmed that the farther the planets were from the sun the more slowly they moved in orbit. His conclusion was that some force or power in the sun was moving the planets and that this force or power diminished with distance. For the first time the sun was seen as a controlling factor in the planetary motions.

Kepler's originality and insight did not end there. Given the records of the motion of the planet Mars by Brahe for analysis, Kepler sought and found a single simple shape for the planet's orbit. This shape turned out to be an ellipse with the sun located at one of its focal points. If this were the actual motion of the planet Mars, then its velocity would have to vary in orbit. Kepler found that the speed of the planet Mars varied in proportion to its distance from the sun just as the speed of the different planets had varied in proportion to their distances. Kepler published these results in 1609 in a book called *Astronomia Nova*. The book contained his first two laws of planetary motions.

1. Each planet describes an elliptical orbit with the sun at one of the focal points of the ellipse.
2. The radius vector (line joining the sun and the planet) sweeps out equal areas in equal times. This is the distance-velocity law.

Ten years later Kepler added a third law of planetary motion: T^2 is proportional to $\bar{R}^3$, where T is the time it takes for the planet to complete its orbit about the sun and $\bar{R}$ is the mean distance of the planet from the sun.

With these laws Kepler had placed heliocentrism in its final form. The sun was indeed central but not in the exact center of every planetary orbit. All the planets, including the earth, moved in orbits about the sun that were elliptical rather than circular. As accurate as this picture turned out to be, however, it was still far from achieving final acceptance.

Galileo and Newton

In the end it took two additional developments to guarantee heliocentrism a permanent place in the scientific world picture. The first development was the invention of the telescope in 1610. An Italian astronomer, Galileo, became aware that such an instrument had been produced in Holland and decided to build one himself to explore the heavens. With his telescope he observed mountains and valleys on the moon suggesting that the heavens were not unlike the earth. He observed the appearance and disappearance of dark spots on the sun, confirming that the heavens can undergo significant changes like earthly objects. These observations were enough to cause the rejection of the Aristotelian cosmology of a perfect heaven once and for all.

But that was just the beginning. Galileo also observed the phases of the planet Venus, which could only be explained if Venus was in orbit about the sun. He discovered four satellites or moons of the planet Jupiter and found that they were in orbit about that planet with velocities that varied with their distance. This showed that the earth was certainly not the center of all orbits, and by analogy it supported the heliocentric argument of Copernicus. Finally, Galileo's discovery of new stars indicated that stars were independent bodies at various distances from the earth. The celestial sphere was gone forever. This meant that the universe had no spatial boundary. Theoretically the universe was now considered to be infinitely large.

Based upon his telescopic discoveries, Galileo became a convinced Copernican and so sought to argue that case to the scientific community. In 1632 he published his *Dialogue on the Two Chief World Systems*, in which he gave a thorough critique of geocentrism from which it never recovered. He then went on to enrich the Copernican argument with his telescopic observations and his own analysis.

With the work of Galileo there was no doubt that Heliocentric Theory held the dominant position in astronomy. What was lacking was a physical theory to explain its main features. That theory was invented by Isaac Newton between 1666 and 1687. In those years Newton developed the first great science, the science of mechanics, which was published in 1687 with the title *The Mathematical Principles of Natural Philosophy*. In his science Newton sought to explain the motions of all material objects in the universe. He started by distinguishing natural states, which could maintain themselves, from coerced states, which required an outside force. Using absolute space as a theoretical reference frame, Newton held that rest and uniform rectilinear motion were natural states that could maintain themselves. Changes in those states, called accelerations, required an unbalanced force.

Unlike Aristotle, who saw circular motion as a natural state for the heavens, Newton saw the somewhat circular orbital motions of the planets as involving a constant acceleration directed toward the center of the orbits, the sun. Since he rejected any distinction between terrestrial and celestial mechanics, Newton felt that the force causing objects to fall in the vicinity of the earth was of the same nature as the force causing the planets to constantly change their direction of motion. This force, which he called gravity, was mathematically formalized in his Law of Universal Gravitation. According to the law, material objects appear to exert an attractive force upon one another that varies proportionally to the product of their masses and inversely with the distance between them squared. Newton was then able to show that his law of gravity, along with his three laws of motion, could explain Kepler's three laws of planetary motion. Thus the reason why the velocity of a planet in orbit varies inversely with its distance from the sun is because the attractive force of gravity between the sun and the planet diminishes with the distance. Heliocentrism now had a physical theory to explain its features.

The Final Word

In the end it was not the discovery of stellar parallax, or the development of a satisfactory explanation for why we have no experience of the double motion of the earth, that led to the complete acceptance of heliocentrism. Rather it was the fact that heliocentrism was consistent with the great new science of mechanics, while geocentrism was not, that ended the competition once and for all. When stellar parallax was finally observed in 1838 by the German astronomer Friedrich Bessel, it was anticlimactic. By then Heliocentric Theory had become a permanent part of the scientific landscape.

1
— Aristotle —
(384–322 B.C.)

Aristotle, a Greek philosopher born in Stagira, in northern Greece, was the son of the court physician to the King of Macedon. While known primarily for his contributions to Western philosophy, Aristotle was also a brilliant naturalist and marine biologist who devoted a great deal of time and effort to the study of the natural world. Aristotle's physical works include the Physics, On the Heavens, On Coming-to-be and Passing-Away, *and the* Meteorology.

His influence on premodern natural science is immeasurable, most notably in the area of astronomy and cosmology. The universe that Aristotle depicts for us is one that is finite, hierarchically and purposefully ordered, in which every thing has its proper place. Most significantly, Aristotle maintains the natural place of the earth to be at the center, hence the geocentric universe. According to Aristotle, the earth remains stationary, while the other celestial bodies rotate around it in concentric circles. These heavenly spheres are made of the incorruptible ether.

Motion is at the heart of Aristotelian physics; according to Aristotle, all motion is caused. This applies no less to the universe as a whole, including the celestial bodies, than it does to the individual object. To avoid an infinite regress of cause of motion, Aristotle postulates the notion of an Unmoved Mover, the discussion of which is found primarily in the Physics *and the* Metaphysics. *This Unmoved*

Mover is of primary importance in the movement of the heavens in that it is the source of (inspiration for) eternal and regular motion of the celestial sphere.

In this reading Aristotle provides the physical foundations for the geocentric theory, including a discussion of the nature of motion, the circular motion of the heavenly bodies, and the natural position of the earth at rest and in the center of the universe. He also discusses the opposing theory of the Pythagoreans that the earth is in motion around a more precious fire. The reading comes from Aristotle's On the Heavens.

THE PHYSICAL FOUNDATION FOR THE GEOCENTRIC UNIVERSE

The question as to the nature of the whole, whether it is infinite in size or limited in its total mass, is a matter for subsequent inquiry. We will now speak of those parts of the whole which are specifically distinct. Let us take this as our starting-point. All natural bodies and magnitudes we hold to be, as such, capable of locomotion; for nature, we say, is their principle of movement. But all movement that is in place, all locomotion, as we term it, is either straight or circular or a combination of these two which are the only simple movements. And the reason is that these two, the straight and the circular line, are the only simple magnitudes. Now revolution about the centre is circular motion, while the upward and downward movements are in a straight line, "upward" meaning motion away from the centre, and "downward" motion towards it. All simple motion, then, must be motion either away from or towards or about the centre. This seems to be in exact accord with what we said above: as body found its completion in three dimensions, so its movement completes itself in three forms.

Bodies are either simple or compounded of such; and by simple bodies I mean those which possess a principle of movement in their own nature, such as fire and earth with their kinds, and whatever is akin to them. Necessarily, then, movements also will be either simple or in some sort compound—simple in the case of the simple bodies, compound in that of the composite—and the motion is according to the prevailing element. Supposing, then, that there is such a thing as simple movement, and that circular movement is simple, and that both movement of a simple body is simple and simple movement is of a simple body (for if it is movement of a compound it will be in virtue of a prevailing element), then there must necessarily be some simple body which moves naturally and in virtue of its own nature with a circular movement. By constraint, of course, it may be brought to move with the motion of something else different from itself, but it cannot so move naturally, since there is one sort of movement natural to each of the simple bodies. Again, if the unnatural movement is the contrary of the natural and a thing can have no more than one contrary, it will follow that circular movement, being a simple motion, must be unnatural, if it is not natural, to the body moved. If then the body whose movement is circular is fire or some other element, its natural motion must be the contrary of the circular motion. But a single thing has a single contrary; and upward and downward motion are the contraries of one another. If, on the other hand, the body moving with this circular motion which is unnatural to it is something different from the elements, there will be some other motion which is natural to it. But this cannot be. For if the natural motion is upward, it will be fire or air, and if downward, water or earth. Further, this circular motion is necessarily primary. For the complete is naturally prior to the incomplete, and the circle is a complete thing. This cannot be said of any straight line:—not of an infinite line; for then it would have a limit and an end: nor of any finite line; for in every case there is something beyond it, since any finite line can be extended. And so, since the prior movement belongs to the body which is naturally prior, and circular movement is prior to straight, and movement in a straight line belongs to simple bodies—fire moving straight upward and earthy bodies straight downward towards the centre—since this is so, it follows that circular movement also must be the movement of some simple body. For the movement of composite bodies is, as we said, determined by that simple body which prevails in the composition. From this it is clear that there is in nature some bodily substance other than the formations we know, prior to them all and more divine than they. Or again, we may take it that all movement is either natural or unnatural, and that the movement which is unnatural to one body is natural to another, as for instance is the case with the upward and downward movements, which are natural and unnatural to fire and earth respectively. It necessarily follows that circular movement, being unnatural to these bodies, is the natural movement of some other. Further, if, on the one hand, circular movement is *natural* to something, it must surely be some simple and primary body which naturally moves with a natural circular motion, as fire moves up and earth down. If, on the other hand, the movement of the rotating bodies about the centre is *unnatural*, it would be remarkable and indeed quite inconceivable that this movement alone should be continuous and eternal, given that it is unnatural. At any rate the evidence of all other cases goes to show that it is the unnatural which quickest passes away. And so, if, as some say, the body so moved is fire, this movement is just as unnatural to it as downward movement; for any one can see that fire moves in a straight line away from the centre. On all these grounds, therefore, we may infer with confidence that there is something beyond the bodies that are about us on this earth, different and separate from them; and that the superior glory of its nature is proportionate to its distance from this world of ours. . . .

Since circular motion is not the contrary of the reverse circular motion, we must consider why there is more than one motion, though we have to pursue our inquiries at a distance—a distance created not so much by our spatial position as by the fact that our senses enable us to perceive very few of the attributes of the heavenly bodies. But let not that deter us. The reason must be sought in the following facts. Everything which has a function exists for its function. The activity of God is immortality, i.e., eternal life. Therefore the movement of God must be eternal. But such is the heaven, viz. a divine body, and for that reason too it is given the circular body whose nature it is to move always in a circle. Why, then, is not the whole body of the heaven of the same character as that part? Because there must be something at rest at the centre of the revolving body; and of that body no part can be at rest, either elsewhere or at the centre. It could do so only if the body's natural movement were towards the centre. But the circular movement is natural, since otherwise it could not be eternal; for nothing unnatural is eternal. The unnatural is subsequent to the natural, being a derangement of the natural which occurs in the course of its generation. Earth then has to exist; for it is earth which is at rest at the centre. (At present we may take this for granted: it will be explained later.) But if earth must exist, so must fire. For, if one of a pair of contraries naturally exists, the other, if it is really contrary, exists also naturally, and has a nature of its own (for the matter of contraries is the same). Also, the positive is prior to its privation (warm, for instance, to cold), and rest and heaviness stand for the privation of lightness and movement. But further, if fire and earth exist, the intermediate bodies must exist also; for each element stands in a contrary relation to every other. (This, again, we will here take for granted and try later to explain.) With these four elements generation clearly is involved, since none of them can be eternal; for contraries interact with one another and destroy one another. Further, it is unreasonable that a movable body should be eternal, if its movement cannot be naturally eternal: and these bodies possess movement. Thus we see that generation is necessarily involved. But if so, there must be at least one other motion; for a single movement of the whole heaven would necessitate an identical relation of the elements of bodies to one another. This matter also will be cleared up in what follows; but for the present so much is clear, that the reason why there is more than one circular body is the necessity of generation, which follows on the presence of fire, which, with that of the other bodies, follows on that of earth; and earth is required because eternal movement in one body necessitates eternal rest in another.

The shape of the heaven is of necessity spherical; for that is the shape most appropriate to its substance and also by nature primary.

First, let us consider generally which shape is primary among planes and solids alike. Every plane figure must be either rectilinear or curvilinear. Now the rectilinear is bounded by more than one line, the curvilinear by one only. But since in any kind the one is naturally prior to the many and the simple to the complex, the circle will be the first of plane figures. Again, if by complete, as previously defined, we mean a thing outside which nothing can be found, and if addition is always possible to the straight line but never to the circular, clearly the line which embraces the circle is complete. If then the complete is prior to the incomplete, it follows on this ground also that the circle is primary among figures. And the sphere holds the same position among solids. For it alone is embraced by a single surface, while rectilinear solids have several. The sphere is among solids what the circle is among plane figures. Further, those who divide bodies into planes and generate them out of planes seem to bear witness to the truth of this. Alone among solids they leave the sphere undivided, as not possessing more than one surface; for the division into surfaces is not just dividing a whole by cutting into its parts, but division into parts different in form. It is clear, then, that the sphere is first of solid figures.

If again, one orders figures according to their numbers, it is most reasonable to arrange them in this way. The circle corresponds to the number one, the triangle, being the sum of two right angles, to the number two. But if one is assigned to the triangle, the circle will not be a figure at all.

Now the first figure belongs to the first body, and the first body is that at the farthest circumference. It follows that the body which revolves with a circular movement must be spherical. The same then will be true of the body continuous with it; for that which is continuous with the spherical is spherical. The same again holds of the bodies between these and the centre. Bodies which are bounded by the spherical and in contact with it must be, as wholes, spherical; and the lower bodies are contiguous with the sphere above them. The sphere then will be spherical throughout; for every body within it is contiguous and continuous with spheres.

Again, since the whole seems—and has been assumed—to revolve in a circle, and since it has been shown that outside the farthest circumference there is neither void nor place, from these grounds also it will follow necessarily that the heaven is spherical. For if it is to be rectilinear in shape, it will follow that there is place and body and void without it. For a rectilinear figure as it revolves never continues in the same room, but where formerly was body, is now none, and where now is none, body will be in a moment because of the changing position of the corners. Similarly, if the world had some other figure with unequal radii, if, for instance, it were lentiform, or oviform, in every case we should have to admit space and void outside the moving body, because the whole body would not always occupy the same room.

Again, if the motion of the heaven is the measure of all movements in virtue of being alone continuous and regular and eternal, and if, in each kind, the measure is the minimum, and the minimum movement is the swiftest, then the movement of the heaven must be the swiftest of all movements. Now of lines which return upon themselves the line which bounds the circle is the shortest; and that movement is the swiftest which follows the shortest line. Therefore, if the heaven moves in a circle and moves more swiftly than anything else, it must necessarily be spherical.

Corroborative evidence may be drawn from the bodies whose position is about the centre. If earth is enclosed by water, water by air, air by fire, and these similarly by the upper bodies—which while not continuous are yet contiguous with them—and if the surface of water is spherical, and that which is continuous with or embraces the spherical must itself be spherical, then on these grounds also it is clear that the heavens are spherical. But the surface of water is seen to be spherical if we take as our starting-point the fact that water naturally tends to collect in the more hollow places—and the more hollow are those nearer the centre. Draw from the centre the lines *AB, AC*, and let them be joined by the straight line *BC*. The line *AD*, drawn to the base of the triangle, will be shorter than either of the radii. Therefore the place in which it terminates will be more hollow. The water then will collect there until equality is established. But the line *AE* is equal to the radii. Thus water lies at the ends of the radii, and there will it rest; but the line which connects the extremities of the radii is circular: therefore the surface of the water *BEC* is spherical.

It is plain from the foregoing that the universe is spherical. It is plain further, that it is so accurately turned that no manufactured thing nor anything else within the range of our observation can even approach it. For the matter of which these are composed does not admit of anything like the same regularity and finish as the substance of the enveloping body; since with each step away from earth the matter manifestly becomes finer in the same proportion as water is finer than earth.

Now there are two ways of moving along a circle, from *A* to *B* or from *A* to *C,* and we have already explained that these movements are not contrary to one another. But nothing which concerns the eternal can be a matter of chance or spontaneity, and the heaven and its circular motion are eternal. We must therefore ask why this motion takes one direction and not the other. Either this is itself a principle or there is a principle behind it. It may seem evidence of excessive folly or excessive zeal to try to provide an explanation of some things, or of everything, admitting no exception. The criticism, however, is not always just: one should first consider what reason there is for speaking, and also what kind of certainty is looked for, whether human merely or of a more cogent kind. When any one shall succeed in finding proofs of greater precision, gratitude will be due to him for the discovery, but at present we must be content with what seems to be the case. If nature always follows the best course possible, and, just as upward movement is the superior form of rectilinear movement, since the upper region is more divine than the lower, so forward movement is superior to backward, then front and back exhibits, like right and left, as we said before and as the difficulty just stated itself suggests, the distinction of prior and posterior, which provides a reason and so solves our difficulty. Supposing that nature is ordered in the best way possible, this may stand as reason of the fact mentioned. For it is best to move with a movement simple and unceasing, and, further, in the superior of two possible directions.

We have next to show that the movement of the heaven is regular and not irregular. This applies to the first heaven and the first movement; for the lower spheres exhibit a composition of several movements into one. If the movement is uneven, clearly there will be acceleration, maximum speed, and retardation, since these appear in all irregular motions. The maximum may occur either at the starting-point or at the goal or between the two; and we expect natural motion to reach its maximum at the goal, unnatural motion at the starting-point, and missiles midway between the two. But circular movement, having no beginning or limit or middle without qualification, has neither whence nor whither nor middle; for in time it is eternal, and in length it returns upon itself without a break. If then its movement has no maximum, it can have no irregularity, since irregularity is produced by retardation and acceleration. Further, since everything that is moved is moved by something, the cause of the irregularity of movement must lie either in the mover or in the moved or in both. For if the mover moved not always with the same force, or if the moved were altered and did not remain the same, or if both were to change, the result might well be an irregular movement in the moved. But none of these possibilities can occur in the case of the heavens. As to that which is moved, we have shown that it is primary and simple and ungenerated and indestructible and generally unchanging; and it is far more reasonable to ascribe those attributes to the mover. It is the primary that moves the primary, the simple the simple, the indestructible and ungenerated that which is indestructible and ungenerated. Since then that which is moved, being a body, is nevertheless unchanging, how should the mover, which is incorporeal, be changed?

For if irregularity occurs, there must be change either in the movement as a whole, from fast to slow and slow to fast, or in its parts. That there is no irregularity in the parts is obvious, since, if there were, some divergence of the stars would have taken place before now in the infinity of time, as one moved slower and another faster; but no alteration of their intervals is ever observed. Nor again is a change in the movement as a whole admissible. Retardation is always due to incapacity, and incapacity is unnatural. The incapacities of animals, age, decay, and the like, are all unnatural, due, it seems, to the fact that the whole animal complex is made up of materials which differ in respect of their proper places, and no single part occupies its own place. If therefore that which is primary contains nothing unnatural, being simple and unmixed and in its proper place and having no contrary, then it has no place for incapacity, nor, consequently, for retardation or (since acceleration involves retardation) for acceleration. Again, it is unreasonable that the mover should first show incapacity for an infinite time, and capacity afterwards for another infinity. For clearly nothing which, like incapacity, is unnatural ever continues for an infinity of time; nor does the unnatural endure as long as the natural, or any form of incapacity as long as the capacity. But if the movement is retarded it must necessarily be retarded for an infinite time. Equally impossible is perpetual acceleration or perpetual retardation. For such movement would be infinite and indefinite; but every movement, in our view, proceeds from one point to another and is definite in character. Again, suppose one assumes a minimum time in less than which the heaven could not complete its movement. For, as a given walk or a given exercise on the harp cannot take any and every time, but every performance has its definite minimum time which is unsurpassable, so, one might suppose, the movement of the heaven could not be completed in any and every time. But in that case perpetual acceleration is impossible (and, equally, perpetual retardation; for the argument holds of both and each), if we may take acceleration to proceed by identical or increasing additions of speed and for an infinite time. The remaining possibility is to say that the movement exhibits an alternation of slower and faster; but this is a mere fiction and quite unreasonable.

Further, irregularity of this kind would be particularly unlikely to pass unobserved, since contrast makes observation easy.

That there is one heaven, then, only, and that it is ungenerated and eternal, and further that its movement is regular, has now been sufficiently explained.

We have next to speak of the stars, as they are called, of their composition, shape, and movements. It would be most reasonable and consequent upon what has been said that each of the stars should be composed of that substance in which their path lies, since as we said, there is an element whose natural movement is circular. In so saying we are only following the same line of thought as those who say that the stars are fiery because they believe the upper body to be fire, the presumption being that a thing is composed of the same stuff as that in which it is situated. The warmth and light which proceed from them are caused by the friction set up in the air by their motion. Movement tends to create one in wood, stone, and iron; and with even more reason should it have that effect on air, a substance which is closer to fire than these. An example is that of missiles, which as they move are themselves fired so strongly that leaden balls are melted and if they are fired the surrounding air must be similarly affected. Now while the missiles are heated by reason of their motion in air, which is turned into fire by the agitation produced by their movement, the upper bodies are carried on a moving sphere, so that, though they are not themselves fired, yet the air underneath the sphere of the revolving body is necessarily heated by its motion, and particularly in that part where the sun is attached to it. Hence warmth increases as the sun gets nearer or higher or overhead. Of the fact, then, that the stars are neither fiery nor move in fire, enough has been said.

Since changes evidently occur not only in the position of stars but also in that of the whole heaven, there are three possibilities: either both are at rest, or both are in motion, or the one is at rest and the other in motion.

That both should be at rest is impossible; for, if the earth is at rest, the hypothesis does not account for the phenomena; and we take it as granted that the earth is at rest. It remains either that both are moved, or that the one is moved and the other at rest.

On the view, first, that both are in motion, we have the absurdity that the stars and the circles move with the same speed, i.e., that the pace of every star is that of the circle in which it moves. For star and circle are seen to come back to the same place at the same moment; from which it follows that the star has reversed the circle and the circle has completed its own movement, i.e., traversed its own circumference, at one and the same moment. But it is unreasonable that the pace of each star should be exactly proportioned to the size of its circle. That the pace of each circle should be proportionate to its size is not absurd but inevitable; but that the same should be true of the movement of the stars contained in the circles is quite unreasonable. For if the star which moves on the greater circle is necessarily swifter, clearly if the stars shifted their position so as to exchange circles, the slower would become swifter and the swifter slower. But this would show that their movement was not their own, but due to the circles. If, on the other hand, the arrangement was a chance combination, the coincidence in every case of a greater circle with a swifter movement of the star contained in it is unreasonable. In one of two cases it might not inconceivably fall out so, but to imagine it in every case alike is a mere fiction. Besides, chance has no place in that which is natural, and what happens everywhere and in every case is no matter of chance.

The same absurdity is equally plain if it is supposed that the circles stand still and that it is the stars themselves which move. For it will follow that the outer stars are the swifter, and that the pace of the stars corresponds to the size of circles.

Since, then, we cannot reasonably suppose either that both are in motion or that the star alone moves, it remains that the circles should move, while the stars are at rest and move with the circles to which they are attached. Only on this supposition are we involved in no absurd consequence. For, in the first place, the quicker movement of the larger circle is reasonable when all the circles are attached to the same centre. Whenever bodies are moving with their proper motion, the larger moves quicker. It is the same here with the revolving bodies; for the arc intercepted by two radii will be larger in the larger circle, and hence it is reasonable that the revolution of the larger circle should take the same time as that of the smaller. And secondly, the fact that the heavens do not break in pieces follows not only from this but also from the proof already given of the continuity of the whole.

Again, since the stars are spherical, as our opponents assert and we may consistently admit, inasmuch as we construct them out of the spherical body, and since the spherical body has two movements proper to itself, namely rolling and spinning, it follows that if the stars have a movement of their own, it will be one of these. But neither is observed. Suppose them to *spin*. They would then stay where they were, and not change their place, as, by observation and general consent, they do. Further, it would be reasonable for them all to exhibit the same movement; but the only star which appears to possess this movement is the sun, at sunrise or sunset, and this appearance is due not to the sun itself but to the distance from which we observe it. The visual ray being excessively prolonged becomes weak and wavering. The same reason probably accounts for the apparent twinkling of the fixed stars and the absence of twinkling in the planets. The planets are near, so that the visual ray reaches them in its full vigour, but when it comes to the fixed stars it is quivering because of the distance and its excessive extension; and its tremor produces an appearance of movement in the star; for it makes no difference whether movement is set up in the ray or in the object of vision.

On the other hand, it is also clear that the stars do not *roll*. For rolling involves rotation; but the "face," as it is called, of the moon is always seen. Therefore, since any movement of their own which the stars possessed would presumably be one proper to themselves, and no such movement is observed in them, clearly they have no movement of their own.

There is, further, the absurdity that nature has bestowed upon them no organ appropriate to such movement. For nature leaves nothing to chance, and would not, while caring for animals, overlook things so precious. Indeed, nature seems deliberately to have stripped them of everything which makes self-originated progression possible, and to have removed them as far as possible from things which have organs of movement. This is just why it seems reasonable that the whole heaven and every star should be spherical. For while of all shapes the sphere is the most convenient for movement in one place, making possible, as it does, the swiftest and most self-contained motion, for forward movement it is the most unsuitable, least of all resembling shapes which are self-moved, in that it has no dependent or projecting part, as a rectilinear figure has, and is in fact as far as possible removed in shape from ambulatory bodies. Since, therefore, the heavens have to move in one place, and the stars are not required to move themselves forward, it is reasonable that both should be spherical—a shape which best suits the movement of the one and the immobility of the other. . . .

It remains to speak of the earth, of its position, of the question whether it is at rest or in motion, and of its shape.

As to its *position* there is some difference of opinion. Most people—all, in fact, who regard the whole heaven as finite—say it lies at the centre. But the Italian philosophers known as Pythagoreans take the contrary view. At the centre, they say, is fire, and the earth is one of the stars, creating night and day by its circular motion about the centre. They further construct another earth in opposition to ours to which they give the name counter-earth. In all this they are not seeking for theories and causes to account for observed facts, but rather forcing their observations and trying to accommodate them to certain theories and opinions of their own. But there are many others who would agree that it is wrong to give the earth the central position, looking for confirmation rather to theory than to the facts of observation. Their view is that the most precious place befits the most precious thing: but fire, they say, is more precious than earth, and the limit than the intermediate, and the circumference and the centre are limits. Reasoning on this basis they take the view that it is not earth that lies at the centre of the sphere, but rather fire. The Pythagoreans have a further reason. They hold that the most important part of the world, which is the centre, should be most strictly guarded, and name it, or rather the fire which occupies that place, the "Guardhouse of Zeus," as if the word "centre" were quite unequivocal, and the centre of the mathematical figure were always the same with that of the thing or the natural centre. But it is better to conceive of the case of the whole heaven as analogous to that of animals, in which the centre of the animal and that of the body are different. For this reason they have no need to be so disturbed about the world, or to call in a guard for its centre: rather let them look for the centre in the other sense and tell us what it is like and where nature has set it. That centre will be something primary and precious; but to the mere position we should give the last place rather than the first. For the middle is what is defined, and what defines it is the limit, and that which contains or limits is more precious than that which is limited, seeing that the latter is the matter and the former the essence of the system.

As to the position of the earth, then, this is the view which some advance, and the views advanced concerning its *rest or motion* are similar. For here too there is no general agreement. All who deny that the earth lies at the centre think that it revolves about the centre, and not the earth only but, as we said before, the counter-earth as well. Some of them even consider it possible that there are several bodies so moving, which are invisible to us owing to the inter-position of the earth. This, they say, accounts for the fact that eclipses of the moon are more frequent than eclipses of the sun: for in addition to the earth each of these moving bodies can obstruct it. Indeed, as in any case the surface of the earth is not actually a centre but distant from it a full hemisphere, there is no more difficulty, they think, in accounting for the observed facts on their view that we do not dwell at the centre, than on the common view that the earth is in the middle. Even as it is, there is nothing in the observations to suggest that we are removed from the centre by half the diameter of the earth. Others, again, say that the earth, which lies at the centre, is rolled, and thus in motion, about the axis of the whole heaven. So it stands written in the *Timaeus*.

There are similar disputes about the *shape* of the earth. Some think it is spher-ical, others that it is flat and drum-shaped. For evidence they bring the fact that, as the sun rises and sets, the part concealed by the earth shows a straight and not a curved edge, whereas if the earth were spherical the line of section would have to be circular. In this they leave out of account the great distance of the sun from the earth and the great size of the circumference, which, seen from a distance on these apparently small circles appears straight. Such an appearance ought not to make them doubt the circular shape of the earth. But they have another argument. They say that because it is at rest, the earth must necessarily have this shape. For there are many different ways in which the movement or rest of the earth has been conceived.

The difficulty must have occurred to every one. It would indeed be a compla-cent mind that felt no surprise that, while a little bit of earth, let loose in mid-air, moves and will not stay still, and the more there is of it the faster it moves, the whole earth, free in mid-air, should show no movement at all. Yet here is this great weight of earth, and it is at rest. And again, from beneath one of these moving fragments of earth, before it falls, take away the earth, and it will con-tinue its downward movement with nothing to stop it. The difficulty then, has naturally passed into a commonplace of philosophy; and one may well wonder that the solutions offered are not seen to involve greater absurdities than the problem itself.

By these considerations some, like Xenophanes of Colophon, have been led to assert that the earth below us is infinite, [saying that it has "pushed its roots to infinity"] in order to save the trouble of seeking for the cause. Hence the sharp rebuke of Empedocles, in the words "if the deeps of the earth are endless and endless the ample ether—such is the vain tale told by many a tongue, poured from the mouths of those who have seen but little of the whole." Others say the earth rests upon water. This, indeed, is the oldest theory that has been preserved, and is attributed to Thales of Miletus. It was supposed to stay still because it floated like wood and other similar substances, which are so consti-tuted as to rest upon water but not upon air. As if the same account had not to be given of the water which carries the earth as of the earth itself! It is not the nature of water, any more than of earth, to stay in mid-air: it must have something to rest upon. Again, as air is lighter than water, so is water than earth: how then can they think that the naturally lighter substance lies below the heavier? Again, if the earth as a whole is capable of floating upon water, that must obviously be the case with any part of it. But observation shows that this is not the case. Any piece of earth goes to the bottom, the quicker the larger it is. These thinkers seem to push their inquiries some way into the problem, but not so far as they might. It is what we are all inclined to do, to direct our inquiry not by the matter itself, but by the views of our opponents: and even when interrogating oneself one pushes the inquiry only to the point at which one can no longer offer any opposition. Hence a good inquirer will be one who is ready in bringing forward the objections proper to the genus, and that he will be when he has gained an understanding of all the differences.

Anaximenes and Anaxagoras and Democritus give the flatness of the earth as the cause of its staying still. Thus, they say, it does not cut, but covers like a lid, the air beneath it. This seems to be the way of flat-shaped bodies: for even the wind can scarcely move them because of their power of resistance. The same immobility, they say, is produced by the flatness of the surface which the earth presents to the air which underlies it; while the air, not having room enough to change its place because it is underneath the earth, stays there in a mass, like the water in the case of the water-clock. And they adduce an amount of evidence to prove that air, when cut off and at rest, can bear a considerable weight.

Now, first, if the shape of the earth is not flat, its flatness cannot be the cause of its immobility. But in their own account it is rather the size of the earth than its flatness that causes it to remain at rest. For the reason why the air is so closely confined that it cannot find a passage, and therefore stays where it is, is its great amount: and this amount is great because the body which isolates it, the earth, is very large. This result, then, will follow, even if the earth is spherical, so long as it retains its size. So far as their arguments go, the earth will still be at rest.

In general, our quarrel with those who speak of movement in this way cannot be confined to the parts; it concerns the whole universe. One must decide at the outset whether bodies have a natural movement or not, whether there is no natural but only constrained movement. Seeing, however, that we have already decided this matter to the best of our ability, we are entitled to treat our results as representing fact. Bodies, we say, which have no natural movement, have no constrained movement; and where there is no natural and no constrained movement there will be no movement at all. This is a conclusion, the necessity of which we have already decided, and we have seen further that rest also will be inconceivable, since rest, like movement, is either natural or constrained. But if there is any natural movement, constraint will not be the sole principle of motion or of rest. If, then, it is by constraint that the earth now keeps its place, the so-called "whirling" movement by which its parts came together at the centre was also constrained. (The form of causation supposed they all borrow from observations of liquids and of air, in which the larger and heavier bodies always move to the centre of the whirl. This is thought by all those who try to generate the heavens to explain why the earth came together at the centre. They then seek a reason for its staying there; and some say, in the manner explained, that the reason is its size and flatness, others, with Empedocles, that the motion of the heavens, moving about it at a higher speed, prevents movement of the earth, as the water in a cup, when the cup is given a circular motion, though it is often underneath the bronze, is for this same reason prevented from moving with the downward movement which is natural to it.) But suppose both the "whirl" and its flatness (the air beneath being withdrawn) cease to prevent the earth's motion, where will the earth move to then? Its movement to the centre was constrained, and its rest at the centre is due to constraint; but there must be some motion which is natural to it. Will this be upward motion or downward or what? It must have some motion; and if upward and downward motion are alike to it, and the air above the earth does not prevent upward movement, then no more could air below it prevent downward movement. For the same cause must necessarily have the same effect on the same thing.

Further, against Empedocles there is another point which might be made. When the elements were separated off by Hate, what caused the earth to keep its place? Surely the "whirl" cannot have been then also the cause. It is absurd too not to perceive that, while the whirling movement may have been responsible for the original coming together of the parts of earth at the centre, the question remains, why *now* do all heavy bodies move to the earth. For the whirl surely does not come near us. Why, again, does fire move upward? Not, surely, because of the whirl. But if fire is naturally such as to move in a certain direction, clearly the same may be supposed to hold of earth. Again, it cannot be the whirl which determines the heavy and the light. Rather that movement caused the pre-existent heavy and light things to go to the middle and stay on the surface respectively. Thus, before ever the whirl began, heavy and light existed; and what can have been the ground of their distinction, or the manner and direction of their natural movements? In the infinite chaos there can have been neither above nor below, and it is by these that heavy and light are determined.

It is to these causes that most writers pay attention: but there are some, Anaximander, for instance, among the ancients, who say that the earth keeps its place because of its indifference. Motion upward and downward and sideways were all, they thought, equally inappropriate to that which is set at the centre and indifferently related to every extreme point; and to move in contrary directions at the same time was impossible: so it must needs remain still. This view is ingenious but not true. The argument would prove that everything, whatever it be, which is put at the centre, must stay there. Fire, then, will rest at the centre: for the proof turns on no peculiar property of earth. But this does not follow. The observed facts about earth are not only that it remains at the centre, but also that it moves to the centre. The place to which any fragment of earth moves must necessarily be the place to which the whole moves; and in the place to which a thing naturally moves, it will naturally rest. The reason then is not in the fact that the earth is indifferently related to every extreme point: for this would apply to any body, whereas movement to the centre is peculiar to earth. Again it is absurd to look for a reason why the earth remains at the centre and not for a reason why fire remains at the extremity. If the extremity is the natural place of fire, clearly earth must also have a natural place. But suppose that the centre is not its place, and that the reason of it remaining there is this necessity of indifference—on the analogy of the hair which, it is said, however great the tension, will not break under it, if it be evenly distributed, or of the men who, though exceedingly hungry and thirsty, and both equally, yet being equidistant from food and drink, is therefore bound to stay where he is—even so, it still remains to explain why fire stays at the extremities. It is strange, too, to ask about things staying still but not about their motion,—why, I mean, one thing, if nothing stops it, moves up, and another thing to the centre. Again, their statements are not true. It happens, indeed, to be the case that a thing to which movement this way and that is equally inappropriate is obliged to remain at the centre. But so far as their argument goes, instead of remaining there, it will move, only not as a mass but in fragments. For the argument applies equally to fire. Fire, if set at the centre, should stay there, like earth, since it will be indifferently related to every point on the extremity. Nevertheless it will move, as in fact it always does move when nothing stops it, away from the centre to the extremity. It will not, however, move in a mass to a single point on the circumference—the only possible result on the lines of the indifference theory—but rather each corresponding portion of fire to the corresponding part of the extremity, each fourth part, for instance, to a fourth part of the circumference. For since no body is a point, it will have parts. The expansion, when the body increased the place occupied, would be on the same principle as the contraction, in which the place was diminished. Thus, for all the indifference theory shows to the contrary, earth also would have moved in this manner away from the centre, unless the centre had been its natural place.

We have now outlined the views held as to the shape, position, and rest or movement of the earth.

2

— Aristarchus —

(ca. 310–230 B.C.)

Aristarchus of Samos was an ancient Greek mathematician and astronomer and a contemporary of Archimedes. He was the first among the Greeks to clearly formulate a model of the universe based upon the heliocentric hypothesis, in which the earth both rotates and revolves around the sun. This is why he is often referred to as the Copernicus of antiquity. The heliocentric model of Aristarchus can be viewed as a legitimate scientific alternative to the geocentric paradigm. Unlike other nongeocentric-oriented ancient Greek cosmological speculation, it is safe to say that the work of Aristarchus was not based upon mere metaphysical prejudice, but upon an attempt to resolve purely astronomical difficulties such as retrograde motions and variations in planetary brightness. The one surviving book by Aristarchus is On the Sizes and Distances of the Sun and Moon, containing the earliest known attempt to calculate the sizes and distances of these bodies in a mathematical fashion.

In this reading, which comes from Archimedes' The Sand-Reckoner and was reprinted in Thomas Heath's Greek Astronomy, the heliocentric theory of Aristarchus is described. Interesting here is a brief discussion dealing with the size of the universe and with various measurements relating to the diameter of the sun, earth, and moon.

AN EARLY VERSION OF HELIOCENTRISM

There are some, King Gelon, who think that the number of the sand is infinite in multitude; and I mean by the sand not only that which exists about Syracuse and the rest of Sicily, but also that which is found in every region, whether inhabited or uninhabited. Again there are some who, without regarding it as infinite, yet think that no number has been named which is great enough to exceed its multitude. And it is clear that they who hold this view, if they imagined a mass made up of sand as large in size as the mass of the earth, including in it all the seas and the hollows of the earth filled up to a height equal to that of the highest mountain, would be many times further still from recognizing that any number could be expressed which exceeded the multitude of the sand so taken. But I will try to show you, by means of geometrical proofs, which you will be able to follow, that, of the numbers named by me and given in the work which I sent to Zeuxippus, some exceed, not only the number of the mass of sand equal in size to the earth filled up in the way described, but also that of a mass equal in size to the universe. Now you are aware that "universe" is the name given by most astronomers to the sphere the centre of which is the centre of the earth, and the radius of which is equal to the straight line between the centre of the sun and the centre of the earth; this you have seen in the treatises written by astronomers.

But Aristarchus of Samos brought out a book consisting of certain hypotheses, in which the premises lead to the conclusion that the universe is many times greater than that now so called. His hypotheses are that the fixed stars and the sun remain motionless, that the earth revolves about the sun in the circumference of a circle, the sun lying in the middle of the orbit, and that the sphere of the fixed stars, situated about the same centre as the sun, is so great that the circle in which he supposes the earth to revolve bears such a proportion to the distance of the fixed stars as the centre of the sphere bears to its surface.

Now it is easy to see that this is impossible; for, since the centre of the sphere has no magnitude, we cannot conceive it to bear any ratio whatever to the surface of the sphere. We must, however, take Aristarchus to mean this: since we conceive the earth to be, as it were, the centre of the universe, the ratio which the earth bears to what we describe as the "universe" is the same as the ratio which the sphere containing the circle in which he supposes the earth to revolve bears to the sphere of the fixed stars. For he adapts the proofs of the phenomena to a hypothesis of this kind, and in particular he appears to suppose the size of the sphere in which he represents the earth as moving to be equal to what we call the "universe."

I say then, that, even if a sphere were made up of sand to a size as great as Aristarchus supposes the sphere of the fixed stars to be, I shall still be able to prove that, of the numbers named in the "Principles," some exceed in multitude the number of the sand which is equal in size to the sphere referred to, provided that the following assumptions be made:

1. The perimeter of the earth is about 3,000,000 stades and not greater.

It is true that some have tried, as you are, of course, aware, to prove that the said perimeter is about 300,000 stades. But I go farther and, putting the size of the earth at ten times the size that my predecessors thought it, I suppose its perimeter to be about 3,000,000 stades and not greater.

2. The diameter of the earth is greater than the diameter of the moon, and the diameter of the sun is greater than the diameter of the earth.

In this assumption I follow most of the earlier astronomers.

3. The diameter of the sun is about 30 times the diameter of the moon and not greater.

It is true that, of the earlier astronomers, Eudoxus declared it to be about nine times as great, and Phidias, my father, twelve times, while *Aristarchus tried to prove that the diameter of the sun is greater than 18 times, but less than 20 times, the diameter of the moon.* But I go even further than Aristarchus, in order that the truth of my proposition may be established beyond dispute, and I suppose the diameter of the sun to be about 30 times that of the moon and not greater.

4. The diameter of the sun is greater than the side of the chiliagon (a regular polygon of 1000 sides) inscribed in the greatest circle in the sphere of the universe. I make this assumption because Aristarchus discovered that the sun appeared to be about 1/720th part of the circle of the zodiac, and I myself tried, by a method which I will now describe, to find experimentally (by means of a mechanical contrivance), the angle subtended by the sun and having its vertex at the eye.

3
— Claudius Ptolemy —
(A.D. 100–172)

Claudius Ptolemy was a Greek astronomer who spent a good part of his life in the Egyptian city of Alexandria. He recovered much of the data collected by ancient Greek astronomers, using it to develop a mathematical astronomy of his own. Key to Ptolemaic astronomy was a series of mathematical hypotheses concerning the orbits of the sun, moon, planets, and stars that enabled him to satisfactorily account for a sizable number of celestial phenomena. Ptolemy employed these hypotheses to substantiate a geocentric view of the universe. To Ptolemy we owe the introduction of the equant into the repertoire of descriptive astronomy, a device that would later be rejected by Copernicus. Ptolemy authored a number of important scientific works, including Optics, *which dealt with atmospheric refraction, as well as a work entitled* Planetary Hypotheses. *His most important and enduring work, which transmitted the geocentric theory in its most advanced mathematical form to the Middle Ages, was the* Almagest. *This title was derived through a series of mistranslations: from the original Greek to Arabic (about A.D. 800), and to Latin in the thirteenth century. Our chief manuscript of the* Almagest *bears the title* Great Composition, *although elsewhere it is also referred to as* Mathematical Composition.

In this reading, taken from the Almagest, *Ptolemy puts forward his argument in support of geocentrism. Despite the mathematically rich treatment of geocentric theory provided by Ptolemy throughout the* Almagest, *the reader should nevertheless note the extent to which Ptolemy relies heavily upon Aristotelian style arguments and commitments.*

―――――――――

THE CASE FOR GEOCENTRISM

Preface

The true philosophers, Syrus, were, I think, quite right to distinguish the theoretical part of philosophy from the practical. For even if practical philosophy, before it is practical, turns out to be theoretical, nevertheless one can see that there is a great difference between the two: in the first place, it is possible for many people to possess some of the moral virtues even without being taught, whereas it is impossible to achieve theoretical understanding of the universe without instruction; furthermore, one derives most benefit in the first case [practical philosophy] from continuous practice in actual affairs, but in the other [theoretical philosophy] from making progress in the theory. Hence we thought it fitting to guide our actions (under the impulse of our actual ideas [of what is to be done]) in such a way as never to forget, even in ordinary affairs, to strive for a noble and disciplined disposition, but to devote most of our time to intellectual matters, in order to teach theories, which are so many and beautiful, and especially those to which the epithet "mathematical" is particularly applied. For Aristotle divides theoretical philosophy too, very fittingly, into three primary categories, physics, mathematics, and theology. For everything that exists is composed of matter, form, and motion; none of these [three] can be observed in its substratum by itself, without the others: they can only be imagined. Now the first cause of the first motion of the universe, if one considers it simply, can be thought of as an invisible and motionless deity; the division [of theoretical philosophy] concerned with investigating this [can be called] "theology," since this kind of activity, somewhere up in the highest reaches of the universe, can only be imagined, and is completely separated from perceptible reality. The division [of theoretical philosophy] which investigates material and ever-moving nature, and which concerns itself with "white," "hot," "sweet," "soft," and suchlike qualities one may call "physics"; such an order of being is situated (for the most part) amongst corruptible bodies and below the lunar sphere. That division [of theoretical philosophy] which determines the nature involved in forms and motion from place to place, and which serves to investigate shape, number, size, and place, time, and suchlike, one may define as "mathematics." Its subject-matter falls as it were in the middle between the other two, since, firstly, it can be conceived of both with and without the aid of the senses, and, secondly, it is an attribute of all existing things without exception, both mortal and immortal: for those things which are perpetually changing in their inseparable form, it changes with them, while for eternal things which have an aethereal nature, it keeps their unchanging form unchanged.

From all this we concluded: that the first two divisions of theoretical philosophy should rather be called guesswork than knowledge, theology because of its completely invisible and ungraspable nature, physics because of the unstable and unclear nature of matter; hence there is no hope that philosophers will ever be agreed about them; and that only mathematics can provide sure and unshakeable knowledge to its devotees, provided one approaches it rigorously. For its kind of proof proceeds by indisputable methods, namely arithmetic and geometry. Hence we were drawn to the investigation of that part of theoretical philosophy, as far as we were able to the whole of it, but especially to the theory concerning divine and heavenly things. For that alone is devoted to the investigation of the eternally unchanging. For that reason it too can be eternal and unchanging (which is a proper attribute of knowledge) in its own domain, which is neither unclear nor disorderly. Furthermore it can work in the domains of the other [two divisions of theoretical philosophy] no less than they do. For this is the best science to help theology along its way, since it is the only one which can make a good guess at [the nature of] that activity which is unmoved and separated; [it can do this because] it is familiar with the attributes of those beings which are on the one hand perceptible, moving and being moved, but on the other hand eternal and unchanging, [I mean the attributes] having to do with motions and the arrangements of motions. As for physics, mathematics can make a significant contribution. For almost every peculiar attribute of material nature becomes apparent from the peculiarities of its motion from place to place. [Thus one can distinguish] the corruptible from the incorruptible by [whether it undergoes] motion in a straight line or in a circle, and heavy from light, and passive from active, by [whether it moves] towards the centre or away from the centre. With regard to virtuous conduct in practical actions and character, this science, above all things, could make men see clearly; from the constancy, order, symmetry, and calm which are associated with the divine, it makes its followers lovers of this divine beauty, accustoming them and reforming their natures, as it were, to a similar spiritual state.

It is this love of the contemplation of the eternal and unchanging which we constantly strive to increase, by studying those parts of these sciences which have already been mastered by those who approached them in a genuine spirit of enquiry, and by ourselves attempting to contribute as much advancement as has been made possible by the additional time between those people and ourselves. We shall try to note down everything which we think we have discovered up to the present time; we shall do this as concisely as possible and in a manner which can be followed by those who have already made some progress in the field. For the sake of completeness in our treatment we shall set out everything useful for the theory of the heavens in the proper order, but to avoid undue length we shall merely recount what has been adequately established by the ancients. However, those topics which have not been dealt with [by our predecessors] at all, or not as usefully as they might have been, will be discussed at length, to the best of our ability.

On the Order of the Theorems

In the treatise which we propose, then, the first order of business is to grasp the relationship of the earth taken as a whole to the heavens taken as a whole. In the treatment of the individual aspects which follows, we must first discuss the position of the ecliptic and the regions of our part of the inhabited world and also the features differentiating each from the others due to the [varying] latitude at each horizon taken in order. For if the theory of these matters is treated first it will make examination of the rest easier. Secondly, we have to go through the motion of the sun and of the moon, and the phenomena accompanying these [motions]; for it would be impossible to examine the theory of the stars thoroughly without first having a grasp of these matters. Our final task in this way of approach is the theory of the stars. Here too it would be appropriate to deal first with the sphere of the so-called "fixed stars," and follow that by treating the five "planets," as they are called. We shall try to provide proofs in all of these topics by using as starting-points and foundations, as it were, for our search the obvious phenomena, and those observations made by the ancients and in our own times which are reliable. We shall attach the subsequent structure of ideas to this [foundation] by means of proofs using geometrical methods.

The general preliminary discussion covers the following topics: the heaven is spherical in shape, and moves as a sphere; the earth too is sensibly spherical in shape, when taken as a whole; in position it lies in the middle of the heavens very much like its centre; in size and distance it has the ratio of a point to the sphere of the fixed stars; and it has no motion from place to place. We shall briefly discuss each of these points for the sake of reminder.

That the Heavens Move Like a Sphere

It is plausible to suppose that the ancients got their first notions on these topics from the following kind of observations. They saw that the sun, moon, and other stars were carried from east to west along circles which were always parallel to each other, that they began to rise up from below the earth itself, as it were, gradually got up high, then kept on going round in similar fashion and getting lower, until, falling to earth, so to speak, they vanished completely, then, after remaining invisible for some time, again rose afresh and set; and [they saw] that the periods of these [motions], and also the places of rising and setting, were, on the whole, fixed and the same.

What chiefly led them to the concept of a sphere was the revolution of the ever-visible stars, which was observed to be circular, and always taking place about one centre, the same [for all]. For by necessity that point became [for them] the pole of the heavenly sphere: those stars which were closer to it re-volved on smaller circles, those that were farther away described circles ever greater in proportion to their distance, until one reaches the distance of the stars which become invisible. In the case of these, too, they saw that those near the ever-visible stars remained invisible for a short time, while those farther away remained invisible for a long time, again in proportion [to their distance]. The result was that in the beginning they got to the aforementioned notion solely from such considerations; but from then on, in their subsequent investigation, they found that everything else accorded with it, since absolutely all phenomena are in contradiction to the alternative notions which have been propounded.

For if one were to suppose that the stars' motion takes place in a straight line towards infinity, as some people have thought, what device could one conceive of which would cause each of them to appear to begin their motion from the same starting-point every day? How could the stars turn back if their motion is toward infinity? Or, if they did turn back, how could this not be obvious? [On such a hypothesis], they must gradually diminish in size until they disappear, whereas, on the contrary, they are seen to be greater at the very moment of their disappearance, at which time they are gradually obstructed and cut off, as it were, by the earth's surface.

But to suppose that they are kindled as they rise out of the earth and are extinguished again as they fall to earth is a completely absurd hypothesis. For even if we were to concede that the strict order in their size and number, their intervals, positions, and periods could be restored by such a random and chance process; that one whole area of the earth has a kindling nature, and another an extinguishing one, or rather that the same part [of the earth] kindles for one set of observers and extinguishes for another set; and that the same stars are already kindled or extinguished for some observers while they are not yet for others: even if, I say, we were to concede all these ridiculous consequences, what could we say about the ever-visible stars, which neither rise nor set? Those stars which are kindled and extinguished ought to rise and set for observers everywhere, while those which are not kindled and extinguished ought always to be visible for observers everywhere. What cause could we assign for the fact that this is not so? We will surely not say that stars which are kindled and extinguished for some observers never undergo this process for other observers. Yet it is utterly obvious that the same stars rise and set in certain regions [of the earth] and do neither at others.

To sum up, if one assumes any motion whatever, except spherical, for the heavenly bodies, it necessarily follows that their distances, measured from the earth upwards, must vary, wherever and however one supposes the earth itself to be situated. Hence the sizes and the mutual distances of the stars must appear to vary for the same observers during the course of each revolution, since at one time they must be at a greater distance, at another at a lesser. Yet we see that no such variation occurs. For the apparent increase in their sizes at the horizons is caused, not by a decrease in their distances, but by the exhalations of moisture surrounding the earth being interposed between the place from which we observe and the heavenly bodies, just as objects placed in water appear bigger than they are, and the lower they sink, the bigger they appear.

The following considerations also lead us to the concept of the sphericity of the heavens. No other hypothesis but this can explain how sundial constructions produce correct results; furthermore, the motion of the heavenly bodies is the most unhampered and free of all motions, and freest motion belongs among plane figures to the circle and among solid shapes to the sphere; similarly, since of different shapes having an equal boundary those with more angles are greater [in area or volume], the circle is greater than [all other] surfaces, and the sphere greater than [all other] solids; [likewise] the heavens are greater than all other bodies.

Furthermore, one can reach this kind of notion from certain physical considerations. E.g., the aether is, of all bodies, the one with constituent parts which are finest and most like each other; now bodies with parts like each other have surfaces with parts like each other; but the only surfaces with parts like each other are the circular, among planes, and the spherical, among three-dimensional surfaces. And since the aether is not plane, but three-dimensional, it follows that it is spherical in shape. Similarly, nature formed all earthly and corruptible bodies out of shapes which are round but of unlike parts, but all aethereal and divine bodies out of shapes which are of like parts and spherical. For if they were flat or shaped like a discus they would not always display a circular shape to all those observing them simultaneously from different places on earth. For this reason it is plausible that the aether surrounding them, too, being of the same nature, is spherical, and because of the likeness of its parts moves in a circular and uniform fashion.

That the Earth Too, Taken as a Whole, is Sensibly Spherical

That the earth, too, taken as a whole, is sensibly spherical can best be grasped from the following considerations. We can see, again, that the sun, moon, and other stars do not rise and set simultaneously for everyone on earth, but do so earlier for those more towards the east, later for those towards the west. For we find that the phenomena at eclipses, especially lunar eclipses, which take place at the same time [for all observers], are nevertheless not recorded as occurring at the same hour (that is at an equal distance from noon) by all observers. Rather, the hour recorded by the more easterly observers is always later than that recorded by the more westerly. We find that the differences in the hour are proportional to the distances between the places [of observation]. Hence one can reasonably conclude that the earth's surface is spherical, because its evenly curving surface (for so it is when considered as a whole) cuts off [the heavenly bodies] for each set of observers in turn in a regular fashion.

If the earth's shape were any other, this would not happen, as one can see from the following arguments. If it were concave, the stars would be seen rising first by those more towards the west; if it were plane, they would rise and set simultaneously for everyone on earth; if it were triangular or square or any other polygonal shape, by a similar argument, they would rise and set simultaneously for all those living on the same plane surface. Yet it is apparent that nothing like this takes place. Nor could it be cylindrical, with the curved surface in the east-west direction, and the flat sides towards the poles of the universe, which some might suppose more plausible. This is clear from the following: for those living on the curved surface none of the stars would be ever-visible, but either all stars would rise and set for all observers, or the same stars, for an equal [celestial] distance from each of the poles, would always be invisible for all observers. In fact, the further we travel toward the north, the more of the southern stars disappear and the more of the northern stars appear. Hence it is clear that here too the curvature of the earth cuts off [the heavenly bodies] in a regular fashion in a north-south direction, and proves the sphericity [of the earth] in all directions.

There is the further consideration that if we sail towards mountains or elevated places from and to any direction whatever, they are observed to increase gradually in size as if rising up from the sea itself in which they had previously been submerged: this is due to the curvature of the surface of the water.

That the Earth Is in the Middle of the Heavens

Once one has grasped this, if one next considers the position of the earth, one will find that the phenomena associated with it could take place only if we assume that it is in the middle of the heavens, like the centre of a sphere. For if this were not the case, the earth would have to be either

[a] not on the axis [of the universe] but equidistant from both poles, or
[b] on the axis but removed towards one of the poles, or
[c] neither on the axis nor equidistant form both poles.

Against the first of these three positions militate the following arguments. If we imagined [the earth] removed towards the zenith or the nadir of some observer, then, if he were at *sphaera recta*, he would never experience equinox, since the horizon would always divide the heavens into two unequal parts, one above and one below the earth; if he were at *sphaera obliqua*, either, again, equinox would never occur at all, or, [if it did occur,] it would not be at a position halfway between summer and winter solstices, since these intervals would necessarily be unequal, because the equator, which is the greatest of all parallel circles drawn about the poles of the [daily] motion would no longer be bisected by the horizon; instead [the horizon would bisect] one of the circles parallel to the equator, either to the north or to the south of it. Yet absolutely everyone agrees that these intervals are equal everywhere on earth, since [everywhere] the increment of the longest day over the equinoctial day at the summer solstice is equal to the decrement of the shortest day from the equinoctial day at the winter solstice. But if, on the other hand, we imagined the displacement to be towards the east or west of some observer, he would find that the sizes and distances of the stars would not remain constant and unchanged at eastern and western horizons, and that the time-interval from rising to culmination would not be equal to the interval from culmination to setting. This is obviously completely in disaccord with the phenomena.

Against the second position, in which the earth is imagined to lie on the axis removed towards one of the poles, one can make the following objections. If this were so, the plane of the horizon would divide the heavens into a part above the earth and a part below the earth which are unequal and always different for different latitudes, whether one considers the relationship of the same part at two different latitudes or the two parts at the same latitude. Only at *sphaera recta* could the horizon bisect the sphere; at a *sphaera obliqua* situation such that the nearer pole were the ever-visible one, the horizon would always make the part above the earth lesser and the part below the earth greater; hence another phenomenon would be that the great circle of the ecliptic would be divided into unequal parts by the plane of the horizon. Yet it is apparent that this is by no means so. Instead, six zodiacal signs are visible above the earth at all times and places, while the remaining six are invisible; then again [at a later time] the latter are visible in their entirety above the earth, while at the same time the others are not visible. Hence it is obvious that the horizon bisects the zodiac, since the same semi-circles are cut off by it, so as to appear at one time completely above the earth, and at another [completely] below it.

And in general, if the earth were not situated exactly below the [celestial] equator, but were removed towards the north or south in the direction of one of the poles, the result would be that at the equinoxes the shadow of the gnomon at sunrise would no longer form a straight line with its shadow at sunset in a plane parallel to the horizon, not even sensibly. Yet this is a phenomenon which is plainly observed everywhere.

It is immediately clear that the third position enumerated is likewise impossible, since the sorts of objection which we made to the first [two] will both arise in that case.

To sum up, if the earth did not lie in the middle [of the universe], the whole order of things which we observe in the increase and decrease of the length of daylight would be fundamentally upset. Furthermore, eclipses of the moon would not be restricted to situations where the moon is diametrically opposite the sun (whatever part of the heaven [the luminaries are in]), since the earth would often come between them when they were not diametrically opposite, but at intervals of less than a semi-circle.

That the Earth Has the Ratio of a Point to the Heavens

Moreover, the earth has, to the senses, the ratio of a point to the distance of the sphere of the so-called fixed stars. A strong indication of this is the fact that the sizes and distances of the stars, at any given time, appear equal and the same from all parts of the earth everywhere, as observations of the same [celestial] objects from different latitudes are found to have not the least discrepancy from each other. One must also consider the fact that gnomons set up in any part of the earth whatever, and likewise the centres of armillary spheres, operate like the real centre of the earth; that is, the lines of sight [to heavenly bodies] and the paths of shadows caused by them agree as closely with the [mathematical] hypotheses explaining the phenomena as if they actually passed through the real centre-point of the earth.

Another clear indication that this is so is that the planes drawn through the observer's lines of sight at any point [on earth], which we call "horizons," always bisect the whole heavenly sphere. This would not happen if the earth were of perceptible size in relation to the distance of the heavenly bodies; in that case only the plane drawn through the centre of the earth could bisect the sphere, while a plane through any point on the surface of the earth would always make the section [of the heavens] below the earth greater than the section above it.

That the Earth Does Not Have any Motion
from Place to Place, Either

One can show by the same arguments as the preceding that the earth cannot have any motion in the aforementioned directions, or indeed ever move at all from its position at the centre. For the same phenomena would result as would if it had any position other than the central one. Hence I think it is idle to seek for causes for the motion of objects towards the centre, once it has been so clearly established from the actual phenomena that the earth occupies the middle place in the universe, and that all heavy objects are carried towards the earth. The following fact alone would most readily lead one to this notion [that all objects fall towards the centre]. In absolutely all parts of the earth, which, as we said, has been shown to be spherical and in the middle of the universe, the direction and path of the motion (I mean the proper, [natural] motion) of all bodies possessing weight is always and everywhere at right angles to the rigid plane drawn tangent to the point of impact. It is clear from this fact that, if [these falling objects] were not arrested by the surface of the earth, they would certainly reach the centre of the earth itself, since the straight line to the centre is also always at right angles to the plane tangent to the sphere at the point of intersection [of that radius] and the tangent.

Those who think it paradoxical that the earth, having such a great weight, is not supported by anything and yet does not move, seem to me to be making the mistake of judging on the basis of their own experience instead of taking into account the peculiar nature of the universe. They would not, I think, consider such a thing strange once they realised that this great bulk of the earth, when compared with the whole surrounding mass [of the universe], has the ratio of a point to it. For when one looks at it in that way, it will seem quite possible that that which is relatively smallest should be overpowered and pressed in equally from all directions to a position of equilibrium by that which is the greatest of all and of uniform nature. For there is no up and down in the universe with respect to itself, any more than one could imagine such a thing in a sphere: instead the proper and natural motion of the compound bodies in it is as follows: light and rarefied bodies drift outwards towards the circumference, but seem to move in the direction which is "up" for each observer, since the overhead direction for all of us, which is also called "up," points towards the surrounding surface; heavy and dense bodies, on the other hand, are carried towards the middle and the centre, but seem to fall downwards, because, again, the direction which is for all us towards our feet, called "down," also points towards the centre of the earth. These heavy bodies, as one would expect, settle about the centre because of their mutual pressure and resistance, which is equal and uniform from all directions. Hence, too, one can see that it is plausible that the earth, since its total mass is so great compared with the bodies which fall towards it, can remain motionless under the impact of these very small weights (for they strike it from all sides), and receive, as it were, the objects falling on it. If the earth had a single motion in common with other heavy objects, it is obvious that it would be carried down faster than all of them because of its much greater size: living things and individual heavy objects would be left behind, riding on the air, and the earth itself would very soon have fallen completely out of the heavens. But such things are utterly ridiculous merely to think of.

But certain people, [propounding] what they consider a more persuasive view, agree with the above, since they have no argument to bring against it, but think that there could be no evidence to oppose their view if, for instance, they supposed the heavens to remain motionless, and the earth to revolve from west to east about the same axis [as the heavens], making approximately one revolution each day; or if they make both heaven and earth move by any amount whatever, provided, as we said, it is about the same axis, and in such a way as to preserve the overtaking of one by the other. However, they do not realise that, although there is perhaps nothing in the celestial phenomena which would count against that hypothesis, at least from simpler considerations, nevertheless from what would occur here on earth and in the air, one can see that such a notion is quite ridiculous. Let us concede to them [for the sake of argument] that such an unnatural thing could happen as that the most rare and light of matter should either not move at all or should move in a way no different from that of matter with the opposite nature (although things in the air, which are less rare [than the heavens] so obviously move with a more rapid motion than any earthy object); [let us concede that] the densest and heaviest objects have a proper motion of the quick and uniform kind which they suppose (although, again, as all agree, earthy objects are sometimes not readily moved even by an external force). Nevertheless, they would have to admit that the revolving motion of the earth must be the most violent of all motions associated with it, seeing that it makes one revolution in such a short time; the result would be that all objects not actually standing on the earth would appear to have the same motion, opposite to that of the earth: neither clouds nor other flying or thrown objects would ever be seen moving towards the east, since the earth's motion towards the east would always outrun and overtake them, so that all other objects would seem to move in the direction of the west and the rear. But if they said that the air is carried around in the same direction and with the same speed as the earth, the compound objects in the air would nonetheless always seem to be left behind by the motion of both [earth and air]; or if those objects too were carried around, fused, as it were, to the air, then they would never appear to have any motion either in advance or rearwards: they would always appear still, neither wandering about nor changing position, whether they were flying or thrown objects. Yet we quite plainly see that they do undergo all these kinds of motion, in such a way that they are not even slowed down or speeded up at all by any motion of the earth.

4

— Nicholaus Copernicus —
(1473–1543)

Nicholaus Copernicus was a Polish astronomer who was the first to offer a complete heliocentric theory in the modern era. Although he did not make many observations of his own, Copernicus used the existing astronomical literature to argue that a heliocentric theory offered a more coherent cosmology than geocen-

trism. Copernican astronomy was not more accurate than the revised geocentrism then in vogue. A lingering problem for Copernicus and his heliocentric model was the continuing failure to detect stellar parallax. Copernicus persisted in his refusal to publish his magnum opus *until finally convinced to do so at the very end of his life. His primary motivations for withholding* De Revolutionibus *from publication can be attributed to theological misgivings.* On the Revolutions of the Heavenly Spheres *was finally published in 1543 under the supervision of Andrew Osiander, a Lutheran theologian who added an unsigned preface* ad Lectorem *(to the Reader).*

The following reading is from The Commentariolus, *a work that scholars believe to have been written sometime between 1510 and 1514. It was circulated by Copernicus to a limited number of people within the scientific community of his time. Unlike the longer and highly technical presentation of his argument in defense of heliocentrism in* De Revolutionibus, *which represents the more mature and final statement of his heliocentric model, the* Commentariolus *is written in a nontechnical style, hence making it readily accessible to the average reader. It was not published during the lifetime of Copernicus and was not rediscovered until 1878. In what follows Copernicus presents his revolutionary astronomical hypothesis for the first time.*

After a brief critique of the theories of Eudoxus and Ptolemy, Copernicus offers seven postulates for his new astronomy. Among them:

1. *The sun is the center of all things;*
2. *The distance to the stars is immense;*
3. *The apparent stellar, solar, and planetary motions are due to the actual motions of the earth.*

Copernicus argues strongly against the immobility of the earth. He concludes with an analysis of planetary motion in a sun-centered universe touching upon the issue of the retrograde motion of the outer planets, Mars, Jupiter, and Saturn, as well as the motions of Venus and Mercury.

FIRST THOUGHTS ON HELIOCENTRISM

Our predecessors assumed, I observe, a large number of celestial spheres mainly for the purpose of explaining the planets' apparent motion by the principle of uniformity. For they thought it altogether absurd that a heavenly body, which is perfectly spherical, should not always move uniformly. By connecting and combining uniform motions in various ways, they had seen, they could make any body appear to move to any position.

Callippus and Eudoxus, who tried to achieve this result by means of concentric circles, could not thereby account for all the planetary movements, not merely the apparent revolutions of those bodies but also their ascent, as it seems to us, at some times and descent at others, [a pattern] entirely incompatible with [the principle of] concentricity. Therefore for this purpose it seemed better to employ eccentrics and epicycles, [a system] which most scholars finally accepted.

Yet the widespread [planetary theories], advanced by Ptolemy and most other [astronomers], although consistent with the numerical [data], seemed likewise to present no small difficulty. For these theories were not adequate unless they also conceived certain equalizing circles, which made the planet appear to move at all times with uniform velocity neither on its deferent sphere nor about its own [epicycle's] center. Hence this sort of notion seemed neither sufficiently absolute nor sufficiently pleasing to the mind.

Therefore, having become aware of these [defects], I often considered whether there could perhaps be found a more reasonable arrangement of circles, from which every apparent irregularity would be derived while everything in itself would move uniformly, as is required by the rule of perfect motion. After I had attacked this very difficult and almost insoluble problem, the suggestion at length came to me how it could be solved with fewer and much more suitable constructions than were formerly put forward, if some postulates (which are called axioms) were granted me. They follow in this order.

POSTULATES

1. There is no one center of all the celestial orbs or spheres.
2. The center of the earth is the center, not of the universe, but only of gravity and of the lunar sphere.
3. All the spheres encircle the sun, which is as it were in the middle of them all, so that the center of the universe is near the sun.
4. The ratio of the earth's distance from the sun to the height of the firmament is so much smaller than the ratio of the earth's radius to its distance from the sun that the distance between the earth and the sun is imperceptible in comparison with the loftiness of the firmament.
5. Whatever motion appears in the firmament is due, not to it, but to the earth. Accordingly, the earth together with the circumjacent elements performs a complete rotation on its fixed poles in a daily motion, while the firmament and highest heaven abide unchanged.
6. What appear to us as motions of the sun are due, not to its motion, but to the motion of the earth and our sphere, with which we revolve about the sun as [we would with] any other planet. The earth has, then, more than one motion.
7. What appears in the planets as [the alternation of] retrograde and direct motion is due, not to their motion, but to the earth's. The motion of the earth alone, therefore, suffices [to explain] so many apparent irregularities in the heaven.

Having thus propounded the foregoing postulates, I shall endeavor briefly to show to what extent the uniformity of the motions can be saved in a systematic way. Here, however, the mathematical demonstrations intended for my larger work should be omitted for brevity's sake, in my judgment. Nevertheless, in the explanation of the circles themselves, I shall set down here the lengths of the spheres' radii. From these anybody familiar with mathematics will readily perceive how excellently this arrangement of circles agrees with the numerical data and observations.

Accordingly, lest anybody suppose that, with the Pythagoreans, I have asserted the earth's motion gratuitously, he will find strong evidence here too in my exposition of the circles. For, the principle arguments by which the natural philosophers attempt to establish the immobility of the earth rest for the most part on appearances. All these arguments are the first to collapse here, since I undermine the earth's immobility as likewise due to an appearance.

The Order of the Spheres

The celestial spheres embrace one another in the following order. The highest is the immovable sphere of the fixed stars, which contains and gives position to all things. Beneath it is Saturn's, which Jupiter's follows, then Mars'. Below Mars' is the sphere on which we revolve; then Venus'; last is Mercury's. The lunar sphere, however, revolves around the center of the earth and moves with it like an epicycle. In the same order also, one sphere surpasses another in speed of revolution, according as they measure out greater or smaller expanses of circles. Thus Saturn's period ends in the thirtieth year, Jupiter's in the twelfth, Mars' in the third, and the earth's with the annual revolution; Venus completes its revolution in the ninth month, Mercury in the third.

The Apparent Motions of the Sun

The earth has three motions. First, it revolves annually in a Grand Orb about the sun in the order of the signs, always describing equal arcs in equal times. From the Grand Orb's center to the sun's center the distance is $\frac{1}{25}$ of the Grand Orb's radius. This Orb's radius is assumed to have a length imperceptible in comparison with the height of the firmament. Consequently the sun appears to revolve with this motion, as if the earth lay in the center of the universe. This [appearance], however, is caused not by the sun's motion but rather by the earth's. Thus, for example, when the earth is in the Goat, the sun is seen diametrically opposite in the Crab, and so on. Moreover, on account of the aforementioned distance of the sun from the Orb's center, this apparent motion of the sun will be nonuniform, the maximum inequality being $2\frac{1}{6}°$. The sun's direction with reference to the Orb's center is invariably toward a point of the firmament about 10° west of the more brilliant bright star in the head of the Twins. Therefore, when the earth is opposite this point, with the Orb's center lying between them, the sun is then seen at its greatest distance [from the earth].

By this Orb not only is the earth revolved, but also whatever else is associated with the lunar sphere.

The earth's second motion is the daily rotation. This is in the highest degree peculiar to the earth, which it turns on its poles in the order of the signs, that is, eastward. On account of this rotation the entire universe appears to revolve with enormous speed. Thus does the earth rotate together with its circumjacent waters and nearby air.

The third is the motion in declination. For, the axis of the daily rotation is not parallel to the Grand Orb's axis, but is inclined [to it at an angle that intercepts] a portion of a circumference, in our time about 23½ °. Therefore, while the earth's center always remains in the plane of the ecliptic, that is, in the circumference of a circle of the Grand Orb, the earth's poles rotate, both of them describing small circles about centers [lying on a line that moves] parallel to the Grand Orb's axis. The period of this motion also is a year, but not quite, being nearly equal to the Grand Orb's [revolution]. The Grand Orb's axis, however, being invariant with regard to the firmament, is directed toward what are called the poles of the ecliptic. The poles of the daily rotation would always be fixed in like manner at the same points of the heavens by the motion in declination combined with the Orb's motion, if their periods were exactly equal. Now with the long passage of time it has become clear that this alignment of the earth changes with regard to the configuration of the firmament. Hence it is the common opinion that the firmament itself has several motions. But even though the principle involved is not yet sufficiently understood, it is less surprising that all these phenomena can occur on account of the earth's motion. I am not prepared to state to what its poles are attached. I am of course aware that in more mundane matters a magnetized iron needle always points toward a single spot in the universe. It has nevertheless seemed a better view to ascribe the phenomena to a sphere, whose turning governs the movements of the poles. This sphere must doubtless be sublunar.

5
— Nicholaus Copernicus —

In this reading Copernicus presents his mature argument in favor of Heliocentric Theory. It comes from his major work On the Revolutions of the Heavenly Spheres. *The introduction by Andrew Osiander, added without the knowledge or permission of Copernicus, is also included.*

Osiander enjoins the reader to view the Copernican theory as hypothetical but with considerable instrumental value. In the process, he offers a classic definition of instrumentalism.

Copernicus begins by arguing that the circle and the sphere are the geometrical keys to astronomy. He goes on to explain the arguments of the ancients as to why the earth was immobile, then criticizes them as inadequate. Since, as he says, nothing hinders the mobility of the earth, he gives it rotational and translational motions and places the sun at rest in the center of the world.

THE HELIOCENTRIC THEORY

Introduction to the Reader concerning The Hypotheses of This Work

Since the newness of the hypotheses of this work—which sets the earth in motion and puts an immovable sun at the centre of the universe—has already received a great deal of publicity, I have no doubt that certain of the savants have taken grave offense and think it wrong to raise any disturbance among liberal disciplines which have had the right set-up for a long time now. If, however, they are willing to weigh the matter scrupulously, they will find that the author of this work has done nothing which merits blame. For it is the job of the astronomer to use painstaking and skilled observation in gathering together the history of the celestial movements, and then—since he cannot by any line of reasoning reach the true causes of these movements—to think up or construct whatever causes or hypotheses he pleases such that, by the assumption of these causes, those same movements can be calculated from the principles of geometry for the past and for the future too. This artist is markedly outstanding in both of these respects: for it is not necessary that these hypotheses should be true, or even probably; but it is enough if they provide a calculus which fits the observations—unless by some chance there is anyone so ignorant of geometry and optics as to hold the epicycle of Venus as probable and to believe this to be a cause why Venus alternately precedes and follows the sun at an angular distance of up to 40° or more. For who does not see that it necessarily follows from this assumption that the diameter of the planet in its perigee should appear more than four times greater, and the body of the planet more than sixteen times greater, than in its apogee? Nevertheless the experience of all the ages is opposed to that. There are also other things in this discipline which are just as absurd, but it is not necessary to examine them right now. For it is sufficiently clear that this art is absolutely and profoundly ignorant of the causes of the apparent irregular movements. And if it constructs and thinks up causes—and it has certainly thought up a good many—nevertheless it does not think them up in order to persuade anyone of their truth but only in order that they may provide a correct basis for calculation. But since for one and the same movement varying hypotheses are proposed from time to time, as eccentricity or epicycle for the movement of the sun, the astronomer much prefers to take the one which is easiest to grasp. Maybe the philosopher demands probability instead; but neither of them will grasp anything certain or hand it on, unless it has been divinely revealed to him. Therefore let us permit these new hypotheses to make a public appearance among old ones which are themselves no more probable, especially since they are wonderful and easy and bring with them a vast storehouse of learned observations. And as far as hypotheses go, let no one expect anything in the way of certainty from astronomy, since astronomy can offer us nothing certain, lest, if anyone take as true that which has been constructed for another use, he go away from this discipline a bigger fool than when he came to it. Farewell.

Preface and Dedication to Pope Paul III

I can reckon easily enough, Most Holy Father, that as soon as certain people learn that in these books of mine which I have written about the revolutions of the spheres of the world I attribute certain motions to the terrestrial globe, they will immediately shout to have me and my opinion hooted off the stage. For my own works do not please me so much that I do not weigh what judgments others will pronounce concerning them. And although I realize that the conceptions of a philosopher are placed beyond the judgment of the crowd, because it is his loving duty to seek the truth in all things, in so far as God has granted that to human reason; nevertheless I think we should avoid opinions utterly foreign to rightness. And when I considered how absurd this "lecture" would be held by those who know that the opinion that the Earth rests immovable in the middle of the heavens as if their centre had been confirmed by the judgments of many ages—if I were to assert to the contrary that the Earth moves; for a long time I was in great difficulty as to whether I should bring to light my commentaries written to demonstrate the Earth's movement, or whether it would not be better to follow the example of the Pythagoreans and certain others who used to hand down the mysteries of their philosophy not in writing but by word of mouth and only to their relatives and friends—witness the letter of Lysis to Hipparchus. They however seem to me to have done that not, as some judge, out of a jealous unwillingness to communicate their doctrines but in order that things of very great beauty which have been investigated by the loving care of great men should not be scorned by those who find it a bother to expend any great energy on letters—except on the money-making variety—or who are provoked by the exhortations and examples of others to the liberal study of philosophy but on account of their natural stupidity hold the position among philosophers that drones hold among bees. Therefore, when I weighed these things in my mind, the scorn which I had to fear on account of the newness and absurdity of my opinion almost drove me to abandon a work already undertaken.

But my friends made me change my course in spite of my long-continued hesitation and even resistance. First among them was Nicholas Schonberg, Cardinal of Capua, a man distinguished in all branches of learning; next to him was my devoted friend Tiedeman Giese, Bishop of Culm, a man filled with the

greatest zeal for the divine and liberal arts: for he in particular urged me frequently and even spurred me on by added reproaches into publishing this book
and letting come to light a work which I had kept hidden among my things for
not merely nine years, but for almost four times nine years. Not a few other
learned and distinguished men demanded the same thing of me, urging me to
refuse no longer—on account of the fear which I felt—to contribute my work
to the common utility of those who are really interested in mathematics: they
said that the absurder my teaching about the movement of the Earth now seems
to very many persons, the more wonder and thanksgiving will it be the object
of, when after the publication of my commentaries those same persons see the
fog of absurdity dissipated by my luminous demonstrations. Accordingly I was
led by such persuasion and by that hope finally to permit my friends to undertake the publication of a work which they had long sought from me.

But perhaps Your Holiness will not be so much surprised at my giving the
results of my nocturnal study to the light—after having taken such care in
working them out that I did not hesitate to put in writing my conceptions as to
the movement of the Earth—as you will be eager to hear from me what came
into my mind that in opposition to the general opinion of mathematicians and
almost in opposition to common sense I should dare to imagine some movement
of the Earth. And so I am unwilling to hide from Your Holiness that nothing
except my knowledge that mathematicians have not agreed with one another in
their researches moved me to think out a different scheme of drawing up the
movements of the spheres of the world. For in the first place mathematicians
are so uncertain about the movements of the sun and moon that they can neither
demonstrate nor observe the unchanging magnitude of the revolving year. Then
in setting up the solar and lunar movements and those of the other five wandering stars, they do not employ the same principles, assumptions, or demonstrations for the revolutions and apparent movements. For some make use of homocentric circles only, others of eccentric circles and epicycles, by means of which
however they do not fully attain what they seek. For although those who have
put their trust in homocentric circles have shown that various different movements can be composed of such circles, nevertheless they have not been able to
establish anything for certain that would fully correspond to the phenomena.
But even if those who have thought up eccentric circles seem to have been able
for the most part to compute the apparent movements numerically by those
means, they have in the meanwhile admitted a great deal which seems to contradict the first principles of regularity of movement. Moreover, they have not been
able to discover or to infer the chief point of all, i.e., the form of the world and
the certain commensurability of its parts. But they are in exactly the same fix as
someone taking from different places hands, feet, head, and the other limbs—
shaped very beautifully but not with reference to one body and without correspondence to one another—so that such parts made up a monster rather than
a man. And so, in the process of demonstration which they call "method," they
are found either to have omitted something necessary or to have admitted something foreign which by no means pertains to the matter; and they would by no

means have been in this fix, if they had followed sure principles. For if the hypotheses they assumed were not false, everything which followed from the hypotheses would have been verified without fail; and though what I am saying may be obscure right now, nevertheless it will become clearer in the proper place.

Accordingly, when I had meditated upon this lack of certitude in the traditional mathematics concerning the composition of movements of the spheres of the world, I began to be annoyed that the philosophers, who in other respects had made a very careful scrutiny of the least details of the world, had discovered no sure scheme for the movements of the machinery of the world, which has been built for us by the Best and Most Orderly Workman of all. Wherefore I took the trouble to reread all the books by philosophers which I could get hold of, to see if any of them even supposed that the movements of the spheres of the world were different from those laid down by those who taught mathematics in the schools. And as a matter of fact, I found first in Cicero that Nicetas thought that the Earth moved. And afterwards I found in Plutarch that there were some others of the same opinion: I shall copy out his words here, so that they may be known to all:

Some think that the Earth is at rest; but Philolaus the Pythagorean says that it moves around the fire with an obliquely circular motion, like the sun and moon. Herakleides of Pontus and Ekphantus the Pythagorean do not give the Earth any movement of locomotion, but rather a limited movement of rising and setting around its centre, like a wheel.

Therefore I also, having found occasion, began to meditate upon the mobility of the Earth. And although the opinion seemed absurd, nevertheless because I knew that others before me had been granted the liberty of constructing whatever circles they pleased in order to demonstrate astral phenomena, I thought that I too would be readily permitted to test whether or not, by the laying down that the Earth had some movement, demonstrations less shaky than those of my predecessors could be found for the revolutions of the celestial spheres.

And so, having laid down the movements which I attribute to the Earth farther on in the work, I finally discovered by the help of long and numerous observations that if the movements of the other wandering stars are correlated with the circular movement of the Earth, and if the movements are computed in accordance with the revolution of each planet, not only do all their phenomena follow from that but also this correlation binds together so closely the order and magnitudes of all the planets and of their spheres or orbital circles and the heavens themselves that nothing can be shifted around in any part of them without disrupting the remaining parts and the universe as a whole.

Accordingly, in composing my work I adopted the following order: in the first book I describe all the locations of the spheres or orbital circles together with the movements which I attribute to the earth, so that this book contains as it were the general set-up of the universe. But afterwards in the remaining books I correlate all the movements of the other planets and their spheres or orbital circles with the mobility of the Earth, so that it can be gathered from that how far the apparent movements of the remaining planets and their orbital circles can be saved by being correlated with the movements of the Earth. And I have no doubt that talented and learned mathematicians will agree with me, if—as Philosophy demands in the first place—they are willing to give not superficial but profound thought and effort to what I bring forward in this work in demonstrating these things. And in order that the unlearned as well as the learned might see that I was not seeking to flee from the judgment of any man, I preferred to dedicate these results of my nocturnal study to Your Holiness rather than to anyone else; because, even in this remote corner of the earth where I live, you are held to be most eminent both in the dignity of your order and in your love of letters and even of mathematics; hence, by the authority of your judgment you can easily provide a guard against the bites of slanderers, despite the proverb that there is no medicine for the bite of a sycophant.

But if perchance there are certain "idle talkers" who take it upon themselves to pronounce judgment, although wholly ignorant of mathematics, and if by shamelessly distorting the sense of some passage in Holy Writ to suit their purpose, they dare to reprehend and to attack my work; they worry me so little that I shall even scorn their judgments as foolhardy. For it is not unknown that Lactantius, otherwise a distinguished writer but hardly a mathematician, speaks in an utterly childish fashion concerning the shape of the Earth, when he laughs at those who have affirmed that the Earth has the form of a globe. And so the studious need not be surprised if people like that laugh at us. Mathematics is written for mathematicians; and among them, if I am not mistaken, my labours will be seen to contribute something to the ecclesiastical commonwealth, the principate of which Your Holiness now holds. For not many years ago under Leo X when the Lateran Council was considering the question of reforming the Ecclesiastical Calendar, no decision was reached, for the sole reason that the magnitude of the year and the months and the movements of the sun and moon had not yet been measured with sufficient accuracy. From that time on I gave attention to making more exact observations of these things and was encouraged to do so by that most distinguished man, Paul, bishop of Fossombrone, who had been present at those deliberations. But what have I accomplished in this matter I leave to the judgment of Your Holiness in particular and to that of all other learned mathematicians. And so as not to appear to Your Holiness to make more promises concerning the utility of this book than I can fulfill, I now pass on to the body of the work.

Book One

Among the many and varied literary and artistic studies upon which the natural talents of man are nourished, I think that those above all should be embraced and pursued with the most loving care which have to do with things that are very beautiful and very worthy of knowledge. Such studies are those which deal with the godlike circular movements of the world and the course of the stars, their magnitudes, distances, risings and settings, and the causes of the other appearances in the heavens; and which finally explicate the whole form. For what could be more beautiful than the heavens which contain all beautiful things? Their very names make this clear: *Caelum* (heavens) by naming that which is beautifully carved; and *Mundus* (world), purity and elegance. Many philosophers have called the world a visible god on account of its extraordinary excellence. So if the worth of the arts were measured by the matter with which they deal, this art—which some call astronomy, others astrology, and many of the ancients the consummation of mathematics—would be by far the most outstanding. This art which is as it were the head of all the liberal arts and the one most worthy of a free man leans upon nearly all the other branches of mathematics. Arithmetic, geometry, optics, geodesy, mechanics, and whatever others, all offer themselves in its service. And since a property of all good arts is to draw the mind of man away from the vices and direct it to better things, these arts can do that more plentifully, over and above the unbelievable pleasure of mind [which they furnish]. For who, after applying himself to things which he sees established in the best order and directed by divine ruling, would not through diligent contemplation of them and through a certain habituation be awakened to that which is best and would not wonder at the Artificer of all things, in Whom is all happiness and every good? For the divine Psalmist surely did not say gratuitously that he took pleasure in the workings of god and rejoiced in the works of His hands, unless by means of these things as by some sort of vehicle we are transported to the contemplation of the highest Good.

Now as regards the utility and ornament which they confer upon a commonwealth—to pass over the innumerable advantages they give to private citizens—Plato makes an extremely good point, for in the seventh book of the *Laws* he says that this study should be pursued in especial, that through it the orderly arrangement of days into months and years and the determination of the times for solemnities and sacrifices should keep the state alive and watchful; and he says that if anyone denies that this study is necessary for a man who is going to take up any of the highest branches of learning, then such a person is thinking foolishly; and he thinks that it is impossible for anyone to become godlike or be called so who has no necessary knowledge of the sun, moon, and the other stars.

However, this more divine than human science, which inquires into the highest things, is not lacking in difficulties. And in particular we see that as regards its principles and assumptions, which the Greeks call "hypotheses," many of those who undertook to deal with them were not in accord and hence did not employ the same methods of calculation. In addition, the courses of the planets and the revolution of the stars cannot be determined by exact calculations and reduced to perfect knowledge unless, through the passage of time and with the help of many prior observations, they can, so to speak, be handed down to posterity. For even if Claud Ptolemy of Alexandria, who stands far in front of all the others on account of his wonderful care and industry, with the help of more than forty years of observations brought this art to such a high point that there seemed to be nothing left which he had not touched upon; nevertheless we see that very many things are not in accord with the movements which should follow from his doctrine but rather with movements which were discovered later and were unknown to him. Whence even Plutarch in speaking of the revolving solar year says, "So far the movement of the stars has overcome the ingenuity of the mathematicians." Now to take the year itself as my example, I believe it is well known how many different opinions there are about it, so that many people have given up hope of making an exact determination of it. Similarly, in the case of the other planets I shall try—with the help of God, without Whom we can do nothing—to make a more detailed inquiry concerning them, since the greater the interval of time between us and the founders of this art— whose discoveries we can compare with the new ones made by us—the more means we have of supporting our own theory. Furthermore, I confess that I shall expound many things differently from my predecessors—although with their aid, for it was they who first opened the road of inquiry into these things.

1. The World Is Spherical

In the beginning we should remark that the world is globe-shaped; whether because this figure is the most perfect of all, as it is an integral whole and needs no joints; or because this figure is the one having the greatest volume and thus is especially suitable for that which is going to comprehend and conserve all things; or even because the separate parts of the world, i.e., the sun, moon, and stars are viewed under such a form; or because everything in the world tends to be delimited by this form, as is apparent in the case of drops of water and other liquid bodies, when they become delimited of themselves. And so no one would hesitate to say that this form belongs to the heavenly bodies.

2. The Earth Is Spherical Too

The Earth is globe-shaped too, since on every side it rests upon its centre. But it is not perceived straightway to be a perfect sphere, on account of the great height of its mountains and the lowness of its valleys, though they modify its universal roundness to only a very small extent.

That is made clear in this way. For when people journey northward from anywhere, the northern vertex of the axis of daily revolution gradually moves overhead, and the other moves downward to the same extent; and many stars situated to the north are seen not to set, and many to the south are seen not to rise any more. So Italy does not see Canopus, which is visible to Egypt. And Italy sees the last star of Fluvius, which is not visible to this region situated in a more frigid zone. Conversely, for people who travel southward, the second group of stars becomes higher in the sky; while those become lower which for us are high up.

Moreover, the inclinations of the poles have everywhere the same ratio with places at equal distances from the poles of the Earth and that happens in no other figure except the spherical. Whence it is manifest that the Earth itself is contained between the vertices and is therefore a globe.

Add to this the fact that the inhabitants of the East do not perceive the evening eclipses of the sun and moon; nor the inhabitants of the West, the morning eclipses; while of those who live in the middle region—some see them earlier and some later.

Furthermore, voyagers perceive that the waters too are fixed within this figure; for example, when land is not visible from the deck of a ship, it may be seen from the top of the mast, and conversely, if something shining is attached to the top of the mast, it appears to those remaining on the shore to come down gradually, as the ship moves from the land, until finally it becomes hidden, as if setting.

Moreover, it is admitted that water, which by its nature flows, always seeks lower places—the same way as earth—and does not climb up the shore any farther than the convexity of the shore allows. That is why the land is so much higher where it rises up from the ocean.

3. How Land and Water Make Up a Single Globe

And so the ocean encircling the land pours forth its waters everywhere and fills up the deeper hollows with them. Accordingly it was necessary for there to be less water than land, so as not to have the whole earth soaked with water—since both of them tend toward the same centre on account of their weight—and so as to leave some portions of land—such as the islands discernible here and there—for the preservation of living creatures. For what is the continent itself and the *orbis terrarum* except an island which is larger than the rest? We should not listen to certain Peripatetics who maintain that there is ten times more water than land and who arrive at that conclusion because in the transmutation of the elements the liquefaction of one part of earth results in ten parts of water. And they say that land has emerged for a certain distance because, having hollow spaces inside, it does not balance everywhere with respect to weight and so the centre of gravity is different from the centre of magnitude. But they fall into error through ignorance of geometry; for they do not know that there cannot be seven times more water than land and some part of the land still remain dry, unless the land abandon its centre of gravity and give place to the waters as being heavier. For spheres are to one another as the cubes of their diameters. If therefore there were seven parts of water and one part of land, the diameter of the land could not be greater than the radius of the globe of the waters. So it is even less possible that the water should be ten times greater. It can be gathered that there is no difference between the centres of magnitude and of gravity of the Earth from the fact that the convexity of the land spreading out from the ocean does not swell continuously, for in that case it would repulse the sea-waters as much as possible and would not in any way allow interior seas and huge gulfs to break through. Moreover, from the seashore outward the depth of the abyss would not stop increasing, and so no island or reef or any spot of land would be met with by people voyaging out very far. Now it is well known that there is not quite the distance of two miles—at practically the centre of the *orbis terrarum*—between the Egyptian and the Red Sea. And on the contrary, Ptolemy in his *Cosmography* extends inhabitable lands as far as the median circle, and he leaves that part of the Earth as unknown, where the moderns have added Cathay and other vast regions as far as 60° longitude, so that inhabited land extends in longitude farther than the rest of the ocean does. And if you add to these the islands discovered in our time under the princes of Spain and Portugal and especially America—named after the ship's captain who discovered her—which they consider a second *orbis terrarum* on account of her so far unmeasured magnitude—besides many other islands heretofore unknown, we would not be greatly surprised if there were antiphodes or antichthones. For reasons of geometry compel us to believe that America is situated diametrically opposite to the India of the Ganges.

And from all that I think it is manifest that the land and the water rest upon one centre of gravity; that this is the same as the centre of magnitude of the land, since land is the heavier; that parts of land which are as it were yawning are filled with water; and that accordingly there is little water in comparison with the land, even if more of the surface appears to be covered by water.

Now it is necessary that the land and the surrounding waters have the figure which the shadow of the Earth casts, for it eclipses the moon by projecting a perfect circle upon it. Therefore the Earth is not a plane, as Empedocles and Anaximenes opined; or a tympanoid, as Leucippus; or a scaphoid, as Heracleitus; or hollowed out in any other way as Democritus; or again a cylinder, as Anaximander; and it is not infinite in its lower part, with the density increasing rootwards, as Xenophanes thought; but it is perfectly round, as the philosophers perceived.

4. The Movement of the Celestial Bodies Is Regular, Circular, and Everlasting—Or Else Compounded of Circular Movements

After this we will recall that the movement of the celestial bodies is circular. For the motion of a sphere is to turn in a circle; by this very act expressing its form, in the most simple body, where beginning and end cannot be discovered or distinguished from one another, while it moves through the same parts in itself.

But there are many movements on account of the multitude of spheres or orbital circles. The most obvious of all is the daily revolution—which the Greeks call νυχθήμερον; i.e., having the temporal span of a day and a night. By means of this movement the whole world—with the exception of the Earth—is supposed to be borne from east to west. This movement is taken as the common measure of all movements, since we measure even time itself principally by the number of days.

Next, we see other as it were antagonistic revolutions; i.e., from west to east, on the part of the sun, moon, and the wandering stars. In this way the sun gives us the year, the moon the months—the most common periods of time; and each of the other five planets follows its own cycle. Nevertheless these movements are manifoldly different from the first movement. First, in that they do not revolve around the same poles as the first movement but follow the oblique ecliptic; next, in that they do not seem to move in their circuit regularly. For the sun and moon are caught moving at times more slowly and at times more quickly. And we perceive the five wandering stars sometimes even to retrograde and to come to a stop between these two movements. And though the sun always proceeds straight ahead along its route, they wander in various ways, straying sometimes towards the south, and at other times towards the north—whence they are called "planets." Add to this the fact that sometimes they are nearer the Earth—and are then said to be at their perigee—and at other times are farther away—and are said to be at their apogee.

We must however confess that these movements are circular or are composed of many circular movements, in that they maintain these irregularities in accordance with a constant law and with fixed periodic returns: and that could not take place, if they were not circular. For it is only the circle which can bring back what is past and over with; and in this way, for example, the sun by a movement composed of circular movements brings back to us the inequality of days and nights and the four seasons of the year. Many movements are recognized in that movement, since it is impossible that a simple heavenly body should be moved irregularly by a single sphere. For that would have to take place either on account of the inconstancy of the motor virtue—whether by reason of an extrinsic cause or its intrinsic nature—or on account of the inequality between it and the moved body. But since the mind shudders at either of these suppositions, and since it is quite unfitting to suppose that such a state of affairs exists among things which are established in the best system, it is agreed that their regular movements appear to us as irregular, whether on account of their circles having different poles or even because the earth is not at the centre of the circles in which they revolve. And so for us watching from the Earth, it happens that the transits of the planets, on account of being at unequal distances from the Earth, appear greater when they are nearer than when they are farther away, as has been shown in optics: thus in the case of equal arcs of an orbital circle which are seen at different distances there will appear to be unequal movements in equal times. For this reason I think it necessary above all that we should note carefully what the relation of the Earth to the heavens is, so as not—when we wish to scrutinize the highest things—to be ignorant of those which are nearest to us, and so as not—by the same error—to attribute to the celestial bodies what belongs to the Earth

5. Does the Earth Have a Circular Movement? And of Its Place

Now that it has been shown that the Earth too has the form of a globe, I think we must see whether or not a movement follows upon its form and what the place of the Earth is in the universe. For without doing that it will not be possible to find a sure reason for the movements appearing in the heavens.

Although there are so many authorities for saying that the Earth rests in the centre of the world that people think the contrary supposition inopinable and even ridiculous; if however we consider the thing attentively, we will see that the question has not yet been decided and accordingly is by no means to be scorned. For every apparent change in place occurs on account of the movement either of the thing seen or of the spectator, or on account of the necessarily unequal movement of both. For no movement is perceptible relatively to things moved equally in the same directions—I mean relatively to the thing seen and the spectator. Now it is from the Earth that the celestial circuit is beheld and presented to our sight. Therefore, if some movement should belong to the Earth it will appear, in the parts of the universe which are outside, as the same movement but in the opposite direction, as though the things outside were passing over. And the daily revolution in especial is such a movement. For the daily revolution appears to carry the whole universe along, with the exception of the Earth and the things around it. And if you admit that the heavens possess none of this movement but that the Earth turns from west to east, you will find—if you make a serious examination—that as regards the apparent rising and setting of the sun, moon, and stars the case is so. And since it is the heavens which contain and embrace all things as the place common to the universe, it will not be clear at once why movement should not be assigned to the contained rather than to the container, to the thing placed rather than to the thing providing the place.

As a matter of fact, the Pythagoreans Herakleides and Ekphantus were of this opinion and so was Hicetas the Syracusan in Cicero; they made the Earth to revolve at the centre of the world. For they believed that the stars set by reason of the interposition of the Earth and that with cessation of that they rose again. Now upon this assumption there follow other things, and a no smaller problem concerning the place of the Earth, though it is taken for granted and believed by nearly all that the Earth is the centre of the world. For if anyone denies that the Earth occupies the midpoint or centre of the world yet does not admit that the distance [between the two] is great enough to be compared with [the distance to] the sphere of the fixed stars but is considerable and quite apparent in relation to the orbital circles of the sun and the planets; and if for that reason he thought that their movements appeared irregular because they are organized around a different centre from the centre of the Earth, he might perhaps be able to bring forward a perfectly sound reason for movement which appears irregular. For the fact that the wandering stars are seen to be sometimes nearer the Earth and at other times farther away necessarily argues that the centre of the Earth is not the centre of their circles. It is not yet clear whether the Earth draws near to them and moves away or they draw near to the Earth and move away.

And so it would not be very surprising if someone attributed some other movement of the earth in addition to the daily revolution. As a matter of fact, Philolaus the Pythagorean—no ordinary mathematician, whom Plato's biographers say Plato went to Italy for the sake of seeing—is supposed to have held that the Earth moved in a circle and wandered in some other movements and was one of the planets.

Many however have believed that they could show by geometrical reasoning that the Earth is in the middle of the world; that it has the proportionality of a point in relation to the immensity of the heavens, occupies the central position, and for this reason is immovable, because, when the universe moves, the centre remains unmoved and the things which are closest to the centre are moved most slowly.

6. On the Immensity of the Heavens in Relation to the Magnitude of the Earth

It can be understood that this great mass which is the Earth is not comparable with the magnitude of the heavens, from the fact that the boundary circles—for that is the translation of the Greek ὁρίζοντες—cut the whole celestial sphere into two halves; for that could not take place if the magnitude of the Earth in comparison with the heavens, or its distance from the centre of the world, were considerable. For the circle bisecting a sphere goes through the centre of the sphere, and is the greatest circle which it is possible to circumscribe.

Now let the horizon be the circle ABCD, and let the Earth, where our point of view is, be E, the centre of the horizon by which the visible stars are separated from those which are not visible. Now with a dioptra or horoscope or level placed at E, the beginning of Cancer is seen to rise at

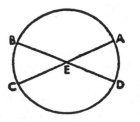

point C; and at the same moment the beginning of Capricorn appears to set at A. Therefore, since AEC is in a straight line with the dioptra, it is clear that this line is a diameter of the ecliptic, because the six signs bound a semicircle, whose centre E is the same as that of the horizon. But when a revolution has taken place and the beginning of Capricorn arises at B, then the setting of Cancer will be visible at D, and BED will be a straight line and a diameter of the ecliptic. But it has already been seen that the line AEC is a diameter of the same circle; therefore, at their common section, point E will be their centre. So in this way the horizon always bisects the ecliptic, which is a great circle of the sphere. But on a sphere, if a circle bisects one of the great circles, then the circle bisecting is a great circle. Therefore the horizon is a great circle; and its centre is the same as that of the ecliptic, as far as appearance goes; although nevertheless the line passing through the centre of the Earth and the line touching to the surface are necessarily different; but on account of their immensity in comparison with the Earth they are like parallel lines, which on account of the great distance between the termini appear to be one line, when the space contained between them is in no perceptible ratio to their length, as has been shown in optics.

From this argument it is certainly clear enough that the heavens are immense in comparison with the Earth and present the aspect of an infinite magnitude, and that in the judgment of sense-perception the Earth is to the heavens as a point to a body and as a finite to an infinite magnitude. But we see that nothing more than that has been shown, and it does not follow that the Earth must rest at the centre of the world. And we should be even more surprised if such a vast world should wheel completely around during the space of twenty-four hours rather than that its least part, the Earth, should. For saying that the centre is immovable and that those things which are closest to the centre are moved least does not argue that the Earth rests at the centre of the world. That is no different from saying that the heavens revolve but the poles are at rest and those things which are closest to the poles are moved least. In this way Cynosura [the pole star] is seen to move much more slowly than Aquila or Canicula because, being very near to the pole, it describes a smaller circle, since they are all on a single sphere, the movement of which stops at its axis and which does not allow any of its parts to have movements which are equal to one another. And nevertheless the revolution of the whole brings them round in equal times but not over equal spaces.

The argument which maintains that the Earth, as a part of the celestial sphere and as sharing in the same form and movement, moves very little because very near to its centre advances to the following position: therefore the Earth will move, as being a body and not a centre, and will describe in the same time arcs similar to, but smaller than, the arcs of the celestical circle. It is clearer than daylight how false that is; for there would necessarily always be noon at one place and midnight at another, and so the daily risings and settings could not take place, since the movement of the whole and the part would be one and inseparable.

But the ratio between things separated by diversity of nature is so entirely different that those which describe a smaller circle turn more quickly than those which describe a greater circle. In this way Saturn, the highest of the wandering stars, completes its revolution in thirty years, and the moon which is without doubt the closest to the Earth completes its circuit in a month, and finally the Earth itself will be considered to complete a circular movement in the space of a day and a night. So this same problem concerning the daily revolution comes up again. And also the question about the place of the Earth becomes even less certain on account of what was just said. For that demonstration proves nothing except that the heavens are of an indefinite magnitude with respect to the Earth. But it is not at all clear how far this immensity stretches out. On the contrary, since the minimal and indivisible corpuscles, which are called atoms, are not perceptible to sense, they do not, when taken in twos or in some small number, constitute a visible body; but they can be taken in such a large quantity that there will at last be enough to form a visible magnitude. So it is as regards the place of the earth; for although it is not at the centre of the world, nevertheless the distance is as nothing, particularly in comparison with the sphere of the fixed stars.

7. Why the Ancients Thought the Earth Was at Rest at the Middle of the World as Its Centre

Wherefore for other reasons the ancient philosophers have tried to affirm that the Earth is at rest at the middle of the world, and as principal cause they put forward heaviness and lightness. For Earth is the heaviest element; and all things of any weight are borne towards it and strive to move towards the very centre of it.

For since the Earth is a globe towards which from every direction heavy things by their own nature are borne at right angles to its surface, the heavy things would fall on one another at the centre if they were not held back at the surface; since a straight line making right angles with a plane surface where it touches a sphere leads to the centre. And those things which are borne toward the centre seem to follow along in order to be at rest at the centre. All the more then will the Earth be at rest at the centre; and, as being the receptacle for falling bodies, it will remain immovable because of its weight.

They strive similarly to prove this by reason of movement and its nature. For Aristotle says that the movement of a body which is one and simple is simple, and the simple movements are the rectilinear and the circular. And of rectilinear movements, one is upward, and the other is downward. As a consequence, every simple movement is either toward the centre, i.e., downward, or away from the centre, i.e., upward, or around the centre, i.e., circular. Now it belongs to earth and water, which are considered heavy, to be borne downward, i.e., to seek the centre: for air and fire, which are endowed with lightness, move upward, i.e., away from the centre. It seems fitting to grant rectilinear movement to these four elements and to give the heavenly bodies a circular movement around the centre. So Aristotle. Therefore, said Ptolemy of Alexandria, if the Earth moved, even if only by its daily rotation, the contrary of what was said above would necessarily take place. For this movement which would traverse the total circuit of the Earth in twenty-four hours would necessarily be very headlong and of an unsurpassable velocity. Now things which are suddenly and violently whirled around are seen to be utterly unfitted for reuniting, and the more unified are seen to become dispersed, unless some constant force constrains them to stick together. And a long time ago, he says, the scattered Earth would have passed beyond the heavens, as is certainly ridiculous; and *a fortiori* so would all the living creatures and all the other separate masses which could by no means remain unshaken. Moreover, freely falling bodies would not arrive at the places appointed them, and certainly not along the perpendicular line which they assume so quickly. And we would see clouds and other things floating in the air always borne toward the west.

8. Answer to the Aforesaid Reasons and Their Inadequacy

For these and similar reasons they say that the Earth remains at rest at the middle of the world and that there is no doubt about this. But if someone opines that the Earth revolves, he will also say that the movement is natural and not violent. Now things which are according to nature produce effects contrary to those which are violent. For things to which force or violence is applied get broken up and are unable to subsist for a long time. But things which are caused by nature are in a right condition and are kept in their best organization. Therefore Ptolemy had no reason to fear that the Earth and all things on the Earth would be scattered in a revolution caused by the efficacy of nature, which is greatly different from that of art or from that which can result from the genius of man. But why didn't he feel anxiety about the world instead, whose movement must necessarily be of greater velocity, the greater the heavens are than the Earth? Or have the heavens become so immense, because an unspeakably vehement motion has pulled them away from the centre, and because the heavens would fall if they came to rest anywhere else?

Surely if this reasoning were tenable, the magnitude of the heavens would extend infinitely. For the farther the movement is borne upward by the vehement force, the faster will the movement be, on account of the ever-increasing circumference which must be traversed every twenty-four hours: and conversely, the immensity of the sky would increase with the increase in movement. In this way, the velocity would make the magnitude increase infinitely, and the magnitude the velocity. And in accordance with the axiom of physics that *that which is infinite cannot be traversed or moved in any way*, then the heavens will necessarily come to rest.

But they say that beyond the heavens there isn't any body or place or void or anything at all; and accordingly it is not possible for the heavens to move outward: in that case it is rather surprising that something can be held together by nothing. But if the heavens were infinite and were finite only with respect to a hollow space inside, then it will be said with more truth that there is nothing outside the heavens, since anything which occupied any space would be in them; but the heavens will remain immobile. For movement is the most powerful reason wherewith they try to conclude that the universe is finite.

But let us leave to the philosophers of nature the dispute as to whether the world is finite or infinite, and let us hold as certain that the Earth is held together between its two poles and terminates in a spherical surface. Why therefore should we hesitate any longer to grant to it the movement which accords naturally with its form, rather than put the whole world in a commotion—the world whose limits we do not and cannot know? And why not admit that the appearance of daily revolution belongs to the heavens but the reality belongs to the Earth? And things are as when Aeneas said in Virgil: "We sail out of the harbor, and the land and the cities move away." As a matter of fact, when a ship floats on over a tranquil sea, all the things outside seem to the voyagers to be moving in a movement which is the image of their own, and they think on the contrary that they themselves and all the things with them are at rest. So it can easily happen in the case of the movement of the Earth, that the whole world should be believed to be moving in a circle. Then what would we say about the clouds and the other things floating in the air or falling or rising up, except that not only the Earth and the watery element with which it is conjoined are moved in this way but also no small part of the air and whatever other things have a similar kinship with the Earth? Whether because the neighbouring air, which is mixed with earthly and watery matter, obeys the same nature as the Earth or because the movement of the air is an acquired one, in which it participates without resistance on account of the contiguity and perpetual rotation of the Earth. Conversely, it is no less astonishing for them to say that the highest region of the air follows the celestial movement, as is shown by those stars which appear suddenly—I mean those called "comets" or "bearded stars" by the Greeks. For that place is assigned for their generation; and like all the other stars they rise and set. We can say that that part of the air is deprived of terrestrial motion on account of its great distance from the Earth. Hence the air which is nearest to the Earth and the things floating in it will appear tranquil, unless they are driven to and fro by the wind or some other force, as happens. For how is the wind in the air different from a current in the sea?

But we must confess that in comparison with the world the movement of falling and of rising bodies is twofold and is in general compounded of rectilinear and the circular. As regards things which move downward on account of their weight because they have very much earth in them, doubtless their parts possess the same nature as the whole, and it is for the same reason that fiery bodies are drawn upward with force. For even this earthly fire feeds principally on earthly matter; and they define flame as glowing smoke. Now it is a property of fire to make that which it invades to expand; and it does this with such force that it can be stopped by no means or contrivance from breaking prison and completing its job. Now expanding movement moves away from the centre to the circumference; and so if some part of the Earth caught on fire, it would be borne away from the centre and upward. Accordingly, as they say, a simple body possesses a simple movement—this is first verified in the case of circular movement—as long as the simple body remain in its unity in its natural place. In this place, in fact, its movement is none other than the circular, which remains entirely in itself, as though at rest. Rectilinear movement, however, is added to those bodies which journey away from their natural place or are shoved out of it or are outside it somehow. But nothing is more repugnant to the order of the whole and to the form of the world than for anything to be outside of its place. Therefore rectilinear movement belongs only to bodies which are not in the right condition and are not perfectly conformed to their nature—when they are separated from their whole and abandon its unity. Furthermore, bodies which are moved upward or downward do not possess a simple, uniform, and regular movement—even without taking into account circular movement. For they cannot be in equilibrium with their lightness or their force of weight. And those which fall downward possess a slow movement at the beginning but increase their velocity as they fall. And conversely we note that this earthly fire—and we have experience of no other—when carried high up immediately dies down, as if through the acknowledged agency of the violence of earthly matter.

Now circular movement always goes on regularly, for it has an unfailing cause; but [in rectilinear movement] the acceleration stops, because, when the bodies have reached their own place, they are no longer heavy or light, and so the movement ends. Therefore, since circular movement belongs to wholes and rectilinear to parts, we can say that the circular movement stands with the recti-

linear, as does animal with sick. And the fact that Aristotle divided simple movement into three genera: away from the centre, toward the centre, and around
the centre, will be considered merely as an act of reason, just as we distinguish
between line, point, and surface, though none of them can subsist without the
others or without body.

In addition, there is the fact that the state of immobility is regarded as more
noble and godlike than that of change and instability, which for that reason
should belong to the Earth rather than to the world. I add that it seems rather
absurd to ascribe movement to the container or to that which provides the place
and not rather to that which is contained and has a place; i.e., the Earth. And
lastly, since it is clear that the wandering stars, are sometimes nearer and sometimes farther away from the Earth, then the movement of one and the same
body around the centre—and they mean the centre of the Earth—will be both
away from the centre and toward the centre. Therefore it is necessary that movement around the centre should be taken more generally; and it should be
enough if each movement is in accord with its own centre. You see therefore
that for all these reasons it is more probably that the Earth moves than that it is
at rest—especially in the case of the daily revolution, as it is the Earth's very
own. And I think that is enough as regards the first part of the question.

9. Whether Many Movements Can Be Attributed to The Earth, And Concerning The Centre of The World

Therefore, since nothing hinders the mobility of the Earth, I think we should
now see whether more than one movement belongs to it, so that it can be
regarded as one of the wandering stars. For the apparent irregular movement of
the planets and their variable distances from the Earth—which cannot be understood as occurring in circles homocentric with the Earth—make it clear that the
Earth is not the centre of their circular movements. Therefore, since there are
many centres, it is not foolhardy to doubt whether the centre of gravity of the
Earth rather than some other is the centre of the world. I myself think that
gravity or heaviness is nothing except a certain natural appetency implanted in
the parts by the divine providence of the universal Artisan, in order that they
should unite with one another in their oneness and wholeness and come together in the form of a globe. It is believable that this affect is present in the
sun, moon, and the other bright planets and that through its efficacy they remain
in the spherical figure in which they are visible, though they nevertheless accomplish their circular movements in many different ways. Therefore if the Earth
too possesses movements different from the one around its centre, then they
will necessarily be movements which similarly appear on the outside in the
many bodies; and we find the yearly revolution among these movements. For if
the annual revolution were changed from being solar to being terrestrial, and
immobility were granted to the sun, the rising and settings of the signs and of
the fixed stars—whereby they become morning or evening stars—will appear

in the same way; and it will be seen that the stoppings, retrogressions, and progressions of the wandering stars are not their own, but are a movement of the Earth and that they borrow the appearances of this movement. Lastly, the sun will be regarded as occupying the centre of the world. And the ratio of order in which these bodies succeed one another and the harmony of the whole world teaches us their truth, if only—as they say—we would look at the thing with both eyes.

10. On the Order of the Celestial Orbital Circles

I know of no one who doubts that the heavens of the fixed stars is the highest up of all visible things. We see that the ancient philosophers wished to take the order of the planets according to the magnitude of their revolutions, for the reason that among things which are moved with equal speed those which are the more distant seem to be borne along more slowly, as Euclid proves in his *Optics*. And so they think that the moon traverses its circle in the shortest period of time because being next to the Earth, it revolves in the smallest circle. But they think that Saturn, which completes the longest circuit in the longest period of time, is the highest. Beneath Saturn, Jupiter. After Jupiter, Mars.

There are different opinions about Venus and Mercury, in that they do not have the full range of angular elongations from the sun that the others do. Wherefore some place them above the sun, as Timaeus does in Plato; some, beneath the sun, as Ptolemy and a good many moderns. Alpetragius makes Venus higher than the sun and Mercury lower. Accordingly, as the followers of Plato suppose that all the planets—which are otherwise dark bodies—shine with light received from the sun, they think that if the planets were below the sun, they would on account of their slight distance from the sun be viewed as only half—or at any rate as only partly—spherical. For the light which they receive is reflected by them upward for the most part, i.e., towards the sun, as we see in the case of the new moon or the old. Moreover, they say that necessarily the sun would sometimes be obscured through their interposition and that its light would be eclipsed in proportion to their magnitude; and as that has never appeared to take place, they think that these planets cannot by any means be below the sun.

On the contrary, those who place Venus and Mercury below the sun claim as a reason the amplitude of the space which they find between the sun and the moon. For they find that the greatest distance between the Earth and the moon, i.e., 64 ⅙ units, whereof the radius of the Earth is one, is contained almost 18 times in the least distance between the sun and the Earth. This distance is 1160 such units, and therefore the distance between the sun and the moon is 1096 such units. And then, in order for such a vast space not to remain empty, they find that the intervals between the perigees and apogees—according to which they reason out the thickness of the spheres—add up to approximately the same sum: in such fashion that the apogee of the moon may be succeeded by the perigee of Mercury, that the apogee of Mercury may be followed by the perigee of Venus, and that finally the apogee of Venus may nearly touch the perigee of the sun. In fact they calculate that the interval between the perigee and the apogee of Mercury contains approximately 177½ of the aforesaid units and that the remaining space is nearly filled by the 910 units of the interval between the perigee and apogee of Venus. Therefore they do not admit that these planets have a certain opacity, like that of the moon; but that they shine either by their own proper light or because their entire bodies are impregnated with sunlight, and that accordingly they do not obscure the sun, because it is an extremely rare occurrence for them to be interposed between our sight and the sun, as they usually withdraw [from the sun] latitudinally. In addition, there is the fact that they are small bodies in comparison with the sun, since Venus even though larger than Mercury can cover scarcely one one-hundredth part of the sun, as al-Battani the Harranite maintains, who holds that the diameter of the sun is ten times greater, and therefore it would not be easy to see such a little speck in the midst of such beaming light. Averroes, however, in his paraphrase of Ptolemy records having seen something blackish, when he observed the conjunction of the sun and Mercury which he had computed. And so they judge that these two planets move below the solar circle.

But how uncertain and shaky this reasoning is, is clear from the fact that though the shortest distance of the moon is 38 units whereof the radius of the Earth is one unit—according to Ptolemy, but more than 49 such units by a truer evaluation, as will be shown below—nevertheless we do not know that this great space contains anything except air, or if you prefer, what they call the fiery element.

Moreover, there is the fact that the diameter of the epicycle of Venus—by reason of which Venus has an angular digression of approximately 45° on either side of the sun—would have to be six times greater than the distance from the centre of the Earth to its perigee, as will be shown in the proper place. Then what will they say is contained in all this space, which is so great as to take in the Earth, air, ether, moon and Mercury, and which moreover the vast epicycle of Venus would occupy if it revolved around an immobile Earth?

Furthermore, how unconvincing is Ptolemy's argument that the sun must occupy the middle position between those planets which have the full range of angular elongation from the sun and those which do not is clear from the fact that the moon's full range of angular elongation proves its falsity.

But what cause will those who place Venus below the sun, and Mercury next, or separate them in some other order—what cause will they allege why these planets do not also make longitudinal circuits separate and independent of the sun, like the other planets—if indeed the ratio of speed or slowness does not falsify their order? Therefore it will be necessary either for the Earth not to be the centre to which the order of the planets and their orbital circles is referred, or for there to be no sure reason for their order and for it not to be apparent why the highest place is due to Saturn rather than to Jupiter or some other planet. Wherefore I judge that what Martianus Capella—who wrote the *Encyclopedia*—and some other Latins took to be the case is by no means to be despised.

For they hold that Venus and Mercury circle around the sun as a centre; and they hold that for this reason Venus and Mercury do not have any further elongation from the sun than the convexity of their orbital circles permits; for they do not make a circle around the earth as do the others, but have perigee and apogee interchangeable [in the sphere of the fixed stars]. Now what do they mean except that the centre of their spheres is around the sun? Thus the orbital circle of Mercury will be enclosed within the orbital circle of Venus—which would have to be more than twice as large—and will find adequate room for itself within that amplitude. Therefore if anyone should take this as an occasion to refer Saturn, Jupiter, and Mars also to this same centre, provided he understands the magnitude of those orbital circles to be such as to comprehend and encircle the Earth remaining within them, he would not be in error, as the table of ratios of their movements makes clear. For it is manifest that the planets are always nearer the Earth at the time of their evening rising, i.e., when they are opposite to the sun and the Earth is in the middle between them and the sun. But they are farthest away from the Earth at the time of their evening setting, i.e., when they are occulted in the neighbourhood of the sun, namely, when we have the sun between them and the Earth. All that shows clearly enough that their centre is more directly related to the sun and is the same as that to which Venus and Mercury refer their revolutions. But as they all have one common centre, it is necessary that the space left between the convex orbital circle of Venus and the concave orbital circle of Mars should be viewed as an orbital circle or sphere homocentric with them in respect to both surfaces, and that it should receive the Earth and its satellite the moon and whatever is contained beneath the lunar globe. For we can by no means separate the moon from the Earth, as the moon is incontestably very near to the Earth—especially since we find in this expanse a place for the moon which is proper enough and sufficiently large. Therefore we are not ashamed to maintain that this totality—which the moon embraces—and the centre of the Earth too traverse that great orbital circle among the other wandering stars in an annual revolution around the sun; and that the centre of the world is around the sun. I also say that the sun remains forever immobile and that whatever apparent movement belongs to it can be verified of the mobility of the Earth; that the magnitude of the world is such that, although the distance from the sun to the Earth in relation to whatsoever planetary sphere you please possesses magnitude which is sufficiently manifest in proportion to these dimensions, this distance, as compared with the sphere of the fixed stars, is imperceptible. I find it much more easy to grant that than to unhinge the understanding by an almost infinite multitude of spheres—as those who keep the earth at the centre of the world are forced to do. But we should rather follow the wisdom of nature, which, as it takes very great care not to have produced anything superfluous or useless, often prefers to endow one thing with many effects. And though all these things are difficult, almost inconceivable, and quite contrary to the opinion of the multitude, nevertheless in what follows we will with God's help make them clearer than day—at least for those who are not ignorant of the art of mathematics.

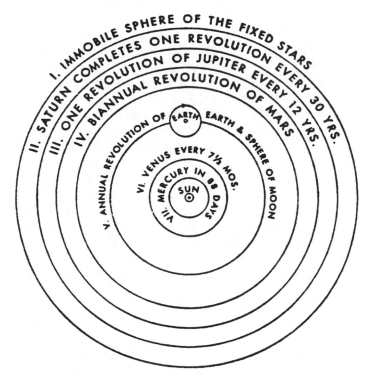

Therefore if the first law is still safe—for no one will bring forward a better one than that the magnitude of the orbital circles should be measured by the magnitude of time—then the order of the spheres will follow in this way—beginning with the highest: the first and highest of all is the sphere of the fixed stars, which comprehends itself and all things, and is accordingly immovable. In fact it is the place of the universe, i.e., it is that to which the movement and position of all the other stars are referred. For in the deduction of terrestrial movement, we will however give the cause why there are appearances such as to make people believe that even the sphere of the fixed stars somehow moves. Saturn, the first of the wandering stars, follows; it completes its circuit in 30 years. After it comes Jupiter moving in a 12-year period of revolution. Then Mars, which completes a revolution every 2 years. The place fourth in order is occupied by the annual revolution in which we said the Earth together with the orbital circle of the moon as an epicycle is comprehended. In the fifth place, Venus, which completes its revolution in 7½ months. The sixth and final place is occupied by Mercury, which completes its revolution in a period of 88 days. In the center of all rests the sun. For who would place this lamp of a very beautiful temple in another or better place than this wherefrom it can illuminate everything at the same time? As a matter of fact, not unhappily do some call it the lantern; others, the mind and still others, the pilot of the world. Trismegistus calls it a "visible god"; Sophocles' Electra, "that which gazes upon all things." And so the sun, as if resting on a kingly throne, governs the family of stars which wheel around. Moreover, the Earth is by no means cheated of the services of the moon; but, as Aristotle says in the *De Animalibus*, the earth has the closest kinship with the moon. The Earth moreover is fertilized by the sun and conceives offspring every year.

Therefore in this ordering we find that the world has a wonderful commensurability and that there is a sure bond of harmony for the movement and magnitude of the orbital circles such as cannot be found in any other way. For now the careful observer can note why progression and retrogradation appear greater in Jupiter than in Saturn and smaller than in Mars; and in turn greater in Venus than in Mercury. And why these reciprocal events appear more often in Saturn than in Jupiter, and even less often in Mars and Venus than in Mercury. In addition, why when Saturn, Jupiter, and Mars are in opposition [to the mean position of the sun] they are nearer to the Earth than at the time of their occultation and their reappearance. And especially why at the times when Mars is in opposition to the sun, it seems to equal Jupiter in magnitude and to be distinguished from Jupiter only by a reddish color, but when discovered through careful observation by means of a sextant is found with difficulty among the stars of second magnitude? All these things proceed from the same cause, which resides in the movement of the Earth.

But that there are no such appearances among the fixed stars argues that they are at an immense height away, which makes the circle of annual movement or its image disappear from before our eyes since every visible thing has a certain distance beyond which it is no longer seen, as is shown in optics. For the brilliance of their lights shows that there is a very great distance between Saturn the highest of the planets and the sphere of the fixed stars. It is by this mark in particular that they are distinguished from the planets, as it is proper to have the greatest difference between the moved and the unmoved. How exceedingly fine is the godlike work of the Best and Greatest Artist!

11. A Demonstration of the Threefold Movement of the Earth

Therefore since so much and such great testimony on the part of the planets is consonant with the mobility of the Earth, we shall now give a summary of its movements, insofar as the appearances can be shown forth by its movement as by an hypothesis. We must allow a threefold movement altogether.

The first—which we said the Greeks call νυχθημέρινος—is the proper circuit of day and night, which goes around the axis of the earth from west to east—as the world is held to move in the opposite direction—and describes the equator or the equinoctial circle—which some, imitating the Greek expression [10b] ισημέρινος, call the equidial.

The second is the annual movement of the centre, which describes the circle of the [zodiacal] signs around the sun similarly from west to east, i.e., towards the signs which follow [from Aries to Taurus] and moves along between Venus and Mars, as we said, together with the bodies accompanying it. So it happens that the sun itself seems to traverse the ecliptic with a similar movement. In this way, for example, when the centre of the Earth is traversing Capricorn, the sun seems to be crossing Cancer; and when Aquarius, Leo, and so on, as we were saying.

It has to be understood that the equator and the axis of the Earth have a variable inclination with the circle and the plane of the ecliptic. For if they remained fixed and only followed the movement of the centre simply, no inequality of days and nights would be apparent, but it would always be the summer solstice or the winter solstice or the equinox, or summer or winter, or some other season of the year always remaining the same. There follows then the third movement, which is the declination: it is also an annual revolution but one towards the signs which precede [from Aries to Pisces], or westwards, i.e., turning back counter to the movement of the centre; and as a consequence of these two movements which are nearly equal to one another but in opposite directions, it follows that the axis of the Earth and the greatest of the parallel circles on it, the equator, always look towards approximately the same quarter of the world, just as if they remained immobile. The sun in the meanwhile is seen to move along the oblique ecliptic with that movement with which the centre of the earth moves, just as if the centre of the earth were the centre of the world—provided you remember that the distance between the sun and the earth in comparison with the sphere of the fixed stars is imperceptible to us.

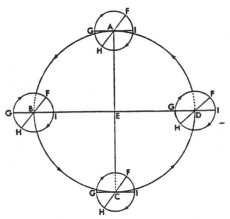

Since these things are such that they need to be presented to sight rather than merely to be talked about, let us draw the circle ABCD, which will represent the annual circuit of the centre of the earth in the plane of the ecliptic, and let E be the sun around its centre. I will cut this circle into four equal parts by means of the diameters AEC and BED. Let the point A be the beginning of Cancer; B of Libra; E of Capricorn; and D of Aries. Now let us put the centre of the earth first at A, around which we shall describe the terrestrial equator FGHI, but not in the same plane [as the ecliptic] except that the diameter GAI is the common section of the circles, i.e., of the equator and the ecliptic. Also let the diameter FAH be drawn at right angles to GAI; and let F be the limit of the greatest southward declination [of the equator], and H of the northward declination. With this set-up, the Earth-dweller will see the sun—which is at the centre E—at the point of the winter solstice in Capricorn—which is caused by the greatest northward declination at H being turned toward the sun; since the inclination of the equator with respect to line *AE* describes by means of the daily revolution the winter tropic, which is parallel to the equator at the distance comprehended by the angle of inclination *EAH*. Now let the centre of the Earth proceed from west to east; and let *F*, the limit of greatest declination, have just as great a movement from east to west, until at *B* both of them have traversed quadrants of circles. Meanwhile, on account of the equality of the revolutions, angle *EAI* will always remain equal to angle *AEB*; the diameters will always stay parallel to one another—*FAH* to *FBH* and *GAI* to *GBI*; and the equator will remain parallel to the equator. And by reason of the cause spoken of many times already, these lines will appear in the immensity of the sky as the same. Therefore from point *B* the beginning of Libra, *E* will appear to be in Aries, and the common section of the two circles [of the ecliptic and the equator] will fall upon line *GBIE*, in respect to which the daily revolution has no declination; but every declination will be on one side or the other of this line. And so the sun will be soon in the spring equinox. Let the centre of the Earth advance under the same conditions; and when it has completed a semicircle at *C*, the sun will appear to be entering Cancer. But since *F* the southward declination of the equator is now turned toward the sun, the result is that the sun is seen in the north, traversing the summer tropic in accordance with angle of inclination *ECF*. Again, when *F* moves on through the third quadrant of the circle, the common section *GI* will fall on line *ED*; whence the sun, seen in Libra, will appear to have reached the autumn equinox. But then as, in the same progressive movement, *HF* gradually turns in the direction of the sun, it will make the situation at the beginning return, which was our point of departure.

6
— T y c h o B r a h e —
(1546–1601)

Tycho Brahe was a Danish nobleman and the greatest astronomer prior to the invention of the telescope. His early astronomical work was done mainly on the site of the astronomical complex he had built for himself on the Danish island of Hveen. The two main buildings of this complex were appropriately named: Uraniborg, or Castle of the Heavens, and Stjerneborg, or Star Castle. Among the most famous of Brahe's early astronomical enterprises were his observations of a bright new star or supernova in the constellation of Cassiopeia in 1572, and his observations, five years later, of the comet of 1577. In both cases Brahe's measurements implied that neither of these phenomena could be sublunar, thus putting them at odds with the established Aristotelian view of the nature of the heavens.

Brahe's precise charting of the locations of stellar and planetary bodies, and his corrections of prominent observational errors of the past, were of great value to succeeding astronomers, especially his research assistant, Johannes Kepler, who would follow Brahe as the royal mathematician at the Emperor Rudolph's court in Prague in 1602.

The great reduction in observational margins of error achieved by Brahe was not the result of chance. He allowed for errors due to faulty instrumentation and corrected for errors arising from atmospheric refraction, or the bending of incoming light. Instrumental efficiency was improved by Brahe with various technical innovations, such as that of restricting instrumental movement, thus ensuring greater stability. Also not to be underestimated is the fact that Brahe's astronomical results were as systematic as they were precise. Brahe observed the heavens continuously, taking care, unlike his predecessors, to document the motions and positions of celestial bodies on a continual basis.

In this reading Brahe describes his observations of the supernova of 1572 and reflects upon its significance. It comes from a treatise he wrote called His Astronomical Coniectur of the New and Much Admired Star Which Appeared in the Year 1572. *Of interest here is Brahe's discussion of the substance out of which the supernova is made, and his conjecture as to where in the heavens it might be situated. The reading was translated by Lisa Dolling.*

THE SUPERNOVA OF 1572

Therefore, concerning the matter of this adventitious Star, that I may first give you my opinion of, I think it was celestial, not differing from the matter of the other stars, but yet in this it did admit of some diversity—that it was not exalted to such a perfection, nor solid composition of the parts, as appears in the everlasting and continuing stars; and therefore it had no perpetual duration, as these have, but was subject in process of time to dissolution; for as much as this Star could not consist of any elementary matter, (such) that cannot be carried into the highest part of the air, nor can obtain there any firm place of abiding. Besides, this Star did at the first in its magnitude exceed the whole Globe of the Earth, and was three hundred times bigger than the whole circumference thereof, and therefore what sublunary matter could be sufficient to the conformation of it.

But some may say, how or whence could it be framed of Celestial matter? I answer that the Heavens did afford it themselves, in like manner as the Earth, the Sea, and the Air; if at any time they exhibit some strange sight, do produce it out of their own proper substance. For although the Heaven itself be thin and pervious, giving way to the motion of the Stars without any hindrance, yet it is not altogether incorporeal, for then it should be infinite and without place. Therefore the very matter of Heaven, though it be subtle, and possible to the courses of the Planets, yet being compacted and condensed into one Globe, and being illustrated by the light of the Sun, might give form and fashion to this Star. Which because it had not its beginning from the common order of nature, therefore it could not have a continual duration equal to the rest, and in like manner, new and monstrous generations arising and compounded out of the Elements cannot long endure. And albeit the large vastness of the Celestial world may afford sufficient matter for the conformation of any adventitious Star, yet there is nowhere more plenty then near unto Via lactea or the Milky Way, which I suppose to be a certain heavenly substance not differing from the matter of the other Stars, but diffused and spread abroad, yet not distinctly conglobated in one body, as the Stars are; and hence I conjecture it came to pass, that this Star appeared in the edge of the Milky Way, and had the same substance as the Galaxia has. Besides, there is discerned a certain mark or scar as it were in that part of the Galaxia, wherein this Star was seated, as in a clear night when the Milky Way is not veiled with clouds we may easily perceive. Which mark or scar I never saw before this Star did arise, neither did I ever read of it. But howsoever, the substance of the Milky Zone is able to supply matter for the framing of this Star, which because it had not attained so excellent a consummation, and solid existence, as the genuine and natural Stars have, therefore it was subject to dissolution and dissipation, either by its own nature, or by the multiplicity of the beams of the Sun and other Stars.

Neither is Aristotle here to be allowed of, who disapproving of the opinions of others, does himself bring in no less absurdities, while he makes the Galaxia to be a certain sublunary concretion attracted and formed out of the Stars which are above it; so that it becomes a Meteor, in the highest part of the Air, not unlike to the Comets, which he (grounding one absurdity upon another) supposes to be generated there. For if it were so, the Milky Way would not have continued in the same form, place, and Magnitude, as it has done from the beginning of the world. And besides, other Stars would attain unto the like Luminous concretion. And moreover, this Galaxia of Aristotle, would then admit of a Parallax, and according to the optical consideration, by the shining of the fixed Stars through it, it would beget a strange refraction, differing from that which is occasioned by the vapors that are seen about the horizon, which seldom rises to the twentieth degree of Altitude, when this proceeding from the Via lactea would reach to the greatest height. All which, Aristotle rather guessed at, grounding it upon conjecture rather than on the doctrine of the Mathematics and optics, and therefore it is no marvel, if he has endeavored, to banish those seldom appearing Comets out of the heaven, and to equal them to sublunary Meteors, whereby he had thrust down the Galaxia beneath the Moon, and has made it participant of a sublunary nature. Hence it is, that Aristotle and other Philosophers, have joined the description and explication of the Galaxia, together with the Comets, because they knew not, the affinity which is between, having only learned by experience, or by the relation of ancient writers, that these beamy Stars have their original and beginning near to the Milky Way. Neither can it be a solecism, in that I affirm, that this new Star was framed of

Celestial matter, being the same whereof the Galaxia and other Stars do consist, yet not so well compacted. . . .

Moreover, forasmuch as this Star was placed in the eight Sphere, above the Orbs of the Planets, it seems that the predictions issuing from it, do not only concern one peculiar tract of Land, but all the Nations of the world; and therefore it will be the longer before the effects will be declared by succeeding events. Which, since they shall not begin until some years after the apparition, so they shall continue for a long time afterward. And if we may take leave to conjecture by Astrological computation of time, concerning the first beginning of that which is portended, we may guess it will be nine years after the great conjunction, whereof this Star was the Prodemus or fore-runner. If therefore we frame our Astrological direction by the place of this Conjunction, which was in the one and twentieth degree of Aquarius, the events of this Star shall begin to show themselves, nine years after this conjunction. And when this is finished, in the year of Christ 1583, and in the latter end of the Month of April, the confirmation and end of this Equinoxital progression to the place of the new Star, will fall out in the year 1592, when the third Septinary of years after the first appearing of the Star shall be accomplished.

7
— Tycho Brahe —

In this reading Brahe considers the comet of 1577 and its ramifications for the existence of the solid celestial spheres of Aristotle, a position he himself had once espoused. He concludes that his observations of the comet represent substantial evidence against the existence of the Aristotelian spheres. Brahe concludes the brief passage with a reflection on his own hybrid model of the universe.

The reading, originally written in 1598, comes from Tycho Brahe's Description of His Instruments and Scientific Work.

OBSERVATIONAL EVIDENCE AGAINST
THE ARISTOTELIAN COSMOLOGY

We also prepared a special book on the immense comet that appeared five years later [in 1577]. In this we discuss it fully, including in the discussion our own observations and determinations as well as the opinions of others. We add a few pamphlets on this same subject, which elucidate this cometary problem more fully, and it was our plan to include all this in the first part of the second volume of the *Progymnasmata*. In the second part we shall, God willing, deal with the remaining six smaller comets that we observed with equal care in some of the subsequent years. Although all this has not yet been quite completed, the more important parts, and most of the demonstration, has been prepared. For the constant stars have not left us with sufficient time to dwell too long on these fading and quickly passing celestial bodies. Yet I hope that I shall soon, with the help of the gracious God, complete the second part of the second volume also. In this volume I shall clearly demonstrate that all the comets observed by me moved in the ethereal regions of the world and never in the air below the moon, as Aristotle and his followers have tried without reason to make us believe for so many centuries; and the demonstration will be clearest for some of the comets, while for others it will be according to the opportunity I had. The reasons why I treat the comets in the second volume of the *Progymnasmata* before I set about the other five planets, which I intend to discuss in the third volume, are given in the same place in the preface. But the principal reason is that the results pertaining to the comets, the true ethereal nature of which I prove conclusively, show that the entire sky is transparent and clear, and cannot contain any solid and real spheres. For the comets as a rule follow orbits of a kind that no celestial sphere whatever would permit, and consequently it is a settled thing that there is nothing unreasonable in the hypothesis invented by us (the Tychonic System), since we have found that there is no such thing as penetration of spheres and limits of distance, as the solid spheres do not really exist.

With regard to that which we have until now accomplished in Astronomy, and to that which has yet to be done, this brief account must now suffice.

8

— Johannes Kepler —

(1571–1630)

Johannes Kepler was a German-born astronomer and physicist, and an adherent of the Copernican system. He is regarded as one of the founders of modern astronomy. In 1600, Kepler went to Prague, where he became assistant to Tycho Brahe and was assigned the task of working out the orbit of Mars. Upon Brahe's death two years later, Kepler inherited the former's twenty-year archive of astronomical observations, enabling him to formulate the first two fundamental laws of planetary motion. A third law was added nine years later. Together, these three laws provided the foundation for all subsequent work on the solar system.

In this reading from the New Astronomy *(1609), Kepler presents his view of a motor force emanating from a central sun that controls the motion of the planets. Within the framework of this discussion he speculates about the genesis of this force, its kinship to light, and the relationship between a planet's distance from the sun and its orbital speed. Kepler offers a metaphysics of light in which he speaks of the "divinity" of the sun, and suggests that light might be the vehicle for the motive force that the sun imposes on the plants.*

Kepler ends the reading by comparing the Tychonic and Copernican models of the universe and by drawing an analogy between the solar forces and magnetism.

THE SUN AS THE SOURCE OF PLANETARY MOTIONS

The Power That Moves the Planets Resides in the Body of the Sun

It was demonstrated in the previous chapter that the elapsed times of a planet on equal parts of the eccentric circle (or on equal distances in the aethereal air) are in the same ratio as the distances of those spaces from the point whence the eccentricity is reckoned; or, more simply, to the extent that a planet is farther from the point which is taken as the centre of the world, it is less strongly urged to move about that point. It is therefore necessary that the cause of this weakening is either in the very body of the planet, in a motive force placed therein, or right at the supposed centre of the world.

Now it is an axiom in natural philosophy of the most common and general application that of those things which can occur at the same time and in the same manner, and which are always subject to like measurements, either one is the cause of the other or both are effects of the same cause. Just so, in this instance, the intension and remission of motion is always in the same ratio as the approach and recession from the centre of the world. Thus, either that weakening will be the cause of the star's motion away from the centre of the world, or the motion away will be the cause of the weakening, or both will have some cause in common. But it would be impossible for anyone to think up some third concurrent thing which would be the cause of these two, and in the following chapters it will become clear that we have no need of feigning any such cause, since the two are sufficient in themselves.

Further, it is not in accord with nature that strength or weakness in longitudinal motion should be the cause of distance from the centre. For distance from the centre is prior both in thought and in nature to motion over an interval. Indeed, motion over an interval is never independent of distance from the centre, since it requires a space in which to be performed, while distance from the centre can be conceived without motion. Therefore, distance will be the cause of intensity of motion, and a greater or lesser distance will result in a greater or lesser amount of time.

And since distance is a kind of relation whose essence resides in end points, while of relation itself, without respect to end points, there can be no efficient cause, it therefore follows (as has been said) that the cause of the variation of intensity of motion inheres in one or the other of the end points.

Now the body of a planet is never by itself made heavier in receding, nor lighter in approaching.

Moreover, that an animal force, which the motion of the heavens suggests is seated in the mobile body of the planet, undergoes intension and remission so many times without ever becoming tired or growing old,—this will surely be absurd to say. Also, it is impossible to understand how this animal force could carry its body through the spaces of the world, since there are no solid orbs, as Tycho Brahe has proved. And on the other hand, a round body lacks such aids as wings or feet, by the moving of which the soul might carry its body through the aethereal air as birds do in the atmosphere, by some kind of pressure upon, and counter-pressure from, that air.

Therefore, the only remaining possibility is that the cause of this intensification and weakening resides in the other endpoint, namely, in that point which is taken to be the centre of the world, from which the distances are measured.

So now, if the distance of the centre of the world from the body of a planet governs its slowness, and approach governs its speeding up, it is a necessary consequence that the source of motive power is at that supposed centre of the world. And with this laid down, the manner in which the cause operates is also clear. For it gives us to understand that the planets are moved rather in the manner of the steelyard or lever. For if the planet is moved with greater difficulty (and hence more slowly) by the power at the centre when it is farther from the centre, it is just as if I had said that where the weight is farther from the fulcrum, it is thereby rendered heavier, not of itself, but by the power of the arm supporting it at that distance. And this is true, both of the steelyard or lever, and of the motion of the planets: that the weakening of power is in the ratio of the distances.

But which body is it that is at the centre? Is there none, as for Copernicus when he is computing, and for Tycho in part? Is it the earth, as for Ptolemy and for Tycho in part? Or finally, is it the sun itself, as I, and Copernicus when he is speculating, would have it? I there supposed as one of the principles . . . that a planet is moved less vigorously when it recedes from the point whence the eccentricity is computed.

From this principle I presented a probable argument that the sun is at that point and at the centre of the world (or the earth for Ptolemy) rather than its being some other point occupied by no body. Allow me, then, to recall that same probable argument, its principles now demonstrated, in the present chapter. Then, as you may remember, I demonstrated . . . that the phenomena at either end of the night follow beautifully if the oppositions of Mars are reckoned according to the sun's apparent position. If this is done, then we likewise set up the eccentricity and the distances from the very centre of the sun's body, so that the sun itself again comes to be at the centre of the world (for Copernicus), or at least at the centre of the planetary system (for Tycho). But of these two arguments, one depends upon physical probability, and the other proceeds from possibility to actuality. And so in the third place I have demonstrated from the observations that we cannot avoid referring Mars to the apparent position of the sun, and drawing the line of apsides, which bisects the eccentric, directly through the sun's body, unless perhaps we wish to allow an eccentric such as will by no means be in accord with the parallax of the annual orb. . . .

Therefore, with the sun belonging in the centre of the system, the source of motive power, from what has now been demonstrated, belongs in the sun, since it too has now been located in the centre of the world.

But indeed, if this very thing which I have just demonstrated *a posteriori* (from the observations) by a rather long deduction, if, I say, I had taken this as something to be demonstrated *a priori* (from the worthiness and eminence of the sun), so that the source of the world's life (which is visible in the motion of the heavens) is the same as the source of the light which forms the adornment of the entire machine, and which is also the source of the heat by which everything grows, I think I would deserve an equal hearing.

Tycho Brahe himself, or anyone who prefers to follow his general hypothesis of the second inequality, should consider by how close a likeness to the truth this physically elegant combination has for the most part been accepted (since for him, too, this substitution of the apparent position of the sun brings the sun back to the centre of the planetary system) yet to some extent recoils from his hypothesis.

For it is obvious from what has been said that only one of the following can be true: either the power residing in the sun, which moves all the planets, by the same action moves the earth as well; or the sun, together with the planets linked to it through its motive force, is borne about the earth by some power which is seated in the earth.

Now Tycho himself destroyed the notion of real orbs, and I in turn have in this third part irrefutably demonstrated that there is an equant in the theory of the sun or earth. From this it follows that the motion of the sun itself (if it is moved) is intensified and remitted according as it is nearer or farther from the earth, and hence that the sun is moved by the earth. But if, on the other hand, the earth is in motion, it too will be moved by the sun with greater or less velocity according as it is nearer or farther from it, while the power in the body of the sun remains perpetually constant. Between these two possibilities, therefore, there is no intermediate.

I myself agree with Copernicus, and allow that the earth is one of the planets.

Now it is true that the same objection may be raised against Copernicus concerning the moon, that I raised against Tycho concerning the five planets; namely, that it appears absurd for the moon to be moved by the earth, and to be associated with it and bound to it as well, so that it too, as a secondary planet, is swept around the sun by the sun. Nevertheless, I prefer to allow one moon, akin to the earth in its corporeal disposition (as I have shown in the *Optics*) to be moved by a power seated in the earth but extended towards the sun . . . than to ascribe to that same earth as well the motion of the sun and of all the planets bound to it.

But let us carry on to a consideration of this motive power residing in the sun, and let us now again observe its very close kinship with light.

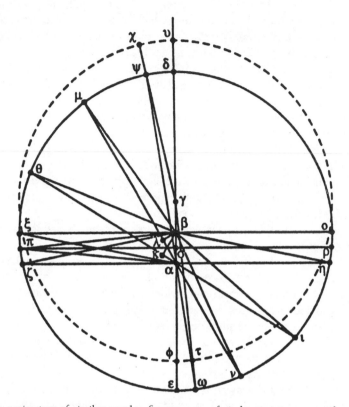

Since the perimeters of similar regular figures, even of circles, are to one another as their semidiameters, therefore as αδ is to αε, so is the circumference of the circle described about α through δ to the circumference of the circle described about the same point α through ε. But as αδ is to αε, so (inversely) is the strength of the power at ε to the strength of the power at δ. . . . Therefore, as the circle at δ is to the smaller circle at ε, so, inversely, is the power at ε to the power at δ that is, the power is weaker to the extent that it is more spread out, and stronger to the extent that it is more concentrated. Hence we may understand that there is the same power in the whole circumference of the circle through δ as there is in the circumference of the smaller circle through ε. This is shown to be true of light in exactly the same way in the Astronomiae pars optica, chapter 1. Therefore, in all respects and in all its attributes, the motive power from the sun coincides with light.

And although this light of the sun cannot be the moving power itself, I leave it to others to see whether light may perhaps be so constituted as to be, as it were, a kind of instrument or vehicle, of which the moving power makes use.

This seems gainsaid by the following: first, light is hindered by the opaque, and therefore if the moving power had light as a vehicle, darkness would result in the movable bodies being at rest; again, light spreads spherically in straight lines, while the moving power, though spreading in straight lines, does so circularly; that is, it is exerted in but one region of the world, from east to west, and not the opposite, not at the poles, and so on. But we shall be able to reply plausibly to these objections in the chapters immediately following.

Finally, since there is just as much power in a larger and more distant circle as there is in a smaller and closer one, nothing of this power is lost in travelling from its source, nothing is scattered between the source and the movable body. The emission, then, in the same manner as light, is immaterial, unlike odours, which are accompanied by a diminution of substance, and unlike heat from a hot furnace, or anything similar which fills the intervening space. The remaining possibility, then, is that, just as light, which lights the whole earth, is an immaterial *species* of that fire which is in the body of the sun, so this power which enfolds and bears the bodies of the planets, is an immaterial *species* residing in the sun itself, which is of inestimable strength, seeing that it is the primary agent of every motion in the universe.

Since, therefore, this *species* of the power, exactly as the *species* of light (for which see the *Astronomiae pars optica* ch. 1), cannot be considered as dispersed throughout the intermediate space between the source and the mobile body, but is seen as collected in the body in proportion to the amount of the circumference it occupies, this power (or *species*) will therefore not be any geometrical body, but is like a variety of surface, just as light is. To generalize this, the *species* which proceed immaterially from things are not by that procession extended through the dimensions of a body, although they arise from a body (as this one does from the body of the sun). Instead, they proceed according to that very law of emission: they do not possess their own boundaries, but just as the surfaces of illuminated things cause light to be considered as surfaces in certain respects, because they receive and terminate its emission, so the bodies of things that are moved suggest that this moving power be considered as if a sort of geometrical body, because their whole masses terminate or receive this emission of the motive *species*, so that the *species* can exist or subsist nowhere in the world but in the bodies of the mobile things themselves. And, exactly like light, between the source and the movable thing it is in a state of becoming, rather than of being.

Moreover, at the same time, a reply can be made here to a possible objection. For it was said above that this motive power is extended throughout the space of the world, in some places more spread out and in others more concentrated, and that the intensification and remission of the motions of the planets are consequent upon this variation. Now, however, it has been said that this power is an immaterial *species* of its source, and never inheres in anything except a mobile subject, such as the body of a planet. But to lack matter and yet to be subject to geometrical dimensions appear to be contradictory. This implies that it is poured out throughout the whole world, and yet does not exist anywhere but where there is something movable.

The reply is this: although the motive power is not anything material, nevertheless, since it is destined to carry matter (namely, the body of a planet), it is not free from geometrical laws, at least on account of this material action of carrying things about. Nor is there need for more, for we see that those motions are carried out in space and time, and that this power arises and is poured out from the source through the space of the world, all of which are geometrical entities. So this power should indeed be subject to other geometrical necessities.

But lest I appear to philosophize with excessive insolence, I shall propose to the reader the clearly authentic example of light, since it also makes its nest in the sun, thence to break forth into the whole world as a companion to this motive power. Who, I ask, will say that light is something material? Nevertheless, it carries out its operations with respect to place, suffers alteration, is reflected and refracted, and assumes quantities so as to be dense or rare, and to be capable of being taken as a surface wherever it falls upon something illuminable. Now just as it is said in optics, that light does not exist in the intermediate space between the source and the illuminable, this is equally true of the motive power. Moreover, although light itself does indeed flow forth in no time, while this power creates motion in time, nonetheless the way in which both do so is the same, if you consider them correctly. Light manifests those things which are proper to it instantaneously, but requires time to effect those which are associated with matter. It illuminates a surface in a moment, because here matter need not undergo any alteration, for all illumination takes place according to surfaces, or at least as if a property of surfaces and not as a property of corporeality as such. On the other hand, light bleaches colours in time, since here it acts upon matter *qua* matter, making it hot and expelling the contrary cold which is embedded in the body's matter and is not on its surface. In precisely the same manner, this moving power perpetually and without any interval of time is present from the sun wherever there is a suitable movable body, for it receives nothing from the movable body to cause it to be there. On the other hand, it causes motion in time, since the movable body is material.

Or if it seems better, frame the comparison in this manner: light is constituted for illumination, and it is just as certain that power is constituted for motion. Light does everything that can be done to achieve the greatest illumination; nonetheless, it does not happen that colour is most greatly illuminated. For colour intermingles its own peculiar *species* with the illumination of light, thus forming some third entity. In like manner, there is no retardation in the moving power to prevent the planet's having as much speed as it has itself, but it does not follow that the planet's speed is that great, since something intervening prevents that, namely, some sort of matter possessed by the surrounding aether, or the disposition of the movable body itself to rest (others might say, "weight," but I do not entirely approve of that, except, indeed, where the earth is concerned). It is the tempering effect of these, together with the weakening of the motive power, that determines a planet's periodic time.

The Sun Is a Magnetic Body, and Rotates in Its Space

Concerning that power that is closely attached to, and draws, the bodies of the planets, we have already said how it is formed, how it is akin to light, and what it is in its metaphysical being. Next, we shall contemplate the deeper nature of its source, shown by the outflowing *species* (or archetype). For it may appear that there lies hidden in the body of the sun a sort of divinity, which may be compared to our soul, from which flows that *species* driving the planets around, just as from the soul of someone throwing pebbles a *species* of motion comes to inhere in the pebbles thrown by him, even when he who threw them removes his hand from them. And to those who proceed soberly, other reflections will soon be provided.

The power that is extended from the sun to the planets moves them in a circular course around the immovable body of the sun. This cannot happen, or be conceived in thought, in any other way than this, that the power traverses the same path along which it carries the other planets. This has been observed to some extent in catapults and other violent motions. Thus, Fracastoro and others, relying on a story told by the most ancient Egyptians, spoke with little probability when they said that some of the planets perchance would have their orbits deflected gradually beyond the poles of the world, and thus afterwards would move in a path opposite to the rest and to their modern course. For it is much more likely that the bodies of the planets are always borne in that direction in which the power emanating from the sun tends.

But this *species* is immaterial, proceeding from its body out to this distance without the passing of any time, and is in all other respects like light. Therefore, it is not only required by the nature of the *species*, but likely in itself owing to this kinship with light, that along with the particles of its body or source it too is divided up, and when any particle of the solar body moves towards some part of the world, the particle of the immaterial *species* that from the beginning of creation corresponded to that particle of the body also always moves towards the same part. If this were not so, it would not be a *species*, and would come down from the body in curved rather than straight lines.

Since the *species* is moved in a circular course, in order thereby to confer motion upon the planets, the body of the sun, or source, must move with it, not, of course, from space to space in the world—for I have said, with Copernicus, that the body of the sun remains in the centre of the world—but upon its centre or axis, both immobile, its parts moving from place to place, while the whole body remains in the same place.

In order that the force of the analogical argument may be that much more evident, I would like you to recall, reader, the demonstration in optics that vision occurs through the emanation of small sparks of light toward the eye from the surfaces of the seen object. Now imagine that some orator in a great crowd of men, encircling him in an orb, turns his face, or his whole body along with it, once around. Those of the audience to whom he turns his eyes directly will also see his eyes, but those who stand behind him then lack the view of his eyes. But when he turns himself around, he turns his eyes around to everyone in the orb. Therefore, in a very short interval of time, all get a glimpse of his eyes. This they get by the arrival of a small spark of light or *species* of colour descending from the eyes of the orator to the eyes of the spectators. Thus by turning his eyes around in the small space in which his head is located, he carries around along with it the rays of the small spark of light in the very large orb in which the eyes of the spectators all around are situated. For unless the small spark of light went around, his spectators would not be recipients of his eyes' glance. Here you see clearly that the immaterial *species* of light either is moved around or stands still depending upon whether that of which it is the *species* either moves or stands still.

Therefore, since the *species* of the source, or the power moving the planets, rotates about the centre of the world, I conclude with good reason, following this example, that that of which it is the *species*, the sun, also rotates.

However, the same thing is also shown by the following argument. Motion which is local and subjected to time cannot inhere in a bare immaterial *species*, since such a *species* is incapable of receiving an applied motion unless the received motion is nontemporal, just as the power is immaterial. Also, although it has been proved that this moving power rotates, it cannot be allowed to have infinite speed (for then it would seem that infinite speed would also have to be imposed upon the bodies), and therefore it completes its rotation in some period of time. Therefore, it cannot carry out this motion by itself, and it is as a consequence necessary that it is moved only because the body upon which it depends is moved.

By the same argument, it appears to be a correct conclusion that there does not exist within the boundaries of the solar body anything immaterial by whose rotation the *species* descending from that immaterial something also rotates. For again, local motion which takes time cannot correctly be attributed to anything immaterial. It therefore remains that the body of the sun itself rotates in the manner described above, indicating the poles of the zodiac by the poles of its rotation (by extension to the fixed stars of the line from the centre of the body through the poles), and indicating the ecliptic by the greatest circle of its body, thus furnishing a natural cause for these astronomical entities.

Further, we see that the individual planets are not carried along with equal swiftness at every distance from the sun, nor is the speed of all of them at their various distances equal. For Saturn takes 30 years, Jupiter 12, Mars 23 months, earth 12, Venus eight and one half, and Mercury three. Nevertheless, it follows from what has been said that every orb of power emanating from the sun (in the space embraced by the lowest, Mercury, as well as that embraced by the highest, Saturn) is twisted around with a whirl equal to that which spins the solar body, with an equal period. (There is nothing absurd in this statement, for the emanating power is immaterial, and by its own nature would be capable of infinite speed if it were possible to impress a motion upon it from elsewhere, for then it could be impeded neither by weight, which it lacks, nor by the obstruction of the corporeal medium.) It is consequently clear that the planets are not so constituted as to emulate the swiftness of the motive power. For Saturn is less receptive than Jupiter, since its returns are slower, while the orb of power at the path of Saturn returns with the same swiftness as the orb of power at the path of Jupiter, and so on in order, all the way to Mercury, which, by example of the superior planets, doubtless moves more slowly than the power that pulls it. It is therefore necessary that the nature of the planetary globes is material, from an inherent property, arising from the origin of things, to be inclined to rest or to the privation of motion. When the tension between these things leads to a fight, the power is more overcome by that planet which is placed in a weaker power, and is moved more slowly by it, and is less overcome by a planet which is closer to the sun.

This analogy shows that there is in all planets, even in the lowest, Mercury, a material force of disengaging itself somewhat from the orb of the sun's power.

From this it is concluded that the rotation of the solar body anticipates considerably the periodic times of all the planets; therefore, it must rotate in its space at least once in a third of a year.

However, in my *Mysterium cosmographicum* I pointed out that there is about the same ratio between the semidiameters of the sun's body and the orb of Mercury as there is between the semidiameters of the body of the earth and the orb of the moon. Hence, you may plausibly conclude that the period of the orb of Mercury would have the same ratio to the period of the body of the sun as the period of the orb of the moon has to the period of the body of the earth. And the semidiameter of the orb of the moon is sixty times the semidiameter of the body of the earth, while the period of the orb of the moon (or the month) is a little less than thirty times the period of the body of the earth (or day), and thus the ratio of the distances is double the ratio of the periodic times. Therefore, if the doubled ratio also holds for the sun and Mercury, since the diameter of the sun's body is about one sixtieth of the diameter of Mercury's orb, the time of rotation of the solar globe will be one thirtieth of 88 days, which is the period of Mercury's orb. Hence it is likely that the sun rotates in about three days.

You may, on the other hand, prefer to prescribe the sun's diurnal period in such a way that the diurnal rotation of the earth is dispensed by the diurnal rotation of the sun, by some sort of magnetic force. I would certainly not object. Such a rapid rotation appears not to be alien to that body in which lies the first impulse for all motion.

This opinion (on the rotation of the solar body as the cause of the motion for the other planets) is beautifully confirmed by the example of the earth and the moon. For the chief, monthly motion of the moon . . . takes its origin entirely from the earth (for what the sun is for the rest of the planets there, the earth is for the moon in this demonstration). Consider, therefore, how our earth occasions the motion of the moon: while this our earth, and its immaterial *species* along with it, rotates twenty-nine and one half times about its axis, at the moon this *species* has the capability of driving it only once around in the same time, in (of course) the same direction in which the earth leads it.

Here, by the way, is a marvel: in any given time the centre of the moon traverses twice as long a line about the centre of the earth as any place on the surface of the earth beneath the great circle of the equator. For if equal spaces were measured out in equal times, the moon ought to return in sixty days, since the size of its orb is sixty times the size of the earth's globe.

This is surely because there is so much force in the immaterial *species* of the earth, while the lunar body is doubtless of great rarity and weak resistance. Thus, to remove your bewilderment, consider that on the principles we have supposed it would necessarily follow that if the moon's material force had no resistance to the motion impressed from outside by the earth, the moon would be carried at exactly the same speed as the earth's immaterial *species*, that is, with the earth itself, and would complete its circuit in 24 hours, in which the earth also completes its circuit. For even if the tenuity of the earth's *species* is great at the distance of 60 semidiameters, the ratio of one to nothing is still the same as the ratio of sixty to nothing. Hence the immaterial *species* of the earth would win out completely, if the moon did not resist.

Here, one might inquire of me, what sort of body I consider the sun to be, from which this motive *species* descends. I would ask him to proceed under the guidance of a further analogy, and urge him to inspect more closely the example of the magnet brought up a little earlier, whose power resides in the entire body of the magnet when it grows in mass, or when by being divided it is diminished. So in the sun the moving power appears so much stronger that it seems likely that its body is of all [those in the world] the most dense.

And the power of attracting iron is spread out in an orb from the magnet so that there exists a certain orb within which iron is attracted, but more strongly so as the iron comes nearer into the embrace of that orb. In exactly the same way the power moving the planets is propagated from the sun in an orb, and is weaker in the more remote parts of the orb.

The magnet, however, does not attract with all its parts, but has filaments (so to speak) or straight fibres (seat of the motor power) extended throughout its length, so that if a little strip of iron is placed in a middle position between the heads of the magnet at the side, the magnet does not attract it, but only directs it parallel to its own fibres. Thus it is credible that there is in the sun no force whatever attracting the planets, as there is in the magnet (for then they would approach the sun until they were quite joined with it), but only a directing force, and consequently that it has circular fibres all set up in the same direction, which are indicated by the zodiac circle.

Therefore, as the sun forever turns itself, the motive force or the outflowing of the *species* from the sun's magnetic fibres, diffused through all the distances of the planets, also rotates in an orb, and does so in the same time as the sun, just as when a magnet is moved about, the magnetic power is also moved, and the iron along with it, following the magnetic force.

The example of the magnet I have hit upon is a very pretty one, and entirely suited to the subject; indeed, it is little short of being the very truth. So why should I speak of the magnet as if it were an example? For, by the demonstration of the Englishman William Gilbert, the earth itself is a big magnet, and is said by the same author, a defender of Copernicus, to rotate once a day, just as I conjecture about the sun. And because of that rotation, and because it has magnetic fibres intersecting the line of its motion at right angles, those fibres lie in various circles about the poles of the earth parallel to its motion. I am therefore absolutely within my rights to state that the moon is carried along by the rotation of the earth and the motion of its magnetic power, only thirty times slower.

I know that the filaments of the earth, and its motion, indicate the equator, while the circuit of the moon is generally related to the zodiac. . . . With this one exception, everything fits: the earth is intimately related to the lunar period, just as the sun is to that of the other planets. And just as the planets are eccentric with respect to the sun, so is the moon with respect to the earth. So it is certain that the earth is looked upon by the moon's mover as its pole star (so to speak), just as the sun is looked upon by the movers belonging to the rest of the planets. . . . It is therefore plausible, since the earth moves the moon through its *species* and is a magnetic body, while the sun moves the planets similarly through an emitted *species*, that the sun is likewise a magnetic body.

9
— Galileo Galilei —
(1564–1642)

Galileo Galilei was an Italian astronomer and physicist who made significant contributions to both fields. His work in mechanics set the stage for Newton and the development of the science of mechanics. The law of falling bodies still bears his name. His work in mechanics appears in his book Discourse Concerning Two New Sciences, *which was published in 1638.*

In astronomy Galileo was the first to develop and use a telescope to observe the heavens. His discoveries, which included the moons of Jupiter, sunspots, and new stars, were revolutionary and contributed substantially to the acceptance of heliocentrism. These early observations appeared in the Sidereal *or* Starry Messenger, *which was published in 1610.*

In this reading Galileo describes his telescopic observations and their significance. The first part comes from the Starry Messenger, *the second from the* Letter on Sunspots, *written in 1613. In the first he describes the mountainous character and brightness of the moon, as well as the construction of the telescope, its powers of magnification, and a method for measuring the brightness and sizes of stars. Finally he offers details of the discovery of four moons of Jupiter.*

In the second part of the reading Galileo describes the discovery of the phases of Venus and offers this discovery as corroboration of Heliocentric Theory.

TELESCOPIC OBSERVATIONS IN SUPPORT OF COPERNICUS

Astronomical Message

Containing and Explaining Observations Recently Made, With the Benefit of a New Spyglass, About the Face of the Moon, the Milky Way, and Nebulous Stars, about Innumerable Fixed Stars and also Four Planets hitherto never seen, and named MEDICEAN STARS.

In this short treatise I propose great things for inspection and contemplation by every explorer of Nature. Great, I say, because of the excellence of the things themselves, because of their newness, unheard of through the ages, and also because of the instrument with the benefit of which they make themselves manifest to our sight.

Certainly it is a great thing to add to the countless multitude of fixed stars visible hitherto by natural means and expose to our eyes innumerable others never seen before, which exceed tenfold the number of old and known ones.

It is most beautiful and pleasing to the eye to look upon the lunar body, distant from us about sixty terrestrial diameters, from so near as if it were distant by only two of these measures, so that the diameter of the same Moon appears as if it were thirty times, the surface nine-hundred times, and the solid body about twenty-seven thousand times larger than when observed only with the naked eye. Anyone will then understand with the certainty of the senses that the Moon is by no means endowed with a smooth and polished surface, but is rough and uneven and, just as the face of the Earth itself, crowded everywhere with vast prominences, deep chasms, and convolutions.

Moreover, it seems of no small importance to have put an end to the debate about the Galaxy or Milky Way and to have made manifest its essence to the senses as well as the intellect; and it will be pleasing and most glorious to demonstrate clearly that the substance of those stars called nebulous up to now by all astronomers is very different from what has hitherto been thought.

But what greatly exceeds all admiration, and what especially impelled us to give notice to all astronomers and philosophers, is this, that we have discovered four wandering stars, known or observed by no one before us. These, like Venus and Mercury around the Sun, have their periods around a certain star notable among the number of known ones, and now precede, now follow, him, never digressing from him beyond certain limits. All these things were discovered and observed a few days ago by means of a glass contrived by me after I had been inspired by divine grace.

Perhaps more excellent things will be discovered in time, either by me or by others, with the help of a similar instrument, the form and construction of which, and the occasion of whose invention, I shall first mention briefly, and then I shall review the history of the observations made by me.

About 10 months ago a rumor came to our ears that a spyglass had been made by a certain Dutchman by means of which visible objects, although far removed from the eye of the observer, were distinctly perceived as though nearby. About this truly wonderful effect some accounts were spread abroad, to which some gave credence while others denied them. The rumor was confirmed to me a few days later by a letter from Paris from the noble Frenchman Jacques Badovere. This finally caused me to apply myself totally to investigating the principles and figuring out the means by which I might arrive at the invention of a similar instrument, which I achieved shortly afterward on the basis of the science of refraction. And first I prepared a lead tube in whose ends I fitted two glasses, both plane on one side while the other side of one was spherically convex and of the other concave. Then, applying my eye to the concave glass, I saw objects satisfactorily large and close. Indeed, they appeared three times closer and nine times larger than when observed with natural vision only. Afterward I made another more perfect one for myself that showed objects more than sixty times larger. Finally, sparing no labor or expense, I progressed so far that I constructed for myself an instrument so excellent that things seen through it appear about a thousand times larger and more than thirty times closer than when observed with the natural faculty only. It would be entirely superfluous to enumerate how many and how great the advantages of this instrument are on land and at sea. But having dismissed earthly things, I applied myself to explorations of the heavens. And first I looked at the Moon from so close that it was scarcely two terrestrial diameters distant. Next, with incredible delight I frequently observed the stars, fixed as well as wandering, and as I saw their huge number I began to think of, and at least discovered, a method whereby I could measure the distances between them. In this matter, it behooves all those who wish to make such observations to be forewarned. For it is necessary first that they prepare a most accurate glass that shows objects brightly, distinctly, and not veiled by any obscurity, and second that it multiply them at least four hundred times and show them twenty times closer. For if it is not an instrument such as that, one will try in vain to see all the things observed in the heavens by us and enumerated below. Indeed, in order that anyone may, with little trouble, make himself more certain about the magnification of the instrument, let him draw two circles or two squares on paper, one of which is four hundred times larger than the other, which will be the case when the larger diameter is twenty times the length of the other diameter. He will then observe from afar both sheets fixed to the same wall, the smaller one with one eye applied to the glass and the larger one with the other, naked eye. This can easily be done with both eyes open at the same time. Both figures will then appear of the same size if the instrument multiplies objects according to the desired proportion. After such an instrument has been prepared, the method of measuring distances is to be investigated, which is achieved by the following procedure. For the sake of easy comprehension, let *ABCD* be the tube and *E* the eye of the observer. When there are no glasses in the tube, the rays proceed to the object *FG* along the straight lines *ECF* and *EDG*, but with the glasses put in they proceed along the refracted lines *ECH* and *EDI*. They are indeed squeezed together and where before, free, they were directed to the object *FG*, now they only grasp the part *HI*. Then, having found the ratio of the distance EH to the line HI, the size of the angle subtended at the eye by the object HI is found from the table of sines, and we will find this angle to contain only some minutes, and if over the glass CD we fit plates perforated some with larger and some with smaller holes, putting now this plate and now that one over it as needed, we form at will angles subtending more or fewer minutes. By this means we can conveniently measure the spaces between stars separated from each other by several minutes with an error of less than one or two minutes. Let it suffice for the present, however, to have touched on this so lightly and to have, so to speak, tasted it only with our lips, for on another occasion we shall publish a complete theory of this instrument. Now let us review the observations made by us during the past 2 months, inviting all lovers of true philosophy to the start of truly great contemplation.

Let us speak first about the face of the Moon that is turned toward our sight, which, for the sake of easy understanding, I divided into two parts, namely a brighter one and a darker one. The brighter part appears to surround and pervade the entire hemisphere, but the darker part, like some cloud, stains its very face and renders it spotted. Indeed, these darkish and rather large spots are obvious to everyone, and every age has seen them. For this reason we shall call them the large or ancient spots, in contrast with other spots, smaller in size and occurring with such frequency that they besprinkle the entire lunar surface, but especially the brighter part. These were, in fact, observed by no one before us. By oft-repeated observations of them we have been led to the conclusion that we certainly see the surface of the Moon to be not smooth, even, and perfectly spherical, as the great crowd of philosophers have believed about this and other heavenly bodies, but, on the contrary, to be uneven, rough, and crowded with depressions and bulges. And it is like the face of the Earth itself, which is marked here and there with chains of mountains and depths of valleys. The observations from which this is inferred are as follows.

On the fourth or fifth day after conjunction, when the Moon displays herself to us with brilliant horns, the boundary dividing the bright from the dark part does not form a uniformly oval line, as would happen in a perfectly spherical solid, but is marked by an uneven, rough, and very sinuous line, as the figure shows. For several, as it were, bright excrescences extend beyond the border between light and darkness into the dark part, and on the other hand little dark parts enter into the light. Indeed, a great number of small darkish spots, entirely separated from the dark part, are distributed everywhere over almost the entire region already bathed by the light of the Sun, except, at any rate, for that part affected by the large and ancient spots. We noticed, moreover, that all these small spots just mentioned always agree in this, that they have a dark part on the side toward the Sun while on the side opposite the Sun they are crowned with brighter borders like shining ridges. And we have an almost entirely similar sight on Earth, around sunrise, when the valleys are not yet bathed in light but the surrounding mountains facing the Sun are already seen shining with light. And just as the shadows of the earthly valleys are diminished as the Sun climbs higher, so those lunar spots lose their darkness as the luminous part grows.

Not only are the boundaries between light and dark on the Moon perceived to be uneven and sinuous, but, what causes even greater wonder, is that very many bright points appear within the dark part of the Moon, entirely separated and removed from the illuminated region and located  no small distance from it. Gradually, after a small period of time, these are increased in size and brightness. Indeed, after 2 or 3 hours they are joined with the rest of the bright part, which has now become larger. In the meantime, more and more bright points light up, as if they are sprouting, in the dark part, grow, and are connected at length with that bright surface as it extends farther in this direction. An example of this is shown in the same figure. Now, on Earth, before sunrise, aren't the peaks of the highest mountains illuminated by the Sun's rays while shadows still cover the plain? Doesn't light grow, after a little while, until the middle and larger parts of the same mountains are illuminated, and finally, when the Sun has risen, aren't the illuminations of plains and hills joined together? These differences between prominences and depressions in the Moon, however, seem to exceed the terrestial roughness greatly, as we shall demonstrate below. Meanwhile, I would by no means be silent about something deserving notice, observed by me while the Moon was rushing toward first quadrature, the appearance of which is also shown in the above figure. For toward the lower horn, a vast dark gulf projected into the bright part. As I observed this for a long time, I saw it very dark. Finally, after about 2 hours, a bit below the middle of this cavity a certain bright peak began to rise and, gradually growing, it assumed a triangular shape, still entirely removed and separated from the bright face. Presently three other small points began to shine around it until, as the Moon was about to set, this enlarged triangular shape, now made larger, joined together with the rest of the bright part, and like a huge promontory, surrounded by the three bright peaks already mentioned, it broke out into the dark gulf. Also, in the tips of both the upper and lower horns, some bright points emerged, entirely separated from the rest of the light, as shown in the same figure. And there was a great abundance of dark spots in both horns, especially in the lower one. Of these, those closer to the boundary between light and dark appeared larger and darker while those farther away appeared less dark and more diluted. But as we have mentioned above, the dark part of the spot always faces the direction of the Sun and the brighter border surrounds the dark spot on the side turned away from the Sun and facing the dark part of the Moon. This lunar surface, which is decorated with spots like the dark blue eyes in the tail of a peacock, is rendered similar to those small glass vessels which, plunged into cold water while still warm, crack and acquire a wavy surface, after which they are commonly called ice-glasses. The large (and ancient) spots of the Moon, however, when broken up in a similar manner, are not seen to be filled with depressions and prominences, but rather to be even and uniform, for they are only here and there sprinkled with some brighter little places. Thus, if anyone wanted to resuscitate the old opinion of the Pythagoreans that the Moon is, as it were, another Earth, its brighter part would represent the land surface while its darker part would more appropriately represent the water surface. Indeed, for me there has never been any doubt that when the terrestrial globe, bathed in sunlight, is observed from a distance, the land surface will present itself brighter to the view and the water surface darker. Moreover, in the Moon the large spots are seen to be lower than the brighter areas, for in her waxing as well as waning, on the border between light and dark, there is always a prominence here or there around these large spots, next to the brighter part, as we have taken care to show in the figures; and the edges of the said spots are not only lower, but more uniform and not broken by creases or roughnesses. Indeed, the brighter part stands out very much near the ancient spots, so that both before the first and near the second quadrature some huge projections arise around a certain spot in the upper, northern part of the Moon, both above and below it, as the adjoining figures show.

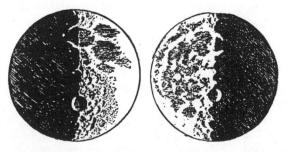

Before the second quadrature this same spot is seen walled around by some darker edges which, like a ridge of very high mountains turned away from the Sun, appear darker; and where they face the Sun they are brighter. The opposite of this occurs in valleys whose part away from the Sun appears brighter, while the part situated toward the Sun is dark and shady. Then, when the bright surface has decreased in size, as soon as almost this entire spot is covered in darkness, brighter ridges of mountains rise loftily out of the darkness. The following figures clearly demonstrate this double appearance.

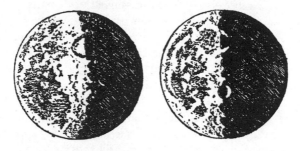

There is another thing that I noticed not without some admiration and that I may not omit. The area around the middle of the Moon is occupied by a certain cavity larger than all others and of a perfectly round figure. I observed this near both quadratures, and I have portrayed it as far as possible in the second figure above. It offers the same aspect to shadow and illumination as a region similar to Bohemia would offer on Earth, if it were enclosed on all sides by very high mountains, placed around the periphery in a perfect circle. For on the Moon it is surrounded by such lofty ranges that its side bordering on the dark part of the Moon is observed bathed in sunlight before the dividing line between light and shadow reaches the middle of the diameter of that circle. But in the manner of the other spots, its shaded part faces the Sun while its bright part is situated toward the dark part of the Moon, which, I advise for the third time, is to be esteemed as a very strong argument for the roughnesses and unevennesses scattered over the entire brighter region of the Moon. Its darker spots are always those that border on the boundary between light and dark, while the farther ones appear both smaller and less dark, so that finally, when the Moon is at opposition and full, the darkness of the depressions differs from the brightness of the prominences by a modest and quite small degree.

These things we have reviewed are observed in the brighter regions of the Moon. In the large spots, however, such a difference between depressions and prominences is not seen to be the same, as we are driven to conclude by necessity in the brighter part on account of the change of shapes caused by the changing illumination of the Sun's rays as it regards the Moon from many different positions. In the large spots there are some darkish areas, as we have shown in the figures, but yet those always have the same appearance, and their darkness is not increased or abated. Rather, they appear, with a very slight difference, now a little darker, now a little lighter, as the Sun's rays fall on them more or less obliquely. Moreover, they join with nearby parts of the spots in a gentle bond, their boundaries mingling and running together. Things happen differently, however, in the spots occupying the brighter part of the Moon, for like sheer cliffs sprinkled with rough and jagged rocks, these are divided by a line which sharply separates shadow from light. Moreover, in those larger spots certain other brighter areas—indeed, some very bright ones—are seen. But the appearance of these and the darker ones is always the same, with no change in shape, light, or shadow. It is thus known for certain and beyond doubt that they appear this way because of a real dissimilarity of parts and not merely because of inequalities in the shapes of their parts and shadows moving diversely because of the varying illumination by the Sun. This does happen beautifully in the other, smaller, spots occupying the brighter part of the Moon; day by day these are altered, increased, diminished, and destroyed, since they only derive from the shadows of rising prominences.

But I sense that many people are affected by great doubts in this matter and are so occupied by the grave difficulty that they are driven to call into doubt the conclusion already explained and confirmed by so many appearances. For if that part of the Moon's surface which more brilliantly reflects the solar rays is filled with innumerable contortions, that is, elevations and depressions, why is it that in the waxing Moon the limb facing west, in the waning Moon the eastern limb, and in the full Moon the entire periphery are seen not uneven, rough, and sinuous, but exactly round and circular, and not jagged with prominences and depressions? And especially because the entire edge consists of the brighter lunar substance which, we have said, is entirely bumpy and covered with depressions, for none of the large (ancient) spots reach the very edge, but all are seen to be clustered far from the periphery. Since these appearances present an opportunity for such serious doubt, I shall put forward a double cause for them and therefore a double explanation of the doubt. First, if the prominences and depressions in the lunar body were spread only along the single circular periphery outlining the hemisphere seen by us, then the Moon could indeed, nay it would have to, show itself to us in the shape of, as it were, a toothed wheel, that is, bumpy and bounded by a sinuous outline. If, however, there were not just one chain of prominences distributed only along a single circumference, but rather very many rows of mountains with their clefts and sinuosities were arranged about the outer circuit of the Moon—and these not only in the visible hemisphere but also in the one turned away from us (yet near the boundary between the hemispheres)—then the eye, observing from afar, could by no means perceive the distinction between the prominences and depressions. For the interruptions in the mountains arranged in the same circle or the same chain are hidden by the interposition of row upon row of other prominences, and especially if the eye of the observer is located on the same line with the peaks of those prominences. Thus on Earth the ridges of many mountains close together appear to be arranged in a flat surface if the observer is far away and situated at the same altitude. So (also) in a billowy sea the high tips of the waves appear stretched out in the same plane, even though between the waves there are very many troughs and gulfs so deep that not only the keels but also the upper decks, masts, and sails of tall ships are hidden. Since, therefore, in the Moon itself and around its perimeter there is a complex arrangement of prominences and depressions, and the eye, observing from afar, is located in about the same plane as their peaks, it should be surprising to no one that, with the visual rays skimming them, they show themselves in an even and not at all wavy line. To this reason another can be added, namely, that, just as around the Earth, there is around the lunar body a certain orb of denser substance than the rest of the ether, able to receive and reflect a ray of the Sun, although not endowed with so much opacity that it can inhibit the passage of vision (especially when it is not illuminated). That orb, illuminated by the solar rays, renders and represents the lunar body in the figure of a larger sphere, and, if it were thicker, it could limit our sight so as not to reach the actual body of the Moon. And it is indeed thicker around the periphery of the Moon; not absolutely thicker I say, but thicker as presented to our (visual) rays that intersect it obliquely. Therefore it can inhibit our vision and, especially when it is luminous, hide the periphery of the Moon exposed to the Sun. This is seen clearly in the adjoining figure, in which the lunar body *ABC* is surrounded by the vaporous orb *DEG*.

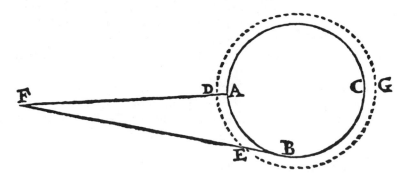

The eye at *F* reaches the middle parts of the Moon, as at A, through the shallower vapors DA, and toward its extreme parts an abundance of deeper vapors, EB, blocks our sight from its boundary. An indication of this is that the part of the Moon bathed in light appears greater in circumference than the remaining dark orb. And someone will perhaps find this cause reasonable to explain why the larger spots of the Moon are nowhere seen to extend to the outer edge, although it is to be expected that some of them would also be found near it. It seems plausible, then, that they are inconspicuous because they are hidden under thicker and brighter vapors.

From the appearances already explained, I think it is sufficiently clear that the brighter surface of the Moon is sprinkled all over with prominences and depressions. It remains for us to speak of their magnitudes, demonstrating that the terrestrial roughnesses are far smaller than the lunar ones. I say smaller, speaking absolutely, not merely in the proportion to the sizes of their globes. This is clearly shown in the following manner.

As has often been observed by me, with the Moon in various aspects to the Sun, some peaks within the dark part of the Moon appear drenched in light, although very far from the boundary line of the light. Comparing their distance from that boundary line to the entire lunar diameter, I found that this interval sometimes exceeds the twentieth part of the diameter. Assuming this, imagine the lunar globe, whose great circle is *CAF*, whose center is *E*, and whose diameter is *CF*, which is the Earth's diameter as 2 to 7. And since according to the most exact observations the terrestrial diameter contains 7000 Italian miles, *CF* will be 2000 miles, *CE* 1000, and the twentieth part of the whole of *CF* will be

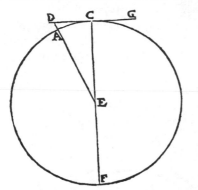

100 miles. Now let *CF* be the diameter of the great circle dividing the luminous from the dark part of the Moon (because of the very great distance of the Sun from the Moon this circle does not differ sensibly from a great circle), and let *A* be distant from point *C* one-twentieth part of it. Draw the semidiameter *EA*, which, when extended, intersects the tangent *GCD* (which represents a ray of light) at *D*. The arc *CA* or the straight line *CD* will therefore be 100 parts of the 1000 represented by *CE*, and the sum of the squares of *CD* and *CE* is 1,010,000, which is equal to the square of *ED*. The whole of ED will therefore be more than 1004, and AD more than 4 parts of the 1000 represented by CE. Therefore the height AD on the Moon, which represents some peak reaching all the way up to the Sun's rays GCD and removed from the boundary line C by the distance CD, is higher than 4 Italian miles. But on Earth no mountains exist that reach even to a perpendicular height of 1 mile.

It is evident, therefore, that the lunar prominences are loftier than the terrestrial ones.

In this place I wish to explain the cause of another lunar phenomenon worthy of notice. This phenomenon was observed by us not recently but rather many years ago, shown to some close friends and pupils, explained, and given a causal demonstration. But since the observation of it is made easier and more noticeable with the aid of the glasses, I thought it not unsuitable to be repeated here, and especially so that the relationship and similarity between the Moon and Earth may appear more clearly.

When, both before and after conjunction, the Moon is found not far from the Sun, she offers to our sight not only that part of her globe that is adorned with shining horns, but also a certain thin, faint periphery that is seen to outline the circle of the dark part (that is, the part turned away from the Sun) and to separate it from the darker field of the ether itself. But if we examine the matter more closely, we will see not only the extreme edge of the dark part shining with a faint brightness, but the entire face of the Moon—that part, namely, that does not yet feel the brightness of the Sun—made white by some not inconsiderable light. At first glance, however, only a slender shining circumference appears on account of the darker parts of the sky bordering it, while, on the contrary, the rest of the surface appears darker because the nearness of the shining horns makes our sight dark. But if one chooses a place for oneself so that those bright horns are concealed by a roof or a chimney or another obstacle between one's sight and the Moon (but positioned far away from the eye), the remaining part of the lunar globe is left exposed to one's view, and then one will discover that this region of the Moon, although deprived of sunlight, also shines with a considerable light, and especially when the chill of the night has already increased through the absence of the Sun. For in a darker field the same light appears brighter. It is moreover ascertained that this secondary brightness (as I call it) of the Moon is greater the less distant the Moon is from the Sun, for as she becomes more distant from him it is decreased more and more so that after the first quadarature and before the second it is found weak and very doubtful, even though it is seen in a darker sky, while at the sextile and smaller elongations it shines in a wonderful way although in the twilight. Indeed, it shines so much that with the aid of a precise glass the large spots can be distinguished in her. This marvelous brightness has caused no small astonishment to those applying themselves to philosophy, and some have put forward one reason and some another as the cause to be assigned to it. Some have said that it is the intrinsic and natural brightness of the Moon herself, others that it is imparted to it by Venus, or by all the stars, and yet others have said that it is imparted by the Sun who penetrates the Moon's vast mass with his rays. But such inventions are refuted with little difficulty and demonstrated to be false. For if this kind of light were either the Moon's own or gathered from the stars, she would retain it and show it especially during eclipses when she is placed in a very dark sky. This is not borne out by experience, however, for the light that appears in the Moon during an eclipse is much weaker, somewhat reddish, and almost coppery, while this light is brighter and whiter. The light that appears during an eclipse is, moreover, changeable and movable, for it wanders across the lunar face so that the part closer to the edge of the circle of the Earth's shadow is always seen brighter while the rest is darker. From this we understand with complete certainty that this light comes about because of the proximity of the solar rays falling upon some denser region which surrounds the Moon on all sides. Because of this contact a certain dawn light is spread over nearby areas of the Moon, just as on Earth twilight is spread in the morning and evening. We will treat this matter at greater length in a book on the system of the world. To declare, on the other hand, that this light is imparted by Venus is so childish as to be unworthy of an answer. For who is so ignorant as not to know that near conjunction and within the sextile aspect it is entirely impossible for the part of the Moon turned away from the Sun to be seen from Venus? But it is equally inconceivable that this light is due to the Sun, who with his light penetrates and fills the solid body of the Moon. For it would never be diminished, since a hemisphere of the Moon is always illuminated by the Sun except at the moment of a lunar eclipse. Yet the light is diminished when the Moon hastens toward quadrature and is entirely dimmed when she has gone beyond quadrature. Since, therefore, this secondary light is not intrinsic and proper to the Moon, and is borrowed neither from any star nor from the Sun, and since in the vastness of the world no other body therefore remains except the Earth, I ask what are we to think? What are we to propose——that the lunar body or some other dark and gloomy body is bathed by light from the Earth? But what is so surprising about that? In an equal and grateful exchange the Earth pays back the Moon with light equal to that which she receives from the Moon almost all the time in the deepest darkness of the night. Let us demonstrate the matter more clearly. At conjunction, when she occupies a place between the Sun and the Earth, the Moon is flooded by solar rays on her upper hemisphere that is turned away from the Earth. But the lower hemisphere turned toward the Earth is covered in darkness, and therefore it in no way illuminates the terrestrial surface. As the Moon gradually recedes from the Sun, some part of the inferior hemisphere turned toward us is soon illuminated and she turns somewhat white but thin horns toward us and lightly illuminates the Earth. The illumination of the Sun grows on the Moon as she approaches quadrature, and on Earth the reflection of her light increases. The brightness of the Moon is extended further, beyond a semicircle, and lights up our clear nights. Finally, the entire face of the Moon that regards the Earth is illuminated with a very bright light from the opposed Sun, and the Earth's surface shines far and wide, perfused by lunar splendor. Afterward, when the Moon is waning, she emits weaker rays toward us, and the Earth is weakly illuminated; and as the Moon hastens toward conjunction, dark night comes over the Earth. In this sequence, then, in alternate succession, the lunar light bestows upon us her monthly illuminations, now brighter, now weaker. But the favor is repaid by the Earth in like manner, for when the Moon is found near the Sun around conjunction, she faces the entire surface of the hemisphere of the Earth exposed to the Sun and illuminated by vigorous rays, and receives reflected light from her. And therefore, because of this reflection, the inferior hemisphere of the Moon, although destitute of solar light, appears of considerable brightness. When the Moon is removed from the Sun by a quadrant, she only sees the illuminated half of the terrestrial hemisphere, that is, the western one, for the other, the eastern half, is darkened by night. The Moon is therefore less brightly illuminated by the Earth, and her secondary light accordingly appears more feeble to us. For if you place the Moon at opposition to the Sun, she will face the hemisphere of the interposed Earth that is entirely dark and steeped in the shadow of night. If therefore, such an opposition were an eclipse, the Moon would receive absolutely no illumination, being deprived alike of solar and terrestrial radiation. In its various aspects to the Sun and Earth, the Moon receives more or less light by reflection from the Earth as she faces a larger or smaller part of the illuminated terrestrial hemisphere. For the relative positions of those two globes are always such that at those times when the Earth is most illuminated by the Moon the Moon is least illuminated by the Earth, and vice versa. Let these few things said here about this matter suffice. We will say more in our *System of the World*, where with very many arguments and experiments a very strong reflection of solar light from the Earth is demonstrated to those who claim that the Earth is to be excluded from the dance of the stars, especially because she is devoid of motion and light. For we will demonstrate that she is movable and surpasses the Moon in brightness, and that she is not the dump heap of the filth and dregs of the universe, and we will confirm this with innumerable arguments from nature.

Up to this point we have discussed the observations made of the lunar body. We will now report briefly on what has been observed by us thus far concerning the fixed stars. And first, it is worthy of notice that when they are observed by means of the spyglass, stars, fixed as well as wandering, are seen not to be magnified in size in the same proportion in which other objects, and also the Moon herself, are increased. In the stars, the increase appears much smaller so that you may believe that a glass capable of multiplying other objects, for example, by a ratio of 100 hardly multiplies stars by a ratio of 4 or 5. The reason for this is that when the stars are observed with the naked eye, they do not show themselves according to their simple and, so to speak, naked size, but rather surrounded by a certain brightness and crowned by twinkling rays, especially as the night advances. Because of this they appear much larger than if they were stripped of these extraneous rays, for the visual angle is determined not by the primary body of the star but by the widely surrounding brilliance. You will perhaps understand this more clearly from this: that stars emerging in the first twilight at sunset, even if they are of the first magnitude, appear very small, and Venus herself, when she presents herself to our view in broad daylight, is perceived so small that she hardly appears to equal a little star of the sixth magnitude. Things are different for other objects and the Moon herself, which, whether she is observed at midday or in the deepest darkness, appears always of the same size to us. Stars are therefore seen unshorn in the midst of darkness, but daylight can shear them of their hair—and not only daylight but also a thin little cloud that is interposed between the star and the eye of the observer. The same effect is also achieved by dark veils and colored glasses, by the opposition and interposition of which the surrounding brightness will desert the stars. The spyglass likewise does the same thing: for first it takes away the borrowed and accidental brightness from the stars and thereupon it enlarges their simple globes (if indeed their figures are globular), and therefore they appear increased by a much smaller ratio, for stars of the fifth or sixth magnitude seen through the spyglass are shown as of the first magnitude.

The difference between the appearance of planets and fixed stars also seems worthy of notice. For the planets present entirely smooth and exactly circular globes that appear as little moons, entirely covered with light, while the fixed stars are not seen bounded by circular outlines but rather as pulsating all around with certain bright rays. With the glass they appear in the same shape as when they are observed with natural vision, but so much larger that a little star of the fifth or sixth magnitude appears to equal the Dog Star, which is the largest of all fixed stars. Indeed, with the glass you will detect below stars of the sixth magnitude such a crowd of others that escape natural sight that it is hardly believable. For you may see more than six further gradations of magnitude. The largest of these, which we may designate as the seventh magnitude, or the first magnitude of the invisible ones, appear larger and brighter with the help of the glass than stars of the second magnitude seen with natural vision. But in order that you may see one or two illustrations of the almost inconceivable crowd of them, and from their example form a judgment about the rest of them, I decided to reproduce two star groups. In the first I had decided to depict the entire constellation of Orion, but overwhelmed by the enormous multitude of stars and a lack of time, I put off this assault until another occasion. For there are more than five hundred new stars around the old ones spread over a space of 1 or 2 degrees. For this reason, to the three in Orion's belt and the six in his sword that were observed long ago, I have added eighty others seen recently, and I have retained their separations as accurately as possible. For the sake of distinction, we have depicted the known or ancient ones larger and outlined by double lines, and the other inconspicuous ones smaller and outlined by single lines. We have also preserved the distinction in size as much as possible. In the second example we have depicted the six stars of the Bull called the Pleiades (I

The Belt and Sword of Orion

The Pleiades

say six since the seventh almost never appears) contained within very narrow limits in the heavens. Near these lie more than forty other invisible stars, none of which is farther removed from the aforementioned six than scarcely half a degree. We have marked down only thirty-six of these, preserving their mutual distances, sizes, and the distinction between old and new ones, as in the case of Orion.

What was observed by us in the third place is the nature or matter of the Milky Way itself, which, with the air of the spyglass, may be observed so well that all the disputes that for so many generations have vexed philosophers are destroyed by visible certainty, and we are liberated from wordy arguments. For the Galaxy is nothing else than a congeries of innumerable stars distributed in clusters. To whatever region of it you direct your spyglass, an immense number of stars immediately offer themselves to view, of which very many appear rather large and very conspicuous but the multitude of small ones is truly unfathomable.

And since that milky luster, like whitish clouds, is seen not only in the Milky Way, but dispersed through the ether, many similarly colored patches shine weakly; if you direct a glass to any of them, you will meet with a dense crowd of stars. Moreover—and what is even more remarkable—the stars that have been called "nebulous" by every single astronomer up to this day are swarms of small stars placed exceedingly closely together. While each individual one escapes our sight because of its smallness or its very great distance from us, from the commingling of their rays arises that brightness ascribed up to now to a denser part of the heavens capable of reflecting the rays of the stars or Sun. We have observed some of these, and we wanted to reproduce the asterisms of two of them.

In the first there is the nebula called Orion's Head, in which we have counted twenty-one stars.

The second figure contains the nebula called Praesepe, which is not a single star but a mass of more than forty little stars. In addition to the ass-colts we have marked down thirty-six stars, arranged as follows.

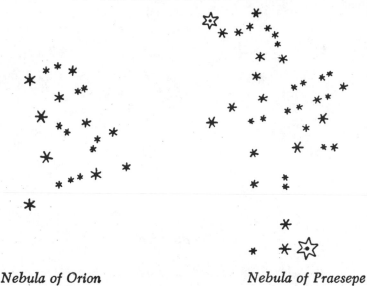

Nebula of Orion   *Nebula of Praesepe*

We have briefly explained our observations thus far about the Moon, the fixed stars, and the Milky Way. It remains for us to reveal and make known what appears to be most important in the present matter: four planets never seen from the beginning of the world right up to our day, the occasion of their discovery and observation, their positions, and the observations made over the past 2 months concerning their behavior and changes. And I call on all astronomers to devote themselves to investigating and determining their periods. Because of the shortness of time, it has not been possible for us to achieve this so far. We advise them again, however, that they will need a very accurate glass like the one we have described at the beginning of this account, lest they undertake such an investigation in vain.

Accordingly, on the seventh day of January of the present year 1610 at the first hour of the night, when I inspected the celestial constellations through a spyglass, Jupiter presented himself. And since I had prepared for myself a superlative instrument, I saw (which earlier had not happened because of the weakness of the other instruments) that three little stars were positioned near him—small but yet very bright. Although I believe them to be among the number of fixed stars, they nevertheless intrigued me because they appeared to be arranged exactly along a straight line and parallel to the ecliptic, and to be brighter than others of equal size. And their disposition among themselves and with respect to Jupiter was as follows:

East   *   *   ◯   *   **West**

That is, two stars were near him on the east and one on the west; the more eastern one and the western one appeared a bit larger than the remaining one.

I was not in the least concerned with their distances from Jupiter, for, as we said above, at first I believed them to be fixed stars. But when, on the eighth, I returned to the same observation, guided by I know not what fate, I found a very different arrangement. For all three little stars were to the west of Jupiter and closer to each other than the previous night, and separated by equal intervals, as shown in the adjoining sketch. Even though at this point I had by no means turned my thought to the mutual motions of these stars, yet I was aroused

East ○ ✳ ✳ ✳ *West*

by the question of how Jupiter could be to the east of all the said fixed stars when the day before he had been to the west of two of them. I was afraid, therefore, that perhaps, contrary to the astronomical computations, his motion was direct and that, by his proper motion, he had bypassed those stars. For this reason I waited eagerly for the next night. But I was disappointed in my hope, for the sky was everywhere covered with clouds.

Then, on the tenth, the stars appeared in this position with regard to Jupiter. Only two stars were near him, both to the east. The third, as I thought, was

East ✳ ✳ ○ *West*

hidden behind Jupiter. As before, they were in the same straight line with Jupiter and exactly aligned along the zodiac. When I saw this, and since I knew that such changes could in no way be assigned to Jupiter, and since I knew, moreover, that the observed stars were always the same ones (for no others, either preceding or following Jupiter, were present along the zodiac for a great distance), now, moving from doubt to astonishment, I found that the observed change was not in Jupiter but in the said stars. And therefore I decided that henceforth they should be observed more accurately and diligently. . . .

These are the observations of the four Medicean planets recently, and for the first time, discovered by me. From them, although it is not yet possible to calculate their periods, something worthy of notice may at least be said. And first, since they sometimes follow and at other times precede Jupiter by similar intervals, and are removed from him toward the east as well as the west by only very narrow limits, and accompany him equally in retrograde and direct motion, no one can doubt that they complete their revolutions about him while, in the meantime, all together they complete a 12-year period about the center of the world. Moreover, they whirl around in unequal circles, which is clearly deduced from the fact that at the greatest separations from Jupiter two planets could never be seen united while, on the other hand, near Jupiter two, three, and occasionally all four planets are found crowded together at the same time. It is further seen that the revolutions of the planets describing smaller circles around Jupiter are faster. For the stars closer to Jupiter are often seen to the east when the previous day they appeared to the west and vice versa, while from a careful examination of its previously accurately noted returns, the planet traversing the largest orb appears to have a semimonthly period. We have moreover an excellent and splendid argument for taking away the scruples of those who, while tolerating with equanimity the revolution of the planets around the Sun in the Copernican system, are so disturbed by the attendance of one Moon around the Earth while the two together complete the annual orb around the Sun that they conclude that this constitution of the universe must be overthrown as impossible. For here we have only one planet revolving around another while both run through a great circle around the Sun: but our vision offers us four stars wandering around Jupiter like the Moon around the Earth while all together with Jupiter traverse a great circle around the Sun in the space of 12 years. Finally, we must not neglect the reason why it happens that the Medicean stars, while completing their very small revolutions around Jupiter, are themselves now and then seen twice as large. We can in no way seek the cause in terrestrial vapors, for the stars appear larger and smaller when the sizes of Jupiter and nearby fixed stars are seen completely unchanged. It seems inconceivable moreover, that they approach and recede from the Earth by such a degree around the perigees and apogees of their orbits as to cause such large changes. For smaller circular motions can in no way be responsible, while an oval motion (which in this case would have to be almost straight) appears to be both inconceivable and by no account harmonious with the appearances. I gladly offer what occurs to me in this matter and submit it to the judgment and censure of right-thinking men. It is well known that because of the interposition of terrestrial vapors the sun and Moon appear larger but the fixed stars and planets smaller. For this reason, near the horizon the luminaries appear larger but the stars (and planets) smaller and generally inconspicuous, and they are diminished even more if the same vapors are perfused by light. For that reason the stars (and planets) appear very small by day and during twilight, but not the Moon, as we have already stated above. From what we have said above as well as from those things that will be discussed more amply in our system, it is moreover certain that not only the Earth but also the Moon has its surrounding vaporous orb. And we can accordingly make the same judgment about the remaining planets, so that it does not appear inconceivable to put around Jupiter an orb denser than the rest of the ether around which the Medicean planets are led like the Moon around the sphere of the elements. And at apogee, by the interposition of this orb, they are smaller, but when at perigee, because of the absence or attenuation of this orb, they appear larger. Lack of time prevents me from proceeding further. The fair reader may expect more about these matters soon.

Letter on Sunspots

Next Apelles suggests that sunspot observations afford a method by which he can determine whether Venus and Mercury revolve about the sun or between the earth and the sun. I am astonished that nothing has reached his ears—or if anything has, that he has not capitalized upon it—of a very elegant, palpable, and convenient method of determining this, discovered by me about two years ago and communicated to so many people that by now it has become notorious. This is the fact that Venus changes shape precisely as does the moon, and if Apelles will now look through his telescope he will see Venus to be perfectly circular in shape and very small (though indeed it was smaller yet when it (recently) emerged as evening star). He may then go on observing it, and he will see that as it reaches its maximum departure from the sun it will be semicircular. Thence it will pass into a horned shape, gradually becoming thinner as it once more approaches the sun. Around conjunction it will appear as does the moon when two or three days old, but the size of its visible circle will have much increased. Indeed, when Venus emerges (from behind the sun) to appear as evening star, its apparent diameter is only one-sixth as great as its evening disappearance (in front of the sun) or its emergence as morning star (several days thereafter), and hence its disk appears forty times as large on the latter occasions.

These things leave no room for doubt about the orbit of Venus. With absolute necessity we shall conclude, in agreement with the theories of the Pythagoreans and of Copernicus, that Venus revolves about the sun just as do all the other planets. Hence it is not necessary to wait for transits and occultations of Venus to make certain of so obvious a conclusion. No longer need we employ arguments that allow any answer, however feeble, from persons whose philosophy is badly upset by this new arrangement of the universe. For these opponents, unless constrained by some stronger argument, would say that Venus either shines with its own light or is of a substance that may be penetrated by the sun's rays, so that it may be lighted not only on its surface but also throughout its depth. They take heart to shield themselves with this argument because there have not been wanting philosophers and mathematicians who have actually believed this—meaning no offense to Apelles, who says otherwise. Indeed, Copernicus himself was forced to admit the possibility and even the necessity of one of these two ideas, as otherwise he could give no reason for Venus failing to appear horned when beneath the sun. As a matter of fact nothing else could be said before the telescope came along to show us that Venus is naturally and actually dark like the moon, and like the moon has phases.

10
— Johannes Kepler —

In this reading Kepler defends the Copernican system, explaining its advantages over that of Ptolemy, Aristotle, and Brahe. He also provides a historical overview of the development of Heliocentric Theory and gives further evidence of its validity. Kepler argues, against instrumentalism, that the goal of astronomy is not just to save appearances but also to know the true structure of the cosmos.

In his support of Copernicus, Kepler severely criticizes the Aristotelian argument for a central earth and offers eighteen separate arguments for the immobility of the sun and the mobility of the earth. Extending his support for Copernicus beyond the merits of his system, Kepler provides a very interesting comparison between the reasoning of Copernicus and that of the ancients, and a comparison between reasoning of Copernicus and that of Brahe.

The reading comes from Kepler's The Epitome of Copernican Philosophy.

THE SUPERIORITY OF THE COPERNICAN SYSTEM

What are the hypotheses or principles wherewith Copernican astronomy saves the appearance in the proper movements of the planets?

They are principally: (1) that the sun is located at the centre of the sphere of the fixed stars—or approximately at the centre—and is immovable in place; (2) that the single planets move really around the sun in their single systems, which are compounded of many perfect circles revolved in an absolutely uniform movement; (3) that the Earth is one of the planets, so that by its mean annual movement around the sun it describes its orbital circle between the orbital circles of Mars and of Venus; (4) that the ratio of its orbital circle to the diameter of the sphere of the fixed stars is imperceptible to sense and therefore, as it were, exceeds measurements; (5) that the sphere of the moon is arranged around the Earth as its centre, so that the annual movement around the sun—and so the movement from place to place—is common to the whole sphere of the moon and to the Earth.

Do you judge that these principles should be held to in this Epitome?

Since astronomy has two ends, to save the appearances and to contemplate the true form of the edifice of the world . . . there is no need of all these principles in order to attain the first end: but some can be changed and others can be omitted; however, the second principle must necessarily be corrected: and even though most of these principles are necessary for the second end, nevertheless they are not yet sufficient.

Which of these principles can be changed or omitted and the appearances still be saved?

Tycho Brahe demonstrates the appearances with the first and third principles changed: for he, like the ancients, places the Earth immobile, at the centre of the world; but the sun—which even for him is the centre of the orbital circles of the five planets—and the system of all the spheres he makes to go around the Earth in the common annual movement, while at the same time in this common system any planet completes its proper movements. Moreover, he omits the fourth principle altogether and exhibits the sphere of the fixed stars as not much greater than the sphere of Saturn.

What in turn do you substitute for the second principle and what else do you add to the true form of the dwelling of the world or to what belongs to the nature of the heavens?

Even though the true movements are to be left singly to the single planets, nevertheless these movements do not move by themselves nor by the revolutions of spheres—for there are no solid spheres—but the sun in the centre of the world, revolving around the centre of its body and around its axis, by this revolution becomes the cause of the single planets going around.

Further, even though the planets are really eccentric to the centre of the sun: nevertheless there are no other smaller circles called epicycles, which by their revolution vary the intervals between the planet and the sun; but the bodies themselves of the planets, by an inborn force (*vi insite*), furnish the occasion for this variation. . . .

On the Place of the Sun at the Centre of the World

By what arguments do you affirm that the sun is situated at the centre of the world?

The very ancient Pythagoreans and the Italian philosophers supply us with some of those arguments in Aristotle (*On the Heavens*, Book, II, Chapter 13); and these arguments are drawn from the dignity of the sun and that of the place, and from the sun's office of vivification and illumination in the world.

State the first argument from dignity.

This is the reasoning of the Pythagoreans according to Aristotle: the more worthy place is due to the most worthy and most precious body. Now the sun—for which they use the word "fire," as sects purposely hiding their teachings—is worthier than the Earth and is the most worthy and most precious body in the whole world, as was shown a little before. But the surface and centre, or mid-point, are the two extremities of a sphere. Therefore one of these places is due to the sun. But not the surface; for that which is the principal body in the whole world should watch over all the bodies; but the centre is suited for this function, and so they used to call it the Watchtower of Jupiter. And so it is not proper that the Earth should be in the middle. For this place belongs to the sun, while the Earth is borne around the centre of its yearly movement.

What answer does Aristotle make to this argument?

1. He says that they assume something which is not granted, namely that the centre of magnitude, i.e. of the sphere, and the centre of the things, i.e. of the body of the world, and so of nature, i.e. of informing or vivifying, are the same.

But just as in animals the centre of vivification and the centre of the body are not the same—for the heart is inside but is not equally distant from the surface—we should think in the same way about the heavens, and we should not fear for the safety of the whole universe or place a guard at the centre; rather, we should ask what sort of body the heart of the world of the centre or vivification is and in what place in the world it is situated.

2. He tried to show the dissimilarity between the midpart of the nature and the midpart of place. For the midpart of nature, or the most worthy and precious body, has the proportionality of a beginning. But in the midpart of place is the last, in quantity considered metaphysically, rather than the first or the beginning. For that which is the midpart of quantity, i.e., is the farthest in, is bounded or circumscribed. But the limits are that which bounds or circumscribes. Now that which goes around on the outside, and limits and encloses, is of greater excellence and worth than that which is on the inside and is bounded: for matter is among those things which are bounded, limited, and contained; but form, or the essence of any creature, is of the number of those things which limit, circumscribe, and comprehend. He thinks that he has proved in this way that not so much the midpart of the world as the extremity belongs to the sun, or as he understood it, to the fire of the Pythagoreans.

How do you rebut this refutation of Aristotle's?

1. Even if it be true that not in all creatures and least in animals is the principal part of the whole creature at the centre of the whole mass: however, since we are arguing about the world, nothing is more probable than this. For the figure of the world is spherical, and that of an animal is not. For animals need organs extending outside themselves, with which they stand upon the ground, and upon which they may move, and with which they may take within themselves the food, drink, forms of things, and sounds received from outside. The world on the contrary, is alone, having nothing outside, resting on itself immobile as a whole; and it alone is all things. And so there is no reason why the heart of the world should be elsewhere than in the center in order that what it is, viz., the heart, might be equally distant from the farthest parts of the world, that is to say, by an interval everywhere equal.

2. Furthermore, as regards his telling us to ask what sort of body the principal part of the whole universe is: he is confused by that riddle of the Pythagoreans and believes that they claim that this element is principal. He is not wrong however in telling us to do that. And accordingly we, following the advice of Aristotle, have picked out the sun; and neither the Pythagoreans in their mystical sense nor Aristotle himself are against us. And when we ask in what place in the world the sun is situated, Copernicus, as being skilled in the knowledge of the heavens, shows us that the sun is in the midpart. The others who exhibit its place as elsewhere are not forced to do this by astronomical arguments but by certain others of a metaphysical character drawn from the consideration of the Earth and its place. Both we and they set a value upon these arguments; and

they themselves too by means of these arguments do not show but seek the place of the sun. So if when seeking the place of the sun in the world, we find that it is the centre of the world; we are doing just as Aristotle; and his refutation does not apply to us.

3. As regards the fact that Aristotle, directly contradicting the Pythagoreans ascribes vileness to the center, he does that contrary to the nature of figures and contrary to their geometrical or metaphysical consideration. . . .

Prove by means of the office of the sun that the centre is due to it.

That has already been partly done in rebutting the Aristotlean refutation. For (1) if the whole world, which is spherical, is equally in need of the light of the sun and its heat, then it would be best for the sun to be at the midpart, whence light and heat may be distributed to all the regions of the world. And that takes place more uniformly and rightly, with the sun resting at the centre than with the sun moving around the centre. For if the sun approached certain regions for the sake of warming them, it would draw away from the opposite regions and would cause alternations while it itself remained perfectly simple. And it is surprising that some people use jokingly the similitude of light at the centre of the lamp, as it is a very apt similitude, lease fitted to satirize this opinion but suited rather to painting the power of this argument.

(2) But a special argument is woven together concerning light, which presupposes fitness, not necessity. Imagine the sphere of the fixed stars as a concave mirror: you know that the eye placed at the centre of such a mirror gazes upon itself everywhere: and if there is a light at the centre, it is everywhere reflected at right angles from the concave surface and the reflected rays come together again at the centre. And in fact that can occur at no other point in the concave mirror except at the centre. Therefore, since the sun is the source of light and eye of the world, the centre is due to it in order that the sun—as the Father in the divine symbolizing—may contemplate itself in the whole concave surface— which is the symbol of God the Son—and take pleasure in the imago of itself, and illuminate itself by shining and inflame itself by warming. These melodious little verses apply to the sun:

> Thou who dost gaze at thy face
> and dost everywhere leap back
> from the navel of the upper air
> O gushing up of the gleams flowing
> through the glass emptiness, Sun,
> who dost again swallow thy reflections.

Nevertheless Copernicus did not place the sun exactly at the center of the world?

It was the intention of Copernicus to show that this node common to all the planetary systems—of which node we shall speak below—is as far distant from the centre of the sun as the ancients made the eccentricity of the sun to be. He established this node as the centre of the world, and was compelled to do so by no astronomical demonstration but on account of fitness alone, in order that this node and, as it were, the common centre of the mobile spheres would not differ from the very centre of the world. But if anyone else, in applying this same fitness, wished to control that we should rather fear to make the sun differ from the centre of the world, and that it was sufficient that this node of the region of the moving planets should be situated very near, even if not exactly at the centre—anyone who wished to make this contention, I say, would have raised no disturbance in Copernican astronomy. So, firstly, the last arguments concerning the place of the sun at the centre are nevertheless unaffected by this opinion of Copernicus concerning the distance of this node from the sun. But secondly we must not agree to the opinion of Copernicus that this node is distant from the centre of the sun. For the common node of the region of the mobile planets is in the sun, as will be proved below; and so by some probable arguments either the one or the other point is set down at the centre of the sphere of the fixed stars, and by the same arguments the other point is brought to the same place, even with the approval of Copernicus.

On the Order of the Movable Spheres

How are the planets divided among themselves?

Into the primary and the secondary. The primary planets are those whose bodies are borne around the sun, as will be shown below: the secondary planets are those whose own circles are arranged not around the sun but around one of the primary planets and who also share in the movement of the primary planet around the sun. Saturn is believed to have two such secondary planets and to draw them around with itself: they come into sight now and then with the help of a telescope. Jupiter has four such planets around itself: D, E, F, H. The Earth (B) has one (C) called the moon. It is not yet clear in the case of Mars, Venus, and Mercury whether they too have such a companion or satellite.

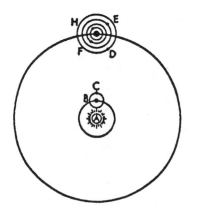

Then how many planets are to be considered in the doctrine on schemata?

No more than seven: the six so-called primary planets: (1) Saturn, (2) Jupiter, (3) Mars, (4) the Earth—the sun to eyesight, (5) Venus, (6) Mercury, and (7) only one of the secondary planets, the moon, because it alone revolves around our home, the Earth; the other secondary planets do not concern us who inhabit the Earth, and we cannot, behold them without excellent telescopes.

In what order are the planets laid out: are they in the same heaven or in different heavens?

Eyesight places them all in that farthest and highest sphere of the fixed stars and opines that they move among the fixed stars. But reason persuades men of all times and of all sects that the case is different. For if the centres of all the planets were in the same sphere and since we see that to sight they are fairly often in conjunction with one another: accordingly one planet would impede the other, and their movements could not be regular and perpetual.

But the reasoning of Copernicus and ancient Aristarchus, which relies upon observations, proves that the regions of the single planets are separated by very great intervals from one another and from the fixed stars.

What is the difference here between the reasoning of Copernicus and that of the ancients?

1. The reasoning of the ancients is merely probable, but the demonstration of Copernicus, arising from his principles, brings necessity.

2. They teach only that there is not more than one planet in any one sphere: Copernicus further adds how great a distance any planet must necessarily be above another.

3. Now the ancients built up one heaven upon another, like layers in a wall, or, to use a closer analogy, like onion skins: the inner supports the outer; for they thought that all intervals had to be filled by spheres and that the higher sphere must be set down as being only as great as the lower sphere of a known magnitude allows; and that is only a material conformation. Copernicus, having measured by his observation the intervals between the single spheres, showed that there is such a great distance between two planetary spheres, that it is unbelievable that it should be filled with spheres. And so this lay-out of his urges the speculative mind to spurn matter and the contiguity of spheres and to look towards the investigation of the formal lay-out or archetype, with reference to which the intervals were made.

4. The ancients, with their material structure, were forced to make the planetary or mobile world many parts greater than Copernicus was forced to do with his formal lay-out. But Copernicus, on the contrary, made the region of the mobile planets not very large, while he made the motionless sphere of the fixed stars immense. The ancients do not make it much greater than the sphere of Saturn.

5. The ancients do not explain and confirm as they desire the reason for their lay-out; Copernicus establishes his lay-out excellently by reasons . . . .

On the Annual Movement of the Earth

Accordingly this philosophy of Copernicus makes the Earth one of the planets and sets it revolving among the stars: I ask what, besides what has been said, is required for the easier perception of the teaching and the arguments.

Since the annual movement of the Earth becomes necessary, because it has been postulated that the centre of the sun is at rest at the centre of the world, and since this movement is caused by the revolution of the sun in that space and clearly removes the truth of the stopping and retrogradation of the planets and explains it as a mere deception of sight: we must distinguish carefully the following questions: (1) Whether the sun sticks to the centre of the world. (2) Whether all the five spheres of the planets and the middle sphere of the Earth are drawn up around the sun, so that the sun is in the embrace of all. (3) Whether the sun occupies the very centre of the whole planetary system, or whether it stands outside of that. (4) Whether this centre of the system and the sun in it revolve in an annual movement, or whether the Earth has an annual movement through the sections opposite to those in which the sun is thought to be moving at any time.

You have proved above that (1) the sun is at the centre of the sphere of the fixed stars. Now prove also that (2) it is within the embrace of the planetary spheres.

That the sun is at the centre of the planetary revolutions is proved first from an accident of this movement, namely, the appearance of stoppings and retrogradings, which are deceptions of sight, or even because the planets seem to be faster when progressing than they really are.

For to begin with the lower planets, now for a long time during the many ages which followed Ptolemy—let us say nothing at present of ancient Aristarchus—it was perceived by the authors Martianus Capella, Campanus, and others, that it is not possible for the sun, Venus, and Mercury to have the same period of time, namely, a year, unless they have the same sphere and unless the sun is at the centre of the two spheres of Venus and Mercury and these planets revolve around the sun: for that reason, when these planets seem to retrograde, they are not really retrograding but are advancing in the same direction in the sphere of the fixed stars but are going around the sun. And that is more consonant with the nature of celestial things.

A few years ago Galileo confirmed this argument by a very clear demonstration: by means of a telescope he disclosed the illumination of Venus. When Venus is progressing and is in the neighbourhood of the sun, it has a round figure; when retrograding, a horn-shaped figure. For from this it is proven with the utmost certainty that its illumination comes from the sun, that when Venus appears round and progresses straight ahead, it is above the sun; but that when it is horn-shaped and retrograding, it is below the sun, and that it thus revolves around the sun. Let the demonstration of this thing by reason of light be joined to the demonstrations of the illuminations of the moon. In the case of Mercury,

Marius brings forward similar things by the aid of the telescope: the feebleness of its light was recognized as the planet came down to the Earth: and that is a sign that the form (*speciem*) of the illumination is changed and the light has become weaker in the horn, so that it moves the eye less when near than when far away. And that would be absurd without this weakening in the horn; because elsewhere things which are nearer appear greater than if they had drawn away farther. Now as regards the three upper planets, Aristarchus, Copernicus, and Tycho Brahe demonstrate that if we set them in order around the sun and put the sun as, so to speak, the common centre of the five

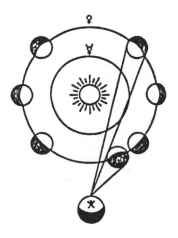

planets, so that the movement of the sun, whether true or apparent, affects all the spheres of the five planets, we are freed—as before in the case of Venus and Mercury, from two eccentric circles, so now in the upper planets—(1) from three epicycles; (2) from the blind unbelievable harmony of their real movement with the movement of the sun; (3) and just as above in the case of Venus and Mercury, they have no real stations and retrogradations with respect to the sun, around which they revolve; (4) thus also very many complications in the latitudinal movement are removed from the doctrine of the schemata; (5) and finally the reasons are disclosed for the difference which makes the five planets become stationary and retrograding, but never the sun and the moon; and (6) why Saturn, the highest of the upper planets, has the least arc of retrogradation; Jupiter, the middle one, the middle arc; and Mars, the nearest one, the greatest arc. All these things will be explained below in Book VI. But the ancient astronomers were totally ignorant of the reasons for these appearances.

But even the secondary planets bear some witness to this thing. For Marius found that in the world of Jupiter the restitutions of the jovial satellites around Jupiter is never regular with respect to the lines which we cast out from the centre of the Earth to Jupiter; but that they are regular, if they are compared with the lines drawn through Jupiter from the centre of the sun. And that argues strongly that the orbit of Jupiter is arranged around the sun and that the distance of the sun from the centre of the orbit of Jupiter is sure and somehow fixed; but that throughout the year the Earth varies its distances from this centre.

How many sects of astronomers are there with respect to this theory, from which the second argument is drawn?

Three: the first, commonly known by the name of the ancients, nevertheless has Ptolemy as its coryphaeus; the second and third are ascribed to the moderns. Though the second, named after Copernicus, is the oldest, Tycho Brahe is the founder of the third.

Accordingly Ptolemy treats the single wandering stars only separately, and he ascribes the apparent causes of all the movements, retrogradations, and stations, to their separate spheres. None the less in the case of each planet he sets down one unchanging sphere which completes its period with respect to the movement of the sun: but Ptolemy does not explain from what causes that comes about—unless the Latin writers, hypnotized by their complete ignorance of the rays, attribute some obscure force to the constant rays of the sun.

The remaining two founders compare the planets with one another; and the things which are found to be common to their movements are deduced from the same common cause. But this common cause, which makes the planets seem

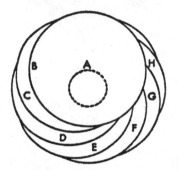

stationary and retrograding in some fixed configuration of the planet and the sun, is still attributed by Brahe to the real movement of all the planetary spheres, but it is completely removed by Copernicus from the spheres of the planets. For Brahe teaches that all the five spheres of the primary planets are bound together at some common point, which is not far distant from the centre of each sphere—as if here all the spheres were described in the common circular table B—and that this, so to speak, common node or knot really revolves with the sun during the year, and that very near to this node the sun—in the small circle made with dots—makes all the spheres revolve along with itself; and so to speak dislocates them from their own regions in the world—in the manner wherein bolters, grasping a sieve by one part of its rim, turn it with their hands and shake it; hence, for example the position of the whole planetary system during the month of June is along circle B, during August along C, during October along D, during December along E, during February along F, during March along G, whence once more along B: meanwhile the planet, which has not been disturbed at all by this dislocation of its sphere, completes its own circle on the sphere around its, as it were, fixed centre. But as regards the time of one year, Copernicus leaves the centres of the spheres absolutely fixed and also the centre of the sun fixed in the neighbourhood of the aforesaid centres. But he ascribes to the Earth, and thus to our eyesight, an annual movement around the sun: hence, since our eyesight thinks that itself is at rest, the sun seems to move with an annual movement and all the five planets seem now to stop, now to go in the opposite direction, now to advance forward very fast.

11
— Galileo Galilei —

In this reading, taken from his Dialogue Concerning the Two Chief World Systems *(1632), Galileo puts forth his mature argument in support of the Copernican theory. The dialogue, which takes place over the course of four days, purports to be a conversation between three principal characters: Salviati, Simplicio, and Sagredo. Salviati endorses the positions of Galileo, while Simplicio is a defender of the Aristotelian view. Sagredo can be taken to represent the intelligent and unprejudiced reader, a group Galileo hopes would also include Church authorities, who should rightly be persuaded by the overall rationale and evidence in favor of Heliocentric Theory.*

The reading has the interlocutors discussing the issue of the earth's motion. They consider such issues as the consistency of this hypothesis with common sense, an aesthetic principle of simplicity, probability, and the existing scholastic physics stemming from Aristotle. After considering seven possible objections to the motion of the earth, Galileo intends the reader to view the Copernican heliocentric theory as the best overall explanation of the phenomena. Galileo offers not a rigorous proof of the Copernican theory, but a series of persuasive arguments that all point toward the same conclusion.

THE COHERENCE OF THE COPERNICAN THEORY

The Second Day

SALVIATI: Then let the beginning of our reflections be the consideration that whatever motion comes to be attributed to the earth must necessarily remain imperceptible to us and as if nonexistent, so long as we look only at terrestrial objects; for as inhabitants of the earth, we consequently participate in the same motion. But on the other hand it is indeed just as necessary that it display itself very generally in all other visible bodies and objects which, being separated from the earth, do not take part in this movement. So the true method of investigating whether any motion can be attributed to the earth, and if so what it may be, is to observe and consider whether bodies separated from the earth exhibit some appearance of motion which belongs equally to all. For a motion which is perceived only, for example, in the moon, and which does not affect Venus or Jupiter or the other stars, cannot in any way be the earth's or anything but the moon's.

Now there is one motion which is most general and supreme over all, and it is that by which the sun, moon, and all other planets and fixed stars—in a word, the whole universe, the earth alone excepted—appear to be moved as a unit from east to west in the space of twenty-four hours. This, in so far as first appearances are concerned, may just as logically belong to the earth alone as to the rest of the universe, since the same appearances would prevail as much in the one situation as in the other. Thus it is that Aristotle and Ptolemy, who thoroughly understood this consideration, in their attempt to prove the earth immovable do not argue against any other motion than this diurnal one, though Aristotle does drop a hint against another motion ascribed to it by an ancient writer, of which we shall speak in the proper place.

SAGREDO: I am quite convinced of the force of your argument, but it raises a question for me from which I do not know how to free myself, and it is this: Copernicus attributed to the earth another motion than the diurnal. By the rule just affirmed, this ought to remain imperceptible to all observations on the earth, but be visible in the rest of the universe. It seems to me that one may deduce as a necessary consequence either that he was grossly mistaken in assigning to the earth a motion corresponding to no appearance in the heavens generally, or that if the correspondent motion does exist, then Ptolemy was equally at fault in not explaining it away, as he explained away the other.

SALVIATI: This is very reasonably questioned, and when we come to treat of the other movement you will see how greatly Copernicus surpassed Ptolemy in acuteness and penetration of mind by seeing what the latter did not—I mean the wonderful correspondence with which such a movement is reflected in all the other heavenly bodies. But let us postpone this for the present and return to the first consideration, with respect to which I shall set forth, commencing with the most general things, those reasons which seem to favor the earth's motion, so that we may then hear their refutation from Simplicio.

First, let us consider only the immense bulk of the starry sphere in contrast with the smallness of the terrestrial globe, which is contained in the former so many millions of times. Now if we think of the velocity of motion required to make a complete rotation in a single day and night, I cannot persuade myself that anyone could be found who would think it the more reasonable and credible thing that it was the celestial sphere which did the turning, and the terrestrial globe which remained fixed.

SAGREDO: If, throughout the whole variety of effects that could exist in nature as dependent upon these motions, all the same consequences followed indifferently to a hairsbreadth from both positions, still my first general impression of them would be this: I should think that anyone who considered it more reasonable for the whole universe to move in order to let the earth remain fixed would be more irrational than one who should climb to the top of your cupola just to get a view of the city and its environs, and then demand that the whole countryside should revolve around him so that he would not have to take the trouble to turn his head. Doubtless there are many and great advantages to be drawn from the new theory and not from the previous one (which to my mind is comparable with or even surpasses the above in absurdity), making the former more credible than the latter. But perhaps Aristotle,

Ptolemy, and Simplicio ought to marshal their advantages against us and set them forth, too, if such there are; otherwise it will be clear to me that there are none and cannot be any.

SALVIATI: Despite much thinking about it, I have not been able to find any difference, so it seems to me I have found that there can be no difference; hence I think it vain to seek one further. For consider: Motion, in so far as it is and acts as motion, to that extent exists relatively to things that lack it; and among things which all share equally in any motion, it does not act, and is as if it did not exist. Thus the goods with which a ship is laden leaving Venice, pass by Corfu, by Crete, by Cyprus and go to Aleppo. Venice, Corfu, Crete, etc. stand still and do not move with the ship; but as to the sacks, boxes, and bundles with which the boat is laden and with respect to the ship itself, the motion from Venice to Syria is as nothing, and in no way alters their relation among themselves. This is so because it is common to all of them and all share equally in it. If, from the cargo in the ship, a sack were shifted from a chest one single inch, this alone would be more of a movement for it than the two-thousand-mile journey made by all of them together.

SIMPLICIO: This is good, sound doctrine, and entirely Peripatetic.

SALVIATI: I should have thought it somewhat older. And I question whether Aristotle entirely understood it when selecting it from some good school of thought, and whether he has not, by altering it in his writings, made it a source of confusion among those who wish to maintain everything he said. When he wrote that everything which is moved is moved upon something immovable, I think he only made equivocal the saying that whatever moves, moves with respect to something motionless. This proposition suffers no difficulties at all, whereas the other has many.

SAGREDO: Please do not break the thread, but continue with the argument already begun.

SALVIATI: It is obvious, then, that motion which is common to many moving things is idle and inconsequential to the relation of these movables among themselves, nothing being changed among them, and that it is operative only in the relation that they have with other bodies lacking that motion, among which their location is changed. Now, having divided the universe into two parts, one of which is necessarily movable and the other motionless, it is the same thing to make the earth alone move, and to move all the rest of the universe, so far as concerns any result which may depend upon such movement. For the action of such a movement is only in the relation between the celestial bodies and the earth, which relation alone is changed. Now if precisely the same effect follows whether the earth is made to move and the rest of the universe stay still, or the earth alone remains fixed while the whole universe shares one motion, who is going to believe that nature (which by general agreement does not act by means of many things when it can do so by means of few) has chosen to make an immense number of extremely large bodies move with inconceivable velocities, to achieve what could have been done by a moderate movement of one single body around its own center?

SIMPLICIO: I do not quite understand how this very great motion is as nothing for the sun, the moon, the other planets, and the innumerable host of the fixed stars. Why do you say it is nothing for the sun to pass from one meridian to the other, rise above this horizon and sink beneath that, causing now the day and now the night; and for the moon, the other planets, and the fixed stars to vary similarly?

SALVIATI: Every one of these variations which you recite to me is nothing except in relation to the earth. To see that this is true, remove the earth; nothing remains in the universe of rising and setting of the sun and moon, nor of horizons and meridians, nor day and night, and in a word from this movement there will never originate any changes in the moon or sun or any stars you please, fixed or moving. All these changes are in relation to the earth, all of them meaning nothing except that the sun shows itself now over China, then to Persia, afterward to Egypt, to Greece, to France, to Spain, to America, etc. And the same holds for the moon and the rest of the heavenly bodies, this effect taking place in exactly the same way if, without embroiling the biggest part of the universe, the terrestrial globe is made to revolve upon itself.

And let us redouble the difficulty with another very great one, which is this. If this great motion is attributed to the heavens, it has to be made in the opposite direction from the specific motion of all the planetary orbs, of which each one incontrovertibly has its own motion from west to east, this being very gentle and moderate, and must then be made to rush the other way; that is, from east to west, with this very rapid diurnal motion. Whereas by making the earth itself move, the contrariety of motions is removed, and the single motion from west to east accommodates all the observations and satisfies them all completely.

SIMPLICIO: As to the contrariety of motions, that would matter little, since Aristotle demonstrates that circular motions are not contrary to one another, and their opposition cannot be called true contrariety.

SALVIATI: Does Aristotle demonstrate that, or does he just say it because it suits certain designs of his? If, as he himself declares, contraries are those things which mutually destroy each other, I cannot see how two movable bodies meeting each other along a circular line conflict any less than if they had met along a straight line.

SAGREDO: Please stop a moment. Tell me, Simplicio, when two knights meet tilting in an open field, or two whole squadrons, or two fleets at sea go to attack and smash and sink each other, would you call their encounters contrary to one another?

SIMPLICIO: I should say they were contrary.

SAGREDO: Then why are two circular motions not contrary? Being made upon the surface of the land or sea, which as you know is spherical, these motions become circular. Do you know what circular motions are not contrary to each other, Simplicio? They are those of two circles which touch from the outside; one being turned, the other naturally moves the opposite way. But if one circle should be inside the other, it is impossible that their motions should be made in opposite directions without their resisting each other.

SALVIATI: "Contrary" or "not contrary," these are quibbles about words, but I know that with facts it is a much simpler and more natural thing to keep everything with a single motion than to introduce two, whether one wants to call them contrary or opposite. But I do not assume the introduction of two to be impossible, nor do I pretend to draw a necessary proof from this; merely a greater probability. The improbability is shown for a third time in the relative disruption of the order which we surely see existing among those heavenly bodies whose circulation is not doubtful, but most certain. This order is such that the greater orbits complete their revolutions in longer times, and the lesser in shorter; thus Saturn, describing a greater circle than the other planets, completes it in thirty years; Jupiter revolves in its smaller one in twelve years, Mars in two; the moon covers its much smaller circle in a single month. And we see no less sensibly that of the satellites of Jupiter (*stelle Medicee*), the closest one to that planet makes its revolution in a very short time, that is in about forty-two hours; the next, in three and a half days; the third in seven days and the most distant in sixteen. And this very harmonious trend will not be a bit altered if the earth is made to move on itself in twenty-four hours. But if the earth is desired to remain motionless, it is necessary, after passing from the brief period of the moon to the other consecutively larger ones, and ultimately to that of Mars in two years, and the greater one of Jupiter in twelve, and from this to the still larger one of Saturn, whose period is thirty years—it is necessary, I say, to pass on beyond to another incomparably larger sphere, and make this one finish an entire revolution in twenty-four hours. Now this is the minimum disorder that can be introduced, for if one wished to pass from Saturn's sphere to the stellar, and make the latter so much greater than Saturn's that it would proportionally be suited to a very slow motion of many thousands of years, a much greater leap would be required to pass beyond that to a still larger one and then make that revolve in twenty-four hours. But by giving mobility to the earth, order becomes very well observed among the periods; from the very slow sphere of Saturn one passes on to the entirely immovable fixed stars, and manages to escape a fourth difficulty necessitated by supposing the stellar sphere to be movable. This difficulty is the immense disparity between the motions of the stars, some of which would be moving very rapidly in vast circles, and others very slowly in little tiny circles, according as they are located farther from or closer to the poles. This is indeed a nuisance, for just as we see that all those bodies whose motion is undoubted move in large circles, so it would not seem to have been good judgment to arrange bodies in such a way that they must move circularly at immense distances from the center, and then make them move in little tiny circles.

Not only will the size of the circles and consequently the velocities of motion of these stars be very diverse from the orbits and motions of some others, but (and this shall be the fifth difficulty) the same stars will keep changing their circles and their velocities, since those which two thousand years ago were on the celestial equator, and which consequently described great circles with their motion, are found in our time to be many degrees distant, and must be made slower in motion and reduced to moving in smaller circles. Indeed, it is not impossible that a time will come when some of the stars which in the past have always been moving will be reduced, by reaching the pole, to holding fast, and then after that time will start moving once more; whereas all those stars which certainly do move describe, as I said, very large circles in their orbits and are unchangeably preserved in them.

For anyone who reasons soundly, the unlikelihood is increased—and this is the sixth difficulty—by the incomprehensibility of what is called the "solidity" of that very vast sphere in whose depths are firmly fixed so many stars which, without changing place in the least among themselves, come to be carried around so harmoniously with such a disparity of motions. If, however, the heavens are fluid (as may much more reasonably be believed) so that each star roves around in it by itself, what law will regulate their motion so that as seen from the earth they shall appear as if made into a single sphere? For this to happen, it seems to me that it is as much more effective and convenient to make them immovable than to have them roam around, as it is easier to count the myriad tiles set in a courtyard than to number the troop of children running around on them.

Finally, for the seventh objection, if we attribute the diurnal rotation to the highest heaven, then this has to be made of such strength and power as to carry with it the innumerable host of fixed stars, all of them vast bodies and much larger than the earth, as well as to carry along the planetary orbs despite the fact that the two move naturally in opposite ways. Besides this, one must grant that the element of fire and the greater part of the air are likewise hurried along, and that only the little body of the earth remains defiant and resistant to such power. This seems to me to be most difficult; I do not understand why the earth, a suspended body balanced on its center and indifferent to motion or to rest, placed in and surrounded by an enclosing fluid, should not give in to such force and be carried around too. We encounter no such objections if we give the motion to the earth, a small and trifling body in comparison with the universe, and hence unable to do it any violence.

12

— Isaac Newton —

(1642–1727)

Isaac Newton was an English physicist and mathematician who is considered by many to have been the most influential scientist of the past millennium. He discovered the binomial theorem, invented the calculus, and made significant contributions to the science of optics.

Newton's most significant contribution, however, was his development of the science of mechanics, often called the first great science. This science of mechanics was published in 1687 under the title "The Mathematical Principles of Natural Philosophy," or simply the "Principia." It is the single most important and influential science book ever written, in which Newton states, as axioms, his three laws of motion and his famous law of universal gravitation.

In the Opticks (1704), Newton proposed a corpuscular theory of light but also used the concept of periodicity (or wave motion) with instrumental success. In this work, he showed that he had great experimental skill to go along with his theoretical genius.

In the reading that follows, Newton incorporates the achievements of Copernicus, Kepler, and Galileo in support of Heliocentric Theory into his own systematic physical theory of the universe. He thus solidifies and extends the explanatory framework of heliocentrism. Along the way Newton demonstrates the consistency between his own science of mechanics and Kepler's three laws of planetary motion. He also shows how the forces exerted upon planets by a central sun obey an inverse square law, thus suggesting the universality of his law of gravity.

The reading is taken from Newton's "System of the World," which makes up Book III of his Mathematical Principles of Natural Philosophy. It includes references to earlier parts of the Mathematical Principles that do not appear in the present volume.

THE PHYSICAL FOUNDATIONS OF HELIOCENTRISM

Centripetal Forces Are Directed to the Individual Centres of the Planets

That there are centripetal forces actually directed to the bodies of the sun, of the earth, and other planets, I thus infer.

The moon revolves about our earth, and by radii drawn to its centre describes areas nearly proportional to the times in which they are described, as is evident from its velocity compared with its apparent diameter; for its motion is slower when its diameter is less (and therefore its distance greater), and its motion is swifter when its diameter is greater.

The revolutions of the satellites of Jupiter about that planet are more regular; for they describe circles concentric with Jupiter by uniform motions, as exactly as our senses can perceive.

And so the satellites of Saturn are revolved about this planet with motions nearly circular and uniform, scarcely disturbed by any eccentricity hitherto observed.

That Venus and Mercury are revolved about the sun, is demonstrable from their moon-like appearances; when they shine with a full face, they are in those parts of their orbits which in respect of the earth lie beyond the sun; when they appear half full, they are in those parts which lie over against the sun; when horned, in those parts which lie between the earth and the sun; and sometimes they pass over the sun's disk, when directly interposed between the earth and the sun.

And Venus, with a motion almost uniform, describes an orbit nearly circular and concentric with the sun.

But Mercury, with a more eccentric motion, makes remarkable approaches to the sun, and goes off again by turns; but it is always swifter as it is near to the sun, and therefore by a radius drawn to the sun still describes areas proportional to the times.

Lastly, that the earth describes about the sun, or the sun about the earth, by a radius from the one to the other, areas exactly proportional to the times is demonstrable from the apparent diameter of the sun compared with its apparent motion.

These are astronomical experiments; from which it follows, by Prop. I, II, III, in the first Book of our *Principles*, and their Corollaries, that there are centripetal forces actually directed (either accurately or without considerable error) to the centres of the earth, of Jupiter, of Saturn, and of the sun. In Mercury, Venus, Mars, and the lesser planets, where experiments are wanting, the arguments from analogy must be allowed in their place.

Centripetal Forces Decrease Inversely as the Square of the Distances from the Centres of the Planets

That those forces decrease as the inverse square of the distances from the centre of every planet, appears by Cor. VI, Prop. IV, Book 1; for the periodic times of the satellites of Jupiter are one to another, as the ³⁄₂th power of their distances from the centre of this planet.

This proportion has been long ago observed in those satellites; and Mr. Flamsteed, who had often measured their distances from Jupiter by the micrometer, and by the eclipses of the satellites, wrote to me, that it holds to all the accuracy that possibly can be discerned by our senses. And he sent me the dimensions of their orbits taken by the micrometer, and reduced to the mean distance of Jupiter from the earth, or from the sun, together with the times of their revolutions. . . .

The Superior Planets Revolve about the Sun, and the Radii Drawn to the Sun Describe Areas Proportional to the Times

That Mars is revolved about the sun is demonstrated from the phases which it shows, and the proportion of its apparent diameters; for from its appearing full

The greatest elongation of the satellites from the centre of Jupiter as seen from the sun			The periodic times of their revolutions			
′	″	″	d	h	m	s
1st 1 48		or 108	1	18	28	36
2d 3 01		or 181	3	13	17	54
3d 4 46		or 286	7	03	59	36
4th 8 13½		or 493½	16	18	5	13

near conjunction with the sun, and gibbous in its quadratures, it is certain that it surrounds the sun.

And since its diameter appears about five times greater when in opposition to the sun than when in conjunction therewith, and its distance from the earth is inversely as its apparent diameter, that distance will be about five times less when in opposition to than when in conjunction with the sun; but in both cases its distance from the sun will be nearly the same with the distance which is inferred from its gibbous appearance in the quadratures. And as it encompasses the sun at almost equal distances, but in respect of the earth is very unequally distant, so by radii drawn to the sun it describes areas nearly uniform; but by radii drawn to the earth, it is sometimes swift, sometimes stationary, and sometimes retrograde.

That Jupiter, in a higher orbit than Mars, is likewise revolved about the sun, with a motion nearly uniform, as well in distance as in the areas described, I infer thus.

Mr. Flamsteed assured me, by letters, that all the eclipses of the innermost satellite which hitherto have been well observed do agree with his theory so nearly, as never to differ therefrom by two minutes of time; that in the outermost the error is little greater; in the outermost but one, scarcely three times greater; that in the innermost but one the difference is indeed much greater, yet so as to agree as nearly with his computations as the moon does with the common tables; and that he computes those eclipses only from the mean motions corrected by the equation of light discovered and introduced by Mr. Römer. Supposing, then, that the theory differs by a less error than that of 2′ from the motion of the outmost satellite as hitherto described, and taking as the periodic time $16^d. 18^h. 5^m. 13^s.$ to 2′ in time, so is the whole circle of 360° to the arc 1′ 48″, the error of Mr. Flamsteed's computation, reduced to the satellite's orbit, will be less than 1′ 48″, that is, the longitude of the satellite, as seen from the centre of Jupiter, will be determined with a less error than 1′ 48″. But when the satellite is in the middle of the shadow, that longitude is the same with the heliocentric longitude of Jupiter; and, therefore, the hypothesis which Mr. Flamsteed follows, viz., the Copernican, as improved by Kepler, and (as to the motion of Jupiter) lately corrected by himself, rightly represents that longitude within a less error than 1′ 48″; but by this longitude, together with the geocentric longitude, which is always easily found, the distance of Jupiter from the sun is determined; which must, therefore, be the very same with that which the hypothesis exhibits. For that greatest error of 1′ 48″ that can happen in the heliocentric longitude is almost insensible, and quite to be neglected, and perhaps may arise from some yet undiscovered eccentricity of the satellite; but since both longitude and distance are rightly determined, it follows of necessity that Jupiter, by radii drawn to the sun, describes areas so conditioned as the hypothesis requires, that is, proportional to the times.

And the same thing may be concluded of Saturn from his satellite, by the observations of Mr. Huygens and Dr. Halley; though a longer series of observations is yet wanting to confirm the thing, and to bring it under a sufficiently exact computation.

The Force Which Controls the Superior Planets is Not Directed to the Earth, but to the Sun

For if Jupiter was viewed from the sun, it would never appear retrograde nor stationary, as it is seen sometimes from the earth, but always to go forwards with a motion nearly uniform. And from the very great inequality of its apparent geocentric motion, we infer that the force by which Jupiter is turned out of a rectilinear course, and made to revolve in an orbit, is not directed to the centre of the earth. The same argument holds good for Mars and for Saturn. Another centre of these forces is therefore to be looked for (by Book 1, Prop. II and III, and the Corollaries of the latter), about which the areas described by radii intervening may be uniform; and that this is the sun, we have proved already in Mars and Saturn nearly, but accurately enough in Jupiter. It may be alleged that the sun and planets are impelled by some other force equally and in the direction of parallel lines; but by such a force (by Cor. VI of the Laws of Motion) no change would happen in the situation of the planets one to another, nor any sensible effect follow: but our business is with the causes of sensible effects. Let us, therefore, neglect every such force as imaginary and precarious, and of no use in the phenomena of the heavens; and the whole remaining force by which Jupiter is impelled will be directed to the centre of the sun.

The Circumsolar Force Decreases in All Planetary Spaces Inversely as the Square of the Distances from the Sun

The distances of the planets from the sun come out the same, whether, with Tycho, we place the earth in the centre of the system, or the sun with Copernicus; and we have already proved that these distances are true in Jupiter.

Kepler and Boulliau have, with great care, determined the distances of the planets from the sun; and hence it is that their tables agree best with the heavens. And in all the planets, in Jupiter and Mars, in Saturn and the earth, as well as in Venus and Mercury, the cubes of their distances are as the squares of their periodic times; and therefore the centripetal circumsolar force throughout all the planetary regions decreases as the inverse square of the distances from the sun. In examining this proportion, we are to use the mean distances, or the transverse semiaxes of the orbits, and to neglect those little fractions, which, in defining the orbits, may have arisen from the insensible errors of observation, or may be ascribed to other causes which we shall afterwards explain. And thus we shall always find the said proportion to hold exactly; for the distances of Saturn, Jupiter, Mars, the earth, Venus, and Mercury, from the sun, obtained from the observations of astronomers, are, according to the computation of Kepler, as the numbers 951000, 519650, 152350, 100000, 72400, 38806; by the computation of Boulliau, as the numbers 954198, 522520, 152350, 100000, 72398, 38585; and from the periodic times they come out 953806, 520116, 152399, 100000, 72333, 38710. Their distances, according to Kepler and Boulliau, scarcely differ by any sensible quantity, and where they differ most the distances calculated from the periodic times fall in between them . . . .

All the Planets Revolve Around the Sun

From comparing the forces of the planets one with another, we have above seen that the circumsolar does more than a thousand times exceed all the rest; but by the action of a force so great it is unavoidable that all bodies within, nay, and far beyond, the bounds of the planetary system must descend directly to the sun, unless by other motions they are impelled towards other parts: nor is our earth to be excluded from the number of such bodies, for certainly the moon is a body of the same nature with the planets, and subject to the same attractions with the other planets, seeing it is by the circumterrestrial force that it is retained in its orbit. But that the earth and moon are equally attracted towards the sun, we have above proved; we have likewise before proved that all bodies are subject to the said common laws of attraction. Nay, supposing any of those bodies to be deprived of its circular motion about the sun, by having its distance from the sun, we may find in what space of time it would in its descent arrive at the sun; namely, in half that periodic time in which the body might be revolved at one-half of its former distance; or in a space of time that is to the periodic time of the planet as 1 to $4\sqrt{2}$; as that Venus in its descent would arrive at the sun in the space of 40 days, Jupiter in the space of two years and one month, and the earth and moon together in the space of 66 days and 19 hours. But, since no such thing happens, it must needs be, that those bodies are moved towards other parts, nor is every motion sufficient for this purpose. To hinder such a descent, a due proportion of velocity is required. And hence depends the force of the argument drawn from the retardation of the motions of the planets. Unless the circumsolar force decreased as the square of their increasing slowness, the excess thereof would force those bodies to descend to the sun; for instance, if the motion (other things being equal) was retarded by one-half, the planet would be held in its orbit by one-fourth of the former circumsolar force, and by the excess of the other three-fourths would descend to the sun. And therefore the planets (Saturn, Jupiter, Mars, Venus, and Mercury) are not really retarded in their perigees, nor become really stationary, or regressive with slow motions. All these are but apparent, and the absolute motions, by which the planets continue to revolve in their orbits, are always direct, and nearly uniform. But that such motions are performed about the sun, we have already proved; and therefore the sun, as the centre of the absolute motions, is quiescent. For we can by no means allow quiescence to the earth, lest the planets in their perigees should indeed be truly retarded, and become truly stationary and regressive, and so for want of motion should descend to the sun. But further; since the planets (Venus, Mars, Jupiter, and the rest) by radii drawn to the sun describe regular orbits, and areas (as we have shown) nearly and to sense proportional to the times, it follows that the sun is moved with no notable force, unless perhaps with such as all the planets are equally moved with, according to their several quantities of matter, in parallel lines, and so the whole system is transferred in right lines. Reject that translation of the whole system, and the sun will be almost quiescent in the centre thereof. If the sun was revolved about the earth, and carried the other planets round about itself, the earth ought to attract the sun with a great force, but the circumsolar planets with no force producing any sensible effect, which is contrary to Cor. III, Prop. LXV, Book I. Add to this, that if hitherto the earth, because of the gravitation of its parts, has been placed by most authors in the lowermost region of the universe; now, for better reason, the sun possessed of a centripetal force exceeding our terrestrial gravitation a thousand times and more, ought to be depressed into the lowermost place, and to be held for the centre of the system. And thus the true disposition of the whole system will be more fully and more exactly understood.

The Common Centre of Gravity of the Sun and All the Planets Is at Rest and the Sun Moves with a Very Slow Motion. Explanation of the Solar Motion

Because the fixed stars are quiescent one in respect of another, we may consider the sun, earth, and planets, as one system of bodies carried hither and thither by various motions among themselves; and the common centre of gravity of all (by Cor. IV of the Laws of Motion) will either be quiescent, or move uniformly forwards in a right line: in which case the whole system will likewise move uniformly forwards in right lines. But this is an hypothesis hardly to be admitted; and, therefore, setting it aside, that common centre will be quiescent: and from it the sun is never far removed. The common centre of gravity of the sun and Jupiter falls on the surface of the sun; and though all the planets were placed towards the same parts from the sun with Jupiter, the common centre of the sun and all of them would scarcely recede twice as far from the sun's center;

and, therefore, though the sun, according to the different situations of the planets, is variously agitated, and always wandering to and fro with a slow motion of libration, yet it never recedes one entire diameter of its own body from the quiescent centre of the whole system. But from the weights of the sun and planets above determined, and the situation of all among themselves, their common centre of gravity may be found; and, this being given, the sun's place at any supposed time may be obtained.

Nevertheless, the Planets Revolve in Ellipses Having Foci at the Centre of the Sun; and the Radii Drawn to the Sun Describe Areas Proportional to the Times

About the sun, thus librated, the other planets are revolved in elliptic orbits, and by radii drawn to the sun, describe areas nearly proportional to the times, as is explained in Prop. LXV, Book I. If the sun were quiescent, and the other planets did not act mutually one upon another, their orbits would be elliptic, and the areas exactly proportional to the times (by Prop. XI. and Cor., Prop. LXVIII, Book I). But the actions of the planets among themselves compared with the actions of the sun on the planets, are of no moment, and produce no sensible errors. And those errors are less in revolutions about the sun agitated in the manner but now described than if those revolutions were made about the sun quiescent (by Prop. LXVI, Book I, and Cor., Prop. LXVIII, Book I), especially if the focus of every orbit is placed in the common centre of gravity of all the lower included planets; viz., the focus of the orbit of Mercury in the centre of the sun; the focus of the orbit of Venus in the common centre of gravity of Mercury and the sun; the focus of the orbit of the earth in the common centre of gravity of Venus, Mercury, and the sun; and so of the rest. And by this means the foci of the orbits of all the planets, except Saturn, will not be sensibly removed from the centre of the sun, nor will the focus of the orbit of Saturn recede sensibly from the common centre of gravity of Jupiter and the sun. And therefore astronomers are not far from the truth, when they reckon the sun's centre the common focus of all the planetary orbits. In Saturn itself the error thence arising does not exceed 1′ 45″. And if its orbit, by placing the focus thereof in the common centre of gravity of Jupiter and the sun, shall happen to agree with the phenomena, from thence all that we have said will be further confirmed. . . .

On the Distance of the Stars

Thus I have given an account of the system of the planets. As to the fixed stars, the smallness of their annual parallax proves them to be removed to immense distances from the system of the planets: that this parallax is less than one minute is most certain; and from this it follows that the distance of the fixed stars is above 360 times greater than the distance of Saturn from the sun. Those who consider the earth one of the planets, and the sun one of the fixed stars, may remove the fixed stars to yet greater distances by the following arguments. From the annual motion of the earth there would happen an apparent transposition of the fixed stars, one in respect of another, almost equal to their double parallax; but the greater and nearer stars, in respect of the more remote, which are only seen by the telescope, have not hitherto been observed to have the least motion. If we should suppose that motion to be only less than 20″, the distance of the nearer fixed stars would exceed the mean distance of Saturn by above 2000 times. Again, the disk of Saturn, which is only 17″ or 18″ in diameter, receives only about 1/2,100,000,000 of the sun's light; for so much less is that disk than the whole spherical surface of the orb of Saturn. Now if we suppose Saturn to reflect about ¼ of this light, the whole light reflected from its illuminated hemisphere will be about 1/4,200,000,000 of the whole light emitted from the sun's hemisphere; and, therefore, since light is rarefied inversely as the square of the distance from the luminous body, if the sun was $10000\sqrt{42}$ times more distant than Saturn, it would yet appear as lucid as Saturn now does without its ring, that is, somewhat more lucid than a fixed star of the first magnitude. Let us, therefore, suppose that the distance from which the sun would shine as a fixed star exceeds that of Saturn by about 100,000 times, and its apparent diameter will be $7^{v}\ 16^{vi}$ and its parallax arising from the annual motion of the earth 13^{iv}: and so great will be the distance, the apparent diameter, and the parallax of the fixed stars of the first magnitude, in bulk and light equal to our sun. Some may, perhaps, imagine that a great part of the light of the fixed stars is intercepted and lost in its passage through so vast spaces, and upon that account pretend to place the fixed stars at nearer distances; but at this rate the remoter stars could be scarcely seen. Suppose, for example, that ¾ of the light is lost in its passage from the nearest fixed stars to us; then ¾ will be lost twice in its passage through a double space, thrice through a triple, and so forth. And, therefore, the fixed stars that are at a double distance will be 16 times more obscure, viz., 4 times more obscure on account of the diminished apparent diameter; and, again, 4 times more on account of the lost light. And, by the same argument, the fixed stars at a triple distance will be 9·4·4, or 144 times more obscure; and those at a quadruple distance will be 16·4·4·4, or 1024 times more obscure; but so great a diminution of light is no ways consistent with the phenomena and with that hypothesis which places the fixed stars at different distances. . . .

Some Propositions from Newton, *Principia*

Proposition I, Theorem I

That the forces by which the circumjovial planets are continually drawn off from rectilinear motions, and retained in their proper orbits, tend to Jupiter's center;

*and are inversely as the squares of the distances of the place of those planets
from that centre.*

The former part of the Proposition is manifest from Phen. V, and Prop. II, Book
I; the latter from Phen. IV, and Cor. VI, Prop. IV, of the same Book. But this
part of the Proposition is, with great accuracy, demonstrable from the quiescence
of the aphelion points; for a very small aberration from the proportion according
to the inverse square of the distances would (by cor. I, Prop. XLV, Book I)
produce a motion of the apsides sensible enough in every single revolution, and
in many of them enormously great.

Proposition V, Theorem V

*That the circumjovial planets gravitate towards Jupiter; the circumsaturnal to-
wards Saturn; the circumsolar towards the sun; and by the forces of their gravity
are drawn off from rectilinear motions, and retained in curvilinear orbits.*

For the revolutions of the circumjovial planets about Jupiter, of the circumsaturnal about Saturn, and of Mercury and Venus, and the other circumsolar planets,
about the sun, are appearances of the same sort with the revolution of the moon
about the earth; and therefore, by Rule 2, must be owing to the same sort of
causes; especially since it has been demonstrated, that the forces upon which
those revolutions depend tend to the centres of Jupiter, of Saturn, and of the
sun; and that those forces, in receding from Jupiter, from Saturn, and from the
sun, decrease in the same proportion, and according to the same law, as the
force of gravity does in receding from the earth.

Corollary I. There is, therefore, a power of gravity tending to all the planets;
for, doubtless, Venus, Mercury, and the rest, are bodies of the same sort with
Jupiter and Saturn. And since all attraction (by Law III) is mutual, Jupiter will
therefore gravitate towards all his own satellites, Saturn towards his, the earth
towards the moon, and the sun towards all the primary planets.

Corollary II. The force of gravity which tends to any one planet is inversely
as the square of the distance of places from that planet's centre.

Corollary III. All the planets do gravitate towards one another, by Cor. I and
II. And hence it is that Jupiter and Saturn, when near their conjunction, by their
mutual attractions sensibly disturb each other's motions. So the sun disturbs the
motions of the moon; and both sun and moon disturb our sea, as we shall
hereafter explain.

Scholium

The force which retains the celestial bodies in their orbits has been hitherto
called centripetal force; but it being now made plain that it can be no other than
a gravitating force, we shall hereafter call it gravity. For the cause of that centripetal force which retains the moon in its orbit will extend itself to all the planets,
by Rule 1, 2, and 4.

Proposition VI, Theorem VI

That all bodies gravitate towards every planet; and that the weights of bodies towards any one planet, at equal distances from the centre of the planet, are proportional to the quantities of matter which they severally contain.

It has been, now for a long time, observed by others, that all sorts of heavy bodies (allowance being made for the inequality of retardation which they suffer from a small power of resistance in the air) descend to the earth *from equal heights* in equal times; and that equality of times we may distinguish to a great accuracy, by the help of pendulums. I tried experiments with gold, silver, lead, glass, sand, common salt, wood, water, and wheat. I provided two wooden boxes, round and equal: I filled the one with wood, and suspended an equal weight of gold (as exactly as I could) in the centre of oscillation of the other. The boxes, hanging by equal threads of 11 feet, made a couple of pendulums perfectly equal in weight and figure, and equally receiving the resistance of the air. And, placing the one by the other, I observed them to play together forwards and backwards, for a long time, with equal vibrations. And therefore the quantity of matter in the gold (by Cor. I and VI, Prop. XXIV, Book II) was to the quantity of matter in the wood as the action of the motive force (or *vis motrix*) upon all the gold to the action of the same upon all the wood; that is, as the weight of the one to the weight of the other: and the like happened in the other bodies. By these experiments, in bodies of the same weight, I could manifestly have discovered a difference of matter less than the thousandth part of the whole, had any such been. But, without all doubt, the nature of gravity towards the planets is the same as towards the earth. For, should we imagine our terrestrial bodies taken to the orbit of the moon, and there, together with the moon, deprived of all motion, to be let go, so as to fall together towards the earth, it is certain from what we have demonstrated before, that, in equal times, they would describe equal spaces with the moon, and of consequence are to the moon, in quantity of matter, as their weights to its weight. Moreover, since the satellites of Jupiter perform their revolutions in times which observe the ³⁄₂th power of the proportion of their distances from Jupiter's centre, their accelerative gravities towards Jupiter will be inversely as the squares of their distances from Jupiter's centre; that is, equal, at equal distances. And, therefore, these satellites, if supposed to fall *towards Jupiter* from equal heights, would describe equal spaces in equal times, in like manner as heavy bodies do on our earth. And by the same argument, if the circumsolar planets were supposed to be let fall at equal distances from the sun, they would, in their descent towards the sun, describe equal spaces in equal times. But forces which equally accelerate unequal bodies must be as those bodies: that is to say, the weights of the planets *towards the sun* must be as their quantities of matter. Further, that the weights of Jupiter and of his satellites towards the sun are proportional to the several quantities of their matter, appears from the exceedingly regular motions of the satellites (by Cor. III, Prop. LXV, Book I). For if some of those bodies were more strongly attracted to the sun in proportion to their quantity of matter than others, the motions of the satellites would be disturbed by that inequality of attraction (by Cor. II, Prop. LXV, Book I). If, at equal distances from the sun, any satellite, in proportion to the quantity of its matter, did gravitate towards the sun with a force greater than Jupiter in proportion to his, according to any given proportion, suppose of d to e; then the distance between the centres of the sun and of the satellite's orbit would be always greater than the distance between the centres of the sun and of Jupiter, nearly as the square root of that proportion: as by some computations I have found. And if the satellite did gravitate towards the sun with a force, less in the proportion of e to d, the distance of the centre of the satellite's orbit from the sun would be less than the distance of the centre of Jupiter from the sun as the square root of the same proportion. Therefore if, at equal distances from the sun, the accelerative gravity of any satellite towards the sun were greater or less than the accelerative gravity of Jupiter towards the sun but by 1/1000 part of the whole gravity, the distance of the centre of the satellite's orbit from the sun would be greater or less than the distance of Jupiter from the sun by one 1/2000 part of the whole distance; that is, by a fifth part of the distance of the utmost satellite from the centre of Jupiter; an eccentricity of the orbit which would be very sensible. But the orbits of the satellites are concentric to Jupiter, and therefore the accelerative gravities of Jupiter, and of all its satellites towards the sun, are equal among themselves. And by the same argument, the weights of Saturn and of his satellites towards the sun, at equal distances from the sun, are as their several quantities of matter; and the weights of the moon and of the earth towards the sun are either none, or accurately proportional to the masses of matter which they contain. But some weight they have, by Cor. I and III, Prop. V.

But further; the weights of all the parts of every planet towards any other planet are one to another as the matter in the several parts; for if some parts did gravitate more, others less, than for the quantity of their matter, then the whole planet, according to the sort of parts with which it most abounds, would gravitate more or less than in proportion to the quantity of matter in the whole. Nor is it of any moment whether these parts are external or internal; for if, for example, we should imagine the terrestrial bodies with us to be raised to the orbit of the moon, to be there compared with its body; if the weights of such bodies were to the weights of the external parts of the moon as the quantities of matter in the one and in the other respectively, but to the weights of the internal parts in a greater or less proportion, then likewise the weights of those bodies would be to the weight of the whole moon in a greater or less proportion, against what we have shown above.

Corollary I. Hence the weights of bodies do not depend upon their forms and textures; for if the weights could be altered with the forms, they would be greater or less, according to the variety of forms, in equal matter; altogether against experience.

Corollary II. Universally, all bodies about the earth gravitate towards the earth; and the weights of all, at equal distances from the earth's centre, are as the quantities of matter which they severally contain. This is the quality of all bodies within the reach of our experiments; and therefore (by Rule 3) to be affirmed of all bodies whatsoever. If the ether, or any other body, were either altogether void of gravity, or were to gravitate less in proportion to its quantity of matter, then because (according to Aristotle, Descartes, and others) there is no difference between that and other bodies but in *mere* form of matter, by a successive change from form to form, it might be changed at last into a body of the same condition with those which gravitate most in proportion to their quantity of matter; and, on the other hand, the heaviest bodies, acquiring the first form of that body, might by degrees quite lose their gravity. And therefore the weights would depend upon the forms of bodies, and with those forms, might be changed: contrary to what was proved in the preceding Corollary.

Corollary III. All spaces are not equally full; for if all spaces were equally full, then the specific gravity of the fluid which fills the region of the air, on account of the extreme density of the matter, would fall nothing short of the specific gravity of quicksilver, or gold, or any other the most dense body; and, therefore, neither gold, nor any other body, could descend in air; for bodies do not descend in fluids, unless they are specifically heavier than the fluids. And if the quantity of matter in a given space can, by any rarefaction, be diminished, what should hinder a diminution to infinity?

Corollary IV. If all the solid particles of all bodies are of the same density, and cannot be rarefied without pores, then a void, space, or vacuum must be granted. By bodies of the same density, I mean those whose inertias are in the proportion of their bulks.

Corollary V. The power of gravity is of a different nature from the power of magnetism; for the magnetic attraction is not as the matter attracted. Some bodies are attracted more by the magnet; others less; most bodies not at all. The power of magnetism in one and the same body may be increased and diminished; and is sometimes far stronger, for the quantity of matter, than the power of gravity; and in receding from the magnet decreases not as the square but almost as the cube of the distance, as nearly as I could judge from some rude observations.

Proposition VII, Theorem VII

That there is a power of gravity pertaining to all bodies, proportional to the several quantities of matter which they contain.

That all the planets gravitate one towards another, we have proved before; as well as that the force of gravity towards every one of them, considered apart, is inversely as the square of the distance of places from the centre of the planet. And thence (by Prop. LXIX, Book I, and its Corollaries) it follows, that the gravity tending towards all the planets is proportional to the matter which they contain.

Moreover, since all the parts of any planet A gravitate towards any other planet B; and the gravity of every part is to the gravity of the whole as the matter of the part to the matter of the whole, and (by Law III) to every action corresponds an equal reaction; therefore the planet B will, on the other hand, gravitate towards all the parts of planet A; and its gravity towards any one part will be to the gravity towards the whole as the matter of the part to the matter of the whole. Q.E.D.

Corollary I. Therefore the force of gravity towards any whole planet arises from, and is compounded of, the forces of gravity towards all its parts. Magnetic and electric attractions afford us examples of this; for all attraction towards the whole arises from the attractions towards the several parts. The thing may be easily understood in gravity, if we consider a greater planet, as formed of a number of lesser planets, meeting together in one globe; for *hence it would appear* that the force of the whole must arise from the forces of the component parts. If it is objected, that, according to this law, all bodies with us must gravitate one towards another, whereas no such gravitation anywhere appears, I answer, that since the gravitation towards these bodies is to the gravitation towards the whole earth as these bodies are to the whole earth, the gravitation towards them must be far less than to fall under the observation of our senses.

Corollary II. The force of gravity towards the several equal particles of any body is inversely as the square of the distance of places from the particles; as appears from Cor. III. Prop. LXXIV, Book I.

Proposition XIII, Theorem XIII

The planets move in ellipses which have their common focus in the centre of the sun, and, by radii drawn to that centre, they describe areas proportional to the times of description.

We have discoursed above on these motions from the Phenomena. Now that we know the principles on which they depend, from those principles we deduce the motions of the heavens *a priori.* Because the weights of the planets towards the sun are inversely as the squares of their distances from the sun's centre, if the sun were at rest, and the other planets did not act one upon another, their orbits would be ellipses, having the sun in their common focus, and they would describe areas proportional to the times *of description,* by Prop. I and XI, and Cor. I, Prop. XIII, Book I. But the actions of the planets one upon another are so very small, that they may be neglected; and by Prop. LXVI, Book I, they disturb the motions of the planets around the sun in motion, less than if those motions were performed about the sun at rest.

It is true, that the action of Jupiter upon Saturn is not to be neglected; for the force of gravity towards Jupiter is to the force of gravity towards the sun (at equal distances, Cor. II, Prop. VIII) as 1 to 1067; and therefore in the conjunction of Jupiter and Saturn, because the distance of Saturn from Jupiter is to the distance of Saturn from the sun almost as 4 to 9, the gravity of Saturn towards Jupiter will be to the gravity of Saturn towards the sun as 81 to 16·1067; or, as 1 to about 211. And hence arises a perturbation of the orbit of Saturn in every conjunction of this planet with Jupiter, so sensible, that astronomers are puzzled with it. As the planet is differently situated in these conjunctions, its eccentricity is sometimes augmented, sometimes diminished; its aphelion is sometimes carried forwards, sometimes backwards, and its mean motion is by turns accelerated and retarded; yet the whole error in its motion about the sun, though arising from so great a force, may be almost avoided (except in the mean motion) by placing the lower focus of its orbit in the common centre of gravity of Jupiter and the sun (according to Prop. LXVII, Book I), and therefore that error, when it is greatest, scarcely exceeds two minutes; and the greatest error in the mean motion scarcely exceeds two minutes yearly. But in the conjunction of Jupiter and Saturn, the accelerative forces of gravity of the sun towards Saturn, of Jupiter towards Saturn, and of Jupiter towards the sun, are almost as 16, 81, and 16·81·3021/25 or 156,609; and therefore the difference of the forces of gravity of the sun towards Saturn, and of Jupiter towards Saturn, is to the force of gravity of Jupiter towards the sun as 65 to 156609, or as 1 to 2409. But the greatest power of Saturn to disturb the motion of Jupiter is proportional to this difference; and therefore the perturbation of the orbit of Jupiter is much less than that of Saturn's. The perturbations of the other orbits are yet far less, except that the orbit of the earth is sensibly disturbed by the moon. The common centre of gravity of the earth and moon moves in an ellipse about the sun in the focus thereof, and, by a radius drawn to the sun, describes areas proportional to the times of description. But the earth in the meantime by a menstrual motion is revolved about this common centre.

13
— John Herschel —
(1792–1871)

John Herschel was a British astronomer and physicist and the son of the famous astronomer William Herschel. His extensive accomplishments in astronomy included a mapping of the southern sky between 1834 and 1838. In 1849 Herschel published Outlines of Astronomy, *which offered a comprehensive account of the state of astronomy at that time. In it Herschel describes Friedrich Bessel's discovery of stellar parallax in 1838, a discovery that removed the final obstacle standing in the way of the acceptance of a moving earth theory. The passage that follows is from that book and includes the description of Bessel's discovery.*

THE DISCOVERY OF STELLAR PARALLAX

A short time previous to the publication of this important result, the detection of a sensible and measurable amount of parallax in the star N° 61 Cygni of

Flamsteed's catalogue of stars was announced by the celebrated astronomer of Königsberg, the late M. Bessel. This is a small and inconspicuous star, hardly exceeding the sixth magnitude, but which had been pointed out for especial observation by the remarkable circumstance of its being affected by a *proper motion*, i.e. a regular and continually progressive annual displacement among the surrounding stars to the extent of more than 5" per annum, a quantity so very much exceeding the average of similar minute annual displacements which many other stars exhibit, as to lead to a suspicion of its being actually nearer to our system. It is not a little remarkable that a similar presumption of proximity exists also in the case of á Centauri, whose unusually large proper motion of nearly 4" per annum is stated by Professor Henderson to have been the motive which induced him to subject his observations of that star to that severe discussion which led to the detection of its parallax. M. Bessel's observations of 61 Cygni were commenced in August, 1837, immediately on the establishment at the Königsberg observatory of a magnificent heliometer, the workmanship of the celebrated optician Fraunhofer of Munich, an instrument especially fitted for the system of observation adopted; which being totally different from that of direct meridional observation, more refined in its conception, and susceptible of far greater accuracy in its practical application, we must now explain. . . .

The star examined by Bessel has two such neighbors, both very minute, and therefore probably very distant, most favorably situated, the one (s) at a distance of 7' 42" the other (s' at 11' 46" from the large star, and so situated, that their directions from that star make nearly a right angle with each other. The effect of parallax therefore would necessarily cause the two distances S s and S s' to vary so as to attain their maximum and minimum values alternately at three-monthly intervals, and this is what was actually observed to take place, the one distance being always most rapidly on the increase or decrease when the other was stationary (the uniform effect of proper motion being understood of course to be always duly accounted for). This alternation, though so small in amount as to indicate, as a final result, a parallax, or rather a difference of parallaxes between the large and small stars of hardly more than one-third of a second, was maintained with such regularity as to leave no room for reasonable doubt as to its cause, and having been confirmed by the further continuance of these observations, and quite recently by the exact coincidence between the result thus obtained, and that deduced by M. Peters from observations of the same star at the observatory of Pulkova, is considered on all hands as fully established.

SELECTED BIBLIOGRAPHY

Berry, A. *A Short History of Astronomy: From the Earliest Times through the Nineteenth Century.* New York: Dover, 1961.

Crowe, M. J. *Theories of the World from Antiquity to the Copernican Revolution.* New York: Dover, 1990.

Dijksterhuis, E. J. *Mechanization of the World Picture.* Princeton: Princeton University Press, 1986.

Duhem, P. *To Save the Phenomena*. Chicago: University of Chicago Press, 1969.

Gingerich, O. *The Eye of Heaven: Ptolemy, Copernicus, Kepler*. New York: American Institute of Physics, 1993.

Koyré, A. *The Astronomical Revolution: Copernicus—Kepler—Borelli*. New York: Dover, 1973.

———. *From the Closed World to the Infinite Universe*. Baltimore: Johns Hopkins University Press, 1957.

Kragh, H. *Cosmology and Controversy: The Historical Development of Two Theories of the Universe*. Princeton: Princeton University Press, 1996.

Kuhn, T. *The Copernican Revolution*. Cambridge: Harvard University Press, 1957.

Penderson, O. *Early Physics and Astronomy*. Revised Edition. Cambridge: Cambridge University Press, 1993.

Stephenson, B. *Kepler's Physical Astronomy*. Princeton: Princeton University Press, 1994.

PART II

ELECTROMAGNETIC FIELD THEORY

INTRODUCTION

///

The acceptance of Electromagnetic Field Theory was the result of the convergence of developments in the sciences of electricity, magnetism, and optics. These developments began in earnest at the beginning of the seventeenth century and culminated with the work of James Clerk Maxwell in the middle of the nineteenth century.

Development of Electric and Magnetic Theory to Faraday

The ancients are said to have known of several phenomena that we would now call electrical: lightning; the ability of the torpedo fish to stun its prey; and the ability of amber to attract small objects when rubbed. Associated with this last phenomenon, which we may call the amber effect, is the phenomenon of attraction exhibited between lodestones, or natural magnets, and iron.

The ancient Greeks, to whom these phenomena were known, were quite willing to speculate on their causes. Since one characteristic of Greek natural philosophy was an abhorrence of action at a distance, they sought an intervening unobservable entity to account for the phenomena. The most important speculation concerning such an entity was given by the atomists, who suggested that some emission must leave the amber and the lodestone and travel to the attracted objects. Eventually this view was refined to the point that the amber and the lodestone were said to be surrounded by an aura, supposedly of limited range, and for the most part at rest. This aura was called effluvium and it did its "attracting" by contact.

The atomists' explanation, which was mechanistic in nature, was resurrected at the dawn of modern science in the fourteenth and fifteenth centuries. In 1600 the Englishman William Gilbert published a treatise on magnetism, *De magnete*, which sought to distinguish magnetic effects from the amber effect. To this end he built an instrument called a versorium, which consisted of a metallic pointer positioned on a pivot so that it could turn freely in one plane. Many materials once rubbed would cause the pointer to turn, and these materials he called "electrics" (from the Greek word for amber, *elektron*). Gilbert's explanation of the electric effect also involved an effluvium, which was supposedly emitted from the material when rubbed, but which did not agitate or move the surrounding air. It was instead thought to be a steady material "cloud" that hovered about the rubbed electric and "attracted" objects by direct contact. Magnetic bodies were able to emit effluvia continuously and without rubbing.

By the latter part of the seventeenth century electricity came to be looked upon as a property of things rather than as a material substance. This property, which was possessed by electrics when rubbed, was called "electric virtue." Nicolo Cabeo, an Italian Jesuit, found that the electric virtue can cause repulsion as well as attraction. The Englishman Stephen Gray in the early part of the eighteenth century found that some materials would conduct the electric virtue and some would not. Building on their efforts, the Frenchman Charles Dufay proposed a new hypothesis that there were two distinct electricities, vitreous and resinous. A body possessing vitreous electricity would repel those possessing resinous electricity. The attraction and repulsion occurred in the context of the intermingling of effluvia vortices.

By the middle of the eighteenth century the concern about the actual mechanism that produced the attraction or repulsion had died down, but the concern for the property of electricity continued to grow. Dufay's hypothesis was developed into a two-fluid theory in which all objects initially possessed equal amounts of both fluids. Upon rubbing, for example, amber and fur together, the amber was said to acquire resinous fluid from the fur while the fur acquired vitreous fluid from the amber. Thus scientists for the first time could account for the electrical property without postulating its creation ex nihilo.

As the two-fluid theory developed in Europe, Benjamin Franklin was working on a one-fluid theory in America. Franklin suggested that every body initially possessed a normal amount of electrical fluid. This fluid consisted of material particles that repelled one another but attracted nonelectrical matter. In the process of electrification an object would either gain some fluid, in which case it would be positively charged, or lose some, in which case it would be negatively charged. It follows from this that the total quantity of electricity (electrical fluid) in an isolated system would remain constant. Thus Franklin seems to have been the first to suggest a conservation principle for electricity.

Franklin explained attraction and repulsion by assuming that ordinary matter is a kind of sponge for electrical fluid. When saturation is reached, additional electricity must lie on or near the surface. Although the electrical particles strongly repel each other, they have an equally strong attraction for ordinary matter. This is why neutral objects do not affect one another. Two objects that have an excess of electrical fluid will repel each other. Two objects, one with an excess and one with a deficiency of the fluid, will attract one another, as would two objects with a deficiency. When it was discovered that two "negatively" charged bodies repel one another, this theory was put in doubt. As to the mechanism whereby the attractions and repulsions take place, Franklin remained in the tradition of the effluvia theories. However, in attempting to explain the effects of a Leyden jar he found himself forced to talk about a kind of action at a distance.[1]

In 1759 Franz Aepinus of St. Petersburg in Russia generalized upon the work of Franklin by showing that all nonconductors were impermeable to the electrical fluid. Most significantly, air, which is a nonconductor, was shown to provide the same kind of electrical shield as the glass had done in the Leyden jar. Aepinus concluded that the electrical fluid must not extend beyond the charged bodies themselves. Since he denied the existence of electric effluvia surrounding charged bodies, the only alternative open to him was to consider that charged bodies attracted and repelled one another across the intervening air, at a distance.

Electrical experimenters, freed from the confinement of an effluvium theory, now addressed themselves to the problem of finding the force law that governed the attraction and repulsion of electrically charged objects. Their job was aided considerably by the fact that electrical phenomena were analogous in character to gravitational phenomena. This analogy suggested the obvious hypothesis that the electrical force varied inversely with the square of the distance between the centers of the charged objects. In 1760 Daniel Bernoulli found this to be approximately the case in his experiments with electrified metal disks. About ten years later Joseph Priestley found that a small object hanging from a silk thread inside an electrified vessel experienced no electrical force. Since Newton had shown that an object within a spherical shell would experience no net gravitational force at any point within that shell, Priestley concluded "the attraction of electricity is subject to the same laws with that of gravitation." Shortly thereafter Henry Cavendish improved upon Priestley's experiment by placing an object within a spherical, completely enclosed, electrified vessel and showing that the object experienced no net electrical force. With this experiment Cavendish had shown that electrical force was precisely analogous to gravitational force and that an inverse square law for electricity would obtain.

Unfortunately Cavendish, because he was a very shy man, withheld the results of his experiments. The demonstration of the inverse square relationship and the law of electrostatics was accomplished publicly by Charles Augustin Coulomb in 1785. Coulomb built a torsion balance for his experiments, consisting of a thin metal fiber from which was suspended a light nonconducting rod that was free to move in a horizontal circle. At the ends of the rod were small pith balls, with a larger pith ball rigidly mounted near one of the smaller ones. When the two pith balls were given charges of the same kind of electricity, they would repel

---

[1]The Leyden jar is a glass jar coated with metal on the inside and the outside. When the inside metal coat is charged, the outside coat also appears charged despite the fact that no charge has been transmitted through the glass. Franklin explained the existence of the charge on the outer coat by induction at a distance. The excess of electrical fluid in the inner coat repelled the fluid in the outer coat to its outer surface, thus giving the appearance of being charged to the outer world. Experiments involving the Leyden jar as well as other experiments led Franklin to speak in terms of an action-at-a-distance theory of electrical repulsion and attraction: "Every particle of matter electrified is repelled by every other particle equally electrified."

each other and the horizontal rod would twist the metal fiber through a certain angle. From his work on elasticity Coulomb knew that the force twisting the metal fiber was proportional to the angle through which the twist had proceeded. He could then calculate the force by observing the angle of rotation of the horizontal rod. By varying the distance between the center of the pith balls and then measuring the force between them, Coulomb was able to confirm the inverse square relation.

He then went on to conjecture that the force exerted by one charge on another was equal to the product of the charges and sought to confirm this conjecture by further experiments. In the most significant of these tests he measured the force of repulsion between two pith balls in the apparatus previously described. Then he placed a pith ball of similar shape and size in contact with the mounted pith ball, draining away half its charge. The force between the two original pith balls was found to be half as great, thus indicating that the force of electrical repulsion (and attraction) was directly proportional to the product of the charges. In its final form, then, Coulomb's Law of Electrostatic Force was

$$F = \frac{C_{q_1 q_2}}{r^2}$$

where C is a constant of proportionality similar to the gravitational constant. Final and more refined confirmation of Coulomb's Law came in the early part of the nineteenth century through the work of K. F. Gauss.

For thirty years following establishment of the Law of Electrostatic Force, the work done in electricity, as well as in magnetism, was carried out within the framework of an action-at-a-distance theory. However, during that time an extremely important development in the history of electricity occurred that was to trigger a century of incredible progress. In 1800 Alessandro Volta, building on the efforts of his fellow Italian Luigi Galvani, produced the first electrochemical or voltaic cell. With this cell he was able to support a constant movement of electricity in a conductor, i.e., to support an electric current. The significance of this achievement became known in 1819 when Hans Christian Oersted of Denmark discovered that an electric current will cause a compass needle (a magnet), placed in its neighborhood, to deflect, just as the magnetic properties of the earth would do. But more important the direction of the force exerted by the current on the magnet was quite unexpected. Oersted found that when the needle was placed in the vicinity of the current-carrying wire it does indeed experience a force, but this force is perpendicular to the line joining the wire and the needle. Further, the force on the needle seemed to circulate about the current-carrying wire, pushing the needle in one direction when it was placed above the wire, and in the opposite direction when it was placed below.

As might be expected, Oersted rejected an action-at-a-distance explanation for this phenomena. For him there were two electrical fluids that conflicted with one another when they came in contact. It was this conflict, which existed in the space surrounding the wire, that would act on magnetic poles while being indifferent to nonmagnetic bodies. On the whole Oersted's explanation is rather vague. He apparently was uncertain of the nature of the "current" that existed in the wire conductor. It is also unclear as to whether he thought that the electric conflict in the space surrounding the conductor was the conflict of two material fluids that had left the wire, or the "electrical" disturbance of a medium that existed between the wire and the magnet. Be that as it may, he found himself forced to look for an answer in a direction that turned out to be most significant.

The man who brought a great deal of clarity at this time to electrical and magnetic science was André Marie Ampère, a Frenchman, and a contemporary of Oersted. First Ampère showed that frictional electricity (electricity produced by rubbing) and voltaic electricity (electricity produced in a conductor by a voltaic cell) are essentially one phenomenon. The difference is that the latter is a continuous flow of the former—it is a current. Further, it is only the current that produce magnetic effects. Ampère reduced magnetism to electricity by asserting that the electric current is more fundamental than any magnetic fluid or magnetic particles. Thus all magnets must owe their properties to small, closed electric circuits within them in which a current is constantly present.

Ampère's most important work concerned the action of two current-carrying wires on one another. He began this work as early as 1819 and published his collected results in 1825. Ampère had reasoned that if an electric current affects a magnet, and a magnet's power comes from internal electric currents, then two electric currents should affect one another. He found that indeed to be the case. Two parallel wires with their currents in the same direction attract one another while the two wires with currents in opposite directions repel one another. Here we have parallel objects (the two wires) in which electricity is flowing along their lengths, and yet the forces they are experiencing are perpendicular to their lengths. As with Oersted's experiment, these experiments of Ampère are difficult to explain in terms of a simple action-at-a-distance theory.

Unfortunately, Ampère was unwilling to speculate as to the cause of this phenomena. Instead, belonging to the analytical school, he sought to save the appearances with mathematics. At that time he wished only to "explain" this phenomena in terms of forces between pairs of particles, and he was not concerned about speculation as to the nature of these forces.

Young, Fresnel, and the Luminiferous Ether

While paralleling the development of electrical science through the seventeenth and eighteenth centuries, optical science found itself at the beginning of the nineteenth century with a considerable amount of experimental and observational data, and with two opposing theories to explain this data, the wave and particle theories. Both theories had been formulated in the middle of the seventeenth century, one by Isaac Newton, the other by Christiaan Huygens. Huygens, basing his theory predominantly on the phenomenon of refraction, suggested that light consisted of vibratory motions in a fluid medium. Newton, in his

Opticks, recognized the instrumental value of the wave theory of light and maintained it as a subsidiary hypothesis. But when it came to a question of the real nature of light, he could not go along with this theory. Water and sound waves bend around corners, but light seemingly did not. Therefore, Newton rejected the wave theory in favor of a particle emission theory, according to which particles of light are emitted by the luminous source and travel in straight lines, obeying the laws of motion and collision. With Newton's name in support, the particle theory held the dominant position in optics until the beginning of the nineteenth century, despite the fact that the wave theory seemed to be superior in explaining the known phenomena.

The wave theory began to supplant the particle theory in 1800 when Dr. Thomas Young, an Englishman, published a paper entitled "Sound and Light." Young attempted to diminish the objections to the wave theory by applying the analogy of sound to light where it would be most advantageous. In addition, he emphasized certain difficulties of the particle theory that had been overlooked by optical scientists. For example, Young wondered how light particles emanating from the intensely hot sun could have the same velocity as particles emanating from the friction of two pebbles. Further, he branded particle "explanations" of simultaneous reflection and refraction, and the colors of thin films, as extremely weak and obscure.

A year later Young decided that the evidence for the wave theory had become overwhelming and concluded that light was probably the undulation of an elastic medium. Among the phenomena he pointed to in support of this conclusion were the following:

1. The velocity of light in the same medium is always equal.
2. All reflections are accompanied by a partial refraction.
3. Diffraction is better explained by a wave theory (in this phenomenon light does seem to bend around corners).
4. The colors of thin plates can be completely explained by the wave theory but not by the particle theory.

Young called the elastic medium for light waves the luminiferous ether, which, he felt, pervaded the entire universe.

From this medium hypothesis Young was able to make two very important predictions. Because he felt that material bodies had an attraction for the ether, it followed that light would travel more slowly in a more dense medium. This was in direct conflict with the particle theory that predicted that light would travel faster in a more dense medium. More important, however, the medium theory suggested that waves of different origins should interfere with one another if their paths crossed.[2]

[2]A wave is any disturbance that propagates with time from one region of space to another. The most obvious examples of wave motion are disturbances in water and those disturbances in air known as sound. If a pebble is dropped into a still pond, ripples will spread from the point of impact to the shore. The water is not significantly displaced horizontally, but the disturbance is. The water particles are displaced vertically under the influence of the disturbance. Sources of sound

In 1803 Young gave his first experimental demonstrations of the interference of light. In these experiments he was able to measure the wavelength of the light being used with some degree of accuracy. Since the speed of light was known to be very large, Young could see that the frequency of vibration of light waves was also going to be very large. This meant that the luminiferous ether would have to possess enormous elasticity.

In 1807 Young published the results of his most significant experiment, namely, the double slit experiment.[3] On the basis of the results of this experiment Young asked the rhetorical question: How could two sources of light produce darkness? Surely if light were a stream of emitted particles, then two sources would not produce darkness. But if light were a wave traveling in a medium, then waves from different sources could destructively interfere and produce darkness.

Following the double slit experiment, the arguments for the wave theory seemed quite imposing; the pendulum had clearly shifted in its direction. However, in 1809 the phenomenon of polarization,[4] totally inexplicable at that time by either theory, once again appeared on the scientific scene. When the Frenchman Dominique Arago showed that two beams of light, polarized in planes perpendicular to one another, do not interfere, Young became worried about his wave hypothesis. The difficulty stemmed from the sound-light analogy. Sound waves were known to be longitudinal, in that the medium vibrated in the direction of propagation. If light waves were also assumed to be longitudinal, then no explanation of polarization seemed possible. This problem bothered Young until 1817, when he finally admitted that some form of transverse vibrations were required to explain polarization. But he tried to derive these transverse vibrations from the longitudinal vibrations that he still deemed primary.

Meanwhile, in France, Augustin Fresnel was repeating much of the work of Young. His major contribution was the use of a principle of Huygens's as a synthesizing agent for all optical explanations. The principle was simply this: that the regions of the original wave could be considered sources of secondary

---

produce disturbances in air. The particles of air are displaced in the direction of the disturbance but their displacement is back and forth about their original position. When more than one wave is produced in a medium like water or air and these waves (disturbances) enter the same region of the medium at the same time, interference can take place: the total influence of the waves on the medium will be in some sense a combination of the influences of each individual wave.

[3]The double slit experiment consisted of a beam of monochromatic light sent to a screen in which there were two thin slits placed close together. The light proceeded through the slits to another screen upon which an interference pattern appeared. If the intensity of the beam was strong enough and if the distance between the two screens was large compared to the separation of the slits, then the pattern consisted of clear and distinct alternating bright and dark bands. Since the interference pattern extends beyond the projection of the slits, this experiment demonstrates conclusively that light does bend around corners.

[4]The two refracted beams from the crystal Iceland Spar are said to be polarized. A slice of tourmaline will transmit one of the refracted beams and not the other in one position and will transmit the second and not the first if rotated 90 degrees. At that time neither the wave nor the particle theory could explain this phenomenon.

waves that spread out in all directions from those regions. In this way diffraction, the bending of light around corners, could be satisfactorily explained. In fact all known optical phenomena could be explained with the exception of polarization.

Between 1816 and 1826 Fresnel worked on the problem of polarization, sometimes alone, sometimes in consultation with Young, and sometimes with the direct help of Arago. In 1826 he and Arago published the results of their experimentation with polarization. From these results Fresnel concluded that the oscillatory motion of light waves is transverse to the line of progress of the wave. Polarization had become explainable in terms of the wave theory.

With the work of Fresnel the wave theory of light was established for the remainder of the nineteenth century. Even the experiments of Foucault and Fizeau, performed in 1850, showing that light indeed moves slower in a more dense medium, were anticlimactic. But waves by themselves are simply mathematical constructions, and if light is to have a role in nature there must be a physical medium for it. In order to carry the rapid vibrations of light waves, the luminiferous ether had to possess enormous elasticity, and to allow high velocities it had to be of very low density. The vibrations of elastic fluids known at that time were all longitudinal. Only solids were able to transmit transverse waves. This seemed to leave only one possible answer as to the nature of the ether. *The luminiferous ether, pervading all space, and penetrating almost all substances, must not only be highly elastic but absolutely solid.* Despite the conceptual and physical difficulties of this hypothesis, the work of Young and Fresnel seemed to have established the conclusion that light is a wave form traveling in an apparently imponderable medium.

Ampère had observed the work of his fellow Frenchman, Fresnel, with great interest. As far as he was concerned Fresnel had demonstrated the existence of the luminiferous ether with his experiments. This apparently encouraged Ampère finally to theorize on his own. Since the accepted theory of electricity at that time was the two-fluid theory, Ampère saw the chance for a synthesis of electricity and optics. He suggested that the luminiferous ether was composed of the two electrical fluids, and that these fluids mutually saturate one another under normal conditions. A current in a wire would disturb this equilibrium and the resultant magnetic effects would occur.

Unfortunately, the mathematical physicists who followed Ampère did not have to accept or even consider his hypothesis. They could work simply with his electrodynamical equations. In so doing, however, they retained the language of action-at-a-distance theories and their instrumental successes lent credence to these theories. Thus action-at-a-distance theories became competitors of the ether and fluid theories in a second-hand way. If predictive power was all that was desired, then action-at-a-distance theories with their simplicity and mathematical susceptibility had a definite advantage. But if coherent explanations were desired, there was a need to speculate about an ethereal medium.

Faraday and the Beginnings of Field Theory

By 1830 the members of the scientific community that had been studying the phenomena of electricity, magnetism, and optics had come to realize that they possessed some of the pieces of a very complex puzzle but that they had no idea how to begin to put them together. The man who was to collect the remaining pieces and begin to put them together was Michael Faraday. Faraday knew little mathematics and used none in his work, but he was surely one of the greatest experimental scientists of his or any age. His lengthy and involved experiments with the materials of nature seemed to give him a greater appreciation for physical hypotheses than the mathematical physicists who used "susceptibility to mathematical analysis" as the chief criteria for evaluating said hypotheses.

Faraday believed in the unity of the forces of nature, and throughout his life he sought to reduce them to one or two primary forces. Knowing from Oersted's experiments that an electric current acts on a magnet, he reasoned that a magnet should somehow act upon or produce a current. His initial assumption was that a magnet would generate a continuous current in a circuit, but his attempts to justify this by experiment all failed. These experiments, however, did produce the discovery of an important new phenomenon: a wire free to revolve around a magnetic pole would do so when a current was passed through it. Similarly, the pole could be made to rotate around the current-carrying wire. This electromagnetic rotation further emphasized Oersted's discovery of the distinctive nature of the electromagnetic force. In addition, it convinced Faraday that electromagnetic phenomena involved something more than simple action at a distance along straight lines. In order now to make sense of the phenomena he had discovered, he developed the notion of magnetic lines of force extending in curved paths through space. A simple but ingenious demonstration showed how this new concept could account for the circular force and for attraction and repulsion. A straight current-carrying wire produces circular forces about the wire at its center. But when the wire is bent to form a loop, the lines are bent through it, causing the loop to behave as a bar magnet.

Faraday produced a further and even more striking demonstration of the appropriateness of his "lines of force" hypothesis by placing iron filings on a piece of paper along a diameter of the loop, and then perpendicular to a bar magnet at its midpoint. As a result, the iron filings lined up along the "lines of force" so that the patterns could easily be seen. The patterns from the loop and the magnet were essentially the same. Faraday concluded that magnetic action occurred along lines of force and that these lines were very often curved. He could not, however, conceive of curved lines of force without the conditions of a physical medium in space. Since magnets act as well when placed in a vacuum, this "physical medium in space" could not be composed of ponderable matter.

This, then, was Faraday's view as to the nature of magnetism. The magnet itself is somehow the source of the phenomena but it cannot act alone. It requires the surrounding medium in order for it to have magnetic properties. Without a medium the power or "energy" possessed by the magnet could not be released. That is, the power of the magnet lay in the medium through which the lines of force passed and not in the magnet. The lines of magnetic force maintained their physical reality. They were strains in space produced by the magnets in some way. The notion of a conditioned space or a conditionable space is the beginning of the concept of the electromagnetic field.

Maxwell's Electromagnetic Field Theory

As impressive as the physical evidence was for a field theory to account for electromagnetic phenomena, it had a strong competitor. Wilhelm Weber, a contemporary of Faraday, had developed a noninstantaneous action-at-a-distance theory to account for electromagnetic phenomena that had a distinct advantage over the field theory. Weber's theory had a mathematical structure that allowed him to make precise quantitative predictions that could then be tested by experiment.

The Scottish physicist James Clerk Maxwell objected to Weber's theory for philosophical as well as commonsense reasons. First, electromagnetic forces do not act along the lines joining the particles involved and thus would be very difficult to picture as the result of an action at a distance. Also, there is no physical concept of energy in Weber's theory. The action appears to be noninstantaneous but we are not told what is propagated in time, why it is propagated in time, how it is propagated in time, or what state it is in while being propagated.

In the first of three key memoirs, written in 1856, Maxwell rejected Weber's theory. Intuitively he found great physical value in Faraday's "lines of force in a surrounding medium" hypothesis. He then took it upon himself to mathematize those lines of force. In so doing Maxwell was able to show that electric and magnetic phenomena can be considered in terms of a medium with an appropriate mathematical structure that rivaled Weber's theory.

In the second memoir Maxwell developed a mechanical theory of the electromagnetic field: he viewed the medium as a mechanism, a series of vortices, subject to the laws of mechanics. With this mechanical medium theory he was now able to account for all existing electric and magnetic phenomena. But that was not all. Maxwell was able to calculate the velocity of a disturbance in this medium—the velocity of an electromagnetic wave. This velocity turned out to be essentially the same as the velocity of a light wave as measured by Fizeau in 1849. Maxwell concluded on the basis of this coincidence that light must be an electromagnetic wave.

Maxwell ended the second memoir with the conviction that he had successfully united the sciences of electricity and magnetism with the science of optics. He was also more convinced than ever that a medium theory was correct. However, because of the difficulties in developing a mechanical theory of the luminiferous ether and because of the apparent unobservable-in-principle character of the electromagnetic field, he was very uncomfortable with his theory of vortices as a physical hypothesis.

In his third memoir, written in 1864, Maxwell declared that the theory of vortices had instrumental value only and that it should not be considered a serious physical hypothesis. He then proceeded to remove the mechanical medium from the electromagnetic field theory by considering the field as an abstract dynamical system transporting energy. With this dynamical analogy he was able to develop the general equations of the electromagnetic field (Maxwell's equations). From these equations he could then derive all the known laws of electricity, magnetism, and optics in terms of a medium or field theory. In addition, these equations contained the extraordinary definition of an electromagnetic wave as a transverse variation in electric and magnetic field strengths, and the prediction of the existence of electromagnetic waves other than light. So impressive was Maxwell's theory that the existence of electromagnetic waves other than light was a foregone conclusion.

Maxwell ended his third memoir with a definite commitment to the reality of the electromagnetic field. But its exact nature still eluded him. Although he speaks as if it is made of matter, he is quick to distinguish the matter of the field from what he called gross matter. Electromagnetic waves were undulations of an ethereal substance and not of gross matter. But this ethereal substance could be called matter only in the sense that it is subject to dynamical analysis. However, in considering the energy that is stored in and transmitted through the field, Maxwell wanted to be taken literally. All energy is the same as mechanical energy, he felt, and thus it needed a physical substratum to support it.

Maxwell ended his work with the view that the electromagnetic field is real and physical in the sense that it can effect gross matter. Further, he felt the field filled all of space but was not identical to it. However, because the field was not made of gross matter it was an unobservable-in-principle reality.

Theoretical and Experimental Support for Maxwell's Theory

Prior to the experimental support that Maxwell's theory received from Hertz, it received considerable theoretical support from J. H. Poynting. Poynting developed an expression that indicates the flux of energy at any point in the electromagnetic field in terms of the electric and magnetic field strengths at that point.

$$\vec{S} = \vec{E} \times \vec{H}$$

In the above expression the vector S is called Poynting's Vector and gives the intensity (the amount of energy crossing a unit area in a unit time) of an electromagnetic wave.[5] Thus, the propagation of E and H, which, of course, is an

[5]Vector quantities stipulate both magnitude and direction. Poynting's Vector would have a direction perpendicular to both $\vec{E}$ and $\vec{H}$.

electromagnetic wave, is accompanied by the propagation of energy. With the Poynting expression the discussion of the relationship between the energy of the field and the material objects that the field interacts with could be accomplished. The validity and efficacy of this expression further supported Maxwell's contention that an "action at a distance in time" theory would violate the physical principle of energy conservation.

Such theoretical support, despite its significance, is usually overlooked because of the experimental support provided for Maxwell's theory by Heinrich Hertz. The results of Hertz's most important work were published in two memoirs, both of which appeared in 1888. These memoirs appear as Chapters 7 and 8 of his translated work *Electric Waves*. Basically the apparatus used by Hertz is rather simple. It consisted of two wires both having spark gaps. One, the transmitter, was connected to an induction coil that could build up large electric charges of opposite signs at either end of the spark gap. When the opposite charges became strong enough, the insulating ability of the air would break down and a current in the form of a spark would jump across the gap. It was known to Hertz through Helmholtz that this spark had an oscillatory character that "gradually" diminished to zero. According to Maxwell's theory, a changing electric field produces a magnetic field and vice versa, with these disturbances propagated in time from the source. Hertz thus reasoned that this "spark" should transmit an electromagnetic wave with its accompanying energy. To detect this wave he used another wire also with a spark gap. If currents are induced in the receiver wire by the electromagnetic wave, sparks should be observed at the gap. Hertz observed such sparks in the receiver wire, which indicated a cause-effect relationship between the transmitter and the receiver. But this was not proof that the energy was propagated in time through the field. In order to show this, Hertz set up stationary electromagnetic waves in the laboratory by reflecting the transmitted waves off a metallic surface. Stationary or standing waves are characterized by stationary nodes (or zero points) and antinodes. Hertz was able to detect these nodes and antinodes with his receiver wire. From wave theory it was known that the distance between successive nodes, or antinodes, is one half the wavelength of the waves that are interfering. Hertz found that the wavelength of the wave was 9.6 meters. He was also able to calculate the frequency of the waves from information about the induction coil. The frequency turned out to be approximately 3×10^7 cycles/sec. From the simple equation relating the velocity of the wave to its frequency and wavelength, Hertz was able to calculate the velocity of propagation of this electromagnetic wave to be 2.88×10^8 m/sec. This was approximately the velocity of light as measured in air, as well as the velocity of propagation of Maxwell's electromagnetic waves as determined theoretically.

It can be said with certainty that Hertz's experiments had shown that electromagnetic action occurs in time rather than instantaneously. But they further showed that Maxwell's electromagnetic wave theory provided an intelligible and accurate physical interpretation of electromagnetic action. A ballistic transmission theory would have to explain physically how the stationary effect is achieved by moving particles. Such a physical explanation would obviously be difficult, if not impossible, to produce. Thus, Maxwell's electromagnetic field theory became quite convincing on the basis of Hertz's experiments.

Concluding Remarks

In the hundred years that have followed Hertz's experiments, we have seen the second great revolution in physical theory, a revolution that has led to the development of the Theory of Relativity and quantum mechanics. Yet despite these extraordinary developments, Electromagnetic Field Theory, with some modification, has maintained a prominent position in physics. Einstein himself pointed this out in 1938:

> In the beginning the field concept was no more than a means of facilitating the understanding of phenomena from the mechanical point of view. In the new field language it is the description of the field between the two charges, and not the charges themselves, which is essential for an understanding of their actions. . . . It was realized that something of great importance had happened in physics. *A new reality was created for which there was no place in the mechanical description.* Slowly and by a struggle the field concept established for itself a leading place in physics and has remained one of the *basic physical concepts. The electromagnetic field is, for the modern physicist, as real as the chair on which he sits.*[6]

At the start of the twenty-first century the role of Electromagnetic Field Theory as a major paradigm in physical science is stronger than ever.

1
— William Gilbert —
(1544–1603)

William Gilbert was an English physician and physicist who was a pioneer in the study of magnetism. He developed two methods for producing magnets; the first involved the stroking of iron by natural magnets or loadstones, and the second involved hammering pieces of iron while they were aligned in the Earth's magnetic field. In 1600, he published the results of his experiments with magnets in a book entitled De Magnete. *It is considered to be a classic of experimental science and one of the first great works of the modern scientific era.*

[6]Albert Einstein and L. Infeld, *The Evolution of Physics* (New York: Simon and Schuster, 1938), 151.

In this reading, taken from De Magnete, *Gilbert describes the properties of magnets. He compares the loadstone, or natural magnet, to the earth itself, suggesting that it too has two poles, a north and a south. By a simple demonstration he shows that like poles repel and unlike poles attract.*

Gilbert also describes the properties of amber, and indeed other "electrics," and shows that the cause of these properties is different from the cause of the properties of magnets. His reference to the "electrical effuvia of the earth" suggests that he was thinking of a medium theory to explain electrical and magnetic phenomena.

THE PROPERTIES OF MAGNETS

The Loadstone Possesses Parts Differing in Their Natural Powers, and Has Poles Conspicuous for Their Properties.

The many qualities exhibited by the loadstone itself, qualities hitherto recognized yet not well investigated, are to be pointed out in the first place, to the end the student may understand the powers of the loadstone and of iron, and not be confused through want of knowledge at the threshold of the arguments and demonstrations. In the heavens, astronomers give to each moving sphere two poles; thus do we find two natural poles of excelling importance even in our terrestrial globe, constant points related to the movement of its daily revolution, to wit, one pole pointing to Arctos (Ursa) and the north; the other looking toward the opposite part of the heavens. In like manner the loadstone has from nature its two poles, a northern and a southern; fixed, definite points in the stone, which are the primary termini of the movements and effects, and the limits and regulators of the several actions and properties. It is to be understood, however, that not from a mathematical point does the force of the stone emanate, but from the parts themselves; and all these parts in the whole—while they belong to the whole—the nearer they are to the poles of the stone the stronger virtues do they acquire and pour out on other bodies. These poles look toward the poles of the earth, and move toward them, and are subject to them. The magnetic poles may be found in every loadstone, whether strong and powerful (male, as the term was in antiquity) or faint, weak, and female; whether its shape is due to design or to chance, and whether it be long, or flat, or four-square, or three-cornered, or polished; whether it be rough, broken-off, or unpolished: the loadstone ever has and ever shows its poles. . . .

One Loadstone Appears to Attract Another in the Natural Position; But in the Opposite Position Repels It and Brings It to Rights.

First we have to describe in popular language the potent and familiar properties of the stone; afterward, very many subtle properties, as yet recondite and unknown, being involved in obscurities, are to be unfolded; and the causes of all these (nature's secrets being unlocked) are in their place to be demonstrated in fitting words and with the aid of apparatus. The fact is trite and familiar, that the loadstone attracts iron; in the same way, too, one loadstone attracts another. Take the stone on which you have designated the poles, N. and S., and put it in its vessel so that it may float; let the poles lie just in the plane of the horizon, or at least in a plane not very oblique to it; take in your hand another stone the poles of which are also known; and hold it so that its south pole shall lie toward the north pole of the floating stone, and near it alongside, the floating loadstone will straightway follow the other (provided it be within the range and dominion of its powers), nor does it cease to move nor does it quit the other till it clings to it, unless by moving your hand away, you manage skillfully to prevent the conjunction. In like manner, if you oppose the north pole of the stone in your hand to the south pole of the floating one, they come together and follow each other. For opposite poles attract opposite poles. But, now, if in the same way you present N. to N. or S. to S., one stone repels the other; and as though a helmsman were bearing on the rudder it is off like a vessel making all sail, nor stands nor stays as long as the other stone pursues. One stone also will range the other, turn the other around, bring it to right about and make it come to agreement with itself. But when the two come together and are conjoined in nature's order, they cohere firmly. For example, if you present the north pole of the stone in your hand to the Tropic of Capricorn (for so we may distinguish with mathematical circles the round stone or terrella, just as we do the globe itself) or to any point between the equator and the south pole: immediately the floating stone turns round and so places itself that its south pole touches the north pole of the other and is most closely joined to it. In the same way you will get like effect at the other side of the equator by presenting pole to pole; and thus by art and contrivance we exhibit attraction and repulsion, and motion in a circle toward the concordant position, and the same movements to avoid hostile meetings. Furthermore, in one same stone we are thus able to demon-

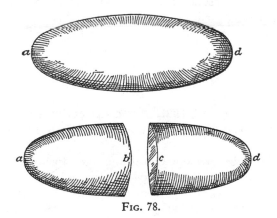

Fig. 78.

strate all this: but also we are able to show how the self-same part of one stone may by division become either north or south. Take the oblong stone *ad* (Fig. 78) in which *a* is the north pole and *d* the south. Cut the stone in two equal parts, and put part *a* in a vessel and let it float in water.

You will find that *a* the north point, will turn to the south as before; and in like manner the point *d* will move to the north, in the divided stone, as before division. But *b* and *c*, before connected, now separated from each other, are not what they were before. *b* is now south while *c* is north. *b* attracts *c*, longing for union and for restoration of the original continuity. They are two stones made out of one, and on that account the *c* of one turning toward the *b* of the other, they are mutually attracted, and being freed from all impediments and from their own weight, borne as they are on the surface of the water, they come together and into conjunction. But if you bring the part or point *a* up to *c* of the other, they repel one another and turn away; for by such a position of the parts nature is crossed and the form of the stone is perverted: but nature observes strictly the laws it has imposed upon bodies: hence the flight of one part from the undue position of the other, and hence the discord unless everything is arranged exactly according to nature. And nature will not suffer an unjust and inequitable peace, or agreement, but makes war and employs force to make bodies acquiesce fairly and justly. Hence, when rightly arranged the parts attract each other, i.e., both stones, the weaker and the stronger, come together and with all their might tend to union: a fact manifest in all loadstones, and not, as Pliny supposed, only in those from Ethiopia. The Ethiopic stones if strong, and those brought from China, which are all powerful stones, show the effect most quickly and most plainly, attract with most force in the parts nighest the pole, and keep turning till pole looks straight on pole. The pole of a stone has strongest attraction for that part of another stone which answers to it (the *adverse* as it is called); e.g., the north pole of one has strongest attraction for, has the most vigorous pull on, the south part of another; so too it attracts iron more powerfully, and iron clings to it more firmly, whether previously magnetized or not. Thus it has been settled by nature, not without reason, that the parts nigher the pole shall have the greatest attractive force; and that in the pole itself shall be the seat, the throne as it were, of a high and splendid power; and that magnetic bodies brought near thereto shall be attracted most powerfully and relinquished with most reluctance. So, too, the poles are readiest to spurn and drive away what is presented to them amiss, and what is inconformable and foreign. . . .

Of Magnetic Coition; and, First, of the Attraction Exerted by Amber, or More Properly the Attachment of Bodies to Amber

Great has ever been the fame of the loadstone and of amber in the writings of the learned: many philosophers cite the loadstone and also amber whenever, in explaining mysteries, their minds become obfuscated and reason can no farther go. Over-inquisitive theologians, too, seek to light up God's mysteries and things beyond man's understanding by means of the loadstone as a sort of Delphic sword and as an illustration of all sorts of things. Medical men also (at the bidding of Galen), in proving that purgative medicines exercise attraction through likeness of substance and kinships of juices (a silly error and gratuitous!), bring in as a witness the loadstone, a substance of great authority and of noteworthy efficiency, and a body of no common order. Thus in very many affairs persons who plead for a cause the merits of which they cannot set forth, bring in as masked advocates the loadstone and amber. But all these, besides sharing the general misapprehension, are ignorant that the causes of the loadstone's movements are very different from those which give to amber its properties; hence they easily fall into errors, and by their own imaginings are led farther and farther astray. For in other bodies is seen a considerable power of attraction, differing from that of the loadstone,—in amber, for example. Of this substance a few words must be said, to show the nature of the attachment of bodies to it, and to point out the vast difference between this and the magnetic actions; for men still continue in ignorance, and deem that inclination of bodies to amber to be an attraction, and comparable to the magnetic coition. The Greeks call this substance 'ἤλεκτρον, because, when heated by rubbing, it attracts to itself chaff; whence it is also called 'ἅρπαξ and from its golden color, χρυσοφορον. But the Moors call it *carabe*, because they used to offer it in sacrifices and in the worship of the gods; for in Arabic *carab* means oblation, not *rapiens paleas* (snatching chaff), as Scaliger would have it, quoting from the Arabic or Persian of Abohali (Hali Abbas). Many call this substance *ambra* (amber), especially that which is brought from India and Ethiopia. The Latin name *succinum* appears to be formed from *succus*, juice. The Sudavienses or Sudini call the substance *geniter*, as though *genitum terra* (produced by the earth). The erroneous opinion of the ancients as to its nature and source being exploded, it is certain that amber comes for the most part from the sea: it is gathered on the coast after heavy storms, in nets and through other means, by peasants, as by the Sudini of Prussia; it is also sometime found on the coast of our own Britain. But it seems to be produced in the earth and at considerable depth below its surface; like the rest of the bitumens; then to be washed out by the sea-waves, and to gain consistency under the action of the sea and the saltness of its waters. For at first it was a soft and viscous matter, and hence contains, buried in its mass forevermore (*aeternis sepulchris relucentes*), but still (shining) visible, flies, grubs, midges, and ants. The ancients as well as moderns tell (and their report is confirmed by experience) that amber attracts straws and chaff. The same is done by jet, a stone taken out of the earth in Britain, Germany, and many other regions: it is a hard concretion of black bitumen,—a sort of transformation of bitumen to stone. Many modern authors have written about amber and jet as attracting chaff and about other facts unknown to the generality, or have copied from other writers: with the results of their labors booksellers' shops are crammed full. Our generation has produced many volumes about recondite, abstruse, and occult causes and wonders, and in all of them amber and jet are represented as attracting chaff; but never a proof from experiments, never a demonstration do you find in them. The writers deal only in words that involve in thicker darkness subject-matter; they treat the subject esoterically, miracle-mongeringly, abstrusely, reconditely, mystically. Hence such philosophy bears no fruit; for it rests simply on a few Greek or unusual terms—just as our barbers toss off a few Latin words in the hearing of the ignorant rabble in token of their learning, and thus win reputation—bears no fruit, because few of the philosophers themselves are investigators, or have any first-hand acquaintance with things; most of them are indolent and untrained, add nothing to knowledge by their writings, and are blind to the things that might throw a light upon their reasonings. For not only do amber and (gagates or) jet, as they suppose attract light corpuscles (substances): the same is done by diamond, sapphire, carbuncle, iris stone, opal, amethyst, vincentina, English gen (Bristol stone, *bristola*), beryl, rock crystal. Like powers of attracting are possessed by glass, especially clear, brilliant glass; by artificial gems made of (paste) glass or rock crystal, antimony glass, many fluor-spars, and belemnites. Sulphur also attracts, and likewise mastich, and sealing-wax (of lac), hard resin, orpiment (weakly). Feeble power of attraction is also possessed in favoring dry atmosphere by sal gemma [native chloride of sodium], mica, rock alum. This we may observe when in mid-winter the atmosphere is very cold, clear, and thin; when the electrical effluvia of the earth offers less impediment, and electric bodies are harder: of all this later. These several bodies (electric) not only draw to themselves straws and chaff, but all metals, wood, leaves, stones, earths, even water and oil; in short, whatever things appeal to our senses or are solid: yet we are told that it attracts nothing but chaff and twigs. Hence Alexander Aphrodiseus incorrectly declares the question of amber to be unsolvable, because that amber does attract chaff, yet not the leaves of basil; but such stories are false, disgracefully inaccurate. Now in order clearly to understand by experience how such attraction takes place, and what those substances may be that so attract other bodies (and in the case of many of these electrical substances, though the bodies influenced by them lean toward them, yet because of the feebleness of the attraction they are not drawn clean up to them, but are easily made to rise), make yourself a rotating-needle (electroscope—*versorium*), of any sort of metal, three or four fingers long, pretty light, and poised on a sharp point after the manner of a magnetic pointer. Bring near to one end of it a piece of amber or a gem, lightly rubbed, polished and shining: at once the instrument revolves.

Several objects are seen to attract not only natural objects, but things artificially prepared, or manufactured, or formed by mixture. Nor is this a rare property possessed by one object or two (as is commonly supposed), but evidently

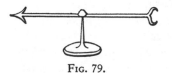

belongs to a multitude of objects, both simple and compound, e.g., sealing-wax and other unctuous mixtures. But why this inclination and what these forces,—on which points a few writers have given a very small amount of information, while the common run of philosophers give us nothing,—these questions must be considered fully.

FIG. 79.

2
— Charles Coulomb —
(1736–1806)

Charles Coulomb, a French physicist, discovered the laws of electric and mag-
netic force. In 1784, he invented a torsion balance capable of detecting very
weak forces with which he conducted a series of very delicate experiments. These
experiments led to the development of a law of electrostatic force that paralleled
Newton's Law of Universal Gravitation. The unit of electric charge is named in
his honor.

In this reading Coulomb describes the experiments that led to his law of elec-
trostatic force. He shows clearly from his "trials" that the force exerted between
two electrified bodies varies inversely with the square of the distance between
them.

The reading comes from a memoir presented to the French Academy of Science
in 1785.

THE LAW OF ELECTRIC FORCE

The repulsive force between two small spheres charged with the same sort of
electricity is in the inverse ration of the squares of the distances between the
centers of the two spheres.

Experiment

We electrify a small conductor, (Fig. 81, 4) which is simply a pin with a large
head insulated by sinking its point into the end of a rod of Spanish wax; we
introduce this pen through the hole *m* and with it touch the ball *t*, which is in
contact with the ball *a*; on withdrawing the pin the two balls are electrified with
electricity of the same sort and they repel each other to a distance which is
measured by looking past the suspension wire and the center of the ball *a* to
the corresponding division of the circle *zoQ*; then by turning the index of the
micrometer in the sense *pno* we twist the suspension wire *lp* and exert a force
proportional to the angle of torsion which tends to bring the ball *a* nearer to
the ball *t*. We observe in this way the distance through which different angles
of torsion bring the ball *a* toward the ball *t*, and by comparing the forces of
torsion with the corresponding distances of the two balls we determine the law
of repulsion. I shall here only present some trials which are easy to repeat and
which will at once make evident the law of repulsion.

First Trial. Having electrified the two balls by means of the pin head while

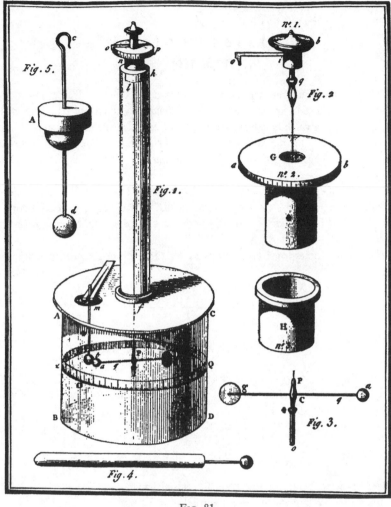

FIG. 81.

the index of the micrometer points to *o*, the ball *a* of the needle is separated from the ball *t* by 36 degrees.

Second Trial. By twisting the suspension wire through 126 degrees as shown by the pointer *o* of the micrometer, the two balls approach each other and stand 18 degrees apart.

Third Trial. By twisting the suspension wire through 567 degrees the two balls approach to a distance of 8 degrees and a half.

Explanation and Result of This Experiment

Before the balls have been electrified they touch, and the center of the ball *a* suspended by the needle is not separated from the point where the torsion of the suspension wire is zero by more than half the diameters of the two balls. It must be mentioned that the silver wire *lp* which formed this suspension was twenty-eight inches long and was so fine that a foot of it weighed only 1/16 grain. By calculating the force which is needed to twist this wire by acting on the point *a* four inches away from the wire *lp* or from the center of suspension, I have found by using the formulas explained in a memoir on the laws of the force of torsion of metallic wires, printed in the Volume of the Academy for 1784, that to twist this wire through 360 degrees the force that was needed when applied at the point *a* so as to act on the lever *an* four inches long was only 1/340 grains: so that since the forces of torsion, as is proved in that memoir, are as the angles of torsion, the least repulsive force between the two balls would separate them sensibly from each other.

We found in our first experiment, in which the index of the micrometer is set on the point *o*, that the balls are separated by 36 degrees, which produces a force of torsion of 36° = 1/3400 of a grain; in the second trial the distance between the balls is 18 degrees, but as the micrometer has been turned through 126 degrees it results that at a distance of 18 degrees the repulsive force was equivalent to 144 degrees; so at half the first distance the repulsion of the balls is quadruple.

In the third trial the suspension wire was twisted through 567 degrees and the two balls are separated by only 8 degrees and a half. The total torsion was consequently 576 degrees, four times that of the second trial, and the distance of the two balls in this third trial lacked only one-half degree of being reduced to half of that at which it stood in the second trial. It results then from these three trials that the repulsive action which the two balls exert on each other when they are electrified similarly is in the inverse ratio of the square of the distances.

3
— Hans Christian Oersted —
(1777–1851)

Hans Christian Oersted was a Danish physicist who discovered the extraordinary relationship between an electric current and a magnet. Oersted was appointed professor of physics at the University of Copenhagen in 1806 and later became director of the Polytechnic Institute in Copenhagen.

In the following reading Oersted describes the experiment that led to his discovery that a current-carrying wire will cause magnets to deflect in directions perpendicular to the length of the wire. Because of the directions of the force exerted on the magnet, Oersted rules out attraction and repulsion and thus ap-

parently action at a distance as an explanation for this phenomenon. He uses the concept of electric conflict for his explanation, which seems to presuppose a medium. The final paragraph indicates Oersted's extraordinary insight into the workings of nature.

The reading was published in pamphlet form in 1820.

THE EFFECT OF A CURRENT OF ELECTRICITY ON A MAGNETIC NEEDLE

The first experiments respecting the subject which I mean at present to explain, were made by me last winter, while lecturing on electricity, galvanism, and magnetism, in the University. It seemed demonstrated by these experiments that the magnetic needle was moved from its position by the galvanic apparatus, but that the galvanic circle must be complete, and not open, which last method was tried in vain some years ago by very celebrated philosophers. But as these experiments were made with a feeble apparatus, and were not, therefore, sufficiently conclusive, considering the importance of the subject, I associated myself with my friend Esmarck to repeat and extend them by means of a very powerful galvanic battery, provided by us in common. Mr. Wleugel, a Knight of the Order of Danneborg, and at the head of the Pilots, was present at, and assisted in, the experiments. There were present likewise Mr. Hauch, a man very well skilled in the Natural Sciences, Mr. Reinhardt, Professor of Natural History, Mr. Jacobsen, Professor of Medicine, and that very skillful chemist, Mr. Zeise, Doctor of Philosophy. I had often made experiments by myself; but every fact which I had observed was repeated in the presence of these gentlemen.

The galvanic apparatus which we employed consists of 20 copper troughs, the length and height of each of which was 12 inches; but the breadth scarcely exceeded 2½ inches. Every trough is supplied with two plates of copper, so bent that they could carry a copper red, which supports the zinc plate in the water of the next trough. The water of the troughs contain $1/60^{th}$ of its weight of sulphuric acid, and an equal quantity of nitric acid. The portion of each zinc plate sunk in the water is a square whose side is about 10 inches in length. A smaller apparatus will answer provided it be strong enough to heat a metallic wire red hot.

The opposite ends of the galvanic battery were joined by a metallic wire, which for shortness sake, we shall call the *uniting conductor*, or the *uniting wire*. To the effect which takes place in this conductor and in the surrounding space, we shall give the name of the *conflict of electricity*.

Let the straight part of this wire be placed horizontally above the magnetic needle, properly suspended, and parallel to it. If necessary, the uniting wire is bent so as to assume a proper position for the experiment. Things being in this state, the needle will be moved, and the end of it next to the negative side of the battery will go westward.

If the distance of the uniting wire does not exceed three-quarters of an inch from the needle, the declination of the needle makes an angle of about 45°. If the distance is increased, the angle diminishes proportionally. The declination likewise varies with the power of the battery.

The uniting wire may change its place, either towards the east or west, provided it continue parallel to the needle, without any other change of the effect than in respect to its quantity. Hence the effect cannot be ascribed to attraction; for the same pole of the magnetic needle, which approaches the uniting wire, while placed on its east side, ought to recede from it when on the west side, if these declinations depended on attractions and repulsions. The uniting conductor may consist of several wires, or metallic ribbons, connected together. The nature of the metal does not alter the effect, but merely the quantity. Wires of platinum, gold, silver, brass, iron, ribbons of lead and tin, a mass of mercury, were employed with equal success. The conductor does not lose its effect, though interrupted by water, unless the interruption amount to several inches in length.

The effect of the uniting wire passes to the needle through glass, metals, wood, water, resin, stoneware, stones; for it is not taken away by imposing plates of glass, metal, and wood, interposed at once, (do not destroy), and indeed scarcely diminish the effect. The disc of the electrophorus, plates of porphyry, a stone-ware vessel, even filled with water, were interposed with the same result. We found the effects unchanged when the needle was included in a brass box filled with water. It is needless to observe that the transmission of effects through all these matters has never before been observed in electricity and galvanism. The effects, therefore, which take place in the conflict of electricity are very different from the effects of either of the electricities.

If the uniting wire be placed in a horizontal plane under the magnetic needle, all the effects are the same as when it is above the needle, only they are in the opposite direction; for the pole of the magnetic needle (next the negative end) of the battery declines to the east.

That these facts may be the more easily retained, we may use this formula— the pole above which the *negative* electricity enters is turned to the *west*; under which, to the *east*.

If the uniting wire is so turned in a horizontal plane as to form a gradually increasing angle with the magnetic meridian, the declination of the needle *increases*, if the motion of the wire is towards the place of the disturbed needle; but it *diminishes* if the wire moves further from that place.

When the uniting wire is situated in the same horizontal plane in which the needle moves by means of the counterpoise, and parallel to it, no declination is produced either to the east or west; but an *inclination* takes place, so that the pole, next which the negative electricity enters the wire, is *depressed* when the wire is situated on the *west* side, and *elevated* when situated on the *east* side.

If the uniting wire be placed perpendicularly to the plane of the magnetic meridian, whether above or below it, the needle remains at rest, unless it be very *near the pole; in that case the pole is elevated when the entrance is from the west* of the wire, and *depressed*, when from the *east* side.

When the uniting wire is placed perpendicularly opposite to the pole of the magnetic needle, and the upper extremity of the wire receives the negative electricity, the pole is moved towards the east; but when the wire is opposite to a point between the pole and the middle of the needle, the pole is most towards the west. When the upper end of the wire received positive electricity, the phenomena are reversed.

If the uniting wire is bent so as to form two legs parallel to each other, it repels or attracts the magnetic poles according to the different conditions of the case. Suppose the wire placed opposite to either pole of the needle, so that the plane of the parallel legs is perpendicular to the magnetic meridian, and let the eastern leg be united with the negative end, the western leg with the positive end of the battery: in that case the nearest pole will be repelled either to the east or west, according to the position of the plane of the legs. The eastmost leg being united with the positive, and the westmost with the negative side of the battery, the nearest pole will be attracted. When the plane of the legs is placed perpendicular to the place between the pole and the middle of the needle, the same effects recur, but reversed.

A brass needle, suspended like a magnetic needle, is not moved by the effect of the uniting wire. Likewise needles of glass and of gum lac remain unacted on.

We may now make a few observations towards explaining these phenomena.

The electric conflict acts only on the magnetic particles of matter. All non-magnetic bodies, or rather their magnetic particles, resist the passage of this conflict. Hence they can be moved by the impetus of the contending powers.

It is sufficiently evident from the preceding facts that the electric conflict is not confined to the conductor, but dispersed pretty widely in the circumjacent space.

From the preceding facts we may likewise collect that this conflict performs circles; for without this condition, it seems impossible that the one part of the uniting wire, when placed below the magnetic pole, should drive it towards the east, and when placed above it towards the west; for it is the nature of a circle that the motions in opposite parts should have an opposite direction. Besides, a motion in circles, joined with a progressive motion, according to the length of the conductor, ought to form a conchoidal or spiral line; but this, unless I am mistaken, contributes nothing to explain the phenomena hitherto observed.

All the effects on the north pole above-mentioned are easily understood by supposing that negative electricity moves in a spiral line bent towards the right, and propels the north pole, but does not act on the south pole. The effects on the south pole are explained in a similar manner, if we ascribe to positive electricity a contrary motion and power of acting on the south pole, but not upon the north. The agreement of this law with nature will be better seen by a repetition of the experiments than by a long explanation. The mode of judging of the experiments will be much facilitated if the course of the electricities in the uniting wire be pointed out by marks or figures.

I shall merely add to the above that I have demonstrated in a book published five years ago that heat and light consist of the conflict of the electricities. From the observations now stated, we may conclude that a circular motion likewise occurs in these effects. This I think will contribute very much to illustrate the phenomena to which the appellation of polarization of light has been given.

4
— André Marie Ampère —
(1775–1836)

André Marie Ampère was a French physicist and a pioneer in the field of electro-dynamics. Impressed by the discovery of Oersted, he performed experiments in-volving current-carrying wires. In 1827, as a result of his experiments, he was able to develop a law describing the forces that are exerted by current-carrying wires on one another. The unit of electric current is named in his honor.

In this reading Ampère recognizes the problems associated with the hypothesis of a subtle medium that carries electrodynamical forces. He recommends that the science of electrodynamics drop such a hypothesis and concentrate only on empirical laws. Interestingly, he cites Newton's unwillingness to offer a causal explanation of gravity as his inspiration.

Although a positivist with regard to electrodynamical phenomena at this time, Ampère later concluded that the phenomena associated with light had demon-strated the existence of a medium to carry its transmission.

The selection is from a memoir read to the Royal Academy of Sciences in 1825.

A POSITIVIST APPROACH TO ELECTROMAGNETISM

The new era in the history of science marked by the works of Newton, is not only the age of man's most important discovery in the causes of natural phenom-ena, it is also the age in which the human spirit has opened a new highway into the sciences which have natural phenomena as their object of study.

Until Newton, the causes of natural phenomena had been sought almost ex-clusively in the impulsion of an unknown fluid which entrained particles of materials in the same direction as its own particles; wherever rotational motion occurred, a vortex in the same direction was imagined.

Newton taught us that motion of this kind, like all motions in nature, must be reducible by calculation to forces acting between two material particles along the straight line between them such that the action of one upon the other is equal and opposite to that which the latter has upon the former and consequently, assuming the two particles to be permanently associated, that no motion whatsoever can result from their interaction. It is this law, now confirmed by every observation and every calculation, which he represented in the three axioms at the beginning of the *Philosophiae naturalis principia mathematica*. But it was not enough to rise to the conception; the law had to be found which governs the variation of these forces with the positions of the particle between which they act, or, what amounts to the same thing, the value of these forces had to be expressed by a formula.

Newton was far from thinking that this law could be discovered from abstract considerations, however plausible they might be. He established that such laws must be deduced from observed facts, or preferably, from empirical laws, like those of Kepler, which are only the generalized results of very many facts.

To observe first the facts, varying the conditions as much as possible, to accompany this with precise measurement, in order to deduce general laws based solely on experience, and to deduce therefrom, independently of all hypothesis regarding the nature of forces which produce the phenomena, the mathematical value of these forces, that is to say, to derive the formula which represents them, such was the road which Newton followed. This was the approach generally adopted by the learned men of France to whom physics owes the immense progress which has been made in recent times, and similarly it has guided me in all my research into electrodynamic phenomena. I have relied solely on experimentation to establish the laws of the phenomena and from them I have derived the formula which alone can represent the forces which are produced; I have not investigated the possible cause of these forces, convinced that all research of this nature must proceed from pure experimental knowledge of the laws and from the value, determined solely by deduction from these laws, of the individual forces in the direction which is, of necessity, that of a straight line drawn through the material points between which the forces act. That is why I shall refrain from discussing any ideas which I might have on the nature of the cause of the forces produced by voltaic conductors, though this is contained in the notes which accompany the "Exposé somaire des nouvelles expériences electromagnétiques faites par plusieurs physiciens depuis le mois de mars 1821," which I read at the public session of the Académie des Sciences, 8 April 1822; my remarks can be seen in these notes on page 215 of my collection of "Observations in Electrodynamics." It does not appear that this approach, the only one which can lead to results which are free of all hypothesis, is preferred by physicists in the rest of Europe like it is by Frenchmen; the famous scientist who first saw the poles of a magnet transported by the action of a conductor in directions perpendicular to those of the wire, concluded that electrical matter revolved about it and pushed the poles along with it, just as Descartes made "the matter of his vortices" revolve in the direction of planetary revolution. Guided by Newtonian philosophy, I have reduced the phenomenon observed by M. Oerstedt, as has been done for all similar natural phenomena, to forces acting along a straight line joining the two particles between which the actions are exerted;

and if I have established that the same arrangement, or the same movement of electricity, which exists in the conductor is present also round the particles of the magnets, it is certainly not to explain their action by impulsion as with a vortex, but to calculate, according to my formula, the resultant forces acting between the particles of a magnet and those of a conductor, or of another magnet, along the lines joining the particles in pairs which are considered to be interacting, and to show that the results of the calculation are completely verified by (1) the experiments of M. Pouillet and my own into the precise determination of the conditions which must exist for a moving conductor to remain in equilibrium when acted upon, whether by another conductor, or by a magnet, and (2) by the agreement between these results and the laws which Coulomb and M. Biot have deduced by their experiments, the former relating to the interaction of two magnets, and the latter to the interaction between a magnet and a conductor.

The principal advantage of formulae which are derived in this way from general facts gained from sufficient observations for their certitude to be incontestable, is that they remain independent, not only of the hypotheses which may have aided in the quest for these formulae, but also independent of those which may later be adopted instead. The expression for universal attraction from the laws of Kepler is completely independent of the hypotheses which some writers have advanced to justify the mechanical cause to which they would ascribe it. The theory of heat is founded on general facts which have been obtained by direct observation; the equation deduced from these facts, being confirmed by the agreement between the results of calculation and of experiment, must be equally accepted as representative of the true laws of heat propagation by those who attribute it to the radiation of calorific molecules as by these who take the view that the phenomenon is caused by the vibration of a diffuse fluid in space; it is only necessary for the former to show how the equations result from their way of looking at heat for the others to derive it from general formulae for vibratory motion; doing so does not add anything to the certitude of the equation, but only substantiates the respective hypotheses. The physicist who refrains from committing himself in this respect, acknowledges the heat equation to be an exact representation of the facts without concerning himself with the manner in which it can result from one or other of the explanations of which we are speaking; and if new phenomena and new calculations should demonstrate that the effects of heat can in fact only be explained in a system of vibrations, the great physicist who first produced the equation and who created the methods of integration to apply it in his research, is still just as much the author of the mathematical theory of heat, as Newton is still the author of the theory of planetary motion, even though the theory was not as completely demonstrated by his works as his successors have been able to do in theirs.

It is the same with the formula by which I represented electrodynamic action. Whatever the physical cause to which the phenomena produced by this action might be ascribed, the formula which I have obtained will always remain the true statement of the facts. If it should later be derived from one of the consideration by which so many other phenomena have been explained, such as attraction in inverse ratio to the square of the distance, considerations which disregard any appreciable distance between particles between which forces are exerted, the vibration of a fluid in space, etc., another step forward will have been made in this field of physics; but this inquiry, in which I myself am no longer occupied, though I fully recognize its importance, will change nothing in the results of my work, since to be in agreement with the facts, the hypothesis which is eventually adopted must always be in accord with the formula which fully represents them.

From the time when I noticed that two voltaic conductors interact, now attracting each other, now repelling each other, ever since I distinguished and described the actions which they exert in the various positions where they can be in relation to each other, and after I had established that the action exerted by a straight conductor is equal to that exerted by a sinuous conductor whenever the latter only deviates slightly from the direction of the former and both terminate at the same points, I have been seeking to express the value of the attractive or repellent force between two elements, or infinitesimal parts, of conducting wires by a formula so as to be able to derive by the known methods of integration the action which takes place between two portions of conductors of the shape in question in any given conditions.

The impossibility of conducting direct experiments on infinitesimal portions of a voltaic circuit makes it necessary to proceed from observations of conductors of finite dimension and to satisfy two conditions, namely that the observations be capable of great precision and that they be appropriate to the determination of the interaction between two infinitesimal portions of wires. It is possible to proceed in either of two ways: one is first to measure values of the mutual action of two portions of finite dimension with the greatest possible exactitude, by placing them successively, one in relation to the other, at different distances and in different positions, for it is evident that the interaction does not depend solely on distance, and then to advance a hypothesis as to the value of the mutual action of two infinitesimal portions, to derive the value of the action which must result for the test conductors of finite dimension, and to modify the hypothesis until the calculated results are in accord with those of observation. It is this procedure which I first proposed to follow, as explained in detail in the paper which I read at the Académie des Sciences 9 October 1820, though it leads to the truth only by the indirect route of hypothesis, it is no less valuable because of that since it is often the only way open in investigation of this kind. A member of this Académie whose works have covered the whole range of physics has aptly expressed this in the "Notice on the Magnetization of Metals by Electricity in Motion," which he read 2 April 1821, saying that prediction of this kind was the aim of practically all physical research.

However, the same end can be reached more directly in the way which I have since followed: it consists in establishing by experiment that a moving conductor remains exactly in equilibrium between equal forces, or between equal rotational moments, these forces and these moments being produced by portions of fixed conductors of arbitrary shape and dimensions without equilibrium being disturbed in the conditions of the experiment, and in determining directly therefrom by calculation what the value of the mutual action of the two infinitesimal portions must be for equilibrium to be, in fact, independent of all variations of shape and dimension compatible with the conditions.

This procedure can only be adopted when the nature of the action being studied is such that cases of equilibrium which are independent of the shape of the body are possible; it is therefore of much more restricted application that the first method which I discussed; but since voltaic conductors do permit equilibrium of this kind, it is natural to prefer the simpler and more direct method which is capable of great exactitude if ordinary precautions are taken for the experiments. There is, however, in connection with the action of conductors, a much more important reason for employing it in the determination of the forces which produce their action; it is the extreme difficulty associated with experiments where it is proposed, for example, to measure the forces by the number of oscillations of the body which is subjected to the actions. This difficulty is due to the fact that when a fixed conductor is made to act upon the moving portion of a circuit, the pieces of apparatus which are necessary for connection to the battery act on the moving portion at the same time as the fixed conductor, thus altering the results of the experiments. I believe, however, that I have succeeded in overcoming this difficulty in a suitable apparatus for measuring the mutual action of two conductors, one fixed and the other moving, by the number of oscillations in the latter for various shapes of the fixed conductor. I shall describe this apparatus in the course of this paper.

It is true that the same obstacles do not arise when the action of a conducting wire on a magnet is measured in the same way: but this method cannot be employed when it is a question of determining the forces which two conductors exert upon each other, the question which must be our first consideration in the investigation of the new phenomena. It is evident that if the action of a conductor on a magnet is due to some other cause than that which produces the effect between two conductors, experiments performed with respect to a conductor and magnet can add nothing to the study of two conductors; if magnets only owe their properties to electric currents, which encircle each of their particles, it is necessary, in order to draw definite conclusions as to the action of the conducting wire on these currents, to be sure that these currents are of the same intensity near to the surface of the magnet as within it, or else to know the law governing the variation of intensity; whether the planes of the currents are everywhere perpendicular to the axis of a bar magnet, as I at first supposed, or whether the mutual action of the currents of the magnet itself inclines them more to the axis when at a greater distance from this axis, which is what I have since concluded from the difference which is noticeable between the position of the poles on a magnet and the position of the points which are endowed with the same properties in a conductor of which one part is helically wound.

5
— Isaac Newton —

In this reading Newton criticizes the wave theory of light and with it the concept of a medium to carry those waves. It is clear that he senses the problems associated with such a medium, assuming all the while that the medium would have to be mechanical. Instead he argues for a particle theory of light. In the final paragraphs of the reading Newton makes some extraordinary comments.

1. *He appeals to the authority of the ancient atomists in support of his rejection of the medium hypothesis.*
2. *He reiterates his conception of scientific methodology as arguing from effects to causes and not "feigning hypotheses."*
3. *He states that the very first cause is surely not mechanical.*

Although chronologically this reading appeared in print after Huygens's Treatise on Light, the sequence tells the story of development more clearly. Both the particle and the wave theories were afloat within the scientific community during the last half of the seventeenth century.

This reading comes from Newton's Opticks *(1704).*

THE PARTICLE THEORY OF LIGHT

Are not all hypotheses erroneous in which light is supposed to consist in pression or motion, propagated through a fluid medium? For in all these hypotheses the phenomena of light have been hitherto explained by supposing that they arise from new modifications of the rays; which is an erroneous supposition.

If light consisted only in pression propagated without actual motion, it could not be able to agitate and heat the bodies which refract and reflect it. If it consisted in motion propagated to all distances in an instant, it would require an infinite force every moment, in every shining particle, to generate that motion. And if it consisted in pression or motion, propagated either in an instant or in time, it would bend into the shadow. For pression or motion cannot be propagated in a fluid in right lines, beyond an obstacle which stops part of the motion, but will bend and spread every way into the quiescent medium which lies beyond the obstacle. Gravity tends downwards, but the pressure of water arising from gravity tends every way with equal force, and is propagated as readily, and with as much force sideways as downwards, and through crooked passages as through straight ones. The waves on the surface of stagnating water, passing by the sides of a broad obstacle which stops part of them, bend afterwards and dilate themselves gradually into the quiet water behind the obstacle. The waves, pulses or vibrations of the air, wherein sounds consist, bend manifestly, though not so much as the waves of water. For a bell or a cannon may be heard beyond a hill which intercepts the sight of the sounding body, and sounds are propagated as readily through crooked pipes as through straight ones. But light is never known to follow crooked passages nor to bend into the shadow. For the fixed stars by the interposition of any of the planets cease to be seen. And so do the parts of the sun by the interposition of the Moon, Mercury or Venus. The rays which pass very near to the edges to any body are bent a little by the action of the body, as we shewed above; but this bending is not towards but from the shadow, and is performed only in the passage of the ray by the body, and at a very small distance from it. So soon as the ray is past the body, it goes right on.

To explain the unusual refraction of island crystal by pression or motion propagated, has not hitherto been attempted (to my knowledge) except by Huygens, who for that end supposed two several vibrating mediums within that crystal. But when he tried the refractions in two successive pieces of that crystal, and found them such as is mentioned above, he confessed himself at a loss for explaining them. For pressions or motions, propagated from a shining body through an uniform medium, must be on all sides alike; whereas by those experiments it appears that the rays of light have different properties in their different sides. He suspected that the pulses of aether in passing through the first crystal might receive certain new modifications, which might determine them to be propagated in this or that medium within the second crystal, according to the position of that crystal. But what modifications those might be he could not say, nor think of anything satisfactory in that point. And if he had known that the unusual refraction depends not on new modifications, but on the original and unchangeable dispositions of the rays, he would have found it as difficult to explain how these dispositions, which he supposed to be impressed on the rays by the first crystal, could be in them before their incidence on that crystal, and in general, how all rays emitted by shining bodies can have those dispositions in them from the beginning. To me, at least, this seems inexplicable, if light be nothing else than pression or motion propagated through aether.

And it is as difficult to explain by these hypotheses how rays can be alternately in fits of easy reflexion and easy transmission, unless perhaps one might suppose that there are in all space two aetheral vibrating mediums, and that the vibrations of one of them constitute light, and the vibrations of the other are swifter, and as often as they overtake the vibrations of the first, put them into those fits. But how two aethers can be diffused through all space, one of which acts upon the other, and by consequence is reacted upon, without retarding, shattering, dispersing and confounding one another's motions, is inconceivable. And against filling the heavens with fluid mediums, unless they be exceeding rare, a great objection arises from the regular and very lasting motions of the planets and comets in all manner of courses through the heavens. For thence it is manifest that the heavens are void of all sensible resistance, and by consequence of all sensible matter.

For the resisting power of fluid mediums arises partly from the attrition of the parts of the medium, and partly from the *vis inertiae* of the matter. That part of the resistance of a spherical body which arises from the attrition of the parts of the medium is very nearly as the diameter, or, at the most, as the *factum* of the diameter, and the velocity of the spherical body together. And that part of the resistance which arises from the *vis inertiae* of the matter is as the square of that *factum*. And by this difference the two sorts of resistance may be distinguished from one another in any medium; and these being distinguished, it will be found that almost all resistance of bodies of a competent magnitude moving in air, water, quick-silver, and such like fluids with a competent velocity, arises from the *vis inertiae* of the parts of the fluid.

Now, that part of the resisting power of any medium which arises from the tenacity, friction or attrition of the parts of the medium, may be diminished by dividing the matter into smaller parts, and making the parts more smooth and slippery; but that part of the resistance which arises from the *vis inertiae* is proportional to the density of the matter, and cannot be diminished by dividing the matter into smaller parts, nor by any other means than by decreasing the density of the medium. And for these reasons the density of fluid mediums is very nearly proportional to their resistance. Liquors which differ not much in density as water, spirit of wine, spirit of turpentine, hot oil, differ not much in resistance. Water is thirteen or fourteen times lighter than quick-silver and by consequence thirteen or fourteen times rarer, and its resistance is less than that of quick-silver in the same proportion, or thereabouts, as I have found by experiments made with pendulums. The open air in which we breathe is eight or nine hundred times lighter than water, and by consequence eight or nine hundred times rarer, and accordingly its resistance is less than that of water in the same proportion, or thereabouts, as I have also found by experiments made with pendulums. And in thinner air the resistance is still less, and at length, by rarefying the air, becomes insensible. For small feathers falling in the open air meet with great resistance, but in a tall glass well emptied of air, they fall as fast as lead or gold, as I have seen tried several times. Whence the resistance seems still to decrease in proportion to the density of the fluid. For I do not find by any experiments that bodies moving in quick-silver, water or air meet with any other sensible resistance than what arises from the density and tenacity of those sensible fluids, as they would do if the pores of those fluids, and all other spaces, were filled with a dense and subtile fluid. Now, if the resistance in a vessel well emptied of air was but a hundred times less than in the open air, it would be about a million of times less than in quick-silver. But it seems to be much less in such a vessel, and still much less in the heavens, at the height of three or four hundred miles from the Earth, or above. For Mr. Boyle has shewed that air may be rarified above ten thousand times in vessels of glass; and the heavens are much emptier of air than any vacuum we can make below. For since the air is compressed by the weight of the incumbent atmosphere and the density of air is proportional to the force compressing it, it follows by computation that, at the height of about seven and a half English miles from the Earth, the air is four times rarer than at the surface of the Earth; and at the height of 15 miles it is sixteen times rarer than that at the surface of the Earth; and at the height of 22½, 30, or 38 miles, it is respectively 64, 256, or 1,024 times rarer, or thereabouts; and at the height of 76, 152, or 228 miles, it is about 1,000,000, 1,000,000,000,000, or 1,000,000,000,000,000,000 times rarer; and so on.

Heat promotes fluidity very much by diminishing the tenacity of bodies. It makes many bodies fluid which are not fluid in cold, and increases the fluidity of tenacious liquids, as of oil, balsam, and honey, and thereby decreases their resistance. But it decreases not the resistance of water considerably, as it would do if any considerable part of the resistance of water arose from the attrition or tenacity of its parts. And, therefore, the resistance of water arises principally and almost entirely from the *vis inertiae* of its matter; and by consequence, if the heavens were as dense as water, they would not have much less resistance than water; if as dense as quick-silver, they would not have much less resistance than quick-silver; if absolutely dense, or full of matter without any vacuum, let the matter be never so subtile and fluid, they would have a greater resistance than quick-silver. A solid globe in such a medium would lose above half its motion in moving three times the length of its diameter, and a globe not solid (such as are the planets), would be retarded sooner. And, therefore, to make way for the regular and lasting motions of the planets and comets, it's necessary to empty the heavens of all matter, except perhaps some very thin vapours, steams, or effluvia, arising from the atmospheres of the Earth, planets, and comets, and from such an exceedingly rare aethereal medium as we described above. A dense fluid can be of no use for explaining the phenomena of Nature, the motions of the planets and comets being better explained without it. It serves only to disturb and retard the motions of those great bodies, and make the frame of Nature languish; and in the pores of bodies it serves only to stop the vibrating motions of their parts, wherein their heat and activity consists. And as it is of no use, and hinders the operations of Nature, and makes her languish, so there is no evidence for its existence; and therefore it ought to be rejected. And if it be rejected, the hypotheses that light consists in pression or motion, propagated through such a medium, are rejected with it.

And, for rejecting such a medium, we have the authority of those the oldest and most celebrated philosophers of Greece and Phoenicia, who made a vacuum, and atoms, and the gravity of atoms, the first principles of their philosophy; tacitly attributing gravity to some other cause than dense matter. Later philosophers banish the consideration of such a cause out of natural philosophy, feigning hypotheses for explaining all things mechanically, and referring other causes to metaphysics; whereas the main business of natural philosophy is to argue from phenomena without feigning hypotheses, and to deduce causes from effects, till we come to the very first cause, which certainly is not mechanical; and not only to unfold the mechanism of the world, but chiefly to resolve these and such like questions. . . .

Are not the rays of light very small bodies emitted from shining substances? For such bodies will pass through uniform mediums in right lines without bending into the shadow, which is the nature of the rays of light. They will also be capable of several properties, and be able to conserve their properties unchanged in passing through several mediums, which is another condition of the rays of light. Pellucid substances act upon the rays of light at a distance in refracting, reflecting, and inflecting them, and the rays mutually agitate the parts of those substances at a distance for heating them; and this action and reaction at a distance very much resembles an attractive force between bodies. If refraction be performed by attraction of the rays, the sines of incidence must be to the sines of refraction in a given proportion, as we shewed in our principles of philosophy. And this rule is true by experience. The rays of light in going out of glass into a vacuum, they are bent towards the glass; and if they fall too obliquely on the vacuum, they are bent backwards into the glass, and totally reflected; and this reflexion cannot be ascribed to the resistance of an absolute vacuum, but must be caused by the power of the glass attracting the rays at their going out of it into the vacuum, and bringing them back. For if the farther surface of the glass be moistened with water or clear oil, or liquid and clear honey, the rays which would otherwise be reflected will go into the water, oil, or honey; and, therefore, are not reflected before they arrive at the farther surface of the glass, and begin to go out of it. If they go out of it into the water, oil, or honey, they go on, because the attraction of the glass is almost balanced and rendered ineffectual by the contrary attraction of the liquor. But if they go out of it into a vacuum which has no attraction to balance that of the glass, the attraction of the glass either bends and refracts them, or brings them back and reflects them. And this is still more evident by laying together two prisms of glass, or two object-glasses of very long telescopes, the one plane, the other a little convex and so compressing them that they do not fully touch, nor are too far asunder. For the light which falls upon the farther surface of the first glass where the interval between the glasses is not above the ten hundred thousandth part of an inch, will go through that surface, and through the air or vacuum between the glasses, and enter into the second glass, as was explained in the first, fourth, and eighth Observations of the first part of the second book. But if the second glass be taken away, the light which goes out of the second surface of the first glass into the air or vacuum, will not go on forwards, but turns back into the first glass, and is reflected; and, therefore, it is drawn back by the power of the first glass, there being nothing else to turn it back. Nothing more is requisite for producing all the variety of colours, and degrees of refrangibility, than that the rays of light be bodies of different sizes, the least of which may take violet the weakest and darkest of the colours, and be more easily diverted by refracting surfaces from the right course; and the rest, as they are bigger and bigger, may make the stronger and more lucid colors (blue, green, yellow, and red) and be more and more difficultly diverted. Nothing more is requisite for putting the rays of light into fits of easy reflexion and easy transmission, than that they be small bodies which by their attractive powers, or some other force, stir up vibrations in what they act upon, which vibrations, being swifter than the rays, overtake them successively, and agitate them so as by turns to increase and decrease their velocities, and thereby put them into those fits. And, lastly, the unusual refraction of island crystal looks very much as if it were performed by some kind of attractive virtue lodged in certain sides both of the rays, and of the particles of the crystal. For were it not for some kind of disposition or virtue lodged in some sides of the particles of the crystal, and not in their other sides, and which includes and bends the rays towards the coast of unusual refraction, the rays which fall perpendicularly on the crystal would not be refracted towards that coast rather than towards any other coast, both at their incidence and at their emergence, so as to emerge perpendicularly by a contrary situation of the coast of unusual refraction at the second surface; the crystal acting upon the rays after they have passed through it, and are emerging into the air; or, if you please, into a vacuum. And since the crystal by this disposition or virtue does not act upon the rays, unless when one of their sides of unusual refraction looks towards that coast, this argues a virtue or disposition in those sides of the rays which answers to, and sympathizes with, that virtue or disposition of the crystal as the poles of two magnets answer to one another. And as magnetism may be intended and remitted, and is found only in the magnet and in iron, so this virtue of refracting the perpendicular rays is greater in island crystal, less in crystal of the rock, and is not yet found in other bodies. I do not say that this virtue is magnetical: it seems to be of another kind. I only say that whatever it be, it's difficult to conceive how the rays of light, unless they be bodies, can have a permanent virtue in two of their sides which is not in their other sides, and this without any regard to their position to the space or medium through which they pass.

6

— Christiaan Huygens —

(1629–1695)

Christiaan Huygens was a Dutch physicist and astronomer who made significant contributions to both fields. With a telescope that he designed and constructed, Huygens discovered the rings of Saturn and its largest moon, Titan. He also designed a pendulum clock that allowed for an accurate measurement of time.

Huygens's greatest contribution to science was his wave theory of light, which he developed in 1678. For Huygens light was a disturbance in an ether made of microscopic particles. He imagined that each point on a wave front acted as a source of other wavelets. From this hypothesis, Huygens was able to account for the laws of reflection and refraction and he predicted that light would travel more slowly in a more dense medium. In this matter he opposed Newton, who held to a particle theory of light that predicted that light would travel more rapidly in a denser medium.

In this reading Huygens describes and defends his wave theory of light. In the process Huygens shows, contra Descartes, that light travels at a finite rate of speed. He uses the work of Olaf Römer with the moons of Jupiter to support his case.

In addressing the question of a medium for light waves, Huygens decides that the medium, though mechanical, is not air. He concludes that it is composed of particles perfectly hard but also quite elastic.

This reading is taken from Huygens's Treatise on Light *(1690).*

THE WAVE THEORY OF LIGHT

As happens in all the sciences in which geometry is applied to matter, the demonstrations concerning Optics are founded on truth drawn from experience. Such are: that the rays of light are propagated in straight lines; that the angles of reflexion and of incidence are equal; and that in refraction the ray is bent according to the law of sines, now so well-known, and which is no less certain than the preceding laws.

The majority of those who have written touching the various parts of Optics have contented themselves with presuming these truths. But some, more inquiring, have desired to investigate the origin and the causes, considering these to be in themselves wonderful effects of nature. In which they advanced some ingenious things, but not, however, such that the most intelligent folk do not wish for better and more satisfactory explanations. Wherefore I here desire to propound what I have meditated on the subject, so as to contribute as much as I can to the explanation of this department of natural science, which not without reason, is reputed to be one of its most difficult parts. I recognize myself to be much indebted to those who were the first to begin to dissipate the strange obscurity in which these things were enveloped, and to give us hope that they might be explained by intelligible reasoning. But, on the other hand, I am astonished also that even here these have often been willing to offer, as assured and demonstrative, reasonings which were far from conclusive. For I do not find that any one has yet given a probable explanation of the first and most notable phenomena of light, namely, why it is not propagated except in straight lines, and how visible rays, coming from an infinitude of diverse places, cross one another without hindering one another in any way. I shall therefore essay in this book, to give, in accordance with the principles accepted in the philosophy of the present day, some clearer and more probable reasons, firstly, of these properties of light propagated rectilinearly; secondly, of light which is reflected on meeting other bodies. Then I shall explain the phenomena of those rays which are said to suffer refraction on passing through transparent bodies of different sorts; and in this part I shall also explain the effects of the refraction of the air by the different densities of the atmosphere.

Thereafter, I shall examine the causes of the strange refraction of a certain kind of crystal which is brought from Iceland. And, finally, I shall treat of the various shapes of transparent and reflecting bodies by which rays are collected at a point or are turned aside in various ways. From this it will be seen with what facility, following our new theory, we find not only the ellipses, hyperbolas, and other curves which M. Descartes has ingeniously invented for this purpose; but also those which the surface of a glass lens ought to possess when its other surface is given as spherical or plane, or of any other figure that may be.

It is inconceivable to doubt that light consists in the motion of some sort of matter. For whether one considers its production, one sees that here upon the earth it is chiefly engendered by fire and flame which contain without doubt bodies that are in rapid motion, since they dissolve and melt many other bodies, even the most solid; or whether one considers its effects, one sees that when light is collected, as by concave mirrors, it has the property of burning as a fire does, that is to say, it disunites the particles of bodies. This is assuredly the mark of motion, at least in the true philosophy, in which one conceives the causes of all natural effects in terms of mechanical motions. This, in my opinion, we must necessarily do, or else renounce all hopes of ever comprehending anything in physics.

And as, according to this philosophy, one holds as certain that the sensation of sight is excited only by the impression of some movement of a kind of matter which acts on the nerves at the back of our eyes, there is here yet one reason more for believing that light consists in a movement of the matter which exists between us and the luminous body.

Further, when one considers the extreme speed with which light spreads on every side, and how, when it comes from different regions, even from those directly opposite, the rays traverse one another without hindrance, one may well understand that when we see a luminous object, it cannot be by any transport of matter coming to us from this object, in the way in which a shot or an arrow traverses the air; for assuredly that would too greatly impugn these two properties of light, especially the second of them. It is then in some other way that light spreads; and that which can lead us to comprehend it is the knowledge which we have of the spreading of sound in the air.

We know that by means of the air, which is an invisible and impalpable body, sound spreads around the spot where it has been produced by a movement which is passed on successively from one part of the air to another; and that the spreading of this movement, taking place equally rapidly on all sides, ought to form spherical surfaces ever enlarging and which strike our ears. Now there is no doubt at all that light also comes from the luminous body to our eyes by some movement impressed on the matter which is between the two; since, as we have already seen, it cannot be by the transport of a body which passes from one to the other. If, in addition, light takes time for its passage—which we are now going to examine—it will follow that this movement, impressed on the intervening matter, is successive; and consequently it spreads, as sound does, by spherical surfaces and waves; for I call them waves from their resemblance to those which are seen to be formed in water when a stone is thrown into it, and which present a successive spreading as circles, though these arise from another cause, and are only in a flat surface.

To see then whether the spreading of light takes time, let us consider first whether there are any facts of experience which can convince us to the contrary. As to those which can be made here on the earth, by striking lights at great distances, although they prove that light takes no sensible time to pass over these distances, one may say with good reason that they are too small, and that the only conclusion to be drawn from them is that the passage of light is extremely rapid. M. Descartes, who was of opinion that it is instantaneous, founded his views, not without reason, upon a better basis of experience, drawn from the eclipses of the moon; which, nevertheless, as I shall show, is not at all convincing. I will set it forth, in a way a little different from his, in order to make the conclusion more comprehensible.

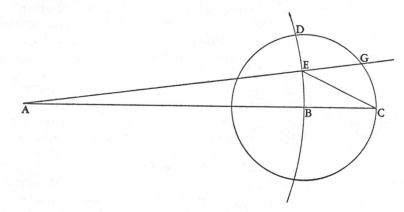

Let A be the place of the sun, BD a part of the orbit or annual path of the earth: ABC a straight line which I suppose to meet the orbit of the moon, which is represented by the circle CD, at C.

Now if light requires time, for example one hour, to traverse the space which is between the earth and the moon, it will follow that the earth having arrived at B, the shadow which it casts, or the interruption of the light, will not yet have arrived at the point C, but will only arrive there an hour after. It will then be one hour after, reckoning from the moment when the earth was at B, that the moon, arriving at C, will be obscured; but this obscuration or interruption of the light will not reach the earth till after another hour. Let us suppose that the earth in these two hours will have arrived at E. The earth then, being at E, will see the eclipsed moon at C, which it left an hour before, and at the same time will see the sun at A. For it being immovable, as I supposed with Copernicus, and the light moving always in straight lines, it must always appear where it is. But one has always observed, we are told, that the eclipsed moon appears at the point of the ecliptic opposite to the sun; and yet here it would appear in arrear of that point by an amount equal to the angle GEC, the supplement of

AEC. This, however, is contrary to experience, since the angle GEC would be very sensible, and about 33 degrees. Now according to our computation, which is given in the treatise on the causes of the phenomena of Saturn, the distance BA between the earth and the sun is about twelve thousand diameters of the earth, and hence four hundred times greater than BC the distance of the moon, which is 30 diameters. Then the angle ECB will be nearly four hundred times greater than BAE, which is five minutes; namely, the path which the earth travels in two hours along its orbit; and thus the angle BCE will be nearly 33 degrees; and likewise the angle CEG, which is greater by five minutes.

But it must be noted that the speed of light in this argument has been assumed such that it takes a time of one hour to make the passage from here to the moon. If one supposes that for this it requires only one minute of time, then it is manifest that the angle CEG will only be 33 minutes; and if it requires only ten seconds of time, the angle will be less than six minutes. And then it will not be easy to perceive anything of it in observations of the eclipse; nor, consequently, will it be permissible to deduce from it that the movement of light is instantaneous.

It is true that we are here supposing a strange velocity that would be a hundred thousand times greater than that of sound. For sound, according to what I have observed, travels about 350 meters in the time of one second, or in about one beat of the pulse. But this supposition ought not to seem to be an impossibility; since it is not a question of the transport of a body with so great a speed, but of a successive movement which is passed on from some bodies to others. I have then made no difficulty, in meditating on these things, in supposing that the emanation of light is accomplished with time, seeing that in this way all its phenomena can be explained, and that in following the contrary opinion everything is incomprehensible. For it has always seemed to me that even M. Descartes, whose aim has been to treat all the subjects of physics intelligibly and who assuredly has succeeded in this better than anyone before him, has said nothing that is not full of difficulties, or even inconceivable, in dealing with light and its properties.

But that which I employed only as a hypothesis, has recently received great seemingness as an established truth by the ingenious proof of Mr. Römer which I am going here to relate, expecting him himself to give all that is needed for its confirmation. It is founded, as is the preceding argument, upon celestial observations, and proves not only that light takes time for its passage, but also demonstrates how much time it takes, and that its velocity is even at least six times greater than that which I have just stated.

For this he makes use of the eclipses suffered by the little planets which revolve around Jupiter, and which often enter his shadow; and see what is his reasoning. Let A be the sun, BCDE the annual orbit of the earth, F, Jupiter, GN the orbit of the nearest of his satellites, for it is this one which is more apt for this investigation than any of the other three because of the quickness of its revolution. Let G be this satellite entering into the shadow of Jupiter, H the same satellite emerging from the shadow.

Let it be then supposed, the earth being at B some time before the last quadrature, that one has seen the said satellite emerge from the shadow; it must needs be, if the earth remains at the same place, that after 42½ hours, one would

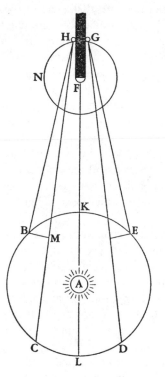

again see a similar emergency, because that is the time in which it makes the round of its orbit, and when it would come again into opposition to the sun. And if the earth, for instance, were to remain always at B during 30 revolutions of this satellite, one would see it again emerge from the shadow after 30 times 42½ hours. But the earth having been carried along during this time to C, increasing thus its distance from Jupiter, it follows that if light requires time for its passage the illumination of the little planet will be perceived later at C than it would have been at B, and that there must be to this time of 30 times 42½ hours that which the light has required to traverse the space MC, the difference of the spaces CH, BH. Similarly, at the other quadrature when the earth has come to E from D while approaching toward Jupiter, the immersions of the satellite ought to be observed at E earlier than they would have been seen if the earth had remained at D.

Now in quantities of observations of these eclipses, made during ten consecutive years, these differences have been found to be very considerable, such as ten minutes and more; and from them it has been concluded that in order to traverse the whole diameter of the annual orbit KL, which is double the distance from here to the sun, light requires about 22 minutes of time.

The movement of Jupiter in his orbit while the earth passed from B to C, or from D to E, is included in this calculation; and this makes it evident that one cannot attribute the retardation of these illuminations or the anticipation of the eclipses, either to any irregularity occurring in the movement of the little planet or to its eccentricity.

If one considers the vast size of the diameter KL, which according to me is some 24 thousand diameters of the earth, one will acknowledge the extreme velocity of light. For, supposing that KL is no more than 22 thousand of these diameters, it appears that being traversed in 22 minutes this makes the speed a thousand diameters in one minute, that is 16⅔ diameters in one second or in one beat of the pulse, which makes more than 11 hundred times a hundred thousand toises; since the diameter of the earth contains 2,865 leagues, reckoned at 25 to the degree, and each league is 2,282 toises, according to the exact measurement which Mr. Picard made by order of the King in 1669. But sound, as I have said above, only travels 180 toises in the same time of one second, hence the velocity of light is more than six hundred thousand times greater than that of sound. This, however, is quite another thing from being instantaneous, since there is all the difference between a finite thing and an infinite. Now the successive movement of light being confirmed in this way, it follows, as I have said, that it spreads by spherical waves, like the movement of sound.

But if the one resembles the other in this respect, they differ in many other things; to wit, in the first production of the movement which causes them; in the matter in which the movement spreads; and in the manner in which it is propagated. As to that which occurs in the production of sound, one knows that it is occasioned by the agitation undergone by an entire body, or by a considerable part of one, which shakes all the contiguous air. But the movement of the light must originate as from each point of the luminous object, else we should not be able to perceive all the different parts of that object, as will be more evident in that which follows. And I do not believe that this movement can be better explained than by supposing that all those of the luminous bodies which are liquid, such as flames, and apparently the sun and the stars, are composed of particles which float in a much more subtle medium which agitates them with great rapidity, and makes them strike against the particles of the ether which surrounds them, and which are much smaller than they. But I hold also that in luminous solids such as charcoal or metal made red-hot in the fire, this same movement is caused by the violent agitation of the particles of the metal or of the wood; those of them which are on the surface striking similarly against the ethereal matter. The agitation, moreover, of the particles which engender the light ought to be much more prompt and more rapid than is that of the bodies which cause sound, since we do not see that the tremors of a body which is giving out a sound are capable of giving rise to light, even as the movement of the hand in the air is not capable of producing sound.

Now if one examines what this matter may be in which the movement coming from the luminous body is propagated, which I call ethereal matter, one will see that it is not the same that serves for the propagation of sound. For one finds that the latter is really that which we feel and which we breathe, and which being removed from any place still leaves there the other kind of matter that serves to convey light. This may be proved by shutting up a sounding body in a glass vessel from which the air is withdrawn by the machine which Mr. Boyle has given us, and with which he has performed so many beautiful experiments. But in doing this of which I speak, care must be taken to place the sounding body on cotton or on feathers, in such a way that it cannot communicate its tremors either to the glass vessel which encloses it, or to the machine; a precaution which has hitherto been neglected. For then after having exhausted all the air one hears no sound from the metal, though it is struck.

One sees here not only that our air, which does not penetrate through glass, is the matter by which sound spreads; but also that it is not the same air but another kind of matter in which light spreads; since if the air is removed from the vessel the light does not cease to traverse it as before.

And this last point is demonstrated even more clearly by the celebrated experiment of Torricelli, in which the tube of glass from which the quicksilver has withdrawn itself, remaining void of air, transmits light just the same as when air is in it. For this proves that a matter different from air exists in this tube, and that this matter must have penetrated the glass or the quicksilver, either one or the other, though they are both impenetrable to the air. And when, in the same experiment, one makes the vacuum after putting a little water above the quicksilver, one concludes equally that the said matter passes through glass or water, or through both.

As regards the different modes in which I have said the movements of sound and of light are communicated, one may sufficiently comprehend how this occurs in the case of sound if one considers that the air is of such a nature that it can be compressed and reduced to a much smaller space than that which it ordinarily occupies. And in proportion as it is compressed the more does it exert an effort to regain its volume; for this property along with its penetrability, which remains notwithstanding its compression, seems to prove that it is made up of small bodies which float about and which are agitated very rapidly in the ethereal matter composed of much smaller parts. So that the cause of the spreading of sound is the effort which these little bodies make in collisions with one another, to regain freedom when they are a little more squeezed together in the circuit of these waves than elsewhere.

But the extreme velocity of light, and other properties which it has, cannot admit of such a propagation of motion, and I am about to show here the way in which I conceive it must occur. For this, it is needful to explain the property which hard bodies must possess to transmit movement from one to another.

When one takes a number of spheres of equal size, made of some very hard substance, and arranges them in a straight line, so that they touch one another, one finds, on striking with a similar sphere against the first of these spheres, that the motion passes as in an instant to the last of them, which separates itself from the row, without one's being able to perceive that the others have been stirred. And even that one which was used to strike remains motionless with them. Whence one sees that the movement passes with an extreme velocity which is the greater, the greater the hardness of the substance of the spheres.

But it is still certain that this progression of motion is not instantaneous, but successive, and therefore must take time. For if the movement, or the disposition to movement, if you will have it so, did not pass successively through all the spheres, they would all acquire the movement at the same time, and hence would all advance together; which does not happen. For the last one leaves the whole row and acquires the speed of the one which was pushed. Moreover there are experiments which demonstrate that all bodies which we reckon of the hardest kind, such as quenched steel, glass, and agate, act as springs and bend somehow, not only when extended as rods but also when they are in the form of spheres or of other shapes. That is to say, they yield a little in themselves at the place where they are stuck, and immediately regain their former figure. For I have found that on striking with a ball of glass or of agate against a large and quite thick piece of the same substance which had a flat surface, slightly soiled with breath or in some other way, there remained round marks, of smaller or larger size according as the blow had been weak or strong. This makes it evident that these substances yield where they meet, and spring back: and for this time must be required.

Now in applying this kind of movement to that which produces light there is nothing to hinder us from estimating the particles of the ether to be of a substance as nearly approaching to perfect hardness and possessing a springiness as prompt as we choose. It is not necessary to examine here the causes of this hardness, or of that springiness, the consideration of which would lead us too far from our subject. I will say, however, in passing that we may conceive from the particles of the ether, notwithstanding their smallness, are in turn composed of other parts and that their springiness consists in the very rapid movement of a subtle matter which penetrates them from every side and constrains their structure to assume such a disposition as to give to this fluid matter the most overt and easy passage possible. This accords with the explanation which M. Descartes gives for the spring, though I do not, like him suppose the pores to be in the form of round hollow canals. And it must not be thought that in this there is anything absurd or impossible, it being on the contrary quite credible that it is this infinite series of different sizes of corpuscles, having different degrees of velocity, of which Nature makes use to produce so many marvelous effects.

But though we shall ignore the true cause of springiness we still see that there are many bodies which possess this property; and thus there is nothing strange in supposing that it exists also in little invisible bodies like the particles of the ether. Also, if one wishes to seek for any other way in which the movement of light is successively communicated, one will find none which agrees better, with uniform progression, as seems to be necessary, than the property of springiness; because if this movement should grow slower in proportion as it is shared over a greater quantity of matter in moving away from the source of the light, it could not conserve this great velocity over great distances. But by supposing springiness in the ethereal matter, its particles will have the property of equally rapid restitution whether they are pushed strongly or feebly; and thus the propagation of light will always go on with an equal velocity.

And it must be known that, although the particles of the ether are not ranged thus in straight lines, as in our row of spheres, but confusedly, so that one of them touches several others, this does not hinder them from transmitting their movement and from spreading it always forward. As to this, it is to be remarked that there is a law of motion serving for this propagation, and verifiable by experiment. It is that when a sphere, such as A here, touches several other similar spheres CCC, if it is struck by another sphere B in such a way as to exert an impulse against all the spheres CCC which touch it, it transmits to them the whole of its movement, and remains after that motionless like the sphere B. And without supposing that the ethereal particles are of spherical form (for I see indeed no need to suppose them so) one may well understand that this property of communicating an impulse does not fail to contribute to the aforesaid propagation of movement.

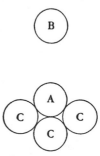

Equality of size seems to be more necessary, because otherwise there ought to be some reflexion of movement backwards when it passes from a smaller particle to a larger one, according to the Laws of Percussion which I published some years ago.

However, one will see hereafter that we have to suppose such an equality not so much as a necessity for the propagation of light as for rendering that propagation easier and more powerful; for it is not beyond the limits of probability that the particles of the ether have been made equal for a purpose so important as that of light, at least in that vast space which is beyond the region of atmosphere and which seems to serve only to transmit the light of the sun and stars.

I have then shown in what manner one may conceive light to spread successively, by spherical waves, and how it is possible that this spreading is accomplished with as great a velocity as that which experiments and celestial observa-

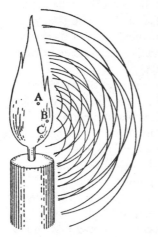

tions demand. Whence it may be further remarked that, although the particles are supposed to be in continual movement (for there are many reasons for this), the successive propagation of the waves cannot be hindered by this; because the propagation consists nowise in the transport of those particles but merely in a small agitation which they cannot help communicating to those surrounding, notwithstanding any movement which may act on them causing them to be changing positions amongst themselves.

But we must consider still more particularly the origin of these waves and the manner in which they spread. And, first, it follows from what has been said of the production of light, that each little region of a luminous body, such as the sun, a candle, or a burning coal, generates its own waves of which that region is the centre. Thus, in the flame of a candle, having distinguished the points A, B, C, concentric circles described about each of these points represent the waves which come from them. And one must imagine the same about every point of the surface and of the part within the flame.

But as the percussions at the centres of these waves possess no regular succession, it must not be supposed that the waves themselves follow one another at equal distances: and if the distances marked in the figure appear to be such, it is rather to mark the progression of one and the same wave at equal intervals of time than to represent several of them issuing from one and the same center.

After all, this prodigious quantity of waves which traverse one another without confusion and without effacing one another must not be deemed inconceivable; it being certain that one and the same particle of matter can serve for many waves coming from different sides or even from contrary directions, not only if it is struck by blows which follow one another closely but even for those which act on it at the same instant. It can do so because the spreading of the movement is successive. This may be proved by the row of equal spheres of hard matter,

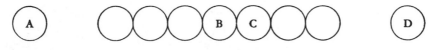

spoken of above. If against this row there are pushed from two opposite sides at the same time two similar spheres A and D, one will see each of them rebound with the same velocity which it had in striking, yet the whole row will remain in its place, although the movement has passed along its whole length twice over. And if these contrary movements happen to meet one another at the middle sphere, B, or at some other such as C, that sphere will yield and act as a spring at both sides, and so will serve at the same instant to transmit these two movements.

But what may at first appear full strange and even incredible is that the undulations produced by such small movements and corpuscles should spread to such immense distances; as for example, from the sun or from the stars to us. For the force of these waves must grow feeble in proportion as they move away from their origin, so that the action of each one in particular will without doubt become incapable of making itself felt to our sight. But one will cease to be astonished by considering how at a great distance from the luminous body an infinitude of waves, though they have issued from different points of this body, unite together in such a way that they sensibly compose one single wave only, which, consequently, ought to have enough force to make itself felt. Thus, this infinite number of waves which originate at the same instant from all points of a fixed star, big it may be as the sun, make practically only one single wave which may well have force enough to produce an impression on our eyes. Moreover, from each luminous point there may come many thousands of waves in the smallest imaginable time, by the frequent percussion of the corpuscles which strike the ether at these points: which further contributes to rendering their action more sensible.

There is the further consideration in the emanation of these waves, that each particle of matter, in which a wave spreads, ought not to communicate its motion only to the next particle which is in the straight line drawn from the luminous point, but that it also imparts some of it necessarily to all the others which touch it and which oppose themselves to its movement. So it arises that around each particle there is made a wave of which that particle is the centre. Thus, if DCF is a wave emanating from the luminous point A, which is its centre, the particle B, one of those comprised within the sphere DCF, will have made its particular or partial wave KCL, which will touch the wave DCF at C at the same moment that the principal wave emanating from the point A has arrived at DCF; and it is clear that it will be only the region C of the wave KCL which will touch the wave DCF, to wit, that which is in the straight line drawn through AB. Similarly the other particles of the sphere DCF, such as *bb*, *dd*, etc. will each make its own wave. But each of these waves can be infinitely feeble only as compared with the wave DCF, to the composition of which all the others contribute by the part of their surface which is most distant from the centre A.

One sees, in addition, that the wave DCF is determined by the distance attained in a certain space of time by the movement which started from the point

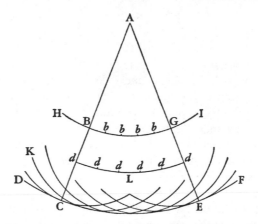

A; there being no movement beyond this wave, though there will be in the space which it encloses, namely, in parts of the particular waves, those parts which do not touch the sphere DCF. And all this ought not seem fraught with too much minuteness or subtlety, since we shall see in the sequel that all the properties of light, and everything pertaining to its reflexion and its refraction, can be explained in principle by this means. This is a matter which has been quite unknown to those who hitherto have begun to consider the waves of light, amongst whom are Mr. Hooke in his *Micrographia*, and Father Pardies, who, in a treatise of which he let me see a portion, and which he was unable to complete as he died shortly afterward, had undertaken to prove by these waves the effects of reflexion and refraction. But the chief foundation, which consists in the remark I have just made, was lacking in his demonstrations; and for the rest he had opinions very different from mine, as maybe will appear some day if his writing has been preserved.

To come to the properties of light. We remark first that each portion of a wave ought to spread in such a way that its extremities lie always between the same straight lines drawn from the luminous point. Thus, the portion BG of the wave, having the luminous point A as its centre, will spread into the arc CE bounded by the straight lines ABC, AGE. For although the particular waves produced by the particles comprised within the space CAE spread also outside this space, they yet do not concur at the same instant to compose a wave which terminates the movement, as they do precisely at the circumference CE, which is their common tangent.

And hence one sees the reason why light, at least if its rays are not reflected or broken spreads only by straight lines, so that it illuminates no object except when the path from its source to that object is open along such lines. For if, for example, there were an opening BG, limited by opaque bodies BH, GI, the wave of light which issues from the point A will always be terminated by the straight lines AC, AE, as has just been shown; the parts of the partial waves which spread outside the space ACE being too feeble to produce light there.

Now, however small we make the opening BG, there is always the same rea-son causing the light there to pass between straight lines; since this opening is always large enough to contain a great number of particles of the ethereal matter, which are of an inconceivable smallness; so that it appears that each little portion of the wave necessarily advances following the straight line which comes from the luminous point. Thus, then, we may take the rays of light as if they were straight lines.

It appears, moreover, by what has been remarked touching the feebleness of the particular waves, that it is not needful that all the particles of the ether should be equal amongst themselves, though equality is more apt for the propa-gation of the movement. For it is true that inequality will cause a particle, by pushing against another larger one, to strive to recoil with a part of its move-ment; but it will thereby merely generate backwards towards the luminous point some partial waves incapable of causing light, and not a wave compounded of many as CE was.

Another property of waves of light, and one of the most marvelous, is that when some of them come from different or even from opposing sides, they produce their effect across one another without any hindrance. Whence also it comes about that a number of spectators may view different objects at the same time through the same opening, and that two persons can at the same time see one another's eyes. Now according to the explanation which has been given of the action of light, how the waves do not destroy nor interrupt one another when they cross one another, these effects which I have just mentioned are easily conceived. But in my judgement they are not at all easy to explain according to the views of M. Descartes, who makes light to consist in a continuous pressure merely tending to movement. For this pressure not being able to act from two opposite sides at the same time against bodies which have no inclination to approach one another, it is impossible so to understand what I have been saying about two persons mutually seeing one another's eyes, or how two torches can illuminate one another.

7

— Thomas Young —

(1773–1829)

Thomas Young was a British physiologist, physician, and physicist who firmly established the wave theory of light and thus settled the dispute between Newton and Huygens. This he accomplished in the years 1800–1804 by a series of extraordinary experiments that demonstrated the interference of beams of light.

He thought that if light beams could both constructively and destructively inter-
fere, light must be a wave phenomenon in an ethereal medium. A man of excep-
tional talents, Young also made a major contribution to the deciphering of the
hieroglyphic inscription on the Rosetta Stone.

In these readings, Young describes the development of his thought on the na-
ture of light. Through his experiments, including those that reveal the phenomena
of interference and polarization, Young concludes that light is a transverse wave
in a medium he calls the luminiferous ether. It should be noted how indebted
Young was to Newton and how he used Newton's work, both theoretical and
experimental, to support his own conclusions. Yet in the end Young replaced the
material particle theory of light, which had the blessing of Newton, with a wave
theory. This result was extremely important in the history of physical theory.

These materials were published between 1801 and 1823.

THE VINDICATION OF THE WAVE THEORY OF LIGHT

Of the Analogy between Light and Sound

Ever since the publication of Sir Isaac Newton's incomparable writings, his doctrines of the emanation of particles of light from lucid substances, and of the
formal pre-existence of coloured rays in white light, have been almost universally admitted in this country, and but little opposed in others. Leonard Euler
indeed, in several of his works, has advanced some powerful objections against
them, but not sufficiently powerful to justify the dogmatical reprobation with
which he treats them; and he has left that system of an ethereal vibration, which
after Huygens and some others he adopted, equally liable to be attacked on
many weak sides. Without pretending to decide positively on the controversy,
it is conceived that some considerations may be brought forwards, which may
tend to diminish the weight of objections to a theory similar to the Huygenian.
There are also one or two difficulties in the Newtonian system, which have been
little observed. The first is, the uniform velocity with which light is supposed to
be projected from all luminous bodies, in consequence of heat, or otherwise.
How happens it that, whether the projecting force is the slightest transmission
of electricity, the friction of two pebbles, the lowest degree of visible ignition,
the white heat of a wind furnace, or the intense heat of the sun itself, these
wonderful corpuscles are always propelled with one uniform velocity? For, if
they differed in velocity, that difference ought to produce a different refraction.
But a still more insuperable difficulty seems to occur, in the partial reflection
from every refracting surface. Why, of the same kind of rays, in every circumstance precisely similar, some should always be reflected, and others transmitted,
appears in this system to be wholly inexplicable. That a medium resembling, in

many properties, that which has been denominated ether, does really exist, is undeniably proved by the phaenomena of electricity; and the arguments against the existence of such an ether throughout the universe, have been pretty sufficiently answered by Euler. The rapid transmission of the electrical shock shows that the electric medium is possessed of an elasticity as great as is necessary to be supposed for the propagation of light. Whether the electric ether is to be considered as the same with the luminous ether, if such a fluid exists, may perhaps at some future time be discovered by experiment; hitherto I have not been able to observe that the refractive power of a fluid undergoes any change by electricity. The uniformity of the motion of light in the same medium, which is a difficulty in the Newtonian theory, favours the admission of the Huygenian; as all impressions are known to be transmitted through an elastic fluid with the same velocity. It has been already shown, that sound, in all probability, has very little tendency to diverge: in a medium so highly elastic as the luminous ether must be supposed to be, the tendency to diverge may be considered as infinitely small, and the grand objection to the system of vibration will be removed. It is not absolutely certain, that the white line visible in all directions on the edge of a knife, in the experiments of Newton and of Mr. Jordan, was not partly occasioned by the tendency of light to diverge. Euler's hypothesis, of the transmission of light by an agitation of the particles of the refracting media themselves, is liable to strong objections; according to this supposition, the refraction of the rays of light, on entering the atmosphere from the pure ether which he describes, ought to be a million times greater than it is. For explaining phaenomena of partial and total reflection, refraction, and inflection, nothing more is necessary than to suppose all refracting media to retain, by their attraction, a greater or less quantity of the luminous ether, so as to make its density greater than that which it possesses in a vacuum, without increasing its elasticity; and that light is a propagation of an impulse communicated to this ether by luminous bodies: whether this impulse is produced by a partial emanation of the ether, or by vibrations of the particles of the body, and whether these vibrations are, as Euler supposed, of various and irregular magnitudes, or whether they are uniform, and comparatively large, remains to be hereafter determined. Now, as the direction of an impulse, transmitted through a fluid, depends on that of the particles in synchronous motion, to which it is always perpendicular, whatever alters the direction of the pulse, will inflect the ray of light. If a smaller elastic body strike against a larger one, it is well known that the smaller is reflected more or less powerfully, according to the difference of their magnitudes: thus, there is always a reflection when the rays of light pass from a rarer to a denser stratum of ether; and frequently an echo when a sound strikes against a cloud. A greater body, striking a smaller one, propels it, without losing all its motion: thus, the particles of a denser stratum of ether do not impart the whole of their motion to a rarer, but, in their effort to proceed, they are recalled by the attraction of the refracting substance with equal force; and thus a reflection is always secondarily produced, when the rays of light pass from a denser to a rarer stratum. . . . It has already been conjectured by Euler, that the colours of light consist in the different frequency of the vibrations of the luminous ether: it does not appear that he has supported this opinion by any argument; but it is strongly confirmed, by the analogy between the colours of a thin plate and the sounds of a series of organ-pipes. The phenomena of the colours of thin plates require, in the Newtonian system, a very complicated supposition, of an ether, anticipating by its motion the velocity of the corpuscles of light, and thus producing the fits of transmission and reflection; and even this supposition does not much assist the explanation. It appears from the accurate analysis of the phenomena which Newton has given, and which has by no means been superseded by any later observations, that the same colour recurs whenever the thickness answers to the terms of an arithmetical progression. Now this is precisely similar to the production of the same sound, by means of an uniform blast, from organ-pipes which are different multiples of the same length. Supposing white light to be a continued impulse or stream of luminous ether, it may be conceived to act on the plates as a blast of air does on the organ-pipes, and to produce vibrations regulated in frequency by the length of the lines which are terminated by the two refracting surfaces. It may be objected that, to complete the analogy, there should be tubes to answer to the organ-pipes: but the tube of an organ-pipe is only necessary to prevent the divergence of the impression, and in light there is little or no tendency to diverge; and indeed, in the case of a resonant passage, the air is not prevented from becoming sonorous by the liberty of lateral motion. It would seem that the determination of a portion of the track of a ray of light through any homogeneous stratum of ether is sufficient to establish a length as a basis for colorific vibrations. In inflections the length of the track of a ray of light through the inflecting atmosphere may determine its vibrations: but, in this case, as it is probable that there is a reflection from every part of the surface of the surrounding atmosphere, contributing to the appearance of the white line in every direction, in the experiments already mentioned, so it is possible that there may be some second reflection at the immediate surface of the body itself, and that, by mutual reflections between these two surfaces, something like the anguiform motion suspected by Newton may really take place; and then the analogy to the colours of thin plates will be still stronger. A mixture of vibrations, of all possible frequencies, may easily destroy the peculiar nature of each, and concur in a general effect of white light. The greatest difficulty in this system is, to explain the different degree of refraction of differently coloured light, and the separation of white light in refraction; yet, considering how imperfect the theory of elastic fluids still remains, it cannot be expected that every circumstance should at once be clearly elucidated. It may hereafter be considered how far the excellent experiments of Count Rumford, which tend very greatly to weaken the evidence of the modern doctrine of heat, may be more or less favourable to one or the other system of light and colours. It does not appear that any comparative experiments have been made on the inflection of light by substances possessed of different refractive powers; undoubtedly some very interesting conclusions might be expected from the inquiry.

An Elastic Medium for Light Waves

Sir, in the Supplement to the Encyclopaedia Britannica are inserted several excellent articles by Professor Robison of Edinburgh; one of them appears to require some public notice on my part, and I consider your valuable Journal as the most eligible channel for such a communication, especially as you have lately done me the honour of reprinting the paper which gave rise to the Professor's animadversions. But in the first place I shall beg leave to recall the attention of your readers, by a summary enumeration, to the principal positions which I have in that paper endeavoured to establish.

Sound, as transmitted through the atmosphere, consists in an undulatory motion of the particles of the air. This is generally admitted; but as the contrary has even very lately been asserted, it is not superfluous to have decisive evidence of the fact. Professor Robison's experiment with a stopcock furnishes an argument nearly similar. . . .

Light is probably the undulation of an elastic medium,
A. Because its velocity in the same medium is always equal.
B. Because all refractions are attended with a partial reflection.
C. Because there is no reason to expect that such a vibration should diverge equally in all directions, and because it is probable that it does diverge in a small degree in every direction.
D. Because the dispersion of differently coloured rays is no more incompatible with this system than with the common opinion, which only assigns for it the nominal cause of different elective attractions.
E. Because refraction and reflection in general are equally explicable on both suppositions.
F. Because inflection is as well, and it may be added, even much better explained by this theory.
G. Because all the phenomena of the colours of thin plates, which are in reality totally unintelligible on the common hypothesis, admit a very complete and simple explanation by this supposition. The analogy which is here superficially indicated, will probably soon be made public more in detail; and will also be extended to the colours of thick plates, and to the fringes produced by inflection, affording, from Newton's own elaborate experiments, a most convincing argument in favour of this system.

A Complete Theory of Light

Although the invention of plausible hypotheses, independent of any connexion with experimental observations, can be of very little use in the promotion of natural knowledge; yet the discovery of simple and uniform principles, by which a great number of apparently heterogeneous phenomena are reduced to coherent and universal laws, must every be allowed to be of considerable importance towards the improvement of the human intellect.

The object of the present dissertation is not so much to propose any opinions which are absolutely new, as to refer some theories, which have been already advanced, to their original inventors, to support them by additional evidence, and to apply them to a great number of diversified facts, which have hitherto been buried in obscurity. Nor is it absolutely necessary in this instance to produce a single new experiment; for of experiments there is already an ample store, which are so much the more unexceptionable, as they must have been conducted without the least partiality for the system by which they will be explained; yet some facts, hitherto unobserved, will be brought forwards, in order to show the perfect agreement of that system with the multifarious phenomena of nature.

The optical observations of Newton are yet unrivalled; and, excepting some casual inaccuracies, they only rise in our estimation as we compare them with later attempts to improve on them. A further consideration of the colours of thin plates, as they are described in the second book of Newton's Optics, has converted that prepossession which I before entertained for the undulatory system of light, into a very strong conviction of its truth and sufficiency; a conviction which has been since most strikingly confirmed by an analysis of the colours of striated substances. The phenomena of thin plates are indeed so singular, that their general complexion is not without great difficulty reconcileable to any theory, however complicated, that has hitherto been applied to them; and some of the principal circumstances have never been explained by the most gratuitous assumptions; but it will appear, that the minutest particulars of these phenomena are not only perfectly consistent with the theory which will now be detailed, but that they are all the necessary consequences of that theory, without any auxiliary suppositions; and this by inferences so simple, that they become particular corollaries, which scarcely require a distinct enumeration.

A more extensive examination of Newton's various writings has shown me that he was in reality the first that suggested such a theory as I shall endeavour to maintain; that his own opinions varied less from his theory than is now almost universally supposed; and that a variety of arguments have been advanced, as if to confute him, which may be found nearly in a similar form in his own works; and this by no less a mathematician than Leonard Euler, whose system of light, as far as it is worthy of notice, either was, or might have been, wholly borrowed from Newton, Hooke, Huygens, and Malebranche.

Those who are attached, as they may be with the greatest justice, to every doctrine which is stamped with the Newtonian approbation, will probably be disposed to bestow on these considerations so much more of their attention, as they appear to coincide more nearly with Newton's own opinions. For this reason, after having briefly stated each particular position of my theory, I shall collect, from Newton's various writings, such passages as seem to be the most favourable to its admission; and although I shall quote some papers which may be thought to have been partly retracted at the publication of the Optics, yet I shall borrow nothing from them that can be supposed to militate against his maturer judgment.

Hypothesis I

A luminiferous ether pervades the universe, rare and elastic in a high degree.

PASSAGES FROM NEWTON

"The hypothesis certainly has a much greater affinity with his own," that is, Dr. Hooke's, "hypothesis, that he seems to be aware of; the vibrations of the ether being as useful and necessary in this as in his."

"To proceed to the hypothesis: first, it is to be supposed therein, that there is an ethereal medium, much of the same constitution with air, but far rarer, subtler, and more strongly elastic. It is not to be supposed that this medium is one uniform matter, but compounded, partly of the main phlegmatic body of ether, partly of other various ethereal spirits, much after the manner that air is compounded of the phlegmatic body of air, intermixed with various vapours and exhalations: for the electric and magnetic effluvia, and gravitating principle, seems to argue such variety."

"Is not the heat (of the warm room) conveyed through the vacuum by the vibrations of a much subtler medium than air?—And is not this medium the same with that medium by which light is refracted and reflected, and by whose vibrations light communicates heat to bodies, and is put into fits of easy reflection, and easy transmission? And do not the vibrations of this medium in hot bodies contribute to the intenseness and duration of their heat? And do not hot bodies communicate their heat to contiguous cold ones, by the vibrations of this medium propagated from them into the cold ones? And is not this medium exceedingly more rare and subtle than the air, and exceedingly more elastic and active? And doth it not readily pervade all bodies? And is it not, by its elastic force, expanded through all the heavens?—May not planets and comets, and all gross bodies, perform their motions in this ethereal medium?—And may not its resistance be so small as to be inconsiderable? For instance, if this ether (for so I will call it) should be supposed 700,000 times more elastic than our air, and above 700,000 times more rare, its resistance would be about 600,000,000 times less than that of water. And so small a resistance would scarce make any sensible alteration in the motions of the planets in ten thousand years. If any one would ask how a medium can be so rare, let him tell me how an electric body can by friction emit an exhalation so rare and subtle, and yet so potent?—And how the effluvia of a magnet can pass through a plate of glass without resistance, and yet turn a magnetic needle beyond the glass?"

Hypothesis II

Undulations are excited in this ether whenever a body becomes luminous.

Scholium. I use the word undulation, in preference to vibration, because vibration is generally understood as implying a motion which is contained alternately backwards and forwards, by a combination of the momentum of the body with an accelerating force, and which is naturally more or less permanent; but an undulation is supposed to consist in a vibratory motion, transmitted successively through different parts of a medium, without any tendency in each particle to continue its motion, except in consequence of the transmission of succeeding undulations, from a distinct vibrating body; as, in the air, the vibrations of a chord produce the undulations constituting sound.

PASSAGES FROM NEWTON

"Were I to assume an hypothesis, it should be this, if propounded more generally, so as not to determine what light is further than that it is something or other capable of exciting vibrations in the ether: for thus it will become so general and comprehensive of other hypotheses, as to leave little room for new ones to be invented."

"In the second place, it is to be supposed that the ether is a vibrating medium like air, only the vibrations far more swift and minute; those of air, made by a man's ordinary voice, succeeding one another at more than half a foot, or a foot distance; but those of ether at a less distance than the hundred thousandth part of an inch. And, as in air, the vibrations are some larger than others, but yet all equally swift, (for in a ring of bells the sound of every tone is heard at two or three miles distance in the same order that the bells are struck,) so, I suppose, the ethereal vibrations differ in bigness, but not in swiftness. Now, these vibrations, beside their use in reflection and refraction, may be supposed the chief means by which the parts of fermenting or putrefying substances, fluid liquors, or melted, burning, or other hot bodies, continue in motion."

"When a ray of light falls upon the surface of any pellucid body, and is there refracted or reflected, may not waves of vibrations, or tremors, be thereby excited in the refracting or reflecting medium? And are not these vibrations propagated from the point of incidence to great distances? And do they not overtake the rays of light, and by overtaking them successively, do not they put them into the fits of easy reflection and easy transmission described above?"

"Light is in fits of easy reflection and easy transmission, before its incidence on transparent bodies. And probably it is put into such fits at its first emission from luminous bodies, and continues in them during all its progress."

Hypothesis III

The Sensation of different Colours depends on the different frequency of Vibrations excited by Light in the Retina.

PASSAGES FROM NEWTON

"The objector's hypothesis, as to the fundamental part of it, is not against me. That fundamental supposition is, that the parts of bodies, when briskly agitated, do excite vibrations in the ether, which are propagated every way from those bodies in straight lines, and cause a sensation of light by beating and dashing against the bottom of the eye, something after the manner that vibrations in the air cause a sensation of sound by beating against the organs of hearing. Now, the most free and natural application of this hypothesis to the solution of phenomena I take to be this—that the agitated parts of bodies, according to their several sizes, figures, and motions, do excite vibrations in the ether of various depths or bignesses, which, being promiscuously propagated through that medium to our eyes, effect in us a sensation of light of a white colour; but if by any means those of unequal bignesses be separated from one another, the largest beget a sensation of a red colour, the least or shortest of a deep violet, and the intermediate ones of intermediate colours; much after the manner that bodies, according to their several sizes, shapes, and motions excite vibrations in the air of various bignesses, which, according to those bignesses, make several tones in sound: that the largest vibrations are best able to overcome the resistance of a refracting superficies, and so break through it with least refraction; whence the vibrations of several bignesses, that is, the rays of several colours, which are blended together in light, must be parted from one another by refraction, and so cause the phenomena of prisms and other refracting substances; and that it depends on the thickness of a thin transparent plate, according to the number of vibrations, interceding the two superficies, they may be reflected or transmitted for many successive thicknesses. And, since the vibrations which make blue and violet are supposed shorter than those which make red and yellow, they must be reflected at a less thickness of the plate; which is sufficient to explicate all the ordinary phenomena of those plates or bubbles, and also of all natural bodies, whose parts are like so many fragments of such plates. These seem to be the most plain, genuine, and necessary conditions of this hypothesis; and they agree so justly with my theory, that, if the animadversor think fit to apply them, he need not, on that account, apprehend a divorce from it; but yet, how he will defend it from other difficulties I know not."

"To explain colours, I suppose, that as bodies of various sizes, densities, or sensations, do by percussion or other action excite sounds of various tones, and consequently vibrations in the air of different bigness; so the rays of light, by impinging on the stiff refracting superficies, excite vibrations in the ether, of various bigness; the biggest, strongest, or most potent rays, the largest vibrations; and others shorter, according to their bigness, strength, or power: and therefore the ends of the capillamenta of the optic nerve, which pave or face the retina, being such refracting superficies, when the rays impinge upon them, they must there excite these vibrations, which vibrations (like those of sound in a trunk or trumpet) will run along the aqueous pores or crystalline pith of the capillamenta, through the optic nerves, into the sensorium; and there, I suppose, affect the sense with various colours, according to their bigness and mixture; the biggest with the strongest colours, reds and yellows; the least with the weakest, blues and violets; the middle with green, and a confusion of all with white—much after the manner that, in the sense of hearing, nature makes use of aerial vibrations of several bignesses to generate sounds of divers tones, for the analogy of nature is to be observed."

"Considering the lastingness of the motions excited in the bottom of the eye by light, are they not of a vibrating nature? Do not the most refrangible rays excite the shortest vibrations, the least refrangible the largest? May not the harmony and discord of colours arise from the proportions of the vibrations propagated through the fibres of the optic nerve into the brain, as the harmony and discord of sounds arise from the proportions of the vibrations of the air?"

Scholium. Since, for the reason here assigned by Newton, it is probable that the motion of the retina is rather of a vibratory than of an undulatory nature, the frequency of the vibrations must be dependent on the constitution of this substance. Now, as it is almost impossible to conceive each sensitive point of the retina to contain an infinite number of particles, each capable of vibrating in perfect unison with every possible undulation, it becomes necessary to suppose the number limited, for instance, to the three principal colours, red, yellow, and blue, of which the undulations are related in magnitude nearly as the number 8, 7, and 6; and that each of the particles is capable of being put in motion less or more forcibly by undulations differing less or more from a perfect unison; for instance, the undulations of green light being nearly in the ratio of 6½, will affect equally the particles in unison with yellow and blue, and produce the same effect as a light composed of those two species; and each sensitive filament of the nerve may consist of three portions, one for each principal colour. Allowing this statement, it appears that any attempt to produce a musical effect from colours must be unsuccessful, or at least that nothing more than a very simple melody could be imitated by them; for the period, which in fact constitutes the harmony of any concord, being a multiple of the periods of the single undulations, would in this case be wholly without the limits of sympathy of the retina, and would lose its effect, in the same manner as the harmony of a third or fourth is destroyed by depressing it to the lowest notes of the audible scale. In hearing, there seems to be no permanent vibration of any part of the organ.

Hypothesis IV

All material Bodies have an Attraction for the ethereal Medium, by means of which it is accumulated within their Substance, and for a small Distance around them, in a state of greater Density, but not of greater Elasticity.

It has been shown that the three former hypotheses, which may be called essential, are literally parts of the more complicated Newtonian system. This fourth hypothesis differs perhaps in some degree from any that have been proposed by former authors, and is diametrically opposite to that of Newton; but both being in themselves equally probable, the opposition is merely accidental, and it is only to be inquired which is the best capable of explaining the phenomena. Other suppositions might perhaps be substituted for this, and therefore I do not consider it as fundamental, yet it appears to be the simplest and best of any that have occurred to me.

Proposition I

All impulses are propagated in a homogenous elastic Medium with an equable Velocity.

Every experiment relative to sound coincides with the observation already quoted from Newton, that all undulations are propagated through the air with equal velocity; and this is further confirmed by calculations. If the impulse be so great as materially to disturb the density of the medium, it will be no longer homogeneous; but, as far as concerns our senses, the quantity of motion may be considered as infinitely small. It is surprising that Euler, although aware of the matter of fact, should still have maintained that the more frequent undulations are more rapidly propagated. It is possible that the actual velocity of the particles of the luminiferous ether may bear a much less proportion to the velocity of the undulations than in sound, for light may be excited by the motion of a body moving at the rate of only one mile in the time that light moves a hundred millions.

Scholium 1. It has been demonstrated that in different mediums the velocity varies in the subduplicate ratio of the force directly, and of the density inversely.

Scholium 2. It is obvious, from the phenomena of elastic bodies, and of sounds, that the undulations may cross each other without interruption; but there is no necessity that the various colours of white light should intermix their undulations, for, supposing the vibrations of the retina to continue but a five hundredth of a second after their excitement, a million undulations of each of a million colours may arrive in distinct succession within this interval of time, and produce the same sensible effect, as if all the colours arrived precisely at the same instant.

Proposition II

An Undulation conceived to originate from the Vibration of a single Particle, must expand through a homogeneous Medium in a spherical Form, but with different quantities of Motion in different Parts.

For, since every impulse, considered as positive or negative, is propagated with a constant velocity, each part of the undulation must in equal times have passed through equal distances from the vibrating point. And, supposing the vibrating particle, in the course of its motion, to proceed forwards to a small distance in a given direction, the principal strength of the undulation will naturally be straight before it; behind it, the motion will be equal, in a contrary direction; and at right angles to the line of vibration, the undulation will be evanescent.

Now, in order that such an undulation may continue its progress to any considerable distance, there must be in each part of it a tendency to preserve its own motion in a right line from the centre; for if the excess of force at any part were communicated to the neighbouring particles, there can be no reason why it should not very soon be equalized throughout, or, in other words, become wholly extinct, since the motions in contrary directions would naturally destroy each other. The origin of sound from the vibration of a chord is evidently of this nature; on the contrary, in a circular wave of water, every part is at the same instant either elevated or depressed. It may be difficult to show mathematically the mode in which this inequality of force is preserved, but the inference from the matter of fact appears to be unavoidable; and while the science of hydrodynamics is so imperfect that we cannot even solve the simple problem of the time required to empty a vessel by a given aperture, it cannot be expected that we should be able to account perfectly for so complicated a series of phenomena as those of elastic fluids. The theory of Huygens, indeed, explains the circumstances in a manner tolerably satisfactory. He supposes every particle of the medium to propagate a distinct undulation in all directions, and that the general effect is only perceptible where a portion of each undulation conspires in direction at the same instant; and it is easy to show that such a general undulation would in all cases proceed rectilinearly, with proportionate force; but, upon this supposition, it seems to follow, that a greater quantity of force must be lost by the divergence of the partial undulations than appears to be consistent with the propagation of the effect to any considerable distance; yet it is obvious that some such limitation of the motion must naturally be expected to take place, for, if the intensity of the motion of any particular part, instead of continuing to be propagated straight forwards, were supposed to affect the intensity of a neighbouring part of the undulation, an impulse must then have travelled from an internal to an external circle in an oblique direction, in the same time as in the direction of the radius, and consequently with a greater velocity, against the first proposition. In the case of water, the velocity is by no means so rigidly limited as in that of an elastic medium. Yet it is not necessary to suppose, nor is it indeed probable, that there is absolutely not the least lateral communication of the force of the undulation, but that, in highly elastic mediums, this communication is almost insensible. In the air, if a chord be perfectly insulated, so as to propagate exactly such vibrations as have been described, they will in fact be much less forcible than if the chord be placed in the neighbourhood of a sounding-board, and probably in some measure because of this lateral communication of motions of an opposite tendency. And the different intensity of different parts of the same circular undulation may be observed, by holding a common tuning-fork at arm's length, while sounding, and turning it, from a plane directed to the ear, into a position perpendicular to that plane. . . .

Proposition VIII

When two Undulations, from different Origins, coincide either perfectly or very nearly in Direction, their joint effect is a Combination of the Motions belonging to each.

Since every particle of the medium is affected by each undulation, wherever the directions coincide, the undulations can proceed no otherwise than by uniting their motions, so that the joint motion may be the sum or difference of the separate motions, accordingly as similar or dissimilar parts of the undulations are coincident.

I have, on a former occasion, insisted at large on the application of this principle to harmonics; and it will appear to be of still more extensive utility in explaining the phenomena of colours. The undulations which are now to be compared are those of equal frequency. When the two series coincide exactly in point of time, it is obvious that the united velocity of the particular motions must be greatest, and in effect at least, double the separate velocities; and also, that it must be smallest, and if the undulations are of equal strength, totally destroyed, when the time of the greatest direct motion belonging to one undulation coincides with that of the greatest retrograde motion of the other. In intermediate states, the joint undulations will be of intermediate strength; but by what laws this intermediate strength must vary, cannot be determined without further data. It is well known that a similar cause produces in sound, that effect which is called a beat; two series of undulations of nearly equal magnitude cooperating and destroying each other alternately, as they coincide more or less perfectly in the times of performing their respective motions. . . .

Proposition IX

Radiant Light consists in undulations of the luminiferous Ether.

This proposition is the general conclusion from all the preceding, and it is conceived that they conspire to prove it in as satisfactory a manner as can possibly be expected from the nature of the subject. It is clearly granted by Newton, that there are undulations, yet he denies that they constitute light; but it is shown in the three first corollaries of the last proposition, that all cases of the increase or diminution of light are referable to an increase or diminution of such undulations, and that all the affections to which the undulations would be liable, are distinctly visible in the phenomena of light; it may therefore be very logically inferred, that the undulations are light.

A few detached remarks will serve to obviate some objections which may be raised against this theory.

1. Newton has advanced the singular refraction of the Iceland crystal, as an argument that the particles of light must be projected corpuscles; since he thinks it probable that the different sides of these particles must be differently attracted by the crystal, and since Huygens has confessed his inability to account in a satisfactory manner for all the phenomena. But, contrarily to what might have been expected from Newton's usual accuracy and candour, he has laid down a new law for the refraction, without giving a reason for rejecting that of Huygens, which Mr. Haüy has found to be more accurate than Newton's; and, without attempting to deduce from his own system any explanation of the more universal and striking effects of doubling spars, he has omitted to observe that Huygen's most elegant and ingenious theory perfectly accords with these general effects, in all particulars, and of course derives from them additional pretensions

to truth; this he omits, in order to point out a difficulty, for which only a verbal solution can be found in his own theory, and which will probably long remain explained by any other.

2. Mr. Michell has made some experiments, which appear to show that the rays of light have an actual momentum, by means of which a motion is produced when they fall on a thin plate of copper delicately suspended. (Priestley's Optics.) But, taking for granted the exact perpendicularity of the plate, and the absence of any ascending current of air, yet since, in every such experiment, a greater quantity of heat must be communicated to the air at the surface on which the light falls than at the opposite surface, the excess of expansion must necessarily produce an excess of pressure on the first surface, and a very perceptible recession of the plate in the direction of the light. Mr. Bennet has repeated the experiment, with a much more sensible apparatus, and also in the absence of air; and very justly infers from its total failure, an argument in favour of the undulatory system of light. For, granting the utmost imaginable subtility of the corpuscles of light, their effects might naturally be expected to bear some proportion to the effects of the much less rapid motions of the electrical fluid, which are so very easily perceptible, even in their weakest states.

3. There are some phenomena of the light of solar phosphori, which at first sight might seem to favour the corpuscular system; for instance, its remaining many months as if in a latent state, and its subsequent re-emission by the action of heat. But, on further consideration, there is no difficulty in supposing the particles of the phosphori which have been made to vibrate by the action of light, to have this action abruptly suspended by the intervention of cold, whether as contracting the bulk of the substance or otherwise; and again, after the restraint is removed, to proceed in their motion, as a spring would do which had been held fast for a time in an intermediate stage of its vibration; nor is it impossible that heat itself may, in some circumstances, become in a similar manner latent. But the affections of heat may perhaps hereafter be rendered more intelligible to us; at present, it seems highly probable that light differs from heat only in the frequency of its undulations or vibrations; those undulations which are within certain limits, with respect to frequency, being capable of affecting the optic nerve, and constituting light; and those which are slower and probably stronger, constituting heat only; that light and heat occur to us, each in two predicaments, the vibratory or permanent, and the undulatory or transient state; vibratory light being the minute motion of ignited bodies, or of solar phosphori, and undulatory or radiant light the motion of the ethereal medium excited by these vibrations; vibratory heat being a motion to which all material substances are liable, and which is more or less permanent; and undulatory heat that motion of the same ethereal medium, which has been shown by Mr. King (Morsels of Criticism, 1786, p. 99), and Mr. Pictet (*Essais de Physique*, 1790), to be as capable of reflection as light, and by Dr. Herschel to be capable of separate refraction. How much more readily heat is communicated by the free access of colder substances, than either by radiation or by transmission through a quiescent medium, has been shown by the valuable experiments of

Count Rumford. It is easy to conceive that some substances permeable to light, may be unfit for the transmission of heat, in the same manner as particular substances may transmit some kinds of light, while they are opaque with respect to others.

On the whole, it appears, that the few optical phenomena which admit of explanation by the corpuscular system, are equally consistent with this theory; that many others, which have long been known, but never understood, become by these means perfectly intelligible; and that several new facts are found to be thus only reducible to a perfect analogy with other facts, and to the simple principles of the undulatory system. It is presumed, that henceforth the second and third books of Newton's Optics will be considered as more fully understood than the first has hitherto been; but, if it should appear to impartial judges, that additional evidence is wanting for the establishment of the theory, it will be easy to enter more minutely into the details of various experiments, and to show the insuperable difficulties attending the Newtonian doctrines, which, without necessity, it would be tedious and invidious to enumerate. The merits of their author in natural philosophy are great beyond all contest or comparison: his optical discovery of the composition of white light would alone have immortalized his name; and the very arguments which tend to overthrow his system, give the strongest proofs of the admirable accuracy of his experiments.

Sufficient and decisive as these arguments appear, it cannot be superfluous to seek for further confirmation; which may with considerable confidence be expected, from an experiment very ingeniously suggested by Professor Robison, on the refraction of the light returning to us from the opposite margins of Saturn's ring: for, on the corpuscular theory, the right must be considerably distorted when viewed through an achromatic prism: a similar distortion ought also to be observed in the disc of Jupiter; but, if it be found that an equal deviation is produced in the whole light reflected from these planets, there can scarcely be any remaining hope to explain the affections of light by a comparison with the motions of projectiles.

The Interference of Light

Whatever opinion may be entertained of the theory of light and colours which I have lately had the honour of submitting to the Royal Society, it must at any rate be allowed that it has given birth to the discovery of a simple and general law, capable of explaining a number of the phenomena of coloured light, which, without this law, would remain insulated and unintelligible. The law is, that "wherever two portions of the same light arrive at the eye by different routes, either exactly or very nearly in the same direction, the light becomes most intense when the difference of the routes is any multiple of a certain length, and least intense in the intermediate state of the interfering portions; and this length is different for light of different colours."

I have already shown in detail, the sufficiency of this law for explaining all the phenomena described in the second and third books of Newton's Optics, as well as some others not mentioned by Newton. But it is still more satisfactory to observe its conformity to other facts, which constitute new and distinct classes of phenomena, and which could scarcely have agreed so well with any anterior law, if that law had been erroneous or imaginary: these are the colours of fibres, and the colours of mixed plates.

As I was observing the appearance of the fine parallel lines of light which are seen upon the margin of an object held near the eye, so as to intercept the greater part of the light of a distant luminous object, and which are produced by the fringes caused by the inflection of light already known, I observed that they were sometimes accompanied by coloured fringes, much broader and more distinct; and I soon found that these broader fringes were occasioned by the accidental interposition of a hair. In order to make them more distinct, I employed a horse-hair, but they were then no longer visible. With a fibre of wool, on the contrary, they became very large and conspicuous; and, with a single silk-worm's thread, their magnitude was so much increased, that two or three of them seemed to occupy the whole field of view. They appeared to extend on each side of the candle, in the same order as the colours of thin plates, seen by transmitted light. It occurred to me that their cause must be sought in the interference of two portions of light, one reflected from the fibre, the other bending round its opposite side, and at last coinciding nearly in direction with the former portion; that, accordingly as both portions deviated more from a rectilinear direction, the difference of the length of their paths would become gradually greater and greater, and would consequently produce the appearances of colour usual in such cases; that supposing them to be inflected at right angles, the difference would amount nearly to the diameter of the fibre, and, that this difference must consequently be smaller as the fibre became smaller; and, the number of fringes in a right angle becoming smaller, that their angular distance would consequently become greater, and the whole appearance would be dilated. It was easy to calculate, that for the light least inflected, the difference of the paths would be to the diameter of the fibre, very nearly as the deviation of the ray, at any point, from the rectilinear direction, to its distance from the fibre.

I therefore made a rectangular hole in a card, and bent its ends so as to support a hair parallel to the sides of the hole; then, upon applying the eye near the hole, the hair of course appeared dilated by indistinct vision into a surface, of which the breadth was determined by the distance of the hair and the magnitude of the hole, independently of the temporary aperture of the pupil. When the hair approached so near to the direction of the margin of a candle that the inflected light was sufficiently copious to produce a sensible effect, the fringes began to appear; and it was easy to estimate the proportion of their breadth to the apparent breadth of the hair, across the image of which they extended. I found that six of the brightest red fringes, nearly at equal distances, occupied the whole of that image. The breadth of the aperture was 66/1000, and its distance from the hair 8/10 of an inch: the diameter of the hair was less than 1/500 of an inch: as nearly as I could ascertain, it was 1/600. Hence, we have

11/1000 for the deviation of the first red fringe at the distance 8/10; and, as 8/10:11/1000::1/600:11/480000 or 1/43646 for the difference of the routes of the red light where it was most intense. The measure deduced from Newton's experiments is 1/39800. I thought this coincidence, with only an error of one-ninth of so minute a quantity, sufficiently perfect to warrant completely the explanation of the phenomenon, and even to render a repetition of the experiment unnecessary; for there are several circumstances which make it difficult to calculate much more precisely what ought to be the result of the measurement.

When a number of fibres of the same kind, for instance, a uniform lock of wool, are held near to the eye, we see an appearance of halos surrounding a distant candle; but their brilliancy, and even their existence, depends on the uniformity of the dimensions of the fibres; and they are larger as the fibres are smaller. It is obvious that they are the immediate consequences of the coincidence of a number of fringes of the same size, which, as the fibres are arranged in all imaginable directions, must necessarily surround the luminous object at equal distances on all sides, and constitute circular fringes.

There can be little doubt that the coloured atmospherical halos are of the same kind: their appearance must depend on the existence of a number of particles of water, of equal dimensions, and in a proper position, with respect to the luminary and to the eye. As there is no natural limit to the magnitude of the spherules of water, we may expect these halos to vary without limit in their diameters; and accordingly, Mr. Jordan has observed that their dimensions are exceedingly various, and has remarked that they frequently change during the time of observation.

I first noticed the colours of mixed plates, in looking at a candle through two pieces of plate-glass, with a little moisture between them. I observed an appearance of fringes resembling the common colours of thin plates; and, upon looking for the fringes by reflection, I found that these new fringes were always in the same direction as the other fringes, but many times larger. By examining the glasses with a magnifier, I perceived that wherever these fringes were visible, the moisture was intermixed with portions of air, producing an appearance similar to dew. I then supposed that the origin of the colours was the same as that of the colours of halos; but, on a more minute examination, I found that the magnitude of the portions of air and water was by no means uniform, and that the explanation was therefore inadmissible. It was, however, easy to find two portions of light sufficient for the production of these fringes; for, the light transmitted through the water, moving in it with a velocity different from that of the light passing through the interstices filled only with air, the two portions would interfere with each other, and produce effects of colour according to the general law. The ratio of the velocities in water and in air is that of 3 to 4; the fringes ought therefore to appear where the thickness is 6 times as great as that which corresponds to the same colour in the common case of thin plates; and, upon making the expermient with a plane glass and a lens slightly convex, I found the sixth dark circle actually of the same diameter as the first in the new fringes. The colours are also very easily produced, when butter or tallow is substituted for water; and the rings then become smaller, on account of the greater refractive density of the oils: but, when water is added, so as to fill up the interstices of the oil, the rings are very much enlarged; for here the difference only of the velocities in water and in oil is to be considered, and this is much smaller than the difference between air and water. All these circumstances are sufficient to satisfy us with respect to the truth of the explanation; and it is still more confirmed by the effect of inclining the plates to the direction of the light; for then, instead of dilating, like the colours of thin plates, these rings contract: and this is the obvious consequence of an increase of the length of the paths of the light, which now traverses both mediums obliquely; and the effect is every where the same as that of a thicker plate.

It must, however, be observed, that the colours are not produced in the whole light that is transmitted through the mediums: a small portion only of each pencil, passing through the water contiguous to the edges of the particle, is sufficiently coincident with the light transmitted by the neighbouring portions of air, to produce the necessary interference; and it is easy to show that, on account of the natural concavity of the surface of each portion of the fluid adhering to the two pieces of glass, a considerable portion of the light which is beginning to pass through the water will be dissipated laterally by reflection at its entrance, and that much of the light passing through the air will be scattered by refraction at the second surface. For these reasons, the fringes are seen when the plates are not directly interposed between the eye and the luminous object; and on account of the absence of foreign light, even more distinctly than when they are in the same right line with that object. And if we remove the plates to a considerable distance out of this line, the rings are still visible, and become larger than before; for here the actual route of the light passing through the air, is longer than that of the light passing more obliquely through the water, and the difference in the times of passage is lessened. It is however impossible to be quite confident with respect to the causes of these minute variations, without some means of ascertaining accurately the forms of the dissipating surfaces.

In applying the general law of interference to these colours, as well as to those of thin plates already known, I must confess that it is impossible to avoid another supposition, which is a part of the undulatory theory, that is, that the velocity of light is the greater, the rarer the medium; and that there is also a condition annexed to the explanation of the colours of thin plates. Which involves another part of the same theory, that is, that where one of the portions of light has been reflected at the surface of a rarer medium, it must be supposed to be retarded one half of the appropriate interval, for instance in the central black spot of a soap-bubble, where the actual lengths of the paths very nearly coincide, but the effect is the same as if one of the portions had been so retarded as to destroy the other. For considering the nature of this circumstance, I ventured to predict, that if the two reflections were of the same kind, made at the surfaces of a thin plate, of a density intermediate between the densities of the mediums containing it, the effect would be reversed, and the central spot, instead of black, would become white; and I have now the pleasure of stating, that I have fully verified this prediction, by interposing a drop of oil of sassafras between a prism of flint-glass and a lens of crown-glass: the central spot seen by reflected light was white, and surrounded by a dark ring. It was however necessary to use some force, in order to produce a contact sufficiently intimate; and the white spot differed, even at last, in the same degree from perfect whiteness, as the black spot usually does from perfect blackness.

The colours of mixed plates suggested to me an idea which appears to lead to an explanation of the dispersion of colours by refraction, more simple and satisfactory than that which I advanced in the last Bakerian Lecture. We may suppose that every refractive medium transmits the undulations constituting light in two separate portions, one passing through its ultimate particles, and the other through its pores; and that these portions re-unite continually, after each successive separation, the one having preceded the other by a very minute but constant interval, depending on the regular arrangement of the particles of a homogeneous medium. Now, if these two portions were always equal, each point of the undulations resulting from their re-union would always be found half-way between the places of the corresponding point in the separate portions; but supposing the preceding portion to be the smaller, the newly combined undulation will be less advanced than if both had been equal, and the difference of its place will depend, not only on the difference of the length of the two routes, which will be constant for all the undulations; but also on the law and magnitude of those undulations; so that the larger undulations will be somewhat further advanced after each re-union than the smaller ones, and, the same operation recurring at every particle of the medium, the whole progress of the larger undulations will be more rapid than that of the smaller; hence the deviation, in consequence of the retardation of the motion of light in a denser medium, will of course be greater for the smaller than for the larger undulations. Assuming the law of the harmonic curve for the motions of the particles, we might without much difficulty reduce this conjecture to a comparison with experiment; but it would be necessary, in order to warrant our conclusions, to be provided with very accurate measures of the refractive and dispersive powers of various substances, for rays of all descriptions.

The Phenomena of Polarisation

We are led from the facts which have been enumerated in the 11th section, respecting "uniaxal and biaxal crystals," to the remarkable coincidence between the discoveries of Dr. Brewster respecting crystal with two axes, and a theory which had been published a few years earlier in order to illustrate the propagation of an undulation in a medium compressed or dilated in a given direction only, and to prove that such an undulation must necessarily assume a spheroidal form upon the mechanical principles of the Huygenian theory. As every contribution to the investigation of so difficult a subject may chance to be of some value, it will be worth while to copy this demonstration here, from the *Quarterly Review* for Nov. 1809, Vol. II. P. 345.

However satisfactorily such a mode of viewing the extraordinary refraction may be applied to the subsequent discoveries relating to the effects of heat and compression, there is another train of ideas, which arises more immediately from the phenomena of polarisation, and which might lead us to a more distinct notion of the separation of the pencil into two or more portions, though it does not seem to comprehend so entirely the phenomena depending on spheroidal undulations.

We may begin this mode of considering the subject in the words which have already been employed in the article *Chromatics, supra*, pp. 334, 335. "If we assume as a mathematical postulate, in the undulatory theory, without attempting to demonstrate its physical foundation, that a transverse motion may be propagated in a direct line, we may derive from this assumption a tolerable illustration of the subdivision of polarised light by reflection in an oblique plane. Supposing polarisation to depend on a transverse motion in the given plane; when a ray completely polarised is subjected to simple reflection in a different plane, at a surface which is destitute of any polarising action, and which may be said to afford a neutral reflection, the polar motion may be conceived to be reflected as any other motion would be reflected at a perfectly smooth surface, the new plane of the motion being always the image of the former plane; and the effect of refraction will be nearly of a similar nature. But when the surface exhibits a new polarising influence, and the beams of light are divided by it into two portions, the intensity of each may be calculated, by supposing the polar motion to be resolved instead of being reflected, the simple velocities of the two portions being as the cosines of the angle, formed by the new planes of motion with the old, and the energies, which are the true measure of the intensity, as the squares of the sines. We are thus insensibly led to confound the intensity of the supposed polar motion with that of the light itself; since it was observed by Malus, that the relative intensity of the two portions, into which light is divided under such circumstances, is indicated by the proportion of the squares of the cosine and sine of the inclination of the planes of polarisation. The imaginary transverse motion must also necessarily be alternate, partly from the nature of a continuous medium, and partly from the observed fact, that there is no distinction between the polarisations produced by causes precisely opposed to each other in the same plane." Another analogous hint is found in the *Philosophical Transactions for 1818, supra, p. 373*: "Supposing the experiments to be perfectly represented by [Dr. Brewster's] general law, it will follow that the tint exhibited depends not on the difference of refracted densities in the direction of the ray transmitted, but on the greatest difference of refractive densities in directions perpendicular to that of the ray. These two conditions lead to the same result, where the effect of one axis only is considered, but they vary materially where two axes are supposed to be combined. . . . There can be little doubt that the direction of the polarisation, in such cases, must be determined by that of the greatest and least of the refractive densities in question"; a "supposition," which Dr. Brewster finds "quite correct."

We may add again to these hints the consideration, that when simple pressure or extension in the direction of any given axis produces a spheroidal undulation in a medium before homogeneous, this state is always accompanied by the condition, that a ray describing the axis, while the densities in all transverse directions remain equal, undergoes no subdivision, but that a ray moving in the plane of the equator, to which the perpendiculars are the axis and another equatorial diameter, undergoes the greatest possible separation into parts that are respectively polarised in the planes passing through these directions.

From these phenomena we are led to be strongly impressed with the analogy of the properties of sound, as investigated cursorily by Mr. Wheatstone, and in a more elaborate manner by the multiplied experiments of Mr. Savart, which have shown that, in many cases, the elementary motions of the substances transmitting sound are transverse to the direction in which the sound is propagated, and that they remain in general parallel to the original impulse.

The next transition carries us from the *mathematical postulate* here mentioned to the *physical condition* assumed by Mr. Fresnel, that the relative situation of the particles of the etherial medium with respect to each other, is such as to produce an elastic force tending to bring back a line of particles, which has been displaced, towards its original situation by the resistance of the particles *surrounding the line*, and at the same time to impel these particles in its own direction, and in that direction only, or principally, while the aggregate effect is propagated in concentric surfaces.

This hypothesis of Mr. Fresnel is at least very ingenious, and may lead us to some satisfactory computations: but it is attended by one circumstance which is perfectly *appalling* in its consequences. The substances on which Mr. Savart made his experiments were *solids* only; and it is only to solids that such a lateral resistance has ever been attributed: so that if we adopted the distinctions laid down by the reviver of the undulatory system himself, in his *Lectures*, it might be inferred that the luminiferous ether, pervading all space, and penetrating almost all substances, is not only highly elastic, but absolutely solid!!! The passage in question is this:

"The immediate cause of solidity, as distinguished from liquidity, is the lateral adhesion of the particles to each other, to which the degree of hardness or solidity is always proportional. This adhesion prevents any change of the relative situation of the particles, so that they cannot be withdrawn from their places without experiencing a considerable resistance from the force of cohesion, while those of liquids may remain equally in contact with the neighbouring particles, notwithstanding their change of form. When a perfect solid is extended or compressed, the particles, being retained in their situations by the force of laterial adhesion, can only approach directly to each other, or be withdrawn further from each other; and the resistance is nearly the same, as if the same substance, in a fluid state, were enclosed in an unalterable vessel, and forcibly compressed or dilated. Thus the resistance of ice to extension or compression is found by experiment to differ very little from that of water contained in a vesssel; and the same effect may be produced even when the solidity is not the most perfect that the substance admits; for *the immediate resistance of iron or steel to flexure is the same, whether it may be harder or softer.* It often happens; however, that the magnitude of the lateral adhesion is so much limited, as to allow a capability of extension or compression, and it may yet retain a power of restoring the bodies to their original form by its reaction. This force may even be the principal or the only source of the body's elasticity: thus when a piece of elastic gum is extended, the mean distance of the particles is not materially increased . . . and the change of form is rather to be attributed to a displacement of the particles than to their separation to a greater distance from each other, and the resistance must be derived from the lateral adhesion only: some other substances also, approaching more nearly to the nature of liquids, may be extended to many times their original length, with a resistance continually increasing; and in such cases there can scarcely be any material changes of the specific gravity of these substances. Professor Robison has mentioned the juice of bryony as affording a remarkable instance of such viscidity.

"It is probable that the immediate cause of the lateral adhesion of solids is a symmetrical arrangement of their constituent parts; it is certain that almost all bodies are disposed, in becoming solid, to assume the form of crystals, which evidently indicates the existence of such an arrangement; and all the hardest bodies in nature are of a crystalline form. It appears, therefore, consistent both with reason and with experience to suppose, that a crystallization more or less perfect is the universal cause of solidity. We may imagine, that when the particles of matter are disposed without any order, they can afford no strong resistance of a motion in any direction; but when they are regularly placed in certain situations with respect to each other, any change of form must displace them in such a manner, as to increase the distance of a whole rank at once; and hence they may be enabled to co-operate in resisting such a change. Any inequality of tension in a particular part of a solid, is also probably so far the cause of hardness, as it tends to increase the strength of union of any part of a series of particles which must be displaced by a change of form."

It must, however, be admitted, that this passage by no means contains a demonstration of the total incapability of fluids to transmit any impressions by lateral adhesion, and the hypothesis remains completely open for discussion, notwithstanding the apparent difficulties attending it; which have appeared to bring us very near to the case stated in the same lectures as a possible one, that there may be independent worlds, some existing in different parts of space, *others pervading each other unseen and unknown in the same space.* We may perhaps accommodate the hypothesis of Mr. Fresnel to the phenomena of the ordinary and extraordinary refraction, by considering the undulations as propagated through the given medium in two different ways; some by the divergence of the elementary motions in the direction of the ray, and others by their remaining parallel to the direction of the impulse or of the polarisation: the former must be supposed to furnish the spheroidal, the latter the spherical refraction. It would indeed follow that the velocity of the spherical undulation ought to vary by innumerable degrees, within certain limits, according to the direction of the supposed elementary motion: while in fact the actual velocity of the spherical undulations seems to be uniformly equal to the velocity in the direction of the axis: but this objection may be obviated by supposing the surface so constituted, that for some unknown reason the parallel elementary motion can only be propagated in the regular manner when it takes place in the direction of the axis, or when it is made to assume that direction: a condition not very simple or natural, but by no means inconceivable; unless we saw any reason to consider the adhesion as a constant force, independent of the direction, and equal to the least or greatest elasticity, or unless it were possible to derive the phenomena of two supposed axes of polarisation, which Mr. Fresnel has explained on the hypothesis of two spheroids, from the supposition of two spherical undulations propagating oblique elementary motions in the direction of the actual polarisation as already determined for these crystals.

If these conjectures should be found to afford a single step, in an investigation so transcendently delicate, it will be best to pause on them for a time, and to wait for further aid from a new supply of experiments and observations.

8

— Augustin Fresnel —

(1788–1827)

and Dominique Arago

(1786–1853)

Augustin Fresnel was a French physicist who made significant contributions to the development of the wave theory of light. Studying polarized light, Fresnel theorized that light must be a transverse wave, not a longitudinal wave as with sound.

Dominique Arago, also a French physicist, collaborated with Fresnel in the study of the polarization of light. In addition Arago suggested a crucial experiment that would settle the debate between the wave and particle theories. He proposed that the speed of light be measured in both air and water. Newton (particle theory) and Huygens (wave theory) had made opposing predictions concerning the velocity of light in water (a medium more dense than air). The experiment was eventually performed by Foucault in 1850 with the results supporting the wave theory, but by that time the results were anticlimactic.

In this reading, Fresnel and Arago describe the interference of polarized light, and although their work is crucial to the determination that light is a transverse wave, they claim that the results of their experiments can be stated independently of all hypotheses. It should also be noted that in this reading Fresnel and Arago express great respect for the work of Thomas Young. This paper was published in the "Annales de Chimie et de Physique" in 1819.

THE TRANSVERSE NATURE OF LIGHT WAVES

Before reporting the experiments which are the subject of this memoir, it will perhaps be worth while to recall some of the beautiful results that Dr. Thomas Young has already obtained by studying, with the rare sagacity which characterizes him, the influence which, in certain circumstances, the rays of light exert on one another.

1. Two rays of homogeneous light coming from the same source, which reach a certain point in space by two different routes which are slightly unequal, enhance each other or destroy each other, forming on the screen which receives them a bright or dark point according as the difference of their routes has one value or another.

2. Two rays enhance each other always when they have passed over equal distances. If we find that they enhance each other again when the difference of the two distances is equal to a certain quantity d, they enhance each other again for all differences which are contained in the series 2d, 3d, 4d, etc. The intermediate values 0 + 1/2d, d + 1/2d, 2d + 1/2d, etc. determine the cases in which the rays mutually annul each other.

3. The quantity d has not the same value for all the homogeneous rays: in the air it is equal to 67/100000 millimeters for the extreme red rays of the spectrum and to 42/100000 only for the violet rays. The values corresponding to the other colors are between these limits.

The periodic colors of colored rings or of halos, etc. appear to depend on the influence that the rays which are first separated exert on one another when they again coincide: in any case, in order that the laws which we have stated should conform to these various phenomena we must assume that the difference in path is not the only cause of the action of two rays at the point where they cross, except when they are both of them always in the same medium; and that if there is any difference between the refractive powers or the thicknesses of the transparent bodies through which the rays severally pass, that will produce an effect equivalent to a difference in path. There has been elsewhere reported a direct experiment tried by M. Arago which gives the same results, and from which there can also be drawn this conclusion, that a transparent body diminishes the velocity of the light which traverses it in the ratio of the sine of the angle of incidence to the sine of the angle of refraction: so that in all the phenomena of interference two different media produce equal effects when their thicknesses are in the inverse ratio of their indices of refraction. These considerations lead us, as we can see, to a new method for measuring small differences of refrangibility.

During the trials that we made together to test the degree of precision of which this method is capable, one of us (M. Arago) conceived that it would be interesting to see if the actions which the ordinary rays habitually exert on one another would not be modified when the two luminous beams were made to interfere only after they had previously been polarized.

It is known that if we illuminate a narrow strip of a body by light which comes from a radiant point, its shadow is bordered on the outside by a set of fringes formed by the interference of the direct light and of rays which have been bent in the neighborhood of the opaque body; and that a part of the same light which enters the geometrical shadow past the two edges of the body, gives rise to fringes of the same sort. We found, to start with, very easily that these two systems of fringes are absolutely similar whether the incident light has not been modified in any way or has reached the body only after having been previously polarized. Thus rays polarized in the same sense influence each other when they meet in the same way as natural rays.

It still remained to try whether two rays originally polarized at right angles, or to use an accepted expression, in opposite senses, would produce phenomena of the same sort when they meet in the interior of the geometrical shadow of an opaque body.

To try this we placed before the radiating source sometimes a rhomboid of calcspar, sometimes an achromatized prism of rock crystal and we obtained thus two luminous points. From each of these proceeded a divergent beam: these two beams, as we see, were polarized in opposite senses. A metallic cylinder was then placed between the two radiant points so as to correspond precisely with the middle of the interval which separated them. With this arrangement a part of the polarized rays of the first beam penetrated from the right into the space behind the cylinder; and a part of the rays of the second beam, polarized in the opposite sense entered it from the left. Some rays of these two groups met each other near the line which joined the center of the cylinder and the middle of the straight line which passed through the two radiant points. These rays had passed over paths which were equal or only slightly different: it seemed then that they ought to have formed fringes; but we did not see the least trace of them even with a lens, in a word, the rays had crossed without influencing each other. The only system of fringes that we perceive in this experiment, came from the interference of the rays which penetrated into the shadow by the two edges of the cylinder, having started from each radiant point considered separately. Those that we tried to produce by crossing rays polarized in opposite senses would have evidently occupied a position between these.

The crystal which we used separated the images only slightly, and the two rays, the ordinary and extraordinary, traversed it through almost equal thicknesses. However, we had already too often noticed, when we made similar experiments, how much the smallest difference in the velocity of the rays, in the thickness, or in the refracting power of the media that they traversed, modified sensibly the phenomena of interference, not to be convinced of the necessity of repeating our test while avoiding all the doubtful circumstances to which we have called attention. Each of us sought to find the way to do this.

M. Fresnel proposed first for this two different methods. The principle of interference shows that rays coming from two different foci, originating from the same source, will form when they cross dark and bright bands, without its being necessary to introduce in their path any opaque body.

To settle the question it is therefore sufficient to try if the two images formed by placing a rhomboid of calcspar in the path of the rays from a luminous point would give a similar result; but since, according to the theory of double refraction, the extra-ordinary ray moves in the calcspar more quickly than the ordinary ray, we must artificially compensate for this excess of velocity before allowing the rays to cross. To do this, utilizing an experiment of M. Arago, which has been cited, M. Fresnel placed in the path of the extraordinary beam a glass plate whose thickness had been determined by calculation in such a way that by traversing it with perpendicular incidence this beam lost almost all the advance which it had made in the crystal on the ordinary beam; further, by slightly tilting the plate we could obtain in this respect an exact compensation. In spite of this, the crossing of the two beams, polarized in opposite senses, did not give the bands.

In another experiment, to compensate the difference of velocity of the two rays, M. Fresnel let both of them fall on an unsilvered sheet of plate glass, whose thickness had been calculated in such a way that the extraordinary ray when reflected perpendicularly at the second face, lost by passing twice through the glass more than it gained in traversing the crystal; a gradual change of inclination led finally to a perfect compensation: nevertheless, under any incidence, the ordinary rays reflected at the upper surface of the glass gave no perceptible bands when they met the rays reflected from the second surface.

M. Fresnel avoided the defect of the preceding experiment, of depending on theoretical considerations, and furthermore preserved all the intensity of the incident light, by the following procedure. He cut a rhomboid of calcspar through the middle and placed the two parts one before the other in such a way that the principal sections were perpendicular: in this situation the ordinary beam of the first crystal experienced extraordinary refraction in the second; and reciprocally, the beam which first followed the extraordinary route experienced next ordinary refraction. In looking through this apparatus there was seen only a double image of the luminous point; each beam had experienced the two sorts of refraction; the sums of the paths traversed by each of them in the two crystals at once ought then to be equal, since by hypothesis these crystals had both the same thickness; all was then compensated in respect of the velocities and the routes traversed; and nevertheless the two systems of rays polarized in opposite senses did not produce any perceptible fringes by interference. We may add, that for fear that the two pieces of the rhomboid had not exactly the same thickness he took pains in each test to make slight and gradual changes of the angle at which the incident rays encountered the second crystal.

The method that M. Arago proposed, on his part, for trying the same experiment was independent of double refraction. It has been known for a long time that if in a thin plate we cut two fine slits near each other and if we illuminate them by light from a single luminous point, there are formed behind the plate very bright fringes resulting from the action that the rays from the slit on the right exert on the rays of the other slit in the points where they meet. To polarize in opposite senses the rays coming from these two openings M. Arago had first proposed to use a thin piece of agate, to cut it through the middle and to place each half before one of the slits in such a way that the portions of the agate which at first were contiguous, would be at right angles to each other. This arrangement should evidently produce the expected effect: but since he did not have at the moment a suitable agate, M. Arago proposed to substitute for it two piles of plates, and in order to preserve the thinness that was necessary for the success of the experiment, to make them of sheets of mica.

To do this we chose fifteen of these sheets, as transparent as possible, and superposed them. Then with a sharp instrument this single pile was divided through the middle. It is clear that the two partial piles which were made by this process should have closely the same thickness, at least in the parts which were at first contiguous, even if the sheets composing the pile were sensibly prismatic. These piles polarized almost completely the light which traversed them when the incidence measured from the surface was 30°. It is precisely at this angle that each of them was placed behind one of the slits of the copper plate.

When the two planes of incidence were parallel, that is, when the two piles were inclined in the same sense, from above downward for example, we clearly saw the bands formed by the interference of the two polarized rays, just as when two ordinary rays of light act on each other; but if, by turning one of the piles about the incident ray, the two planes of incidence became perpendicular; if while the first pile always remained inclined from above downward, the second was inclined, for example, from left to right, the two emergent beams, then polarized at right angles or in opposite senses, no longer formed any perceptible bands when they encountered each other.

The precautions that we had taken to give the same thickness to the two piles show clearly enough that when we placed them behind the slits we took pains to have them traversed by the light in those parts which were contiguous before the pile was cut. It has otherwise been seen, and this circumstance removes all the difficulties which could have been suggested in this regard, that the fringes showed themselves as in ordinary light when the rays were polarized in the same sense; we may nevertheless add that a slow and gradual change in the inclination of one of the piles never made the bands appear, when the planes of incidence were at right angles.

On the same day on which we tried the system of the two piles we made an experiment suggested by M. Fresnel, really less direct than the preceding one but also easier to carry out, which also demonstrates the impossibility of producing fringes by the crossing of luminous rays at right angles.

We placed behind the plate of copper, pierced with its two slits, a thin plate of selenite, for example: since this crystal is doubly refracting, there come from each slit two rays polarized in opposite senses; now if the rays of one sort can act on the rays of the opposite sort we ought to see with this apparatus three distinct systems of fringes. The ordinary rays from the right combined with the ordinary rays from the left should give a first system corresponding exactly to the middle of the interval between the two slits; the bands formed by the interference of the two extraordinary beams should occupy the same place as the others, and should increase their intensity but should not be distinguishable from them. As to those which result from the action of the ordinary rays on the right and the extraordinary rays on the left, and reciprocally, they should be placed on the right and on the left of the central fringes and so much further apart as the plate employed is thicker: for we have seen that a difference of velocity changes the position of the fringes as well as a difference of route. Now since the fringes in the middle are the only ones visible, even when the interposed plate is so thin that the two other systems should not be far removed, we must conclude that rays of different names or polarized in opposite senses do not affect each other.

To further confirm this conclusion, let us suppose that we cut in two our plate of selenite; that one of the halves is set up at the first slit; that the other is placed behind the other slit and that the axes, instead of being parallel, as they were when the plate was uncut, are now perpendicular. With this arrangement the ordinary ray coming from the slit on the right will be polarized in the same sense as the extraordinary ray coming from the slit on the left, and reciprocally. These rays will then form fringes; but since their velocities in the crystal are not equal these fringes do not correspond to the center of the interval between the two openings; only the ordinary or the extraordinary rays from one of the slits, by meeting the rays of the same name coming from the other slit, can give central images; but since from the particular arrangement that has been made of the two pieces of crystal these rays are polarized in opposite senses they ought not to affect each other, so we see only the first two systems of fringes separated by an interval which is uniformly white.

If, without changing any other arrangements of the preceding experiment, we set the two plates of selenite in such a way that their axes, instead of being perpendicular, make an angle of 45°, we then perceive all at once three systems of fringes; for each pencil from the right acts on the two pencils on the left and reciprocally, since their planes of polarization are no longer perpendicular. We should here remark that the system in the middle is the most intense and results from the perfect superposition of the bands formed by the interference of the beams of the same name.

Let us take up again the arrangement with the piles and suppose that the planes of incidence are perpendicular and that the beams transmitted through the two slits are polarized in opposite senses. Let us further place between the sheet of copper and the eye a double refracting crystal, whose principal section makes an angle of 45° with the planes of incidence. From the known laws of double refraction the rays transmitted by the piles will each be divided in the crystal into two rays of the same intensity and polarized in two perpendicular directions, one of which is precisely that of the principal section. One might then expect to observe in this experiment a series of fringes produced by the action of the ordinary beam on the right and the ordinary beam on the left, and a second similar series coming from the interference of the two extraordinary beams; nevertheless, we do not perceive the slightest trace of it and the four beams when they cross give only continuous light.

This experiment, which originated with M. Arago, proves that two rays which have been originally polarized in opposite senses can then be brought to the same plane of polarization without thereby acquiring the power of affecting each other.

In order that two rays polarized in opposite senses and then brought to the same state of polarization can mutually affect each other, it is necessary that they should have first started from the same plane of polarization. This results from the experiment devised by M. Fresnel, which we proceed to describe.

We place in a pencil of polarized light coming from a radiant point a plate of selenite cut parallel to the axis, and covered with a thin sheet of copper pierced with two openings; the incidence is perpendicular and the axis of the plate makes an angle of 45° with the original plane of polarization. As in all the analogous experiments we observe the shadow of the sheet with a lens; but this time we place in addition in front of its focus a rhomboid of calcspar which gives a sensible double refraction and whose principal section makes in its turn with that of the plate an angle of 45°. Then we see in each image three systems of fringes; the one of them corresponds exactly to the middle of the shadow; the other systems are to the left and right of the first.

Let us now examine how these three systems of fringes arise in one of the two images, in the ordinary image, for example.

The beams polarized in the same sense which pass through the two slits are each divided when they traverse the plate of selenite into two beams polarized in opposite senses. Since the double refraction of the plate is insensible, the ordinary and extraordinary parts of each beam follow the same route but with different velocities.

One of these double beams, that of the slit on the right, for example, is divided when it traverses the rhomboid into four beams, two ordinary and two extraordinary; but in fact we only see two, since the component parts of the beams of the same name coincide. It is further evident, from the known laws of double refraction and the positions which we have given to the plate of selenite and to the rhomboid, that when the ordinary beam leaves this last crystal, it is composed of half of the ray which was ordinary in the plate and of half of the extraordinary ray; and that the two other halves of these same rays pass to the extraordinary image, which we have agreed to neglect for the present. The beam coming from the slit on the left behaves in the same way. We see, in a word, that after having traversed the two crystals in this new apparatus, the ordinary beams coming from the slit on the right or from that on the left are both of them composed of a portion of light which has always followed the ordinary route in the two crystals and of a second portion which at first was extraordinary.

Those rays coming from the two slits which, when they traverse the plate of selenite and the rhomboid, always follow the ordinary route traverse equal paths with the same velocity and therefore, when they reunite, should give rise to central fringes. The same is true of the rays which, while extraordinary in the plate of selenite, have become at the same time ordinary by the action of the rhomboid; the fringes of the middle of the shadow therefore result from the superposition of two different systems.

As to the portion of light on the right, which is extraordinary, for example, in the plate of selenite and becomes ordinary by traversing the rhomboid, it will have traversed a path equal to that of the portion of the beam on the left, which is always refracted ordinarily; but as these rays in the plate have slightly unequal velocities, the points where they form sensible fringes when they meet, instead of corresponding to the middle of the interval between the two slits, will be on the right, that is to say, on the side opposite to the ray which is for a moment extraordinary and so moves more slowly. There comes finally for the last combination the interference of the part of the beam from the right, which is ordinary in the two crystals, with the portion of the beam from the left which is extraordinary in the plate and ordinary in the rhomboid and which therefore gives rise to bands situated on the left of the center.

We have now explained the passage of the rays which take part in the formation of the three systems of fringes in the apparatus in question; and we may notice that the systems on the right and left result from the interference of rays first polarized in opposite senses in the plate of selenite and brought back finally to a similar polarization by the action of the rhomboid. Two rays polarized in opposite senses and brought back to the same plane of polarization can give fringes when they meet; but that this may happen it is indispensable that they have been first polarized in the same sense.

We have left out of consideration hitherto the mutual action of the two pencils which undergo extraordinary refraction in the rhomboid. These pencils furnish also three systems of fringes, but they are separated from the former one. If, while everything remains in the same condition, we substitute for the rhomboid a plate of selenite or of rock crystal, which does not give two distinct images, the six systems, instead of producing three by superposition, are reduced to that of the middle one. This remarkable result demonstrates: 1. That fringes resulting from the interference of ordinary rays are complements to the fringes produced by the extraordinary rays; and, 2. That these two systems are so placed that a bright fringe of the first system corresponds to a dark fringe of the second and reciprocally; without these two conditions we would perceive nothing but a uniform and continuous light on the two sides of the central fringes. We encounter here the difference of a half wave, as in the phenomena of colored rings.

The experiments that we have reported lead us then definitely to the following conclusions:

1. In the same condition in which two rays of ordinary light seem to destroy each other mutually, two rays polarized at right angles or in opposite senses exert on each other no appreciable action;
2. The rays of light polarized in one sense act on one another like natural rays: so that in these two sorts of light the phenomena of interference are absolutely the same;
3. Two rays originally polarized in opposite senses can be brought to the same plane of polarization without acquiring thereby the power of affecting each other.

4. Two rays polarized in opposite senses and brought to similar states of polarization affect each other like natural rays, if they come from a pencil originally polarized in one sense;

5. In the phenomena of interference produced by rays which have experienced double refraction, the place of the fringes is not determined uniquely by the difference of the paths and of the velocities; and in certain circumstances which we have pointed out we must take account in addition of a difference of half a wave.

All these laws are deduced, as we have seen, from direct experiment. We might reach them more simply still by starting from the phenomena which are presented by crystalline plates; but then we should have had to admit that the tints with which these plates are colored when they are illuminated by a beam of polarized light arise from the interference of several systems of waves. The demonstrations which we have reported have the advantage of establishing the same laws independently of all hypothesis.

9

— Michael Faraday —

(1791–1867)

Michael Faraday was a British chemist and physicist who invented the electric motor, the dynamo, and the transformer. His greatest contributions were to the sciences of electricity and magnetism. In the 1830s Faraday developed the concept of lines of force that could be used to map out a magnetic field. This concept proved useful in the design of many experiments, but more important it had a great influence on Maxwell's development of Electromagnetic Field Theory.

It is curious that Faraday, one of the greatest experimentalists in the history of science, did not use mathematics to any significant extent in his work. In a sense, Maxwell mathematized the ideas of Faraday.

In this reading Faraday describes his experiments involving the induction of electric currents. Faraday discovers that it is a change in current in one coil that produces a momentary current in another coil. This result played a major role in the ultimate development of his lines-of-force hypothesis. The reading comes from a paper given by Faraday to the Royal Society in November 1831.

ELECTROMAGNETIC INDUCTION

1. The power which electricity of tension possesses of causing an opposite electrical state in its vicinity has been expressed by the general term Induction;

which, as it has been received into scientific language, may also, with propriety, be used in the same general sense to express the power which electrical currents may possess of inducing any particular state upon matter in their immediate neighborhood, otherwise indifferent. It is with this meaning that I propose using it in the present paper.

2. Certain effects of the induction of electrical currents have already been recognized and described; as those of magnetization; Ampère's experiments of bringing a copper disc near to a flat spiral; his repetition with electro-magnets of Arago's extraordinary experiments, and perhaps a few others. Still it appeared unlikely that these could be all the effects which induction by currents could produce; especially as, upon dispensing with iron, almost the whole of them disappear, whilst yet an infinity of bodies, exhibiting definite phenomena of induction with electricity of tension, still remain to be acted upon by the induction of electricity in motion.

3. Further: Whether Ampère's beautiful theory were adopted, or any other, or whatever reservation were mentally made, still it appeared very extraordinary, that as every electric current was accompanied by a corresponding intensity of magnetic action at right angles to the current, good conductors of electricity, when placed within the sphere of this action, should not have any current induced through them, so some sensible effect produced equivalent in force to such a current.

4. These considerations, with their consequence, the hope of obtaining electricity from ordinary magnetism, have stimulated me at various times to investigate experimentally the inductive effect of electric currents. I lately arrived at positive results; and not only had my hopes fulfilled, but obtained a key which appeared to me to open out a full explanation of Arago's magnetic phenomena, and also to discover a new state, which may probably have great influence in some of the most important effects of electric currents.

5. These results I propose describing, not as they were obtained, but in such a manner as to give the most concise view of the whole.

Induction of Electric Currents

6. About twenty-six feet of copper wire one-twentieth of an inch in diameter were wound round a cylinder of wood as a helix, the different spires of which were prevented from touching a thin interposed twine. This helix was covered with calico, and then a second wire applied in the same manner. In this way twelve helices were superposed, each containing an average length of wire of twenty-seven feet, and all in the same direction. The first, third, fifth, seventh, ninth, and eleventh of these helices were connected at their extremities end to end, so as to form one helix; the others were connected in a similar manner; and thus two principal helices were produced, closely interposed, having the same direction, not touching anywhere, and each containing one hundred and fifty-five feet in length of wire.

7. One of these helices was connected with a galvanometer, the other with a voltaic battery of ten pairs of plates four inches square, with double coppers and well charged; yet not the slightest sensible deflection of the galvanometer needle could be observed.

8. A similar compound helix, consisting of six lengths of copper and six of soft iron wire, was constructed. The resulting iron helix contained two hundred and fourteen feet of wire, the resulting copper helix two hundred and eight feet; but whether the current from the trough was passed through the copper or the iron helix, no effect upon the other could be perceived at the galvanometer.

9. In these and many similar experiments no difference in action of any kind appeared between iron and other metals.

10. Two hundred and three feet of copper wire in one length were coiled round a large block of wood; other two hundred and three feet of similar wire were interposed as a spiral between the turns of the first coil, and metallic contact everywhere prevented by twine. One of these helices was connected with a galvanometer, and the other with a battery of one hundred pairs of plates four inches square, with double coppers, and well charged. When the contact was made, there was a sudden and very slight effect at the galvanometer, and there was also a similar slight effect when the contact with the battery was broken. But whilst the voltaic current was continuing to pass through the one helix, no galvanometrical appearance nor any effect like induction upon the other helix could be perceived, although the active power of the battery was proved to be great, by its heating the whole of its own helix, and by the brilliancy of the discharge when made through charcoal.

11. Repetition of the experiments with a battery of one hundred and twenty pairs of plates produced no other effects; but it was ascertained, both at this and the former time, that the slight deflection of the needle occurring at the moment of completing the connexion, was always in one direction, and that the equally slight deflection produced when the contact was broken, was in the other direction; and also, that these effects occurred when the first helices were used.

12. The results which I had by this time obtained with magnets led me to believe that the battery current through one wire, did, in reality, induce a similar current through the other wire, but that it continued for an instant only, and partook more of the nature of the electrical wave passed through from the shock of a common Leyden jar that of the current from a voltaic battery, and therefore might magnetise a steel needle, although it scarcely affected the galvanometer.

13. This expectation was confirmed; for on substituting a small hollow helix, formed round a glass tube, for the galvanometer, introducing a steel needle, making contact as before between the battery and the inducing wire and then removing the needle before the battery contact was broken, it was found magnetised.

14. When the battery contact was first made, then an unmagnetised needle introduced into the small indicating helix and lastly the battery contact broken, the needle was found magnetised to an equal degree apparently as before; but the poles were of the contrary kind.

15. The same effects took place on using the large compound helices first described.

16. When the unmagnetised needle was put into the indicating helix, before contact of the inducing wire with the battery, and remained there until the contact was broken, it exhibited little or no magnetism; the first effect having been nearly neutralised by the second. The force of the induced current upon making contact was found always to exceed that of the induced current at breaking of contact; and if therefore the contact was made and broken many times in succession, whilst the needle remained in the indicating helix, it at last came out not unmagnetised, but a needle magnetised as if the induced current upon making contact had acted alone on it. This effect may be due to the accumulation (as it is called) at the poles of the unconnected pile, rendering the current upon first making contact more powerful than when it is afterwards, at the moment of breaking contact.

17. If the circuit between the helix or wire under induction and the galvanometer or indicating spiral was not rendered complete *before* the connexion between the battery and the inducing wire was completed or broken, then no effects were perceived at the galvanometer. Thus, if the battery communications were first made, and then the wire under induction connected with the indicating helix, no magnetising power was there exhibited. But still retaining the latter communications, when those with the battery were broken, was formed in the helix, but of the second kind, i.e. with poles indicating a current in the same direction to that belonging to the battery current, or to that always induced by that current at its cessation.

18. In the preceding experiments the wires were placed near to each other, and the contact of the inducing one with the battery made when the inductive effect was required; but as the particular action might be supposed to be exerted only at the moments of making and breaking contact, the induction was produced in another way. Several feet of copper wire were stretched in wide zigzag forms, representing the letter W, on one surface of a broad board; a second wire was stretched in precisely similar forms on a second board, so that when brought near the first, the wires should everywhere touch, except that a sheet of thick paper was interposed. One of these wires was connected with the galvanometer, and the other with a voltaic battery. The first wire was then moved towards the second, and as it approached, the needle was deflected. Being then removed, the needle was deflected in the opposite direction. By first making the wires approach and then recede, simultaneously with the vibrations of the needle, the latter soon became very extensive; but when the wires ceased to move from or towards each other, the galvanometer needle soon came to its usual position.

19. As the wires approximated, the induced current was in the *contrary* direction to the inducing current. As the wires receded, the induced current was in the *same* direction as the inducing current. When the wires remained stationary, there was no induced current.

20. When a small voltaic arrangement was introduced into the circuit between the galvanometer and its helix or wire, so as to cause a permanent deflection of 30° or 40°, and then the battery of one hundred pairs of plates connected with the inducing wire, there was an instantaneous action as before; but the galvanometer-needle immediately resumed and retained its place unaltered, notwithstanding the continued contact of the inducing wire with the trough such as the case in whichever way the contacts were made.

21. Hence it would appear that collateral currents, either in the same or in opposite directions, exert no permanent inducing power on each other, affecting their quantity or tension.

22. I could obtain no evidence by the tongue, by spark, or by heating fine wire or charcoal, of the electricity passing through the wire under induction; neither could I obtain any chemical effects, though the contacts with metallic and other solutions were made and broken alternately with those of the battery, so that the second effect of induction should not oppose or neutralize the first.

23. This deficiency of effect is not because the induced current of electricity cannot pass fluids, but probably because of its brief duration and feeble intensity; for on introducing two large copper plates into the circuit on the induced side, the plates being immersed in brine, but prevented from touching each other by an interposed cloth, the effect at the indicating galvanometer or helix occurred as before. The induced electricity could also pass through as voltaic trough. When, however, the quantity of interposed fluid was reduced to a drop, the galvanometer gave no indication.

24. Attempts to obtain similar effects by the use of wires conveying ordinary electricity were doubtful in the results. A compound helix similar to that already described, containing eight elementary helices was used. Four of the helices had their similar ends bound together by wire, and the two general terminations thus produced connected with the small magnetising helix containing an unmagnetised needle. The other four helices were similarly arranged, but their ends connected with a Leyden jar. On passing the discharge, the needle was found to be a magnet; but it appeared probable that a part of the electricity of the jar had passed off to the small helix, and so magnetised the needle. There was indeed no reason to expect that the electricity of a jar possessing as it does great tension, would not diffuse itself through all the metallic matter interposed between the coatings.

25. Still it does not follow that the discharge of ordinary electricity through a wire does not produce analagous phenomena to those arising from voltaic electricity; but as it appears impossible to separate the effects produced at the moment when the discharge begins to pass, from the equal and contrary effects produced when it ceases to pass, inasmuch as with ordinary electricity these periods are simultaneous, so there can be scarcely any hope that in this form of the experiment they can be perceived.

26. Hence it is evident that currents of voltaic electricity present phenomena of induction somewhat analagous to those produced by electricity of tension, although, as will be seen hereafter, many differences exist between them. The result is the production of other currents, (but which are only momentary,) parallel, or tending to parallelism, with the inducing current. By reference to the poles of the needle formed in the indicating helix and to the deflections of the galvanometer needle it was found in all cases that the induced current, produced by the first action of the inducing current was in the contrary direction to the latter, but that the current produced by the cessation of the inducing current was in the same direction. For the purpose of avoiding periphrasis, I propose to call this action of the current from the voltaic battery *volta-electric induction*. The properties of the second wire, after induction has developed the first current, and whilst the electricity from the battery continues to flow through its inducing neighbour, constitute a peculiar electric condition, the consideration of which will be resumed hereafter. All these results have been obtained with a voltaic apparatus consisting of a single pair of plates.

10
— Michael Faraday —
(1791–1867)

In these readings, Faraday describes the development of the concept of a magnetic field and the mapping of its lines of forces. He compares magnetic force with gravitational force and decides that gravitational force may be explained by action at a distance because the force is always along the line separating the two masses. Since magnetic force is not analogous to gravitational force Faraday concludes that he must view magnetic lines of force as having a physical existence. However, he notes, although these lines of force have a physical existence, they can exist in a vacuum as well as where there is matter. In the final reading Faraday describes the essentials of an electromagnetic field theory. These materials were originally published in the Philosophical Magazine *in 1852.*

THE CONCEPT OF AN ELECTROMAGNETIC FIELD

I have recently been engaged in describing and defining the lines of magnetic force, i.e. those lines which are indicated in a general manner by the disposition of iron filings or small magnetic needles, around or between magnets; and I have shown, I hope satisfactorily, how these lines may be taken as exact representants of the magnetic power, both as to disposition and amount; also how they may be recognized by a moving wire in a manner altogether different in principle from the indications given by a magnetic needle, and in numerous cases with great and peculiar advantages. The definition then given had no reference to the physical nature of the force at the place of action, and will apply with equal accuracy whatever that may be; and this being very thoroughly understood, I am now about to leave the strict line of reasoning for a time, and enter upon a few speculations respecting the physical character of the lines of force, and the manner in which they may be supposed to be continued through space. We are obliged to enter into such speculations with regard to numerous natural powers, and indeed, that of gravity is the only instance where they are apparently shut out.

It is not to be supposed for a moment that speculations of this kind are useless, or necessarily hurtful, in natural philosophy. They should ever be held as doubtful, and liable to error and to change; but they are wonderful aids in the hands of the experimentalist and mathematician. For not only are they useful in rendering the vague idea more clear for the time, giving it something like a definite shape, that it may be submitted to experiment and calculation; but they lead on, by deduction and correction, to the discovery of new phenomena, and so cause an increase and advance of real physical truth, which, unlike the hypothesis that led to it, becomes fundamental knowledge not subject to change. Who is not aware of the remarkable progress in the development of the nature of light and radiation in modern times, and the extent to which that progress has been aided by the hypotheses both of emission and undulation? Such considerations form my excuse for entering now and then upon speculations; but though I value them highly when cautiously advanced, I consider it as an essential character of a sound mind to hold them in doubt; scarcely giving them the character of opinions, but esteeming them merely as probabilities and possibilities, and making a very broad distinction between them and the facts and laws of nature.

In the numerous cases of force acting at a distance, the philosopher has gradually learned that it is by no means sufficient to rest satisfied with the mere fact, and has therefore directed his attention to the manner in which the force is transmitted across the intervening space; and even when he can learn nothing sure of the manner, he is still able to make clear distinctions in different cases, by what may be called the affections of the lines of power; and thus, by these and other means, to make distinctions in the nature of the lines of force of different kinds of power as compared with each other and therefore between the powers to which they belong. In the action of gravity, for instance, the line of force is a straight line as far as we can test it by the resultant phenomena. It cannot be deflected, or even affected, in its course. Neither is the action in one line at all influenced, either in direction or amount, by a like action in another line; i.e. one particle gravitating toward another particle has exactly the same amount of force in the same direction, whether it gravitates to the one alone or towards myriads of other like particles, exerting in the latter case upon each one of them a force equal to that which it can exert upon the single one when alone: the results of course can combine, but the direction and amount of force between any two given particles remain unchanged. So gravity presents us with the simplest case of attraction; and appearing to have no relation to any physical process by which the power of the particles is carried on between them, seems to be a pure case of attraction or action at a distance, and offers therefore the simplest type of the cases which may be like it in that respect. My object is to consider how far magnetism, is such an action at a distance; or how far it may partake of the nature of other powers, the lines of which depend, for the communication of force, upon intermediate physical agencies.

There is one question in relation to gravity, which, if we could ascertain or touch it, would greatly enlighten us. It is, whether gravitation requires time. If it did, it would show undeniably that a physical agency existed in the course of the line of force. It seems equally impossible to prove or disprove this point; since there is no capability of suspending, changing, or annihilating the power (gravity), or annihilating the matter in which the power resides.

When we turn to radiation phenomena, then we obtain the highest proof, that though nothing ponderable passes, yet the lines of force have a physical existence independent, in a manner, of the body radiating, or of the body receiving the rays. They may be turned aside in their course, and then deviate from a straight into a bent or a curved line. They may be affected in their nature so as to be turned on their axis, or else to have different properties impressed on different sides. Their sum of power is limited; so that if the force, as it issues from its source, is directed on to or determined upon a given set of particles, or in a given direction, it cannot be in any degree directed upon other particles, or into another direction, without being proportionately removed from the first. The lines have no dependence upon a second or reacting body, as in gravitation; and they require time for their propagation. In all these things they are in marked contrast with the lines of gravitating force.

When we turn to the electric force, we are presented with a very remarkable general condition intermediate between the conditions of the two former cases. The power (and its lines) here requires the presence of two or more acting particles or masses, as in the case of gravity; and cannot exist with one only, as in the case of light. But though two particles are requisite, they must be in an *antithetical* condition in respect of each other, and not, as in the case of gravity, alike in relation to the force. The power is now dual; there it was simple. Requiring two or more particles like gravity, it is unlike gravity in that the power is limited. One electro-particle cannot affect a second, third and fourth, as much as it does the first; to act upon the latter its power must be proportionately removed from the former, and this limitation appears to exist as a necessity in the dual character of the force; for the two states, or places, or directions of force must be equal to each other.

With the electric force we have both the static and dynamic state. I use these words merely as names, without pretending to have a clear notion of the physical condition which they seem meaningly to imply. Whether there are two fluids or one, or any fluid of electricity, or such a thing as may be rightly called a current, I do not know; still there are well-established electric conditions and effects which the words *static, dynamic,* and *current* are generally employed to express; and with this reservation they express them as well as any other. The lines of the force of the static condition of electricity are present in all cases of induction. They terminate at the surfaces of the conductors under induction, or at the particles of non-conductors, which, being electrified, are in that condition. They are subject to infection in their course, and may be compressed or rarefied by bodies of different inductive capacities, but they are in those cases affected by the intervening matter, and it is not certain how the line of electric force would exist in relation to a perfect vacuum, i.e. whether it would be a straight line, as that of gravity is assumed to be, or curved in such manner as to show something like physical existence separate from the mere distant actions of the surfaces or particles bounding or terminating the induction. No condition of *quality* or *polarity* has as yet been discovered in the line of static electric force; nor has any relation of *time* been established in respect of it.

The lines of force of dynamic electricity are either limited in their extent, as in the lowering by discharge, or otherwise of the inductive condition of static electricity; or endless and continuous, as closed curves in the case of a voltaic circuit. Being definite in their amount for a given source, they can still be expanded, contracted, and deflected almost to any extent, according to the nature and size of the media through which they pass, and to which they have a direct relation. It is probable that matter is always essentially present; but the hypothetical aether may perhaps be admitted here as well as elsewhere. No condition of quality or polarity has as yet been recognised in them. In respect of *time,* it has been found, in the case of a Leyden discharge, that time is necessary even with the best conduction; indeed there is reason to think it is as necessary there as in the cases dependent on bad conducting media, as for instance, in the lightning flash.

Three great distinctions at least may be taken among these cases of the exertion of force at a distance; that of gravitation, where propagation of the force by physical lines through the intermediate space is not supposed to exist; that of radiation, where the propagation does exist, and where the propagating line or ray, once produced, has existence independent either of its source, or termination; and that of electricity, where the propagating process has intermediate existence, like a ray, but at the same time depends upon both extremities of the line of force, or upon conditions (as in the connected voltaic pile) equivalent to such extremities. Magnetic action at a distance has to be compared with these. It may be unlike any of them; for who shall say we are aware of all the physical methods or forms under which force is communicated? It has been assumed, however, by some, to be a pure case of force at a distance, and so like that of gravity; whilst others have considered it as better represented by the idea of streams of power. The question at present appears to be, whether the lines of magnetic force have or have not a physical existence; and if they have, whether such physical existence has a static or dynamic form.

The lines of magnetic force have not as yet been affected in their *qualities,* i.e. nothing analogous to the polarization of a ray of light or heat has been impressed on them. A relation between them and the rays of light when polarized has been discovered; but it is not of such a nature as to give proof as yet, either that the lines of magnetic force have a separate existence, or that they have not; though I think the facts are in favour of the former supposition. The investigation is an open one, and very important.

No relation of *time* to the lines of magnetic force has as yet been discovered. That iron requires *time* for its magnetization is well known. Plücker says the same is the case for bismuth, but I have not been able to obtain the effect showing this result. If that were the case, then mere space with its aether ought to have a similar relation for it comes between bismuth and iron; and such a result would go far to show that the lines of magnetic force have a separate physical existence. At present such results as we have cannot be accepted as in any degree proving the point of *time;* though if that point were proved, they would most probably come under it. It may be as well to state, that in the case also of the moving wire or conductor, time is required. There seems no hope of touching the investigation by any method like those we are able to apply to a ray of light, or to the current of the Leyden discharge; but the mere statement of the problem may help towards its solution.

If an *action* in *curved* lines or directions could be proved to exist in the case of the lines of magnetic force, it would also prove their physical existence external to the magnet on which they might depend; just as the same proof applies in the case of static electric induction. But the simple disposition of the lines, as they are shown by iron particles, cannot as yet be brought in proof of such a curvature, because they may be dependent upon the presence of these particles and their mutual action on each other and the magnets; and it is possible that attractions and repulsions in right lines might produce the same arrangement. The results therefore obtained by the moving wire; are more likely to supply data fitted to elucidate this point, when they are extended, and the true magnetic relation of the moving wire to the space which it occupies is fully ascertained.

The *amount* of the lines of magnetic force, or the force which they represent, is clearly limited, and therefore quite unlike the force of gravity in that respect; and this is true, even though the force of a magnet in free space must be conceived of as extending to incalculable distances. This limitation in amount of force appears to be intimately dependent upon the dual nature of the power, and is accompanied by a displacement or removeability of it from one object to another, utterly unlike anything which occurs in gravitation. The lines of force abutting on one end or pole of a magnet may be changed in their direction almost at pleasure, though the original seats of their further parts may otherwise remain the same. For, by bringing fresh terminals of power into presence, a new disposition of force upon them may be occasioned; but though these may be made, either in part or entirely, to receive the external power, and thus alter its direction, no change in the amount of the force is thus produced. And this is the case in strict experiments, whether the new bodies introduced are soft iron or magnets. In this respect, therefore, the lines of magnetic force and of electric force agree. Results of this kind are well shown in some recent experiments on the effect of iron, when passing by a copper wire in the magnetic field of a horseshoe magnet, and also the action of iron and magnets on each other.

It is evident, I think, that the experimental data are as yet insufficient for a full comparison of the various lines of power. They do not enable us to conclude, with much assurance, whether the magnetic lines of force are analogous to those of gravitation, or direct actions at a distance; or whether, having a physical existence, they are more like in their nature to those of electric induction or the electric current. I incline at present to the latter view more than to the former, and will proceed to the further and future elucidation of the subject.

I think I have understood that the mathematical expression of the laws of magnetic action at a distance is the same as that of the laws of static electric actions; and it has been assumed at times that the supposition of north and south magnetism, spread over the poles or respective ends of a magnet, would account for all its external actions on other magnets or bodies. In either the static or dynamic view, or in any other view like them, the exertion of the magnetic forces outwards, at the poles or ends of the magnet, must be an essential condition. Then, with a given bar-magnet, can these forces exist without a mutual relation of the two, or else a relation to contrary magnetic forces of equal amount originating in other sources? I do not believe they can, because, as I have shown in recent researches, the sum of the lines of force is equal for any section across them taken anywhere externally between the poles. Besides that, there are many other experimental facts which show the relation and connexion of the forces at one pole to those at the other; and there is also the analogy with static electrical induction, where the one electricity cannot exist without relation to, equality with, and dependence on the other. Every dual power appears subject to this law as a law of necessity. If the opposite magnetic forces could be independent of each other, then it is evident that *a charge* with one magnetism only is possible; but such a possibility is negatived by every known experiment and fact.

But supposing this necessary relation, which constitutes polarity, to exist, then how is it sustained or permitted in the case of an independent bar-magnet situated in free space? It appears to me, that the outer forces at the poles can only have relation to each other by *curved* lines of force through the surrounding space; and I cannot conceive curved lines of force without the conditions of a physical existence in that intermediate space. If they exist, it is not by a succession of particles, as in the case of static electric induction, but by the condition of space free from such material particles. A magnet placed in the middle of the best vacuum we can produce, and whether that vacuum be formed in a space previously occupied by paramagnetic or diamagnetic bodies, acts as well upon a needle as if it were surrounded by air, water or glass; and therefore these lines exist in such a vacuum as well as where there is matter.

On a former occasion certain lines about a bar-magnet were described and defined (being those which are depicted to the eye by the use of iron filings sprinkled in the neighbourhood of the magnet), and were recommended as expressing accurately the nature, condition, direction, and amount of the force in any given region either within or outside of the bar. At that time the lines were considered in the abstract. Without departing from or unsettling anything then said, the inquiry is now entered upon of the possible and probable *physical existence* of such lines. Those who wish to reconsider the different points belonging to these parts of magnetic science may refer to two papers in the first part of the Phil. Trans. for 1852 for data concerning the *representative lines* of force, and to a paper in the Phil. Mag. 4th Series, 1852, vol. iii. p. 401, for the argument respecting the *physical* lines of force.

Many powers act manifestly at a distance; their physical nature is incomprehensible to us; still we may learn much that is real and positive about them, and amongst other things something of the condition of the space between the body acting and that acted upon, or between the two mutually acting bodies. Such powers are presented to us by the phenomena of gravity, light, electricity, magnetism, &c. These when examined will be found to present remarkable differences in relation to their respective lines of forces; and at the same time that they establish the existence of real physical lines in some cases, will facilitate the consideration of the question as applied especially to magnetism.

When two bodies, *a, b,* gravitate towards each other, the line in which they act is a straight line, for such is the line which either would follow if free to move. The attractive force is not altered, either in *direction* or *amount,* if a third body is made to act by gravitation or otherwise upon either or both of the two first. A balanced cylinder of brass gravitates to the earth with a weight exactly the same, whether it is left like a pendulum freely to hang towards it, or whether it is drawn aside by other attraction or by tension, whatever the amount of the latter may be. A new gravitating force may be exerted upon *a*, but that does not in the least affect the amount of power which it exerts towards *b*. We have no evidence that *time* enters in any way into the exercise of this power, whatever the distance between the acting bodies, as that from the sun to the earth, or from star to star. We can hardly conceive of this force in one particle by itself; it is when two or more are present that we comprehend it: yet in gaining this idea we perceive no difference in the character of the power in the different particles; all of the same kind are *equal, mutual*, and *alike*. In the case of gravitation, no effect which sustains the idea of an independent or physical line of force is presented to us; and as far as we are at present know, the line of gravitation is merely an ideal line representing the direction in which the power is exerted.

Take the Sun in relation to another force which it exerts upon the earth, namely its illuminating or warming power. In this case rays (which are lines of force) pass across the intermediate space; but then we may affect these lines by different media applied to them in their course. We may alter their direction either by reflection or refraction, we may make them pursue curved or angular courses. We may cut them off at their origin and then search for and find them before they have attained their object. They have a relation to time, and occupy 8 minutes in coming from the sun to the earth: so that they may exist independently either of their source or their final home, and have in fact a clear distinct physical existence. They are in extreme contrast with the lines of gravitating power in this respect; as they are also in respect of their condition at their terminations. The two bodies terminating a line of gravitating force are alike in their actions in every respect, and so the line joining them has like relations in both directions. The two bodies at the terminals of a ray are utterly unlike in action; one is a source, the other a destroyer of the line; and the line itself has the relation of a stream flowing in one direction. In these two cases of gravity and radiation, the difference between an abstract and a physical line of force is immediately manifest.

Turning to the case of Static Electricity we find here attractions (and other actions) at a distance as in the former cases; but when we come to compare the attraction with that of gravity, very striking distinctions are presented which immediately affect the question of a physical line of force. In the first place, when we examine the bodies bounding or terminating the lines of attraction, we find them as before, mutually and equally concerned in the action; but they are not alike: on the contrary, though each is endued with a force which speaking generally is of the like nature, still they are in such contrast that their actions on a third body in a state like either of them are precisely the reverse of each other—what the one attracts the other repels; and the force makes itself evident as one of those manifestations of power endued with a dual and antithetical condition. Now with all such dual powers, attraction cannot occur unless the two conditions of force are present and in face of each other through the lines of force. Another essential limitation is that these two conditions must be exactly equal in amount, not merely to produce the effects of attraction, but in every other case; for it is impossible so to arrange things that there shall be present or be evolved more electric power of the one kind than of the other. Another limitation is that they must be in physical relation to each other; and that when a positive and a negative electrified surface are thus associated, we cannot cut off this relation except by transferring the forces of these surfaces to equal amounts of the contrary forces provided elsewhere. Another limitation is that the power is definite in amount. If a ball *a* be charged with 10 of positive electricity, it may be made to act with that amount of power on another ball *b* charged with 10 of negative electricity; but if 5 of its power be taken up by a third ball *c* charged with negative electricity, then it can only act with 5 of power on ball *a*, and that ball must find or evolve 5 of positive power elsewhere: this is quite unlike what occurs with gravity, a power that presents us with nothing dual in its character. Finally, the electric force acts in curved lines. If a ball be electrified positively and insulated in the air, and a round metallic plate be placed about 12 or 15 inches off, facing it and uninsulated, the latter will be found, by the necessity mentioned above, in a negative condition; but it is not negative only on the side facing the ball, but on the other or outer face also, as may be shown by a carrier applied there, or by a strip of gold or silver leaf hung against that outer face. Now the power affecting this face does not pass through the uninsulated plate, for the thinnest gold leaf is able to stop the inductive action, but round the edges of the face, and therefore acts in curved lines. All these points indicate the existence of physical lines of electric force:—the absolutely essential relation of positive and negative surfaces to each other, and their dependence on each other contrasted with the known mobility of the forces, admit of no other conclusion. The action also in curved lines must depend upon a physical line of force. And there is a third important character of the force leading to the same result, namely its affection by media having different specific inductive capacities.

When we pass to Dynamic Electricity the evidence of physical lines of force is far more patent. A voltaic battery having its extremities connected by a conducting medium, has what has been expressively called a current of force running round the circuit, but this current is an axis of power having equal and contrary forces in opposite directions. It consists of lines of force which are compressed or expanded according to the transverse action of the conductor, which changes in direction with the form of the conductor, which are found in every part of the conductor, and can be taken out from any place by channels properly appointed for the purpose; and nobody doubts that they are physical lines of force.

Finally as regards a Magnet, which is the object of the present discourse. A magnet presents a system of forces perfect in itself, and able, therefore, to exist by its own mutual relations. It has the dual and antithetic character belonging to both static and dynamic electricity; and this is made manifest by what are called its polarities, i.e. by the opposite powers of like kind found at and towards its extremities. These powers are found to be absolutely equal to each other; one cannot be changed in any degree as to amount without an equal change of the other, and this is true when the opposite polarities of a magnet are not related to each other, but to the polarities of other magnets. The polarities, or the *northness* and *southness* of a magnet are not only related to each other, through or within the magnet itself, but they are also related externally to opposite polarities (in the manner of static electric induction), or they cannot exist: and this external relation involves and necessitates an exactly equal amount of the new opposite polarities to which those of the magnet are related. So that if the force of a magnet *a* is related to that of another magnet *b*, it cannot act on a third magnet *c* without being taken off from *b*, to an amount proportional to its action on *c*. The lines of magnetic force are shown by the moving wire to exist both within and outside of the magnet; also they are shown to be closed curves passing in one part of their course through the magnet; and the amount of those within the magnet at its equator is exactly equal in force to the amount in any section including the whole of those on the outside. The lines of force outside a magnet can be affected in their direction by the use of various media placed in their course. A magnet can in no way be procured having only one magnetism, or even the smallest excess of northness or southness one over the other. When the polarities of a magnet are not related externally to the forces of other magnets, then they are related to each other: i.e. the northness and southness of an isolated magnet and externally dependent on and sustained by each other.

Now all these facts, and many more, point to the existence of physical lines of force external to the magnets as well as within. They exist in curved as well as in straight lines; for if we conceive of an isolated straight bar-magnet, or more especially of a round disc of steel magnetized regularly, so that its magnetic axis shall be in one diameter, it is evident that the polarities must be related to each other externally by curved lines of force; for no straight line can at the same time touch two points having northness and southness. Curved lines of force can, as I think, only consist with physical lines of force.

The phenomena exhibited by the moving wire confirm the same conclusion. As the wire moves across the lines of force, a current of electricity passes or tends to pass through it, there being no such current before the wire is moved. The wire when quiescent has no such current, and when it moves it need not pass into places where the magnetic force is greater or less. It may travel in such a course that if a magnetic needle were carried through the same course it would be entirely unaffected magnetically, i.e., it would be a matter of absolute indifference to the needle whether it were moving or still. Matters may be so arranged that the wire when still shall have the same diamagnetic force as the medium surrounding the magnet, and so in no way cause disturbance of the lines of force passing through both; and yet when the wire moves, a current of electricity shall be generated in it. The mere fact of motion cannot have produced this current: there must have been a state or condition around the magnet and sustained by it, within the range of which the wire was placed; and this state shows the physical constitution of the lines of magnetic force.

What this state is, or upon what it depends, cannot as yet be declared. It may depend upon the aether, as a ray of light does, and an association has already been shown between light and magnetism. It may depend upon a state of tension, or a state of vibration, or perhaps some other state analogous to the electric current, to which the magnetic forces are so intimately related. Whether it of necessity requires matter for its sustentation will depend upon what is understood by the term matter. If that is to be confined to ponderable or gravitating substance, then matter is not essential to the physical lines of magnetic force any more than to a ray of light or heat; but if in the assumption of an aether we admit it to be a species of matter, then the lines of force may depend upon some function of it. Experimentally mere space is magnetic; but then the idea of such mere space must include that of the aether, when one is talking on that belief; or if hereafter any other conception of the state or condition of space rise up, it must be admitted into the view of that, which just now in relation to experiment is called mere space. On the other hand it is, I think, an ascertained fact, that ponderable matter is not essential to the existence of physical lines of magnetic force.

11
— James Clerk Maxwell —
(1831–1879)

James Clerk Maxwell was a British physicist and one of the most important figures in the history of science. He authored the kinetic theory of gases and made contributions to optics and astronomy. But it was in the field of electromagnetism that Maxwell made his greatest contribution to science. Using a mechanical model, Maxwell mathematized the field concept of Faraday. From the equations that now bear his name, he was able to derive the velocity of a disturbance traveling through the field, which was very close to the velocity of light as measured experimentally.

In this reading, Maxwell describes the essence of his Electromagnetic Field Theory. He starts with a critique of the action-at-a-distance theory of Weber and Neumann, suggesting that the mechanical difficulties involved make it unacceptable as an ultimate explanation. Maxwell turns instead to a field concept in which the space between bodies is filled with an ethereal substance, distinct from "gross matter," which carries electric and magnetic action. He notes that the science of optics has come to a similar conclusion concerning the existence of an ethereal substance.

In the final page of the reading Maxwell asserts that the energy of the field is mechanical energy and that the field and/or ether is a substratum that can contain and transmit energy, which when transferred to gross matter is transferred as mechanical energy.

The reading comes from a memoir entitled a "A Dynamical Theory of the Electromagnetic Field" (1865).

THE THEORY OF THE ELECTROMAGNETIC FIELD

1. The most obvious mechanical phenomenon in electrical and magnetical experiments is the mutual action by which bodies in certain states set each other in motion while still at a sensible distance from each other. The first step, therefore, in reducing these phenomena into scientific form, is to ascertain the magnitude and direction of the force acting between the bodies, and when it is found that this force depends in a certain way upon the relative position of the bodies and on their electric or magnetic condition, it seems at first sight natural to explain the facts by assuming the existence of something either at rest or in motion in each body, constituting its electric or magnetic state, and capable of acting at a distance according to mathematical laws.

In this way mathematical theories of statical electricity, of magnetism, of the mechanical action between conductors carrying currents, and of the induction of currents have been formed. In these theories the force acting between the two bodies is treated with reference only to the condition of the bodies and their relative position, and without any express consideration of the surrounding medium.

These theories assume, more or less explicitly, the existence of substances the particles of which have the property of acting on one another at a distance by attraction or repulsion. The most complete development of a theory of this kind is that of M. W. Weber, who has made the same theory include electrostatic and electromagnetic phenomena.

In doing so, however, he has found it necessary to assume that the force between two electric particles depends on their relative velocity, as well as on their distance.

This theory, as developed by M.M.W. Weber and C. Neumann, is exceedingly ingenious, and wonderfully comprehensive in its application to the phenomena of statical electricity, electromagnetic attractions, induction of currents and diamagnetic phenomena; and it comes to us with the more authority, as it has served to guide the speculations of one who has made so great an advance in the practical part of electric science, both by introducing a consistent system of units in electrical measurement, and by actually determining electrical quantities with an accuracy hitherto unknown.

2. The mechanical difficulties, however, which are involved in the assumption of particles acting at a distance with forces which depend on their velocities are such as to prevent me from considering this theory as an ultimate one though it may have been, and may yet be useful in leading to the coordination of phenomena.

I have therefore preferred to seek an explanation of the fact in another direction, by supposing them to be produced by actions which go on in the surrounding medium as well as in the excited bodies, and endeavouring to explain the action between distant bodies without assuming the existence of forces capable of acting directly at sensible distances.

3. The theory I propose may therefore be called a theory of the *Electromagnetic Field*, because it has to do with the space in the neighbourhood of the electric or magnetic bodies, and it may be called a Dynamical Theory, because it assumes that in that space there is matter in motion, by which the observed electromagnetic phenomena are produced.

4. The electromagnetic field is that part of space which contains and surrounds bodies in electric or magnetic conditions.

It may be filled with any kind of matter, or we may endeavour to render it empty of all gross matter, as in the case of Geissler's tubes and other so called vacua.

There is always, however, enough of matter left to receive and transmit the undulations of light and heat, and it is because the transmission of these radiations is not greatly altered when transparent bodies of measurable density are substituted for the so-called vacuum, that we are obliged to admit that the undulation are those of an aethereal substance, and not of the gross matter, the presence of which merely modifies in some way the motion of the æther.

We have therefore some reason to believe, from the phenomena of light and heat, that there is an aethereal medium filling space and permeating bodies, capable of being set in motion and of transmitting that motion from one part to another, and of communicating that motion to gross matter so as to heat it and affect it in various ways.

5. Now the energy communicated to the body in heating it must have formerly existed in the moving medium, for the undulations had left the source of heat some time before they reached the body, and during that time the energy must have been half in the form of the medium and half in the form of elastic resilience. From these considerations Professor W. Thomson has argued, that the medium must have a density capable of comparison with that of gross matter, and has even assigned an inferior limit to that density.

6. We may therefore receive, as a datum derived from a branch of science independent of that with which we have to deal, the existence of a pervading medium, of small but real density, capable of being set in motion, and of transmitting motion from one part to another with great, but not infinite, velocity.

Hence the parts of this medium must be so connected that the motion of one part depends in some way on the motion of the rest; and at the same time these connexions must be capable of a certain kind of elastic yielding, since the communication of motion is not instantaneous, but occupies time.

The medium is therefore capable of receiving and storing up two kinds of energy, namely, the "actual" energy depending on the motions of its parts, and "potential" energy, consisting of the work which the medium will do in recovering from displacement in virtue of its elasticity.

The propagation of undulations consists in the continual transformation of one of these forms of energy into the other alternately, and at any instant the amount of energy in the whole medium is equally divided, so that half is energy of motion, and half is elastic resilience.

7. A medium having such a constitution may be capable of other kinds of motion and displacement than those which produce the phenomena of light and heat, and some of these may be of such a kind that they may be evidenced to our senses by the phenomena they produce.

8. Now we know that the luminiferous medium is in certain cases acted on by magnetism, for Faraday discovered that when a plane polarized ray traverses a transparent diamagnetic medium in the direction of the lines of magnetic force produced by magnets or currents in the neighbourhood, the plane of polarization is caused to rotate.

This rotation is always in the direction in which positive electricity must be carried round the diamagnetic body in order to produce the actual magnetization of the field.

M. Verdet has since discovered that if a paramagnetic body, such as solution of perchloride of iron in ether, be substituted for the diamagnetic body, the rotation is in the opposite direction.

Now Professor W. Thomson has pointed out that no distribution of forces acting between the parts of a medium whose only motion is that of the luminous vibrations, is sufficient to account for the phenomena, but that we must admit the existence of a motion in the medium depending on the magnetization, in addition to the vibratory motion which constitutes light.

It is true that the rotation by magnetism of the plane of polarization has been observed only in media of considerable density; but the properties of the magnetic field are not so much altered by the substitution of one medium for another, or for a vacuum, as to allow us to suppose that the sense medium does anything more than merely modify the motion of the ether. We have therefore warrantable grounds for inquiring whether there may not be a motion of the ethereal medium going on whenever magnetic effects are observed, and we have some reason to suppose that this motion is one of rotation, having the direction of the magnetic force as its axis.

9. We may now consider another phenomenon observed in the electromagnetic field. When a body is moved across the lines of magnetic force it experiences what is called an electromotive force; the two extremities of the body tend to become oppositely electrified, and an electric current tends to flow through the body. When the electromotive force is sufficiently powerful, and is made to act on certain compound bodies, it decomposes them and causes one of their components to pass towards one extremity of the body, and the other in the opposite direction.

Here we have evidence of a force causing an electric current in spite of resistance; electrifying the extremities of a body in opposite ways, a condition which is sustained only by the action of the electromotive force, and which as soon as that force is removed, tends, with an equal and opposite force, to produce a counter current through the body and to restore the original electrical state of the body; and finally, if strong enough, tearing to pieces chemical compounds and carrying their components in opposite directions, while their natural tendency is to combine, and to combine with a force which can generate an electromotive force in the reverse direction.

This, then, is a force acting on a body caused by its motion through the electromagnetic field, or by changes occurring in that field itself; and the effect of the force is either to produce a current and heat the body, or to decompose the body, or, when it can do neither, to put the body in a state of electric polarization,—a state of constraint in which opposite extremities are oppositely electrified and from which the body tends to relieve itself as soon as the disturbing force is removed.

10. According to the theory which I propose to explain, this "electromotive force" is the force called into play during the communication of motion from one part of the medium to another, and it is by means of this force that the motion of one part causes motion in another part. When electromotive force acts on a conducting circuit, it produces a current, which, as it meets with resistance, occasions a continual transformation of electrical energy into heat, which is incapable of being restored again to the form of electrical energy by any reversal of the process.

11. But when electromotive force acts on a dielectric it produces a state of polarization of its parts similar in distribution to the polarity of the parts of a mass of iron under the influence of a magnet, and like the magnetic polarization, capable of being described as a state in which every particle has it opposite poles in opposite conditions.

In a dielectric under the action of electromotive force, we may conceive that the electricity in each molecule is so displaced that one side is rendered positively and the other negatively electrical, but that the electricity remains entirely connected with the molecule, and does not pass from one molecule to another. The effect of this action on the whole dielectric mass is to produce a general displacement of electricity in a certain direction. This displacement does not amount to a current, because when it has attained to a certain value it remains constant, but it is the commencement of a current, and its variations constitute currents in the positive or the negative direction according as the displacement is increasing or decreasing. In the interior of the dielectric there is no indication of electrification, because the electrification of the surface of any molecule is neutralized by the opposite electrification of the surface of the molecules in contact with it; but at the founding surface of the dielectric, where the electrification is not neutralized, we find the phenomena which indicate positive or negative electrification.

The relation between the electromotive force and the amount of electric displacement it produces depends on the nature of the dielectric, the same electromotive force producing generally a greater electric displacement in solid dielectrics, such as glass or sulphur, than in air.

12. Here, then, we perceive another effect of electromotive force, namely, electric displacement, which according to our theory is a kind of elastic yielding to the action of the force, similar to that which takes place in structures and machines owing to the want of perfect rigidity of the connexions.

13. The practical investigation of the inductive capacity of dielectrics is rendered difficult on account of two disturbing phenomena. The first is the conductivity of the dielectric, which, though in many cases exceedingly small, is not altogether insensible. The second is the phenomenon called electric absorption, in virtue of which, when the dielectric is exposed to electromotive force, the electric displacement gradually increases, and when the electromotive force is removed, the dielectric does not instantly return to its primitive state, but only discharges a portion of its electrification, and when left to itself gradually acquires electrification on its surface as the interior gradually becomes depolarized. Almost all solid dielectrics exhibit this phenomenon, which gives rise to the residual charge in the Leyden jar, and to several phenomena of electric cables described by Mr. F. Jenkin.

14. We have here two other kinds of yielding besides the yielding of the perfect dielectric, which we have compared to a perfectly elastic body. The yielding due to conductivity may be compared to that of a viscous fluid (that is to say, a fluid having great internal friction), or a soft solid on which the smallest force produces a permanent alteration of figure increasing with the time during which the force acts. The yielding due to electric absorption may be compared to that of a cellular elastic body containing a thick fluid in its cavities. Such a body, when subjected to pressure, is compressed by degrees on account of the gradual yielding of the thick fluid; and when the pressure is removed it does not at once recover its figure, because the elasticity of the substance of the body has gradually to overcome the tenacity of the fluid before it can regain complete equilibrium.

Several solid bodies in which no such structure as we have supposed can be found, seem to possess a mechanical property of this kind; and it seems probable that the same substances, if dielectrics, may possess the analogous electrical property, and if magnetic, may have corresponding properties relating to the acquisition, retention, and loss of magnetic polarity.

15. It appears therefore that certain phenomena in electricity and magnetism lead to the same conclusion as those of optics, namely, that there is an aetheral medium pervading all bodies, and modified only in degree by their presence; that the parts of this medium are capable of being set in motion by electric currents and magnets; that this motion is communicated from one part of the medium to another by forces arising from the connexions of those parts; that under the action of these forces there is a certain yielding depending on the elasticity of these connexions; and that therefore energy in two different forms may exist in the medium, the one form being the actual energy of motion of its parts, and the other being the potential energy stored up in the connexions, in virtue of their elasticity.

16. Thus, then we are led to the conception of a complicated mechanism capable of a vast variety of motion, but at the same time so connected that the motion of one part depends, according to definite relations, on the motion of other parts, these motions being communicated by forces arising from the relative displacement of the connected parts, in virtue of their elasticity. Such a mechanism must be subject to the general laws of Dynamics, and we ought to be able to work out all the consequences of its motion, provided we know the form of the relation between the motions of the parts.

17. We know that when an electric current is established in a conducting circuit, the neighbouring part of the field is characterized by certain magnetic properties, and that if two circuits are in the field, the magnetic properties of the field due to the currents are combined. Thus each part of the field is in connexion with both currents, and the two currents are put in connexion with each other in virtue of their connexion with the magnetization of the field. The first result of this connexion that I propose to examine, is the induction of one current by another, and by the motion of conductors in the field.

The second result, which is deduced from this, is the mechanical action between conductors carrying currents. The phenomenon of the induction of currents has been deduced from their mechanical action by Helmholtz and Thomson. I have followed the reverse order, and deduced the mechanical action from the laws of induction. I have then described experimental methods of determining the quantities L, M, N, on which these phenomena depend.

18. I then apply the phenomena of induction and attraction of currents to the exploration of the electromagnetic field, and the laying down systems of lines of magnetic force which indicate its magnetic properties. By exploring the same field with a magnet, I shew the distribution of its equipotential magnetic surfaces, cutting the lines of force at right angles.

In order to bring these results within the power of symbolical calculation, I then express them in the form of the General Equations of the Electromagnetic Field. These equations express—

A. The relation between electric displacement, true conduction, and the total current, compounded of both.

B. The relation between the lines of magnetic force and the inductive coefficients of a circuit, as already deduced from the laws of induction.

C. The relation between the strength of a current and its magnetic effects, according to the electromagnetic system of measurement.

D. The value of the electromotive force in a body, as arising from the motion of the body in the field, the alteration of the field itself, and the variation of electric potential from one part of the field to another.

E. The relation between electric displacement, and the electromotive force which produces it.

F. The relation between an electric current, and the electromotive force which produces it.

G. The relation between the amount of free electricity at any point, and the electric displacements in the neighbourhood.

H. The relation between the increase or diminution of free electricity and the electric currents in the neighbourhood.

There are twenty of these equations in all, involving twenty variable quantities.

19. I then express in terms of these quantities the intrinsic energy of the Electromagnetic Field as depending partly on its magnetic and partly on its electric polarization at every point.

From this I determine the mechanical force acting, 1st on a moveable conductor carrying an electric current; 2ndly, on a magnetic pole; 3rdly, on an electrified body.

The last result, namely, the mechanical force acting on an electrified body, gives rise to an independent method of electrical measurement founded on its electrostatic effects. The relation between the units employed in the two methods is shewn to depend on what I have called the "electric elasticity" of the medium, and to be a velocity, which has been experimentally determined by M.M.W. Weber and Kohlrausch.

I then shew how to calculate the electrostatic capacity of a condenser, and the specific inductive capacity of a dielectric.

The case of a condenser composed of parallel layers of substances of different electric resistances and inductive capacities is next examined, and it is shewn that the phenomenon called electric absorption will generally occur, that is, the condenser, when suddenly discharged, will after a short time shew signs of a *residual* charge.

20. The general equations are next applied to the case of a magnetic disturbance propagated through a non-conducting field, and it is shewn that the only disturbances which can be so propagated are those which are transverse to the direction of propagation, and that the velocity of propagation is the velocity *v*, found from experiments such as those of Weber, which expresses the number of electrostatic units of electricity which are contained in one electromagnetic unit.

This velocity is so nearly that of light, that it seems we have strong reason to conclude that light itself (including radiant heat, and other radiations if any) is an electromagnetic disturbance in the form of waves propagated through the electromagnetic field according to electromagnetic laws. If so, the agreement between the elasticity of the medium as calculated from the rapid alternations of luminous vibrations, and as found by the slow processes of electrical experiments, shews how perfect and regular the elastic properties of the medium must be when not encumbered with any matter denser than air. If the same character of the elasticity is retained in dense transparent bodies, it appears that the square of the index of refraction is equal to the product of the specific dielectric capacity and the specific magnetic capacity. Conducting media are shewn to absorb such radiations rapidly, and therefore to be generally opaque.

The conception of the propagation of transverse magnetic disturbances to the exclusion of normal ones is distinctly set forth by Professor Faraday in his "Thoughts on Ray Vibrations." The electromagnetic theory of light, as proposed by him is the same in substance as that which I have begun to develop in this paper, except that in 1846 there were no data to calculate the velocity of propagation.

The general equations are then applied to the calculation of the coefficients of mutual induction of two circular currents and the coefficient of self-induction in a coil. The want of uniformity of the current in the different parts of the section of a wire at the commencement of the current is investigated, I believe for the first time, and the consequent correction of the coefficient of self-induction is found.

These results are applied to the calculation of the self-induction of the coil used in the experiments of the Committee of the British Association on Standards of Electric Resistance, and the value compared with that deduced from the experiments. . . .

73. I have on a former occasion attempted to describe a particular kind of motion and a particular kind of strain, so arranged as to account for the phenomena. In the present paper I avoid any hypothesis of this kind; and in using such words as electric momentum and electric elasticity in reference to the known phenomena of the induction of currents and the polarization of dielectrics, I wish merely to direct the mind of the reader to mechanical phenomena which will assist him in understanding the electrical ones. All such phases in the present paper are to be considered as illustrative, not as explanatory.

74. In speaking of the energy of the field, however, I wish to be understood literally. All energy is the same as mechanical energy, whether it exists in the form of motion or in that of elasticity, or in any other form. The energy in electromagnetic phenomena is mechanical energy. The only question is, Where does it reside? On the old theories it resides in the electrified bodies conducting circuits, and magnets, in the form of an unknown quality called potential energy, or the power of producing certain effects at a distance. On our theory it resides in the electromagnetic field, in the space surrounding the electrified and magnetic bodies, as well as in those bodies themselves, and is in two different forms, which may be described without hypothesis as magnetic polarization and electric polarization, or, according to a very probable hypothesis, as the motion and the strain of one and the same medium.

75. The conclusions arrived at in the present paper are independent of this hypothesis, being deduced from experimental facts of three kinds:

1. The induction of electric currents by the increase or diminution of neighbouring currents according to the changes in the lines of force passing through the circuit.
2. The distribution of magnetic intensity according to the variations of a magnetic potential.
3. The induction (or influence) of statical electricity through dielectrics.
4. We may now proceed to demonstrate from these principles the existence and laws of the mechanical forces which act upon electric currents, magnets, and electrified bodies placed in the electromagnetic field.

12
— James Clerk Maxwell —

In this reading, Maxwell gives his argument that light is an electromagnetic wave. In so doing he unites the fields of electricity and magnetism with the field of optics, with all known laws of these sciences derivable from Maxwell's equations. The key to this unification is the comparison of the measured velocity of light to the calculated velocity of propagation of an electromagnetic disturbance in the electromagnetic field.

The reading is from Maxwell's A Treatise on Electricity and Magnetism, *originally published in 1873.*

THE ELECTROMAGNETIC THEORY OF LIGHT

In several parts of this treatise an attempt has been made to explain electromagnetic phenomena by means of mechanical action transmitted from one body to another by means of a medium occupying the space between them. The undulatory theory of light also assumes the existence of a medium. We have now to show that the properties of the electromagnetic medium are identical with those of the luminiferous medium.

To fill all space with a new medium whenever any new phenomenon is to be explained is by no means philosophical, but if the study of two different branches of science has independently suggested the idea of a medium and if the properties which must be attributed to the medium in order to account for electromagnetic phenomena are of the same kind as those which we attribute to the luminiferous medium in order to account for the phenomena of light, the evidence for the physical existence of the medium will be considerably strengthened.

But the properties of bodies are capable of quantitative measurement. We therefore obtain the numerical value of some property of the medium, such as the velocity with which a disturbance is propagated through it, which can be calculated from electromagnetic experiments, and also observed directly in the case of light. If it should be found that the velocity of propagation of electromagnetic disturbances is the same as the velocity of light, and this not only in air, but in other transparent media, we shall have strong reasons for believing that light is an electromagnetic phenomenon, and the combination of the optical with the electrical evidence will produce a conviction of the reality of the medium similar to that which we obtain, in the case of other kinds of matter, from the combined evidence of the senses.

When the light is emitted, a certain amount of energy is expended by the luminous body, and if the light is absorbed by another body, this body becomes heated, shewing that it has received energy from without. During the interval of time after the light left the first body and before it reached the second, it must have existed as energy in the intervening space.

According to the theory of emission, the transmission of energy is effected by the actual transference of light-corpuscles from the luminous to the illuminated body, carrying with them their kinetic energy, together with any other kind of energy of which they may be the receptacles.

According to the theory of undulation, there is a material medium which fills the space between the two bodies, and it is by the action of contiguous parts of this medium that the energy is passed on, from one portion to the next, till it reaches the illuminated body.

The luminiferous medium is therefore, during the passage of light through it, a receptacle of energy. In the undulatory theory, as developed by Huygens, Fresnel, Young, Green, &c, this energy is supposed to be partly potential and partly kinetic. The potential energy is supposed to be due to the distortion of the elementary portions of the medium. We must therefore regard the medium as elastic. The kinetic energy is supposed to be due to the vibratory motion of the medium. We must therefore regard the medium as having a finite density.

In the theory of electricity and magnetism adopted in this treatise, two forms of energy are recognised, the electrostatic and the electrokinetic, and these are supposed to have their seat, not merely in the electrified or magnetized bodies, but in every part of the surrounding space, where electric or magnetic force is observed to act. Hence our theory agrees with the undulatory theory in assuming the existence of a medium which is capable of becoming a receptacle of two forms of energy. . . .

The quantity V, which expresses the velocity of propagation of electromagnetic disturbances in a non-conducting medium is, equal to $1/\sqrt{K\mu}$.

If the medium is air, and if we adopt the electrostatic system of measurement, $K = 1$ and $\mu = 1/v^2$, so that $V = v$, or the velocity of propagation is numerically equal to the number of electrostatic units of electricity in one electromagnetic unit. If we adopt the electromagnetic system, $K = 1/v^2$ and $\mu = 1$, so that the equation $V = v$ is still true.

On the theory that light is an electromagnetic disturbance, propagated in the same medium through which other electromagnetic actions are transmitted, V must be the velocity of light, a quantity the value of which has been estimated by several methods. On the other hand, v is the number of electrostatic units of electricity in one electromagnetic unit, and the methods of determining this quantity have been described in the last chapter. They are quite independent of the methods of finding the velocity of light. Hence the agreement or disagreement of the values of V and of v furnishes a test of the electromagnetic theory of light.

In the following table, the principal results of direct observation of the velocity of light, either through the air or through the planetary spaces, are compared with the principal results of the comparison of the electric units:

Velocity of Light (metres per second)		Ratio of electric Units (metres per second)	
Fizeau	314000000	Weber	310740000
Aberration, &c Sun's		Maxwell	288000000
Parallax	308000000		
Foucault	298360000	Thomson	282000000

It is manifest that the velocity of light and the ratio of the units are quantities of the same order of magnitude. Neither of them can be said to be determined as yet with such a degree of accuracy as to enable us to assert that the one is greater or less than the other. It is to be hoped that, by further experiment, the relation between the magnitudes of the two quantities may be more accurately determined.

In the meantime our theory, which asserts that these two quantities are equal, and assigns a physical reason for this equality, is certainly not contradicted by the comparison of these results such as they are.

In other media than air, the velocity V is inversely proportional to the square root of the product of the dielectric and the magnetic inductive capacities. According to the undulatory theory, the velocity of light in different media is inversely proportional to their indices of refraction.

There are no transparent media for which the magnetic capacity differs from that of air more than by a very small fraction. Hence the principal part of the difference between these media must depend on their dielectric capacity. According to our theory, therefore, the dielectric capacity of a transparent medium should be equal to the square of its index of refraction.

13
— James Clerk Maxwell —

In this reading, Maxwell offers a substantial philosophical analysis of the concept of action at a distance. Skillfully weaving theoretical reflection with experimental results, he argues that an undetectable medium hypothesis is far richer than any literal action-at-a-distance hypothesis. Maxwell goes to great lengths to credit Faraday not only with the conceptual insight of this medium but also with the important experimental work that confirms its existence and nature. Maxwell ends the passage with some grand metaphysical and theological commentary on this undetectable medium. His comment on the "already discovered properties of that which has often been called a vacuum, or nothing at all" captures the metaphysical difficulty of this medium hypothesis. The reading comes from a paper entitled "On Action at a Distance."

THE MEDIUM FOR ELECTROMAGNETIC ACTION

I have no new discovery to bring before you this evening. I must ask you to go over very old ground, and to turn your attention to a question which has been raised again and again ever since men began to think.

The question is that of the transmission of force. We see that two bodies at a distance from each other exert a mutual influence on each other's motion. Does this mutual action depend on the existence of some third thing, some medium of communication, occupying the space between the bodies, or do the bodies act on each other immediately, without the intervention of anything else?

The mode in which Faraday was accustomed to look at phenomena of this kind differs from that adopted by many other modern inquirers, and my special aim will be to enable you to place yourselves at Faraday's point of view, and to point out the scientific value of that conception of *lines of force* which, in his hands, became the key to the science of electricity.

When we observe one body acting on another at a distance, before we assume that this action is direct and immediate, we generally inquire whether there is any material connection between the two bodies; and if we find strings or rods, or mechanism of any kind, capable of accounting for the observed action between the bodies, we prefer to explain the action by means of these intermediate connections, rather than to admit the notion of direct action at a distance.

Thus, when we ring a bell by means of a wire, the successive parts of the wire are first tightened and then moved, till at last the bell is rung at a distance by a process in which all the intermediate particles of the wire have taken part one after the other. We may ring a bell at a distance in other ways, as by forcing air into a long tube, at the other end of which is a cylinder with a piston which is made to fly out and strike the bell. We may also use a wire; but instead of pulling it, we may connect it at one end with a voltaic battery, and at the other with an electro-magnet, and thus ring the bell by electricity.

Here are three different ways of ringing a bell. They all agree, however, in the circumstance that between the ringer and the bell there is an unbroken line of communication, and that at every point of this line some physical process goes on by which the action is transmitted from one end to the other. The process of transmission is not instantaneous, but gradual; so that there is an interval of time after the impulse has been given to one extremity of the line of communication, during which the impulse is on its way, but has not reached the other end.

It is clear, therefore, that in many cases the action between bodies at a distance may be accounted for by a series of actions between each successive pair of a series of bodies which occupy the intermediate space; and it is asked, by the advocates of mediate action, whether, in those cases in which we cannot perceive the intermediate agency, it is not more philosophical to admit the existence of a medium which we cannot at present perceive, than to assert that a body can act at a place where it is not.

To a person ignorant of the properties of air, the transmission of force by means of that invisible medium would appear as unaccountable as any other example of action at a distance, and yet in this case we can explain the whole process and determine the rate at which the action is passed on from one portion to another of the medium.

Why then should we not admit that the familiar mode of communicating motion by pushing and pulling with our hands is the type and exemplification of all action between bodies, even in case in which we can observe nothing between the bodies which appears to take part in the action?

Here for instance is a kind of attraction with which Professor Guthrie has made us familiar. A disk is set in vibration, and is then brought near a light suspended body, which immediately begins to move towards the disk, as if drawn towards it by an invisible cord. What is this cord? Sir. W. Thomson has pointed out that in a moving fluid the pressure is least where the velocity is greatest. Hence the pressure of the air on the suspended body is less on the side nearest the disk than on the opposite side, the body yields to the greater pressure, and moves toward the disk.

The disk, therefore, does not act where it is not. It sets the air next [to] it in motion by pushing it, this motion is communicated to more and more distant portions of the air in turn, and thus the pressures on opposite sides of the suspended body are rendered unequal, and it moves towards the disk in consequence of the excess of pressure. The force is therefore a force of the old school—a case of *vis a tergo*—a shove from behind.

The advocates of the doctrine of action at a distance, however, have not been put to silence by such arguments. What right, say they, have we to assert that a body cannot act where it is not? Do we not see an instance of action at a distance in the case of a magnet, which acts on another magnet not only at a distance, but with the most complete indifference to the nature of the matter which occupies the intervening space? If the action depends on something occupying the space between the two magnets, it cannot surely be a matter of indifference whether this space is filled with air or not, or whether wood, glass, or copper, be placed between the magnets.

Besides this, Newton's law of gravitation, which every astronomical observation only tends to establish more firmly, asserts not only that the heavenly bodies act on one another across immense intervals of space, but that two portions of matter, the one buried a thousand miles deep in the interior of the earth, and the other a hundred thousand miles deep in the body of the sun, act on one another with precisely the same force as if the strata beneath which each is buried had been non-existent. If any medium takes part in transmitting this action, it must surely make some difference whether the space between the bodies contains nothing but this medium, or whether it is occupied by the dense matter of the earth or of the sun.

But the advocates of direct action at a distance are not content with instances of this kind, in which the phenomena, even at first sight, appear to favour their doctrine. They push their operations into the enemy's camp, and maintain that even when the action is apparently the pressure of contiguous portions of matter, the contiguity is only apparent—that a space *always* intervenes between the bodies which act on each other. They assert, in short, that so far from action at a distance being impossible, it is the only kind of action which even occurs, and that the favourite old *vis a tergo* of the schools has no existence in nature, and exists only in the imagination of schoolmen.

The best way to prove that when one body pushes another it does not touch it, is to measure the distance between them. Here are two glass lenses, one of which is pressed against the other by means of weight. By means of the electric light we may obtain on the screen an image of the place where the one lens presses against the other. A series of coloured rings is formed on the screen. These rings were first observed and first explained by Newton. The particular colour of any ring depends on the distance between the surfaces of the pieces of glass. Newton formed a table of the colours corresponding to different distances, so that by comparing the colour of any ring with Newton's table, we may ascertain the distance between the surfaces at that ring. The colours are arranged in rings because the surfaces are spherical, and therefore the interval between the surfaces depends on the distance from the line joining the centres of the spheres. The central spot of the rings indicates the place where the lenses are nearest together, and each successive ring corresponds to an increase of about the 4000th part of a millimetre in the distance of the surfaces.

The lenses are now pressed together with a force equal to the weight of an ounce; but there is still a measurable interval between them, even at the place where they are nearest together. They are not in optical contact. To prove this, I apply a greater weight. A new colour appears at the central spot, and the diameters of all the rings increase. This shews that the surfaces are now nearer than at first, but they are not yet in optical contact, for if they were, the central spot would be black. I therefore increase the weights so as to press the lenses into optical contact.

But what we call optical contact is not real contact. Optical contact indicates only that the distance between the surfaces is much less than a wavelenght of light. To shew that surfaces are not in real contact, I remove the weights. The rings contract and several of them vanish at the centre. Now it is possible to bring two pieces of glass so close together, that they will not tend to separate at all, but adhere together so firmly, that when torn asunder the glass will break, not at the surface of contact, but at some other place. The glasses must then be many degrees nearer than when in mere optical contact.

Thus we have shewn that bodies begin to press against each other whilst still at a measurable distance, and that even when pressed together with great force they are not in absolute contact, but may be brought nearer still, and that by many degrees.

Why, then, say the advocates of direct action, should we continue to maintain the doctrine, founded only on the rough experience of a pre-scientific age, that matter cannot act where it is not, instead of admitting that all the facts from which our ancestors concluded that contact is essential to action were in reality cases of action at a distance, the distance being too small to be measured by their imperfect means of observation?

If we are ever to discover the laws of nature, we must do so by obtaining the most accurate acquaintance with the facts of nature, and not be dressing up in philosophical language the loose opinions of men who had not knowledge of the facts which throw most light on these laws. And as for those who introduce aetherial, or other media, to account for these actions, without any direct evidence of the existence of such media, or any clear understanding of how the media do their work, and who fill all space three and four times over with aethers of different sorts, why the less these men talk about their philosophical scruples about admitting action at a distance the better.

If the progress of science were regulated by Newton's first law of motion, it would be easy to cultivate opinions in the advance of the age. We should only have to compare the science of to-day with that of fifty years ago; and by producing, in the geometrical sense, the line of progress, we should obtain the science of fifty years hence.

The progress of science in Newton's time consisted in getting rid of the celestial machinery with which generations of astronomers had encumbered the heavens, and thus "sweeping cobwebs off the sky."

Though the planets had already got rid of their crystal spheres, they were still swimming in the vortices of Descartes. Magnets were surrounded by effluvia, and electrified bodies by atmospheres, the properties of which resembled in no respect those of ordinary effluvia and atmospheres.

When Newton demonstrated that the force which acts on each of the heavenly bodies depends on its relative position with respect to the other bodies, the new theory met with violent opposition from the advanced philosophers of the day, who described the doctrine of gravitation as a return to the exploded method of explaining everything by occult causes, attractive virtues, and the like.

Newton himself, with that wise moderation which is characteristic of all his speculations, answered that he made no pretence of explaining the mechanism by which the heavenly bodies act on each other. To determine the mode in which their mutual action depends on their relative position was a great step in science, and this step Newton asserted that he had made. To explain the process by which this action is effected was a quite distinct step, and this step Newton, in his *Principia*, does not attempt to make.

But so far was Newton from asserting that bodies really do act on one another at a distance, independently of anything between them, that in a letter to Bentley, which has been quoted by Faraday in this place, he says:—

"It is inconceivable that inanimate brute matter should, without the mediation of something else, which is not material, operate upon and affect other matter without mutual contact, as it must do if gravitation, in the sense of Epicurus, be essential and inherent in it. . . . That gravity should be innate, inherent, and essential to matter, so that one body can act upon another at a distance, through a vacuum, without the mediation of anything else, by and through which their action and force may be conveyed from one to another, is to me so great an absurdity, that I believe no man who has in philosophical matters a competent faculty of thinking can ever fall into it."

Accordingly, we find in his *Optical Queries*, and in his letters to Boyle, that Newton had very early made the attempt to account for gravitation by means of the pressure of a medium, and that the reason he did not publish these investigations "proceeded from hence only, that he found he was not able, from experiment and observations, to give a satisfactory account of this medium, and the manner of its operation in producing the chief phenomena of nature."

The doctrine of direct action at a distance cannot claim for its author the discoverer of universal gravitation. It was first asserted by Roger Cotes, in his preface to the *Principia*, which he edited during Newton's life. According to Cotes, it is by experience that we learn that all bodies gravitate. We do not learn in any other way that they are extended, moveable, or solid.Gravitation, therefore, has as much right to be considered an essential property of matter as extension, mobility, or impenetrability.

And when the Newtonian philosophy gained ground in Europe, it was the opinion of Cotes rather than that of Newton that became most prevalent, till at last Boscovich propounded his theory, that matter is a congeries of mathematical points, each endowed with the power of attracting or repelling the others according to fixed laws. In his world, matter is unextended, and contact is impossible. He did not forget, however, to endow his mathematical points with inertia. In this some of the modern representatives of his school have thought that he "had not quite got so far as the strict modern view of 'matter' as being but an expression for modes or manifestations of 'force.'"

But if we leave out of account for the present the development of the ideas of science, and confine our attention to the extension of its boundaries, we shall see that it was most essential that Newton's method should be extended to every branch of science to which it was applicable—that we should investigate the forces with which bodies act on each other in the first place, before attempting to explain how that force is transmitted. No men could be better fitted to apply themselves exclusively to the first part of the problem, than those who consider the second part quite unnecessary.

Accordingly Cavendish, Coulomb, and Poisson, the founders of the exact sciences of electricity and magnetism, paid no regard to those old notions of "magnetic effluvia" and "electric atmospheres," which had been put forth in the previous century, but turned their undivided attention to the determination of the law of force, according to which electrified and magnetized bodies attract or repel each other. In this way the true laws of these actions were discovered, and this was done by men who never doubted that the action took place at a distance, without the intervention of any medium, and who would have regarded the discovery of such a medium as complicating rather than as explaining the undoubted phenomena of attraction.

We have now arrived at the great discovery by Oersted of the connection between electricity and magnetism. Oersted found that an electric current acts on a magnetic pole, but that it neither attracts it nor repels it, but causes it to move round the current. He expressed this by saying that "the electric conflict acts in a revolving manner."

The most obvious deduction from this new fact was that the action of the current on the magnet is not a push-and-pull force, but a rotatory force, and accordingly many minds were set a-speculating on vortices and streams of aether whirling round the current.

But Ampère, by a combination of mathematical skill with experimental ingenuity, first proved that two electric currents act on one another, and then analysed this action into the resultant of a system of push-and-pull forces between the elementary parts of these currents.

The formula of Ampère, however, is of extreme complexity, as compared with Newton's law of gravitation, and many attempts have been made to resolve it into something of greater apparent simplicity.

I have no wish to lead you into a discussion of any of these attempts to improve a mathematical formula. Let us turn to the independent method of investigation employed by Faraday in those researches in electricity and magnetism which have made this Institution one of the most venerable shrines of science.

No man ever more conscientiously and systematically laboured to improve all his powers of mind than did Faraday from the very beginning of his scientific career. But whereas the general course of scientific method then consisted in the application of the ideas of mathematics and astronomy to each new investigation in turn, Faraday seems to have had no opportunity of acquiring a technical knowledge of mathematics, and his knowledge of astronomy was mainly derived from books.

Hence, though he had a profound respect for the great discovery of Newton, he regarded the attraction of gravitation as a sort of sacred mystery, which, as he was not an astronomer, he had no right to gainsay or to doubt, his duty being to believe it in the exact form in which it was delivered to him. Such a dead faith was not likely to lead him to explain new phenomena by means of direct attractions.

Besides this, the treatises of Poisson and Ampère are of so technical a form, that to derive any assistance from them the student must have been thoroughly trained in mathematics, and it is very doubtful if such a training can be begun with advantage in mature years.

Thus Faraday, with his penetrating intellect, his devotion to science, and his opportunities for experiments, was debarred from following the course of thought which had led to the achievements of the French philosophers, and was obliged to explain the phenomena to himself by means of a symbolism which he could understand, instead of adopting what had hitherto been the only tongue of the learned.

This new symbolism consisted of those lines of force extending themselves in every direction from electrified and magnetic bodies, which Faraday in his mind's eye saw as distinctly as the solid bodies from which they emanated.

The idea of lines of force and their exhibition by means of iron filings was nothing new. They had been observed repeatedly, and investigated mathematically as an interesting curiosity of science. But let us hear Faraday himself, as he introduces to his reader the method which in his hands became so powerful.

"It would be a voluntary and unnecessary abandonment of most valuable aid if an experimentalist, who chooses to consider magnetic power as represented by lines of magnetic force, were to deny himself the use of iron filings. By their employment he may make many conditions of the power, even in complicated cases, visible to the eye at once, may trace the varying direction of the lines of force and determine the relative polarity, may observe in which direction the power is increasing or diminishing, and in the complex systems may determine the neutral points, or places where there is neither polarity nor power, even when they occur in the midst of powerful magnets. By their use probable results may be seen at once, and many a valuable suggestion gained for future leading experiments."

In this experiment each filing becomes a little magnet. The poles of opposite names belonging to different filings attract each other and stick together, and more filings attach themselves to the exposed poles, that is, to the ends of the row of filing. In this way the filings, instead of forming a confused system of dots over the paper, draw together, filing to filing, till long fibres of filings are formed, which indicate by their direction the lines of force in every part of the field.

The mathematicians saw in this experiment nothing but a method of exhibiting at one view the direction in different places of the resultant of two forces, one directed to each pole of the magnet; a somewhat complicated result of the simple law of force.

But Faraday, by a series of steps as remarkable for their geometrical definiteness as for their speculative ingenuity, imparted to this conception of these lines of force a clearness and precision far in advance of that with which the mathematicians could then invest their own formulae.

In the first place, Faraday's lines of force are not to be considered merely as individuals, but as forming a system, drawn in space in a definite manner so that the number of lines which pass through an area, say of one square inch, indicates the intensity of the force acting through the area. Thus the lines of force become definite in number. The strength of a magnetic pole is measured by the number of lines which proceed from it; the electro-tonic state of a circuit is measured by the number of lines which pass through it.

In the second place, each individual line has a continuous existence in space and time. When a piece of steel becomes a magnet, or when an electric current begins to flow, the lines of force do not start into existence each in its own place, but as the strength increases new lines are developed within the magnet or current, and gradually grow outwards, so that the whole system expands from within, like Newton's rings in our former experiment. Thus every line of force preserves its identity during the whole course of its existence, though its shape and size may be altered to any extent.

I have no time to describe the methods by which every question relating to the forces acting on magnets or on currents, or to the induction of currents in conducting circuits, may be solved by the consideration of Faraday's lines of force. In this place they can never be forgotten. By means of this new symbolism, Faraday defined with mathematical precision the whole theory of electromagnetism, in language free from mathematical technicalities, and applicable to the most complicated as well as the simplest cases. But Faraday did not stop here. He went on from the conception of geometrical lines of force to that of physical lines of force. He observed that the motion which the magnetic or electric force tends to produce is invariably such as to shorten the lines of force and to allow them to spread out laterally from each other. He thus perceived in the medium a state of stress, consisting of a tension, like that of a rope, in the direction of the lines of force, combined with a pressure in all directions at right angles to them.

This is quite a new conception of action at a distance, reducing it to a phenomenon of the same kind as that action at a distance which is exerted by means of the tension of ropes and the pressure of rods. When the muscles of our bodies are excited by that stimulus which we are able in some unknown way to apply to them, the fibres tend to shorten themselves and at the same time to expand laterally. A state of stress is produced in the muscle, and the limb moves. This explanation of muscular action is by no means complete. It gives no account of the cause of the excitement of the state of stress, nor does it even investigate those forces of cohesion which enable the muscles to support this stress. Nevertheless, the simple fact, that it substitutes a kind of action which extends continuously along a material substance for one of which we know only a cause and an effect at a distance from each other, induces us to accept it as a real addition to our knowledge of animal mechanics.

For similar reasons we may regard Faraday's conception of a state of stress in the electro-magnetic field as a method of explaining action at a distance by means of the continuous transmission of force, even though we do not know how the state of stress is produced.

But one of Faraday's most pregnant discoveries, that of the magnetic rotation of polarised light, enables us to proceed a step farther. The phenomenon, when analysed into its simplest elements, may be described thus:—Of two circularly polarised rays of light, precisely similar in configuration, but rotating in opposite directions, that ray is propagated with the greater velocity which rotates in the same direction as the electricity of the magnetizing current.

It follows from this, as Sir W. Thomson has shewn by strict dynamical reasoning, that the medium when under the action of magnetic force must be in a state of rotation—that is to say, that small portions of the medium which we may call molecular vortices, are rotating, each on its own axis, the direction of this axis being that of the magnetic force.

Here, then, we have an explanation of the tendency of the lines of magnetic forces to spread out laterally and to shorten themselves. It arises from the centrifugal force of the molecular vortices.

The mode in which electromotive force acts in starting and stopping the vortices is more abstruse, though it is of course consistent with dynamical principles.

We have thus found that there are several different kinds of work to be done by the electro-magnetic medium if it exists. We have also seen that magnetism has an intimate relation to light, and we know that there is a theory of light which supposes it to consist of the vibrations of a medium. How is this luminiferous medium related to our electro-magnetic medium?

It fortunately happens that electro-magnetic measurements have been made from which we can calculate by dynamical principles the velocity of progagation of small magnetic disturbances in the supposed electro-magnetic medium.

This velocity is very great, from 288 to 314 millions of metres per second, according to different experiments. Now the velocity of light, according to Foucault's experiments, is 298 millions of metres per second. In fact, the different determination of either velocity differ from each other more than the estimated velocity of light does from the estimated velocity of propagation of small electro-magnetic disturbance. But if the luminiferous and the electro-magnetic media occupy the same place, and transmit disturbances with the same velocity, what reason have we to distinguish the one from the other? By considering them as the same, we avoid at least the reproach of filling space twice over with different kinds of aether.

Besides this, the only kind of electro-magnetic disturbances which can be propagated through a non-conducting medium is a disturbance transverse to the direction of propagation, agreeing in this respect with what we know of that disturbance which we call light. Hence, for all we know, light also may be an electromagnetic disturbance in a non-conducting medium. If we admit this, the electro-magnetic theory of light will agree in every respect with the undulatory theory, and the work of Thomas Young and Fresnel will be established on a firmer basis than ever, when joined with that of Cavendish and Coulomb by the key-stone of the combined sciences of light and electricity—Faraday's great discovery of the electro-magnetic rotation of light.

The vast interplanetary and interstellar regions will no longer be regarded as waste places in the universe, which the Creator has not seen fit to fill with the symbols of the manifold order of His kingdom. We shall find them to be already full of this wonderful medium; so full, that no human power can remove it from the smallest portion of space, or produce the slightest flaw in its infinite continuity. It extends unbroken from star to star; and when a molecule of hydrogen vibrates in the dog-star, the medium receives the impulses of these vibrations; and after carrying them in its immense bosom for three years, delivers them in due course, regular order, and full tale into the spectroscope of Mr. Huggins, at Tulse Hill.

But the medium has other functions and operations besides bearing light from man to man, and from world to world, and giving evidence of the absolute unity of the metric system of the universe. Its minute parts may have rotatory as well as vibratory motions, and the axes of rotation form those lines of magnetic force which extend in unbroken continuity into regions which no eye has seen, and which, by their action on our magnets, are telling us in language not yet interpreted, what is going on in the hidden underworld from minute to minute and from century to century.

And these lines must not be regarded as mere mathematical abstractions. They are the directions in which the medium is exerting a tension like that of a rope, or rather, like that of our own muscles. The tension of the medium in the direction of the earth's magnetic force is in this country one grain weight on eight square feet. In some of Dr. Joule's experiments, the medium has exerted a tension of 200 lbs. weight per square inch.

But the medium, in virtue of the very same elasticity by which it is able to transmit the undulations of light, is also able to act as a spring. When properly wound up, it exerts a tension, different from the magnetic tension, by which it draws oppositely electrified bodies together, produces effects through the length of telegraph wires, and when of sufficient intensity, leads to the rupture and explosion called lightning.

These are some of the already discovered properties of that which has often been called vacuum, or nothing at all. They enable us to resolve several kinds of action at a distance into actions between contiguous parts of a continuous substance. Whether this resolution is of the nature of explication or complication, I must leave to the metaphysicians.

14
— Heinrich Hertz —
1857–1894

Heinrich Hertz, a German physicist, was the first person to demonstrate the existence of electromagnetic waves other than light. His experiment showing that electromagnetic waves interfere and can be made to produce standing waves was a crucial bit of evidence in support of electromagnetic field theory.

In this reading, Hertz describes his experiments that led to the production of electromagnetic waves other than light. In those experiments he demonstrated that electromagnetic action is propagated at a finite rate of speed. His experiment in which he produced standing electromagnetic waves is a classic. Interestingly, he ends the description of his experiments by saying that their demonstrative power is independent of any particular theory. However, he does admit that his experiments amount to so many reasons in favor of the electromagnetic field theory of Faraday and Maxwell. The reading comes from his book entitled Electric Waves.

THE PRODUCTION OF ELECTROMAGNETIC WAVES

On Electromagnetic Waves in Air and Their Reflection

I have recently endeavoured to prove by experiment that electromagnetic actions are propagated through air with finite velocity. The inferences upon which that proof rested appear to me to be perfectly valid; but they are deduced in a complicated manner from complicated facts, and perhaps for this reason will not quite carry conviction to any one who is not already prepossessed in favour of the views therein adopted. In this respect the demonstration there given may be fitly supplemented by a consideration of the phenomena now to be described, for these exhibit the propagation of induction through the air by wave-motion in a visible and almost tangible form. These new phenomena also admit of a direct measurement of the wave-length in air. The fact that the wave-lengths thus obtained by direct measurement only differ slightly from the previous indirect determinations (using the same apparatus), may be regarded as an indication that the earlier demonstration was in the main correct.

In experimenting upon the action between a rectilinear oscillation and a secondary conductor I had often observed phenomena which seemed to point to a reflection of the induction action from the walls of the building. For example, feeble sparks frequently appeared when the secondary conductor was so situated that any direct action was quite impossible, as was evident from simple geometrical considerations of symmetry; and this most frequently occurred in the neighbourhood of solid walls. In especial, I continually encountered the following phenomenon:—In examining the sparks in the secondary conductor at great distances from the primary conductor, when the sparks were already exceedingly feeble, I observed that in most positions of the secondary conductor the sparks became appreciably stronger when I approached a solid wall, but again disappeared almost suddenly close to the wall. It seemed to me that the simplest way of explaining this was to assume that the electromagnetic action, spreading outwards in the form of waves, was reflected from the walls and that the reflected waves reinforced the advancing waves at certain distances, and weakened them at other distances, stationary waves in air being produced by the interference of the two systems. As I made the conditions more and more favourable for reflection the phenomenon appeared more and more distinct, and the explanation of it given above more probable. But without dwelling upon these preliminary trials I proceed at once to describe the principal experiments.

The physics lecture-room in which these experiments were carried out is about 15 metres long, 14 metres broad, and 6 metres high. Parallel to the two longer walls there are two rows of iron pillars, each of which rows behaves much like a solid wall towards the electromagnetic action, so that the parts of the room which lie outside these cannot be taken into consideration. Thus only the central space, 15 metres long, 8.5 metres broad, and 6 metres high, remained for the purpose of experiment. From this space I have the hanging parts of the gas-pipes and the chandeliers removed, so that it contained nothing except wooden tables and benches which could not well be removed. No objectionable effects were to be feared from these, and none were observed. The front wall of the room, from which the reflection was to take place, was a massive sandstone wall in which were two doorways, and a good many gas-pipes extended into it. In order to give the wall more of the nature of a conducting surface a sheet of zinc 4 metres high and 2 metres broad was fastened on to it; this was connected by wires with the gas-pipes and with a neighbouring water-pipe, and especial care was taken that any electricity that might accumulate at the upper and lower ends of the sheet should be able to flow away as freely as possible.

The primary conductor was set up opposite the middle of this wall at a distance of 13 metres from it, and was therefore 2 metres away from the opposite wall. It was the same conductor that had already been used in the experiments on the rate of propagation. The direction of the conducting wire was now vertical; hence the forces which have here to be considered oscillate up and down

in a vertical direction. The middle point of the primary conductor was 2.5 metres above the level floor; the observations were also carried out at the same
distance above the floor, a gangway for the observer being built up with tables
and boards at a suitable height. We shall denote as the normal a straight line
drawn from the centre of the primary conductor perpendicularly to the reflecting
surface. Our experiments are restricted to the neighbourhood of this normal;
experiments at greater angles of incidence would be complicated by having to
take into consideration the varying polarisation of the waves. Any vertical plane
parallel to the normal will be called a plane of oscillation, and any plane perpendicular to the normal will be called a wave-plane.

The secondary conductor was the circle of 35 cm. radius, which had also
been used before. It was mounted so as to revolve about an axis passing through
its centre and perpendicular to its plane. In the experiments the axis was horizontal; it was mounted in a wooden frame, so that both circle and axis could
be rotated about a vertical axis. For the most part it does well enough for the
observer to hold the circle, mounted in an insulating wooden frame, in his hand,
and then to bring it as may be most convenient into the various positions. But,
inasmuch as the body of the observer always exercises a slight influence, the
observations thus obtained must be controlled by others obtained from greater
distances. The sparks too are strong enough to be seen in the dark several metres
off; but in a well-lit room practically nothing can be seen, even at close quarters,
of the phenomena which are about to be described.

After we have made these preparations the most striking phenomenon that
we encounter is the following: We place the secondary circle with its centre on
the normal and its place in the plane of oscillation, and spark-gap first towards
the wall and then away from it. Generally the sparks differ greatly in the two
positions. If the experiment is arranged at a distance of about 0.8 metre from
the wall the sparks are much stronger when the spark-gap is turned towards the
wall. The length of the sparks can be so regulated that a continuous stream of
sparks passes over when the spark-gap is turned towards the wall, whereas no
sparks whatever pass over in the opposite position. If we repeat the experiment
at a distance of 3 metres from the wall we find, on the contrary, a continuous
stream of sparks pass over when the spark-gap is turned away from the wall,
whereas the sparks disappear when the spark-gap is turned towards the wall. If
we proceed further to a distance of 5.5 metres, a fresh reversal has taken place;
the sparks on the side toward the wall are stronger than the sparks on the
opposite side. Finally, at a distance of 8 metres from the wall, we find that
another reversal has been executed; the sparking is stronger on the side remote
from the wall, but the difference is no longer so noticeable. Nor does any further
reversal occur; for it is prevented by the preponderating strength of the direct
action and by the complicated forces which exist in the neighbourhood of the
primary oscillation. Our figure (the scale in which indicates the distances from
the wall) shows at, I., II., III., IV., the secondary circle in those positions in
which the sparks were most strongly developed. The alternating character of the
conditions of the space is clearly exhibited.

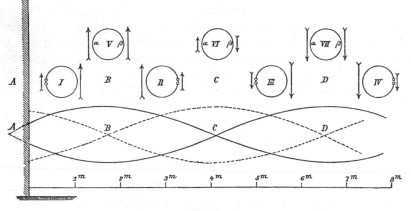

Fig. 26.

At distances lying between those mentioned both sets of sparks under consideration were of equal strength, and in the immediate neighbourhood of the wall too the distinction between them diminishes. We may therefore denote these points—namely, the points A, B, C, D in the figure—as being nodal points in a certain sense. Still we must not consider the distance between any one of these points and the next as being the half wave-length. For if *all* the electrical disturbances change their direction in passing through one of these points, then the phenomena in the secondary circle should repeat themselves without reversal; for in the spark-length there is nothing which corresponds to a change of direction in the oscillation. We should rather conclude from these experiments that in passing through any one of these points one part of the action undergoes reversal, while another part does not. On the other hand, it is allowable to assume that double the distance between any two of the points corresponds to the half wave-length, so that these points each indicate the end of a quarter wave-length. And, indeed, on the basis of this assumption and of the fundamental view just expressed, we shall arrive at a complete explanation of the phenomenon.

For let us suppose that a vertical wave of electric force proceeds towards the wall, is reflected with slightly diminished intensity, and so gives rise to stationary waves. If the wall were a perfect conductor a node would form at its very surface. For inside a conductor or at its boundary the electric force must always be vanishingly small. Now our wall cannot be regarded as a perfect conductor. For, in the first place, it is only metallic in part, and the part which is metallic is not very extensive. Hence at its surface the force will still have a certain value, and this in the sense of the advancing wave. The node, which would be formed at the wall itself if it were perfectly conducting, must therefore lie really somewhat behind the surface of the wall, say at the point A in the figure. If double the distance A B, that is the distance A C, corresponds to the half wave-length, then the geometrical relations of the stationary wave are of the kind which are represented in the usual symbolic fashion by the continuous wave-line in the figure. The forces acting on both sides of the circle in the positions I, II, III, and IV are correctly represented for any given instant in magnitude and direction by the arrows at the sides. If, then, in the neigbourhood of a node the spark-gap is turned towards the node, we have in the circle a stronger force acting under favourable conditions against a weaker force, which acts under unfavourable conditions. But if the spark-gap is turned away from the node, the stronger force now acts under unfavourable conditions against a weaker force, which in this case is acting under favourable conditions. And whether in this latter case the one preponderates or the other, the sparks must necessarily be weaker than in the former case. Thus the change of sign of our phenomenon every quarter wave-length is explained.

Our explanation carries with it a means of further testing its correctness. If it is correct, then the change of sign, at the points B and D should occur in a manner quite different from the change of sign at C. At *V*, *VI*, and *VII* in the figure the circle and the acting forces in these positions are represented, and it is easily seen that if at *B* or *D* we transfer the spark-gap from the one position to the other by rotating the circle within itself, the oscillation changes its direction relatively to a fixed direction within the circle; during this rotation the sparks must therefore become zero either once or an uneven number of times. On the other hand, if the same operation is performed at C, the direction of the oscillation does not change; and therefore the sparks must either not disappear at all, or else they must disappear an even number of times. Now when we actually make the experiment, what we observe is this:—At B the intensity of the sparks diminishes as soon as we remove the spark-gap from a, becomes zero at the highest point, and again increases to its original value when we come to B. Similarly at D. At C on the other hand, the sparks persist without change during the rotation, or, if anything, are somewhat stronger at the highest and lowest points than at those which we have been considering. Furthermore, it strikes the observer that the change of sign ensues after a much smaller displacement at C than at B and D, so that in this respect also there is a contrast between the change at C and that at B and D.

The representation of the electric wave which we have thus sketched can be verified in yet another way, and a very direct one. Instead of placing the plane of our circle in the plane of oscillation, let us place it in the wave-plane; the electric force is now equally strong at all parts of the circle, and for similar positions of the sparks their intensity is simply proportional to this electric force. As might be expected, the sparks are now zero at the highest and lowest points of the circle at all distances, and are strongest at the points along the normal in a horizontal plane. Let us then bring the spark-gap into one of these latter positions, and move slowly away from the wall. This is what we observe:—Just at the conducting metallic surface there are no sparks, but they make their appearance at a very small distance from it; they increase rapidly, are comparatively strong at B, and then again diminish. At C again they are exceedingly feeble, but become stronger as we proceed further. They do not, however again diminish, but continue to increase in strength, because we are now approaching the primary oscillation. If we were to illustrate the strength of the sparks along the interval A D by a curve carrying positive and negative signs, we should obtain almost exactly the curve which has been sketched. And perhaps it would have been better to start from this experiment. But it is not really so striking as the first one described; and furthermore, a periodic change of sign seems to be a clearer proof of wave-motion than a periodic waxing and waning.

We are now quite certain that we have recognised nodes of the electric wave at A and C, and antinodes at B and D. We might, however, in another sense call B and D nodes, for these points are nodes of a stationary wave of magnetic force, which, according to theory, accompanies the electric wave and is displaced a quarter wave-length relatively to it. This statement can be illustrated experimentally as follows:—We again place our circle in the plane of oscillation, but now bring the spark-gap to the highest point. In this position the electric force, if it were homogeneous over the whole extent of the secondary circle, could induce no sparks. It only produces an effect in so far as its magnitude varies in various parts of the circle, and its integral taken around the circle is not zero. This integral is proportional to the number of lines of magnetic force which flow backwards and forwards through the circle. In this sense, we may say that in this position, the sparks measure the magnetic force, which is perpendicular to the plane of the circle. But now we find that in this position near the wall there is vigorous sparking which rapidly diminishes, disappears at B, increases again up to C, then again decreases to a marked minimum at D, after which it continuously increases as we approach the primary oscillation. Representing the strength of these sparks as ordinates with positive and negative signs, we obtain approximately the dotted line of our figure, which thus represents correctly the magnetic wave. The phenomenon which we first described can also be explained as resulting from the co-operation of the electric and the magnetic force. The former changes sign at the points A and C, the latter at the points B and D; thus one part of the action changes sign at each of these points while the other retains its sign; hence the resulting action (as the product) changes sign at each of the points. Clearly this explanation only differs in mode of expression, and not in meaning, from the one first given. . . .

I have described the present set of experiments, as also the first set on the propagation of induction, without paying special regard to any particular theory; and, indeed the demonstrative power of the experiments is independent of any particular theory. Nevertheless, it is clear that the experiments amount to so many reasons in favour of that theory of electromagnetic phenomena which was first developed by Maxwell from Faraday's views. It also appears to me that the hypothesis as to the nature of light which is connected with that theory now forces itself upon the mind with still stronger reason than heretofore. Certainly it is a fascinating idea that the processes in air which we have been investigating represent to us on a million-fold larger scale the same processes which go on in the neighbourhood of a Fresnel mirror or between the glass plates used for exhibiting Newton's rings.

That Maxwell's theory, in spite of all internal evidence of probability, cannot dispense with such confirmation as it has already received, and may yet receive, is proved—if indeed proof be needed—by the fact that electric action is not propagated along wires of good conductivity with approximately the same velocity as through air. Hitherto it has been inferred from all theories, Maxwell's included, that the velocity along wires should be the same as that of light. I hope in time to be able to investigate and report upon the causes of this conflict between theory and experiment.

SELECTED BIBLIOGRAPHY

Buchwald, J. *From Maxwell to Microphysics*. Chicago: University of Chicago Press, 1985.

———. *The Rise of the Wave Theory of Light*. Chicago: University of Chicago Press, 1989.

Harman, P. *Energy, Force and Matter: The Conceptual Development of Nineteenth Century Physics*. Cambridge: Cambridge University Press, 1982.

Jackson, J. D. *Classical Electrodynamics*. 2nd edition. New York: John Wiley and Sons, 1975.

Meyer, H. W. *History of Electricity and Magnetism*. Cambridge: MIT Press, 1971.

Meyer-Arendt, J. R. *Introduction to Classical and Modern Optics*. Englewood Cliffs, N.J.: Prentice-Hall, 1972.

Reitz, J. R., and F. J. Milford. *Foundations of Electromagnetic Theory*. New York: Addison-Wesley, 1967.

Roller, D. E., and D.H. Roller. *Development of the Concept of Electric Charge: Electricity from the Greeks to Coulomb*. Cambridge: Harvard University Press, 1954.

Whitaker, E. *A History of the Theories of Aether and Electricity*. 2 vols. New York: Humanities Press, 1973.

Young, T. *A Course of Lectures on Natural Philosophy*. London: Taylor and Walton, 1845.

PART III

THE THEORY OF RELATIVITY

//

INTRODUCTION

////////////////////////////////////

There are, of course, two theories of relativity, the special and the general. The Special Theory of Relativity states that the laws of nature are the same in every Galilean or inertial reference frame and that the velocity of light is the same in every such reference frame. This means that the structure of space and the flow of time are relative to an observer measuring them.

The General Theory of Relativity states that the laws of nature are the same in any reference frame. This leads to a theory of gravity in which gravity is related to the structure of the space-time continuum, and that structure is related to the distribution of mass and energy in the continuum. Both theories were formulated by the physicist Albert Einstein, the Special Theory in 1905 and the General Theory between 1911 and 1915.

Newton's Universe

The basic concepts of the theories of relativity are definitely counterintuitive. Intuitively we tend to view space, time, and motion as Newton outlined in his science of mechanics. But curiously, Newton's concepts were felt to be counterintuitive to many in the seventeenth century who still viewed the physical universe as Aristotle had in the fourth century B.C. For Aristotle there were natural states for bodies that required no force or cause to be maintained. These states were:

1. The up and down motions of light and heavy objects.
2. The states of rest for those bodies in their natural places.
3. The circular motions of the more perfect heavens.

Any other state of motion, for example, projectile motion, required a cause for its duration. Aristotle used the center of the earth as the reference point or reference frame for all motion. In his cosmology the earth was spherical, immobile, and geometrically in the center of a spherical universe.

Because of the Copernican Revolution, Newton came to view the universe as indefinitely if not infinitely large. The theoretical breaking of Aristotle's celestial sphere by the work of Brahe and Galileo suggested a space for the universe that had no boundary and no center. Using this concept of an unchanging, permanent space as the arena for the activities of the matter of the universe, Newton decided to begin his science of mechanics by suggesting that there were the following two natural states for material objects:

1. Rest.
2. Uniform rectilinear motion (motion in a straight line at a constant rate of speed).

No force or cause was needed to maintain a natural state. A force was needed to change a natural state. From these natural states of motion Newton went on to develop the axioms of his science of mechanics, which have come to be known as his laws of motion:

1. Every body persists in its state of rest or uniform rectilinear motion unless acted upon by an unbalanced external force. This is called the Law of Inertia.
2. An unbalanced force produces a change in a natural state, that is, it produces an acceleration.
3. For every action there is an equal and opposite reaction.

It should be noted that for Newton motion meant velocity and velocity was a vector quantity. Thus a change in velocity (an acceleration) could be a change in speed, or a change in direction, or both.

Newton realized that making theoretical sense of these laws required certain metaphysical assumptions about space and time. To that end, he began the *Principia* with definitions of what he called absolute space and absolute time. Absolute space was said to be

a. distinct and separate from matter;
b. always and everywhere similar, that is, structurally undifferentiated and homogeneous;
c. permanently immobile and thus an absolute reference frame for motion;
d. causally inert, such that space did not affect bodies.

Absolute time was said to be

a. imperceptible, flowing without relation to anything external and thus distinct and separable from motion, with its own motion, such that bodies endure in time whether moving or not;
b. isochronous, flowing equably, evenly, uniformly, with one part of time equal to the next;
c. synchronous, everywhere the same, thus two events would either be simultaneous or not in an absolute sense.

Because absolute space was homogenous, it had to be Euclidean in nature (parallel lines never met) and infinite in extent. Because absolute space and time were homogeneous and causally inert, the laws of mechanics were thought to be universal.

As theoretically valuable as these concepts were, and as commonsensible and intuitive as they seemed to be, there was an overriding problem. They were unobservable or unmeasurable in-principle realities. The best that Newton could do was to offer a thought experiment that supposedly showed there was such a thing as real motion. By inference, then, absolute space and time must be real. The thought experiment involved a pail half-filled with water, hanging by a rope in empty space. If the rope were twisted and started to unravel, the pail would be moving with respect to the water but the water level would be flat. When the rope became completely unraveled and twisted in the opposite direction, the pail would come to rest but the water would still be moving within the pail. In both cases there would be relative motion between the pail and the water, but in the second case, Newton reasoned, the water should move up the sides of the pail in whirlpool fashion. This dynamical effect would indicate that the water must *really* be moving in the second case. Real motion would mean an objective reference frame, which must be an absolute space.

Although there were theoretical difficulties with this thought experiment it was sufficient, along with the success of the new science of mechanics, to guarantee the acceptance of the concepts of absolute space and time until the twentieth century.

The Problem of Reference Frames

Since absolute space was not detectable it could not be used to measure motions. A reference frame was required, but its choice seemed to be arbitrary. Newton, as Galileo before him, noticed that reference frames were not indistinguishable one from another. If a point on the earth's surface were chosen as a reference frame, the only force that appeared was gravity. But if a point on a rotating disk were chosen, strange motions would appear, requiring strange forces. The frames in which only the force of gravity appeared were called Galilean or inertial reference frames. There were, theoretically, an infinite number of such frames (any reference frame moving at a constant rate of speed relative to the earth's surface would be Galilean). Mechanical experiments done in such frames showed that free bodies moved in straight lines at constant speeds, abiding by Newton's first law of motion. In fact, mechanical phenomena when referred to these frames could be organized into simple laws.

Non-Galilean reference frames when viewed from Galilean frames were found to be accelerating. Mechanical laws in non-Galilean frames were hopelessly complex because free bodies would describe capricious curves at varying velocities. Further, Newton's law of gravity would not be able to account for the motion of falling bodies or the motions of the planets if these motions were viewed from non-Galilean frames. For these reasons classical science selected Galilean frames as standard and referred all the laws of physics to them. In fact, Galilean or inertial reference frames came to be defined as ones in which the laws of classical mechanics (Newton's three laws of motion and the law of universal gravitation) hold true.

Because the Copernican revolution had resulted in the view that the earth was in motion in a somewhat circular orbit about the sun, classical science decided that a point on the earth's surface would only approximate a Galilean reference frame. True Galilean frames must then be ones with respect to which the stars appear to be fixed.

In objectivizing space Newton assumed that Galilean reference frames are either at rest in absolute space or moving at a constant velocity with respect to absolute space. Non-Galilean frames would be accelerating with respect to Galilean frames and thus were thought to be accelerating with respect to absolute space. Because the laws of mechanics are the same in every Galilean reference frame, classical science developed the following principle of relativity: *It is impossible for an observer situated in a Galilean reference frame to ascertain by any mechanical experiment whether he is at rest in space or in a state of uniform rectilinear motion through space.* Assuming that the earth is moving through absolute space relatively uniformly, there apparently would be no way of detecting that motion by a mechanical experiment.

Maxwell, Electromagnetism, and the Ether

As we have seen, the nineteenth century saw the development and ultimately the unification of the sciences of electricity, magnetism, and optics. The discovery of the interference of light beams by Thomas Young seemed to confirm the wave theory of light, and if light is a wave there must be a medium to carry it. Although further investigation showed that this medium, called the ether, must possess apparently incompatible mechanical properties, its existence had to be maintained if the wave theory was to be retained.

When Maxwell, in 1865, showed that the velocity of a disturbance in what he called the electromagnetic field, determined theoretically, was essentially the same as the velocity of light in air, measured experimentally, he naturally concluded that light must be an electromagnetic wave. This was the final step in the unification of the sciences of electricity, magnetism, and optics. From his set of equations he was then able to derive all the known laws of those sciences. But Maxwell's equations indicated that electromagnetic waves will move at one specific speed with respect to an observer in a vacuum regardless of how fast the source of the wave is moving. If we assume that an ether must be present in that vacuum in order for there to be waves at all, this means that Maxwell's equations hold for a privileged reference frame that is at rest with respect to the ether. The result is that an optical or electromagnetic experiment should reveal motion through the ether, and since the ether was assumed to be at rest in absolute space such an experiment should reveal absolute motion.

The Michelson-Morley Experiment

One of the first people to attempt an optical experiment designed to reveal motion through the ether was the American experimenter Albert Michelson. In

1881 Michelson built an instrument called an interferometer that was designed to detect and measure the earth's motion through the ether.

In setting up the experiment Michelson assumed that there was a stationary ether that was at rest with respect to the fixed stars and that carried light waves. He also assumed that since the earth was moving with respect to the fixed stars it must be moving through the ether. From the reference frame of the earth there should then be an ether wind blowing past the earth. Since atmospheric wind adds or subtracts from the velocity of sound waves, Michelson reasoned that the ether wind should have the same effect on light waves.

With these assumptions in mind Michelson set up the interferometer to split a beam of monochromatic light into two beams, sending them in perpendicular directions. One beam was sent parallel to the earth's motion through the ether, while the other beam was sent transverse to the earth's motion. The beams were sent through equivalent optical distances and then brought together. Because of the influence of the ether wind, Michelson reasoned, the beams should be out of phase when reunited. The parallel beam should be about 1/25 of a wavelength in advance of the transverse beam. Thus the two beams should interfere with one another and the interference pattern produced would amount to a detection of the ether wind, or conversely the earth's motion through the ether. In fact the pattern should allow for a measurement of the earth's velocity through the ether. But when the experiment was done no interference pattern appeared.

Perhaps because of its extraordinary result very little attention was paid to the experiment. It was assumed that some flaw would be detected in the instrument or that the results would be found to be improperly interpreted. Convinced that his experiment had been sound, Michelson decided to repeat it. In 1887, in Cleveland, he and Edward Morley built a new and more refined interferometer. This time the experiment was performed at a variety of venues and at different times during the year. But the results were always the same—no ether wind was detected. This time, however, the results became accepted and the experiment of 1887 became known as the Michelson-Morley experiment.

Fitzgerald and Lorentz

Once the results of the Michelson-Morley experiment were accepted, attempts were made to explain them within the framework of the ether hypothesis. Two such explanations were quickly rejected because they contradicted well-documented phenomena. The first, the suggestion that the velocity of light is determined by the velocity of the source, is at odds with the observed behavior of binary star systems. The second, the hypothesis that the earth drags the ether along with it as it hurtles through space, is contradicted by the observation of stellar aberration, that is, telescopes must be tilted in the direction of the motion of the earth to make stellar observations. If the ether were dragged along by the earth, there would be no need to tilt telescopes at all.

In 1889 the Irish physicist G. F. Fitzgerald offered the curious suggestion that perhaps material bodies physically contract as they push through the ether. Thus the interferometer arm in the direction of the motion of the earth through the ether would contract just the amount necessary to guarantee that the two beams of light remain in phase. The suggestion was curious for two reasons: the ether seemed much too subtle a medium to physically compress solid metal, and the contraction had to be just the extent necessary to guarantee that the beams remain in phase.

Three years later H. A. Lorentz arrived independently at the same contraction hypothesis. He concluded from the Michelson-Morley experiment that any observer in any Galilean reference frame will measure the velocity of light to be the same value. From this conclusion he wrote a new set of transformation equations relating the description of events given by two observers in Galilean reference frames moving relative to one another. For slow velocities these transformation equations reduce to the classical transformation equations. But at great velocities two extraordinary phenomena occur, length contraction and time dilation. According to the transformation equations, the amount of length contraction and time dilation was just sufficient to produce the null result of the Michelson-Morley experiment.

But the Lorentz hypothesis had two major problems. Although Lorentz felt that length contraction was the result of a physical compression by the all-present ether, he had no physical explanation for the dilation of time. Further, corroborative experimental results for length contraction did not exist. When optical bodies contract, double refraction should result, and when metal contracts, its electrical conductivity should change. These results are observed when said bodies are cooled (and thus contract). But double refraction and electrical conductivity changes are not observed because of motion through the ether.

In addition to these experimental problems, this Lorentz-Fitzgerald contraction hypothesis had a major theoretical difficulty. The transformation equations stated that all bodies would contract equally as they move through the ether. But if this contraction were actually the result of a physical compression, it was difficult to imagine that the compression would have the exact same effect on all bodies regardless of their hardness or softness.

Despite these problems, the notion that the earth moves through a stagnant ether with a definite velocity was generally accepted by the end of the nineteenth century. This velocity is eternally unknowable because of precise adjustments in the lengths of bodies and the passage of time made by nature. There remains one privileged reference frame: the frame at rest in the stagnant ether.

But it was clear that this view was artificial and ad hoc. As the twentieth century began, the very foundations of science were in question and a radical change was necessary.

Special Relativity

It was to this situation that a young German physicist named Albert Einstein responded by initiating one of the greatest conceptual revolutions in human history. To Einstein the problems of electrodynamics had arisen because of the retention of the stagnant ether as an absolute reference frame. The negative results of experiments like Michelson-Morley proved that velocity through the stagnant ether or through absolute space was a physically meaningless idea. This meant that not only would mechanical experiments fail to reveal absolute motion but so would experiments of any kind, including optical and electromagnetic.

Einstein's insight, simply put, was that when it comes to motion the relative is the real—the motion of a material object is only meaningful relative to some other material object. He was led to this conclusion by Ernst Mach's critique of Newton's water pail thought experiment and by the results of certain electrodynamical experiments. Mach had maintained that it was the relative motion of the water with respect to the "fixed" stars and galaxies that would cause the water to move up the sides of the pail and not its motion with respect to absolute space. This view of Mach, that only relative motion is meaningful, seemed to be supported by simple electrodynamical experiments. According to classical electromagnetism, moving charges produce magnetic fields and charges moving through magnetic fields experience magnetic forces. The relative motion of two charged bodies reveal this magnetic force. But if the earth is moving through the ether then every charge on the earth should be producing a magnetic field. Experiments do not detect this magnetic field. The only type of velocity that seemed to have any significance was relative velocity.

Einstein applied this insight to the problems of electrodynamics and in 1905 published his epochal paper entitled "On the Electrodynamics of Moving Bodies." In this paper he draws two consequences from his insight. Since the relative is the real and there is no meaning to absolute motion, the laws of nature that had been formulated for a Galilean reference frame must be the same in every Galilean reference frame. There can be no privileged observer in a privileged reference frame. Further, since the laws of electromagnetism must be the same in every Galilean reference frame, the velocity of light that is a consequence of those laws must be the same in every Galilean reference frame. That is, all observers in Galilean reference frames should measure the velocity of a beam of light to be of the same value regardless of their state of motion relative to one another. These are the two postulates of what is now called the Special Theory of Relativity.

The Lorentz transformation equations flow naturally from the postulates of the Special Theory but their interpretation is quite different from the one Lorentz gave them. For Lorentz material bodies physically contracted as they moved through the ether. For Einstein the contraction as well as time dilation are due to a modification of our space and time measurements because of relative motion. That is, the structure of space and the passage of time are relative to the state of motion of the observer. This implies that length is not something inherent in a body but is rather a relationship between observer and observed. Similarly, the rate of time flow, although appearing to be unchanging for each observer, will nevertheless be different for another observer moving with respect to the first. This latter implication leads to a phenomenon called the relative nature of simultaneity, such that there is no absolute sequence of events. Given two events A and B under certain circumstances, one observer may observe A preceding B, another may observe B preceding A, and a third may see them occurring simultaneously, and each is correct. Each is as "objective" as the other.

It should be noted that space contraction and time dilation only become significant phenomena at very high relative velocities according to the Lorentz transformations. For ordinary velocities the relativistic effects are not observable.

Consequences of Special Relativity for Mechanics

As mentioned, the postulates of Special Relativity required the use of the Lorentz transformation equations when moving from one Galilean reference frame to the next, at least in discussing electrodynamical phenomena, the laws of electrodynamics being invariant to Lorentz transformation equations. However, the laws of classical mechanics were not invariant to such transformations. Einstein concluded from this that the laws of classical mechanics must be incorrect. He then set himself to the task of formulating new laws of mechanics that would be invariant to the Lorentz transformations but that would reduce to the classical laws at "ordinary" relative velocities.

Einstein was able to formulate these laws rather easily and quickly found that they had extraordinary consequences. The first such consequence he discovered was that mass, instead of being a permanent and changeless feature of matter, was a function of relative motion. It was found to increase with an increase in relative velocity. In fact, as the velocity of a body approached the speed of light, its mass approached infinity. Thus the velocity of light became the uppermost velocity in the universe. It would require an infinite force to accelerate a body to the speed of light or beyond.

A second important consequence of the revised laws of mechanics was the identification of mass with energy. This occurred because both mass and energy were now functions of motion, whereas in classical mechanics only energy was. Rest mass came to be seen as a kind of potential or bound energy and, theoretically speaking, the conversion of mass into energy became possible. Einstein was able to derive an equation that expresses the convertibility of mass into energy, an equation that was destined to become the symbol of Relativity Theory: $E = mc^2$.

The invariant velocity of light also led to the development of the concept of the space-time continuum. Because the velocity of light is the same in every Galilean reference frame, there is a separation of any two point events that will have the same value in every Galilean reference frame. This separation, called the Einsteinian interval, is thus an absolute quantity that underlies the relativity of space and time when considered separately. The physicist Herman Minkowski recognized the Einsteinian interval as the expression of the square of a distance in a four-dimensional continuum. The four-dimensional continuum in this case was neither space nor time but a combination of the two. Thus the space-time continuum was a continuum of point events. Bodies that exist for prolonged periods of time would be represented by a continuous sequence of point events, a world line in the space-time continuum.

General Relativity

By 1915 Einstein had decided that, as promising as Special Relativity was, its postulates were too restrictive. Through a combination of insight and intuition he concluded that the laws of physics must be the same in any reference frame whether Galilean or not. As extraordinary as this conclusion was, this General Principle of Covariance, as it came to be called, seemed to follow naturally from the view that there are no privileged observers or privileged reference frames.

Part of Einstein's insight was the realization that the effects produced by gravity are the same as those produced by an acceleration. Under the influence of the gravitational field of the earth, a freely falling body will experience a force that will cause it to accelerate downward at a rate of 32 ft/sec^2. If the same body is placed in an elevator in interstellar space and the elevator is accelerated upward at a rate of 32 ft/sec^2, the body will experience the same "force" with the same result. In the elevator all bodies would "fall" to the floor at the same rate regardless of their mass, just as they do under the influence of gravity. This equivalence between gravitational forces and inertial forces was supported by experiments performed in 1889 by Roland von Eötvös, showing that gravitational mass and inertial mass were essentially equal. Einstein pointed to those experiments as having had a great influence on him.

What followed from the Principle of Equivalence was quite striking. Because accelerations have an effect on space and time measurements and because there is an equivalence of gravitation and acceleration, gravitation should also influence space and time measurements, i.e. the presence of mass in space should influence the structure of space and the passage of time in its vicinity. If this is the case, Einstein concluded, then non-Euclidean geometries were more appropriate in describing the geometrical features of the universe. He then proceeded to develop a set of equations that allowed him to predict the curvature of space on the basis of the configuration of masses in it, and from this how objects would move in such curved spaces. With these equations, which formed the foundation of the General Theory of Relativity, Einstein had essentially developed a new law of universal gravitation. In most cases Einstein's predictions of how objects would move in space were the same as the predictions that followed from Newton's mechanics. However, in some cases the predictions were quite different and Einstein's were more accurate.

Two successes, following quickly upon the publication of General Relativity, guaranteed it a prominent place in physics. The first success concerned a phenomenon associated with the planet Mercury. The path of Mercury's orbit did not remain stationary in space but precessed about the sun. Newtonian gravity could not account for this phenomenon. In 1916 Einstein was able to show that the precession was predicted by the equations of General Relativity.

The second success concerned an extraordinary and striking prediction involving the bending of light rays. General Relativity suggested that large masses would curve space in their vicinity significantly, thus causing light rays to change direction. Since the sun was a large mass it followed that light rays from stars that passed in its vicinity would be significantly bent. In 1919 a team of scientists that included Arthur Eddington did an experiment during a total eclipse of the sun that demonstrated clearly that light rays from a star had been bent as they passed the sun. The magnitude of the bending coincided with that predicted by the equations of General Relativity.

The Final Word

As counterintuitive and as noncommonsensical as the concepts of Relativity appear to be, the theory has survived all experimental tests. Much of this testing has occurred in the last thirty-five years, thanks to the development of new measuring tools that have been provided by the technological revolution. Among the confirmations of General Relativity have been the detection and measurement of the gravitational red shift of light (the Pound-Rebka Experiment), the detection of the lens effect of galaxies (the curving of space-time by galaxies), and the experiments of R. H. Dicke and his colleagues that corroborated, with much greater accuracy, the experiments of Eötvös involving the equivalence of inertial mass and gravitational mass.

But perhaps the most striking and impressive confirmations of General Relativity have been the detection of the expansion of the universe and the indirect determination of the existence of black holes. Because the distortion of space-time affects the motion of masses, which in turn produces further distortions, the equations of General Relativity predict that the size of the universe must be either expanding or contracting. It cannot be static. Einstein was surprised and disturbed by this result. To maintain a static universe, which he felt was the correct view, he introduced into his equations a cosmological constant. When Edwin Hubble later convinced Einstein that the universe was expanding, he called the cosmological constant the biggest blunder of his life. What is important here is that the original equations of General Relativity had predicted that the universe was either expanding or contracting.

The prediction of the theoretical possibility of black holes, based upon the equations of General Relativity, came from the work of German astronomer Karl Schwarzschild. In 1916, Schwarzschild applied General Relativity to the problem of how space warps and time dilates in the vicinity of a spherical star. He was able to show that if the mass of the star was concentrated in a small enough region, the resulting space-time warp will be so great that nothing will be able to escape from the star, not even light. Because light could not escape these extremely dense stars, they came to be known as black holes. In the latter part of the twentieth century a considerable amount of indirect evidence supported the existence of these black holes.

As the twenty-first century begins, it is safe to say that the Theory of Relativity has not only met every test successfully, but has also proven to be incredibly rich in the discovery of new phenomena.

1

— James Clerk Maxwell —

In this reading, Maxwell gives a description of various conceptions of the ether, especially that of Huygens' luminiferous ether as a medium for the propagation of light waves. He reaffirms that light must be a wave process in a medium rather than a substance and shows how the phenomenon of polarization fits in with the wave theory.

Maxwell goes on to identify several properties of the ether, such as elasticity, tenacity, and density, and uses those properties to distinguish the ether from what he calls gross matter. In anticipation of the Michelson-Morley experiment, Maxwell discusses experimental strategies for determining the relative motion of the ether. Finally, Maxwell reaffirms that light must be an electromagnetic phenomenon. This article was written in 1878 for the Encyclopedia Britannica.

THE ETHER

Ether or æther, . . . a material substance of a more subtle kind than visible bodies, supposed to exist in those parts of space which are apparently empty.

The hypothesis of an æther has been maintained by different speculators for very different reasons. To those who maintained the existence of a plenum as a philosophical principle, nature's abhorrence of a vacuum was a sufficient reason for imagining an all-surrounding æther, even though every other argument should be against it. To Descartes, who made extension the sole essential property of matter, and matter a necessary condition of extension, the bare existence of bodies apparently at a distance was a proof of the existence of a continuous medium between them.

But besides these high metaphysical necessities for a medium, there were more mundane uses to be fulfilled by æthers. Æthers were invented for the planets to swim in, to constitute electric atmospheres and magnetic effluvia, to convey sensations from one part of our bodies to another, and so on, till all space had been filled three or four times over with æthers. It is only when we remember the extensive and mischievous influence on science which hypotheses about æthers used formerly to exercise, that we can appreciate the horror of æthers which sober-minded men had during the eighteenth century, and which, probably as a sort of hereditary prejudice, descended even to the late Mr. John Stuart Mill.

The disciples of Newton maintained that in the fact of the mutual gravitation of the heavenly bodies, according to Newton's law, they had a complete quantitative account of their motions; and they endeavoured to follow out the path which Newton had opened up by investigating and measuring the attractions and repulsions of electrified and magnetic bodies, and the cohesive forces in the interior of bodies, without attempting to account for these forces.

Newton himself, however, endeavoured to account for gravitation by differences of pressure in an æther; but he did not publish his theory, "because he was not able from experiment and observation to give a satisfactory account of this medium, and the manner of its operation in producing the chief phenomena of nature."

On the other hand, those who imagined æthers in order to explain phenomena could not specify the nature of the motion of these media, and could not prove that the media, as imagined by them, would produce the effects they were meant to explain. The only æther which has survived is that which was invented by Huygens to explain the propagation of light. The evidence for the existence of the luminiferous æther has accumulated as additional phenomena of light and other radiations have been discovered; and the properties of this medium, as deduced from the phenomena of light, have been found to be precisely those required to explain electromagnetic phenomena.

Function of the æther in the propagation of radiation.—The evidence for the undulatory theory of light will be given in full, under the Article on *Light*, but we may here give a brief summary of it so far as it bears on the existence of the æther.

That light is not itself a substance may be proved from the phenomenon of interference. A beam of light from a single source is divided by certain optical methods into two parts, and these, after travelling by different paths, are made to reunite and fall upon a screen. If either half of the beam is stopped, the other falls on the screen and illuminates it, but if both are allowed to pass, the screen in certain places becomes dark, and thus shews that the two portions of light have destroyed each other.

Now, we cannot suppose that two bodies when put together can annihilate each other; therefore light cannot be a substance. What we have proved is that one portion of light can be the exact opposite of another portion, just as $+a$ is the exact opposite of $-a$, whatever a may be. Among physical quantities we find some which are capable of having their signs reversed, and others which are not. Thus a displacement in one direction is the exact opposite of an equal displacement in the opposite direction. Such quantities are the measures, not of substances, but always of processes taking place in a substance. We therefore conclude that light is not a substance but a process going on in a substance, the process going on in the first portion of light being always the exact opposite of the process going on in the other at the same instant, so that when the two portions are combined no process goes on at all. To determine the nature of the process in which the radiation of light consists, we alter the length of the path of one or both of the two portions of the beam, and we find that the light is extinguished when the difference of the length of the paths is an odd multiple of a certain small distance called a half wave-length. In all other cases there is more or less light; and when the paths are equal, or when their difference is a multiple of a whole wave-length, the screen appears four times as bright as when one portion of the beam falls on it. In the ordinary form of the experiment these different cases are exhibited simultaneously at different points of the screen, so that we see on the screen a set of fringes consisting of dark lines at equal intervals, with bright bands of graduated intensity between them.

If we consider what is going on at different points in the axis of a beam of light at the same instant, we shall find that if the distance between the points is a multiple of a wave-length the same process is going on at the two points at the same instant, but if the distance is an odd multiple of half a wave-length the process going on at one point is the exact opposite of the process going on at the other.

Now, light is known to be propagated with a certain velocity $(3 \cdot 004 \times 10^{10}$ centimetres per second in vacuum, according to Cornu). If, therefore, we suppose a movable point to travel along the ray with this velocity, we shall find the same process going on at every point of the ray as the moving point reaches it. If, lastly, we consider a fixed point in the axis of the beam, we shall observe a rapid alternation of these opposite processes, the interval of time between similar processes being the time light takes to travel a wave-length.

These phenomena may be summed up in the mathematical expression

$$u = A \ cos \ (nt - px + a)$$

which gives u, the phase of the process, at a point whose distance measured from a fixed point in the beam is x, and at a time t.

We have determined nothing as to the nature of the process. It may be a displacement, or a rotation, or an electrical disturbance, or indeed any physical quantity which is capable of assuming negative as well as positive values. Whatever be the nature of the process, if it is capable of being expressed by an equation of this form, the process going on at a fixed point is called a *vibration*; the constant A is called the *amplitude*; the time $2\pi/n_p$ is called the *period*; and $nt - px + a$ is the *phase*.

The configuration at a given instant is called a *wave*, and the distance $2\pi/p$ is called the *wave-length*. The velocity of propagation is n/p. When we contemplate the different parts of the medium as going through the same process in succession, we use the word undulatory to denote this character of the process without in any way restricting its physical nature.

A further insight into the physical nature of the process is obtained from the fact that if the two rays are polarized, and if the plane of polarization of one of them be made to turn round the axis of the ray, then when the two planes of polarization are parallel the phenomena of interference appear as above described. As the plane turns round, the dark and light bands become less distinct, and when the planes of polarization are at right angles, the illumination of the screen becomes uniform, and no trace of interference can be discovered.

Hence the physical process involved in the propagation of light must not only be a directed quantity or vector capable of having its direction reversed, but this vector must be at right angles to the ray, and either in the plane of polarization or perpendicular to it. Fresnel supposed it to be a displacement of the medium perpendicular to the plane of polarization. Maccullagh and Neumann supposed it to be a displacement in the plane of polarization. The comparison of these two theories must be deferred till we come to the phenomena of dense media.

The process may, however, be an electromagnetic one, and as in this case the electric displacement and the magnetic disturbance are perpendicular to each other, either of these may be supposed to be in the plane of polarization.

All that has been said with respect to the radiations which affect our eyes, and which we call light, applies also to those radiations which do not produce a luminous impression on our eyes, for the phenomena of interference have been observed, and the wave-lengths measured, in the case of radiations, which can be detected only by their heating or by their chemical effects. . . .

The æther distinct from gross matter.—When light travels through the atmosphere it is manifest that the medium through which the light is propagated is not the air itself, for in the first place the air cannot transmit transverse vibrations, and the normal vibrations which the air does transmit travel about a million times slower than light. Solid transparent bodies, such as glass and crystals, are no doubt capable of transmitting transverse vibrations, but the velocity of transmission is still hundreds of thousands times less than that with which light is transmitted through these bodies. We are therefore obliged to suppose that the medium through which light is propagated is something distinct from the transparent medium known to us, though it interpenetrates all transparent bodies and probably opaque bodies too.

The velocity of light, however, is different in different transparent media, and we must therefore suppose that these media take some part in the process, and that their particles are vibrating as well as those of the æther, but the energy of the vibrations of the gross particles must be very much smaller than that of the æther, for otherwise a much larger proportion of the incident light would be reflected when a ray passes from vacuum to glass or from glass to vacuum than we find to be the case.

Relative motion of the æther.—We must therefore consider the æther within dense bodies as somewhat loosely connected with the dense bodies, and we have next to inquire whether, when these dense bodies are in motion through the great ocean of æther, they carry along with them the æther they contain, or whether the æther passes through them as the water of the sea passes through the meshes of a net when it is towed along by a boat. If it were possible to determine the velocity of light by observing the time it takes to travel between one station and another on the earth's surface, we might, by comparing the observed velocities in opposite directions, determine the velocity of the æther with respect to these terrestrial stations. All methods, however, by which it is practicable to determine the velocity of light from terrestrial experiments depend on the measurement of the time required for the double journey from one station to the other and back again, and the increase of this time on account of a relative velocity of the æther equal to that of the earth in its orbit would be only about one hundred millionth part of the whole time of transmission, and would therefore be quite insensible.

The theory of the motion of the æther is hardly sufficiently developed to enable us to form a strict mathematical theory of the aberration of light, taking into account the motion of the æther. Professor Stokes, however, has shewn that, on a very probable hypothesis with respect to the motion of the æther, the amount of aberration would not be sensibly affected by that motion.

The only practicable method of determining directly the relative velocity of the æther with respect to the solar system is to compare the values of the velocity of light deduced from the observation of the eclipses of Jupiter's satellites when Jupiter is seen from the earth at nearly opposite points of the ecliptic.

Arago proposed to compare the deviation produced in the light of a star after passing through an achromatic prism when the direction of the ray within the prism formed different angles with the direction of motion of the earth in its orbit. If the æther were moving swiftly through the prism, the deviation might be expected to be different when the direction of the light was the same as that of the æther, and when these directions were opposite.

The present writer arranged the experiment in a more practicable manner by using an ordinary spectroscope, in which a plane mirror was substituted for the slit of the collimator. The cross wires of the observing telescope were illuminated. The light from any point of the wire passed through the object-glass and then through the prisms as a parallel pencil till it fell on the object-glass of the collimator, and came to a focus at the mirror, where it was reflected, and after passing again through the object-glass it formed a pencil passing through each of the prisms parallel to its original direction, so that the object-glass of the observing telescope brought it to a focus coinciding with the point of the cross wires from which it originally proceeded. Since the image coincided with the object, it could not be observed directly, but by diverting the pencil by partial reflection at a plane surface of glass, it was found that the image of the finest spider line could be distinctly seen, though the light which formed the image had passed twice through three prisms of 60°. The apparatus was first turned so that the direction of the light in first passing through the second prism was that of the earth's motion in its orbit. The apparatus was afterwards placed so that the direction of the light was opposite to that of the earth's motion. If the deviation of the ray by the prisms was increased or diminished for this reason in the first journey, it would be diminished or increased in the return journey, and the image would appear on one side of the object. When the apparatus was turned round it would appear on the other side. The experiment was tried at different times of the year, but only negative results were obtained. We cannot, however, conclude absolutely from this experiment that the æther near the surface of the earth is carried along with the earth in its orbit, for it has been shown by Professor Stokes that according to Fresnel's hypothesis the relative velocity of the æther within the prism would be to that of the æther outside inversely as the square of the index of refraction, and that in this case the deviation would not be sensibly altered on account of the motion of the prism through the æther.

Fizeau, however, by observing the change of the plane of polarization of light transmitted obliquely through a series of glass plates, obtained what he supposed to be evidence of a difference in the result when the direction of the ray in space was different, and Angström obtained analogous results by diffraction. The writer is not aware that either of these very difficult experiments has been verified by repetition.

In another experiment of M. Fizeau, which seems entitled to greater confidence, he has observed that the propagation of light in a stream of water takes place with greater velocity in the direction in which the water moves than in the opposite direction, but that the change of velocity is less than that which would be due to the actual velocity of the water, and that the phenomenon does not occur when air is substituted for water. This experiment seems rather to verify Fresnel's theory of the æther; but the whole question of the state of the luminiferous medium near the earth, and of its connexion with gross matter, is very far as yet from being settled by experiment.

Function of the æther in electromagnetic phenomena.—Faraday conjectured that the same medium which is concerned in the propagation of light might also be the agent in electromagnetic phenomena. "For my own part," he says, "considering the relation of a vacuum to the magnetic force, and the general character of magnetic phenomena external to the magnet, I am much more inclined to the notion that in the transmission of the force there is such an action, external to the magnet, than that the effects are merely attraction and repulsion at a distance. Such an action may be a function of the æther; for it is not unlikely that, if there be an æther, it should have other uses than simply the conveyance of radiation." This conjecture has only been strengthened by subsequent investigations.

Electrical energy is of two kinds, electrostatic and electrokinetic. We have reason to believe that the former depends on a property of the medium in virtue of which an electric displacement elicits an electromotive force in the opposite direction, the electromotive force for unit displacement being inversely as the specific induction capacity of the medium.

The electrokinetic energy, on the other hand, is simply the energy of the motion set up in the medium by electric currents and magnets, this motion not being confined to the wires which carry the currents, or to the magnet, but existing in every place where magnetic force can be found.

Electromagnetic Theory of Light.—The properties of the electromagnetic medium are therefore as far as we have gone similar to those of the luminiferous medium, but the best way to compare them is to determine the velocity with which an electromagnetic disturbance would be propagated through the medium. If this should be equal to the velocity of light, we would have strong reason to believe that the two media, occupying as they do the same space, are really identical. The data for making the calculation are furnished by the experiments made in order to compare the electromagnetic with the electrostatic system of units. The velocity of propagation of an electromagnetic disturbance in air, as calculated from different sets of data, does not differ more from the velocity of light in air, as determined by different observers, than the several calculated values of these quantities differ among each other.

If the velocity of propagation of an electromagnetic disturbance is equal to that of light in other transparent media, then in non-magnetic media the specific inductive capacity should be equal to the square of the index of refraction.

Boltzmann has found that this is very accurately true for the gases which he has examined. Liquids and solids exhibit a greater divergence from this relation, but we can hardly expect even an approximate verification when we have to compare the results of our sluggish electrical experiments with the alternations of light, which take place billions of times in a second.

The undulatory theory, in the form which treats the phenomena of light as the motion of an elastic solid, is still encumbered with several difficulties.

The first and most important of these is that the theory indicates the possibility of undulations consisting of vibrations normal to the surface of the wave. The only way of accounting for the fact that the optical phenomena which would arise from these waves do not take place is to assume that the æther is incompressible.

The next is that, whereas the phenomena of reflection are best explained on the hypothesis that the vibrations are perpendicular to the plane of polarization, those of double refraction require us to assume that the vibrations are in that plane.

The third is that, in order to account for the fact that in a double refracting crystal the velocity of rays in any principal plane and polarized in that plane is the same, we must assume certain highly artificial relations among the coefficients of elasticity.

The electromagnetic theory of light satisfies all these requirements by the single hypothesis that the electric displacement is perpendicular to the plane of polarization. No normal displacement can exist, and in doubly refracting crystals the specific dielectric capacity for each principal axis is assumed to be equal to the square of the index of refraction of a ray perpendicular to that axis, and polarized in a plane perpendicular to that axis. Boltzmann has found that these relations are approximately true in the case of crystallized sulphur, a body having three unequal axes. The specific dielectric capacity for these axes are respectively

4.773   3.970   3.811

and the squares of the indices of refraction

4.576   3.886   3.591

Physical constitution of the æther.—What is the ultimate constitution of the æther? is it molecular or continuous?

We know that the æther transmits transverse vibrations to very great distances without sensible loss of energy by dissipation. A molecular medium, moving under such conditions that a group of molecules once near together remain near each other during the whole motion, may be capable of transmitting vibrations without much dissipation of energy, but if the motion is such that the groups of molecules are not merely slightly altered in configuration but entirely broken up, so that their component molecules pass into new types of grouping, then in the passage from one type of grouping to another the energy of regular vibrations will be frittered away into that of the irregular agitation which we call heat.

We cannot therefore suppose the constitution of the æther to be like that of a gas, in which the molecules are always in a state of irregular agitation, for in such a medium a transverse undulation is reduced to less than one five-hundredth of its amplitude in a single wave-length. If the æther is molecular, the grouping of the molecules must remain of the same type, the configuration of the groups being only slightly altered during the motion.

Mr. S. Tolver Preston has supposed that the æther is like a gas whose molecules very rarely interfere with each other, so that their mean path is far greater than any planetary distances. He has not investigated the properties of such a medium with any degree of completeness, but it is easy to see that we might form a theory in which the molecules *never* interfere with each other's motion of translation, but travel in all directions with the velocity of light; and if we further suppose that vibrating bodies have the power of impressing on these molecules some vector property (such as rotation about an axis) which does not interfere with their motion of translation, and which is then carried along by the molecules, and if the alternation of the average value of this vector for all the molecules within an element of volume be the process which we call light, then the equations which express this average will be of the same form as that which expresses the displacement in the ordinary theory.

It is often asserted that the mere fact that a medium is elastic or compressible is a proof that the medium is not continuous, but is composed of separate parts having void spaces between them. But there is nothing inconsistent with experience in supposing elasticity or compressibility to be properties of every portion, however small, into which the medium can be conceived to be divided, in which case the medium would be strictly continuous. A medium, however, though homogeneous, and continuous as regards its density, may be rendered heterogeneous by its motion, as in Sir W. Thomson's hypothesis of vortex, molecules in a perfect liquid.

The æther, if it is the medium of electromagnetic phenomena, is probably molecular, at least in this sense.

Sir W. Thomson has shewn that the magnetic influence on light discovered by Faraday depends on the direction of motion of moving particles, and that it indicates a rotational motion in the medium when magnetized. . . .

Now, it is manifest that this rotation cannot be that of the medium as a whole about an axis, for the magnetic field may be of any breadth, and there is no evidence of any motion the velocity of which increases with the distance from a single fixed line in the field. If there is any motion of rotation, it must be a rotation of very small portions of the medium each about its own axis, so that the medium must be broken up into a number of molecular vortices.

We have as yet no data from which to determine the size or the number of these molecular vortices. We know, however, that the magnetic force in the region in the neighbourhood of a magnet is maintained as long as the steel retains its magnetization, and as we have no reason to believe that a steel magnet would lose all its magnetization by the mere lapse of time, we conclude that the molecular vortices do not require a continual expenditure of work in order to maintain their motion, and that therefore this motion does not necessarily involve dissipation of energy.

No theory of constitution of the æther has yet been invented which will account for such a system of molecular vortices being maintained for an indefinite time without their energy being gradually dissipated into that irregular agitation of the medium which, in ordinary media, is called heat.

Whatever difficulties we may have in forming a consistent idea of the constitution of the æther, there can be no doubt that the interplanetary and interstellar spaces are not empty, but are occupied by a material substance or body, which is certainly the largest, and probably the most uniform body of which we have any knowledge.

Whether this vast homogeneous expanse of isotropic matter is fitted not only to be a medium of physical interaction between distant bodies, and to fulfil other physical functions of which, perhaps, we have as yet no conception, but also, as the authors of the *Unseen Universe* seem to suggest, to constitute the material organism of beings exercising functions of life and mind as high or higher than ours are at present, is a question far transcending the limits of physical speculation.

2
— Albert Michelson —
(1852–1931)

Albert Michelson was an American physicist who designed and built an optical instrument called an interferometer that was capable of making extremely accurate measurements of optical phenomena. In 1881 he used his interferometer in an experiment designed to show the effect on light of the earth's motion through the ether. Unexpectedly, the experiment showed a null result, that the earth's motion through the ether seemed to have no effect on light. Because this result

represented a challenge to classical physics, Michelson, along with Edward Mor-
ley, repeated the experiment in 1887 and confirmed the null result. This experi-
ment, now known as the Michelson-Morley experiment, is one of the most famous
in the history of science.

In this reading, Michelson discusses the physical character that the luminifer-
ous ether must have to allow for the extremely high velocity of light waves. He
considers the question of whether the ether is an ordinary form of matter, a solid
or fluid, and whether it permeates all forms of matter.

Michelson then provides details of two of his famous experiments: his modifi-
cation of Fizeau's experiment, which attempted to detect an increase in the veloc-
ity of light due to the motion of a medium, and his interferometer experiment
with its negative result.

Michelson concludes with the supposition that the phenomena of the physical
universe are manifestations of the various modes of motion of an all-pervading
ethereal substance. The reading was taken from Michelson's Light Waves and
Their Uses, *published in 1903.*

———————

THE ETHER AND OPTICAL EXPERIMENTS

The velocity of light is so enormously greater than anything with which we are
accustomed to deal that the mind has some little difficulty in grasping it. A
bullet travels at the rate of approximately half a mile a second. Sound, in a steel
wire, travels at the rate of three miles a second. From this—if we agree to except
the velocities of the heavenly bodies—there is no intermediate step to the velocity of light, which is about 186,000 miles a second. We can, perhaps, give a
better idea of this velocity by saying that light will travel around the world seven
times between two ticks of a clock.

Now, the velocity of wave propagation can be seen, without the aid of any
mathematical analysis, to depend on the elasticity of the medium and its density;
for we can see that if a medium is highly elastic the disturbance would be
propagated at a great speed. Also, if the medium is dense the propagation would
be slower than if it were rare. It can easily be shown that if the elasticity were
represented by E, and the density by D, . . . the velocity of light is some 60,000
times as great as that of the propagation of sound in a steel wire, must be
60,000 squared times as great as the elasticity of steel. Thus, this medium which
propagates light vibrations would have to have an elasticity of the order of
3,600,000,000 times the elasticity of steel. Or, if the elasticity of the medium
were the same as that of steel, the density would have to be 3,600,000,000
times as small as that of steel, that is to say, roughly speaking, about 50,000
times as small as the density of hydrogen, the lightest known gas. Evidently,
then, a medium which propagates vibrations with such an enormous velocity

must have an enormously high elasticity or abnormally low density. In any case, its properties would be of an entirely different order from the properties of the substances with which we are accustomed to deal, so that it belongs in a category by itself.

Another course of reasoning which leads to this same conclusion—namely, that this medium is not any ordinary form of matter, such as air or gas or steel—is the following: Sound is produced by a bell under a receiver of an air pump. When the air has the same density inside the receiver as outside, the sound reaches the ear of an observer without difficulty. But when the air is gradually pumped out of the receiver, the sound becomes fainter and fainter until it ceases entirely. If the same thing were true of light, and we exhausted a vessel in which a source of light—an incandescent lamp, for example—had been placed, then, after a certain degree of exhaustion was reached, we ought to see the light less clearly than before. We know, however, that the contrary is the case, i.e., that the light is actually brighter and clearer when the exhaustion of the receiver has been carried to the highest possible degree. The probabilities are enormously against the conclusion that light is transmitted by the very small quantity of residual gas. There are other theoretical reasons, into which we will not enter.

Whatever the process of reasoning, we are led to the same result. We know that light vibrations are transverse to the direction of propagation, while sound vibrations are in the direction of propagation. We know also that in the case of a solid body transverse vibrations can be readily transmitted. Thus, if we have a long cylindrical rod and we give one end of it a twist, the twist will travel along from one end to the other. If the medium, instead of being a solid rod, were a tube of liquid, and were twisted at one end, there would be no corresponding transmission of the twist to the other end, for a liquid cannot transmit a torsional strain. Hence this reasoning leads to the conclusion that if the medium which propagates light vibrations has the properties of ordinary matter, it must be considered to be an elastic solid rather than a fluid.

This conclusion was considered one of the most formidable objections to the undulatory theory that light consists of waves. For this medium, notwithstanding the necessity for the assumption that it has the properties of a solid, must yet be of such a nature as to offer little resistance to the motion of a body through it. Take, for example, the motion of the planets around the sun. The resistance of the medium is so small that the earth has been traveling around the sun millions of years without any appreciable increase in the length of the year. Even the vastly lighter and more attenuated comets return to the same point periodically, and the time of such periodical returns has been carefully noted from the earliest historical times, and yet no appreciable increase in it has been detected. We are thus confronted with the apparent inconsistency of a solid body which must at the same time possess in such a marked degree the properties of a perfect fluid as to offer no appreciable resistance to the motion of bodies so very light and extended as the comets. We are, however, not without analogies, for, as was stated in the first lecture, substances such as shoemaker's wax show the properties of an elastic solid when reacting against rapid motions, but act like a liquid under pressures.

In the case of shoemaker's wax both of these contradictory properties are very imperfectly realized, but we can argue from this fact that the medium which we are considering might have the various properties which it must possess in an enormously exaggerated degree. It is, at any rate, not at all inconceivable that such a medium should at the same time possess both properties. We know that the air itself does not possess such properties, and that no matter which we know possesses them in sufficient degree to account for the propagation of light. Hence the conclusion that light vibrations are not propagated by ordinary matter, but by something else. Cogent as these three lines of reasoning may be, it is undoubtedly true that they do not always carry conviction. There is, so far as I am aware, no process of reasoning upon this subject which leads to a result which is free from objection and absolutely conclusive.

But these are not the only paradoxes connected with the medium which transmits light. There was an observation made by Bradley a great many years ago, for quite another purpose. He found that when we observe the position of a star

by means of the telescope, the star seems shifted from its actual position, by a certain small angle called the angle of aberration. He attributed this effect to the motion of the earth in its orbit, and gave an explanation of the phenomenon which is based on the corpuscular theory and is apparently very simple. We will give this explanation, notwithstanding the fact that we know the corpuscular theory to be erroneous.

Let us suppose a raindrop to be falling vertically and an observer to be carrying, say, a gun, the barrel being as nearly vertical as he can hold it. If the observer is not moving and the raindrop falls in the center of the upper end of the barrel, it will fall centrally through the lower end. Suppose, however, that the observer is in motion in the direction *bd* (Fig. 104); the raindrop will still fall exactly vertically, but if the gun advances laterally while the raindrop is within the barrel, it strikes against the side. In order to make the raindrop move centrally along the axis of the barrel, it is evidently necessary to incline the gun at an angle such as *bad*. The gun barrel is now pointing, apparently, in the wrong direction, by an angle whose tangent is the ratio of the velocity of the observer to the velocity of the raindrop.

According to the undulatory theory, the explanation is a trifle more complex; but it can easily be seen that, if the medium we are considering is motionless and the gun barrel represents a telescope, and the waves from the star are moving in the direction *ad*, they will be concentrated at a point which is in the axis of the telescope, unless the latter is in motion. But if the earth carrying the telescope is moving with a velocity something like twenty miles a second,

FIG. 104

and we are observing the stars in a direction approximately at right angles to the direction of that motion, the light from the star will not come to a focus on the axis of the telescope, but will form an image in a new position, so that the telescope appears to be pointing in the wrong direction. In order to bring the image on the axis of the instrument, we must turn the telescope from its position through an angle whose tangent is the ratio of the velocity of the earth in its orbit to the velocity of light. The velocity of light is, as before stated, 186,000 miles a second—200,000 in round numbers—and the velocity of the earth in its orbit is roughly twenty miles a second. Hence the tangent of the angle of aberration would be measured by the ratio of 1 to 10,000. More accurately, this angle is 20″.445. The limit of accuracy of the telescope, as was pointed out in several of the preceding lectures, is about one-tenth of a second; but, by repeating these measurements under a great many variations in the conditions of the problem, this limit may be passed, and it is practically certain that this number is correct to the second decimal place.

When this variation in the apparent position of the stars was discovered, it was accounted for correctly by the assumption that light travels with a finite velocity, and that, by measuring the angle of aberration, and knowing the speed of the earth in its orbit, the velocity of light could be found. This velocity has since been determined much more accurately by experimental means, so that now we use the velocity of light to deduce the velocity of the earth and the radius of its orbit.

The objection to this explanation was, however, raised that if this angle were the ratio of the velocity of the earth in its orbit to the velocity of light, and if we filled a telescope with water, in which the velocity of light is known to be only three-fourths of what it is in air, it would take one and one-third times as long for the light to pass from the center of the objective to the cross-wires, and hence we ought to observe, not the actual angle of aberration, but one which should be one-third greater. The experiment was actually tried. A telescope was filled with water, and observations on various stars were continued throughout the greater part of the year, with the result that almost exactly the same value was found for the angle of aberration.

This result was considered a very serious objection to the undulatory theory until an explanation was found by Fresnel. He proposed that we consider that the medium which transmits the light vibrations is carried along by the motion of the water in the telescope in the direction of the motion of the earth around the sun. Now, if the light waves were carried along with the full velocity of the earth in its orbit, we should be in the same difficulty, or in a more serious difficulty, than before. Fresnel, however, made the further supposition that the velocity of the carrying along of the light waves by the motion of the medium was less than the actual velocity of the medium itself, by a quantity which depended on the index of refraction of the substance. In the case of water the value of this factor is seven-sixteenths.

This, at first sight, seems a rather forced explanation; indeed, at the time it was proposed it was treated with considerable incredulity. An experiment was made by Fizeau, however, to test the point—in my opinion one of the most ingenious experiments that have ever been attempted in the whole domain of physics. The problem is to find the increase in the velocity of light due to a motion of the medium. We have an analogous problem in the case of sound, but in this case it is a very much simpler matter. We know by actual experiment, as we should infer without experiment, that the velocity of sound is increased by the velocity of a wind which carries the air in the same direction, or diminished if the wind moves in the opposite direction. But in the case of light waves the velocity is so enormously great that it would seem, at first sight, altogether out of the question to compare it with any velocity which we might be able to obtain in a transparent medium such as water or glass. The problem consists in finding the change in the velocity of light produced by the greatest velocity we can get— about twenty feet a second—in a column of water through which light waves pass. We thus have to find a difference of the order of twenty feet in 186,000 miles, i.e., of one part in 50,000,000. Besides, we can get only a relatively small column of water to pass light through and still see the light when it returns.

The difficulty is met, however, by taking advantage of the excessive minuteness of light waves themselves. This double length of the water column is something like forty feet. In this forty feet there are, in round numbers, 14,000,000 waves. Hence the difference due to a velocity of twenty feet per second, which is the velocity of the water current, would produce a displacement of the interference fringes (produced by two beams, one of which passes down the column and the other up the column of the moving liquid) of about one-half a fringe, which corresponds to a difference of one-half a light wave in the paths. Reversing the water current should produce a shifting of one-half a fringe in the opposite direction, so that the total shifting would actually be of the order of one interference fringe. But we can easily observe one-tenth of a fringe, or in some cases even less than that. Now, one fringe would be the displacement if water is the medium which transmits the light waves. But this other medium we have been talking about moves, according to Fresnel, with a smaller velocity than the water, and the ratio of the velocity of the medium to the velocity of the water should be a particular fraction, namely, seven-sixteenths. In other words, then, instead of the whole fringe we ought to get a displacement of seven-sixteenths of a fringe by the reversal of the water current. The experiment was actually tried by Fizeau, and the result was that the fringes were shifted by a quantity less than they should have been if water had been the medium; and hence we conclude that the water was not the medium which carried the vibrations.

The arrangement of the apparatus which was used in the experiment is shown in Fig. 105. The light starts from a narrow slit S, is rendered parallel by a lens L, and separated into two pencils by apertures in front of the two tubes TT, which carry the column of water. Both tubes are closed by pieces of the same plane-parallel plate of glass. The light passes through these two tubes and is brought to a focus by the lens in condition to produce interference fringes. The apparatus might have been arranged in this way but for the fact that there would be changes in the position of the interference fringes whenever the density or temperature of the medium changed; and, in particular, whenever the current changes direction there would be produced alterations in length and changes in density; and these exceedingly slight differences are quite sufficient to account for any motion of the fringes. In order to avoid this disturbance, Fresnel had the idea of placing at the focus of the lens the mirror M, so that the two rays return, the one which came through the upper tube going back through the lower, and *vice versa* for the other ray. In this way the two rays pass through identical paths and come together at the same point from which they started. With this arrangement, if there is any shifting of the fringes, it must be due to the reversal of the change in velocity due to the current of water. For one of the two beams, say the upper one, travels with the current in both tubes; the other, starting at the same point, travels against the current in both tubes. Upon reversing the direction of the current of water the circumstances are exactly the reverse: the beam which before traveled with the current now travels against it, etc. The result of the experiment, as before stated, was that there was produced a displacement of less than should have been produced by the motion of the liquid. How much less was not determined. To this extent the experiment was imperfect.

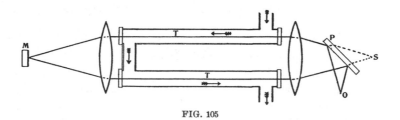

FIG. 105

On this account, and also for the reason that the experiment was regarded as one of the most important in the entire subject of optics, it seemed to me that it was desirable to repeat it in order to determine, not only the fact that the displacement was less than could be accounted for by the motion of the water, but also, if possible, how much less. For this purpose the apparatus was modified in several important points, and is shown in Fig. 106.

It will be noted that the principle of the interferometer has been used to produce interference fringes of considerable breadth without at the same time reducing the intensity of the light. Otherwise, the experiment is essentially the same as that made by Fizeau. The light starts from a bright flame of ordinary gas light, is rendered parallel by the lens, and then falls on the surface, which divides it into two parts, one reflected and one transmitted. The reflected portion goes down one tube, is reflected twice by the total reflection prism P through the other tube, and passes, after necessary reflection, into the observing telescope. The other ray pursues the contrary path, and we see interference fringes in the telescope as before, but enormously brighter and more definite. This arrangement made it possible to make measurements of the displacement of the fringes which were very accurate. The result of the experiment was that the measured displacement was almost exactly seven-sixteenths of what it would

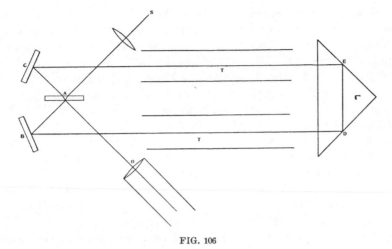

FIG. 106

have been had the medium which transmits the light waves moved with the velocity of the water.

It was at one time proposed to test this problem by utilizing the velocity of the earth in its orbit. Since this velocity is so very much greater than anything we can produce at the earth's surface, it was supposed that such measurements could be made with considerable ease; and they were actually tried in quite a considerable number of different ways and by very eminent men. The fact is, we cannot utilize the velocity of the earth in its orbit for such experiments, for the reason that we have to determine our directions by points outside of the earth, and the only thing we have is the stars, and the stars are displaced by this very element which we want to measure; so the results would be entirely negative. It was pointed out by Lorentz that it is impossible by any measurements made on the surface of the earth to detect any effect of the earth's motion.

Maxwell considered it possible, theoretically at least, to deal with the square of the ratio of the two velocities; that is, the square of 1/10000, or 1/100000000. He further indicated that if we made two measurements of the velocity of light, one in the direction in which the earth is traveling in its orbit, and one in a direction at right angles to this, then the time it takes light to pass over the same length of path is greater in the first case than in the second.

We can easily appreciate the fact that the time is greater in this case, by considering a man rowing in a boat, first in a smooth pond and then in a river. If he rows at the rate of four miles an hour, for example, and the distance between the stations is twelve miles, then it would take him three hours to pull there and three to pull back—six hours in all. This is his time when there is no current. If there is a current, suppose at the rate of one mile an hour, then the time it would take to go from one point to the other, would be, not 12 divided by 4, but 12 divided by 4 + 1, i.e., 2.4 hours. In coming back the time would be 12 divided by 4 − 1, which would be 4 hours, and this added to the other time equals 6.4 instead of 6 hours. It takes him longer, then, to pass back and forth when the medium is in motion than when the medium is at rest. We can understand, then, that it would take light longer to travel back and forth in the direction of the motion of the earth. The difference in the times is, however, so exceedingly small, being of the order of 1 in 100,000,000, that Maxwell considered it practically hopeless to attempt to detect it.

In spite of this apparently hopeless smallness of the quantities to be observed, it was thought that the minuteness of the light waves might again come to our rescue. As a matter of fact, an experiment was devised for detecting this small quantity. The conditions which the apparatus must fulfil are rather complex. The total distance traveled must be as great as possible, something of the order of one hundred million waves, for example. Another condition requires that we be able to interchange the direction without altering the adjustment by even the one hundredth-millionth part. Further, the apparatus must be absolutely free from vibration.

The problem was practically solved by reflecting part of the light back and forth a number of times and then returning it to its starting-point. The other path was at right angles to the first, and over it the light made a similar series of excursions, and was also reflected back to the starting-point. This starting-point was a separating plane in an interferometer, and the two paths at right angles were the two arms of an interferometer. Notwithstanding the very considerable difference in path, which must involve an exceedingly high order of accuracy in the reflecting surfaces and a constancy of temperature in the air between, it was possible to see fringes and to keep them in position for several hours at a time.

These conditions having been fulfilled, the apparatus was mounted on a stone support, about four feet square and one foot thick, and this stone was mounted on a circular disc of wood which floated in a tank of mercury. The resistance to motion is thus exceedingly small, so that by a very slight pressure on the circumference the whole could be kept in slow and continuous rotation. It would take, perhaps, five minutes to make one single turn. With this slight motion there is practically no oscillation; the observer has to follow around and at intervals to observe whether there is any displacement of the fringes.

It was found that there was no displacement of the interference fringes, so that the result of the experiment was negative and would, therefore, show that there is still a difficulty in the theory itself; and this difficulty, I may say, has not yet been satisfactorily explained. I am presenting the case, not so much for solution, but as an illustration of the applicability of light waves to new problems.

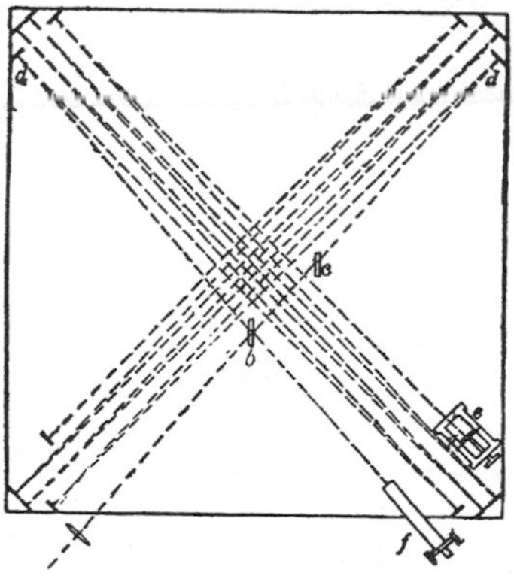

The actual arrangement of the experiment is shown in Fig. 107. A lens makes the rays

FIG. 107

nearly parallel. The dividing surface and the two paths are easily recognized. The telescope was furnished with a micrometer screw to determine the amount of displacement of the fringes, if there were any. The last mirror is mounted on a slide; so these two paths may be made equal to the necessary degree of accuracy—something of the order of one fifty-thousandth of an inch.

Fig. 108 represents the actual apparatus. The stone and the circular disc of wood supporting the stone in the tank filled with mercury are readily recognized; also the dividing surface and the various mirrors.

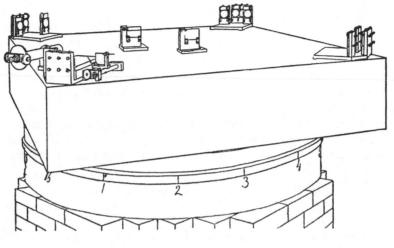

FIG. 108

It was considered that, if this experiment gave a positive result, it would determine the velocity, not merely of the earth in its orbit, but of the earth through the ether. With good reason it is supposed that the sun and all the planets as well are moving through space at a rate of perhaps twenty miles per second in a certain particular direction. The velocity is not very well determined, and it was hoped that with this experiment we could measure this velocity of the whole solar system through space. Since the result of the experiment was negative, this problem is still demanding a solution.

The experiment is to me historically interesting, because it was for the solution of this problem that the interferometer was devised. I think it will be admitted that the problem, by leading to the invention of the interferometer, more than compensated for the fact that this particular experiment gave a negative result.

From all that precedes it appears practically certain that there must be a medium whose proper function it is to transmit light waves. Such a medium is also necessary for the transmission of electrical and magnetic effects. Indeed, it is fairly well established that light is an electro-magnetic disturbance, like that due to a discharge from an induction coil or a condenser. Such electric waves can be reflected and refracted and polarized, and be made to produce vibrations and other changes, just as the light waves can. The only difference between them and the light waves is in the wave length.

This difference may be enormous or quite moderate. For example, a telegraphic wave, which is practically an electromagnetic disturbance, may be as long as one thousand miles. The waves produced by the oscillations of a condenser, like a Leyden jar, may be as short as one hundred feet; the waves produced by a Hertz oscillator may be as short as one-tenth of an inch. Between this and the longest light wave there is not an enormous gap, for the latter has a length of about one-thousandth of an inch. Thus the difference between the Hertz vibrations and the longest light wave is less than the difference between the longest and shortest light waves, for some of the shortest oscillations are only a few millionths of an inch long. Doubtless even this gap will soon be bridged over.

The settlement of the fact that light is a magneto-electric oscillation is in no sense an explanation of the nature of light. It is only a transference of the problem, for the question then arises as to the nature of the medium and of the mechanical actions involved in such a medium which sustains and transmits these electro-magnetic disturbances.

A suggestion which is very attractive on account of its simplicity is that the ether itself is electricity; a much more probable one is that electricity is an ether strain—that a displacement of the ether is equivalent to an electric current. If this is true, we are returning to our elastic-solid theory. I may quote a statement which Lord Kelvin made in reply to a rather skeptical question as to the existence of a medium about which so very little is supposed to be known. The reply was: "Yes, ether is the only form of matter about which we know anything at all." In fact, the moment we begin to inquire into the nature of the ultimate particles of ordinary matter, we are at once enveloped in a sea of conjecture and hypotheses—all of great difficulty and complexity.

One of the most promising of these hypotheses is the "ether vortex theory," which, if true, has the merit of introducing nothing new into the hypotheses already made, but only of specifying the particular form of motion required. The most natural form of such vortex motions with which to deal is that illustrated by ordinary smoke rings, such as are frequently blown from the stack of a locomotive. Such vortex rings may easily be produced by filling with smoke a box which has a circular aperture at one end and a rubber diaphragm at the other, and then tapping the rubber. The friction against the side of the opening, as the puff of smoke passes out, produces a rotary motion, and the result will be smoke rings or vortices.

Investigation shows that these smoke rings possess, to a certain degree, the properties which we are accustomed to associate with atoms, notwithstanding the fact that the medium in which these smoke rings exists is far from ideal. If the medium were ideal, it would be devoid of friction, and then the motion, when once started, would continue indefinitely, and that part of the ether which is differentiated by this motion would ever remain so.

Another peculiarity of the ring is that it cannot be cut—it simply winds around the knife. Of course, in a very short time the motion in a smoke ring ceases in consequence of the viscosity of the air, but it would continue indefinitely in such a frictionless medium as we suppose the ether to be.

There are a number of other analogies which we have not time to enter into—quite a number of details and instances of the interactions of the various atoms which have been investigated. In fact, there are so many analogies that we are tempted to think that the vortex ring is in reality an enlarged image of the atom. The mathematics of the subject is unfortunately very difficult, and this seems to be one of the principal reasons for the slow progress made in the theory.

Suppose that an ether strain corresponds to an electric charge, an ether displacement to the electric current, these ether vortices to the atoms—if we continue these suppositions, we arrive at what may be one of the grandest generalizations of modern science—of which we are tempted to say that it ought to be true even if it is not—namely, that all the phenomena of the physical universe are only different manifestations of the various modes of motions of one all-pervading substance—the ether.

All modern investigation tends toward the elucidation of this problem, and the day seems not far distant when the converging lines from many apparently remote regions of thought will meet on this common ground. Then the nature of the atoms, and the forces called into play in their chemical union; the interactions between these atoms and the non-differentiated ether as manifested in the phenomena of light and electricity; the structures of the molecules and molecular systems of which the atoms are the units; the explanation of cohesion, elasticity, and gravitation—all these will be marshaled into a single compact and consistent body of scientific knowledge.

Summary

1. A number of independent courses of reasoning lead to the conclusion that the medium which propagates light waves is not an ordinary form of matter. Little as we know about it, we may say that our ignorance of ordinary matter is still greater.
2. In all probability, it not only exists where ordinary matter does not, but it also permeates all forms of matter. The motion of a medium such as water is found not to add its full value to the velocity of light moving through it, but only such a fraction of it as is perhaps accounted for on the hypothesis that the ether itself does not partake of this motion.
3. The phenomenon of the aberration of the fixed stars can be accounted for on the hypothesis that the ether does not partake of the earth's motion in its revolution about the sun. All experiments for testing this hypothesis have, however, given negative results, so that the theory may still be said to be in an unsatisfactory condition.

3
— George F. Fitzgerald —
(1851–1901)

George Fitzgerald was an Irish theoretical physicist who offered a contraction hypothesis to explain the null result of the Michelson-Morley experiment. Fitzgerald suggested that a body physically contracts as it pushes its way through the ether in just the right amount necessary to guarantee that the motion will not be detected. This hypothesis was also suggested by H.A. Lorentz and came to be known as the Lorentz-Fitzgerald contraction hypothesis. Although later shown to be mistaken, the hypothesis played a definite role in the development of the Theory of Relativity.

In this reading Fitzgerald proposes that molecular forces are modified in an analogous way to that of electric forces. Molecular changes, brought about by motion relative to the ether, translate into lengthwise contractions. Fitzgerald further suggests an experimental strategy for detecting changes in electrical attractions between permanently electrified bodies due to possible variations in the orbit and rotation of the earth. The text of this reading first appeared in the journal Science *in 1889 under the title "The Ether and the Earth's Atmosphere."*

THE CONTRACTION HYPOTHESIS

I have read with much interest Messrs. Michelson and Morley's wonderfully delicate experiment attempting to decide the important question as to how far the ether is carried along by the earth. Their result seems opposed to other experiments showing that the ether in the air can be carried along only to an inappreciable extent. I would suggest that almost the only hypothesis that can reconcile this opposition is that the length of material bodies changes, according as they are moving through the ether or across it, by an amount depending on the square of the ratio of their velocities to that of light. We know that electric forces are affected by the motion of the electrified bodies relative to the ether, and it seems a not improbable supposition that the molecular forces are affected by the motion, and that the size of a body alters consequently. It would be very important if secular experiments on electrical attractions between permanently electrified bodies, such as in a very delicate quadrant electrometer, were instituted in some of the equatorial parts of the earth to observe whether there is any diurnal and annual variation of attraction—diurnal due to the rotation of the earth being added and subtracted from its orbital velocity, and annual similarly for its orbital velocity and the motion of the solar system.

4
— Hendrik A. Lorentz —
(1853–1928)

Hendrik Lorentz was a Dutch theoretical physicist who played a significant role in the development of the theory of the electron. However, he is probably best known for his contraction hypothesis which he used in an attempt to explain the Michelson-Morley Experiment. In 1904 Lorentz developed transformation equations, based upon the contraction hypothesis, that provided a mathematical description for the results of the experiment. Later Einstein showed that these transformation equations followed from the principles of his Special Theory of Relativity.

In this reading Lorentz gives an account of the null result of the Michelson experiment of 1881 and the Michelson-Morley experiment of 1887. He then proceeds to give a theoretical explanation for the failure to detect the ether. Like Fitzgerald, he suggests that a physical contraction of the interferometer in the direction of the earth's motion through the ether could account for the null result.

The reading is taken from a paper given at Leyden in 1895 entitled "Michelson's Interference Experiment."

THE CONTRACTION HYPOTHESIS

1. As Maxwell first remarked and as follows from a very simple calculation, the time required by a ray of light to travel from a point A to a point B and back to A must vary when the two points together undergo a displacement without carrying the ether with them. The difference is, certainly, a magnitude of second order; but it is sufficiently great to be detected by a sensitive interference method.

The experiment was carried out by Michelson in 1881. His apparatus, a kind of interferometer, had two horizontal arms, P and Q, of equal length and at right angles one to the other. Of the two mutually interfering rays of light the one passed along the arm P and back, the other along the arm Q and back. The whole instrument, including the source of light and the arrangement for taking observations, could be revolved about a vertical axis; and those two positions come especially under consideration in which the arm P or the arm Q lay as nearly as possible in the direction of the Earth's motion. On the basis of Fresnel's theory it was anticipated that when the apparatus was revolved from one of these *principal positions* into the other there would be a displacement of the interference fringes.

But of such a displacement—for the sake of brevity we will call it the Maxwell displacement—conditioned by the change in the times of propagation, no trace was discovered, and accordingly Michelson thought himself justified in concluding that while the Earth is moving, the ether does not remain at rest. The correctness of this inference was soon brought into question, for by an oversight Michelson had taken the change in the phase difference, which was to be expected in accordance with the theory, at twice its proper value. If we make the necessary correction, we arrive at displacements no greater than might be masked by errors of observation.

Subsequently Michelson took up the investigation anew in collaboration with Morley, enhancing the delicacy of the experiment by causing each pencil to be reflected to and fro between a number of mirrors, thereby obtaining the same advantage as if the arms of the earlier apparatus had been considerably lengthened. The mirrors were mounted on a massive stone disc, floating on mercury, and therefore easily revolved. Each pencil now had to travel a total distance of 22 meters, and on Fresnel's theory the displacement to be expected in passing from the one principal position to the other would be 0.4 of the distance between the interference fringes. Nevertheless the rotation produced displacements not exceeding 0.02 of this distance, and these might well be ascribed to errors of observation.

Now, does this result entitle us to assume that the ether takes part in the motion of the Earth, and therefore that the theory of aberration given by Stokes is the correct one? The difficulties which this theory encounters in explaining aberration seem too great for me to share this opinion, and I would rather try to remove the contradiction between Fresnel's theory and Michelson's result. An hypothesis which I brought forward some time ago, and which, as I subsequently learned, has also occurred to Fitzgerald, enables us to do this. The next paragraph will set out this hypothesis.

2. To simplify matters we will assume that we are working with apparatus as employed in the first experiments, and that in the one principal position the arm P lies exactly in the direction of the motion of the Earth. Let v be the velocity of this motion, L the length of either arm, and hence 2L the path traversed by the rays of light. According to the theory, the turning of the apparatus through 90° causes the time in which the one pencil travels along P and back to be longer than the time which the other pencil takes to complete its journey by

$$\frac{Lv^2}{c^3}$$

There would be this same difference if the translation had no influence and the arm P were longer than the arm Q by ½ Lv^2/c^2. Similarly with the second principal position.

Thus we see that the phase differences expected by the theory might also arise if, when the apparatus is revolved, first the one arm and then the other arm were the longer. It follows that the phase differences can be compensated by contrary changes of the dimensions.

If we assume the arm which lies in the direction of the Earth's motion to be shorter than the other by ½ Lv^2/c^2, and, at the same time, that the translation has the influence which Fresnel's theory allows it, then the result of the Michelson experiment is explained completely.

Thus one would have to imagine that the motion of a solid body (such as a brass rod or the stone disc employed in the later experiments) through the resting ether exerts upon the dimensions of that body an influence which varies according to the orientation of the body with respect to the direction of motion. If, for example, the dimensions parallel to this direction were changed in the proportion of 1 to $1 + \delta$, and those perpendicular in the proportion of 1 to $1 + \varepsilon$, then we should have the equation

$$\varepsilon - \delta = \frac{1}{2} \frac{v^2}{c^2} \tag{1}$$

in which the value of one of the quantities δ and ε would remain undetermined. It might be that $\varepsilon = 0$, $\delta = -½ \, v^2/c^2$, but also $\varepsilon = ½ \, v^2/c^2$, $\delta = 0$, or $\varepsilon = ¼ \, v^2/c^2$, and $\delta = -¼ \, v^2/c^2$.

3. Surprising as this hypothesis may appear at first sight, yet we shall have to admit that it is by no means far-fetched, as soon as we assume that molecular forces are also transmitted through the ether, like the electric and magnetic forces of which we are able at the present time to make this assertion definitely. If they are so transmitted, the translation will very probably affect the action between two molecules or atoms in a manner resembling the attraction or repulsion between charged particles. Now, since the form and dimensions of a solid body are ultimately conditioned by the intensity of molecular actions, there cannot fail to be a change of dimensions as well.

From the theoretical side, therefore, there would be no objection to the hypothesis. As regards its experimental proof, we must first of all note that the lengthenings and shortenings in question are extraordinarily small. We have $v^2/c^2 = 10^{-8}$, and thus, if $\varepsilon = 0$, the shortening of the one diameter of the Earth would amount to about 6.5 cm. The length of a meter rod would change, when moved from one principal position into the other, by about 1/200 micron. One could hardly hope for success in trying to perceive such small quantities except by means of an interference method. We should have to operate with two perpendicular rods, and with two mutually interfering pencils of light, allowing the one to travel to and fro along the first rod, and the other along the second rod. But in this way we should come back once more to the Michelson experiment, and revolving the apparatus we should perceive no displacement of the fringes. Reversing a previous remark, we might now say that the displacement produced by the alterations of length is compensated by the Maxwell displacement.

4. It is worth noticing that we are led to just the same changes of dimensions as have been presumed above if we, *firstly*, without taking molecular movement into consideration, assume that in a solid body left to itself the forces, attractions or repulsions, acting upon any molecule maintain one another in equilibrium, and, *secondly*—though to be sure, there is no reason for doing so—if we apply to these molecular forces the law which in another place we deduced for electrostatic actions. For if we now understand by S_1 and S_2 not, as formerly, two systems of charged particles, but two systems of molecules—the second at rest and the first moving with a velocity v in the direction of the axis of x—between the dimensions of which the relationship subsists as previously stated; and if we assume that in both systems the x components of the forces are the same, while the y and z components differ from one another by the factor $\sqrt{1 - v^2/c^2}$, then it is clear that the forces in S_1, will be in equilibrium whenever they are so in S_2. If therefore S_2 is the state of equilibrium of a solid body at rest, then the molecules in S_1 have precisely those positions in which they can persist under the influence of translation. The displacement would naturally bring about this disposition of the molecules of its own accord, and thus effect a shortening in the direction of motion in the proportion of 1 to $\sqrt{1 - v^2/c^2}$, in accordance with the forumlae given in the above-mentioned paragraph. This leads to the values

$$\delta = -\tfrac{1}{2}\frac{v^2}{c^2}, \; \varepsilon = 0$$

in agreement with (1).

In reality the molecules of a body are not at rest, but in every "state of equilibrium" there is a stationary movement. What influence this circumstance may have in the phenomenon which we have been considering is a question which we do not here touch upon; in any case the experiments of Michelson and Morley, in consequence of unavoidable errors of observation, afford considerable latitude for the values of δ and ε.

5

— Henri Poincaré —

(1854–1912)

Jules Henri Poincaré was a French mathematician and philosopher. At the University of Paris he was given charge of courses in both mathematics and experimental physics and lectured on mechanics, mathematical physics and astronomy. Poincaré wrote a great number of papers on mathematics and physics, several important books on the philosophy of science, and some popular essays on science. It is because Einstein had such profound respect for his work in mathematics and the philosophy of science that he appears in this volume.

This reading is a combination of three separate selections. The first comes from Poincaré's 1898 paper entitled "La Mesure du Temps," first published in the Revue Metaphysique Morale. *It inquires into the objective meaning of simultaneity. In doing so Poincaré steps beyond the limits of contemporary concern that fixed its gaze on the mere measurement of time. He concludes that any definition of simultaneity must be guided by a specification of maximal simplicity in the articulation of natural laws. Thus the vaunted conventionalism made famous in Poincaré's later and better-known writing is already evident in his consideration of time. The second selection, which contains his rejection of absolute space and absolute time, comes from his comments on classical mechanics in* Science and Hypothesis, *published in 1902. The third selection, which contains Poincaré's call for a new mechanics in which the velocity of light would be a limiting velocity, comes from a keynote address he gave at the World's Fair in St. Louis in 1904.*

A PRELUDE TO RELATIVITY

Simultaneity and the Measure of Time

So long as we do not go outside the domain of consciousness, the notion of time is relatively clear. Not only do we distinguish without difficulty present sensation from the remembrance of past sensations or the anticipation of future sensations, but we know perfectly well what we mean when we say that of two conscious phenomena which we remember, one was anterior to the other; or that, of two foreseen conscious phenomena, one will be anterior to the other.

When we say that two conscious facts are simultaneous, we mean that they profoundly interpenetrate, so that analysis can not separate them without mutilating them.

The order in which we arrange conscious phenomena does not admit of any arbitrariness. It is imposed upon us and of it we can change nothing.

I have only a single observation to add. For an aggregate of sensations to have become a remembrance capable of classification in time, it must have ceased to be actual, we must have lost the sense of its infinite complexity, otherwise it would have remained present. It must, so to speak, have crystallized around a center of associations of ideas which will be a sort of label. It is only when they thus have lost all life that we can classify our memories in time as a botanist arranges dried flowers in his herbarium.

But these labels can only be finite in number. On that score, psychologic time should be discontinuous. Whence comes the feeling that between any two instants there are others? We arrange our recollections in time, but we know that there remain empty compartments. How could that be, if time were not a form preexistent in our mind? How could we know there were empty compartments, if these compartments were revealed to us only by their content?

But that is not all; into this form we wish to put not only the phenomena of our own consciousness, but those of which other consciousness are the theater. But more, we wish to put there physical facts, these I know not what with which we people space and which no consciousness sees directly. This is necessary because without it science could not exist. In a word, psychologic time is given to us and must needs create scientific and physical time. There the difficulty begins, or rather the difficulties, for there are two.

Think of two consciousnesses, which are like two worlds impenetrable one to the other. By what do we strive to put them into the same mold, to measure them by the same standard? Is it not as if one strove to measure length with a gram or weight with a meter? And besides, why do we speak of measuring? We know perhaps that some fact is anterior to some other, but not *by how much* it is anterior.

Therefore two difficulties: (1) Can we transform psychologic time, which is qualitative, into a quantitative time? (2) Can we reduce to one and the same measure facts which transpire in different worlds?

The first difficulty has long been noticed; it has been the subject of long discussions and one may say the question is settled. *We have not a direct intuition of the equality of two intervals of time.* The persons who believe they possess this intuition are dupes of an illusion. When I say, from noon to one the same time passes as from two to three, what meaning has this affirmation?

The least reflection shows that by itself it has none at all. It will only have that which I choose to give it, by a definition which will certainly possess a certain degree of arbitrariness. Psychologists could have done without this definition; physicists and astronomers could not; let us see how they have managed.

To measure time they use the pendulum and they suppose by definition that all the beats of this pendulum are of equal duration. But this is only a first approximation; the temperature, the resistance of the air, the barometric pressure, make the pace of the pendulum vary. If we could escape these sources of error, we should obtain a much closer approximation, but it would still be only an approximation. New cases, hitherto neglected, electric, magnetic or others, would introduce minute perturbations.

In fact, the best chronometers must be corrected from time to time, and the corrections are made by the aid of astronomic observations; arrangements are made so that the sidereal clock marks the same hour when the same star passes the meridian. In other words, it is the sidereal day, that is, the duration of the rotation of the earth, which is the constant unit of time. It is supposed, by a new definition substituted for that based on the beats of the pendulum, that two complete rotations of the earth about its axis have the same duration.

However, the astronomers are still not content with this definition. Many of them think that the tides act as a check on our globe, and that the rotation of the earth is becoming slower and slower. Thus would be explained the apparent acceleration of the motion of the moon, which would seem to be going more rapidly than theory permits because our watch, which is the earth, is going slow.

All this is unimportant, one will say; doubtless our instruments of measurement are imperfect, but it suffices that we can conceive a perfect instrument.

This ideal can not be reached, but it is enough to have conceived it and so to have put rigor into the definition of the unit of time.

The trouble is that there is no rigor in the definition. When we use the pendulum to measure time, what postulate do we implicitly admit? *It is that the duration of two identical phenomena is the same*; or, if you prefer, that the same causes take the same time to produce the same effects.

And at first blush, this is a good definition of the equality of two durations. But take care. Is it impossible that experiment may some day contradict our postulate?

Let me explain myself. I suppose that at a certain place in the world the phenomenon α happens, causing as consequence at the end of a certain time the effect α'. At another place in the world very far away from the first, happens the phenomena β, which causes as consequence the effect β'. The phenomena α and β are simultaneous, as are also the effects α' and β'.

Later, the phenomenon α is reproduced under approximately the same conditions as before, and *simultaneously* the phenomenon β is also reproduced at a very distant place in the world and almost under the same circumstances. The effects α' and β' also take place. Let us suppose that the effect α' happens perceptibly before the effect β'.

If experience made us witness such a sight, our postulate would be contradicted. For experience would tell us that the first duration $\alpha\alpha'$ is equal to the first duration $\beta\beta'$ and that the second duration $\alpha\alpha'$ is less than the second duration $\beta\beta$. On the other hand, our postulate would require that the two durations $\alpha\alpha'$ should be equal to each other, as likewise the two durations $\beta\beta$. The equality and the inequality deduced from experience would be incompatible with the two equalities deduced from the postulate.

Now can we affirm that the hypotheses I have just made are absurd? They are in no wise contrary to the principle of contradiction. Doubtless they could not happen without the principle of sufficient reason seeming violated. But to justify a definition so fundamental I should prefer some other guarantee.

But this is not all. In physical reality one cause does not produce a given effect, but a multitude of distinct causes contribute to produce it, without our having any means of discriminating the part of each of them.

Physicists seek to make this distinction; but they make it only approximately, and, however they progress, they never will make it except approximately. It is approximately true that the motion of the pendulum is due solely to the earth's attraction; but in all rigor every attraction, even of Sirius, acts on the pendulum.

Under these conditions, it is clear that the causes which have produced a certain effect will never be reproduced except approximately. Then we should modify our postulate and our definition. Instead of saying: "The same causes take the same time to produce the same effects," we should say: "Causes almost identical take almost the same time to produce almost the same effects."

Our definition therefore is no longer anything but approximate. Besides, as M. Calinon very justly remarks in a recent memoir:

One of the circumstances of any phenomenon is the velocity of the earth's rotation; if this velocity of rotation varies, it constitutes in the reproduction of this phenomenon a circumstance which no longer remains the same. But to suppose this velocity of rotation constant is to suppose that we know how to measure time.

Our definition is therefore not yet satisfactory; it is certainly not that which the astronomers of whom I spoke above implicitly adopt, when they affirm that the terrestrial rotation is slowing down.

What meaning according to them has this affirmation? We can only understand it by analyzing the proofs they give of their proposition. They say first that the friction of the tides producing heat must destroy *vis viva*. They invoke therefore the principle of *vis viva*, or of the conservation of energy.

They say next that the secular acceleration of the moon, calculated according to Newton's law, would be less than that deduced from observations unless the correction relative to the slowing down of the terrestrial rotation were made. They invoke therefore Newton's law. In other words, they define duration in the following way: time should be so defined that Newton's law and that of *vis viva* may be verified. Newton's law is an experimental truth; as such it is only approximate, which shows that we still have only a definition by approximation.

If now it be supposed that another way of measuring time is adopted, the experiments on which Newton's law is founded would none the less have the same meaning. Only the enunciation of the law would be different, because it would be translated into another language; it would evidently be much less simple. So that the definition implicitly adopted by the astronomers may be summed up thus: Time should be so defined that the equations of mechanics may be as simple as possible. In other words, there is not one way of measuring time more true than another; that which is generally adopted is only more *convenient*. Of two watches, we have no right to say that the one goes true, the other wrong; we can only say that it is advantageous to conform to the indications of the first.

The difficulty which has just occupied us has been, as I have said, often pointed out; among the most recent works in which it is considered, I may mention, besides M. Calinon's little book, the treatise on mechanics of M. Andrade.

The second difficulty has up to the present attracted much less attention; yet it is altogether analogous to the preceding; and even, logically, I should have spoken of it first.

Two psychological phenomena happen in two different consciousnesses; when I say they are simultaneous, what do I mean? When I say that a physical phenomenon, which happens outside of every consciousness, is before or after a psychological phenomenon, what do I mean?

In 1572, Tycho Brahe noticed in the heavens a new star. An immense conflagration had happened in some far distant heavenly body; but it had happened long before; at least two hundred years were necessary for the light from that star to reach our earth. This conflagration therefore happened before the discovery of America. Well, when considering this gigantic phenomenon, which perhaps had no witness, since the satellites of that star were perhaps uninhabited, I say this phenomenon is anterior to the formation of the visual image of the isle of Española in the consciousness of Christopher Columbus, what do I mean?

A little reflection is sufficient to understand that all these affirmations have by themselves no meaning. They can have one only as the outcome of a convention.

We should first ask ourselves how one could have had the idea of putting into the same frame so many worlds impenetrable to each other. We should like to represent to ourselves the external universe, and only by so doing could we feel that we understood it. We know we never can attain this representation: our weakness is too great. But at least we desire the ability to conceive an infinite intelligence for which this representation would be possible, a sort of great consciousness which should see all, and which should classify all *in its time*, as we classify, *in our time*, the little we see.

This hypothesis is indeed, crude and incomplete, because this supreme intelligence would be only a demigod; infinite in one sense, it would be limited in another, since it would have only an imperfect recollection of the past; and it could have no other, since otherwise all recollections would be equally present to it and for it there would be no time. And yet when we speak of time, for all which happens outside of us, do we not unconsciously adopt this hypothesis; do we not put ourselves in the place of this imperfect god; and do not even the atheists put themselves in the place where god would be if he existed?

What I have just said shows us, perhaps, why we have tried to put all physical phenomena into the same frame. But that can not pass for a definition of simultaneity, since this hypothetical intelligence, even if it existed, would be for us impenetrable. It is therefore necessary to seek something else.

The ordinary definitions which are proper for psychologic time would suffice us no better. Two simultaneous psychologic facts are so closely bound together that analysis can not separate without mutilating them. Is it the same with two physical facts? Is not my present nearer my past of yesterday than the present of Sirius?

It has also been said that two facts should be regarded as simultaneous when the order of their succession may be inverted at will. It is evident that this definition would not suit two physical facts which happen far from one another, and that, in what concerns them, we no longer even understand what this reversibility would be; besides, succession itself must first be defined.

Let us then seek to give an account of what is understood by simultaneity or antecedence, and for this let us analyze some examples.

I write a letter; it is afterward read by the friend to whom I have addressed it. There are two facts which have had for their theater two different consciousnesses. In writing this letter I have had the visual image of it, and my friend has had in his turn the same visual image in reading the letter. Though these two facts happen in impenetrable worlds, I do not hesitate to regard the first as anterior to the second, because I believe it is its cause.

I hear thunder, and I conclude there has been an electric discharge; I do not hesitate to consider the physical phenomenon as anterior to the auditory image perceived in my consciousness, because I believe it is its cause.

Behold then the rule we follow, and the only one we can follow: when a phenomenon appears to us as the cause of another, we regard it as anterior. It is therefore by cause that we define time; but most often, when two facts appear to us bound by a constant relation, how do we recognize which is the cause and which the effect? We assume that the anterior fact, the antecedent, is the cause of the other, of the consequent. It is then by time that we define cause. How save ourselves from this *petitio principii*?

We say now *post hoc, ergo propter hoc*; now *propter hoc, ergo post hoc*; shall we escape from this vicious circle?

Let us see, not how we succeed in escaping, for we do not completely succeed, but how we try to escape.

I execute a voluntary act A and I feel afterward a sensation D, which I regard as a consequence of the act A; on the other hand, for whatever reason, I infer that this consequence is not immediate, but that outside my consciousness two facts B and C, which I have not witnessed, have happened, and in such a way that B is the effect of A, that C is the effect of B, and D of C.

But why? If I think I have reason to regard the four facts A, B, C, D, as bound to one another by a causal connection, why range them in the causal order A B C D, and at the same time in the chronologic order A B C D, rather than in any other order?

I clearly see that in the act A I have the feeling of having been active, while in undergoing the sensation D I have that of having been passive. This is why I regard A as the initial cause and D as the ultimate effect; this is why I put A at the beginning of the chain and D at the end; but why put B before C rather than C before B?

If this question is put, the reply ordinarily is: we know that it is B which is the cause of C because we *always* see B happen before C. These two phenomena, when witnessed, happen in a certain order; when analogous phenomena happen without witness, there is no reason to invert this order.

Doubtless, but take care; we never know directly the physical phenomena B and C. What we know are sensations B' and C' produced respectively by B and C. Our consciousness tells us immediately that B' precedes C' and we *suppose* that B and C succeed one another in the same order.

This rule appears in fact very natural, and yet we are often led to depart from it. We hear the sound of the thunder only some seconds after the electric discharge of the cloud. Of two flashes of lightning, the one distant, the other near, can not the first be anterior to the second, even though the sound of the second comes to us before that of the first?

Another difficulty; have we really the right to speak of the cause of a phenomenon? If all the parts of the universe are interchained in a certain measure, any one phenomenon will not be the effect of a single cause, but the resultant of causes infinitely numerous; it is, one often says, the consequence of the state of the universe a moment before. How enunciate rules applicable to circumstances so complex? And yet it is only thus that these rules can be general and rigorous.

Not to lose ourselves in this infinite complexity let us make a simpler hypothesis. Consider three stars, for example, the sun, Jupiter and Saturn; but, for greater simplicity, regard them as reduced to material points and isolated from the rest of the world. The positions and the velocities of three bodies at a given instant suffice to determine their positions and velocities at the following instant, and consequently at any instant. Their positions at the instant t determine their positions at the instant $t + h$ as well as their positions at the instant $t - h$.

Even more; the position of Jupiter at the instant t, together with that of Saturn at the instant $t + a$, determines the position of Jupiter at any instant and that of Saturn at any instant.

The aggregate of positions occupied by Jupiter at the instant $t + e$ and Saturn at the instant $t + a + e$ is bound to the aggregate of positions occupied by Jupiter at the instant t and Saturn at the instant $t + a$, by laws as precise as that of Newton, though more complicated. Then why not regard one of these aggregates as the cause of the other, which would lead to considering as simultaneous the instant t of Jupiter and the instant $t + a$ of Saturn?

In answer there can only be reasons, very strong, it is true, of convenience and simplicity.

But let us pass to examples less artificial; to understand the definition implicitly supposed by the savants, let us watch them at work and look for the rules by which they investigate simultaneity.

I will take two simple examples, the measurement of the velocity of light and the determination of longitude.

When an astronomer tells me that some stellar phenomenon, which his telescope reveals to him at this moment, happened nevertheless fifty years ago, I seek his meaning, and to that end I shall ask him first how he knows it, that is, how he has measured the velocity of light.

He has begun by *supposing* that light has a constant velocity, and in particular that its velocity is the same in all directions. That is a postulate without which no measurement of this velocity could be attempted. This postulate could never be verified directly by experiment; it might be contradicted by it if the results of different measurements were not concordant. We should think ourselves fortunate that this contradiction has not happened and that the slight discordances which may happen can be readily explained.

The postulate, at all events, resembling the principle of sufficient reason, has been accepted by everybody; what I wish to emphasize is that it furnishes us with a new rule for the investigation of simultaneity, entirely different from that which we have enunciated above.

This postulate assumed, let us see how the velocity of light has been measured. You know that Roemer used eclipses of the satellites of Jupiter, and sought how much the event fell behind its prediction. But how is this prediction made? It is by the aid of astronomic laws, for instance Newton's law.

Could not the observed facts be just as well explained if we attributed to the velocity of light a little different value from that adopted, and supposed Newton's law only approximate? Only this would lead to replacing Newton's law by another more complicated. So for the velocity of light a value is adopted, such that the astronomic laws compatible with this value may be as simple as possible. When navigators or geographers determine a longitude, they have to solve just the problem we are discussing; they must, without being at Paris, calculate Paris time. How do they accomplish it? They carry a chronometer set for Paris. The qualitative problem of simultaneity is made to depend upon the quantitative problem of the measurement of time. I need not take up the difficulties relative to this latter problem, since above I have emphasized them at length.

Or else they observe an astronomic phenomenon, such as an eclipse of the moon, and they suppose that this phenomenon is perceived simultaneously from all points of the earth. That is not altogether true, since the propagation of light is not instantaneous; if absolute exactitude were desired, there would be a correction to make according to a complicated rule.

Or else finally they use the telegraph. It is clear first that the reception of the signal at Berlin, for instance, is after the sending of this same signal from Paris. This is the rule of cause and effect analyzed above. But how much after? In general, the duration of the transmission is neglected and the two events are regarded as simultaneous. But, to be rigorous, a little correction would still have to be made by a complicated calculation; in practise it is not made, because it would be well within the errors of observation; its theoretic necessity is none the less from our point of view, which is that of a rigorous definition. From this discussion, I wish to emphasize two things: (1) The rules applied are exceedingly various. (2) It is difficult to separate the qualitative problem of simultaneity from the quantitative problem of the measurement of time; no matter whether a chronometer is used, or whether account must be taken of a velocity of transmission, as that of light, because such a velocity could not be measured without *measuring* a time.

To conclude: We have not a direct intuition of simultaneity, nor of the equality of two durations. If we think we have this intuition, this is an illusion. We replace it by the aid of certain rules which we apply almost always without taking count of them.

But what is the nature of these rules? No general rule, no rigorous rule; a multitude of little rules applicable to each particular case.

These rules are not imposed upon us and we might amuse ourselves in inventing others; but they could not be cast aside without greatly complicating the enunciation of the laws of physics, mechanics and astronomy.

We therefore choose these rules, not because they are true, but because they are the most convenient, and we may recapitulate them as follows: "The simultaneity of two events, or the order of their succession, the equality of two durations, are to be so defined that the enunciation of the natural laws may be as simple as possible. In other words, all these rules, all these definitions are only the fruit of an unconscious opportunism."

The Classical Mechanics

The English teach mechanics as an experimental science; on the Continent it is taught always more or less as a deductive and a priori science. The English are right, no doubt. How is it that the other method has been persisted in for so long; how is it that Continental scientists who have tried to escape from the practice of their predecessors have in most cases been unsuccessful? On the other hand, if the principles of mechanics are only of experimental origin, are they not merely approximate and provisory? May we not be some day compelled by new experiments to modify or even to abandon them? These are the questions which naturally arise, and the difficulty of solution is largely due to the fact that treatises on mechanics do not clearly distinguish between what is experiment, what is mathematical reasoning, what is convention, and what is hypothesis. This is not all.

1. There is no absolute space, and we only conceive of relative motion; and yet in most cases mechanical facts are enunciated as if there is an absolute space to which they can be referred.
2. There is no absolute time. When we say that two periods are equal, the statement has no meaning, and can only acquire a meaning by a convention.
3. Not only have we no direct intuition of the equality of two periods, but we have not even direct intuition of the simultaneity of two events occurring in two different places. I have explained this in an article entitled "Mesure du Temps."
4. Finally, is not our Euclidean geometry in itself only a kind of convention of language? Mechanical facts might be enunciated with reference to a non-Euclidean space which would be less convenient but quite as legitimate as our ordinary space; the enunciation would become more complicated, but it still would be possible.

Thus, absolute space, absolute time, and even geometry are not conditions which are imposed on mechanics. All these things no more existed before mechanics than the French language can be logically said to have existed before the truths which are expressed in French. We might endeavour to enunciate the fundamental law of mechanics in a language independent of all these conventions; and no doubt we should in this way get a clearer idea of those laws in themselves. This is what M. Andrade has tried to do, to some extent at any rate, in his *Leçons de Mécanique physique*. Of course the enunciation of these laws would become much more complicated, because all these conventions have been adopted for the very purpose of abbreviating and simplifying the enunciation.

The Future of Mathematical Physics

Perhaps, too, we shall have to construct an entirely new mechanics that we only succeed in catching a glimpse of, where, inertia increasing with the velocity, the velocity of light would become an impassable limit. The ordinary mechanics, more simple, would remain a first approximation, since it would be true for velocities not too great, so that the old dynamics would still be found under the new. We should not have to regret having believed in the principles, and even, since velocities too great for the old formulas would always be only exceptional, the surest way in practice would be still to act as if we continued to believe in them. They are so useful, it would be necessary to keep a place for them. To determine to exclude them altogether would be to deprive oneself of a precious weapon. I hasten to say in conclusion that we are not yet there, and as yet nothing proves that the principles will not come forth from out the fray victorious and intact.

6
— Albert Einstein —
(1879–1955)

Albert Einstein, a German theoretical physicist, was surely the most important scientist of the twentieth century and one of the two or three most influential scientists of the last millennium. In addition to his original and revolutionary theories of relativity, Einstein made major contributions to quantum physics, thermodynamics, and cosmology. The paper announcing the Special Theory of Relativity appeared in 1905 when Einstein was still a clerk at a Swiss patent office in Bern. In that remarkable year Einstein also published ground-breaking papers on the photoelectric effect and Brownian motion. In 1915 Einstein published his General Theory of Relativity, which, after some stunning predictive success, guaranteed his lofty position in the history of science. Einstein spent most of the rest of his career in a search for a field theory that would unify electromagnetism and gravity, and also in a debate with Niels Bohr and others over the nature of quantum mechanics. He received the Nobel Prize for physics in 1921.

In this reading, Einstein announces the postulates of the Special Theory of Relativity. He uses these postulates to systematically complete the existing electrodynamical theory of Maxwell and Lorentz and paves the way for the General Theory that would appear a decade later. In addition to solving certain outstanding problems remaing from nineteenth-century physics, the postulates also led to a number of counterintuitive results. At the heart of these results was Einstein's profound insight into the unified character of space and time.

In the excerpt that follows, Einstein uses his two postulates to define the concept of simultaneity. He then proceeds to show the impact of this new relativistic framework of thinking upon the length of moving rods and the measurement of time. The reading ends with the theory behind the transformation of equivalent systems of coordinates.

THE POSTULATES OF THE SPECIAL THEORY OF RELATIVITY

It is well known that Maxwell's electrodynamics—as usually understood at present—when applied to moving bodies, leads up to asymmetries that do not seem to attach to the phenomena. Let us recall, for example, the electrodynamic interaction between a magnet and a conductor. The observable phenomenon depends here only on the relative motion of conductor and magnet, while according to the customary conception the two cases, in which, respectively, either the one or the other of the two bodies is the one in motion, are to be strictly differentiated from each other. For if the magnet is in motion and the conductor is at rest, there arises in the surroundings of the magnet an electric field endowed with a certain energy value that produces a current in the places where parts of the conductor are located. But if the magnet is at rest and the conductor is in motion, no electric field arises in the surroundings of the magnet, while in the conductor an electromotive force will arise, to which in itself there does not correspond any energy, but which, provided that the relative motion in the two cases considered is the same, gives rise to electrical currents that have the same magnitude and the same course as those produced by the electric forces in the first-mentioned case.

Examples of a similar kind, and the failure of attempts to detect a motion of the earth relative to the "light medium," lead to the conjecture that not only in mechanics, but in electrodynamics as well, the phenomena do not have any properties corresponding to the concept of absolute rest, but that in all coordinate systems in which the mechanical equations are valid, also the same electrodynamic and optical laws are valid, as has already been shown for quantities of the first order. We shall raise this conjecture (whose content will be called "the principle of relativity" hereafter) to the status of a postulate and shall introduce, in addition, the postulate, only seemingly incompatible with the former one, that in empty space light is always propagated with a definite velocity V which is independent of the state of motion of the emitting body. These two postulates suffice for arriving at a simple and consistent electrodynamics of moving bodies on the basis of Maxwell's theory for bodies at rest. The introduction of a "light ether" will prove superfluous, inasmuch as in accordance with the concept to be developed here, no "space at absolute rest" endowed with special properties will be introduced, nor will a velocity vector be assigned to a point of empty space at which electromagnetic processes are taking place.

Like every other electrodynamics, the theory to be developed is based on the kinematics of the rigid body, since the assertions of each and any theory concern the relations between rigid bodies (coordinate systems), clocks, and electromagnetic processes. Insufficient regard for this circumstance is at the root of the difficulties with which the electrodynamics of moving bodies must presently grapple.

I. Kinematic Part

§1. Definition of Simultaneity

Consider a coordinate system in which the Newtonian mechanical equations are valid. To distinguish it verbally from the coordinate systems that will be introduced later on, and to visualize it more precisely, we will designate this system as the "system at rest."

If a material point is at rest relative to this coordinate system, its position relative to the latter can be determined by means of rigid measuring rods using the methods of Euclidean geometry and can be expressed in Cartesian coordinates.

If we want to describe the *motion* of a material point, we give the values of its coordinates as a function of time. However, we should keep in mind that for such a mathematical description to have physical meaning, we first have to clarify what is to be understood here by "time." We have to bear in mind that all our propositions involving time are always propositions about *simultaneous events*. If, for example, I say that "the train arrives here at 7 o'clock," that means, more or less, "the pointing of the small hand of my clock to 7 and the arrival of the train are simultaneous events."

It might seem that all difficulties involved in the definition of "time" could be overcome by my substituting "position of the small hand of my clock" for "time." Such a definition is indeed sufficient if time has to be defined exclusively for the place at which the clock is located; but the definition becomes insufficient as soon as series of events occurring at different locations have to be linked temporally, or—what amounts to the same—events occurring at places remote from the clock have to be evaluated temporally.

To be sure, we could content ourselves with evaluating the time of the events by stationing an observer with the clock at the coordinate origin, and having him assign the corresponding clock-hand position to each light signal that attests to an event to be evaluated and reaches him through empty space. But as we know from experience, such an assignment has the drawback that it is not independent of the position of the observer equipped with the clock. We arrive at a far more practical arrangement by the following consideration.

If there is a clock at point A of space, then an observer located at A can evaluate the time of the events in the immediate vicinity of A by finding the clock-hand positions that are simultaneous with these events. If there is also a clock at point B—we should add, "a clock of exactly the same constitution as that at A"—then the time of the events in the immediate vicinity of B can likewise be evaluated by an observer located at B. But it is not possible to compare the time of an event at A with one at B without a further stipulation; thus far we have only defined an "A-time" and a "B-time" but not a "time" common to A and B. The latter can now be determined by establishing *by definition* that the "time" needed for the light to travel from A to B is equal to the "time" it needs to travel from B to A. For, suppose a ray of light leaves from A toward B at "A-time" t_A, is reflected from B toward A at "B-time" t_B, and arrives back at A at "A-time" t_A'. The two clocks are synchronous by definition if

$$t_B - t_A = t_A' - t_B.$$

We assume that it is possible for this definition of synchronism to be free of contradictions, and to be so for arbitrarily many points, and that the following relations are therefore generally valid:

1. If the clock in B is synchronous with the clock in A, then the clock in A is synchronous with the clock in B.

2. If the clock in A is synchronous with the clock in B as well as with the clock in C, then the clocks in B and C are also synchronous relative to each other.

With the help of some physical (thought) experiments, we have thus laid down what is to be understood by synchronous clocks at rest that are situated at different places, and have obviously obtained thereby a definition of "synchronous" and of "time." The "time" of an event is the reading obtained simultaneously with the event from a clock at rest that is located at the place of the event and that for all time determinations is in synchrony with a specified clock at rest.

Based on experience, we also postulate that the quantity

$$\frac{\overline{2AB}}{t_A' - t_A} = V$$

is a universal constant (the velocity of light in empty space).

It is essential that we have defined time by means of clocks at rest in a system at rest; because it belongs to the system at rest, we designate the time just defined as "the time of the system at rest."

§2. On the Relativity of Lengths and Times

The considerations that follow are based on the principle of relativity and the principle of the constancy of the velocity of light, two principles that we define as follows:

1. The laws governing the changes of the state of any physical system do not depend on which one of two coordinate systems in uniform translational motion relative to each other these changes of the state are referred to.

2. Each ray of light moves in the coordinate system "at rest" with the definite velocity V independent of whether this ray of light is emitted by a body at rest or a body in motion. Here,

$$\text{velocity} = \frac{\text{light path}}{\text{time interval}},$$

where "time interval" should be understood in the sense of the definition in §1.

Let there be given a rigid rod at rest; its length, measured by a measuring rod that is also at rest, shall be ℓ. We now imagine that the axis of the rod is placed along the X-axis of the coordinate system at rest, and that the rod is then set in uniform parallel translational motion (velocity v) along the X-axis in the direction of increasing x. We now seek to determine the length of the *moving* rod, which we imagine to be obtained by the following two operations:

(a) The observer co-moves with the above-mentioned measuring rod and the rod to be measured, and measures the length of the rod directly, by applying the measuring rod exactly as if the rod to be measured, the observer, and the measuring rod were at rest.

(b) Using clocks at rest that are set up in the system at rest and are synchronous in the sense of §1, the observer determines the points of the system at rest at which the beginning and the end of the rod to be measured are found at some given time t. The distance between these two points, measured by the rod used before, which in the present case is at rest, is also a length, which can be designated as the "length of the rod."

According to the principle of relativity, the length to be found in operation (a), which we shall call "the length of the rod in the moving system," must equal the length ℓ of the rod at rest.

We will determine the length to be found in operation (b), which we shall call "the length of the (moving) rod in the system at rest," on the basis of our two principles, and will find it to be different from ℓ.

The commonly used kinematics tacitly assumes that the lengths determined by the two methods mentioned are exactly identical, or, in other words, that in the time epoch t a moving rigid body is totally replaceable, as far as geometry is concerned, by the *same* body when it is *at rest* in a particular position.

Further, we imagine that the two ends (A and B) of the rod are equipped with clocks that are synchronous with the clocks of the system at rest, i.e., whose readings always correspond to the "time of the system at rest" at the locations they happen to occupy; hence, these clocks are "synchronous in the system at rest."

We further imagine that each clock has an observer co-moving with it, and that these observers apply to the two clocks the criterion for synchronism formulated in §1. Suppose a ray of light starts out from A at time t_A, is reflected from B at time t_B, and arrives back at A at time t'_A. Taking into account the principle of the constancy of the velocity of light, we find that

$$t_B - t_A = \frac{{}^{r}AB}{V - v}$$

and

$$t'_A - t_B = \frac{{}^{r}AB}{V + v},$$

where ${}^{r}AB$ denotes the length of the moving rod, measured in the system at rest. The observers co-moving with the moving rod would thus find that the two clocks do not run synchronously while the observers in the system at rest would declare them synchronous.

Thus we see that we must not ascribe *absolute* meaning to the concept of simultaneity; instead, two events that are simultaneous when observed from some particular coordinate system can no longer be considered simultaneous when observed from a system that is moving relative to that system.

7

— Herman Minkowski —

(1864–1909)

Herman Minkowski was a German physicist who at one time was Einstein's teacher in Zurich. Minkowski was an exceptional mathematician who was able to simplify the equations of Special Relativity, thus facilitating the development of General Relativity. Minkowski was also responsible for the concept of the space-time continuum, which became a central element in Relativity Theory.

In this reading, Minkowski describes the concept of a four-dimensional space-time continuum as the medium for physical phenomena. In the continuum, all four dimensions are to be treated identically. This radical revision of the concepts of space and time follows from the postulates of Special Relativity.

The reading comes from an address entitled "Space and Time" given by Minkowski in 1908 at Cologne.

THE SPACE-TIME CONTINUUM

The views of space and time which I wish to lay before you have sprung from the soil of experimental physics, and therein lies their strength. They are radical. Henceforth space by itself, and time by itself, are doomed to fade away into mere shadows, and only a kind of union of the two will preserve an independent reality.

I

First of all I should like to show how it might be possible, setting out from the accepted mechanics of the present day, along a purely mathematical line of thought, to arrive at changed ideas of space and time. The equations of Newton's mechanics exhibit a two-fold invariance. Their form remains unaltered, firstly, if we subject the underlying system of spatial co-ordinates to any arbitrary *change of position*; secondly, if we change its state of motion, namely, by imparting to it any *uniform translatory motion*; furthermore, the zero point of time is given no part to play. We are accustomed to look upon the axioms of geometry as finished with, when we feel ripe for the axioms of mechanics, and for that reason the two invariances are probably rarely mentioned in the same breath. Each of them by itself signifies, for the differential equations of mechanics, a certain group of transformations. The existence of the first group is looked upon as a fundamental characteristic of space. The second group is preferably treated with disdain, so that we with untroubled minds may overcome the difficulty of never being able to decide, from physical phenomena, whether space, which is supposed to be stationary, may not be after all in a state of uniform translation. Thus the two groups, side by side, lead their lives entirely apart. Their utterly heterogeneous character may have discouraged any attempt to compound them. But it is precisely when they are compounded that the complete group, as a whole, gives us to think.

We will try to visualize the state of things by the graphic method. Let *x*, *y*, *z* be rectangular co-ordinates for space, and let *t* denote time. The objects of our perception invariably include places and times in combination. Nobody has ever noticed a place except at a time, or a time except at a place. But I still respect the dogma that both space and time have independent significance. A point of space at a point of time, that is, a system of values *x*, *y*, *z*, *t*, I will call a *world-point*. The multiplicity of all thinkable *x*, *y*, *z*, *t* systems of values we will christen the *world*. With this most valiant piece of chalk I might project upon the blackboard four world-axes. Since merely one chalky axis, as it is, consists of molecules all a-thrill, and moreover is taking part in the earth's travels in the universe, it already affords us ample scope for abstraction; the somewhat greater abstraction associated with the number four is for the mathematician no infliction. Not to leave a yawning void anywhere, we will imagine that everywhere and everywhen there is something perceptible. To avoid saying "matter" or "electricity" I will use for this something the word "substance." We fix our attention on the substantial point which is at the world-point *x*, *y*, *z*, *t*, and imagine that we are able to recognize this substantial point at any other time. Let the variations *dx*, *dy*, *dz* of the space co-ordinates of this substantial point correspond to a time element *dt*. Then we obtain, as an image, so to speak, of the everlasting career of the substantial point, a curve in the world, a *world-line*, the points of which can be referred unequivocally to the parameter t from $-\infty$ to $+\infty$. The whole universe is seen to resolve itself into similar world-lines, and I would fain anticipate myself by saying that in my opinion physical laws might find their most perfect expression as reciprocal relations between these world-lines.

I will state at once what is the value of c with which we shall finally be dealing. It is the velocity of the propagation of light in empty space. To avoid speaking either of space or of emptiness, we may define this magnitude in another way, as the ratio of the electromagnetic to the electrostatic unit of electricity.

The existence of the invariance of natural laws for the relevant group G_c would have to be taken, then, in this way:—

From the totality of natural phenomena it is possible, by successively enhanced approximations, to derive more and more exactly a system of reference x, y, z, t, space and time, by means of which these phenomena then present themselves in agreement with definite laws. But when this is done, this system of reference is by no means unequivocally determined by the phenomena. *It is still possible to make any change in the system of reference that is in conformity with the transformations of the group G_c, and leave the expression of the laws of nature unaltered.*

For example, in correspondence with the figure described above, we may also designate time t', but then must of necessity, in connexion therewith, define space by the manifold of the three parameters x', y, z, in which case physical laws would be expressed in exactly the same way by means of x', y, z, t' as by means of x, y, z, t. We should then have in the world no longer *space*, but an infinite number of spaces, analogously as there are in three-dimensional space an infinite number of planes. Three-dimensional geometry becomes a chapter in four-dimensional physics. Now you know why I said at the outset that space and time are to fade away into shadows, and only a world in itself will subsist.

II

The question now is, what are the circumstances which force this changed conception of space and time upon us? Does it actually never contradict experience? And finally, is it advantageous for describing phenomena?

Before going into these questions, I must make an important remark. If we have in any way individualized space and time, we have, as a world-line corresponding to a stationary substantial point, a straight line parallel to the axis of t; corresponding to a substantial point in uniform motion, a straight line at an angle to the axis of t; to a substantial point in varying motion, a world-line in some form of a curve. If at any world-point x, y, z, t we take the world-line passing through that point, and find it parallel to any radius vector OA' of the above-mentioned hyperboloidal sheet, we can introduce OA' as a new axis of time, and with the new concepts of space and time thus given, the substance at the world-point concerned appears as at rest. We will now introduce this fundamental axiom:—

The substance at any world-point may always, with the appropriate determination of space and time, be looked upon as at rest.

The axiom signifies that at any world-point the expression

$$c^2dt^2 - dx^2 - dy^2 - dz^2$$

always has a positive value, or, what comes to the same thing, that any velocity v always proves less than c. Accordingly c would stand as the upper limit for all substantial velocities, and that is precisely what would reveal the deeper significance of the magnitude c. In this second form the first impression made by the axiom is not altogether pleasing. But we must bear in mind that a modified form of mechanics, in which the square root of this quadratic differential expression appears, will now make its way, so that cases with a velocity greater than that of light will henceforward play only some such part as that of figures with imaginary co-ordinates in geometry.

Now the impulse and true motive for assuming the group G_c came from the fact that the differential equation for the propagation of light in empty space possesses that group G_c. On the other hand, the concept of rigid bodies has meaning only in mechanics satisfying the group G_∞. If we have a theory of optics with G_c, and if on the other hand there were rigid bodies, it is easy to see that one and the same direction of t would be distinguished by the two hyperboloidal sheets appropriate to G_c and G_∞, and this would have the further consequence, that we should be able, by employing suitable rigid optical instruments in the laboratory, to perceive some alteration in the phenomena when the orientation with respect to the direction of the earth's motion is changed. But all efforts directed towards this goal, in particular the famous interference experiment of Michelson, have had a negative result. To explain this failure, H. A. Lorentz set up an hypothesis, the success of which lies in this very invariance in optics for the group G_c. According to Lorentz any moving body must have undergone a contraction in the direction of its motion, and in fact with a velocity v, a contraction in the ratio

$$1 : \sqrt{1 - v^2/c^2}.$$

This hypothesis sounds extremely fantastical, for the contraction is not to be looked upon as a consequence of resistances in the ether, or anything of that kind, but simply as a gift from above,—as an accompanying circumstance of the circumstance of motion. . . .

In mechanics as reformed in accordance with the world-postulate, the disturbing lack of harmony between Newtonian mechanics and modern electrodynamics disappears of its own accord. Before concluding I will just touch upon the attitude of Newton's law of attraction toward this postulate. I shall assume that when two points of mass m, m_1 describe their world-lines, a motive force vector is exerted by m on m_1, of exactly the same form as that just given in the case of electrons, except that $+ mm_1$ must now take the place of $-ee_1$. We now specially consider the case where the acceleration vector of m is constantly zero. Let us then introduce t in such a way that m is to be taken as at rest, and let only m_1 move under the motive force vector which proceeds from m. If we now modify this given vector in the first place by adding the factor $t^{-1} = \sqrt{1 - v^2/c^2}$, which, to the order of $1/c^2$, is equal to 1, it will be seen that for the positions $x_1, y_1, z_1,$ of m_1 and their variations in time, we should arrive exactly at Kepler's laws again, except that the proper times τ_1 of m_1 would take the place of the times t_1. From this simple remark it may then be seen that the proposed law of attraction combined with the new mechanics is no less well adapted to explain astronomical observations than the Newtonian law of attraction combined with Newtonian mechanics.

The fundamental equations for electromagnetic processes in ponderable bodies also fit in completely with the world-postulate. As I shall show elsewhere, it is not even by any means necessary to abandon the derivation of these fundamental equations from ideas of the electronic theory, as taught by Lorentz, in order to adapt them to the world-postulate.

The validity without exception of the world-postulate, I like to think, is the true nucleus of an electromagnetic image of the world, which, discovered by Lorentz, and further revealed by Einstein, now lies open in the full light of day. In the development of its mathematical consequences there will be ample suggestions for experimental verifications of the postulate, which will suffice to conciliate even those to whom the abandonment of old-established views is unsympathetic or painful, by the idea of a pre-established harmony between pure mathematics and physics.

8
— Albert Einstein —

In this reading Einstein acknowledges the insufficiency of Special Relativity in regard to the actual geometrical interpretation of space and time. Also, by the use of a thought experiment he concludes that the laws of physics must apply to reference frames in any state of motion, not just to Galilean or inertial frames.

Using the perceived equivalence between accelerated reference frames and gravitational fields, Einstein develops a physical description of phenomena in accelerating frames as if they were stationary. To this Einstein adds the Principle of General Covariance, which requires that the general laws of nature must be expressed by equations that hold for all reference frames. The result is the General Theory of Relativity. Einstein points out that one of the extraordinary implications of Equivalence and Covariance is that the path of a light ray will be curved in a gravitational field.

The reading is taken from an article entitled "The Foundation of the General Theory of Relativity," which appeared in the Annals of Physics in 1916.

THE FOUNDATION OF THE GENERAL THEORY OF RELATIVITY

Fundamental Considerations on the Postulate of Relativity

1. Observations on the Special Theory of Relativity

The special theory of relativity is based on the following postulate, which is also satisfied by the mechanics of Galileo and Newton.

If a system of co-ordinates K is chosen so that, in relation to it, physical laws hold good in their simplest form, the *same* laws also hold good in relation to any other system of co-ordinates K′ moving in uniform translation relatively to K. This postulate we call the "special principle of relativity." The word "special" is meant to intimate that the principle is restricted to the case when K′ has a motion of uniform translation relatively to K, but that the equivalence of K′ and K does not extend to the case of non-uniform motion of K′ relatively to K.

Thus the special theory of relativity does not depart from classical mechanics through the postulate of relativity, but through the postulate of the constancy of the velocity of light *in vacuo*, from which, in combination with the special principle of relativity, there follow, in the well-known way, the relativity of simultaneity, the Lorentzian transformation, and the related laws for the behaviour of moving bodies and clocks.

The modification to which the special theory of relativity has subjected the theory of space and time is indeed far-reaching, but one important point has remained unaffected. For the laws of geometry, even according to the special theory of relativity, are to be interpreted directly as laws relating to the possible relative positions of solid bodies at rest; and, in a more general way, the laws of kinematics are to be interpreted as laws which describe the relations of measuring bodies and clocks. To two selected material points of a stationary rigid body there always corresponds a distance of quite definite length, which is independent of the locality and orientation of the body, and is also independent of the time. To two selected positions of the hands of a clock at rest relatively to the privileged system of reference there always corresponds an interval of time of a definite length, which is independent of place and time. We shall soon see that the general theory of relativity cannot adhere to this simple physical interpretation of space and time.

2. The Need for an Extension of the Postulate of Relativity

In classical mechanics, and no less in the special theory of relativity, there is an inherent epistemological defect which was, perhaps for the first time, clearly pointed out by Ernst Mach. We will elucidate it by the following example: —Two fluid bodies of the same size and nature hover freely in space at so great a distance from each other and from all other masses that only those gravitational forces need be taken into account which arise from the interaction of different parts of the same body. Let the distance between the two bodies be invariable, and in neither of the bodies let there be any relative movements of the parts with respect to one another. But let either mass, as judged by an observer at rest relatively to the other mass, rotate with constant angular velocity about the line joining the masses. This is a verifiable relative motion of the two bodies. Now let us imagine that each of the bodies has been surveyed by means of measuring instruments at rest relatively to itself, and let the surface of S_1 prove to be a sphere, and that of S_2 an ellipsoid of revolution. Thereupon we put the question—What is the reason for this difference in the two bodies? No answer can be admitted as epistemologically satisfactory, unless the reason given is an *observable fact of experience*. The law of causality has not the significance of a statement as to the world of experience, except when *observable facts* ultimately appear as causes and effects.

Newtonian mechanics does not give a satisfactory answer to this question. It pronounces as follows:—The laws of mechanics apply to the space R_1, in respect to which the body S_1 is at rest, but not to the space R_2, in respect to which the body S_2 is at rest. But the privileged space R_1 of Galileo, thus introduced, is a merely *factitious* cause, and not a thing that can be observed. It is therefore clear that Newton's mechanics does not really satisfy the requirement of causality in the case under consideration, but only apparently does so, since it makes the factitious cause R_1 responsible for the observable difference in the bodies S_1 and S_2.

The only satisfactory answer must be that the physical system consisting of S_1 and S_2 reveals within itself no imaginable cause to which the differing behaviour of S_1 and S_2 can be referred. The cause must therefore lie *outside* this system. We have to take it that the general laws of motion, which in particular determine the shapes of S_1 and S_2, must be such that the mechanical behaviour of S_1 and S_2 is partly conditioned, in quite essential respects, by distant masses which we have not included in the system under consideration. These distant masses and their motions relative to S_1 and S_2 must then be regarded as the seat of the causes (which must be susceptible to observation) of the different behaviour of our two bodies S_1 and S_2. They take over the role of the factitious cause R_1. Of all imaginable spaces R_1, R_2, etc., in any kind of motion relatively to one another, there is none which we may look upon as privileged *a priori* without reviving the above-mentioned epistemological objection. *The laws of physics must be of such a nature that they apply to systems of reference in any kind of motion.* Along this road we arrive at an extension of the postulate of relativity.

In addition to this weighty argument from the theory of knowledge, there is a well-known physical fact which favours an extension of the theory of relativity. Let K be a Galilean system of reference, i.e. a system relatively to which (at least in the four-dimensional region under consideration) a mass, sufficiently distant from other masses, is moving with uniform motion in a straight line. Let K' be a second system of reference which is moving relatively to K in *uniformly accelerated* translation. Then, relatively to K', a mass sufficiently distant from other masses would have an accelerated motion such that its acceleration and direction of acceleration are independent of the material composition and physical state of the mass.

Does this permit an observer at rest relatively to K' to infer that he is on a "really" accelerated system of reference? The answer is in the negative; for the above-mentioned relation of freely movable masses to K' may be interpreted equally well in the following way. The system of reference K' is unaccelerated, but the space-time territory in question is under the sway of a gravitational field, which generates the accelerated motion of the bodies relatively to K'.

This view is made possible for us by the teaching of experience as to the existence of a field of force, namely the gravitational field, which possesses the remarkable property of imparting the same acceleration to all bodies. The mechanical behaviour of bodies relatively to K' is the same as presents itself to experience in the case of systems which we are wont to regard as "stationary" or as "privileged." Therefore, from the physical standpoint, the assumption readily suggests itself that the systems K and K' may both with equal right be looked upon as "stationary," that is to say, they have an equal title as systems of reference for the physical description of phenomena.

It will be seen from these reflexions that in pursuing the general theory of relativity we shall be led to a theory of gravitation, since we are able to "produce" a gravitational field merely by changing the system of co-ordinates. It will also be obvious that the principle of the constancy of the velocity of light *in vacuo* must be modified, since we easily recognize that the path of a ray of light with respect to K' must in general be curvilinear, if with respect to K light is propagated in a straight line with a definite constant velocity.

3. The Space-Time Continuum. Requirement of General Co-Variance for the Equations Expressing General Laws of Nature

In classical mechanics, as well as in the special theory of relativity, the co-ordinates of space and time have a direct physical meaning. To say that a point-event has the X_1 co-ordinate x_1 means that the projection of the point-event on the axis of X_1, determined by rigid rods and in accordance with the rules of Euclidean geometry, is obtained by measuring off a given rod (the unit of length) x_1 times from the origin of co-ordinates along the axis of X_1. To say that a point-event has the X_4 co-ordinate $x_4 = t$, means that a standard clock, made to measure time in a definite unit period, and which is stationary relatively to the system of co-ordinates and practically coincident in space with the point-event, will have measured off $x_4 = t$ periods at the occurrence of the event.

This view of space and time has always been in the minds of physicists, even if, as a rule, they have been unconscious of it. This is clear from the part which these concepts play in physical measurements; it must also have underlain the reader's reflexions on the preceding paragraph (§2) for him to connect any meaning with what he there read. But we shall now show that we must put it aside and replace it by a more general view, in order to be able to carry through the postulate of general relativity, if the special theory of relativity applies to the special case of the absence of a gravitational field.

In a space which is free of gravitational fields we introduce a Galilean system of reference K (x, y, z, t), and also a system of co-ordinates K' (x', y', z', t') in uniform rotation relatively to K. Let the origins of both systems, as well as their axes of Z, permanently coincide. We shall show that for a space-time measurement in the system K' the above definition of the physical meaning of lengths and times cannot be maintained. For reasons of symmetry it is clear that a circle around the origin in the X, Y plane of K may at the same time be regarded as a circle in the X', Y' plane of K'. We suppose that the circumference and diameter of this circle have been measured with a unit measure infinitely small compared with the radius, and that we have the quotient of the two results. If this experiment were performed with a measuring-rod at rest relatively to the Galilean system K, the quotient would be π. With a measuring-rod at rest relatively to K', the quotient would be greater than π. This is readily understood if we envisage the whole process of measuring from the "stationary" system K, and take into consideration that the measuring-rod applied to the periphery undergoes a Lorentzian contraction, while the one applied along the radius does not. Hence Euclidean geometry does not apply to K'. The notion of co-ordinates defined above, which presupposes the validity of Euclidean geometry, therefore breaks down in relation to the system K'. So, too, we are unable to introduce a time corresponding to physical requirements in K', indicated by clocks at rest relatively to K'. To convince ourselves of this impossibility, let us imagine two clocks of identical constitution placed, one at the origin of co-ordinates, and the other at the circumference of the circle, and both envisaged from the "stationary" system K. By a familiar result of the special theory of relativity, the clock at the circumference—judged from K—goes more slowly than the other, because the former is in motion and the latter at rest. An observer at the common origin of co-ordinates, capable of observing the clock at the circumference by means of light, would therefore see it lagging behind the clock beside him. As he will not make up his mind to let the velocity of light along the path in question depend explicitly on the time, he will interpret his observations as showing that the clock at the circumference "really" goes more slowly than the clock at the origin. So he will be obliged to define time in such a way that the rate of a clock depends upon where the clock may be.

We therefore reach this result:—In the general theory of relativity, space and time cannot be defined in such a way that differences of the spatial co-ordinates can be directly measured by the unit measuring-rod, or differences in the time co-ordinate by a standard clock.

The method hitherto employed for laying co-ordinates into the space-time continuum in a definite manner thus breaks down, and there seems to be no other way which would allow us to adapt systems of co-ordinates to the four-dimensional universe so that we might expect from their application a particularly simple formulation of the laws of nature. So there is nothing for it but to regard all imaginable systems of co-ordinates, on principle, as equally suitable for the description of nature. This comes to requiring that:—

The general laws of nature are to be expressed by equations which hold good for all systems of co-ordinates, that is, are co-variant with respect to any substitutions whatever (generally co-variant).

It is clear that a physical theory which satisfies this postulate will also be suitable for the general postulate of relativity. For the sum of *all* substitutions in any case includes those which correspond to all relative motions of three-dimensional systems of co-ordinates. That this requirement of general co-variance, which takes away from space and time the last remnant of physical objectivity, is a natural one, will be seen from the following reflexion. All our space-time verifications invariably amount to a determination of space-time coincidences. If, for example, events consisted merely in the motion of material points, then ultimately nothing would be observable but the meetings of two or more of these points. Moreover, the results of our measurings are nothing but verifications of such meetings of the material points of our measuring instruments with other material points, coincidences between the hands of a clock and points on the clock dial, and observed point-events happening at the same place at the same time.

The introduction of a system of reference serves no other purpose than to facilitate the description of the totality of such coincidences. We allot to the universe four space-time variables x_1, x_2, x_3, x_4 in such a way that for every point-event there is a corresponding system of values of the variables $x_1 \ldots x_4$. To two coincident point-events there corresponds one system of values of the variables $x_1 \ldots x_4$, i.e. coincidence is characterized by the identity of the co-ordinates. If, in place of the variables $x_1 \ldots x_4$, we introduce functions of them, x'_1, x'_2, x'_3, x'_4, as a new system of co-ordinates, so that the systems of values are made to correspond to one another without ambiguity, the equality of all four co-ordinates in the new system will also serve as an expression for the space-time coincidence of the two point-events. As all our physical experience can be ultimately reduced to such coincidences, there is no immediate reason for preferring certain systems of co-ordinates to others, that is to say, we arrive at the requirement of general co-variance.

9
— Albert Einstein —

In this reading Einstein offers a popular account of the Special and General Theories of Relativity and their major ramifications. Of significant import are the explanations and predictions that flow from General Relativity:

1. *The explanation of the precession of the orbit of Mercury.*
2. *The prediction of the curvature of light rays by the gravitational field of the sun.*
3. *The displacement of spectral lines of light reaching us from large stars as compared with spectral lines of similar terrestrial light sources.*

The reading is taken from Part I, Sections 5–9, 19–20, and 27–29 of the book Relativity: The Special and General Theory, *written by Einstein in 1916.*

THE RAMIFICATIONS OF THE SPECIAL
AND GENERAL THEORIES OF RELATIVITY

FIVE

The Principle of Relativity (in the Restricted Sense)

In order to attain the greatest possible clearness, let us return to our example of the railway carriage supposed to be travelling uniformly. We call its motion a uniform translation ("uniform" because it is of constant velocity and direction, "translation" because although the carriage changes its position relative to the embankment yet it does not rotate in so doing). Let us imagine a raven flying through the air in such a manner that its motion, as observed from the embankment, is uniform and in a straight line. If we were to observe the flying raven from the moving railway carriage, we should find that the motion of the raven would be one of different velocity and direction, but that it would still be uniform and in a straight line. Expressed in an abstract manner we may say: If a mass m is moving uniformly in a straight line with respect to a co-ordinate system K, then it will also be moving uniformly and in a straight line relative to a second co-ordinate system K', provided that the latter is executing a uniform translatory motion with respect to K. In accordance with the discussion contained in the preceding section, it follows that:

If K is a Galileian co-ordinate system, then every other co-ordinate system K' is a Galileian one, when, in relation to K, it is in a condition of uniform motion of translation. Relative to K' the mechanical laws of Galilei-Newton hold good exactly as they do with respect to K.

We advance a step farther in our generalisation when we express the tenet thus: If, relative to K, K' is a uniformly moving co-ordinate system devoid of rotation, then natural phenomena run their course with respect to K' according to exactly the same general laws as with respect to K. This statement is called the *principle of relativity* (in the restricted sense).

As long as one was convinced that all natural phenomena were capable of representation with the help of classical mechanics, there was no need to doubt the validity of this principle of relativity. But in view of the more recent development of electrodynamics and optics it became more and more evident that classical mechanics affords an insufficient foundation for the physical description of all natural phenomena. At this juncture the question of the validity of the principle of relativity became ripe for discussion, and it did not appear impossible that the answer to this question might be in the negative.

Nevertheless, there are two general facts which at the outset speak very much in favour of the validity of the principle of relativity. Even though classical mechanics does not supply us with a sufficiently broad basis for the theoretical presentation of all physical phenomena, still we must grant it a considerable measure of "truth," since it supplies us with the actual motions of the heavenly bodies with a delicacy of detail little short of wonderful. The principle of relativity must therefore apply with great accuracy in the domain of *mechanics*. But that a principle of such broad generality should hold with such exactness in one domain of phenomena, and yet should be invalid for another, is *a priori* not very probable.

We now proceed to the second argument, to which, moreover, we shall return later. If the principle of relativity (in the restricted sense) does not hold, then the Galileian co-ordinate systems K, K', K'', etc., which are moving uniformly relative to each other, will not be *equivalent* for the description of natural phenomena. In this case we should be constrained to believe that natural laws are capable of being formulated in a particularly simple manner, and of course only on condition that, from amongst all possible Galileian co-ordinate systems, we should have chosen *one* (K_0) of a particular state of motion as our body of reference. We should then be justified (because of its merits for the description of natural phenomena) in calling this system "absolutely at rest," and all other Galileian systems K "in motion." If, for instance, our embankment were the system K_0, then our railway carriage would be a system K, relative to which less simple laws would hold than with respect to K_0. This diminished simplicity would be due to the fact that the carriage K would be in motion (i.e. "really") with respect to K_0. In the general laws of nature which have been formulated with reference to K, the magnitude and direction of the velocity of the carriage would necessarily play a part. We should expect, for instance, that the note emitted by an organ-pipe placed with its axis parallel to the direction of travel would be different from that emitted if the axis of the pipe were placed perpendicular to this direction. Now in virtue of its motion in an orbit round the sun, our earth is comparable with a railway carriage travelling with a velocity of about 30 kilometres per second. If the principle of relativity were not valid we should therefore expect that the direction of motion of the earth at any moment would enter into the laws of nature, and also that physical systems in their behaviour would be dependent on the orientation in space with respect to the earth. For owing to the alteration in direction of the velocity of revolution of the earth in the course of a year, the earth cannot be at rest relative to the hypothetical system K_0 throughout the whole year. However, the most careful observations have never revealed such anisotropic properties in terrestrial physical space, i.e. a physical non-equivalence of different directions. This is a very powerful argument in favour of the principle of relativity.

SIX

The Theorem of the Addition of Velocities Employed in Classical Mechanics

Let us suppose our old friend the railway carriage to be travelling along the rails with a constant velocity v, and that a man traverses the length of the carriage in the direction of travel with a velocity w. How quickly or, in other words, with what velocity W does the man advance relative to the embankment during the process? The only possible answer seems to result from the following consideration: If the man were to stand still for a second, he would advance relative to the embankment through a distance v equal numerically to the velocity of the carriage. As a consequence of his walking, however, he traverses an additional distance w relative to the carriage, and hence also relative to the embankment, in this second, the distance w being numerically equal to the velocity with which he is walking. Thus in total he covers the distance $W = v + w$ relative to the embankment in the second considered. We shall see later that this result, which expresses the theorem of the addition of velocities employed in classical mechanics, cannot be maintained; in other words, the law that we have just written down does not hold in reality. For the time being, however, we shall assume its correctness.

SEVEN

The Apparent Incompatibility of the Law of Propagation of Light with the Principle of Relativity

There is hardly a simpler law in physics than that according to which light is propagated in empty space. Every child at school knows, or believes he knows, that this propagation takes place in straight lines with a velocity $c = 300,000$ km./sec. At all events we know with great exactness that this velocity is the same for all colours, because if this were not the case, the minimum of emission would not be observed simultaneously for different colours during the eclipse of a fixed star by its dark neighbour. By means of similar considerations based on observations of double stars, the Dutch astronomer De Sitter was also able to show that the velocity of propagation of light cannot depend on the velocity of motion of the body emitting the light. The assumption that this velocity of propagation is dependent on the direction "in space" is in itself improbable.

In short, let us assume that the simple law of the constancy of the velocity of light c (in vacuum) is justifiably believed by the child at school. Who would imagine that this simple law has plunged the conscientiously thoughtful physicist into the greatest intellectual difficulties? Let us consider how these difficulties arise.

Of course we must refer the process of the propagation of light (and indeed every other process) to a rigid reference-body (co-ordinate system). As such a system let us again choose our embankment. We shall imagine the air above it to have been removed. If a ray of light be sent along the embankment, we see from the above that the tip of the ray will be transmitted with the velocity c relative to the embankment. Now let us suppose that our railway carriage is again travelling along the railway lines with the velocity v, and that its direction is the same as that of the ray of light, but its velocity of course much less. Let us inquire about the velocity of propagation of the ray of light relative to the carriage. It is obvious that we can here apply the consideration of the previous section, since the ray of light plays the part of the man walking along relatively to the carriage. The velocity W of the man relative to the embankment is here replaced by the velocity of light relative to the embankment. w is the required velocity of light with respect to the carriage, and we have

$$w = c - v.$$

The velocity of propagation of a ray of light relative to the carriage thus comes out smaller than c.

But this result comes into conflict with the principle of relativity set forth in Section 5. For, like every other general law of nature, the law of transmission of light *in vacuo* must, according to the principle of relativity, be the same for the railway carriage as reference-body as when the rails are the body of reference. But, from our above consideration, this would appear to be impossible. If every ray of light is propagated relative to the embankment with the velocity c, then for this reason it would appear that another law of propagation of light must necessarily hold with respect to the carriage—a result contradictory to the principle of relativity.

In view of this dilemma there appears to be nothing else for it than to abandon either the principle of relativity or the simple law of the propagation of light *in vacuo*. Those of you who have carefully followed the preceding discussion are almost sure to expect that we should retain the principle of relativity, which appeals so convincingly to the intellect because it is so natural and simple. The law of the propagation of light *in vacuo* would then have to be replaced by a more complicated law conformable to the principle of relativity. The development of theoretical physics shows, however, that we cannot pursue this course. The epoch-making theoretical investigations of H. A. Lorentz on the electrodynamical and optical phenomena connected with moving bodies show that experience in this domain leads conclusively to a theory of electromagnetic phenomena, of which the law of the constancy of the velocity of light *in vacuo* is a necessary consequence. Prominent theoretical physicists were therefore more inclined to reject the principle of relativity, in spite of the fact that no empirical data had been found which were contradictory to this principle.

At this juncture the theory of relativity entered the arena. As a result of an analysis of the physical conceptions of time and space, it became evident that *in reality there is not the least incompatibility between the principle of relativity and the law of propagation of light*, and that by systematically holding fast to both these laws a logically rigid theory could be arrived at. This theory has been called the *special theory of relativity* to distinguish it from the extended theory, with which we shall deal later. In the following pages we shall present the fundamental ideas of the special theory of relativity.

EIGHT

On the Idea of Time in Physics

Lightning has struck the rails on our railway embankment at two places A and B far distant from each other. I make the additional assertion that these two lightning flashes occurred simultaneously. If I ask you whether there is sense in this statement, you will answer my question with a decided "Yes." But if I now approach you with the request to explain to me the sense of the statement more precisely, you find after some consideration that the answer to this question is not so easy as it appears at first sight.

After some time perhaps the following answer would occur to you: "The significance of the statement is clear in itself and needs no further explanation; of course it would require some consideration if I were to be commissioned to determine by observations whether in the actual case the two events took place simultaneously or not." I cannot be satisfied with this answer for the following reason. Supposing that as a result of ingenious consideration an able meteorologist were to discover that the lightning must always strike the places A and B simultaneously, then we should be faced with the task of testing whether or not this theoretical result is in accordance with the reality. We encounter the same difficulty with all physical statements in which the conception "simultaneous" plays a part. The concept does not exist for the physicist until he has the possibility of discovering whether or not it is fulfilled in an actual case. We thus require a definition of simultaneity such that this definition supplies us with the method by means of which, in the present case, he can decide by experiment whether or not both the lightning strokes occurred simultaneously. As long as this requirement is not satisfied, I allow myself to be deceived as a physicist (and of course the same applies if I am not a physicist), when I imagine that I am able to attach a meaning to the statement of simultaneity. (I would ask the reader not to proceed farther until he is fully convinced on this point.)

After thinking the matter over for some time you then offer the following suggestion with which to test simultaneity. By measuring along the rails, the connecting line AB should be measured up and an observer placed at the midpoint M of the distance AB. This observer should be supplied with an arrangement (e.g. two mirrors inclined at 90°) which allows him visually to observe both places A and B at the same time. If the observer perceives the two flashes of lightning at the same time, then they are simultaneous.

I am very pleased with this suggestion, but for all that I cannot regard the matter as quite settled, because I feel constrained to raise the following objection: "Your definition would certainly be right, if only I knew that the light by means of which the observer at M perceives the lightning flashes travels along the length $A \rightarrow M$ with the same velocity as along the length $B \rightarrow M$. But an examination of this supposition would only be possible if we already had at our disposal the means of measuring time. It would thus appear as though we were moving here in a logical circle."

After further consideration you cast a somewhat disdainful glance at me—and rightly so—and you declare: "I maintain my previous definition nevertheless, because in reality it assumes absolutely nothing about light. There is only *one* demand to be made of the definition of simultaneity, namely, that in every real case it must supply us with an empirical decision as to whether or not the conception that has to be defined is fulfilled. That my definition satisfies this demand is indisputable. That light requires the same time to traverse the path $A \to M$ as for the path $B \to M$ is in reality neither a *supposition nor a hypothesis* about the physical nature of light, but a *stipulation* which I can make of my own freewill in order to arrive at a definition of simultaneity."

It is clear that this definition can be used to give an exact meaning not only to *two* events, but to as many events as we care to choose, and independently of the positions of the scenes of the events with respect to the body of reference (here the railway embankment). We are thus led also to a definition of "time" in physics. For this purpose we suppose that clocks of identical construction are placed at the points A, B and C of the railway line (co-ordinate system), and that they are set in such a manner that the positions of their pointers are simultaneously (in the above sense) the same. Under these conditions we understand by the "time" of an event the reading (position of the hands) of that one of these clocks which is in the immediate vicinity (in space) of the event. In this manner a time-value is associated with every event which is essentially capable of observation.

This stipulation contains a further physical hypothesis, the validity of which will hardly be doubted without empirical evidence to the contrary. It has been assumed that all these clocks go *at the same rate* if they are of identical construction. Stated more exactly: When two clocks arranged at rest in different places of a reference-body are set in such a manner that a *particular* position of the pointers of the one clock is *simultaneous* (in the above sense) with the *same* position of the pointers of the other clock, then identical "settings" are always simultaneous (in the sense of the above definition).

NINE

The Relativity of Simultaneity

Up to now our considerations have been referred to a particular body of reference, which we have styled a "railway embankment." We suppose a very long train travelling along the rails with the constant velocity v and in the direction indicated in Fig. 1. People travelling in this train will with advantage use the train as a rigid reference-body (co-ordinate system); they regard all events in reference to the train. Then every event which takes place along the line also takes place at a particular point of the train. Also the definition of simultaneity can be given relative to the train in exactly the same way as with respect to the embankment. As a natural consequence, however, the following question arises:

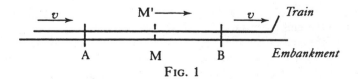

FIG. 1

Are two events (e.g. the two strokes of lightning A and B) which are simultaneous *with reference to the railway embankment* also simultaneous *relatively to the train?* We shall show directly that the answer must be in the negative.

When we say that the lightning strokes A and B are simultaneous with respect to the embankment, we mean: the rays of light emitted at the places A and B, where the lightning occurs, meet each other at the mid-point M of the length $A \rightarrow B$ of the embankment. But the events A and B also correspond to positions A and B on the train. Let M' be the mid-point of the distance $A \rightarrow B$ on the travelling train. Just when the flashes of lightning occur, this point M' naturally coincides with the point M, but it moves towards the right in the diagram with the velocity v of the train. If an observer sitting in the position M' in the train did not possess this velocity, then he would remain permanently at M, and the light rays emitted by the flashes of lightning A and B would reach him simultaneously, i.e. they would meet just where he is situated. Now in reality (considered with reference to the railway embankment) he is hastening towards the beam of light coming from B, whilst he is riding on ahead of the beam of light coming from A. Hence the observer will see the beam of light emitted from B earlier than he will see that emitted from A. Observers who take the railway train as their reference-body must therefore come to the conclusion that the lightning flash B took place earlier than the lightning flash A. We thus arrive at the important result:

Events which are simultaneous with reference to the embankment are not simultaneous with respect to the train, and *vice versa* (relativity of simultaneity). Every reference-body (co-ordinate system) has its own particular time; unless we are told the reference-body to which the statement of time refers, there is no meaning in a statement of the time of an event.

Now before the advent of the theory of relativity it had always tacitly been assumed in physics that the statement of time had an absolute significance, i.e. that it is independent of the state of motion of the body of reference. But we have just seen that this assumption is incompatible with the most natural definition of simultaneity; if we discard this assumption, then the conflict between the law of the propagation of light *in vacuo* and the principle of relativity (developed in Section 7) disappears.

We were led to that conflict by the considerations of Section 6, which are now no longer tenable. In that section we concluded that the man in the carriage, who traverses the distance w *per second* relative to the carriage, traverses the same distance also with respect to the embankment *in each second* of time. But, according to the foregoing considerations, the time required by a particular occurrence with respect to the carriage must not be considered equal to the duration of the same occurrence as judged from the embankment (as reference-body). Hence it cannot be contended that the man in walking travels the distance w relative to the railway line in a time which is equal to one second as judged from the embankment.

Moreover, the considerations of Section 6 are based on yet a second assumption, which, in the light of a strict consideration, appears to be arbitrary, although it was always tacitly made even before the introduction of the theory of relativity.

TEN

On the Relativity of the Conception of Distance

Let us consider two particular points on the train travelling along the embankment with the velocity v, and inquire as to their distance apart. We already know that it is necessary to have a body of reference for the measurement of a distance, with respect to which body the distance can be measured up. It is the simplest plan to use the train itself as reference-body (co-ordinate system). An observer in the train measures the interval by marking off his measuring-rod in a straight line (*e.g.* along the floor of the carriage) as many times as is necessary to take him from the one marked point to the other. Then the number which tells us how often the rod has to be laid down is the required distance.

It is a different matter when the distance has to be judged from the railway line. Here the following method suggests itself. If we call A' and B' the two points on the train whose distance apart is required, then both of these points are moving with the velocity v along the embankment. In the first place we require to determine the points A and B of the embankment which are just being passed by the two points A' and B' at a particular time t—judged from the embankment. These points A and B of the embankment can be determined by applying the definition of time given in Section 8. The distance between these points A and B is then measured by repeated application of the measuring-rod along the embankment.

A priori it is by no means certain that this last measurement will supply us with the same result as the first. Thus the length of the train as measured from the embankment may be different from that obtained by measuring in the train itself. This circumstance leads us to a second objection which must be raised against the apparently obvious consideration of Section 6. Namely, if the man in the carriage covers the distance w in a unit of time—*measured from the train*—then this distance—*as measured from the embankment*—is not necessarily also equal to w.

NINETEEN

The Gravitational Field

"If we pick up a stone and then let it go, why does it fall to the ground?" The usual answer to this question is: "Because it is attracted by the earth." Modern physics formulates the answer rather differently for the following reason. As a result of the more careful study of electromagnetic phenomena, we have come to regard action at a distance as a process impossible without the intervention of some intermediary medium. If, for instance, a magnet attracts a piece of iron, we cannot be content to regard this as meaning that the magnet acts directly on the iron through the intermediate empty space, but we are constrained to imagine—after the manner of Faraday—that the magnet always calls into being something physically real in the space around it, that something being what we call a "magnetic field." In its turn this magnetic field operates on the piece of iron, so that the latter strives to move towards the magnet. We shall not discuss here the justification for this incidental conception, which is indeed a somewhat arbitrary one. We shall only mention that with its aid electromagnetic phenomena can be theoretically represented much more satisfactorily than without it, and this applies particularly to the transmission of electromagnetic waves. The effects of gravitation also are regarded in an analogous manner.

The action of the earth on the stone takes place indirectly. The earth produces in its surroundings a gravitational field, which acts on the stone and produces its motion of fall. As we know from experience, the intensity of the action on a body diminishes according to a quite definite law, as we proceed farther and farther away from the earth. From our point of view this means: The law governing the properties of the gravitational field in space must be a perfectly definite one, in order correctly to represent the diminution of gravitational action with the distance from operative bodies. It is something like this: The body (e.g. the earth) produces a field in its immediate neighbourhood directly; the intensity and direction of the field at points farther removed from the body are thence determined by the law which governs the properties in space of the gravitational fields themselves.

In contrast to electric and magnetic fields, the gravitational field exhibits a most remarkable property, which is of fundamental importance for what follows. Bodies which are moving under the sole influence of a gravitational field receive an acceleration, *which does not in the least depend either on the material or on the physical state of the body*. For instance, a piece of lead and a piece of wood fall in exactly the same manner in a gravitational field (*in vacuo*), when they start off from rest or with the same initial velocity. This law, which holds most accurately, can be expressed in a different form in the light of the following consideration.

According to Newton's law of motion, we have

$$(Force) = (inertial\ mass) \times (acceleration),$$

where the "inertial mass" is a characteristic constant of the accelerated body. If now gravitation is the cause of the acceleration, we then have

$$(Force) = (gravitational\ mass) \times (intensity\ of\ the\ gravitational\ field),$$

where the "gravitational mass" is likewise a characteristic constant for the body. From these two relations follows:

$$(acceleration) = \frac{(gravitational\ mass)}{(inertial\ mass)} \times (intensity\ of\ the\ gravitational\ field).$$

If now, as we find from experience, the acceleration is to be independent of the nature and the condition of the body and always the same for a given gravitational field, then the ratio of the gravitational to the inertial mass must likewise be the same for all bodies. By a suitable choice of units we can thus make this ratio equal to unity. We then have the following law: The *gravitational* mass of a body is equal to its *inertial* mass.

It is true that this important law had hitherto been recorded in mechanics, but it had not been *interpreted*. A satisfactory interpretation can be obtained only if we recognize the following fact: *The same* quality of a body manifests itself according to circumstances as "inertia" or as "weight" (literally "heaviness"). In the following section we shall show to what extent this is actually the case, and how this question is connected with the general postulate of relativity.

TWENTY

The Equality of Inertial and Gravitational Mass as an Argument for the General Postulate of Relativity

We imagine a large portion of empty space, so far removed from stars and other appreciable masses, that we have before us approximately the conditions required by the fundamental law of Galilei. It is then possible to choose a Galilean reference-body for this part of space (world), relative to which points at rest remain at rest and points in motion continue permanently in uniform rectilinear motion. As reference-body let us imagine a spacious chest resembling a room with an observer inside who is equipped with apparatus. Gravitation naturally does not exist for this observer. He must fasten himself with strings to the floor, otherwise the slightest impact against the floor will cause him to rise slowly towards the ceiling of the room.

To the middle of the lid of the chest is fixed externally a hook with rope attached, and now a "being" (what kind of a being is immaterial to us) begins pulling at this with a constant force. The chest together with the observer then begin to move "upwards" with a uniformly accelerated motion. In course of time their velocity will reach unheard-of values—provided that we are viewing all this from another reference-body which is not being pulled with a rope.

But how does the man in the chest regard the process? The acceleration of the chest will be transmitted to him by the reaction of the floor of the chest. He must therefore take up this pressure by means of his legs if he does not wish to be laid out full length on the floor. He is then standing in the chest in exactly the same way as anyone stands in a room of a house on our earth. If he releases a body which he previously had in his hand, the acceleration of the chest will no longer be transmitted to this body, and for this reason the body will approach the floor of the chest with an accelerated relative motion. The observer will further convince himself *that the acceleration of the body towards the floor of the chest is always of the same magnitude, whatever kind of body he may happen to use for the experiment.*

Relying on his knowledge of the gravitational field (as it was discussed in the preceding section), the man in the chest will thus come to the conclusion that he and the chest are in a gravitational field which is constant with regard to time. Of course he will be puzzled for a moment as to why the chest does not fall in this gravitational field. Just then, however, he discovers the hook in the middle of the lid of the chest and the rope which is attached to it, and he consequently comes to the conclusion that the chest is suspended at rest in the gravitational field.

Ought we to smile at the man and say that he errs in his conclusion? I do not believe we ought to if we wish to remain consistent; we must rather admit that his mode of grasping the situation violates neither reason nor known mechanical laws. Even though it is being accelerated with respect to the "Galileian space" first considered, we can nevertheless regard the chest as being at rest. We have thus good grounds for extending the principle of relativity to include bodies of reference which are accelerated with respect to each other, and as a result we have gained a powerful argument for a generalised postulate of relativity.

We must note carefully that the possibility of this mode of interpretation rests on the fundamental property of the gravitational field of giving all bodies the same acceleration, or, what comes to the same thing, on the law of the equality of inertial and gravitational mass. If this natural law did not exist, the man in the accelerated chest would not be able to interpret the behaviour of the bodies around him on the supposition of a gravitational field, and he would not be justified on the grounds of experience in supposing his reference-body to be "at rest."

Suppose that the man in the chest fixes a rope to the inner side of the lid, and that he attaches a body to the free end of the rope. The result of this will be to stretch the rope so that it will hang "vertically" downwards. If we ask for an opinion of the cause of tension in the rope, the man in the chest will say: "The suspended body experiences a downward force in the gravitational field, and this is neutralised by the tension of the rope; what determines the magnitude of the tension of the rope is the *gravitational mass* of the suspended body." On the other hand, an observer who is poised freely in space will interpret the condition of things thus: "The rope must perforce take part in the accelerated motion of the chest, and it transmits this motion to the body attached to it. The tension of the rope is just large enough to effect the acceleration of the body. That which determines the magnitude of the tension of the rope is the *inertial mass* of the body." Guided by this example, we see that our extension of the principle of relativity implies the *necessity* of the law of the equality of inertial and gravitational mass. Thus we have obtained a physical interpretation of this law.

From our consideration of the accelerated chest we see that a general theory of relativity must yield important results on the laws of gravitation. In point of fact, the systematic pursuit of the general idea of relativity has supplied the laws satisfied by the gravitational field. Before proceeding farther, however, I must warn the reader against a misconception suggested by these considerations. A gravitational field exists for the man in the chest, despite the fact that there was no such field for the co-ordinate system first chosen. Now we might easily suppose that the existence of a gravitational field is always only an *apparent* one. We might also think that, regardless of the kind of gravitational field which may be present, we could always choose another reference-body such that *no* gravitational field exists with reference to it. This is by no means true for all gravitational fields, but only for those of quite special form. It is, for instance, impossible to choose a body of reference such that, as judged from it, the gravitational field of the earth (in its entirety) vanishes.

We can now appreciate why that argument is not convincing, which we brought forward against the general principle of relativity. It is certainly true that the observer in the railway carriage experiences a jerk forwards as a result of the application of the brake, and that he recognises in this the non-uniformity of motion (retardation) of the carriage. But he is compelled by nobody to refer this jerk to a "real" acceleration (retardation) of the carriage. He might also interpret his experience thus: "My body of reference (the carriage) remains permanently at rest. With reference to it, however, there exists (during the period of application of the brakes) a gravitational field which is directed forwards and which is variable with respect to time. Under the influence of this field, the embankment together with the earth moves nonuniformly in such a manner that their original velocity in the backwards direction is continuously reduced."

TWENTY-SEVEN

The Space-Time Continuum of the General Theory of Relativity Is Not a Euclidean Continuum

In the first part of this book we were able to make use of space-time co-ordinates which allowed of a simple and direct physical interpretation, and which . . . can be regarded as four-dimensional Cartesian co-ordinates. This was possible on the basis of the law of the constancy of the velocity of light. But . . . the general theory of relativity cannot retain this law. On the contrary, we arrived at the result that according to this latter theory the velocity of light must always depend on the co-ordinates when a gravitational field is present. [Thus] . . . we found that the presence of a gravitational field invalidates the definition of the co-ordinates and the time, which led us to our objective in the special theory of relativity.

In view of the results of these considerations we are led to the conviction that, according to the general principle of relativity, the space-time continuum cannot be regarded as a Euclidean one, but that here we have the general case, corresponding to the marble slab with local variations of temperature, and with which we make acquaintance as an example of a two-dimensional continuum. Just as it was there impossible to construct a Cartesian co-ordinate system from equal rods, so here it is impossible to build up a system (reference-body) from rigid bodies and clocks, which shall be of such a nature that measuring-rods and clocks, arranged rigidly with respect to one another, shall indicate position and time directly. . . .

We refer the four-dimensional space-time continuum in an arbitrary manner to Gauss co-ordinates. We assign to every point of the continuum (event) four numbers, x_1, x_2, x_3, x_4 (co-ordinates), which have not the least direct physical significance, but only serve the purpose of numbering the points of the continuum in a definite but arbitrary manner. This arrangement does not even need to be of such a kind that we must regard x_1, x_2, x_3 as "space" co-ordinates and x_4 as a "time" co-ordinate.

The reader may think that such a description of the world would be quite inadequate. What does it mean to assign to an event the particular co-ordinates x_1, x_2, x_3, x_4, if in themselves these co-ordinates have no significance? More careful consideration shows, however, that this anxiety is unfounded. Let us consider, for instance, a material point with any kind of motion. If this point had only a momentary existence without duration, then it would be described in space-time by a single system of values x_1, x_2, x_3, x_4. Thus its permanent existence must be characterised by an infinitely large number of such systems of values, the co-ordinate values of which are so close together as to give continuity: corresponding to the material point, we thus have a (uni-dimensional) line in the four-dimensional continuum. In the same way, any such lines in our continuum correspond to many points in motion. The only statements having regard to these points which can claim a physical existence are in reality the statements about their encounters. In our mathematical treatment, such an encounter is expressed in the fact that the two lines which represent the motions of the points in question have a particular system of co-ordinate values, x_1, x_2, x_3, x_4, in common. After mature consideration the reader will doubtless admit that in reality such encounters constitute the only actual evidence of a time-space nature with which we meet in physical statements.

When we were describing the motion of a material point relative to a body of reference, we stated nothing more than the encounters of this point with particular points of the reference-body. We can also determine the corresponding values of the time by the observation of encounters of the body with clocks, in conjunction with the observation of the encounter of the hands of clocks with particular points on the dials. It is just the same in the case of space-measurements by means of measuring-rods, as a little consideration will show.

The following statements hold generally: Every physical description resolves itself into a number of statements, each of which refers to the space-time coincidence of two events A and B. In terms of Gaussian co-ordinates, every such statement is expressed by the agreement of their four co-ordinates x_1, x_2, x_3, x_4. Thus in reality, the description of the time-space continuum by means of Gauss co-ordinates completely replaces the description with the aid of a body of reference, without suffering from the defects of the latter mode of description; it is not tied down to the Euclidean character of the continuum which has to be represented.

TWENTY-EIGHT

Exact Formulation of the General Principle of Relativity

We are now in a position to replace the provisional formulation of the general principle of relativity . . . by an exact formulation. The form there used, "All bodies of reference K, K' etc., are equivalent for the description of natural phenomena (formulation of the general laws of nature), whatever may be their state of motion," cannot be maintained, because the use of rigid reference-bodies, in the sense of the method followed in the special theory of relativity, is in general not possible in space-time description. The Gauss co-ordinate system has to take the place of the body of reference. The following statement corresponds to the fundamental idea of the general principle of relativity: "*All Gaussian co-ordinate systems are essentially equivalent for the formulation of the general laws of nature.*"

We can state this general principle of relativity in still another form, which renders it yet more clearly intelligible than it is when in the form of the natural extension of the special principle of relativity. According to the special theory of relativity, the equations which express the general laws of nature pass over into equations of the same form when, by making use of the Lorentz transformation, we replace the space-time variables x, y, z, t, of a (Galileian) reference-body K by the space-time variables x', y', z', t', of a new reference-body K'. According to the general theory of relativity, on the other hand, by application of *arbitrary substitutions* of the Gauss variables x_1, x_2, x_3, x_4, the equations must pass over into equations of the same form; for every transformation (not only the Lorentz transformation) corresponds to the transition of one Gauss co-ordinate system into another.

If we desire to adhere to our "old-time" three-dimensional view of things, then we can characterise the development which is being undergone by the fundamental idea of the general theory of relativity as follows: The special theory of relativity has reference to Galileian domains, i.e., to those in which no gravitational field exists. In this connection a Galileian reference-body serves as body of reference, i.e., a rigid body the state of motion of which is so chosen that the Galileian law of the uniform rectilinear motion of "isolated" material points holds relatively to it.

Certain considerations suggest that we should refer the same Galileian domains to *non-Galileian* reference-bodies also. A gravitational field of a special kind is then present with respect to these bodies.

In gravitational fields there are no such things as rigid bodies with Euclidean properties; thus the fictitious rigid body of reference is of no avail in the general theory of relativity. The motion of clocks is also influenced by gravitational fields, and in such a way that a physical definition of time which is made directly with the aid of clocks has by no means the same degree of plausibility as in the special theory of relativity.

For this reason non-rigid reference-bodies are used, which are as a whole not only moving in any way whatsoever, but which also suffer alterations in form *ad lib.* during their motion. Clocks, for which the law of motion is of any kind, however irregular, serve for the definition of time. We have to imagine each of these clocks fixed at a point on the non-rigid reference-body. These clocks satisfy only the one condition, that the "readings" which are observed simultaneously on adjacent clocks (in space) differ from each other by an indefinitely small amount. This non-rigid reference-body, which might appropriately be termed a "reference-mollusc," is in the main equivalent to a Gaussian four-dimensional co-ordinate system chosen arbitrarily. That which gives the "mollusc" a certain comprehensibility as compared with the Gauss co-ordinate system is the (really unjustified) formal retention of the separate existence of the space co-ordinates as opposed to the time co-ordinate. Every point on the mollusc is treated as a space-point, and every material point which is at rest relatively to it as at rest, so long as the mollusc is considered as reference-body. The general principle of relativity requires that all these molluscs can be used as reference-bodies with equal right and equal success in the formulation of the general laws of nature; the laws themselves must be quite independent of the choice of mollusc.

The great power possessed by the general principle of relativity lies in the comprehensive limitation which is imposed on the laws of nature in consequence of what we have seen above.

TWENTY-NINE

The Solution of the Problem of Gravitation on the Basis of the General Principle of Relativity

If the reader has followed all our previous considerations, he will have no further difficulty in understanding the methods leading to the solution of the problem of gravitation.

We start off from a consideration of a Galileian domain, i.e., a domain in which there is no gravitational field relative to the Galileian reference-body K. The behaviour of measuring-rods and clocks with reference to K is known from the special theory of relativity, likewise the behaviour of "isolated" material points; the latter move uniformly and in straight lines.

Now let us refer this domain to a random Gauss co-ordinate system or to a "mollusc" as reference-body K′. Then with respect to K′ there is a gravitational field G (of a particular kind). We learn the behaviour of measuring-rods and clocks and also of freely-moving material points with reference to K′ simply by mathematical transformation. We interpret this behaviour as the behaviour of measuring-rods, clocks and material points under the influence of the gravitational field G. Hereupon we introduce a hypothesis: that the influence of the gravitational field on measuring-rods, clocks and freely-moving material points continues to take place according to the same laws, even in the case where the prevailing gravitational field is *not* derivable from the Galileian special case, simply by means of a transformation of co-ordinates.

The next step is to investigate the space-time behaviour of the gravitational field *G*, which was derived from the Galileian special case simply by transformation of the co-ordinates. This behaviour is formulated in a law, which is always valid, no matter how the reference-body (mollusc) used in the description may be chosen.

This law is not yet the *general* law of the gravitational field, since the gravitational field under consideration is of a special kind. In order to find out the general law-of-field of gravitation we still require to obtain a generalisation of the law as found above. This can be obtained without caprice, however, by taking into consideration the following demands:

(a) The required generalisation must likewise satisfy the general postulate of relativity.

(b) If there is any matter in the domain under consideration, only its inertial mass, and thus according to Section 15 only its energy is of importance for its effect in exciting a field.

(c) Gravitational field and matter together must satisfy the law of the conservation of energy (and of impulse).

Finally, the general principle of relativity permits us to determine the influence of the gravitational field on the course of all those processes which take place according to known laws when a gravitational field is absent, i.e., which have already been fitted into the frame of the special theory of relativity. In this connection we proceed in principle according to the method which has already been explained for measuring-rods, clocks and freely-moving material points.

The theory of gravitation derived in this way from the general postulate of relativity excels not only in its beauty; nor in removing the defect attaching to classical mechanics . . . nor in interpreting the empirical law of the equality of inertial and gravitational mass; but it has also already explained a result of observation in astronomy, against which classical mechanics is powerless.

If we confine the application of the theory to the case where the gravitational fields can be regarded as being weak, and in which all masses move with respect to the co-ordinate system with velocities which are small compared with the velocity of light, we then obtain as a first approximation the Newtonian theory. Thus the latter theory is obtained here without any particular assumption, whereas Newton had to introduce the hypothesis that the force of attraction between mutually attracting material points is inversely proportional to the square of the distance between them. If we increase the accuracy of the calculation, deviations from the theory of Newton make their appearance, practically all of which must nevertheless escape the test of observation owing to their smallness.

We must draw attention here to one of these deviations. According to Newton's theory, a planet moves round the sun in an ellipse, which would permanently maintain its position with respect to the fixed stars, if we could disregard the motion of the fixed stars themselves and the action of the other planets under consideration. Thus, if we correct the observed motion of the planets for these two influences, and if Newton's theory be strictly correct, we ought to obtain for the orbit of the planet an ellipse, which is fixed with reference to the fixed stars. This deduction, which can be tested with great accuracy, has been confirmed for all the planets save one, with the precision that is capable of being obtained by the delicacy of observation attainable at the present time. The sole exception is Mercury, the planet which lies nearest the sun. Since the time of Leverrier, it has been known that the ellipse corresponding to the orbit of Mercury, after it has been corrected for the influences mentioned above, is not stationary with respect to the fixed stars, but that it rotates exceedingly slowly in the plane of the orbit and in the sense of the orbital motion. The value obtained for this rotary movement of the orbital ellipse was 43 seconds of arc per century, an amount ensured to be correct to within a few seconds of arc. This effect can be explained by means of classical mechanics only on the assumption of hypotheses which have little probability, and which were devised solely for this purpose.

On the basis of the general theory of relativity, it is found that the ellipse of every planet round the sun must necessarily rotate in the manner indicated above; that for all the planets, with the exception of Mercury, this rotation is too small to be detected with the delicacy of observation possible at the present time; but that in the case of Mercury it must amount to 43 seconds of arc per century, a result which is strictly in agreement with observation.

Apart from this one, it has hitherto been possible to make only two deductions from the theory which admit of being tested by observation, to wit, the curvature of light rays by the gravitational field of the sun, and a displacement of the spectral lines of light reaching us from large stars, as compared with the corresponding lines for light produced in an analogous manner terrestrially (*i.e.*, by the same kind of atom). These two deductions from the theory have both been confirmed.

10
— Arthur Eddington —
(1882–1944)

Arthur Eddington was a British astrophysicist who pioneered the study of stellar structure and discovered the relationship between mass and luminosity in stars. It is this relationship that allows for the determination of the mass of a star based upon its intrinsic brightness. Eddington was an early proponent of Relativity Theory and in 1919 supervised the team that observed and measured the bending of light rays from stars as those rays passed close to the sun. This experiment, which had to be done during a total eclipse of the sun, confirmed the prediction of Relativity Theory and guaranteed its acceptance by the scientific community.

In this reading, Eddington gives a description of two expeditions that tested Einstein's prediction of the bending of light in a gravitational field. Observers on these two expeditions, to western Africa and northern Brazil, photographed starlight during the total eclipse of the sun that occurred on May 29, 1919. Eddington describes the photographic equipment used, the techniques employed in the measurement of stellar displacements, and the precautions taken to rule out observational errors. His results are in agreement with Einstein's prediction. The initial results from the Brazilian expedition seemed to favor the Newtonian gravitational theory, but an analysis of observational error brought these results in line with those of Eddington. The final set of seven plates from the Brazilian expedition clearly favor the prediction of Einstein.

This reading was taken from Eddington's book Space, Time and Gravitation, *published in 1920.*

THE BENDING OF LIGHT RAYS

Query 1. Do not Bodies act upon Light at a distance, and by their action bend its Rays, and is not this action (caeteris paribus) strongest at the least distance?
—Newton, Opticks

We come now to the experimental test of the influence of gravitation on light discussed theoretically in the last chapter. It is not the general purpose of this book to enter into details of experiments; and if we followed this plan consistently, we should, as hitherto, summarise the results of the observations in a few lines. But it is this particular test which has turned public attention towards the relativity theory, and there appears to be widespread desire for information. We shall therefore tell the story of the eclipse expeditions in some detail. It will make a break in the long theoretical arguments, and will illustrate the important applications of this theory to practical observations.

It must be understood that there were two questions to answer: firstly, whether light has weight (as suggested by Newton), or is indifferent to gravitation; secondly, if it has weight, is the amount of the deflection in accordance with Einstein's or Newton's laws?

It was already known that light possesses mass or inertia like other forms of electromagnetic energy. This is manifested in the phenomena of radiation-pressure. Some force is required to stop a beam of light by holding an obstacle in its path; a search-light experiences a minute force of recoil just as if it were a machine-gun firing material projectiles. The force, which is predicted by orthodox electromagnetic theory, is exceedingly minute; but delicate experiments have detected it. Probably this inertia of radiation is of great cosmical importance, playing a great part in the equilibrium of the more diffuse stars. Indeed it is probably the agent which has carved the material of the universe into stars of roughly uniform mass. Possibly the tails of comets are a witness to the power of the momentum of sunlight, which drives outwards the smaller or the more absorptive particles.

It is legitimate to speak of a pound of light as we speak of a pound of any other substance. The mass of ordinary quantities of light is however extremely small, and I have calculated that at the low charge of 8*d*. a unit, an Electric Light Company would have to sell light at the rate of £140,000,000 a pound. All the sunlight falling on the earth amounts to 160 tons daily.

It is perhaps not easy to realise how a wave-motion can have inertia, and it is still more difficult to understand what is meant by its having weight. Perhaps this will be better understood if we put the problem in a concrete form. Imagine a hollow body, with radiant heat or light-waves traversing the hollow; the mass of the body will be the sum of the masses of the material and of the radiant energy in the hollow; a greater force will be required to shift it because of the light-waves contained in it. Now let us weigh it with scales or a spring-balance. Will it also weigh heavier on account of the radiation contained, or will the weight be that of the solid material alone? If the former, then clearly from this aspect light has weight; and it is not difficult to deduce the effect of this weight on a freely moving light-beam not enclosed within a hollow.

The effect of weight is that the radiation in the hollow body acquires each second a downward momentum proportional to its mass. This in the long run is transmitted to the material enclosing it. For a free light-wave in space, the added momentum combines with the original momentum, and the total momentum determines the direction of the ray, which is accordingly bent. Newton's theory suggests no means for bringing about the bending, but contents itself with predicting it on general principles. Einstein's theory provides a means, viz. the variation of velocity of the waves.

Hitherto mass and weight have always been found associated in strict proportionality. One very important test had already shown that this proportionality is not confined to material energy. The substance uranium contains a great deal of radio-active energy, presumably of an electromagnetic nature, which it slowly liberates. The mass of this energy must be an appreciable fraction of the whole mass of the substance. But it was shown by experiments with Eötvös torsion-balance that the ratio of weight to mass for uranium is the same as for all other substances; so the energy of radio-activity has weight. Still even this experiment deals only with bound electromagnetic energy, and we are not justified in deducing the properties of the free energy of light.

It is easy to see that a terrestrial experiment has at present no chance of success. If the mass and weight of light are in the same proportion as for matter, the ray of light will be bent just like the trajectory of a material particle. On the earth a rifle bullet, like everything else, drops 16 feet in the first second, 64 feet in two seconds, and so on, below its original line of flight; the rifle must thus be aimed above the target. Light would also drop 16 feet in the first second;

but, since it has travelled 186,000 miles along its course in that time, the bend is inappreciable.

In fact any terrestrial course is described so quickly that gravitation has scarcely had time to accomplish anything.

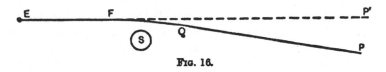

Fig. 16.

The experiment is therefore transferred to the neighbourhood of the sun. There we get a pull of gravitation 27 times more intense than on the earth; and—what is more important—the greater size of the sun permits a much longer trajectory throughout which the gravitation is reasonably powerful. The deflection in this case may amount to something of the order of a second of arc, which for the astronomer is a fairly large quantity.

In Fig. 16 the line *EFQP* shows the track of a ray of light from a distant star *P* which reaches the earth *E*. The main part of the bending of the ray occurs as it passes the sun *S*; and the initial course *PQ* and the final course *FE* are practically straight. Since the light rays enter the observer's eye or telescope in the direction *FE*, this will be the direction in which the star appears. But its true direction from the earth is *QP*, the initial course. So the star appears displaced outwards from its true position by an angle equal to the total deflection of the light.

It must be noticed that this is only true because a star is so remote that its true direction with respect to the earth *E* is indistinguishable from its direction with respect to the point *Q*. For a source of light within the solar system, the apparent displacement of the source is by no means equal to the deflection of the light-ray. It is perhaps curious that the attraction of light by the sun should produce an apparent displacement of the star away from the sun; but the necessity for this is clear.

The bending affects stars seen near the sun, and accordingly the only chance of making the observation is during a total eclipse when the moon cuts off the dazzling light. Even then there is a great deal of light from the sun's corona which stretches far above the disc. It is thus necessary to have rather bright stars near the sun, which will not be lost in the glare of the corona. Further the displacements of these stars can only be measured relatively to other stars, preferably more distant from the sun and less displaced; we need therefore a reasonable number of outer bright stars to serve as reference points.

In a superstitious age a natural philosopher wishing to perform an important experiment would consult an astrologer to ascertain an auspicious moment for the trial. With better reason, an astronomer to-day consulting the stars would announce that the most favourable day of the year for weighing light is May 29. The reason is that the sun in its annual journey round the ecliptic goes through fields of stars of varying richness, but on May 29 it is in the midst of a quite exceptional patch of bright stars—part of the Hyades—by far the best star-field encountered. Now if this problem had been put forward at some other period of history, it might have been necessary to wait some thousands of years for a total eclipse of the sun to happen on the lucky date. But by strange good fortune an eclipse did happen on May 29, 1919. Owing to the curious sequence of eclipses a similar opportunity will recur in 1938; we are in the midst of the most favourable cycle. It is not suggested that it is impossible to make the test at other eclipses; but the work will necessarily be more difficult.

Attention was called to this remarkable opportunity by the Astronomer Royal in March, 1917; and preparations were begun by a Committee of the Royal Society and Royal Astronomical Society for making the observations. Two expeditions were sent to different places on the line of totality to minimise the risk of failure by bad weather. Dr. A.C.D. Crommelin and Mr. C. Davidson went to Sobral in North Brazil; Mr. E. T. Cottingham and the writer went to the Isle of Principe in the Gulf of Guinea, West Africa. The instrumental equipment for both expeditions was prepared at Greenwich Observatory under the care of the Astronomer Royal; and here Mr. Davidson made the arrangements which were the main factor in the success of both parties.

The circumstances of the two expeditions were somewhat different and it is scarcely possible to treat them together. We shall at first follow the fortunes of the Principe observers. They had a telescope of focal length 11 feet 4 inches. On their photographs 1 second of arc (which was about the largest displacement to be measured) corresponds to about 1/1500 inch—by no means an inappreciable quantity. The aperture of the object-glass was 13 inches, but as used it was stopped down to 8 inches to give sharper images. It is necessary, even when the exposure is only a few seconds, to allow for the diurnal motion of the stars across the sky, making the telescope move so as to follow them. But since it is difficult to mount a long and heavy telescope in the necessary manner in a temporary installation in a remote part of the globe, the usual practice at eclipses is to keep the telescope rigid and reflect the stars into it by a coelostat—a plane mirror kept revolving at the right rate by clock-work. This arrangement was adopted by both expeditions.

The observers had rather more than a month on the island to make their preparations. On the day of the eclipse the weather was unfavourable. When totality began the dark disc of the moon surrounded by the corona was visible through cloud, much as the moon often appears through cloud on a night when no stars can be seen. There was nothing for it but to carry out the arranged programme and hope for the best. One observer was kept occupied changing the plates in rapid succession, whilst the other gave the exposures of the required length with a screen held in front of the object-glass to avoid shaking the telescope in any way.

> For in and out, above, about, below
> 'Tis nothing but a Magic *Shadow*-show
> Played in a Box whose candle is the Sun
> Round which we Phantom Figures come and go.

Our shadow-box takes up all our attention. There is a marvellous spectacle above, and, as the photographs afterwards revealed, a wonderful prominence-flame is poised a hundred thousand miles above the surface of the sun. We have no time to snatch a glance at it. We are conscious only of the weird half-light of the landscape and the hush of nature, broken by the calls of the observers, and beat of the metronome ticking out the 302 seconds of totality.

Sixteen photographs were obtained, with exposures ranging from 2 to 20 seconds. The earlier photographs showed no stars, though they portrayed the remarkable prominence; but apparently the cloud lightened somewhat towards the end of totality, and a few images appeared on the later plates. In many cases one or other of the most essential stars was missing through cloud, and no use could be made of them; but one plate was found showing fairly good images of five stars, which were suitable for a determination. This was measured on the spot a few days after the eclipse in a micrometric measuring-machine. The problem was to determine how the apparent positions of the stars, affected by the sun's gravitational field, compared with the normal positions on a photograph taken when the sun was out of the way. Normal photographs for comparison had been taken with the same telescope in England in January. The eclipse photograph and a comparison photograph were placed film to film in the measuring-machine so that corresponding images fell close together, and the small distances were measured in two rectangular directions. From these the relative displacements of the stars could be ascertained. In comparing two plates, various allowances have to be made for refraction, aberration, plate-orientation, etc.; but since these occur equally in determinations of stellar parallax, for which much greater accuracy is required, the necessary procedure is well-known to astronomers.

The results from this plate gave a definite displacement, in good accordance with Einstein's theory and disagreeing with the Newtonian prediction. Although the material was very meagre compared with what had been hoped for, the writer (who it must be admitted was not altogether unbiassed) believed it convincing.

It was not until after the return to England that any further confirmation was forthcoming. Four plates were brought home undeveloped, as they were of a brand which would not stand development in the hot climate. One of these was found to show sufficient stars; and on measurement it also showed the deflection predicted by Einstein, confirming the other plate.

The bugbear of possible systematic error affects all investigations of this kind. How do you know that there is not something in your apparatus responsible for this apparent deflection? Your object-glass has been shaken up by travelling; your have introduced a mirror into your optical system; perhaps the 50° rise of temperature between the climate at the equator and England in winter has done some kind of mischief. To meet this criticism, a different field of stars was photographed at night in Principe and also in England at the same altitude as the eclipse field. If the defection were really instrumental, stars on these plates should show relative displacements of a similar kind to those on the eclipse plates. But on measuring these check-plates no appreciable displacements were found. That seems to be satisfactory evidence that the displacement observed during the eclipse is really due to the sun being in the region, and is not due to differences in instrumental conditions between England and Principe. Indeed the only possible loophole is a difference between the night conditions at Principe when the check-plates were taken, and the day, or rather eclipse, conditions when the eclipse photographs were taken. That seems impossible since the temperature at Principe did not vary more than 1° between day and night.

The problem appeared to be settled almost beyond doubt; and it was with some confidence that we awaited the return of the other expedition from Brazil. The Brazil party had had fine weather and had gained far more extensive material on their plates. They had remained two months after the eclipse to photograph the same region before dawn, when clear of the sun, in order that they might have comparison photographs taken under exactly the same circumstances. One set of photographs was secured with a telescope similar to that used at Principe. In addition they used a longer telescope of 4 inches aperture and 19 feet focal length. The photographs obtained with the former were disappointing. Although the full number of stars expected (about 12) were shown, and numerous plates had been obtained, the definition of the images had been spoiled by some cause, probably distortion of the coelostat-mirror by the heat of the sunshine falling on it. The observers were pessimistic as to the value of these photographs; but they were the first to be measured on return to England, and the results came as a great surprise after the indications of the Principe plates. The measures pointed with all too good agreement to the "half-deflection," that is to say, the Newtonian value which is one-half the amount required by Einstein's theory. It seemed difficult to pit the meagre material of Principe against the wealth of data secured from the clear sky of Sobral. It is true the Sobral images were condemned, but whether so far as to invalidate their testimony on this point was not at first clear; besides the Principe images were not particularly well-defined, and were much enfeebled by cloud. Certain compensating advantages of the latter were better appreciated later. Their strong point was the satisfactory check against systematic error afforded by the photographs of the check-field; there were no check-plates taken at Sobral, and, since it was obvious that the discordance of the two results depended on systematic error and not on the wealth of material, this distinctly favoured the Principe results. Further, at Principe there could be no evil effects from the sun's rays on the mirror, for the sun had withdrawn all too shyly behind the veil of cloud. A further advantage was provided by the check-plates at Principe, which gave an independent determination of the difference of scale of the telescope as used in England and at the eclipse; for the Sobral plates this scale-difference was eliminated by the method of reduction, with the consequence that the results depended on the measurement of a much smaller relative displacement.

There remained a set of seven plates taken at Sobral with the 4-inch lens; their measurement had been delayed by the necessity of modifying a micrometer to hold them, since they were of unusual size. From the first no one entertained any doubt that the final decision must rest with them, since the images were almost ideal, and they were on a larger scale than the other photographs. The use of this instrument must have presented considerable difficulties—the unwieldy length of the telescope, the slower speed of the lens necessitating longer exposures and more accurate driving of the clock-work, the larger scale rendering the focus more sensitive to disturbances—but the observers achieved success, and the perfection of the negatives surpassed anything that could have been hoped for.

These plates were now measured and they gave a final verdict definitely confirming Einstein's value of the deflection, in agreement with the results obtained at Principe.

It will be remembered that Einstein's theory predicts a deflection of $1''.74$ at the edge of the sun, the amount falling off inversely as the distance from the sun's centre. The simple Newtonian deflection is half this, $0''.87$. The final results (reduced to the edge of the sun) obtained at Sobral and Principe with their "probable accidental errors" were

Sobral	$1''.98 \pm 0''.12$,
Principe	$1''.61 \pm 0''.30$.

It is usual to allow a margin of safety of about twice the probable error on either side of the mean. The evidence of the Principe plates is thus just about sufficient to rule out the possibility of the "half-deflection," and the Sobral plates exclude it with practical certainty. The value of the material found at Principe cannot be put higher than about one-sixth of that at Sobral; but it certainly makes it less easy to bring criticism against this confirmation of Einstein's theory seeing that it was obtained independently with two different instruments at different places and with different kinds of checks.

The best check on the results obtained with the 4-inch lens at Sobral is the striking internal accordance of the measures for different stars. The theoretical deflection should vary inversely as the distance from the sun's centre; hence, if we plot the mean radial displacement found for each star separately against the inverse distance, the points should lie on a straight line.

This is shown in Fig. 17 where the broken line shows the theoretical prediction of Einstein, the deviations being within the accidental errors of the determinations. A line of half the slope representing the half-deflection would clearly be inadmissible.

Moreover, values of the deflection were deduced from the measures in right ascension and declination independently. These were in close agreement.

A diagram showing the relative positions of the stars is given in Fig. 18.

The square shows the limits of the plates used at Principe, and the oblique rectangle the limits with the 4-inch lens at Sobral. The centre of the sun moved from S to P in the 2¼ hours interval between totality at the two stations; the sun is here represented for a time about midway between. The stars measured on the Principe plates were Nos. 3, 4, 5, 6, 10, 11; those at Sobral were 11, 10, 6, 5, 4, 2, 3 (in the order of the dots from left to right in Fig. 17). . . . .

It has been objected that although the observations establish a deflection of light in passing the sun equal to that predicted by Einstein, it is not immediately obvious that this deflection must necessarily be attributed to the sun's gravita-

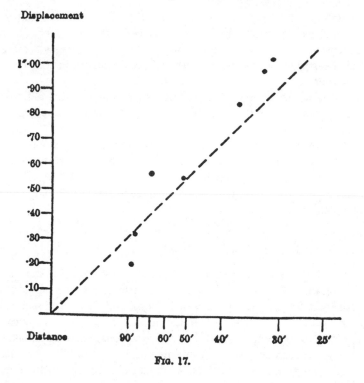

Fᴵɢ. 17.

tional field. It is suggested that it may not be an essential effect of the sun as a massive body, but an accidental effect owing to the circumstance that the sun is surrounded by a corona which acts as a refracting atmosphere. It would be a strange coincidence if this atmosphere imitated the theoretical law in the exact quantitative way shown in Fig. 17; and the suggestion appears to us far-fetched. However the objection can be met in a more direct way. We have already shown that the gravitational effect on light is equivalent to that produced by a refracting medium round the sun and have calculated the necessary refractive index. At a height of 400,000 miles above the surface the refractive index required is 1.0000021. This corresponds to air at l/140 atmosphere, hydrogen at 1/70 atmosphere, helium at 1/20 atmospheric pressure. It seems obvious that there can be no material of this order of density at such a distance from the sun. The pressure on the sun's surface of the columns of material involved would be of the order 10,000 atmospheres; and we know from spectroscopic evidence that there is no pressure of this order. If it is urged that the mass could perhaps be supported by electrical forces, the argument from absorption is even more cogent. The light from the stars photographed during the eclipse has passed through a depth of at least a million miles of material of this order of density—or say the equivalent of 10,000 miles of air at atmospheric density. We know to our cost what absorption the earth's 5 miles of homogeneous atmosphere can effect. And yet at the eclipse the stars appeared on the photographs with their normal bright-

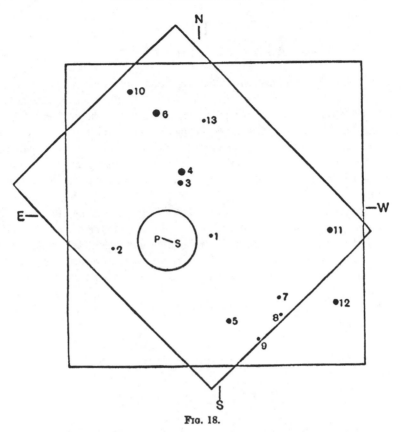

Fig. 18.

ness. If the irrepressible critic insists that the material round the sun may be composed of some new element with properties unlike any material known to us, we may reply that the mechanism of refraction and of absorption is the same, and there is a limit to the possibility of refraction without appreciable absorption. Finally it would be necessary to arrange that the density of the material falls off inversely as the distance from the sun's centre in order to give the required variation of refractive index.

Several comets have been known to approach the sun within the limits of distance here considered. If they had to pass through an atmosphere of the density required to account for the displacement, they would have suffered enormous resistance. Dr. Crommelin has shown that a study of these comets sets an upper limit to the density of the corona, which makes the refractive effect quite negligible.

Those who regard Einstein's law of gravitation as a natural deduction from a theory based on the minimum of hypotheses will be satisfied to find that his remarkable prediction is quantitatively confirmed by observation, and that no unforeseen cause has appeared to invalidate the test.

11

— Albert Einstein —

In this reading Einstein gives his mature reflections on the meaning of the concepts of absolute space, the ether, and the electromagnetic field. He accepts the work of Maxwell and Hertz as representing a rejection of the mechanical view of nature, and the work of Lorentz of ridding the ether of most of its mechanical properties. Special Relativity, Einstein claims, rids the ether of its one remaining property, immobility. However, in a very interesting comment, Einstein declares that Special Relativity does not compel us to deny the existence of a nonmechanical ether.

Einstein goes on to claim that General Relativity requires an ether because there is no part of space without a gravitational potential. The ether both conditions the behavior of inert masses and is conditioned by them. He concludes that the present view of the universe involves a commitment to both a gravitational ether and an electromagnetic field that are conceptually separate but causally related.

The reading is taken from an address given by Einstein in 1920 at the University of Leyden.

ETHER AND RELATIVITY

How does it come about that alongside of the idea of ponderable matter, which is derived by abstraction from everyday life, the physicists set the idea of the existence of another kind of matter, the ether? The explanation is probably to be sought in those phenomena which have given rise to the theory of action at a distance, and in the properties of light which have led to the undulatory theory. Let us devote a little while to the consideration of these two subjects.

Outside of physics we know nothing of action at a distance. When we try to connect cause and effect in the experiences which natural objects afford us, it seems at first as if there were no other mutual actions than those of immediate contact, e.g. the communication of motion by impact, push and pull, heating or inducing combustion by means of a flame, etc. It is true that even in everyday experience weight, which is in a sense action at a distance, plays a very important part. But since in daily experience the weight of bodies meets us as something constant, something not linked to any cause which is variable in time or place, we do not in everyday life speculate as to the cause of gravity, and therefore do not become conscious of its character as action at a distance. It was Newton's theory of gravitation that first assigned a cause for gravity by interpreting it as action at a distance, proceeding from masses. Newton's theory is probably the greatest stride ever made in the effort towards the causal nexus of natural phenomena. And yet this theory evoked a lively sense of discomfort among Newton's contemporaries, because it seemed to be in conflict with the principle springing from the rest of experience, that there can be reciprocal action only through contact, and not through immediate action at a distance.

It is only with reluctance that man's desire for knowledge endures a dualism of this kind. How was unity to be preserved in his comprehension of the forces of nature? Either by trying to look upon contact forces as being themselves distant forces which admittedly are observable only at a very small distance— and this was the road which Newton's followers, who were entirely under the spell of his doctrine, mostly preferred to take; or by assuming that the Newtonian action at a distance is only *apparently* immediate action at a distance, but in truth is conveyed by a medium permeating space, whether by movements or by elastic deformation of this medium. Thus the endeavour toward a unified view of the nature of forces leads to the hypothesis of an ether. This hypothesis, to be sure, did not at first bring with it any advance in the theory of gravitation or in physics generally, so that it became customary to treat Newton's law of force as an axiom not further reducible. But the ether hypothesis was bound always to play some part in physical science, even if at first only a latent part.

When in the first half of the nineteenth century the far-reaching similarity was revealed which subsists between the properties of light and those of elastic waves in ponderable bodies, the ether hypothesis found fresh support. It appeared beyond question that light must be interpreted as a vibratory process in an elastic, inert medium filling up universal space. It also seemed to be a necessary consequence of the fact that light is capable of polarisation that this medium, the ether, must be of the nature of a solid body, because transverse waves are not possible in a fluid, but only in a solid. Thus the physicists were bound to arrive at the theory of the "quasi-rigid" luminiferous ether, the parts of which can carry out no movements relatively to one another except the small movements of deformation which correspond to light-waves.

This theory—also called the theory of the stationary luminiferous ether— moreover found a strong support in an experiment which is also of fundamental importance in the special theory of relativity, the experiment of Fizeau, from which one was obliged to infer that the luminiferous ether does not take part in the movements of bodies. The phenomenon of aberration also favoured the theory of the quasi-rigid ether.

The development of the theory of electricity along the path opened up by Maxwell and Lorentz gave the development of our ideas concerning the ether quite a peculiar and unexpected turn. For Maxwell himself the ether indeed still had properties which were purely mechanical, although of a much more complicated kind than the mechanical properties of tangible solid bodies. But neither Maxwell nor his followers succeeded in elaborating a mechanical model for the ether which might furnish a satisfactory mechanical interpretation of Maxwell's laws of the electro-magnetic field. The laws were clear and simple, the mechanical interpretations clumsy and contradictory. Almost imperceptibly the theoretical physicists adapted themselves to a situation which, from the standpoint of their mechanical programme, was very depressing. They were particularly influenced by the electro-dynamical investigations of Heinrich Hertz. For whereas they previously had required of a conclusive theory that it should content itself with the fundamental concepts which belong exclusively to mechanics (e.g. densities, velocities, deformations, stresses) they gradually accustomed themselves to admitting electric and magnetic force as fundamental concepts side by side with those of mechanics, without requiring a mechanical interpretation for them. Thus the purely mechanical view of nature was gradually abandoned. But this change led to a fundamental dualism which in the long-run was insupportable. A way of escape was now sought in the reverse direction, by reducing the principles of mechanics to those of electricity, and this especially as confidence in the strict validity of the equations of Newton's mechanics was shaken by the experience with β-rays and rapid kathode rays.

This dualism still confronts us in unextenuated form in the theory of Hertz, where matter appears not only as the bearer of velocities, kinetic energy, and mechanical pressures, but also as the bearer of electromagnetic fields. Since such fields also occur *in vacuo*—i.e. in free ether—the ether also appears as bearer of electromagnetic fields. The ether appears indistinguishable in its functions from ordinary matter. Within matter it takes part in the motion of matter and in empty space it has everywhere a velocity; so that the ether has a definitely assigned velocity throughout the whole of space. There is no fundamental difference between Hertz's ether and ponderable matter (which in part subsists in the ether).

The Hertz theory suffered not only from the defect of ascribing to matter and ether, on the one hand mechanical states, and on the other hand electrical states, which do not stand in any conceivable relation to each other; it was also at variance with the result of Fizeau's important experiment on the velocity of the propagation of light in moving fluids, and with other established experimental results.

Such was the state of things when H. A. Lorentz entered upon the scene. He brought theory into harmony with experience by means of a wonderful simplification of theoretical principles. He achieved this, the most important advance in the theory of electricity since Maxwell, by taking from ether its mechanical, and from matter its electromagnetic qualities. As in empty space, so too in the interior of material bodies, the ether, and not matter viewed atomistically, was exclusively the seat of electromagnetic fields. According to Lorentz the elementary particles of matter alone are capable of carrying out movements; their electromagnetic activity is entirely confined to the carrying of electric charges. Thus Lorentz succeeded in reducing all electromagnetic happenings to Maxwell's equations for free space.

As to the mechanical nature of the Lorentzian ether, it may be said of it, in a somewhat playful spirit, that immobility is the only mechanical property of which it has not been deprived by H. A. Lorentz. It may, be added that the whole change in the conception of the ether which the special theory of relativity brought about, consisted in taking away from the ether its last mechanical quality, namely, its immobility. How this is to be understood will forthwith be expounded.

The space-time theory and the kinematics of the special theory of relativity were modelled on the Maxwell-Lorentz theory of the electromagnetic field. This theory therefore satisfies the conditions of the special theory of relativity, but when viewed from the latter it acquires a novel aspect. For if K be a system of co-ordinates relatively to which the Lorentzian ether is at rest, the Maxwell-Lorentz equations are valid primarily with reference to K. But by the special theory of relativity the same equations without any change of meaning also hold in relation to any new system of co-ordinates K′ which is moving in uniform translation relatively to K. Now comes the anxious question:—Why must I in the theory distinguish the K system above all K′ systems, which are physically equivalent to it in all respects, by assuming that the ether is at rest relatively to the K system? For the theoretician such asymmetry in the theoretical structure, with no corresponding asymmetry in the system of experience, is intolerable. If we assume the ether to be at rest relatively to K, but in motion relatively to K′, the physical equivalence of K and K′ seems to me from the logical standpoint, not indeed downright incorrect, but nevertheless unacceptable.

The next position which it was possible to take up in face of this state of things appeared to be the following. The ether does not exist at all. The electromagnetic fields are not states of a medium, and are not bound down to any bearer, but they are independent realities which are not reducible to anything else, exactly like the atoms of ponderable matter. This conception suggests itself the more readily as, according to Lorentz's theory, electromagnetic radiation, like ponderable matter, brings impulse and energy with it, and as, according to the special theory of relativity, both matter and radiation are but special forms of distributed energy, ponderable mass losing its isolation and appearing as a special form of energy.

More careful reflection teaches us, however, that the special theory of relativity does not compel us to deny ether. We may assume the existence of an ether; only we must give up ascribing a definite state of motion to it, i.e. we must by abstraction take from it the last mechanical characteristic which Lorentz had still left it. We shall see later that this point of view, the conceivability of which I shall at once endeavour to make more intelligible by a somewhat halting comparison, is justified by the results of the general theory of relativity.

Think of waves on the surface of water. Here we can describe two entirely different things. Either we may observe how the undulatory surface forming the boundary between water and air alters in the course of time; or else—with the help of small floats, for instance—we can observe how the position of the separate particles of water alters in the course of time. If the existence of such floats for tracking the motion of the particles of a fluid were a fundamental impossibility in physics—if, in fact, nothing else whatever were observable than the shape of the space occupied by the water as it varies in time, we should have no ground for the assumption that water consists of movable particles. But all the same we could characterise it as a medium.

We have something like this in the electromagnetic field. For we may picture the field to ourselves as consisting of lines of force. If we wish to interpret these lines of force to ourselves as something material in the ordinary sense, we are tempted to interpret the dynamic processes as motions of these lines of force, such that each separate line of force is tracked through the course of time. It is well known, however, that this way of regarding the electromagnetic field leads to contradictions.

Generalising we must say this:—There may be supposed to be extended physical objects to which the idea of motion cannot be applied. They may not be thought of as consisting of particles which allow themselves to be separately tracked through time. In Minkowski's idiom this is expressed as follows:—Not every extended conformation in the four-dimensional world can be regarded as composed of world-threads. The special theory of relativity forbids us to assume the ether to consist of particles observable through time, but the hypothesis of ether in itself is not in conflict with the special theory of relativity. Only we must be on our guard against ascribing a state of motion to the ether.

Certainly, from the standpoint of the special theory of relativity, the ether hypothesis appears at first to be an empty hypothesis. In the equations of the electromagnetic field there occur, in addition to the densities of the electric charge, *only* the intensities of the field. The career of electromagnetic processes *in vacuo* appears to be completely determined by these equations, uninfluenced by other physical quantities. The electromagnetic fields appear as ultimate, irreducible realities, and at first it seems superfluous to postulate a homogeneous, isotropic ether-medium, and to envisage electromagnetic fields as states of this medium.

But on the other hand there is a weighty argument to be adduced in favour of the ether hypothesis. To deny the ether is ultimately to assume that empty space has no physical qualities whatever. The fundamental facts of mechanics do not harmonize with this view. For the mechanical behaviour of a corporeal system hovering freely in empty space depends not only on relative positions (distances) and relative velocities, but also on its state of rotation, which physically may be taken as a characteristic not appertaining to the system in itself. In order to be able to look upon the rotation of the system, at least formally, as something real, Newton objectivises space. Since he classes his absolute space together with real things, for him rotation relative to an absolute space is also something real. Newton might no less well have called his absolute space "Ether"; what is essential is merely that besides observable objects, another thing, which is not perceptible, must be looked upon as real, to enable acceleration or rotation to be looked upon as something real.

It is true that Mach tried to avoid having to accept as real something which is not observable by endeavouring to substitute in mechanics a mean acceleration with reference to the totality of the masses in the universe in place of an acceleration with reference to absolute space. But inertial resistance opposed to relative acceleration of distant masses presupposes action at a distance; and as the modern physicist does not believe that he may accept this action at a distance, he comes back once more, if he follows Mach, to the ether, which has to serve as medium for the effects of inertia. But this conception of the ether to which we are led by Mach's way of thinking differs essentially from the ether as conceived by Newton, by Fresnel, and by Lorentz. Mach's ether not only *conditions* the behaviour of inert masses, but *is also conditioned* in its state by them.

Mach's idea finds its full development in the ether of the general theory of relativity. According to this theory the metrical qualities of the continuum of space-time differ in the environment of different points of space-time, and are partly conditioned by the matter existing outside of the territory under consideration. This space-time variability of the reciprocal relations of the standards of space and time, or, perhaps, the recognition of the fact that "empty space" in its physical relation is neither homogeneous nor isotropic, compelling us to describe its state by ten functions (the gravitation potentials $g_{\mu\nu}$), has, I think, finally disposed of the view that space is physically empty. But therewith the conception of the ether has again acquired an intelligible content, although this content differs widely from that of the ether of the mechanical undulatory theory of light. The ether of the general theory of relativity is a medium which is itself devoid of *all* mechanical and kinematical qualities, but helps to determine mechanical (and electromagnetic) events.

What is fundamentally new in the ether of the general theory of relativity as opposed to the ether of Lorentz consists in this, that the state of the former is at every place determined by connections with the matter and the state of the ether in neighbouring places, which are amenable to law in the form of differential equations; whereas the state of the Lorentzian ether in the absence of electromagnetic fields is conditioned by nothing outside itself, and is everywhere the same. The ether of the general theory of relativity is transmuted conceptually into the ether of Lorentz if we substitute constants for the functions of space which describe the former, disregarding the causes which condition its state. Thus we may also say, I think, that the ether of the general theory of relativity is the outcome of the Lorentzian ether, through relativation.

As to the part which the new ether is to play in the physics of the future we are not yet clear. We know that it determines the metrical relations in the space-time continuum, e.g. the configurative possibilities of solid bodies as well as the gravitational fields; but we do not know whether it has an essential share in the structure of the electrical elementary particles constituting matter. Nor do we know whether it is only in the proximity of ponderable masses that its structure differs essentially from that of the Lorentzian ether; whether the geometry of spaces of cosmic extent is approximately Euclidean. But we can assert by reason of the relativistic equations of gravitation that there must be a departure from Euclidean relations, with spaces of cosmic order of magnitude, if there exists a positive mean density, no matter how small, of the matter in the universe. In this case the universe must of necessity be spatially unbounded and of finite magnitude, its magnitude being determined by the value of that mean density.

If we consider the gravitational field and the electromagnetic field from the standpoint of the ether hypothesis, we find a remarkable difference between the

two. There can be no space nor any part of space without gravitational potentials; for these confer upon space its metrical qualities, without which it cannot
be imagined at all. The existence of the gravitational field is inseparably bound
up with the existence of space. On the other hand a part of space may very
well be imagined without an electromagnetic field; thus in contrast with the
gravitational field, the electromagnetic field seems to be only secondarily linked
to the ether, the formal nature of the electromagnetic field being as yet in no
way determined by that of gravitational ether. From the present state of theory
it looks as if the electromagnetic field, as opposed to the gravitational field, rests
upon an entirely new formal *motif*, as though nature might just as well have
endowed the gravitational ether with fields of quite another type, for example,
with fields of a scalar potential, instead of fields of the electromagnetic type.

Since according to our present conceptions the elementary particles of matter
are also, in their essence, nothing else than condensations of the electromagnetic
field, our present view of the universe presents two realities which are completely separated from each other conceptually, although connected causally,
namely, gravitational ether and electromagnetic field, or—as they might also be
called—space and matter.

Of course it would be a great advance if we could succeed in comprehending
the gravitational field and the electromagnetic field together as one unified conformation. Then for the first time the epoch of theoretical physics founded by
Faraday and Maxwell would reach a satisfactory conclusion. The contrast between ether and matter would fade away, and, through the general theory of
relativity, the whole of physics would become a complete system of thought,
like geometry, kinematics, and the theory of gravitation. An exceedingly ingenious attempt in this direction has been made by the mathematician H. Weyl;
but I do not believe that his theory will hold its ground in relation to reality.
Further, in contemplating the immediate future of theoretical physics we ought
not unconditionally to reject the possibility that the facts comprised in the quantum theory may set bounds to the field theory beyond which it cannot pass.

Recapitulating, we may say that according to the general theory of relativity
space is endowed with physical qualities; in this sense, therefore, there exists an
ether. According to the general theory of relativity space without ether is unthinkable; for in such space there not only would be no propagation of light,
but also no possibility of existence for standards of space and time (measuring-
rods and clocks), nor therefore any space-time intervals in the physical sense.
But this ether may not be thought of as endowed with the quality characteristic
of ponderable media, as consisting of parts which may be tracked through time.
The idea of motion may not be applied to it.

12
— Albert Einstein —

In this reading, Einstein offers reflective comments on General Relativity after it had achieved success and notoriety because of its ability to explain the precession of the orbit of Mercury and its successful prediction of the bending of light rays. The reading is taken from "Die Kultur der Gegenwart. Ihre Entwicklung und ihre Ziele," published in the 1925 (revised) edition of Physik.

LATER COMMENTS ON GENERAL RELATIVITY

The special theory of relativity is based on the fundamental idea that certain coordinate systems (inertial systems) are equivalent for the formulation of physical laws; these are those coordinate systems with respect to which the law of inertia and the law of constancy of the velocity of light in vacuum claim validity. Are these systems indeed privileged in nature, or does this privileged status stem from an incomplete understanding of the laws of nature? To be sure, according to Galileo's law of inertia, the inertial systems seem to be privileged over coordinate systems that move in a different manner. But the law of inertia has a serious deficiency that makes the cogency of this argument appear doubtful.

Let us imagine a portion of space that is entirely force-free in the sense of classical mechanics, thus, a space from which gravitating masses are far removed. Then, according to mechanics, there exists an inertial system K, with respect to which a mass M that is left to itself moves rectilinearly and uniformly through the portion of space under consideration. If one now introduces a coordinate system K' that is uniformly accelerated with respect to K, then the mass M, which is left to itself, does not move in a straight line with respect to K' but, rather, along a parabola, akin to the way in which a mass near the surface of the earth moves, relative to it, under the influence of gravity.

Can one draw from this conclusion that K' is (absolutely) accelerated? This conclusion would not be justified. One can just as well view K' as being "at rest" provided that one assumes the presence, with respect to K', of a uniform gravitational field, which is seen as the reason why the bodies move in accelerated fashion with respect to K'.

To be sure, it could be argued against such a conception that one cannot indicate the masses that create this gravitational field. But without violating the fundamental principles of Newtonian mechanics, one can imagine that these masses are practically infinitely distant. Besides, we do not know to what degree of exactness the Newtonian law of gravitation corresponds to the truth.

There is *one circumstance* that speaks forcefully for our conception. All masses fall with the same acceleration with respect to K' independently of their particular physical and chemical nature. Experience shows that the same holds true, indeed, with extraordinary accuracy, with respect to the gravitational field. In light of the remarkable fact that we recognize in the gravitational field a physical state of space that brings about the same behavior of bodies as that which obtains with respect to K', the hypothesis seems completely natural that, with respect to K', there exists a gravitational field that is essentially identical to the gravitational fields generated by masses according to Newton's law.

From this point of view, there is no substantial difference between inertia and gravitation; for it depends on the coordinate system, i.e., on the point of view, whether, at a certain instant, a body is under the exclusive influence of inertia or under the combined influence of inertia and gravitation.

Thus, widely known physical facts lead us to the general relativity principle, i.e., to the conception that the laws of nature are to be formulated in such a way that they hold with respect to arbitrarily moving coordinate systems.

The Theory of the Gravitational Field

From what has been said so far, one sees immediately that the general relativity principle must lead to a theory of the gravitational field. For if one starts out from the gravitation-free inertial system K and introduces a coordinate system K' that moves arbitrarily with respect to the former, then, with respect to K', there exists a precisely known gravitational field, and one can find the general properties of gravitational fields from the general properties of those gravitational fields at which one arrives in this way.

However, one must be careful not to assume that, conversely, every gravitational field can be made to vanish, i.e., can be turned into a gravitation-free region, by means of a suitable choice of the coordinate system. For example, it is impossible to make the gravitational field of the Earth vanish by means of a suitable choice of the coordinate system. In fact, for a region of *finite extension* this is only possible with gravitational fields of a very special kind. But for an infinitely small region the coordinates can always be chosen such that no gravitational field will be present in it. With respect to such an infinitely small region one may then assume that the special theory of relativity is valid. That way the general theory of relativity is connected with the special theory of relativity, and the results of the latter can be utilized for the former.

The Bending of Light Rays

A simple argument shows that a ray of light that propagates rectilinearly and uniformly with respect to the inertial system K must describe a curved trajectory with respect to the coordinate system K' that is in accelerated translation motion. From this we conclude that light rays are bent by gravitational fields, which means, according to Huygens's principle, that the velocity of light in gravitational fields depends on the location. This consequence was confirmed for the first time on the occasion of the solar eclipse in 1919.

Further, it can easily be seen that, according to the general theory of relativity, the gravitational field must have a much more complicated structure than according to Newton's theory. For if one assumes that K' is in uniform rotation with respect to the inertial system K, then the motion of material points relative to K' is such that their acceleration depends not only on their position (centrifugal force) but also on their velocity (Coriolis force).

Further, from the fact of the Lorentz construction, which was derived above as a consequence of the special theory of relativity, one can draw the conclusion that the possibilities for arranging practically rigid bodies with respect to K' are not accurately described by Euclidean geometry and that the rate of identically constructed clocks depends on the location. Thus, a geometry and kinematics that are independent of the rest of physics do not exist in the general theory of relativity, since the behavior of measuring rods and clocks is determined by the gravitational field.

It is thanks to this circumstance that the general theory of relativity entails a much more profound change in the theory of space and time than does the special theory of relativity. For according to the latter, the spatial and temporal coordinates have a direct physical meaning: between two points (x_1, y_1, z_1) and (x_2, y_2, z_2) of the coordinate system, one can lay down a rigid measuring rod of length

$$\sqrt{(x_2 - x_1)^2 + (y_2 - y_1)^2 + (z_2 - z_1)^2},$$

and the time difference $t_2 - t_1$ of two events taking place at the same point of the coordinate system can be measured directly by a clock (of identical construction for all points) set up at this point (or in its immediate vicinity). Such a direct physical meaning cannot be ascribed to the coordinates in the general theory of relativity. To be sure, the totality of events, i.e., of the point events, is arranged in a four-dimensional continuum (space-time) in this theory as well, but the behavior of measuring rods and clocks (the geometry or the metric in general) in this continuum is determined by the gravitational field; the latter is thus a physical state of space that simultaneously determines gravitation, inertia, and the metric. Herein lies the deepening and the unification that the foundation of physics underwent due to the general theory of relativity.

Quantitative Verification

In remarkable contrast to the profound conceptual change that the foundation of physics underwent due to the general theory of relativity, the difference between the quantitatively verifiable assertions of the new and the old theory is slight. Besides the already-mentioned bending of light rays by the gravitational field of the sun, observable only during total solar eclipse, one could mention a tiny rotation of the orbital ellipse of the planet Mercury (40 seconds in 100 years), which can be explained by the general theory of relativity but not by Newton's theory of gravitation. Finally, the general theory of relativity demands a slight displacement of the spectral lines produced on the surface of the sun or the fixed stars with respect to the corresponding spectral lines produced on the surface of the Earth. Observations have lent probability to the existence of this phenomenon but have not yet proved it with certainty.

13
— Albert Einstein —

In this reading Einstein presents an understandable derivation of the famous equation of relativity, $E = MC^2$. The derivation is based upon the law of the conservation of momentum, the pressure of radiation, and the aberration of light.

The reading is taken from two articles written in 1946. The first, "$E = MC^2$," appeared originally in Science Illustrated; *the other, "An Elementary Derivation of the Relationship between Mass and Energy," appeared in the* Technion Journal. *Both have been reprinted in several collections of Einstein's essays.*

$E = MC^2$

The Meaning of the Formula

In order to understand the law of the equivalence of mass and energy, we must go back to two conservation or "balance" principles which, independent of each other, held a high place in pre-relativity physics. These were the principle of the conservation of energy and the principle of the conservation of mass. The first of these, advanced by Leibnitz as long ago as the seventeenth century, was developed in the nineteenth century essentially as a corollary of a principle of mechanics.

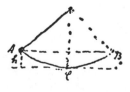

Drawing from
Dr. Einstein's
manuscript.

Consider, for example, a pendulum whose mass swings back and forth between the points A and B. At these points the mass m is higher by the amount h than it is at C, the lowest point of the path (see drawing). At C, on the other hand, the lifting height has disappeared and instead of it the mass has a velocity v. It is as though the lifting height could be converted entirely into velocity, and vice versa. The exact relation would be expressed as $mgh = (m/2)v^2$, with g representing the acceleration of gravity. What is interesting here is that this relation is independent of both the length of the pendulum and the form of the path through which the mass moves.

The significance is that something remains constant throughout the process, and that something is energy. At A and at B it is an energy of position, or "potential" energy; at C it is an energy of motion, or "kinetic" energy. If this conception is correct, then the sum $mgh + m(v^2/2)$ must have the same value for any position of the pendulum, if h is understood to represent the height above C, and v the velocity at that point in the pendulum's path. And such is found to be actually the case. The generalization of this principle gives us the law of the conservation of mechanical energy. But what happens when friction stops the pendulum?

The answer to that was found in the study of heat phenomena. This study, based on the assumption that heat is an indestructible substance which flows from a warmer to a colder object, seemed to give us a principle of the "conservation of heat." On the other hand, from time immemorial it has been known that heat could be produced by friction, as in the fire-making drills of the Indians. The physicists were for long unable to account for this kind of heat "production." Their difficulties were overcome only when it was successfully established that, for any given amount of heat produced by friction, an exactly proportional amount of energy had to be expended. Thus did we arrive at a principle of the "equivalence of work and heat." With our pendulum, for example, mechanical energy is gradually converted by friction into heat.

In such fashion the principles of the conservation of mechanical and thermal energies were merged into one. The physicists were thereupon persuaded that the conservation principle could be further extended to take in chemical and electromagnetic processes—in short, could be applied to all fields. It appeared that in our physical system there was a sum total of energies that remained constant through all changes that might occur.

Now for the principle of the conversation of mass. Mass is defined by the resistance that a body opposes to its acceleration (inert mass). It is also measured by the weight of the body (heavy mass). That these two radically different definitions lead to the same value for the mass of a body is, in itself, an astonishing fact. According to the principle—namely, that masses remain unchanged under any physical or chemical changes—the mass appeared to be the essential (because unvarying) quality of matter. Heating, melting, vaporization, or combining into chemical compounds would not change the total mass.

Physicists accepted this principle up to a few decades ago. But it proved inadequate in the face of the special theory of relativity. It was therefore merged with the energy principle—just as, about 60 years before, the principle of the conservation of mechanical energy had been combined with the principle of the conservation of heat. We might say that the principle of the conservation of energy, having previously swallowed up that of the conservation of heat, now proceeded to swallow that of the conservation of mass—and holds the field alone.

It is customary to express the equivalence of mass and energy (though somewhat inexactly) by the formula $E = mc^2$, in which c represents the velocity of light, about 186,000 miles per second. E is the energy that is contained in a stationary body; m is its mass. The energy that belongs to the mass m is equal to this mass, multiplied by the square of the enormous speed of light—which is to say, a vast amount of energy for every unit of mass.

But if every gram of material contains this tremendous energy, why did it go so long unnoticed? The answer is simple enough: so long as none of the energy is given off externally, it cannot be observed. It is as though a man who is fabulously rich should never spend or give away a cent; no one could tell how rich he was.

Now we can reverse the relation and say that an increase of E in the amount of energy must be accompanied by an increase of E/c^2 in the mass. I can easily supply energy to the mass—for instance, if I heat it by 10 degrees. So why not measure the mass increase, or weight increase, connected with this change? The trouble here is that in the mass increase the enormous factor c^2 occurs in the denominator of the fraction. In such a case the increase is too small to be measured directly; even with the most sensitive balance.

For a mass increase to be measurable, the change of energy per mass unit must be enormously large. We know of only one sphere in which such amounts of energy per mass unit are released: namely, radioactive disintegration. Schematically, the process goes like this: An atom of the mass M splits into two atoms of the mass M' and M", which separate with tremendous kinetic energy. If we imagine these two masses as brought to rest—that is, if we take this energy of motion from them—then, considered together, they are essentially poorer in energy than was the original atom. According to the equivalence principle, the mass sum M' + M" of the disintegration products must also be somewhat smaller than the original mass M of the disintegrating atom—in contradiction to the old principle of the conservation of mass. The relative difference of the two is on the order of 1/10 of one percent.

Now, we cannot actually weigh the atoms individually. However, there are indirect methods for measuring their weights exactly. We can likewise determine the kinetic energies that are transferred to the disintegration products M' and M". Thus it has become possible to test and confirm the equivalence formula. Also, the law permits us to calculate in advance, from precisely determined atom weights, just how much energy will be released with any atom disintegration we have in mind. The law says nothing, of course, as to whether—or how—the disintegration reaction can be brought about.

What takes place can be illustrated with the help of our rich man. The atom M is a rich miser who, during his life, gives away no money (*energy*). But in his will he bequeaths his fortune to his sons M' and M", on condition that they give to the community a small amount, less than one thousandth of the whole estate (*energy or mass*). The sons together have somewhat less than the father had (*the mass sum M' + M" is somewhat smaller than the mass M of the radioactive atom*). But the part given to the community, though relatively small, is still so enormously large (*considered as kinetic energy*) that it brings with it a great threat of evil. Averting that threat has become the most urgent problem of our time.

An Elementary Derivation

The following derivation of the law of equivalence, which has not been published before, has two advantages. Although it makes use of the principle of special relativity, it does not presume the formal machinery of the theory but uses only three previously known laws:

1. The law of the conservation of momentum.
2. The expression for the pressure of radiation; that is, the momentum of a complex of radiation moving in a fixed direction.
3. The well known expression for the aberration of light (influence of the motion of the earth on the apparent location of the fixed stars—Bradley).

We now consider the following system. Let the body B rest freely in space with respect to the system K_0. Two complexes of radiation S, S' each of energy

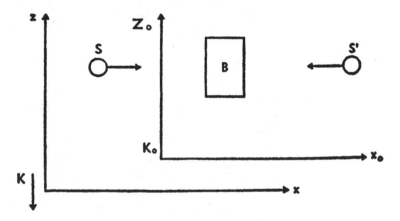

E/2 move in the positive and negative x_0 direction respectively and are eventually absorbed by B. With this absorption the energy of B increases by E. The body B stays at rest with respect to K_0 by reasons of symmetry.

Now we consider this same process with respect to the system K, which moves with respect to K_0 with the constant velocity v in the negative Z_0 direction. With respect to K the description of the process is as follows: The body B moves in the positive Z direction with velocity v. The two complexes of radiation now have directions with respect to K which make an angle α with the x axis. The law of aberration states that in the first approximation $\alpha = c/v$, where c is

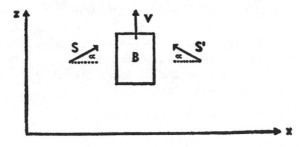

the velocity of light. From the consideration with respect to K_o we know that the velocity v of B remains unchanged by the absorption of S and S'.

Now we apply the law of conservation of momentum with respect to the z direction to our system in the coordinate-frame K.

I. *Before the absorption* let M be the mass of B; Mv is then the expression of the momentum of B (according to classical mechanics). Each of the complexes has the energy of E/2 and hence, by a well known conclusion of Maxwell's theory, it has the momentum E/2c. Rigorously speaking this is the momentum of S with respect to K_o. However, when v is small with respect to c, the momentum with respect to K is the same except for a quantity of second order of magnitude (v^2/c^2 compared to 1). The z-component of this momentum is E/2c sin α or with sufficient accuracy (except for quantities of higher order of magnitude) E/2c α or E/2 · v/c^2. S and S' together therefore have a momentum E v/c^2 in the z direction. The total momentum of the system before absorption is therefore

$$Mv + E/c^2 \cdot v$$

II. *After the absorption* let M' be the mass of B. We anticipate here the possibility that the mass increased with the absorption of the energy E (this is necessary so that the final result of our consideration be consistent). The momentum of the system after absorption is then

$$M'v$$

We now assume the law of the conservation of momentum and apply it with respect to the z direction. This gives the equation

$$Mv + (E/c^2)v = M'v$$

or

$$M' - M = E/c^2$$

This equation expresses the law of the equivalence of energy and mass. The energy increase E is connected with the mass increase E/c^2. Since energy according to the usual definition leaves an additive constant free, we may so choose the latter that

$$E = Mc^2$$

SELECTED BIBLIOGRAPHY

Bose, S. K. *An Introduction to General Relativity*. New Delhi: Wiley Eastern Limited, 1980.

Dirac, P.A.M. *General Theory of Relativity*. Princeton: Princeton University Press, 1996.

Friedman, M. *Foundations of Space-Time Theories: Relativistic Physics and Philosophy of Science*. Princeton: Princeton University Press, 1983.

Merman, W. D. *Space and Time in Special Relativity*. New York: McGraw-Hill, 1968.

Miller, A. I. *Albert Einstein's Special Theory of Relativity: Emergence (1905) and Early Interpretation (1905–1911)*. Reading: Addison-Wesley, 1981.

Miser, C. W., K. S. Thorney, and J. A. Wheeler. *Gravitation*. San Francisco: W. H. Freeman, 1973.

Mook, D. E., and T. Varish. *Inside Relativity*. Princeton: Princeton University Press, 1987.

Pauli, W. *Theory of Relativity*. New York: Dover, 1981

Rindler, W. *Essential Relativity*. 2nd ed. New York: Springer, 1977.

Stachel, J., ed. *Einstein's Miraculous Year: Five Papers That Changed the Face of Physics*. Princeton: Princeton University Press, 1998.

Taylor, E. F., and J. A. Wheeler. *Spacetime Physics*. San Francisco: W. H. Freeman, 1966.

PART IV

QUANTUM THEORY

///

INTRODUCTION

//////////////////////////////////////

Quantum Theory is used primarily to describe phenomena at the atomic and subatomic levels. Its acceptance as such by the scientific community is by and large universal. Nevertheless, as we shall see, the route to gaining such acceptance has not been without great struggle and controversy. Indeed, critics and proponents alike have made the claim that the discovery of quantum physics has caused physicists to lose their grasp on reality. Part of this introduction will consider why that is the case.

The generally accepted interpretation of Quantum Theory was formulated by Niels Bohr, Werner Heisenberg, and Wolfgang Pauli during the early part of the twentieth century at Bohr's laboratory in Copenhagen, Denmark. This account, commonly referred to as the "Copenhagen Interpretation," has been fraught with problems and controversies, many of which continue to be debated today.

Conceptual Background

The conceptual precursor of quantum physics is found in the seventeenth-century conflict regarding the true description of the nature of light. Both the wave theory and the particle theory of light had gained the allegiance of philosophers and scientists alike, as both were able to account for various physical phenomena. For a time, however, the particle theory held the upper hand since it seemed obvious that light did not bend around corners as waves do. It was for this reason that the particle theory had earned the support of Isaac Newton and eventually Pierre Laplace. Nonetheless, the wave theory of light did seem superior in its ability to explain such extraordinary phenomena as refraction and diffraction.

The theoretical problem, of course, lies in the fact that light cannot definitively be both a particle and a wave. Waves are divisible by nature, not localizable by coordinates, and are able to occupy the same space with other waves, whereas particles are finite in extension, identifiable by sharp coordinates, and occupy a region of space exclusively. It was thought to be merely a matter of time before evidence was found to prove once and for all which account of light was true.

That evidence was believed to be found in the early part of the nineteenth century, when Thomas Young showed that two beams of light can both constructively and destructively interfere. Since the wave theory could readily explain this interference phenomena while the particle theory could not, the wave theory assumed the dominant position. The position of the wave theory was enormously strengthened in mid-century when Maxwell showed that light was an electromagnetic wave traveling in the electromagnetic field.

Max Planck and Black-Body Radiation

By 1900 Maxwell's electromagnetic field theory had become well established. However, it still had difficulty explaining some features of electromagnetic energy propagation. For example, Maxwell's theory makes an absurd prediction concerning so-called black-body radiation emitted from a metallic cavity, a prediction that is also in considerable disagreement with experimental results. According to the theory, electromagnetic waves should bounce around in the cavity and form standing waves. Since the smaller wavelengths have a better opportunity to form these standing waves, the intensity of the emitted energy at each frequency should increase as the frequency increases. This prediction is absurd because it suggests that the total power radiated by a black-body cavity tends to infinity. Experimentally, the intensity of radiation at a given frequency did not tend to infinity as the frequency increased, but rather varied according to a bell-shaped curve, approaching zero at the higher frequencies.

In 1900, the German physicist Max Planck, working by trial and error, developed a functional relationship that fit the experimental data. To complete his formula, Planck had to assume that radiated energy was not continuous, as classical physics demanded, but rather that radiation of frequency v could be emitted or absorbed in discrete quantities or bundles of energy of magnitude hv, where h was a constant with the extremely small value of 6.625×10^{-34} joule seconds. This account of radiation contradicted Maxwell's theory, which had assumed that energy was emitted and absorbed continuously.

The notion of a packet or bundle of energy introduced into nature an element of discontinuity that was quite foreign to classical physics. So radical was the idea that Planck himself refused to believe it in the beginning. Despite his reluctance to make his findings public, he eventually did so in December of 1900. In the following year Planck introduced the term *Elementariquanta*, and the era of "quantum" physics had begun.

Einstein and the Photoelectric Effect

Planck's work had left the scientific community with two commonsense problems. First, his discrete quanta of energy had wavelike characteristics associated with them, namely wavelengths and frequencies. Given the incompatibility of particle and wave descriptions, it was not clear what to make of this "dual nature" of energy. Further, the energies of different modes of electromagnetic radiation seemed to have discrete spectra, that is, only certain energy levels seemed to be possible, with no designated values in between. It was not clear at this time whether these were temporary problems caused by Planck's particular mathematical analysis or whether they had a deeper significance.

It became clear in 1905 that these problems did indeed have deeper significance when Einstein found that he was able to explain a phenomenon called the photoelectric effect with Planck's ideas. In general, the photoelectric effect involves the use of light rays to liberate electrons from a metal surface. In an important experiment involving this effect, two metal surfaces were introduced into an electric circuit involving an electromotive force. Since there was a gap between the metal surfaces, the circuit was not closed and there was no current flowing. The surface of lower electric potential was then illuminated by light of a single frequency, liberating electrons from that surface. These electrons traveled to the plate of higher electric potential, thus closing the circuit and producing a current. The results of the experiment were as follows:

1. The electron emissions began as soon as the surface was illuminated. There was no observable time lag.
2. The current in the circuit was proportional to the intensity of the light.
3. The maximum kinetic energy of the liberated electrons was independent of the intensity of the light.
4. The maximum kinetic energy of the liberated electrons was directly proportional to the frequency of the light.

Classical electromagnetic theory could only account for the second of these results. It could not account for the other three.

In order to explain these results Einstein assumed that the energy emitted by the light source was initially localized in a small volume of space and remained localized as it moved to the metallic surface. When these packets of energy reach the surface they would either strike a free electron or they would not. If they did strike an electron they would give it all their energy. Regardless of how low the intensity of light is, some quanta will strike some electrons almost immediately. If the frequency v is such that hv is greater than the binding energy of the electron, then the electron is kicked free and the current begins. Since it was assumed that only one packet can strike an electron at any one time, the maximum kinetic energy of any freed electron would be equal to hv minus the binding energy of the electron.

These packets of energy, which are now called photons, were literally energy particles. However, now light seemed to possess an inherent duality. In its interaction with matter (i.e., the emission and absorption of energy), light acted as a particle, while in its movement from a source to a target, it still acted like a wave. While Einstein was aware of the conceptual difficulties in suggesting that light acts as both a particle and a wave, he seemed to have no choice but to utilize both descriptions. However, because of the tremendous amount of evidence in favor of the electromagnetic wave theory, Einstein chose to interpret his photon theory heuristically, believing that the apparent dualism would eventually be resolved.[1]

Bohr's Theory of the Atom and Correspondence Principle

By 1913, the quantum idea and the concomitant notion of discontinuity had become acceptable theoretical tools for physicists. It was at this time that the Danish physicist Niels Bohr put the idea to use in describing the structure and activity of the atom. Two years earlier, Ernest Rutherford had described the atom as consisting of a relatively heavy nucleus with a positive electric charge, and much smaller electrons that moved in orbit about the nucleus much the way the planets move about the sun. The electrons were negatively charged and were kept in orbit by the attractive electric force exerted by the nucleus.

This analogy between an atom and the solar system broke down rather quickly when it was realized that an atom had the ability to return to its original state when disturbed—something no planetary system following the laws of Newton's mechanics would be able to do. It was in response to this problem that Bohr developed what is known as the "old" Quantum Theory of 1913, which included an extraordinary picture of the hydrogen atom. First, Bohr assumed that electrons would not radiate energy as they moved in orbit, but would only do so in transition from one orbit to the other. (Classical theory would have the electrons continuously radiating energy because they were continuously accelerating.) With this assumption Bohr was able to explain why atoms don't quickly collapse, emitting a continuous spectrum of radiation.

Next, Bohr assumed that the energy of an atom was a function of the orbital energies of the electrons and that radiation would only occur when the electrons moved from one orbit (of higher energy) to another orbit (of lower energy). Using the Planck-Einstein notion that energy can only be emitted in discrete packets—or quanta—of magnitude hv, Bohr concluded that electrons could only exist in orbits corresponding to specific energy levels and these energy levels would differ by whole-number multiples of hv.

If we apply Bohr's theory to the hydrogen atom we find that the atom has a stable state that corresponds to its lowest energy level. At that level the electron occupies its innermost orbit about the proton. When the atom absorbs energy the electron will jump to orbits farther from the proton, but these orbits will be unstable. When the electron falls back to a lower level it will radiate energy in the amount of the difference between the energy levels. This "falling" from a higher energy level (orbit) to a lower one is a "quantum jump," a sudden transi-

---

1. In 1923 the American physicist Arthur Holly Compton provided experimental corroboration of the photon hypothesis. He observed that X-rays, which, like light, are a form of electromagnetic radiation, interact with electrons as if they were particles.

tion from one level to another. Intermediate levels corresponding to intermediate orbits are simply not permitted.

With this theory of the hydrogen atom Bohr could account for the spectrum of hydrogen, which is not continuous, but consists of sharp lines corresponding to the quantum jumps that he had suggested were possible. In fact, the agreement with experiment was truly remarkable. However, Bohr's theory was only useful for atoms with one electron in orbit.[2]

Because Planck's constant (n) is so small, Bohr realized that the quanticization that occurs in the hydrogen atom would never be observed on the macroscopic level. So Bohr offered what was called the Correspondence Principle, which suggested that at a certain level classical physics and quantum physics are equivalent and that quantum considerations should only be used when classical accounts fail. Although the success of this principle was short lived, it proved to be very valuable inasmuch as it was used by Heisenberg in the development of the Uncertainty Principle, one of the cornerstones of Quantum Theory.

Wave and Matrix Mechanics

In 1924, the Frenchman Louis de Broglie carried the development of Quantum Theory to the next level. De Broglie suggested that if photons, or energy particles, have wavelike properties associated with them, then perhaps material particles, like electrons, also have wavelike properties. By mathematical analysis, de Broglie was able to derive an equation for the wavelength of a wave supposedly associated with a particle. The wavelength was equal to Planck's constant divided by the momentum of the particle (h/p).

It turned out that the American physicist Clinton Davisson had already been doing experiments that seemed to confirm de Broglie's hypothesis. Davisson had sent electrons through a crystal lattice and the electrons scattered in what appeared to be a diffraction pattern. After de Broglie's hypothesis was made public, Davisson and a colleague, Leotes Germer, repeated the experiments and found that electrons did form diffraction patterns when sent though the crystals. When the "wavelengths" were measured from these patterns, the results agreed with the wavelengths predicted by the de Broglie equation. Shortly thereafter experiments were done that showed that protons and neutrons also had wavelike properties.

De Broglie felt that the wavelike properties of particles actually came from pilot waves that somehow directed the behavior of the particle. But the nature of these waves remained mysterious. Putting aside that issue, the Austrian physicist Erwin Schrödinger sought to develop a wave equation for these pilot waves. He took the classical wave equation and substituted the de Broglie postulate (h/p) for the wavelength. Schrödinger then applied his wave equation to the

---

2. Bohr's theory was also successful in predicting the spectra of singly ionized helium and doubly ionized lithium.

hydrogen atom and found that he could explain why the atom could only exist at certain energy levels. This was the claim that Niels Bohr had made in 1913 to account for the stability of the hydrogen atom and its characteristic spectrum.

With its success in accounting for certain atomic phenomena, Schrödinger's Wave Mechanics was taken seriously by the scientific community. Attention turned to the question of the nature of the wave functions which were the solutions to the wave equation. Upon reflection Schrödinger became convinced that material reality, at least at the atomic level, was wavelike in nature and that the wave functions described electromagnetic waves in the electromagnetic field.

Unfortunately it was impossible to picture these wave functions in any realistic way. First of all, at times they contained the imaginary number i (the square root of minus one). Second, wave functions of complex systems were functions of more than three dimensions. In 1926 Max Born suggested that the wave functions did not describe real waves but rather described guiding waves in a phantom field. His conclusion was that the square of the amplitude of the wave function in a given region of space gives us the probability of finding the associated particle in that region. This reduction of the wave equation and the wave function to strictly mathematical formalisms allowed Born to resolve many of the difficulties in Schrödinger's view.

As Schrödinger was developing wave mechanics, Heisenberg was developing what came to be known as matrix mechanics. At this stage in his work Heisenberg's approach was purely positivistic. That is to say, he was not interested in describing an underlying physical reality to quantum phenomena and concerned himself solely with the mathematical formalism. In that respect, Heisenberg was only interested in relating what he called "observable quantities." What he ended up with was a mathematical formalism involving matrices that did connect measurable quantities and that was as successful in accounting for quantum phenomena as wave mechanics. Later, Schrödinger would show that wave mechanics and matrix mechanics were mathematically equivalent.

By 1930 Paul Dirac and John von Neumann had perfected the mathematical formalism of quantum mechanics; their work superseded that of both Schrödinger and Heisenberg. In fact, Dirac was able to show that wave and matrix mechanics were special cases of his own theory.

Heisenberg's Uncertainty Principle

As mentioned, there were several attempts to resolve the wave-particle dualism that had arisen in quantum physics. De Broglie had come to see the waves as real electromagnetic waves guiding the particle's motion. Further analysis showed that if this were the case the particle would at times have to be traveling faster than the speed of light, a violation of the Theory of Relativity. Schrödinger tried to resolve the dualism by removing the concept of particle, suggesting that the electron is actually a wave packet. The problem here was that waves disperse while the electron, at least at times, seems to remain as a unit and act in a localized manner. Born's resolution, that waves should not be considered real in any sense but should be used to express probabilities, seemed to be the least problematic.

Born decided to pursue this solution by a thorough analysis of the concept of a wave packet. He discovered that it was impossible by this method to describe the exact position, momentum, and energy of a particle. The properties of wave packets seemed to show that the higher the accuracy with which the position of a particle can be determined at an instant (i.e., the smaller the wave packet), the more uncertain the description of the energy and momentum of the particle. The converse also seemed to be true: the greater the accuracy in describing the energy and momentum of a particle, the more uncertain its position.

In 1927, Heisenberg used Born's analysis to determine what he called the uncertainty relations and to formulate the Uncertainty Principle. Heisenberg found that when we try to determine the position and momentum of a particle at any one instant, there is an element of uncertainty that we cannot overcome. Specifically, the product of the uncertainty in the position of a particle and the uncertainty in its momentum cannot be less than Planck's constant. As the precision in the determination of the position of a particle increases the uncertainty in its momentum increases, and vice versa.

Heisenberg understood this result to indicate that Quantum Theory tells us that there are limits to what is both measurable and knowable. This measurement limitation he called the Uncertainty Principle.[3] The reason for these limits resides in the very act of measurement, since the observer or measurer must "disturb" the object involved in order to make the desired measurement. On the macroscopic level the disturbance is negligible, but when attempting to make measurements involving, for example, electrons, the disturbance is quite significant.

The importance of the Uncertainty Principle cannot be overestimated. Classical physics held that if the exact position and momentum of each particle in a system is known, then the future history of that system can be predicted with perfect accuracy from the known laws of mechanics. If the exact position and momentum cannot be simultaneously known, as dictated by the Uncertainty Principle, then there would necessarily be limits to our predictive ability. Classical determinism is surely called into question by the Uncertainty Principle.

Complementarity and the Copenhagen Interpretation

Bohr's reaction to Heisenberg's Uncertainty Principle was expressed in a paper delivered in September 1927 in Como, Italy, and then given again at the Solvay

3. The Uncertainty Principle not only excludes the possibility of precise simultaneous measurements of position and momentum, but also excludes such measurements for other pairs of properties, such as energy and time, angular momentum and angular positions, and moment of inertia and angular velocity.

Conference in Brussels two months later. It is in these papers that Bohr introduced his philosophically rich notion of *complementarity*, an idea that would become the crux of the Copenhagen Interpretation of Quantum Theory. In the Como paper, Bohr prefaced his remarks with the assertion that he wanted to describe "a certain general point of view . . . which [I] hope will be helpful in order to harmonize the apparently conflicting views taken by different scientists."

What Bohr hoped to harmonize initially were the opposing descriptions of classical and quantum physics. We recall that classical physics, with its continuity, determinism, realistic description, and so on, is directly opposed to Quantum Theory's emphasis on random jumps, probability, and now uncertainty. Thus, the notion of complementarity was introduced by Bohr as a way of describing the relationship between the two world pictures. Rather than merely opposing each other, the classical and quantum descriptions should be seen as *complementing* each other. As such, both were necessary for a full description of reality.

In addition, complementarity provided Bohr with an adequate explanation of wave-particle dualism. Depending on the experimental situation, quantum entities, like photons and electrons, will exhibit either wave characteristics or particle characteristics, but never both at once. The classical assumption that elementary particles have an intrinsic nature which can never change is replaced by the assumption that they can act either like waves or particles, depending on how they are treated by the surrounding environment. Rather than contradicting each other, such diverse modes of behavior should be seen as complementing each other. Only together do they give an adequate description of reality.

Another crucial aspect of the Copenhagen Interpretation was the claim that acknowledgment of the measurement problem as put forward by Heisenberg will lead to the recognition that on the quantum level, when we speak of a phenomenon or event, we must invariably include the observer or agent, whether it be the scientist, the scientist's apparatus, or the measuring instrument, in the description as part of the phenomenon or event. That has led prominent scientists such as John Archibald Wheeler, one-time assistant to Bohr and champion of the Copenhagen Interpretation, to describe the world we live in as a "participatory universe," one where the observer plays a crucial role in the actual existence of the phenomena. In his account of observer-created reality on the quantum level, Wheeler goes so far as to maintain that not only does the observer create current attributes of quantum entities, but also past attributes as well.

All these elements—complementarity, the measurement problem, uncertainty, wave-particle dualism, the probabilistic interpretation of the wave function, and the influence of the observer on the phenomena being observed or measured—together form the core of the Copenhagen Interpretation of quantum physics. Bohr, along with Heisenberg and Wolfgang Pauli, developed this interpretation after much heated, and often philosophically rich, discussion and debate.

The EPR Paradox and the Completeness Debate

As noted, Einstein was never satisfied with Quantum Theory, especially the Copenhagen Interpretation. Because of his more realist/determinist tendencies, Einstein naturally rejected the inherent randomness implied by Quantum Theory. Furthermore, he wholly rejected the notion that reality was somehow observer-created or observer-dependent. Nevertheless, his early attack on the theory (e.g., his photon box experiment) proved to be unsuccessful and he was forced to agree with Bohr that Quantum Theory did in fact correctly describe the outcomes of all conceivable experiments. Undaunted, Einstein decided to change his tactics and attack Quantum Theory as being an incomplete theory of reality.

The most compelling of Einstein's arguments to that end came in a paper written jointly with Boris Podolsky and Nathan Rosen in 1935, known as the EPR paper. The authors tried to show that there are aspects of reality (in the form of information about particles) that do in fact exist independently of any direct act of observation, something not allowed by Quantum Theory as dictated by the measurement problem. As a result of EPR thought experiments, Einstein maintained that certain elements of reality exist in the world that are not described by the quantum formalism.

The EPR paper begins with what appears to be a reasonable definition of physical reality: "If without in any way disturbing a system, we can predict with certainty (i.e., with probability equal to unity) the value of a physical quantity, then there exists an element of physical reality corresponding to this physical quantity." The paper then went on to apply this definition to a thought experiment involving two particles that interact with one another and then fly apart, not interacting with anything else until an experimenter decides to examine one of them. Quantum Theory would allow us to measure precisely the total momentum of the system (the two particles taken together) and the original distance between them. Then at a later time when we decide to measure precisely the momentum of particle A, we can calculate the precise momentum of particle B by subtracting the momentum of A from the total momentum of the system. Of course, we could have chosen to measure the exact position of A, and from that information we could calculate the exact position of B. (Quantum Theory, we recall, would not allow us to measure the precise momentum *and* position of A.) Thus, according to Quantum Theory the state of particle B somehow depends upon the measurement we make of particle A, despite the fact that we have not disturbed B in any way. Further, the information about the measurement performed on particle A must travel instantaneously to B regardless of how far apart they are. To the EPR authors, this appeared to be a "spooky" form of action at a distance that violated the Theory of Relativity.

The EPR authors felt that this thought experiment showed that particle B had intrinsic, classically real properties that existed independent of observation, something denied by the Copenhagen Interpretation. Since such values did exist and since their determination was not provided for by Quantum Theory, the theory must be considered incomplete.

In his response, Bohr rejected the charge of incompleteness. Instead, he argued that Quantum Theory is concerned with the results of the actual measurements of atomic particles at the time of measurement—not any previous history, such as being formerly found in a twin state. Thus the properties of particle B do depend on the act of measurement—albeit the measurement of particle A. There is, then, according to Bohr, some sort of communication between particles A and B such that the measurement has an effect on both. For Bohr the position and momentum of particle B had no objective meaning until they were measured, regardless of any measurement of particle A.

The EPR paper and Bohr's response brought into focus the central problem of Quantum Theory. The implication of the Copenhagen Interpretation is that elements of physical reality are literally defined by the experimental apparatus used to perform measurements on the quantum system. To ask what state a system is in prior to any measurement, according to Quantum Theory, does not make sense.

Hidden Variables Theories

Many scientists were quite uncomfortable with Bohr's response to the EPR paper and accepted the conclusion of Einstein et al. that Quantum Theory was indeed incomplete. Some sought to complete Quantum Theory by introducing a new set of variables that would determine which physical states would be preferred in a quantum process. These variables would be entirely picturable in classical terms but would remain, at least for the moment, hidden; i.e., they would not be detected by laboratory experiments. Unfortunately, hidden variable theorists had to overcome the "impossibility proof" of the mathematician John von Neumann. In 1932, von Neumann showed that if you assume that electrons are classical objects, then their behavior would contradict the predictions of Quantum Theory. His conclusion was that the world cannot ultimately be made of classical objects that possess dynamic attributes of their own. That seemed to eliminate the possibility of ever finding a successful hidden variable theory.

However, in the twenty years following von Neumann's work, serious reflection on his "proof" revealed that hidden variables might not be impossible after all. In 1952, physicist David Bohm, greatly influenced by Einstein, began to reconsider a hidden variable hypothesis first offered by Louis de Broglie. De Broglie had suggested that particles like electrons were guided by real pilot waves in a real electromagnetic field. The hidden variable in that theory was the actual position of the particle. De Broglie dropped this hypothesis when he realized that influences between particles would have to be transmitted at speeds faster than the speed of light. Bohm was not able to overcome this apparent violation of the Theory of Relativity, but he was able to produce a hypothesis in which all the connections between particles were causal. This is accomplished by something called the "quantum potential," through which causal influences are transmitted. The quantum potential organizes every region of space into an indivisible whole and allows for instantaneous connections.

Bell's Theorem

One person who was influenced by Bohm's attempt to resolve the EPR paradox was the Irish physicist John Stewart Bell. In the early 1960s, Bell worked on the problem of the measurements carried out simultaneously on two separated particles. In attempting to establish the theoretical limits to which such results can be correlated, Bell made the following two assumptions: that the quantum behavior is the result of hidden classical variables and that information cannot be transmitted between particles at a rate faster than the speed of light. The latter is known as the locality assumption. In 1964, Bell published a paper that contained the proof of a powerful mathematical theorem establishing strict limits in the correlation of simultaneous two-particle measurement results based upon these assumptions.[4] These limits were called Bell's inequalities.

According to the Copenhagen Interpretation, Quantum Theory should violate Bell's inequalities. In other words, Quantum Theory requires a cooperation between separated particles that would not be permitted by a local theory. Thus it could be inferred that Quantum Theory is not consistent with any local reality. Any model of reality that was to be consistent with the quantum facts must be non-local. To most scientists, Bell's theorem had demonstrated that the world, while phenomenally local, was actually supported by a reality that is non-local. Be that as it may, Bell also showed how experiments could be done to detect this phenomenon of nonlocality.

The Definitive Experiments

Although there was general evidence to suggest that reality indeed was nonlocal, the experiments outlined by Bell had never been performed in the laboratory. In 1972, the first direct tests were performed by Stuart Friedman and John Clauser at the Lawrence Livermore Laboratory in California. Although these experiments produced violations of Bell's inequalities as predicted by Quantum Theory, they were somewhat suspect because of certain auxiliary hypotheses that were used for data analysis.

The most significant experiments designed to test Bell's inequalities were performed in 1981 and 1982 by Alain Aspect and his colleagues at the University

---

4. Bell discovered that all local hidden variable theories are inconsistent with some of the statistically based predictions of quantum mechanics, thereby proving that the premises of the EPR thought experiment were incorrect. By a consideration of widely separated but still physically entangled particles, Bell was able to show that the nonlocal interaction is a prerequisite for reproducing various quantum mechanical predictions.

of Paris in Orsay. Aspect's experiments involved the separation of correlated pairs of photons. When an observer modified the experimental apparatus involving one of the photons, the distant photon was immediately affected, regardless of how distant it was. The Aspect experiments are generally understood to have confirmed that reality is nonlocal. Even if Quantum Theory is eventually superseded by a more comprehensive theory, the nonlocal nature of reality would have to be retained.

Experiments designed to test Bell's inequalities have not stopped with Aspect. In 1997, Nicolas Gisin of the University of Geneva performed an experiment on a twin photon system in which he was able to separate them by a distance of seven miles; previous experiments had achieved a separation of 100 yards or less. Again, the results indicated some sort of communication between the two particles as predicted by Quantum Theory.

The Final Word

Bell's theorem and the subsequent experimentation offer a strong vindication of the Copenhagen Interpretation of Quantum Theory. This interpretation involves not only indeterminism and a reality constituted through experimental interaction, but now also nonlocality. Philosophers who tend to favor Einstein in the Completeness Debate, such as Abner Shimony, reluctantly admit that even if a hidden variable theory turns out to be successful in the future, it would still have to recognize nonlocality as a permanent feature of the physical world.

HISTORICAL AND CONCEPTUAL DEVELOPMENT

/////////////////////////////////

1
— Max Planck —
1858–1947

A prominent German physicist who studied under Hermann von Helmholtz and Gustav Kirchhoff at the University of Berlin, Max Planck is best known for formulating the equation containing the constant that became emblematic of the age of Quantum Theory. Planck's doctoral dissertation at the University of Munich was in the field of Thermodynamics, which, along with his interest in electromagnetic radiation, eventually brought him to the problem of the distribution of energy in the black-body spectrum. He eventually solved the problem by introducing the quantum of action, which stated that rather than being continuous, as previously held, energy is instead radiated and absorbed discontinuously, in small bundles or packets later known as quanta. Planck reported his findings to the German Physical Society in the form of a paper entitled "On the Law of Distribution of Energy in the Normal Spectrum" in 1900, marking both the climax of his work and a turning point in the history of modern physics.

In 1918 Planck was awarded the Nobel Prize for physics and in 1930 was elected president of the Kaiser Wilhelm Institute for the Advancement of Science. At the end of the Second World War the institute was moved to Göttingen and renamed the Max Planck Institute.

In this selection, Planck discusses the conditions under which an accepted theory must be amended in the face of contradictory evidence. In this case, the time-honored Aristotelian precept "natura non facit saltus" is put in jeopardy by the quantum postulate, causing Planck to conclude, "Nature does indeed seem to make jumps and very extraordinary ones." Despite the problematic and oftentimes contradictory results of the physical experimentation, Planck in the end does affirm his faith in scientific realism and maintains the conviction that such a realist attitude is crucial for there to be any advance in knowledge of physical reality.

THE QUANTUM HYPOTHESIS

Experimental research in physics has not experienced, for a long time, such a stormy period, nor has its significance for human culture ever been so generally acknowledged as nowadays. Wireless waves, electrons, Röntgen rays, radioactive phenomena more or less arouse everybody's interest. If we consider only the wider question of how these new and brilliant discoveries have influenced and advanced our understanding of Nature and her laws, their importance does not appear, at first sight, to be commensurate with their brilliance.

Whoever tries to judge the state of present-day physical theories from a detached point of view may easily be led to the opinion that theoretical research is complicated to a certain extent by many new experimental discoveries, some of which were quite unforeseen. He will find that the present time is an unedifying period of aimless groping, in direct contrast to the clearness and certainty characteristic of the recent theoretical epoch, which may, therefore, with some justification, be called the classical epoch. Everywhere, old ideas, firmly rooted, are being displaced, generally accepted theorems are being cast aside and new hypotheses taking their place. Some of these hypotheses are so startling that they put a great strain on our comprehension and on our scientific ideas, and do not appear to inspire confidence in the steady advance of science towards a fixed goal. Modern theoretical physics gives one the impression of an old and honoured building which is falling into decay, with parts tottering one after the other, and its foundations threatening to give way.

No conception could be more erroneous than this. Great fundamental changes are, indeed, taking place in the structure of theoretical physics, but closer examination shows that this is not a case of destruction, but one of perfection and extension, that certain blocks of the building are only removed from their place in order to find a firmer and more suitable position elsewhere, and that the real fundamentals of the theory are to-day as fixed and immutable as ever they have been. After this more general consideration we will examine thoroughly the basis of these remarks.

The first impulse towards a revision and reconstruction of physical theory is nearly always given by the discovery of one or more facts which cannot be fitted into the existing theory. Facts always form a central point about which the most important theories hinge. Nothing is more interesting to the true theorist than a fact which directly contradicts a theory generally accepted up to that time, for this is his particular work.

What is to be done in such a case? Only one thing is certain: The existing theory must be altered in such a way that it is made to agree with the newly discovered fact. But it is often a very difficult and complicated question to decide in what part of the theory the improvement has to be made. A theory is not formed from a single isolated fact, but from a whole series of individual propositions combined together. It resembles a complicated organism, whose separate parts are so intimately connected that any interference in one part must, to some extent, affect other parts, often apparently quite remote, and it is not always easy to realize this fully. Thus, since each consequence of the theory is the result of the co-ordination of several propositions, any erroneous result deduced from the theory can generally be attributed to several of the propositions, and there are almost always numerous ways out of the difficulty. Usually, in the end, the question resolves itself into a conflict between two or three theorems, which were hitherto related in the theory, at least one of which theorems must be discarded on account of the new facts. The dispute often rages for years, and the final settlement means not only the exclusion of one of the theorems considered, but also, and this is specially important, quite naturally the corresponding strengthening and establishing of the remaining accepted propositions.

The most important and remarkable result of all such disputes arising in recent times is that the great general principles of physics have been established. These are the principle of conservation of energy, the principle of conservation of momentum, the principle of least action, and the laws of thermodynamics. These principles have, without exception, held the field and their force has increased appreciably in consequence. On the other hand, the theorems that failed to survive are ones which certainly served as apparent starting-points for all theoretical developments, but only because they were thought to be so self-evident that it was usually considered unnecessary to mention them specifically, or they were completely forgotten. Briefly, it can be said that the latest developments of theoretical physics were vindicated through the triumph of the great principles of physics over certain deeply rooted assumptions and conceptions, which were accepted from habit.

A few such propositions may be mentioned in order to illustrate this. These had been accepted as self-evident foundations of their respective theories without any consideration, but in the light of new discoveries have been proved untenable or, at least, highly improbable, as opposed to the general principles of physics. I mentioned three: the invariability of the chemical atom, the mutual independence of space and time, the continuity of all dynamical effects.

It is naturally not my intention to recount all the weighty arguments which have been directed against the *invariableness of the chemical atom*; I will only quote a single fact which has led to an inevitable conflict between a physical principle and this assumption, which had always hitherto been considered as self evident. The fact is the continual development of heat from a radium compound, the physical principle is conservation of energy, and the conflict finally ended with the complete victory of this principle, though at first many wished to throw doubt on it.

A radium salt, enclosed in a lead chamber of sufficient thickness, continually develops heat at a calculated rate of about 135 calories per grain of radium per hour. Consequently, it always remains, like a furnace, at a higher temperature than its surroundings. The principle of conservation of energy states that the observed heat cannot be created out of nothing, but that there must be a corresponding change as the cause. In the case of a furnace, we have the continual burning, in the case of the radium compound, through lack of any other chemical phenomenon, a variation of the radium atom itself must be assumed. This hypothesis has established itself everywhere, although it appeared daring when looked at from the previously accepted point of view of chemical science.

Strictly considered, there is, indeed, a certain contradiction in the conception of a variable atom, since atoms were originally defined as the invariable ultimate particles of all matter. Accordingly, to be quite accurate, the term "atom" should be reserved for the really invariable elements, as, perhaps, electrons and hydrogen. But apart from the fact that perhaps it can never be established that an invariable element exists in an absolute sense, such an uncertainty of meaning of terms would lead to a terrible confusion in the literature. The present chemical atoms are no longer the atoms of Democritus, but they are accurately determinable numerically by means of another and much more rigid definition. Only these numbers are referred to when atomic change is mentioned, and a misunderstanding in the direction indicated appears quite impossible.

Until recently the *absolute independence of space and time* was considered no less self-evident than the invariableness of the atom. There was a definite physical meaning to the question of the simultaneity of two observations at different places, without any necessity to inquire who had taken the time measurement. Today it is quite the opposite. For a fact which up to now has been repeatedly verified by the most delicate optical and electro-dynamical experiments has brought that simple conception into conflict with the so-called principle of constant velocity of light, which came into favour through Maxwell and Lorentz, and which states that the velocity of propagation of light in empty space is independent of the motion of the waves. This fact is briefly, if not also clearly called the relativity of all motion. If one accepts the relativity as being experimentally proved, one must abandon either the principle of constant velocity of light or the mutual independence of space and time.

Let us take a simple example. A time signal is sent out from a central station such as the Eiffel Tower by means of wireless telegraphy, as proposed in the projected international time service. Then all the surrounding stations which are at the same distance from the central station, receive the signal at the same time and can adjust their clocks accordingly. But such a method of adjusting time is inadmissible if one, mindful of the relativity of all motion, shifts his viewpoint from the earth to the sun, and thus regards the earth in motion. For according to the principle of constant velocity of light, it is clear that those stations which, seen from the central station, lie in the direction of the earth's motion, will receive the signal later than those lying in the opposite direction, for the former move away from the oncoming light waves and must be overtaken by them, which the latter move to meet. This, according to the principle of constant velocity of light, makes it quite impossible to determine time absolutely, i.e. independently of the movement of the observers. The two principles cannot exist side by side. So far, in the conflict, the principle of constant velocity of light has decidedly had the upper hand, and it is very probable that no change will come about, in spite of many ideas recently promulgated.

The third of the above-mentioned propositions deals with the *continuity of all dynamical effects*, formerly an undisputed hypothesis of all physical theories, which was condensed in Aristotle's well-known dogma: *natura non facit saltus*. But even in this stronghold, always respected from ancient times, modern research has made an appreciable breach. In this case, recent discoveries have shown that the proposition is not in agreement with the principles of thermodynamics, and, unless appearances are deceptive, the days of its validity are numbered. Nature certainly seems to move in jerks, indeed of a very definite kind. To illustrate this more clearly, may I present a straightforward comparison.

Let us consider a sheet of water in which strong winds have produced high waves. After the wind has completely died down, the waves will continue for appreciable time and move from one shore to the other. But a certain characteristic change will take place. The energy of the big, long waves will be transformed, to an ever-increasing extent, into the energy of small short waves, particularly when beating against the shore or some other rigid obstacle. This process will continue until finally the waves have become so small, and the movements so fine that they become quite invisible to the eye. This is the well-known transformation of visible motion into heat, molar into molecular movements, ordered movements into disorder; for in the case of ordered movement, many neighbouring molecules have the same velocity, while in disorderly movements, any particular molecule has its own special velocity and direction.

The process described here does not go on indefinitely, but is limited, naturally, by the size of the atom. The movement of a single atom, considered by itself, is always orderly, since all the individual parts of an atom move with a common velocity. The bigger the atom, the less the dissipation of energy that can take place. So far everything is clear, and the classical theory fits in perfectly.

Now let us consider another, quite analogous phenomenon, dealing, not with water waves, but with rays of light and heat. We here assume that rays emitted by a bright body are condensed into a closed space by suitable reflection and are there scattered by reflection from the walls. Here, again, there is a gradual transformation of radiant energy of long waves into shorter ones, of ordered waves into disorderly ones; the big, long waves correspond to the infra-red rays, the small, short ones to the ultra violet rays of the spectrum. According to the classical theory, one would expect all the energy to be concentrated ultimately at the ultra-violet end of the spectrum. In other words, that the infra-red and visible rays are gradually lost and transformed into invisible and chemically active ultra-violet rays.

No trace of such a phenomenon can be discovered in Nature. The transformation sooner or later reaches quite a definite, determinative end, and then the radiation remains completely stable.

Different attempts have been made to bring this fact into line with classical theory, but hitherto these attempts have always shown that the contradiction struck too deeply at the roots of the theory to leave it undisturbed. Nothing remains but to reconsider once more the foundation of the theory. And it has been once more substantiated that the principles of thermo-dynamics are unshakable. For the only way found hitherto which appears to promise a complete solution of the riddle, is to start with the two laws of thermo-dynamics. It joins these, however, with a new peculiar hypothesis, the significance of which in the two cases mentioned above, can be described as follows.

In the water waves, the dissipation of kinetic energy comes to an end on account of the fact that the atom retains its energy in a certain way, such that each atom represents a certain finite quantum of matter which can only move as a whole. Similarly, although the light and heat radiation is of a non-material nature, certain phenomena must occur, which imply that radiation energy is retained in certain finite quanta, and the shorter the wave length, and the quicker the oscillations, the more energy is retained.

Nothing can as yet be said with certainty of the dynamical representation of such quanta. Perhaps one could imagine quanta occurring in this manner, viz.

that any source of radiation can only emit energy after the energy has reached a certain value, as, for example, a rubber tube, into which air is gradually pumped, suddenly bursts and discharges its contents when a certain definite quantity of air has been pumped in.

In all cases, the quantum hypothesis has given rise to this idea, that in Nature, changes occur which are not continuous, but of an explosive nature. I need only mention that this idea has been brought into prominence by the discovery of, and closer research into, radio-active phenomena. The difficulties connected with exact investigations are lessened, since the results obtained on the quantum hypotheses agree better with observation than do those deduced from all previous theories.

But, further, if it is advantageous for a new hypothesis that it proves itself useful in spheres for which it was not intended, the quantum hypothesis has a great deal in its favour. I will here just refer to one point which is particularly striking. When air, hydrogen and helium were liquefied, a rich field of experimental research in low temperature was opened, and has already yielded a series of new and very remarkable results.

To raise the temperature of a piece of copper by one degree, from −250° to −249°, the quantity of heat required is not the same as that necessary to raise it from 0° to 1°, but is about thirty times less. The lower the initial temperature of the copper, the less is the heat necessary, without any assignable limits. This fact not only contradicts our accepted ideas, but also is diametrically opposed to the demands of the classical ideas. For though man had learnt, more than a hundred years ago, to differentiate between temperature and quantity of heat, the kinetic theory of matter led to the deduction that the two quantities, though not exactly proportional, moved more or less parallel to one anther.

The quantum hypothesis has fully explained this difficulty and, at the same time, has given us another result of great importance, namely, that the forces produced in a body by heat vibrations are of exactly the same type as those set up in elastic vibrations. With the help of the quantum hypothesis, the heat energy at different temperatures of a body containing atoms of one sort only can be calculated quantitatively from a knowledge of its elastic properties—the classical theory was very far from accomplishing this. A large number of further questions arise from this, questions which at first sight seem very strange, such as, for example, whether the oscillations of a tuning-fork are absolutely continuous, or whether they follow the nature of quanta. Indeed, in sound waves, the energy quanta are extremely small, on account of the relatively small frequency of the waves; in the case of á, for example, they are of the order of three quadrillionths of work units in absolute mechanical units. The ordinary theory of elasticity would, therefore, need just as little alteration on account of this, as it needs on account of the circumstance that it looks on matter as completely continuous, whereas, strictly speaking, matter is of an atomic, that is, of a quantum nature. But the revolutionary nature of the new development must be evident to everybody, and though the form of the dynamical quanta may for the present remain rather mysterious, the facts available today leave no doubt as to their existence in some form or other. For that which can be measured, exists.

Thus, in the light of modern research, the physical world-picture begins to show an ever-increasing connection between its separate parts and at the same time a certain definite form: the refinement of the parts appeared to have been missing from the earlier less detailed view and must have remained concealed. But, the question can always be repeated afresh: How far does this progress fundamentally satisfy our desire for knowledge? By refining our world-picture, do we attain a clearer understanding of Nature itself? Let us now consider briefly these important questions. Not that I shall say anything essentially new about this matter, with its manifold aspects, but because at present opinions are so divided on some things, and because all who take a deep interest in the real aims of science must take one side or the other.

Thirty-five years ago Hermann von Helmholtz reached the conclusion that our perceptions provide, not a representation of the external world, but at most only an indication thereof. For we have no grounds on which to make any sort of comparison between the actualities of external effects and those of the perceptions provoked by them. All ideas we form of the outer world are ultimately only reflections of our own perceptions. Can we logically set up against our self-consciousness a "Nature" independent of it? Are not all so-called natural laws really nothing more or less than expedient rules with which we associate the run of our perceptions as exactly and conveniently as possible? If that were so, it would follow that not only ordinary commonsense, but also exact natural research, have been fundamentally at fault from the beginning; for it is impossible to deny that the whole of the present-day development of physical knowledge works towards as far-reaching a separation as possible of the phenomena in external Nature from those in human consciousness.

The way out of this awkward difficulty is very soon evident if one continues the argument a step further. Let us assume a physical world-picture has been discovered which satisfies all claims that can be made upon it, and which, therefore, can represent completely all natural laws discovered empirically. Then it can in no way be proved that such a picture in any way represents "actual" Nature. However, there is a converse to this, and far too little emphasis is usually laid on it. Exactly similarly, the still more daring assertion cannot be disproved, that the assumed picture represents quite accurately actual Nature in all points without exception. For, to disprove this, one must be able to speak of actual Nature with certainty, which is acknowledged to be impossible.

Here yawns an enormous vacuum, into which no science can penetrate; and the filling up of this vacuum is the work, not of pure, but of practical reason; it is the work of a healthy view of the world.

However difficult it may be to prove scientifically such a view of the world, one can build so well on it, that it will stand unperturbed by any assaults, so long as it is consistent with itself and in agreement with the observed facts. But one must not imagine that advance is possible, even in the most exact of all natural sciences, without some view of the world, i.e. quite without some hypotheses not capable of proof. The theorem holds also in physics, that one cannot be happy without belief, at least belief in some sort of reality outside us. This undoubting belief points the way to the progressing creative power, it alone provides the necessary point of support in the aimless groping; and only it can uplift the spirit wearied by failure and urge it onwards to fresh efforts. A research worker who is not guided in his work by any hypothesis, however prudently and provisionally formed, renounces from the beginning a deep understanding of his own results. Whoever rejects the belief in the reality of the atom and the electron, or in the electro-magnetic nature of light waves, or in the identity of heat and motion, can most certainly never be convinced by a logical or empirical contradiction. But he must be careful how, from his point of view, he makes any advance in physical knowledge.

Indeed, belief alone is not enough. It is shown in the history of every science how easily it can lead to mistakes and deteriorate into narrow-mindedness and fanaticism. To remain a trust-worthy guide, it must be continually verified by logic and experience, and to this end the only ultimate aid is conscientious and often wearisome and self-denying efforts. He is no scientist who is not at least competent and willing to do the lowliest work, if necessary, whether in the laboratory or in the library, in the open air or at the desk. It is in such severe surroundings that the view of the world is ripened and purified. The significance and meaning of such a process can only be realized by those who have experienced it personally.

2

— Albert Einstein —

In this famous paper of 1905, often considered part of the early or "old" Quantum Theory, Einstein introduces the hypothesis of light quanta (later known as photons). This is the first articulation of the need to use both wave and particle models to describe the same phenomenon—for example, in the behavior of black-body radiation. This realization would provide the groundwork for what would become Quantum Theory's most problematic concept—namely, complementarity, the idea that wave and particle descriptions complement one another and are therefore both required for a full description of the phenomenon in question. This development would cause Einstein a great deal of consternation, as he would refuse to accept Quantum Theory as a complete description of nature. Until the end of his life, Einstein maintained that Quantum Theory would be replaced—or at least supplemented—by a more viable account of reality.

This paper, which originally appeared in The Annals of Physics *in 1905, was titled "On a Heuristic Point of View about the Creation and Conversion of*

Light." The equation that appears at the end of the reading is for the kinetic energy of electrons liberated from a metallic surface by light quanta. In that equation R/N equals Boltzmann's constant, where R equals the ideal gas constant and N equals Avogadro's number. Since B is simply a constant, the part of the equation R/N is equal to Planck's constant.

THE PHOTON

On a Heuristic Point of View Concerning the Production and Transformation of Light

There exists a profound formal difference between the theoretical conceptions physicists have formed about gases and other ponderable bodies, and Maxwell's theory of electromagnetic processes in so-called empty space. While we conceive of the state of a body as being completely determined by the positions and velocities of a very large but nevertheless finite number of atoms and electrons, we use continuous spatial functions to determine the electromagnetic state of a space, so that a finite number of quantities cannot be considered as sufficient for the complete description of the electromagnetic state of a space. According to Maxwell's theory, energy is to be considered as a continuous spatial function for all purely electromagnetic phenomena, hence also for light, while according to the current conceptions of physicists the energy of a ponderable body is to be described as a sum extending over the atoms and electrons. The energy of a ponderable body cannot be broken up into arbitrarily many, arbitrarily small parts, while according to Maxwell's theory (or, more generally, according to any wave theory) the energy of a light ray emitted from a point source of light spreads continuously over a steadily increasing volume.

The wave theory of light, which operates with continuous spatial functions, has proved itself splendidly in describing purely optical phenomena and will probably never be replaced by another theory. One should keep in mind, however, that optical observations apply to time averages and not to momentary values, and it is conceivable that despite the complete confirmation of the theories of diffraction, reflection, refraction, dispersion, etc., by experiment, the theory of light, which operates with continuous spatial functions, may lead to contradictions with experience when it is applied to the phenomena of production and transformation of light.

Indeed, it seems to me that the observations regarding "black-body radiation," photoluminescence, production of cathode rays by ultraviolet light, and other groups of phenomena associated with the production or conversion of light can be understood better if one assumes that the energy of light is discontinuously distributed in space. According to the assumption to be contemplated here, when a light ray is spreading from a point, the energy is not distributed continuously over ever-increasing spaces, but consists of a finite number of energy quanta that are localized in points in space, move without dividing, and can be absorbed or generated only as a whole.

In this paper I wish to communicate my train of thought and present the facts that led me to this course, in the hope that the point of view to be elaborated may prove of use to some researchers in their investigations. . . .

On the Generation of Cathode Rays by Illumination of Solid Bodies

The usual conception, that the energy of light is continuously distributed over the space through which it travels, meets with especially great difficulties when one attempts to explain the photoelectric phenomena; these difficulties are presented in a pioneering work by Mr. Lenard.

According to the conception that the exciting light consists of energy quanta with an energy $(R/N)\beta\nu$, the production of cathode rays by light can be conceived in the following way. The body's surface layer is penetrated by energy quanta whose energy is converted at least partially to kinetic energy of electrons. The simplest possibility is that a light quantum transfers its entire energy to a single electron; we will assume that this can occur. However, we will not exclude the possibility that the electrons absorb only a part of the energy of the light quanta. An electron provided with kinetic energy in the interior of the body will have lost a part of its kinetic energy by the time it reaches the surface. In addition, it will have to be assumed that in leaving the body, each electron has to do some work P (characteristic for the body). The greatest perpendicular velocity on leaving the body will be that of electrons located directly on the surface and excited perpendicular to it. The kinetic energy of such electrons is

$$(R/N)\beta\nu - P. \ . \ . \ .$$

As far as I can see, our conception does not conflict with the properties of the photoelectric effect observed by Mr. Lenard. If each energy quantum of the exciting light transmits its energy to electrons independent of all others, then the velocity distribution of the electrons, i.e., the quality of the cathode rays produced, will be independent of the intensity of the exciting light; on the other hand, under otherwise identical circumstances, the number of electrons leaving the body will be proportional to the intensity of the exciting light.

3
— Niels Bohr —
(1885–1962)

Niels Bohr is considered one of the most important physicists of the twentieth century, second only to Einstein in innovation and influence. Born in Denmark, where his father was professor of physiology at the University of Copenhagen, Bohr distinguished himself in the field of physics at an early age, winning the gold medal of the Royal Danish Academy of Science at twenty two and earning a doctorate from Copenhagen at twenty six. Soon thereafter Bohr went to the Cavendish Laboratory at Cambridge to study with J. J. Thomson, who, just over a decade earlier, had discovered the electron. The following year Bohr traveled to Manchester to work alongside Ernest Rutherford, who was then proposing his nuclear model of the atom. Bohr soon applied Planck's constant to Rutherford's model, suggesting that, similar to the experience of light radiation, energy in an atomic system likewise takes place discontinuously. This allowed Bohr to account for the observed spectrum of hydrogen and hydrogen-like atoms.

In 1916, Bohr was appointed professor of theoretical physics at the University of Copenhagen and in 1920 made director of the Institute for Theoretical Physics (often referred to as the Bohr Institute), which would become one of the leading theoretical centers in Europe. While at the institute, Bohr would provide the foundations for what was to become the "Copenhagen Interpretation" of quantum physics, which grappled with such issues as the "measurement problem"— whereby one must always include the experimental arrangement (e.g., observer, instrument) in the full description of any phenomena—as well as wave-particle dualism, which Bohr resolved by means of his rich notion of "complementarity." In 1922 Bohr was awarded the Nobel Prize for physics.

Upon his arrival in the United States in 1943, Bohr joined the Manhattan Project at Los Alamos in the efforts to develop atomic weapons. However, his sentiments would soon change and he spent a great deal of time and energy lobbying Roosevelt and Churchill on both the danger of atomic weapons as well as the need for some kind of understanding between the West and the Soviet Union. Bohr organized the first Atoms for Peace Conference in Geneva in 1955.

Niels Bohr is known not only for his groundbreaking work in physical theory and efforts toward trying to establish world peace, but also for his deep concern for the philosophical and epistemological consequences inherent in the new Quantum Theory. This, along with his attempts to apply the principles of the theory— especially that of complementarity—to a wide range of disciplines and topics outside physical science, prompted Werner Heisenberg to remark that Bohr "is primarily a philosopher, not a physicist." Nevertheless, Quantum Theory certainly would not have progressed to the extent that it has without the efforts and insights of this brilliant scientist.

In this reading, also part of the "old" or "classical" Quantum Theory, Bohr describes how, while working alongside Ernest Rutherford in his laboratory, he made use of the quantum hypothesis by applying it to the structure of the atom. Thus, according to Bohr, energy radiated from an atomic system occurs discontinuously, and is measured in terms of discrete levels of energy between different "stationary" states. This use of the relatively new notion of "quanta" would open the door to its potential effectiveness and eventual indispensability in explaining physical phenomena at the atomic level.

THE QUANTUM CHARACTER OF THE ATOM

On the Constitution of Atoms and Molecules

In order to explain the results of experiments on scattering of α-rays by matter Professor Rutherford has given a theory of the structure of atoms. According to this theory, the atoms consist of a positively charged nucleus surrounded by a system of electrons kept together by attractive forces from the nucleus; the total negative charge of the electrons is equal to the positive charge of the nucleus. Further, the nucleus is assumed to be the seat of the essential part of the mass of the atom, and to have linear dimensions exceedingly small compared with the linear dimensions of the whole atom. The number of electrons in an atom is deduced to be approximately equal to half the atomic weight. Great interest is to be attributed to this atom-model; for, as Rutherford has shown, the assumption of the existence of nuclei, as those in question, seems to be necessary in order to account for the results of the experiments on large angle scattering of the α-rays.

In an attempt to explain some of the properties of matter on the basis of this atom-model we meet, however, with difficulties of a serious nature arising from the apparent instability of the system of electrons: difficulties purposely avoided in atom-models previously considered, for instance, in the one proposed by Sir J. J. Thomson. According to the theory of the latter the atom consists of a sphere of uniform positive electrification, inside which the electrons move in circular orbits.

The principal difference between the atom-models proposed by Thomson and Rutherford consists in the circumstance that the forces acting on the electrons in the atom-model of Thomson allow of certain configurations and motions of the electrons for which the system is in a stable equilibrium; such configurations, however, apparently do not exist for the second atom-model. The nature of the difference in question will perhaps be most clearly seen by noticing that among the quantities characterizing the first atom a quantity appears—the radius of the positive sphere—of dimensions of a length and of the same order of magnitude as the linear extension of the atom, while such a length does not appear among the quantities characterizing the second atom, viz. the charges and masses of the electrons and the positive nucleus; nor can it be determined solely by help of the latter quantities.

The way of considering a problem of this kind has, however, undergone essential alterations in recent years owing to the development of the theory of the energy radiation, and the direct affirmation of the new assumptions introduced in this theory, found by experiments on very different phenomena such as specific heats, photoelectric effect, Röntgen rays, etc. The result of the discussion of these questions seems to be a general acknowledgement of the inadequacy of the classical electrodynamics in describing the behaviour of systems of atomic size. Whatever the alteration in the laws of motion of the electrons may be, it seems necessary to introduce in the laws in question a quantity foreign to the classical electrodynamics, i.e. Planck's constant, or as it often is called the elementary quantum of action. By the introduction of this quantity the question of the stable configuration of the electrons in the atoms is essentially changed, as this constant is of such dimensions and magnitude that it, together with the mass and charge of the particles, can determine a length of the order of magnitude required.

This paper is an attempt to show that the application of the above ideas to Rutherford's atom-model affords a basis for a theory of the constitution of atoms. It will further be shown that from this theory we are led to a theory of the constitution of molecules.

In the present first part of the paper the mechanism of the binding of electrons by a positive nucleus is discussed in relation to Planck's theory. It will be shown that it is possible from the point of view taken to account in a simple way for the law of the line spectrum of hydrogen. Further, reasons are given for a principal hypothesis on which the considerations contained in the following parts are based. . . .

Binding of Electrons by Positive Nuclei

The inadequacy of the classical electrodynamics in accounting for the properties of atoms from an atom-model such as Rutherford's, will appear very clearly if we consider a simple system consisting of a positively charged nucleus of very small dimensions and an electron describing closed orbits around it. For simplicity, let us assume that the mass of the electron is negligibly small in comparison with that of the nucleus, and further, that the velocity of the electron is small compared with that of light.

Let us at first assume that there is no energy radiation. In this case the electron will describe stationary elliptical orbits. The frequency of revolution ω and the major-axis of the orbit $2a$ will depend on the amount of energy W which must be transferred to the system in order to remove the electron to an infinitely great distance apart from the nucleus. . . .

Let us now, however, take the effect of the energy radiation into account, calculated in the ordinary way from the acceleration of the electron. In this case the electron will no longer describe stationary orbits. W will continuously increase, and the electron will approach the nucleus describing orbits of smaller and smaller dimensions, and with greater and greater frequency; the electron on the average gaining in kinetic energy at the same time as the whole system loses energy. This process will go on until the dimensions of the orbit are of the same order of magnitude as the dimensions of the electron or those of the nucleus. A simple calculation shows that the energy radiated out during the process considered will be enormously great compared with that radiated out by ordinary molecular processes.

It is obvious that the behaviour of such a system will be very different from that of an atomic system occurring in nature. In the first place, the actual atoms in their permanent state seem to have absolutely fixed dimensions and frequencies. Further, if we consider any molecular process, the result seems always to be that after a certain amount of energy characteristic for the systems in question is radiated out, the systems will again settle down in a stable state of equilibrium, in which the distances apart of the particles are of the same order of magnitude as before the process.

Now the essential point in Planck's theory of radiation is that the energy radiation from an atomic system does not take place in the continuous way assumed in the ordinary electrodynamics, but that it, on the contrary, takes place in distinctly separated emissions, the amount of energy radiated out from an atomic vibrator of frequency v in a single emission being equal to $\tau h v$ where τ is an entire number, and h is a universal constant. . . .

In the present paper an attempt has been made to develop a theory of the constitution of atoms and molecules on the basis of the ideas introduced by Planck in order to account for the radiation from a black body, and the theory of the structure of atoms proposed by Rutherford in order to explain the scattering of α particles by matter. Planck's theory deals with the emission and absorption of radiation from an atomic vibrator of a constant frequency, independent of the amount of energy possessed by the system in the moment considered. The assumption of such vibrators, however, involves the assumption of quasi-elastic forces and is inconsistent with Rutherford's theory, according to which all the forces between the particles of an atomic system vary inversely as the square of the distance apart. In order to apply the main results obtained by Planck it is therefore necessary to introduce new assumptions as to the emission and absorption of radiation by an atomic system.

The main assumptions used in the present paper are:

1. That energy radiation is not emitted (or absorbed) in the continuous way assumed in the ordinary electrodynamics, but only during the passing of the systems between different "stationary" states.

2. That the dynamical equilibrium of the systems in the stationary states is governed by the ordinary laws of mechanics, while these laws do not hold for the passing of the systems between the different stationary states.

3. That the radiation emitted during the transition of a system between two stationary states is homogeneous, and that the relation between the frequency v

and the total amount of energy emitted E is given by $E = hv$, where h is Planck's constant.

4. That the different stationary states of a simple system consisting of an electron rotating round a positive nucleus are determined by the condition that the ratio between the total energy, emitted during the formation of the configuration, and the frequency of revolution of the electron is an entire multiple of $h/2$. Assuming that the orbit of the electron is circular, this assumption is equivalent with the assumption that the angular momentum of the electron round the nucleus is equal to an entire multiple of $h/2\pi$.

5. That the "permanent" state of any atomic system—i.e. the state in which the energy emitted is maximum—is determined by the condition that the angular momentum of every electron round the centre of its orbit is equal to $h/2\pi$.

It is shown that, applying these assumptions to Rutherford's atom model, it is possible to account for the laws of Balmer and Rydberg connecting the frequency of the different lines in the line-spectrum of an element. Further, outlines are given of a theory of the constitution of the atoms of the elements and of the formation of molecules of chemical combinations, which on several points is shown to be in approximate agreement with experiments.

4
— Louis de Broglie —
(1892–1987)

Louis de Broglie was a French physicist who taught theoretical physics at the Henri Poincaré Institute at the Sorbonne in Paris. His contribution to Quantum Theory consists primarily in his suggestion that particles, such as electrons, could actually behave like and exhibit properties of *waves. He first published this revolutionary idea in his doctoral thesis of 1924, and went on to receive the Nobel Prize for physics in 1929. The theories contributed by de Broglie laid the foundations for the development of wave mechanics.*

In this reading, taken from a collection of essays entitled Matter and Light: The New Physics, *de Broglie suggests that, just as the wave and particle concepts have to be used simultaneously in the discussion of light, so also must they be used simultaneously in the discussion of matter. Specifically, de Broglie declares that the motion of the material particle is guided by a wave and that a study of the propagation of that wave will reveal to us information about the localization of the particle. This attachment of waves to particles was precisely the starting point for the development of wave mechanics.*

In the end, de Broglie recognizes the eminently changeable nature of scientific theory, declaring that we must remain open to the possibility that "the discovery of a new series of phenomena [may] destroy our finest theories like a house of cards [since] . . . the richness of Nature is always greater than our imagination."

THE WAVE NATURE OF THE ELECTRON

Physicists had for long been wondering whether Light did not consist of minute corpuscles in rapid motion, an idea going back to the philosophers of antiquity, and sustained in the eighteenth century by Newton. After interference phenomena had been discovered by Thomas Young, however, and Augustin Fresnel had completed his important investigation, the assumption that Light had a granular structure was entirely disregarded, and the Wave Theory was unanimously adopted. In this way the physicists of last century came to abandon completely the idea that Light had an atomic structure. But the Atomic Theory, being thus banished from optics, began to achieve great success, not only in Chemistry, where it provided a simple explanation of the laws of definite proportions, but also in pure Physics, where it enabled a fair number of the properties of solids, liquids and gases to be interpreted. Among other things it allowed the great kinetic theory of gases to be formulated, which, in the generalized form of statistical Mechanics, has enabled clear significance to be given to the abstract concepts of thermodynamics. We have seen how decisive evidence in favour of the atomic structure of electricity was also provided by experiments. Thanks to Sir J. J. Thomson, the notion of the corpuscle of electricity was introduced; and the way in which H. A. Lorentz has exploited this idea in his electron Theory is well known.

Some thirty years ago, then, Physics was divided into two camps. On the one hand there was the Physics of Matter, based on the concepts of corpuscles and atoms which were assumed to obey the classical laws of Newtonian Mechanics; on the other hand there was the Physics of radiation, based on the idea of wave propagation in a hypothetical continuous medium: the ether of Light and of electromagnetism. But these two systems of Physics could not remain alien to each other; an amalgamation had to be effected; and this was done by means of a theory of the exchange of energy between Matter and radiation. It was at this point, however, that the difficulties began, for in the attempt to render the two systems of Physics compatible with each other, incorrect and even impossible conclusions were reached with regard to the energy equilibrium between Matter and radiation in an enclosed and thermally isolated region. Some investigators even going so far as to say that Matter would transfer all its energy to radiation, and hence tend towards the temperature of absolute zero. This absurd conclusion had to be avoided at all costs; and by a brilliant piece of intuition, Planck succeeded in doing so. Instead of assuming, as did the classical Wave Theory, that a light-source emits its radiation continuously, he assumed that it emits it in equal and finite quantities—in quanta. The energy of each quantum, still further, was supposed to be proportional to the frequency of the radiation, v, and to be equal to hv, where h is the universal constant since known as Planck's Constant.

The success of Planck's ideas brought with it some serious consequences. For if Light is emitted in quanta, then surely, once radiated, it ought to have a granular structure. Consequently the existence of quanta of radiation brings us back to the corpuscular conception of Light. On the other hand, it can be shown—as has in fact been done by Jeans and H. Poincaré—that if the motion of material particles in a light-source obeyed the laws of classical Mechanics, we could never obtain the correct Law of black body radiation—Planck's Law. It must therefore be admitted that the older dynamics, even as modified by Einstein's Theory of Relativity, cannot explain motion on a very minute scale.

The existence of a corpuscular structure of Light and of other types of radiation has been confirmed by the discovery of the photo-electric effect which, as I have already observed, is easily explained by the assumption that the radiation consists of quanta—hv—capable of transferring their entire energy to an electron in the irradiated substance; and in this way we are brought to the theory of light-quanta which, as we have seen, was advanced in 1905 by Einstein—a theory which amounts to a return to Newton's corpuscular hypothesis, supplemented by the proportionality subsisting between the energy of the corpuscles and the frequency. A number of arguments were adduced by Einstein in support of his view, which was confirmed by Compton's discovery in 1922 of the scattering of X-rays, a phenomenon named after him. At the same time it still remained necessary to retain the Wave Theory to explain the phenomena of diffraction and interference, and no means was apparent to reconcile this Theory with the existence of light-corpuscles.

I have pointed out that in the course of investigation some doubt had been thrown on the validity of small-scale Mechanics. Let us imagine a material point describing a small closed orbit—an orbit returning on itself; then according to classical dynamics there is an infinity of possible movements of this type in accordance with the initial conditions, and the possible values of the energy of the moving material point form a continuous series. Planck, on the other hand, was compelled to assume that only certain privileged movements—*quantized* motion—are possible, or at any rate stable, so that the available values of the energy form a discontinuous series. At first this seemed a very strange idea; soon, however, its truth had to be admitted because it was by its means that Planck arrived at the correct Law of black body radiation and because its usefulness has since been proved in many other spheres. Finally, Bohr founded his famous atomic Theory on this idea of the quantization of atomic motion—a theory so familiar to scientists that I will refrain from summing it up here.

Thus we see once again it had become necessary to assume two contradictory theories of Light, in terms of waves, and of corpuscles, respectively; which it was impossible to understand why, among the infinite number of paths which an electron ought to be able to follow in the atom according to classical ideas, there was only a restricted number which it could pursue in fact. Such were the problems facing physicists at the time when I returned to my studies.

When I began to consider these difficulties I was chiefly struck by two facts. On the one hand the Quantum Theory of Light cannot be considered satisfactory, since it defines the energy of a light-corpuscle by the equation $W = h\nu$, containing the frequency ν. Now a purely corpuscular theory contains nothing that enables us to define a frequency; for this reason alone, therefore, we are compelled, in the case of Light, to introduce the idea of a corpuscle and that of periodicity simultaneously.

On the other hand, determination of the stable motion of electrons in the atom introduces integers; and up to this point the only phenomena involving integers in Physics were those of interference and of normal modes of vibration. This fact suggested to me the idea that electrons too could not be regarded simply as corpuscles, but that periodicity must be assigned to them also.

In this way, then, I obtained the following general idea, in accordance with which I pursued my investigations—that it is necessary in the case of Matter, as well as of radiation generally and of Light in particular, to introduce the idea of the corpuscle and of the wave simultaneously; or in other words, in the one case as well as in the other, we must assume the existence of corpuscles accompanied by waves. But corpuscles and waves cannot be independent of each other; in Bohr's terms, they are two complementary aspects of Reality: and it must consequently be possible to establish a certain parallelism between the motion of a corpuscle and the propagation of its associated wave.

The general formulae establishing the parallelism between waves and corpuscles can be applied to light-corpuscles if we assume that in that case the rest-mass m_0 is infinitely small. If then for any given value of the energy W we make m_0 tend to zero, we find that both ν and V tend to c, and in the limit we obtain the two fundamental formulae on which Einstein erected his Theory of Light-quanta

$$W = h\nu \quad p = \frac{h\nu}{c}.$$

Such were the principal ideas which I had developed during my earlier researches. They showed clearly that it was possible to establish a correspondence between waves and corpuscles of such a kind that the Laws of Mechanics correspond to those of geometrical optics. But we know that in the Wave theory geometrical optics is only an approximation; there are limits to the validity of this approximation, and especially when the phenomena of interference and of diffraction are concerned it is wholly inadequate. This suggests the idea that the older Mechanics too may be no more than an approximation as compared with a more comprehensive Mechanics of an undulatory character. This was what I expressed at the beginning of my researches when I said that a new Mechanics must be formulated, standing in the same relation to the older Mechanics as that in which wave optics stands to geometrical optics. This new Mechanics has since been developed, thanks in particular to the fine work done by Schrö-dinger. It starts from the equations of wave propagation, which are taken as the basis, and rigorously determines the temporal changes of the wave associated with a corpuscle. More particularly, it has succeeded in giving a new and more satisfactory form to the conditions governing the quantization of intra-atomic motion: for, as we have seen, the older conditions of quantization are encoun-tered again if we apply geometrical optics to the waves associated with intra-atomic corpuscles; and there is strictly no justification for this application.

I cannot here trace even briefly the development of the new Mechanics. All that I wish to say is that on examination it has shown itself to be identical with a Mechanics developed independently, first by Heisenberg and later by Born, Jordan, Pauli, Dirac and others. This latter Mechanics—Quantum Mechanics—and Wave Mechanics are, from the mathematical point of view, equivalent to each other.

Here we must confine ourselves to a general consideration of the results ob-tained. To sum up the significance of Wave Mechanics, we can say that a wave must be associated with each particle, and that a study of the propagation of the wave alone can tell us anything about the successive localizations of the corpus-cle in space. In the usual large-scale mechanical phenomena, the localizations predicted lie along a curve which is the trajectory in the classical sense of the term. What, however, happens if the wave is not propagated according to the laws of geometrical optics; if, for example, interference or diffraction occurs? In such a case we can no longer assign to the corpuscle motion in accordance with classical dynamics. So much is certain. But a further question arises: Can we even suppose that at any given moment the corpuscle has an exactly determined position within the wave, and that in the course of its propagation the wave carries the corpuscle with it, as a wave of water would carry a cork? These are difficult questions, and their discussion would carry us too far and actually to the borderland of Philosophy. All that I shall say here is that the general modern tendency is to assume that it is not always possible to assign an exactly defined position within the wave to the corpuscle, that whenever an observation is made enabling us to localize the corpuscle, we are invariably led to attribute to it a position inside the wave, and that the probability that this position is at a given point, M, within the wave is proportional to the square of the amplitude, or the intensity, at M.

What has just been said can also be expressed in the following way. If we take a cloud of corpuscles all associated with the same wave, then the intensity of the wave at any given point is proportional to the density of the cloud of corpuscles at that point, i.e. to the number of corpuscles per unit of volume around that point. This assumption must be made in order to explain how it is that in the case of interference the luminous energy is found concentrated at those points where the intensity of the wave is at a maximum: if it is assumed that the luminous energy is transferred by light-corpuscles, or photons, then it follows that the density of the photons in the wave is proportional to this intensity.

We thus find that in order to describe the properties of Matter, as well as those of Light, we must employ waves and corpuscles simultaneously. We can no longer imagine the electron as being just a minute corpuscle of electricity: we must associate a wave with it. And this wave is not just a fiction: its length can be measured and its interferences calculated in advance. In fact, a whole group of phenomena was in this way predicted before being actually discovered. It is, therefore, on this idea of the dualism in Nature between waves and corpuscles, expressed in a more or less abstract form, that the entire recent development of theoretical Physics has been built up, and that its immediate future development appears likely to be erected.

5
— Niels Bohr —

In this essay, the "father of Quantum Theory" gives a thorough account of the "new" Quantum Theory and introduces the reader to the rich notion of "complementarity." Among other things, Bohr describes the various ways complementarity can be invoked to explain problems previously encountered by physicists, including those discussed by Einstein, de Broglie, Schrödinger, and Born. Bohr also gives a fluid account of what is at the heart of the quantum situation—the so-called "measurement problem"—as well as a clear explanation of the concept of indeterminacy. He concludes the article by alerting us to the fact that in order to accept fully the conclusions of the quantum postulate, we must "be prepared to meet with a renunciation as to visualization in the ordinary sense," something Bohr would repeat often throughout his writings. The reading comes from an article that appeared in a supplement of the journal Nature *in 1928.*

COMPLEMENTARITY AND THE NEW QUANTUM THEORY

The Quantum Postulate and Causality

The Quantum Theory is characterized by the acknowledgment of a fundamental limitation in the classical physical ideas when applied to atomic phenomena. The situation thus created is of a peculiar nature, since our interpretation of the experimental material rests essentially upon the classical concepts. Notwithstanding the difficulties which, hence, are involved in the formulation of the Quantum Theory, it seems, as we shall see, that its essence may be expressed in the so-called quantum postulate, which attributes to any atomic process an essential discontinuity, or rather individuality, completely foreign to the classical theories and symbolized by Planck's quantum of action.

This postulate implies a renunciation as regards the causal space-time co-ordination of atomic processes. Indeed, our usual description of physical phenomena is based entirely on the idea that the phenomena concerned may be observed without disturbing them appreciably. This appears, for example, clearly in the Theory of Relativity, which has been so fruitful for the elucidation of the classical theories. As emphasized by Einstein, every observation or measurement ultimately rests on the coincidence of two independent events at the same space-time point. Just these coincidences will not be affected by any differences which the space-time co-ordination of different observers otherwise may exhibit. Now, the quantum postulate implies that any observation of atomic phenomena will involve an interaction with the agency of observation not to be neglected. Accordingly, an independent reality in the ordinary physical sense can neither be ascribed to the phenomena nor to the agencies of observation. After all, the concept of observation is in so far arbitrary as it depends upon which objects are included in the system to be observed. Ultimately, every observation can, of course, be reduced to our sense perceptions. The circumstance, however, that in interpreting observations use has always to be made of theoretical notions entails that for every particular case it is a question of convenience at which point the concept of observation involving the quantum postulate with its inherent "irrationality" is brought in.

This situation has far-reaching consequences. On one hand, the definition of the state of a physical system, as ordinarily understood, claims the elimination of all external disturbances. But in that case, according to the quantum postulate, any observation will be impossible, and, above all, the concepts of space and time lose their immediate sense. On the other hand, if in order to make observation possible we permit certain interactions with suitable agencies of measurement, not belonging to the system, an unambiguous definition of the state of the system is naturally no longer possible, and there can be no question of causality in the ordinary sense of the word. The very nature of the Quantum Theory thus forces us to regard the space-time co-ordination and the claim of causality, the union of which characterizes the classical theories, as complementary but exclusive features of the description, symbolizing the idealization of observation and definition respectively. Just as the relativity theory has taught us that the convenience of distinguishing sharply between space and time rests solely on the smallness of the velocities ordinarily met with compared to the velocity of light, we learn from the Quantum Theory that the appropriateness of our usual causal space-time description depends entirely upon the small value of the quantum of action as compared to the actions involved in ordinary sense perceptions. Indeed, in the description of atomic phenomena, the quantum postulate presents us with the task of developing a "Complementarity" theory the consistency of which can be judged only by weighing the possibilities of definition and observation.

This view is already clearly brought out by the much-discussed question of the nature of light and the ultimate constituents of matter. As regards light, its propagation in space and time is adequately expressed by the electromagnetic theory. Especially the interference phenomena *in vacuo* and the optical properties of material media are completely governed by the wave theory superposition principle. Nevertheless, the conservation of energy and momentum during the interaction between radiation and matter, as evident in the photo-electric and Compton effect, finds its adequate expression just in the light quantum idea put forward by Einstein. As is well known, the doubts regarding the validity of the superposition principle, on the one hand, and of the conservation laws, on the other, which were suggested by this apparent contradiction, have been definitely disproved through direct experiments. This situation would seem clearly to indicate the impossibility of a causal space-time description of the light phenomena. On one hand, in attempting to trace the laws of the time-spatial propagation of light according to the quantum postulate, we are confined to statistical considerations. On the other hand, the fulfillment of the claim of causality for the individual light processes, characterized by the quantum of action, entails a renunciation as regards the space-time description. Of course, there can be no question of a quite independent application of the ideas of space and time and of causality. The two views of the nature of light are rather to be considered as different attempts at an interpretation of experimental evidence in which the limitation of the classical concepts is expressed in complementary ways.

The problem of the nature of the constituents of matter presents us with an analogous situation. The individuality of the elementary electrical corpuscles is forced upon us by general evidence. Nevertheless, recent experience, above all the discovery of the selective reflection of electrons from metal crystals, requires the use of the wave theory superposition principle in accordance with the original ideas of L. de Broglie. Just as in the case of light, we have consequently in the question of the nature of matter, so far as we adhere to classical concepts, to face an inevitable dilemma which has to be regarded as the very expression of experimental evidence. In fact, here again we are not dealing with contradictory but with complementary pictures of the phenomena, which only together offer a natural generalization of the classical mode of description. In the discussion of these questions, it must be kept in mind that, according to the view taken above, radiation in free space as well as isolated material particles are abstractions, their properties on the Quantum Theory being definable and observable only through their interaction with other systems. Nevertheless, these abstractions are, as we shall see, indispensable for a description of experience in connection with our ordinary space-time view.

The difficulties with which a causal space-time description is confronted in the Quantum Theory, and which have been the subject of repeated discussions, are now placed into the foreground by the recent development of the symbolic methods. An important contribution to the problem of a consistent application of these methods has been made lately by Heisenberg. In particular, he has stressed the peculiar reciprocal uncertainty which affects all measurements of atomic quantities. Before we enter upon his results, it will be advantageous to show how the complementary nature of the description appearing in this uncertainty is unavoidable already in an analysis of the most elementary concepts employed in interpreting experience.

Quantum of Action and Kinematics

The fundamental contrast between the quantum of action and the classical concepts is immediately apparent from the simple formulae which form the common foundation of the theory of light quanta and of the wave theory of material particles. If Planck's constant be denoted by h, as is well known,

$$E\tau = I\lambda = h, \qquad \qquad \dots\dots (1)$$

where E and I are energy and momentum respectively, τ and λ the corresponding period of vibration and wave-length. In these formulae the two notions of light and also of matter enter in sharp contrast. While energy and momentum are associated with the concept of particles, and, hence, may be characterized according to the classical point of view by definite space-time co-ordinates, the period of vibration and wave-length refer to a plane harmonic wave train of unlimited extent in space and time. Only with the aid of the superposition principle does it become possible to attain a connection with the ordinary mode of description. Indeed, a limitation of the extent of the wave-fields in space and time can always be regarded as resulting from the interference of a group of elementary harmonic waves. As shown by de Broglie, the translational velocity of the individuals associated with the waves can be represented by just the so-called group-velocity. Let us denote a plane elementary wave by

$$A \cos 2\pi \ (vt - x\sigma_x - y\sigma_y - z\sigma_z + \delta),$$

where A and δ are constants determining respectively the amplitude and the phase. The quantity $v = 1/\tau$ is the frequency, σ_x, σ_y, σ_z the wave numbers in the direction of the co-ordinate axes, which may be regarded as vector components of the wave number $\sigma = 1/\lambda$ in the directions of propagation. While the wave or phase velocity is given by v/σ, the group-velocity is defined by $dv/d\sigma$. Now according to the relativity theory we have for a particle with the velocity v:

$$I = v/c^2 \ E \text{ and } vdI = dE,$$

where c denotes the velocity of light. Hence by equation (I) the phase velocity is c^2/v and the group-velocity v. The circumstance that the former is in general greater than the velocity of light emphasizes the symbolic character of these considerations. At the same time, the possibility of identifying the velocity of the particle with the group-velocity indicates the field of application of space-time pictures in the Quantum Theory. Here the complementary character of the description appears, since the use of wave-groups is necessarily accompanied by a lack of sharpness in the definition of period and wave-length, and hence also in the definition of the corresponding energy and momentum as given by relation (1).

Rigorously speaking, a limited wave-field can only be obtained by the super-position of a manifold of elementary waves corresponding to all values of v and σ_x, σ_y, σ_z. But the order of magnitude of the mean difference between these values for two elementary waves in the group is given in the most favourable case by the condition

$$\Delta t\ \Delta v = \Delta x\ \Delta\sigma_x = \Delta y\ \Delta\sigma_y = \Delta z\ \Delta\sigma_z = 1$$

where Δt, Δx, Δy, Δz denote the extension of the wave-field in time and in the directions of space corresponding to the co-ordinate axes. These relations—well known from the theory of optical instruments, especially from Rayleigh's investigation of the resolving power of spectral apparatus—express the condition that the wave-trains extinguish each other by interference at the space-time boundary of the wave-field. They may be regarded also as signifying that the group as a whole has no phase in the same sense as the elementary waves. From equation (1) we find thus:

$$\Delta t\ \Delta E = \Delta x\ \Delta I_x = \Delta y\ \Delta I_y = \Delta z\ \Delta I_z = h, \qquad \ldots\ldots (2)$$

as determining the highest possible accuracy in the definition of the energy and momentum of the individuals associated with the wave-field. In general, the conditions for attributing an energy and a momentum value to a wave-field by means of formula (1) are much less favourable. Even if the composition of the wave-group corresponds in the beginning to the relations (2), it will in the course of time be subject to such changes that it becomes less and less suitable for representing an individual. It is this very circumstance which gives rise to the paradoxical character of the problem of the nature of light and of material particles. The limitation in the classical concepts expressed through relation (2), is, besides, closely connected with the limited validity of classical mechanics, which in the wave theory of matter corresponds to the geometrical optics in which the propagation of waves is depicted through "rays." Only in this limit can energy and momentum be unambiguously defined on the basis of space-time pictures. For a general definition of these concepts we are confined to the conservation laws, the rational formulation of which has been a fundamental problem for the symbolical methods to be mentioned below.

In the language of the relativity theory, the content of the relations (2) may be summarized in the statement that according to the Quantum Theory a general reciprocal relation exists between the maximum sharpness of definition of the space-time and energy-momentum vectors associated with the individuals. This circumstance may be regarded as a simple symbolical expression for the complementary nature of the space-time description and the claims of causality. At the same time, however, the general character of this relation makes it possible to a certain extent to reconcile the conservation laws with the space-time co-ordination of observations, the idea of a coincidence of well-defined events in a space-time point being replaced by that of unsharply defined individuals within finite space-time regions.

This circumstance permits us to avoid the well-known paradoxes which are encountered in attempting to describe the scattering of radiation by free electrical particles as well as the collision of two such particles. According to the classical concepts, the description of the scattering requires a finite extent of the radiation in space and time, while in the change of the motion of the electron demanded by the quantum postulate one seemingly is dealing with an instantaneous effect taking place at a definite point in space. Just as in the case of radiation, however, it is impossible to define momentum and energy for an electron without considering a finite space-time region. Furthermore, an application of the conservation laws to the process implies that the accuracy of definition of the energy-momentum vector is the same for the radiation and the electron. In consequence, according to relation (2), the associated space-time regions can be given the same size for both individuals in interaction.

A similar remark applies to the collision between two material particles, although the significance of the quantum postulate for this phenomenon was disregarded before the necessity of the wave concept was realized. Here, this postulate does, indeed, represent the idea of the individuality of the particles which, transcending the space-time description, meets the claim of causality. While the physical content of the light-quantum idea is wholly connected with the conservation theorems for energy and momentum, in the case of the electrical particles the electric charge has to be taken into account in this connection. It is scarcely necessary to mention that for a more detailed description of the interaction between individuals we cannot restrict ourselves to the facts expressed by formulae (1) and (2), but must resort to a procedure which allows us to take into account the coupling of the individuals, characterizing the interaction in question, where just the importance of the electric charge appears. As we shall see, such a procedure necessitates a further departure from visualization in the usual sense.

Measurements in the Quantum Theory

In his investigations already mentioned on the consistency of the quantum-theoretical methods, Heisenberg has given the relation (2) as an expression for the maximum precision with which the space-time co-ordinates and momentum-energy components of a particle can be measured simultaneously. His view was based on the following consideration: On one hand, the co-ordinates of a particle can be measured with any desired degree of accuracy by using, for example, an optical instrument, provided radiation of sufficiently short wave-length is used for illumination. According to the Quantum Theory, however, the scattering of radiation from the object is always connected with a finite change in momentum, which is the larger the smaller the wave-length of the radiation used. The momentum of a particle, on the other hand, can be determined with any desired degree of accuracy by measuring, for example, the Doppler effect of the scattered radiation, provided the wave-length of the radiation is so large that the effect of recoil can be neglected, but then the determination of the space coordinates of the particle becomes correspondingly less accurate.

The essence of this consideration is the inevitability of the quantum postulate in the estimation of the possibilities of measurement. A closer investigation of the possibilities of definition would still seem necessary in order to bring out the general complementary character of the description. Indeed, a discontinuous change of energy and momentum during observation could not prevent us from ascribing accurate values to the space-time co-ordinates, as well as to the momentum-energy components before and after the process. The reciprocal uncertainty which always affects the values of these quantities is, as will be clear from the preceding analysis, essentially an outcome of the limited accuracy with which changes in energy and momentum can be defined, when the wave-fields used for the determination of the space-time co-ordinates of the particle are sufficiently small.

In using an optical instrument for determinations of position, it is necessary to remember that the formation of the image always requires a convergent beam of light. Denoting by λ the wave-length of the radiation used, and by ε the so-called numerical aperture, that is, the sine of half the angle of convergence, the resolving power of a microscope is given by the well-known expression $\lambda/2\varepsilon$. Even if the object is illuminated by parallel light, so that the momentum h/λ of the incident light quantum is known both as regards magnitude and direction, the finite value of the aperture will prevent an exact knowledge of the recoil accompanying the scattering. Also, even if the momentum of the particle were accurately known before the scattering process, our knowledge of the component of momentum parallel to the focal plane after the observation would be affected by an uncertainty amounting to $2\varepsilon h/\lambda$. The product of the least inaccuracies with which the positional co-ordinate and the component of momentum in a definite direction can be ascertained is therefore just given by formula (2). One might perhaps expect that in estimating the accuracy of determining the position, not only the convergence but also the length of the wave-train has to be taken into account, because the particle could change its place during the finite time of illumination. Due to the fact, however, that the exact knowledge of the wave-length is immaterial for the above estimate, it will be realized that for any value of the aperture the wave-train can always be taken so short that a change of position of the particle during the time of observation may be neglected in comparison to the lack of sharpness inherent in the determination of position due to the finite resolving power of the microscope.

In measuring momentum with the aid of the Doppler effect—with due regard to the Compton effect—one will employ a parallel wave-train. For the accuracy, however, with which the change in wave-length of the scattered radiation can be measured the extent of the wave-train in the direction of propagation is essential. If we assume that the directions of the incident and scattered radiation are parallel and opposite, respectively, to the direction of the position co-ordinate and momentum component to be measured, then $c\lambda/2l$ can be taken as a measure of the accuracy in the determination of the velocity, where l denotes the length of the wave-train. For simplicity, we here have regarded the velocity of light as large compared to the velocity of the particle. If m represents the mass of the particle, then the uncertainty attached to the value of the momentum after observation is $cm\lambda/2l$. In this case the magnitude of the recoil, $2h/\lambda$, is sufficiently well defined in order not to give rise to an appreciable uncertainty in the value of the momentum of the particle after observation. Indeed, the general theory of the Compton effect allows us to compute the momentum components in the direction of the radiation before and after the recoil from the wave-lengths of the incident and scattered radiation. Even if the positional co-ordinates of the particle were accurately known in the beginning, our knowledge of the position after observation nevertheless will be affected by an uncertainty. Indeed, on account of the impossibility of attributing a definite instant to the recoil, we know the mean velocity in the direction of observation during the scattering process only with an accuracy $2h/m\lambda$. The uncertainty in the position after observation hence is $2hl/mc\lambda$. Here, too, the product of the inaccuracies in the measurement of position and momentum is thus given by the general formula (2).

Just as in the case of the determination of position, the time of the process of observation for the determination of momentum may be made as short as is desired, if only the wave-length of the radiation used is sufficiently small. The fact that the recoil then gets larger does not, as we have seen, affect the accuracy of measurement. It should further be mentioned, that in referring to the velocity of a particle as we have here done repeatedly, the purpose has only been to obtain a connection with the ordinary space-time description convenient in this case. As it appears already from the considerations of de Broglie mentioned above, the concept of velocity must always in the Quantum Theory be handled with caution. It will also be seen that an unambiguous definition of this concept is excluded by the quantum postulate. This is particularly to be remembered when comparing the results of successive observations. Indeed, the position of an individual at two given moments can be measured with any desired degree of accuracy; but if, from such measurements, we would calculate the velocity of the individual in the ordinary way, it must be clearly realized that we are dealing with an abstraction, from which no unambiguous information concerning the previous or future behaviour of the individual can be obtained.

According to the above considerations regarding the possibilities of definition of the properties of individuals, it will obviously make no difference in the discussion of the accuracy of measurements of position and momentum of a particle if collisions with other material particles are considered instead of scattering of radiation. In both cases, we see that the uncertainty in question equally affects the description of the agency of measurement and of the object. In fact, this uncertainty cannot be avoided in a description of the behaviour of individuals with respect to a co-ordinate system fixed in the ordinary way by means of solid bodies and unperturbable clocks. The experimental devices—opening and closing of apertures, etc.—are seen to permit only conclusions regarding the space-time extension of the associated wave-fields.

Development of Atomic Theory

In tracing observations back to our sensations, once more regard has to be taken to the quantum postulate in connection with the perception of the agency of observation, be it through its direct action upon the eye or by means of suitable auxiliaries such as photographic plates, Wilson clouds, etc. It is easily seen, however, that the resulting additional statistical element will not influence the uncertainty in the description of the object. It might even be conjectured that the arbitrariness in what is regarded as object and what as agency of observation would open up a possibility of avoiding this uncertainty altogether. In connection with the measurement of the position of a particle, one might, for example, ask whether the momentum transmitted by the scattering could not be determined by means of the conservation theorem from a measurement of the change of momentum of the microscope—including light source and photographic plate—during the process of observation. A closer investigation shows, however, that such a measurement is impossible, if at the same time one wants to know the position of the microscope with sufficient accuracy. In fact, it follows from the experiences which have found expression in the wave theory of matter that the position of the centre of gravity of a body and its total momentum can only be defined within the limits of reciprocal accuracy given by relation (2).

Strictly speaking, the idea of observation belongs to the causal space-time way of description. Due to the general character of relation (2), however, this idea can be consistently utilized also in the Quantum Theory, if only the uncertainty expressed through this relation is taken into account. As remarked by Heisenberg, one may even obtain an instructive illustration of the quantum-theoretical description of atomic (microscopic) phenomena by comparing this uncertainty with the uncertainty, due to imperfect measurements, inherently contained in any observation as considered in the ordinary description of natural phenomena. He remarks on that occasion that even in the case of macroscopic phenomena we may say, in a certain sense, that they are created by repeated observations. It must not be forgotten, however, that in the classical theories any succeeding observation permits a prediction of future events with ever-increasing accuracy, because it improves our knowledge of the initial state of the system. According to the Quantum Theory, just the impossibility of neglecting the interaction with the agency of measurement means that every observation introduces a new uncontrollable element. Indeed, it follows from the above considerations that the measurement of the positional co-ordinates of a particle is accompanied not only by a finite change in the dynamical variables, but also the fixation of its position means a complete rupture in the causal description of its dynamical behaviour, while the determination of its momentum always implies a gap in the knowledge of its spatial propagation. Just this situation brings out most strikingly the complementary character of the description of atomic phenomena which appears as an inevitable consequence of the contrast between the quantum postulate and the distinction between object and agency of measurement, inherent in our very idea of observation. . . .

Wave Mechanics and the Quantum Postulate

Already in his first considerations concerning the wave theory of material particles, de Broglie pointed out that the stationary states of an atom may be visualized as an interference effect of the phase wave associated with a bound electron. It is true that this point of view at first did not, as regards quantitative results, lead beyond the earlier methods of Quantum Theory, to the development of which Sommerfeld has contributed so essentially. Schrödinger, however, succeeded in developing a wave-theoretical method which has opened up new aspects, and has proved to be of decisive importance for the great progress in atomic physics during the last years. Indeed, the proper vibrations of the Schrödinger wave equation have been found to furnish a representation of the stationary states of an atom meeting all requirements. The energy of each state is connected with the corresponding period of vibration according to the general quantum relation (1). Furthermore, the number of nodes in the various characteristic vibrations gives a simple interpretation to the concept of quantum number which was already known from the older methods, but at first did not seem to appear in the matrix formulation. In addition, Schrödinger could associate with the solutions of the wave equation a continuous distribution of charge and current which, if applied to a characteristic vibration, represents the electrostatic and magnetic properties of an atom in the corresponding stationary state. Similarly, the superposition of two characteristic solutions corresponds to a continuous vibrating distribution of electrical charge, which on classical electrodynamics would give rise to an emission of radiation, illustrating instructively the consequences of the quantum postulate and the correspondence requirement regarding the transition process between two stationary states formulated in matrix mechanics. Another application of the method of Schrödinger, important for the further development, has been made by Born in his investigation of the problem of collisions between atoms and free electric particles. In this connection he succeeded in obtaining a statistical interpretation of the wave functions, allowing a calculation of the probability of the individual transition processes required by the quantum postulate. This includes a wave-mechanical formulation of the adiabatic principle of Ehrenfest, the fertility of which appears strikingly in the promising investigations of Hund on the problem of the formation of molecules.

In view of these results, Schrödinger has expressed the hope that the development of the wave theory will eventually remove the irrational element expressed by the quantum postulate and open the way for a complete description of atomic phenomena along the line of the classical theories. In support of this view, Schrödinger, in a recent paper, emphasizes the fact that the discontinuous exchange of energy between atoms required by the quantum postulate, from the point of view of the wave theory, is replaced by a simple resonance phenomenon. In particular, the idea of individual stationary states would be an illusion and its applicability only an illustration of the resonance mentioned. It must be kept in mind, however, that just in the resonance problem mentioned we are concerned with a closed system which, according to the view presented here, is not accessible to observation. In fact, wave mechanics, just as the matrix theory, on this view represents a symbolic transcription of the problem of motion of classical mechanics adapted to the requirements of Quantum Theory and only to be interpreted by an explicit use of the quantum postulate. Indeed, the two formulations of the interaction problem might be said to be complementary in the same sense as the wave and particle idea in the description of the free individuals. The apparent contrast in the utilization of the energy concept in the two theories is just connected with this difference in the starting-point.

The fundamental difficulties opposing a space-time description of a system of particles in interaction appear at once from the inevitability of the superposition principle in the description of the behaviour of individual particles. Already for a free particle the knowledge of energy and momentum excludes, as we have seen, the exact knowledge of its space-time co-ordinates. This implies that an immediate utilization of the concept of energy in connection with the classical idea of the potential energy of the system is excluded. In the Schrödinger wave equation these difficulties are avoided by replacing the classical expression of the Hamiltonian by a differential operator by means of the relation

$$p = \sqrt{-1}\ h/2\pi\ \delta/\delta q$$

where p denotes a generalized component of momentum and q the canonically conjugated variable. Here the negative value of the energy is regarded as conjugated to the time. So far, in the wave equation, time and space as well as energy and momentum are utilized in a purely formal way.

The symbolical character of Schrödinger's method appears not only from the circumstance that its simplicity, similarly to that of the matrix theory, depends essentially upon the use of imaginary arithmetic quantities. But above all there can be no question of an immediate connection with our ordinary conceptions because the "geometrical" problem represented by the wave equation is associated with the so-called co-ordinate space, the number of dimensions of which is equal to the number of degrees of freedom of the system, and, hence, in general greater than the number of dimensions of ordinary space. Further, Schrödinger's formulation of the interaction problem, just as the formulation offered by matrix theory, involves a neglect of the finite velocity of propagation of the forces claimed by relativity theory.

On the whole, it would scarcely seem justifiable, in the case of the interaction problem, to demand a visualization by means of ordinary space-time pictures. In fact, all our knowledge concerning the internal properties of atoms is derived from experiments on their radiation or collision reactions, such that the interpretation of experimental facts ultimately depends on the abstractions of radiation in free space, and free material particles. Hence, our whole space-time view of physical phenomena, as well as the definition of energy and momentum, depends ultimately upon these abstractions. In judging the applications of these auxiliary ideas, we should only demand inner consistency, in which connection special regard has to be paid to the possibilities of definition and observation.

In the characteristic vibrations of Schrödinger's wave equation we have, as mentioned, an adequate representation of the stationary states of an atom allowing an unambiguous definition of the energy of the system by means of the general quantum relation (1). This entails, however, that in the interpretation of observations a fundamental renunciation regarding the space-time description is unavoidable. In fact, the consistent application of the concept of stationary states excludes, as we shall see, any specification regarding the behaviour of the separate particles in the atom. In problems where a description of this behaviour is essential, we are bound to use the general solution of the wave equation which is obtained by superposition of characteristic solutions. We meet here with a complementarity of the possibilities of definition quite analogous to that which we have considered earlier in connection with the properties of light and free material particles. Thus, while the definition of energy and momentum of individuals is attached to the idea of a harmonic elementary wave, every space-time feature of the description of phenomena is, as we have seen, based on a consideration of the interferences taking place inside a group of such elementary waves. Also in the present case the agreement between the possibilities of observation and those of definition can be directly shown.

According to the quantum postulate any observation regarding the behaviour of the electron in the atom will be accompanied by a change in the state of the atom. As stressed by Heisenberg, this change will, in the case of atoms in stationary states of low quantum number, consist in general in the ejection of the electron from the atom. A description of the "orbit" of the electron in the atom with the aid of subsequent observations is, hence, impossible in such a case. This is connected with the circumstance that from characteristic vibrations with only a few nodes no wave packages can be built up which would even approximately represent the "motion" of a particle. The complementary nature of the description, however, appears particularly in that the use of observations concerning the behaviour of particles in the atom rests on the possibility of neglecting, during the process of observation, the interaction between the particles, thus regarding them as free. This requires, however, that the duration of the process is short compared with the natural periods of the atom, which again means that the uncertainty in the knowledge of the energy transferred in the process is large compared to the energy differences between neighbouring stationary states.

In judging the possibilities of observation it must, on the whole, be kept in mind that the wave-mechanical solutions can be visualized only in so far as they can be described with the aid of the concept of free particles. Here the difference between classical mechanics and the quantum-theoretical treatment of the problem of interaction appears most strikingly. In the former such a restriction is unnecessary because the "particles" are here endowed with an immediate "reality," independently of their being free or bound. This situation is particularly important in connection with the consistent utilization of Schrödinger's electric density as a measure of the probability for electrons being present within given space regions of the atom. Remembering the restriction mentioned, this interpretation is seen to be a simple consequence of the assumption that the probability of the presence of a free electron is expressed by the electric density associated with the wave-field in a similar way to that by which the probability of the presence of a light quantum is given by the energy density of the radiation. . . .

Reality of Stationary States

In the conception of stationary states we are, as mentioned, concerned with a characteristic application of the quantum postulate. By its very nature this conception means a complete renunciation as regards a time description. From the point of view taken here, just this renunciation forms the necessary condition for an unambiguous definition of the energy of the atom. Moreover, the conception of a stationary state involves, strictly speaking, the exclusion of all interactions with individuals not belonging to the system. The fact that such a closed system is associated with a particular energy value may be considered as an immediate expression for the claim of causality contained in the theorem of conservation of energy. This circumstance justifies the assumption of the supramechanical stability of the stationary states, according to which the atom, before as well as after an external influence, always will be found in a well-defined state, and which forms the basis for the use of the quantum postulate in problems concerning atomic structure. . . .

The application of the conception of stationary states demands that in any observation, say by means of collision or radiation reactions, permitting a distinction between different stationary states, we are entitled to disregard the previous history of the atom. The fact that the symbolical Quantum Theory methods ascribe a particular phase to each stationary state the value of which depends upon the previous history of the atom, would for the first moment seem to contradict the very idea of stationary states. As soon as we are really concerned with a time problem, however, the consideration of a strictly closed system is excluded. The use of simply harmonic proper vibrations in the interpretation of observations means, therefore, only a suitable idealization which in a more rigorous discussion must always be replaced by a group of harmonic vibrations, distributed over a finite frequency interval. Now, as already mentioned, it is a general consequence of the superposition principle that it has no sense to co-ordinate a phase value to the group as a whole, in the same manner as may be done for each elementary wave constituting the group.

This inobservability of the phase, well known from the theory of optical instruments, is brought out in a particularly simple manner in a discussion of the Stern-Gerlach experiment, so important for the investigation of the properties of single atoms. As pointed out by Heisenberg, atoms with different orientation in the field may only be separated if the deviation of the beam is larger than the diffraction at the slit of the de Broglie waves representing the translational motion of the atoms. This condition means, as a simple calculation shows, that the product of the time of passage of the atom through the field, and the uncertainty due to the finite width of the beam of its energy in the field, is at least equal to the quantum of action. This result was considered by Heisenberg as a support of relation (2) as regards the reciprocal uncertainties of energy and time values. It would seem, however, that here we are not simply dealing with a measurement of the energy of the atom at a given time. But since the period of the proper vibrations of the atom in the field is connected with the total energy by relation (1), we realize that the condition for separability mentioned just means the loss of the phase. This circumstance removes also the apparent contradictions, arising in certain problems concerning the coherence of resonance radiation, which have been discussed frequently, and were also considered by Heisenberg.

To consider an atom as a closed system, as we have done above, means to neglect the spontaneous emission of radiation which even in the absence of external influences puts an upper limit to the lifetime of the stationary states. The fact that this neglect is justified in many applications is connected with the circumstance that the coupling between the atom and the radiation field, which is to be expected on classical electrodynamics, is in general very small compared to the coupling between the particles in the atom. It is, in fact, possible in a description of the state of an atom to a considerable extent to neglect the reaction of radiation, thus disregarding the unsharpness in the energy values connected with the lifetime of the stationary states according to relation (2). This is the reason why it is possible to draw conclusions concerning the properties of radiation by using classical electrodynamics. . . .

The Problem of the Elementary Particles

When due regard is taken of the complementary feature required by the quantum postulate, it seems, in fact, possible with the aid of the symbolic methods to build up a consistent theory of atomic phenomena, which may be considered as a rational generalization of the causal space-time description of classical physics. This view does not mean, however, that classical electron theory may be regarded simply as the limiting case of a vanishing quantum of action. Indeed, the connection of the latter theory with experience is based on assumptions which can scarcely be separated from the group of problems of the Quantum Theory. A hint in this direction was already given by the well-known difficulties met with in the attempts to account for the individuality of ultimate electrical particles on general mechanical and electrodynamical principles. In this respect also, the general relativity theory of gravitation has not fulfilled expectations. A satisfactory solution of the problems touched upon would seem to be possible only by means of a rational quantum-theoretical transcription of the general field theory, in which the ultimate quantum of electricity has found its natural position as an expression of the feature of individuality characterizing the Quantum Theory. . . .

Already the formulation of the relativity argument implies essentially the union of the space-time co-ordination and the demand of causality characterizing the classical theories. In the adaptation of the relativity requirement to the quantum postulate, we must therefore be prepared to meet with a renunciation as to visualization in the ordinary sense going still further than in the formulation of the quantum laws considered here. Indeed, we find ourselves here on the very path taken by Einstein of adapting our modes of perception borrowed from the sensations to the gradually deepening knowledge of the laws of Nature. The hindrances met with on this path originate above all in the fact that, so to say, every word in the language refers to our ordinary perception. In the Quantum Theory we meet this difficulty at once in the question of the inevitability of the feature of irrationality characterizing the quantum postulate. I hope, however, that the idea of complementarity is suited to characterize the situation, which bears a deep-going analogy to the general difficulty in the formation of human ideas, inherent in the distinction between subject and object.

6
— N i e l s B o h r —

In this piece, written originally for the volume, dedicated to Albert Einstein as part of the series "Living Philosophers" and later included in Volume 2 of the author's Philosophical Essays, *Niels Bohr provides a fluid account of the various conceptual difficulties encountered in Quantum Theory, with special attention paid to the various thought experiments proposed by Einstein at the Solvay conferences in Brussels while attempting to refute the theory. After providing a thoroughgoing historical overview of the conceptual origins of Quantum Theory—including a word on the contributions of de Broglie—Bohr elucidates the many conceptual problems inherent in Quantum Theory, including indeterminacy, complementarity, and the inseparability of object and means of observation in any experimental arrangement. Here Bohr suggests not only solutions to these problems, but also how these solutions—especially the form of complementarity—might be of great use outside the domain of physical science.*

Moreover, in this rich essay, Bohr chronicles the epistemological debates he had with Einstein, and in each case he includes his various ripostes to Einstein's proposals. Of particular note is his response to the now well-known quip by Einstein that God could not possibly "play dice."

THE DEBATE WITH EINSTEIN

When invited by the Editor of the series "Living Philosophers" to write an article for this volume in which contemporary scientists are honouring the epoch-making contributions of Albert Einstein to the progress of natural philosophy and are acknowledging the indebtedness of our whole generation for the guidance his genius has given us, I thought much of the best way of explaining how much I owe to him for inspiration. In this connection, the many occasions through the years on which I had the privilege to discuss with Einstein epistemological problems raised by the modern development of atomic physics have come back vividly to my mind and I have felt that I could hardly attempt anything better than to give an account of these discussions which have been of greatest value and stimulus to me. I hope also that the account may convey to wider circles an impression of how essential the open-minded exchange of ideas has been for the progress in a field where new experience has time after time demanded a reconsideration of our views.

From the very beginning the main point under debate has been the attitude to take to the departure from customary principles of natural philosophy characteristic of the novel development of physics which was initiated in the first year of this century by Planck's discovery of the universal quantum of action. This discovery, which revealed a feature of atomicity in the laws of nature going far beyond the old doctrine of the limited divisibility of matter, has indeed taught us that the classical theories of physics are idealizations which can be unambiguously applied only in the limit where all actions involved are large compared with the quantum. The question at issue has been whether the renunciation of a causal mode of description of atomic processes involved in the endeavours to cope with the situation should be regarded as a temporary departure from ideals to be ultimately revived or whether we are faced with an irrevocable step towards obtaining the proper harmony between analysis and synthesis of physical phenomena. To describe the background of our discussions and to bring out as clearly as possible the arguments for the contrasting viewpoints, I have felt it necessary to go to a certain length in recalling some main features of the development to which Einstein himself has contributed so decisively.

As is well known, it was the intimate relation, elucidated primarily by Boltz-mann, between the laws of thermodynamics and the statistical regularities exhib-ited by mechanical systems with many degrees of freedom, which guided Planck in his ingenious treatment of the problem of thermal radiation, leading him to his fundamental discovery. While, in his work, Planck was principally concerned with considerations of essentially statistical character and with great caution re-frained from definite conclusions as to the extent to which the existence of the quantum implied a departure from the foundations of mechanics and electrody-namics, Einstein's great original contribution to Quantum Theory (1905) was just the recognition of how physical phenomena like the photo-effect may de-pend directly on individual quantum effects. In these very same years when, in the development of his Theory of Relativity, Einstein laid down a new founda-tion for physical science, he explored with a most daring spirit the novel features of atomicity which pointed beyond the framework of classical physics.

With unfailing intuition Einstein thus was led step by step to the conclusion that any radiation process involves the emission or absorption of individual light quanta or "photons" with energy and momentum

$$E = h\nu \text{ and } P = h\sigma \tag{1}$$

respectively, where h is Planck's constant, while ν and σ are the number of vibrations per unit time and the number of waves per unit length, respectively. Notwithstanding its fertility, the idea of the photon implied a quite unforeseen dilemma, since any simple corpuscular picture of radiation would obviously be irreconcilable with interference effects, which present so essential an aspect of radiative phenomena, and which can be described only in terms of a wave pic-ture. The acuteness of the dilemma is stressed by the fact that the interference effects offer our only means of defining the concepts of frequency and wave-length entering into the very expressions for the energy and momentum of the photon.

In this situation, there could be no question of attempting a causal analysis of radiative phenomena, but only, by a combined use of the contrasting pictures, to estimate probabilities for the occurrence of the individual radiation processes. However, it is most important to realize that the recourse to probability laws under such circumstances is essentially different in aim from the familiar appli-cation of statistical considerations as practical means of accounting for the prop-erties of mechanical systems of great structural complexity. In fact, in Quantum Physics we are presented not with intricacies of this kind, but with the inability of the classical frame of concepts to comprise the peculiar feature of indivisibil-ity, or "individuality," characterizing the elementary processes.

The failure of the theories of classical physics in accounting for atomic phe-nomena was further accentuated by the progress of our knowledge of the struc-ture of atoms. Above all, Rutherford's discovery of the atomic nucleus (1911) revealed at once the inadequacy of classical mechanical and electromagnetic con-cepts to explain the inherent stability of the atom. Here again the Quantum

Theory offered a clue for the elucidation of the situation and especially it was found possible to account for the atomic stability, as well as for the empirical laws governing the spectra of the elements, by assuming that any reaction of the atom resulting in a change of its energy involved a complete transition between two so-called stationary quantum states and that, in particular, the spectra were emitted by a step-like process in which each transition is accompanied by the emission of a monochromatic light quantum of an energy just equal to that of an Einstein photon.

These ideas, which were soon confirmed by the experiments of Franck and Hertz (1914) on the excitation of spectra by impact of electrons on atoms, involved a further renunciation of the causal mode of description, since evidently the interpretation of the spectral laws implies that an atom in an excited state in general will have the possibility of transitions with photon emission to one or another of its lower energy states. In fact, the very idea of stationary states is incompatible with any directive for the choice between such transitions and leaves room only for the notion of the relative probabilities of the individual transition processes. The only guide in estimating such probabilities was the so-called correspondence principle which originated in the search for the closest possible connection between the statistical account of atomic processes and the consequences to be expected from classical theory, which should be valid in the limit where the actions involved in all stages of the analysis of the phenomena are large compared with the universal quantum.

At that time, no general self-consistent Quantum Theory was yet in sight, but the prevailing attitude may perhaps be illustrated by the following passage from a lecture by the writer from 1913:

> I hope that I have expressed myself sufficiently clearly so that you may appreciate the extent to which these considerations conflict with the admirably consistent scheme of conceptions which has been rightly termed the classical theory of electrodynamics. On the other hand, I have tried to convey to you the impression that—just by emphasizing so strongly this conflict—it may also be possible in course of time to establish a certain coherence in the new ideas.

Important progress in the development of Quantum Theory was made by Einstein himself in his famous article on radiative equilibrium in 1917, where he showed that Planck's law for thermal radiation could be simply deduced from assumptions conforming with the basic ideas of the Quantum Theory of atomic constitution. To this purpose, Einstein formulated general statistical rules regarding the occurrence of radiative transitions between stationary states, assuming not only that, when the atom is exposed to a radiation field, absorption as well as emission processes will occur with a probability per unit time proportional to the intensity of the irradiation, but that even in the absence of external disturbances spontaneous emission processes will take place with a rate corresponding to a certain *a priori* probability. Regarding the latter point, Einstein emphasized the fundamental character of the statistical description in a most suggestive way by drawing attention to the analogy between the assumptions regarding the occurrence of the spontaneous radiative transitions and the well-known laws governing transformations of radioactive substances.

In connection with a thorough examination of the exigencies of thermodynamics as regards radiation problems, Einstein stressed the dilemma still further by pointing out that the argumentation implied that any radiation process was "unidirected" in the sense that not only is a momentum corresponding to a photon with the direction of propagation transferred to an atom in the absorption process, but that also the emitting atom will receive an equivalent impulse in the opposite direction, although there can on the wave picture be no question of a preference for a single direction in an emission process. Einstein's own attitude to such startling conclusions is expressed in a passage at the end of the article (*loc. cit.*, pp. 127f.), which may be translated as follows:

> These features of the elementary processes would seem to make the development of a proper quantum treatment of radiation almost unavoidable. The weakness of the theory lies in the fact that on the one hand, no closer connection with the wave concepts is obtainable and that on the other hand, it leaves to chance (*Zufall*) the time and the direction of the elementary processes; nevertheless, I have full confidence in the reliability of the way entered upon.

When I had the great experience of meeting Einstein for the first time during a visit to Berlin in 1920, these fundamental questions formed the theme of our conversations. The discussions, to which I have often reverted in my thoughts, added to all my admiration for Einstein a deep impression of his detached attitude. Certainly, his favoured use of such picturesque phrases as "ghost waves (*Gespensterfelder*) guiding the photons" implied no tendency to mysticism, but illuminated rather a profound humour behind his piercing remarks. Yet, a certain difference in attitude and outlook remained, since, with his mastery for co-ordinating apparently contrasting experience without abandoning continuity and causality, Einstein was perhaps more reluctant to renounce such ideals than someone for whom renunciation in this respect appeared to be the only way open to proceed with the immediate task of co-ordinating the multifarious evidence regarding atomic phenomena, which accumulated from day to day in the exploration of this new field of knowledge.

In the following years, during which the atomic problems attracted the attention of rapidly increasing circles of physicists, the apparent contradictions inherent in Quantum Theory were felt ever more acutely. Illustrative of this situation is the discussion raised by the discovery of the Stern-Gerlach effect in 1922. On the one hand, this effect gave striking support to the idea of stationary states and in particular to the Quantum Theory of the Zeeman effect developed by Sommerfeld; on the other hand, as exposed so clearly by Einstein and Ehrenfest, it presented with unsurmountable difficulties any attempt at forming a picture of the behaviour of atoms in a magnetic field. Similar paradoxes were raised by the discovery by Compton (1924) of the change in wave-length accompanying the scattering of X-rays by electrons. This phenomenon afforded, as is well known, a most direct proof of the adequacy of Einstein's view regarding the transfer of energy and momentum in radiative processes; at the same time, it was equally clear that no simple picture of a corpuscular collision could offer an exhaustive description of the phenomenon. Under the impact of such difficulties, doubts were for a time entertained even regarding the conservation of energy and momentum in the individual radiation processes; a view, however, which very soon had to be abandoned in face of more refined experiments bringing out the correlation between the deflection of the photon and the corresponding electron recoil.

The way to the clarification of the situation was, indeed, first to be paved by the development of a more comprehensive Quantum Theory. A first step towards this goal was the recognition by de Broglie in 1925 that the wave-corpuscle duality was not confined to the properties of radiation, but was equally unavoidable in accounting for the behaviour of material particles. This idea, which was soon convincingly confirmed by experiments on electron interference phenomena, was at once greeted by Einstein, who had already envisaged the deep-going analog between the properties of thermal radiation and of gases in the so-called degenerate state. The new line was pursued with the greatest success by Schrödinger (1926) who, in particular, showed how the stationary states of atomic systems could be represented by the proper solutions of a wave-equation to the establishment of which he was led by the formal analogy, originally traced by Hamilton, between mechanical and optical problems. Still, the paradoxical aspects of Quantum Theory were in no way ameliorated, but even emphasized, by the apparent contradiction between the exigencies of the general superposition principle of the wave description and the feature of individuality of the elementary atomic processes.

At the same time, Heisenberg (1925) had laid the foundation of a rational quantum mechanics, which was rapidly developed through important contributions by Born and Jordan as well as by Dirac. In this theory, a formalism is introduced, in which the kinematical and dynamical variables of classical mechanics are replaced by symbols subjected to a non-commutative algebra. Notwithstanding the renunciation of orbital pictures, Hamilton's canonical equations of mechanics are kept unaltered and Planck's constant enters only in the rules of commutation

$$qp - pq = \sqrt{-1}\,\frac{h}{2\pi} \tag{2}$$

holding for any set of conjugate variables q and p. Through a representation of the symbols by matrices with elements referring to transitions between stationary states, a quantitative formulation of the correspondence principle became for the first time possible. It may here be recalled that an important preliminary step towards this goal was reached through the establishment, especially by contributions of Kramers, of a Quantum Theory of dispersion making basic use of Einstein's general rules for the probability of the occurrence of absorption and emission processes.

This formalism of quantum mechanics was soon proved by Schrödinger to give results identical with those obtainable by the mathematically often more convenient methods of wave theory, and in the following years general methods were gradually established for an essentially statistical description of atomic processes combining the features of individuality and the requirements of the superposition principle, equally characteristic of Quantum Theory. Among the many advances in this period, it may especially be mentioned that the formalism proved capable of incorporating the exclusion principle which governs the states of systems with several electrons, and which already before the advent of quantum mechanics had been derived by Pauli from an analysis of atomic spectra. The quantitative comprehension of a vast amount of empirical evidence could leave no doubt as to the fertility and adequacy of the quantum-mechanical formalism, but its abstract character gave rise to a widespread feeling of uneasiness. An elucidation of the situation should, indeed, demand a thorough examination of the very observational problem in atomic physics.

This phase of the development was, as is well known, initiated in 1927 by Heisenberg, who pointed out that the knowledge obtainable of the state of an atomic system will always involve a peculiar "indeterminacy." Thus, any measurement of the position of an electron by means of some device, like a microscope, making use of high-frequency radiation, will, according to the fundamental relations (1), be connected with a momentum exchange between the electron and the measuring agency, which is the greater the more accurate a position measurement is attempted. In comparing such considerations with the exigencies of the quantum-mechanical formalism, Heisenberg called attention to the fact that the commutation rule (2) imposes a reciprocal limitation on the fixation of two conjugate variables, q and p, expressed by the relation

$$\Delta q \cdot \Delta p \approx h, \tag{3}$$

where Δq and Δp are suitably defined latitudes in the determination of these variables. In pointing to the intimate connection between the statistical description in quantum mechanics and the actual possibilities of measurement, this so-called indeterminacy relation is, as Heisenberg showed, most important for the elucidation of the paradoxes involved in the attempts of analyzing quantum effects with reference to customary physical pictures.

The new progress in atomic physics was commented upon from various sides at the International Physical Congress held in September 1927 at Como in commemoration of Volta. In a lecture on that occasion, I advocated a point of view conveniently termed "Complementarity," suited to embrace the characteristic features of individuality of quantum phenomena, and at the same time to clarify the peculiar aspects of the observational problem in this field of experience. For this purpose, it is decisive to recognize that, *however far the phenomena transcend the scope of classical physical explanation, the account of all evidence must be ex-*

pressed in classical terms. The argument is simply that by the word "experiment" we refer to a situation where we can tell others what we have done and what we have learned and that, therefore, the account of the experimental arrangement and of the results of the observations must be expressed in unambiguous language with suitable application of the terminology of classical physics.

This crucial point, which was to become a main theme of the discussions reported in the following, implies the *impossibility of any sharp separation between the behaviour of atomic objects and the interaction with the measuring instruments which serve to define the conditions under which the phenomena appear.* In fact, the individuality of the typical quantum effects finds its proper expression in the circumstance that any attempt of subdividing the phenomena will demand a change in the experimental arrangement introducing new possibilities of interaction between objects and measuring instruments which in principle cannot be controlled. Consequently, evidence obtained under different experimental conditions cannot be comprehended within a single picture, but must be regarded as *complementary* in the sense that only the totality of the phenomena exhausts the possible information about the objects.

Under these circumstances an essential element of ambiguity is involved in ascribing conventional physical attributes to atomic objects, as is at once evident in the dilemma regarding the corpuscular and wave properties of electrons and photons, where we have to do with contrasting pictures, each referring to an essential aspect of empirical evidence. An illustrative example, of how the apparent paradoxes are removed by an examination of the experimental conditions under which the complementary phenomena appear, is also given by the Compton effect, the consistent description of which at first had presented us with such acute difficulties. Thus, any arrangement suited to study the exchange of energy and momentum between the electron and the photon must involve a latitude in the space-time description of the interaction sufficient for the definition of wave-number and frequency which enter into the relation (1). Conversely, any attempt of locating the collision between the photon and the electron more accurately would, on account of the unavoidable interaction with the fixed scales and clocks defining the space-time reference frame, exclude all closer account as regards the balance of momentum and energy.

As stressed in the lecture, an adequate tool for a complementary way of description is offered precisely by the quantum-mechanical formalism which represents a purely symbolic scheme permitting only predictions, on lines of the correspondence principle, as to results obtainable under conditions specified by means of classical concepts. It must here be remembered that even in the indeterminacy relation (3) we are dealing with an implication of the formalism which defies unambiguous expression in words suited to describe classical physical pictures. Thus, a sentence like "we cannot know both the momentum and the position of an atomic object" raises at once questions as to the physical reality of two such attributes of the object, which can be answered only by referring to the conditions for the unambiguous use of space-time concepts, on the one hand, and dynamical conservation laws on the other hand. While the combination of these concepts into a single picture of a causal chain of events is the essence of classical mechanics, room for regularities beyond the grasp of such a description is just afforded by the circumstance that the study of the complementary phenomena demands mutually exclusive experimental arrangements.

The necessity, in atomic physics, of a renewed examination of the foundation for the unambiguous use of elementary physical ideas recalls in some way the situation that led Einstein to his original revision of the basis for all application of space-time concepts which, by its emphasis on the primordial importance of the observational problem, has lent such unity to our world picture. Notwithstanding all novelty of approach, causal description is upheld in relativity theory within any given frame of reference, but in Quantum Theory the uncontrollable interaction between the objects and the measuring instruments forces us to a renunciation even in such respect. This recognition, however, in no way points to any limitation of the scope of the quantum-mechanical description, and the trend of the whole argumentation presented in the Como lecture was to show that the viewpoint of Complementarity may be regarded as a rational generalization of the very ideal of causality.

At the general discussion in Como, we all missed the presence of Einstein, but soon after, in October 1927, I had the opportunity to meet him in Brussels at the Fifth Physical Conference of the Solvay Institute, which was devoted to the theme "Electrons and Photons." At the Solvay meetings, Einstein had from their beginning been a most prominent figure, and several of us came to the conference with great anticipations to learn his reaction to the latest stage of the development which, to our view, went far in clarifying the problems which he had himself from the outset elicited so ingeniously. During the discussions, where the whole subject was reviewed by contributions from many sides and where also the arguments mentioned in the preceding pages were again presented, Einstein expressed, however, a deep concern over the extent to which causal account in space and time was abandoned in quantum mechanics.

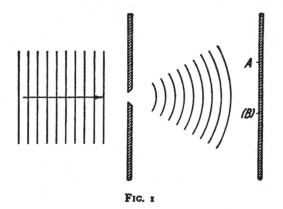

FIG. 1

To illustrate his attitude, Einstein referred at one of the sessions to the simple example, illustrated by Figure 1, of a particle (electron or photon) penetrating through a hole or a narrow slit in a diaphragm placed at some distance before a photographic plate. On account of the diffraction of the wave connected with the motion of the particle and indicated in the figure by the thin lines, it is under such conditions not possible to predict with certainty at what point the electron will arrive at the photographic plate, but only to calculate the probability that, in an experiment, the electron will be found within any given region of the plate. The apparent difficulty, in this description, which Einstein felt so acutely, is the fact that, if in the experiment the electron is recorded at one point A of the plate, then it is out of the question of ever observing an effect of this electron at another point (B), although the laws of ordinary wave propagation offer no room for a correlation between two such events.

Einstein's attitude gave rise to ardent discussions within a small circle, in which Ehrenfest, who through the years had been a close friend of us both, took part in a most active and helpful way, Surely, we all recognized that, in the above example, the situation presents no analogue to the application of statistics in dealing with complicated mechanical systems, but rather recalled the background for Einstein's own early conclusions about the unidirection of individual radiation effects which contrasts so strongly with a simple wave picture. The discussions, however, centered on the question of whether the quantum-mechanical description exhausted the possibilities of accounting for observable phenomena or, as Einstein maintained, the analysis could be carried further and, especially, of whether a fuller description of the phenomena could be obtained by bringing into consideration the detailed balance of energy and momentum in individual processes.

To explain the trend of Einstein's arguments, it may be illustrative here to consider some simple features of the momentum and energy balance in connection with the location of a particle in space and time. For this purpose, we shall examine the simple case of a particle penetrating through a hole in a diaphragm without or with a shutter to open and close the hole, as indicated in Figures 2a and 2b, respectively. The equidistant parallel lines to the left in the figures indicate the train of plane waves corresponding to the state of motion of a particle which, before reaching the diaphragm, has a momentum P related to the wavenumber σ by the second of equations (1). In accordance with the diffraction of the waves when passing through the hole, the state of motion of the particle to the right of the diaphragm is represented by a spherical wave train with a suitably defined angular aperture θ and, in case of Figure 2b, also with a limited radial extension. Consequently, the description of this state involves a certain latitude Δp in the momentum component of the particle parallel to the diaphragm and, in the case of a diaphragm with a shutter, an additional latitude ΔE of the kinetic energy.

Since a measure for the latitude Δq in location of the particle in the plane of the diaphragm is given by the radius a of the hole, and since $\theta \approx 1/\sigma a$, we get, using (1), just $\Delta p \approx \theta P \approx h/\Delta q$, in accordance with the indeterminacy relation (3). This result could, of course, also be obtained directly by noticing that, due to the limited extension of the wave-field at the place of the slit, the component

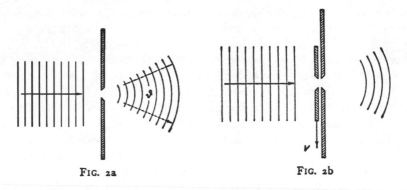

FIG. 2a   FIG. 2b

of the wave-number parallel to the plane of the diaphragm will involve a latitude $\Delta\sigma \approx (1/\alpha) \approx (1/\Delta q)$. Similarly, the spread of the frequencies of the harmonic components in the limited wave-train in Figure 2b is evidently $\Delta v \approx (1/\Delta t)$, where Δt is the time interval during which the shutter leaves the hole open and, thus, represents the latitude in time of the passage of the particle through the diaphragm. From (1), we therefore get

$$\Delta E \cdot \Delta t \approx h, \tag{4}$$

again in accordance with the relation (3) for the two conjugated variables E and t.

From the point of view of the laws of conservation, the origin of such latitudes entering into the description of the state of the particle after passing through the hole may be traced to the possibilities of momentum and energy exchange with the diaphragm or the shutter. In the reference system considered in Figures 2a and 2b, the velocity of the diaphragm may be disregarded and only a change of momentum Δp between the particle and the diaphragm needs to be taken into consideration. The shutter, however, which leaves the hole opened during the time Δt, moves with a considerable velocity $v \approx a/\Delta t$, and a momentum transfer Δp involves therefore an energy exchange with the particle, amounting to

$$v\Delta p \approx (1\Delta t)\, \Delta q\, \Delta p \approx (h/\Delta t),$$

being just of the same order of magnitude as the latitude ΔE given by (4) and, thus, allowing for momentum and energy balance.

The problem raised by Einstein was now to what extent a control of the momentum and energy transfer, involved in a location of the particle in space and time, can be used for a further specification of the state of the particle after passing through the hole. Here, it must be taken into consideration that the position and the motion of the diaphragm and the shutter have so far been assumed to be accurately coordinated with the space-time reference frame. This assumption implies, in the description of the state of these bodies, an essential latitude as to their momentum and energy which need not, of course, noticeably affect the velocities, if the diaphragm and the shutter are sufficiently heavy. However, as soon as we want to know the momentum and energy of these parts of the measuring arrangement with an accuracy sufficient to control the momentum and energy exchange with the particle under investigation, we shall, in accordance with the general indeterminacy relations, lose the possibility of their accurate location in space and time. We have, therefore, to examine how far this circumstance will affect the intended use of the whole arrangement and, as we shall see, this crucial point clearly brings out the complementary character of the phenomena.

Returning for a moment to the case of the simple arrangement indicated in Figure 1, it has so far not been specified to what use it is intended. In fact, it is only on the assumption that the diaphragm and the plate have well-defined positions in space that it is impossible, within the frame of the quantum-mechanical formalism, to make more detailed predictions as to the point of the photographic plate where the particle will be recorded. If, however, we admit a sufficiently large latitude in the knowledge of the position of the diaphragm, it should, in principle, be possible to control the momentum transfer to the diaphragm and, thus, to make more detailed predictions as to the direction of the electron path from the hole to the recording point. As regards the quantum-mechanical description, we have to deal here with a two-body system consisting of the diaphragm as well as of the particle, and it is just with an explicit application of conservation laws to such a system that we are concerned in the Compton effect where, for instance, the observation of the recoil of the electron by means of a cloud chamber allows us to predict in what direction the scattered photon will eventually be observed.

The importance of considerations of this kind was, in the course of the discussions, most interestingly illuminated by the examination of an arrangement where between the diaphragm with the slit and the photographic plate is inserted another diaphragm with two parallel slits, as is shown in Figure 3. If a

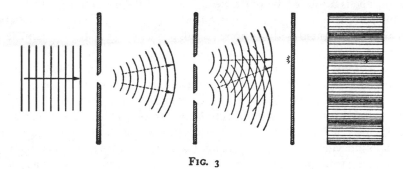

FIG. 3

parallel beam of electrons (or photons) falls from the left on the first diaphragm, we shall, under usual conditions, observe on the plate an interference pattern indicated by the shading of the photographic plate shown in front view to the right of the figure. With intense beams, this pattern is built up by the accumulation of a large number of individual processes, each giving rise to a small spot on the photographic plate, and the distribution of these spots follows a simple law derivable from the wave analysis. The same distribution should also be found in the statistical account of many experiments performed with beams so faint that in a single exposure only one electron (or photon) will arrive at the photographic plate at some spot shown in the figure as a small star. Since, now, as indicated by the broken arrows, the momentum transferred to the first diaphragm ought to be different if the electron was assumed to pass through the upper or the lower slit in the second diaphragm, Einstein suggested that a control of the momentum transfer would permit a closer analysis of the phenomenon and, in particular, make it possible to decide through which of the two slits the electron had passed before arriving at the plate.

A closer examination showed, however, that the suggested control of the momentum transfer would involve a latitude in the knowledge of the position of the diaphragm which would exclude the appearance of the interference phenomena in question. In fact, if ω is the small angle between the conjectured paths of a particle passing through the upper or the lower slit, the difference of momentum transfer in these two cases will, according to (1), be equal to $h\sigma\omega$, and any control of the momentum of the diaphragm with an accuracy sufficient to measure this difference will, due to the indeterminacy relation, involve a minimum latitude of the position of the diaphragm, comparable with $1/\sigma\omega$. If, as in the figure, the diaphragm with the two slits is placed in the middle between the first diaphragm and the photographic plate, it will be seen that the number of fringes per unit length will be just equal to $\sigma\omega$ and, since an uncertainty in the position of the first diaphragm of the amount of $1/\sigma\omega$ will cause an equal uncertainty in the positions of the fringes, it follows that no interference effect can appear. The same result is easily shown to hold for any other placing of the second diaphragm between the first diaphragm and the plate, and would also be obtained if, instead of the first diaphragm, another of these three bodies were used for the control, for the purpose suggested, of the momentum transfer.

This point is of great logical consequence, since it is only the circumstance that we are presented with a choice of *either* tracing the path of a particle *or* observing interference effects, which allows us to escape from the paradoxical necessity of concluding that the behaviour of an electron or a photon should depend on the presence of a slit in the diaphragm through which it could be proved not to pass. We have here to do with a typical example of how the complementary phenomena appear under mutually exclusive experimental arrangements and are just faced with the impossibility, in the analysis of quantum effects, of drawing any sharp separation between an independent behaviour of atomic objects and their interaction with the measuring instruments which serve to define the conditions under which the phenomena occur.

Our talks about the attitude to be taken in face of a novel situation as regards analysis and synthesis of experience touched naturally on many aspects of philosophical thinking, but, in spite of all divergencies of approach and opinion, a most humorous spirit animated the discussions. On his side, Einstein mockingly asked us whether we could really believe that the providential authorities took recourse to dice-playing (" . . . *ob der liebe Gott würfelt*"), to which I replied by pointing at the great caution, already called for by ancient thinkers, in ascribing attributes to Providence in everyday language. I remember also how at the peak of the discussion Ehrenfest, in his affectionate matter of teasing his friends, jokingly hinted at the apparent similarity between Einstein's attitude and that of the opponents of relativity theory; but instantly Ehrenfest added that he would not be able to find relief in his own mind before concord with Einstein was reached.

Einstein's concern and criticism provided a most valuable incentive for us all to reexamine the various aspects of the situation as regards the description of atomic phenomena. To me it was a welcome stimulus to clarify still further the role played by the measuring instruments and, in order to bring into strong relief the mutually exclusive character of the experimental conditions under which the complementary phenomena appear, I tried in those days to sketch various apparatus in a pseudo-realistic style of which the following figures are examples. Thus, for the study of an interference phenomenon of the type indicated in Figure 3, it suggests itself to use an experimental arrangement like that shown in Figure 4, where the solid parts of the apparatus, serving as diaphragms and plate-holder, are firmly bolted to a common support. In such an arrangement, where the knowledge of the relative positions of the diaphragms and the photographic plate is secured by a rigid connection, it is obviously impossible to control the momentum exchanged between the particle and the separate parts of the apparatus. The only way in which, in such an arrangement, we could insure that the particle passed through one of the slits in the second diaphragm is to cover the other slit by a lid, as indicated in the figure; but if the slit is covered, there is of course no question of any interference phenomenon, and on the plate we shall simply observe a continuous distribution as in the case of the single fixed diaphragm in Figure 1.

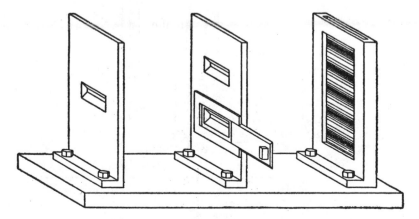

FIG. 4

In the study of phenomena in the account of which we are dealing with detailed momentum balance, certain parts of the whole device must naturally be given the freedom to move independently of others. Such an apparatus is

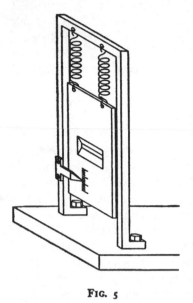

sketched in Figure 5, where a diaphragm with a slit is suspended by weak springs from a solid yoke bolted to the support on which also other immobile parts of the arrangement are to be fastened. The scale on the diaphragm together with the pointer on the bearings of the yoke refer to such study of the motion of the diaphragm, as may be required for an estimate of the momentum transferred to it, permitting one to draw conclusions as to the deflection suffered by the particle in passing through the slit. Since, however, any reading of the scale, in whatever way performed, will involve an uncontrollable change in the momentum of the diaphragm, there will always be, in conformity with the indeterminacy principle, a reciprocal relationship between our knowledge of the position of the slit and the accuracy of the momentum control.

FIG. 5

In the same semi-serious style, Figure 6 represents a part of an arrangement suited for the study of phenomena which, in contrast to those just discussed, involve time coordination explicitly. It consists of a shutter rigidly connected with a robust clock resting on the support

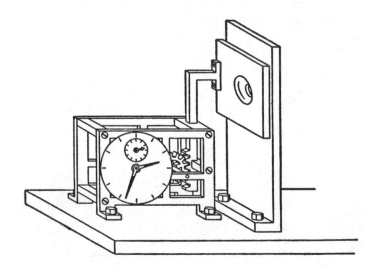

FIG. 6

which carries a diaphragm and on which further parts of similar character, regulated by the same clock-work or by other clocks standardized relatively to it, are also to be fixed. The special aim of the figure is to underline that a clock is a piece of machinery, the working of which can completely be accounted for by ordinary mechanics and will be affected neither by reading of the position of its hands nor by the interaction between its accessories and an atomic particle. In securing the opening of the hole at a definite moment, an apparatus of this type might, for instance, be used for an accurate measurement of the time an electron or a photon takes to come from the diaphragm to some other place, but, evidently, it would leave no possibility of controlling the energy transfer to the shutter with the aim of drawing conclusions as to the energy of the particle which has passed through the diaphragm. If we are interested in such conclusions we must, of course, use an arrangement where the shutter devices can no longer serve as accurate clocks, but where the knowledge of the moment when the hole in the diaphragm is open involves a latitude connected with the accuracy of the energy measurement by the general relation (4).

The contemplation of such more or less practical arrangements and their more or less fictitious use proved most instructive in directing attention to essential features of the problems. The main point here is the distinction between the *objects* under investigation and the *measuring instruments* which serve to define, in classical terms, the conditions under which the phenomena appear. Incidentally, we may remark that, for the illustration of the preceding considerations, it is not relevant that experiments involving an accurate control of the momentum or energy transfer from atomic particles to heavy bodies like diaphragms and shutters would be very difficult to perform, if practicable at all. It is only decisive that, in contrast to the proper measuring instruments, these bodies together with the particles would in such a case constitute the system to which the quantum-mechanical formalism has to be applied. As regards the specification of the conditions for any well-defined application of the formalism, it is moreover essential that the *whole experimental arrangement* be taken into account. In fact, the introduction of any further piece of apparatus, like a mirror, in the way of a particle might imply new interference effects essentially influencing the predictions as regards the results to be eventually recorded.

The extent to which renunciation of the visualization of atomic phenomena is imposed upon us by the impossibility of their subdivision is strikingly illustrated by the following example to which Einstein very early called attention and often has reverted. If a semi-reflecting mirror is placed in the way of a photon, leaving two possibilities for its direction of propagation, the photon may either be recorded on one, and only one, of two photographic plates situated at great distances in the two directions in question, or else we may, by replacing the plates by mirrors, observe effects exhibiting an interference between the two reflected wave-trains. In any attempt of a pictorial representation of the behaviour of the photon we would, thus, meet with the difficulty: to be obliged to say, on the one hand, that the photon always chooses one of the two ways and, on the other hand, that it behaves as if it had passed *both* ways.

It is just arguments of this kind which recall the impossibility of subdividing quantum phenomena and reveal the ambiguity in ascribing customary physical attributes to atomic objects. In particular, it must be realized that—besides in the account of the placing and timing of the instruments forming the experimental arrangement—all unambiguous use of space-time concepts in the description of atomic phenomena is confined to the recording of observations which refer to marks on a photographic plate or to similar practically irreversible amplification effects like the building of a water drop around an ion in a cloud-chamber. Although, of course, the existence of the quantum of action is ultimately responsible for the properties of the materials of which the measuring instruments are built and on which the functioning of the recording devices depends, this circumstance is not relevant for the problems of the adequacy and completeness of the quantum-mechanical description in its aspects here discussed.

These problems were instructively commented upon from different sides at the Solvay meeting, in the same session where Einstein raised his general objections. On that occasion an interesting discussion arose also about how to speak of the appearance of phenomena for which only predictions of statistical character can be made. The question was whether, as to the occurrence of individual effects, we should adopt a terminology proposed by Dirac, that we were concerned with a choice on the part of "nature," or, as suggested by Heisenberg, we should say that we have to do with a choice on the part of the "observer" constructing the measuring instruments and reading their recording. Any such terminology would, however, appear dubious since, on the one hand, it is hardly reasonable to endow nature with volition in the ordinary sense, while, on the other hand, it is certainly not possible for the observer to influence the events which may appear under the conditions he has arranged. To my mind, there is no other alternative than to admit that, in this field of experience, we are dealing with individual phenomena and that our possibilities of handling the measuring instruments allow us only to make a choice between the different complementary types of phenomena we want to study.

The epistemological problems touched upon here were more explicitly dealt with in my contribution to the issue of *Naturwissenschaften* in celebration of Planck's 70th birthday in 1929. In this article, a comparison was also made between the lesson derived from the discovery of the universal quantum of action and the development which has followed the discovery of the finite velocity of light and which, through Einstein's pioneer work, has so greatly clarified basic principles of natural philosophy. In relativity theory, the emphasis on the dependence of all phenomena on the reference frame opened quite new ways of tracing general physical laws of unparalleled scope. In Quantum Theory, it was argued, the logical comprehension of hitherto unsuspected fundamental regularities governing atomic phenomena has demanded the recognition that no sharp separation can be made between an independent behaviour of the objects and their interaction with the measuring instruments which define the reference frame.

In this respect Quantum Theory presents us with a novel situation in physical science, but attention was called to the very close analogy with the situation as regards analysis and synthesis of experience, which we meet in many other fields of human knowledge and interest. As is well known, many of the difficulties in psychology originate in the different placing of the separation lines between object and subject in the analysis of various aspects of psychical experience. Actually, words like "thoughts" and "sentiments," equally indispensable to illustrate the variety and scope of conscious life, are used in a similar complementary way as are space-time coordination and dynamical conservation laws in atomic physics. A precise formulation of such analogies involves, of course, intricacies of terminology, and the writer's position is perhaps best indicated in a passage in the article, hinting at the mutually exclusive relationship which will always exist between the practical use of any word and attempts at its strict definition. The principal aim, however, of these considerations, which were not least inspired by the hope of influencing Einstein's attitude, was to point to perspectives of bringing general epistemological problems into relief by means of a lesson derived from the study of new, but fundamentally simple, physical experience.

At the next meeting with Einstein at the Solvay Conference in 1930, our discussions took quite a dramatic turn. As an objection to the view that a control of the interchange of momentum and energy between the objects and the measuring instruments was excluded if these instruments should serve their purpose of defining the space-time frame of the phenomena, Einstein brought forward the argument that such control should be possible when the exigencies of relativity theory were taken into consideration. In particular, the general relationship between energy and mass, expressed in Einstein's famous formula

$$E = mc^2, \tag{5}$$

should allow, by means of simple weighing, to measure the total energy of any system and, thus, in principle to control the energy transferred to it when it interacts with an atomic object.

As an arrangement suited for such purpose, Einstein proposed the device indicated in Figure 7, consisting of a box with a hole in its side, which could be opened or closed by a shutter moved by means of a clock-work within the box. If, in the beginning, the box contained a certain amount of radiation and the clock was set to open the shutter for a very short interval at a chosen time, it could be achieved that a single photon was released through the hole at a moment known with as great accuracy as desired. Moreover, it would apparently also be possible, by weighing the whole box before and after this event, to measure the energy of the photon with any accuracy wanted, in definite contradiction to the reciprocal

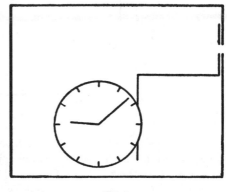

FIG. 7

indeterminacy of time and energy quantities in quantum mechanics.

This argument amounted to a serious challenge and gave rise to a thorough examination of the whole problem. At the outcome of the discussion, to which Einstein himself contributed effectively, it became clear, however, that the argument could not be upheld. In fact, in the consideration of the problem, it was found necessary to look closer into the consequences of the identification of inertial and gravitational mass implied in the application of relation (5). Especially, it was essential to take into account the relationship between the rate of a clock and its position in a gravitational field—well known from the red-shift of the lines in the sun's spectrum—following from Einstein's principle of equivalence between gravity effects and the phenomena observed in accelerated reference frames.

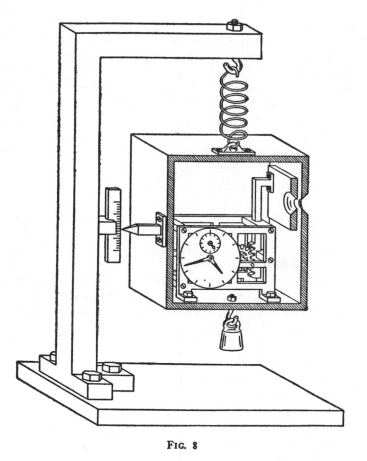

FIG. 8

Our discussion concentrated on the possible application of an apparatus incorporating Einstein's device and drawn in Figure 8 in the same pseudo-realistic style as some of the preceding figures. The box, of which a section is shown in order to exhibit its interior, is suspended in a spring-balance and is furnished with a pointer to read its position on a scale fixed to the balance support. The weighing of the box may thus be performed with any given accuracy Δm by adjusting the balance to its zero position by means of suitable loads. The essential point is now that any determination of this position with a given accuracy Δq will involve a minimum latitude Δp in the control of the momentum of the box connected with Δq by the relation (3). This latitude must obviously again be smaller than the total impulse which, during the whole interval T of the balancing procedure, can be given by the gravitational field to a body with a mass Δm, or

$$\Delta p \approx h/\Delta q < T \cdot g \cdot \Delta m, \tag{6}$$

where g is the gravity constant. The greater the accuracy of the reading q of the pointer, the longer must, consequently, be the balancing interval T, if a given accuracy Δm of the weighing of the box with its content shall be obtained.

Now, according to general relativity theory, a clock, when displaced in the direction of the gravitational force by an amount of Δq, will change its rate in such a way that its reading in the course of a time interval T will differ by an amount ΔT given by the relation

$$\Delta T/T = 1/c^2 \, g\Delta q. \tag{7}$$

By comparing (6) and (7) we see, therefore, that after the weighing procedure there will in our knowledge of the adjustment of the clock be a latitude

$$\Delta T > \frac{h}{c^2 \Delta m}.$$

Together with the formula (5), this relation again leads to

$$\Delta T \cdot \Delta E > h,$$

in accordance with the indeterminacy principle. Consequently, a use of the apparatus as a means of accurately measuring the energy of the photon will prevent us from controlling the moment of its escape.

The discussion, so illustrative of the power and consistency of relativistic arguments, thus emphasized once more the necessity of distinguishing, in the study of atomic phenomena, between the proper measuring instruments which serve to define the reference frame and those parts which are to be regarded as objects under investigation and in the account of which quantum effects cannot be disregarded. Notwithstanding the most suggestive confirmation of the soundness and wide scope of the quantum-mechanical way of description, Einstein nevertheless, in a following conversation with me, expressed a feeling of disquietude as regards the apparent lack of firmly laid down principles for the explanation of nature, in which all could agree. From my viewpoint, however, I could only answer that, in dealing with the task of bringing order into an entirely new field of experience, we could hardly trust in any accustomed principles, however broad, apart from the demand of avoiding logical inconsistencies and, in this respect, the mathematical formalism of quantum mechanics should surely meet all requirements.

The Solvay meeting in 1930 was the last occasion where, in common discussions with Einstein, we could benefit from the stimulating and mediating influence of Ehrenfest, but shortly before his deeply deplored death in 1933 he told me that Einstein was far from satisfied and with his usual acuteness had discerned new aspects of the situation which strengthened his critical attitude. In fact, by further examining the possibilities for the application of a balance arrangement, Einstein had perceived alternative procedures which, even if they did not allow the use he originally intended, might seem to enhance the paradoxes beyond the possibilities of logical solution. Thus, Einstein had pointed out that, after a preliminary weighing of the box with the clock and the subsequent escape of the photon, one was still left with the choice of either repeating the weighing or opening the box and comparing the reading of the clock with the standard time scale. Consequently, we are at this stage still free to choose whether we want to draw conclusions either about the energy of the photon or about the moment when it left the box. Without in any way interfering with the photon between its escape and its later interaction with other suitable measuring instruments, we are, thus, able to make accurate predictions pertaining *either* to the moment of its arrival *or* to the amount of energy liberated by its absorption. Since, however, according to the quantum-mechanical formalism, the specification of the state of an isolated particle cannot involve both a well-defined connection with the time scale and an accurate fixation of the energy, it might thus appear as if this formalism did not offer the means of an adequate description.

Once more Einstein's searching spirit had elicited a peculiar aspect of the situation in Quantum Theory, which in a most striking manner illustrated how far we have here transcended customary explanation of natural phenomena. Still, I could not agree with the trend of his remarks as reported by Ehrenfest. In my opinion, there could be no other way to deem a logically consistent mathematical formalism as inadequate than by demonstrating the departure of its consequences from experience or by proving that its predictions did not exhaust the possibilities of observation, and Einstein's argumentation could be directed to neither of these ends. In fact, we must realize that in the problem in question we are not dealing with a *single* specified experimental arrangement, but are referring to *two* different, mutually exclusive arrangements. In the one, the balance together with another piece of apparatus like a spectrometer is used for the study of the energy transfer by a photon; in the other, a shutter regulated by a standardized clock together with another apparatus of similar kind, accurately timed relatively to the clock, is used for the study of the time of propagation of a photon over a given distance. In both these cases, as also assumed by Einstein, the observable effects are expected to be in complete conformity with the predictions of the theory.

The problem again emphasizes the necessity of considering the *whole* experimental arrangement, the specification of which is imperative for any well-defined application of the quantum-mechanical formalism. Incidentally, it may be added that paradoxes of the kind contemplated by Einstein are encountered also in such simple arrangements as sketched in Figure 5. In fact, after a preliminary measurement of the momentum of the diaphragm, we are in principle offered the choice, when an electron or photon has passed through the slit, either to repeat the momentum measurement or to control the position of the diaphragm and, thus, to make predictions pertaining to alternative subsequent observations. It may also be added that it obviously can make no difference, as regards observable effects obtainable by a definite experimental arrangement, whether our plans of constructing or handling the instruments are fixed beforehand or whether we prefer to postpone the completion of our planning until a later moment when the particle is already on its way from one instrument to another.

In the quantum-mechanical description our freedom of constructing and handling the experimental arrangement finds its proper expression in the possibility of choosing the classically defined parameters entering in any proper application of the formalism. Indeed, in all such respects quantum mechanics exhibits a correspondence with the state of affairs familiar from classical physics, which is as close as possible when considering the individuality inherent in the quantum phenomena. Just in helping to bring out this point so clearly, Einstein's concern had therefore again been a most welcome incitement to explore the essential aspects of the situation.

The next Solvay meeting in 1933 was devoted to the problems of the structure and properties of atomic nuclei, in which field such great advances were made just in that period owing to the experimental discoveries as well as to new fruitful applications of quantum mechanics. It need in this connection hardly be recalled that just the evidence obtained by the study of artificial nuclear transformations gave a most direct test of Einstein's fundamental law regarding the equivalence of mass and energy, which was to prove an evermore important guide for researches in nuclear physics. It may also be mentioned how Einstein's intuitive recognition of the intimate relationship between the law of radioactive transformations and the probability rules governing individual radiation effects was confirmed by the quantum-mechanical explanation of spontaneous nuclear disintegrations. In fact, we are here dealing with a typical example of the statistical mode of description, and the complementary relationship between energy-momentum conservation and time-space coordination is most strikingly exhibited in the well-known paradox of particle penetration through potential barriers.

Einstein himself did not attend this meeting, which took place at a time darkened by the tragic developments in the political world which were to influence his fate so deeply and add so greatly to his burdens in the service of humanity. A few months earlier, on a visit to Princeton where Einstein was then guest of the newly founded Institute for Advanced Study to which he soon after became permanently attached, I had, however, opportunity to talk with him again about the epistemological aspects of atomic physics, but the difference between our ways of approach and expression still presented obstacles to mutual understanding. While, so far, relatively few persons had taken part in the discussions reported in this article, Einstein's critical attitude towards the views on Quantum Theory adhered to by many physicists was soon after brought to public attention through a paper with the title "Can Quantum-Mechanical Description of Physical Reality Be Considered Complete?," published in 1935 by Einstein, Podolsky and Rosen.

The argumentation in this paper is based on a criterion which the authors express in the following sentence: "If, without in any way disturbing a system, we can predict with certainty (i.e., with probability equal to unity) the value of a physical quantity, then there exists an element of physical reality corresponding to this physical quantity." By an elegant exposition of the consequences of the quantum-mechanical formalism as regards the representation of a state of a system, consisting of two parts which have been in interaction for a limited time interval, it is next shown that different quantities, the fixation of which cannot be combined in the representation of one of the partial systems, can nevertheless be predicted by measurements pertaining to the other partial system. According to their criterion, the authors therefore conclude that quantum mechanics does not "provide a complete description of the physical reality," and they express their belief that it should be possible to develop a more adequate account of the phenomena.

Due to the lucidity and apparently incontestable character of the argument, the paper of Einstein, Podolsky and Rosen created a stir among physicists and has played a large role in general philosophical discussion. Certainly the issue is of a very subtle character and suited to emphasize how far, in Quantum Theory, we are beyond the reach of pictorial visualization. It will be seen, however, that we are here dealing with problems of just the same kind as those raised by Einstein in previous discussions, and, in an article which appeared a few months later, I tried to show that from the point of view of complementarity the apparent inconsistencies were completely removed. The trend of the argumentation was in substance the same as that exposed in the foregoing pages, but the aim of recalling the way in which the situation was discussed at that time may be an apology for citing certain passages from my article.

Thus, after referring to the conclusions derived by Einstein, Podolsky and Rosen on the basis of their criterion, I wrote:

Such an argumentation, however, would hardly seem suited to affect the soundness of quantum-mechanical description, which is based on a coherent mathematical formalism covering automatically any procedure of measurement like that indicated. The apparent contradiction in fact discloses only an essential inadequacy of the customary viewpoint of natural philosophy for a rational account of physical phenomena of the type with which we are concerned in quantum mechanics. Indeed the *finite interaction between object and measuring agencies* conditioned by the very existence of the quantum of action entails—because of the impossibility of controlling the reaction of the object on the measuring instruments, if these are to serve their purpose—the necessity of a final renunciation of the classical ideal of causality and a radical revision of our attitude towards the problem of physical reality. In fact, as we shall see, a criterion of reality like that proposed by the named authors contains—however cautious its formulation may appear—an essential ambiguity when it is applied to the actual problems with which we are here concerned.

As regards the special problem treated by Einstein, Podolsky and Rosen, it was next shown that the consequences of the formalism as regards the representation of the state of a system consisting of two interacting atomic objects correspond to the simple arguments mentioned in the preceding in connection with the discussion of the experimental arrangements suited for the study of complementary phenomena. In fact, although any pair q and p of conjugate space and momentum variables obeys the rule of non-commutative multiplication expressed by (2), and can thus only be fixed with reciprocal latitudes given by (3), the difference $q_1 - q_2$ between two space-coordinates referring to the constituents of the system will commute with the sum $p_1 + p_2$ of the corresponding momentum components, as follows directly from the commutability of q_1 with p_2 and q_2 with p_1. Both $q_1 - q_2$ and $p_1 + p_2$ can, therefore, be accurately fixed in a state of the complex system and, consequently, we can predict the values of either q_1 or p_1 if either q_2 or p_2, respectively, is determined by direct measurements. If, for the two parts of the system we take a particle and a diaphragm, like that sketched in Figure 5, we see that the possibilities of specifying the state of the particle by measurements on the diaphragm just correspond to the situation described (above), where it was mentioned that after the particle has passed through the diaphragm, we have in principle the choice of measuring either the position of the diaphragm or its momentum and, in each case, making predictions as to subsequent observations pertaining to the particle. As repeatedly stressed, the principal point here is that such measurements demand mutually exclusive experimental arrangements.

The argumentation of the article was summarized in the following passage:

> From our point of view we now see that the wording of the above-mentioned criterion of physical reality proposed by Einstein, Podolsky and Rosen contains an ambiguity as regards the meaning of the expression "without in any way disturbing a system." Of course there is in a case like that just considered no question of a mechanical disturbance of the system under investigation during the last critical stage of the measuring procedure. But even at this stage there is essentially the question of *an influence on the very conditions which define the possible types of predictions regarding the future behaviour of the system.* Since these conditions constitute an inherent element of the description of any phenomenon to which the term "physical reality" can be properly attached, we see that the argumentation of the mentioned authors does not justify their conclusion that quantum-mechanical description is essentially incomplete. On the contrary, this description, as appears from the preceding discussion, may be characterized as a rational utilization of all possibilities of unambiguous interpretation of

measurements, compatible with the finite and uncontrollable interaction between the objects and the measuring instruments in the field of Quantum Theory. In fact, it is only the mutual exclusion of any two experimental procedures, permitting the unambiguous definition of complementary physical quantities, which provides room for new physical laws, the coexistence of which might at first sight appear irreconcilable with the basic principles of science. It is just this entirely new situation as regards the description of physical phenomena that the notion of *complementarity* aims at characterizing.

Rereading these passages, I am deeply aware of the inefficiency of expression which must have made it very difficult to appreciate the trend of the argumentation aiming to bring out the essential ambiguity involved in a reference to physical attributes of objects when dealing with phenomena where no sharp distinction can be made between the behaviour of the objects themselves and their interaction with the measuring instruments. I hope, however, that the present account of the discussions with Einstein in the foregoing years, which contributed so greatly to make us familiar with the situation in Quantum Physics, may give a clearer impression of the necessity of a radical revision of basic principles for physical explanation in order to restore logical order in this field of experience.

Einstein's own views at that time are presented in an article "Physics and Reality," published in 1936 in the *Journal of the Franklin Institute*. Starting from a most illuminating exposition of the gradual development of the fundamental principles in the theories of classical physics and their relation to the problem of physical reality, Einstein here argues that the quantum-mechanical description is to be considered merely as a means of accounting for the average behaviour of a large number of atomic systems, and his attitude to the belief that it should offer an exhaustive description of the individual phenomena is expressed in the following words: "To believe this is logically possible without contradiction; but it is so very contrary to my scientific instinct that I cannot forego the search for a more complete conception."

Even if such an attitude might seem well balanced in itself, it nevertheless implies a rejection of the whole argumentation exposed in the preceding, aiming to show that, in quantum mechanics, we are not dealing with an arbitrary renunciation of a more detailed analysis of atomic phenomena, but with a recognition that such an analysis is in *principle* excluded. The peculiar individuality of the quantum effects presents us, as regards the comprehension of well-defined evidence, with a novel situation unforeseen in classical physics and irreconcilable with conventional ideas suited for our orientation and adjustment to ordinary experience. It is in this respect that Quantum Theory has called for a renewed revision of the foundation for the unambiguous use of elementary concepts as a further step in the development which, since the advent of relativity theory, has been so characteristic of modern science.

In the following years, the more philosophical aspects of the situation in atomic physics aroused the interest of ever larger circles and were, in particular, discussed at the Second International Congress for the Unity of Science in Copenhagen in July 1936. In a lecture on this occasion, I tried especially to stress the analogy in epistemological respects between the limitation imposed on the causal description in atomic physics and situations met with in other fields of knowledge. A principal purpose of such parallels was to call attention to the necessity in many domains of general human interest of facing problems of a similar kind as those which had arisen in Quantum Theory and thereby to give a more familiar background for the apparently extravagant way of expression which physicists have developed to cope with their acute difficulties.

Besides the complementary features conspicuous in psychology and already touched upon (above), examples of such relationships can also be traced in biology, especially as regards the comparison between mechanistic and vitalistic viewpoints. Just with respect to the observational problem, this last question had previously been the subject of an address to the International Congress on Light Therapy held in Copenhagen in 1932 where it was incidentally pointed out that even the psycho-physical parallelism as envisaged by Leibniz and Spinoza has obtained a wider scope through the development of atomic physics, which forces us to an attitude towards the problem of explanation recalling ancient wisdom, that when searching for harmony in life one must never forget that in the drama of existence we are ourselves both actors and spectators.

Utterances of this kind would naturally in many minds evoke the impression of an underlying mysticism foreign to the spirit of science; at the above-mentioned Congress in 1936 I therefore tried to clear up such misunderstandings and to explain that the only question was an endeavour to clarify the conditions, in each field of knowledge, for the analysis and synthesis of experience. Yet, I am afraid that I had in this respect only little success in convincing my listeners, for whom the dissent among the physicists themselves was naturally a cause of scepticism about the necessity of going so far in renouncing customary demands as regards the explanation of natural phenomena. Not least through a new discussion with Einstein in Princeton in 1937, where we did not get beyond a humourous contest concerning which side Spinoza would have taken if he had lived to see the development of our days, I was strongly reminded of the importance of utmost caution in all questions of terminology and dialectics.

These aspects of the situation were especially discussed at a meeting in Warsaw in 1938, arranged by the International Institute of Intellectual Co-operation of the League of Nations. The preceding years had seen great progress in Quantum Physics owing to a number of fundamental discoveries regarding the constitution and properties of atomic nuclei as well as important developments of the mathematical formalism taking the requirements of relativity theory into account. In the last respect, Dirac's ingenious Quantum Theory of the electron offered a most striking illustration of the power and fertility of the general quantum-mechanical way of description. In the phenomena of creation and annihilation of electron pairs we have in fact to do with new fundamental features of atomicity, which are intimately connected with the non-classical aspects of quantum statistics expressed in the exclusion principle, and which have demanded a still more far-reaching renunciation of explanation in terms of a pictorial representation.

Meanwhile, the discussion of the epistemological problems in atomic physics attracted as much attention as ever and, in commenting on Einstein's views as regards the incompleteness of the quantum-mechanical mode of description, I entered more directly on questions of terminology. In this connection I warned especially against phrases, often found in the physical literature, such as "disturbing of phenomena by observation" or "creating physical attributes to atomic objects by measurements." Such phrases, which may serve to remind of the apparent paradoxes in Quantum Theory, are at the same time apt to cause confusion, since words like "phenomena" and "observations," just as "attributes" and "measurements," are used in a way hardly compatible with common language and practical definition.

As a more appropriate way of expression I advocated the application of the word *phenomenon* exclusively to refer to the observations obtained under specified circumstances, including an account of the whole experimental arrangement. In such terminology, the observational problem is free of any special intricacy since, in actual experiments, all observations are expressed by unambiguous statements referring, for instance, to the registration of the point at which an electron arrives at a photographic plate. Moreover, speaking in such a way is just suited to emphasize that the appropriate physical interpretation of the symbolic quantum-mechanical formalism amounts only to predictions, of determinate or statistical character, pertaining to individual phenomena appearing under conditions defined by classical physical concepts.

7
— Albert Einstein —

In this piece, part of an essay entitled "Reply to Criticism" that appeared at the end of Albert Einstein: Philosopher-Scientist, *Einstein responds to his chief opponents, including Max Born, Wolfgang Pauli, and Niels Bohr, and stands steadfast in his conviction that the statistical character of contemporary Quantum Theory, which his colleagues deemed an essential and indispensable feature of the theory, is due to the incomplete nature of the quantum description of reality. While recognizing the progress the statistical approach has brought to physical theory, Einstein claims that it does not provide what he maintains to be the "programmatic aim of all physics: the complete description of any (individual) real situation." Einstein therefore criticizes the quantum theorist for having a basically positivistic attitude. Whether that charge is justified remains for the reader to decide.*

RESPONSE TO BOHR

I now come to what is probably the most interesting subject which absolutely must be discussed in connection with the detailed arguments of my highly esteemed colleagues Born, Pauli, Heitler, Bohr, and Margenau. They are all firmly convinced that the riddle of the double nature of all corpuscles (corpuscular and undulatory character) has in essence found its final solution in the statistical Quantum Theory. On the strength of the successes of this theory they consider it proved that a theoretically complete description of a system can, in essence, involve only statistical assertions concerning the measurable quantities of this system. They are apparently all of the opinion that Heisenberg's indeterminacy-relation (the correctness of which is, from my own point of view, rightfully regarded as finally demonstrated) is essentially prejudicial in favor of the character of all thinkable reasonable physical theories in the mentioned sense. In what follows I wish to adduce reasons which keep me from falling in line with the opinion of almost all contemporary theoretical physicists. I am, in fact, firmly convinced that the essentially statistical character of contemporary Quantum Theory is solely to be ascribed to the fact that this [theory] operates with an incomplete description of physical systems.

Above all, however, the reader should be convinced that I fully recognize the very important progress which the statistical Quantum Theory has brought to theoretical physics. In the field of *mechanical* problems—i.e. wherever it is possible to consider the interaction of structures and of their parts with sufficient accuracy by postulating a potential energy between material points—[this theory] even now presents a system which, in its closed character, correctly describes the empirical relations between stable phenomena as they were theoretically to be expected. This theory is until now the only one which unites the corpuscular and undulatory dual character of matter in a logically satisfactory fashion; and the (testable) relations, which are contained in it, are, within the natural limits fixed by the indeterminacy-relation, *complete*. The formal relations which are given in this theory—i.e., its entire mathematical formalism—will probably have to be contained, in the form of logical inferences, in every useful future theory.

What does not satisfy me in that theory, from the standpoint of principle, is its attitude towards that which appears to me to be the programmatic aim of all physics: the complete description of any (individual) real situation (as it supposedly exists irrespective of any act of observation or substantiation). Whenever the positivistically inclined modern physicist hears such a formulation his reaction is that of a pitying smile. He says to himself: "There we have the naked formulation of a metaphysical prejudice, empty of content, a prejudice, moreover, the conquest of which constitutes the major epistemological achievement of physicists within the last quarter-century. Has any man ever perceived a 'real physical situation'? How is it possible that a reasonable person could today still believe that he can refute our essential knowledge and understanding by drawing up such a bloodless ghost?" Patience! The above laconic characterization was not meant to convince anyone; it was merely to indicate the point of view around which the following elementary considerations freely group themselves. In doing this I shall proceed as follows: I shall first of all show in simple special cases what seems essential to me, and then I shall make a few remarks about some more general ideas which are involved.

We consider as a physical system, in the first instance, a radioactive atom of definite average decay time, which is practically exactly localized at a point of the co-ordinate system. The radioactive process consists in the emission of a (comparatively light) particle. For the sake of simplicity we neglect the motion of the residual atom after the disintegration-process. Then it is possible for us, following Gamow, to replace the rest of the atom by a space of atomic order of magnitude, surrounded by a closed potential energy barrier which, at a time $t = 0$, encloses the particle to be emitted. The radioactive process thus schematized is then, as is well known, to be described—in the sense of elementary quantum mechanics—by a ψ-function in three dimensions, which at the time $t = 0$ is different from zero only inside of the barrier, but which, for positive times, expands into the outer space. This ψ-function yields the probability that the particle, at some chosen instant, is actually in a chosen part of space (i.e., is actually found there by a measurement of position). On the other hand, the ψ-function does not imply any assertion *concerning the time instant of the disintegration* of the radioactive atom.

Now we raise the question: Can this theoretical description be taken as the *complete* description of the disintegration of a single individual atom? The immediately plausible answer is: No. For one is, first of all, inclined to assume that the individual atom decays at a definite time; however, such a definite time-value is not implied in the description by the ψ-function. If, therefore, the individual atom has a definite disintegration time, then as regards the individual atom its description by means of the ψ-function must be interpreted as an incomplete description. In this case the ψ-function is to be taken as the description, not of a singular system, but of an ideal ensemble of systems. In this case one is driven to the conviction that a complete description of a single system should, after all, be possible; but for such complete description there is no room in the conceptual world of statistical Quantum Theory.

To this the quantum theorist will reply: This consideration stands and falls with the assertion that there actually is such a thing as a definite time of disintegration of the individual atom (an instant of time existing independently of any observation). But this assertion is, from my point of view, not merely arbitrary but actually meaningless. The assertion of the existence of a definite time-instant for the disintegration makes sense only if I can in principle determine this time-instant empirically. Such an assertion, however (which, finally, leads to the attempt to prove the existence of the particle outside of the force barrier), involves a definite disturbance of the system in which we are interested; so that the result of the determination does not permit a conclusion concerning the status of the undisturbed system. The supposition, therefore, that a radioactive atom has a definite disintegration-time is not justified by anything whatsoever; it is, therefore, not demonstrated either that the ψ-function cannot be conceived as a complete description of the individual system. The entire alleged difficulty proceeds from the fact that one postulates something not observable as "real." (This the answer of the quantum theorist.)

What I dislike in this kind of argumentation is the basic positivistic attitude, which from my point of view is untenable, and which seems to me to come to the same thing as Berkeley's principle, *esse est percipi*. "Being" is always something which is mentally constructed by us, that is, something which we freely posit (in the logical sense). The justification of such constructs does not lie in their derivation from what is given by the senses. Such a type of derivation (in the sense of logical deducibility) is nowhere to be had, not even in the domain of pre-scientific thinking. The justification of the constructs, which represent "reality" for us, lies alone in their quality of making intelligible what is sensorily given (the vague character of this expression is here forced upon me by my striving for brevity). Applied to the specifically chosen example this consideration tells us the following:

One may not merely ask: "Does a definite time instant for the transformation of a single atom exist?" but rather: "Is it, within the framework of our theoretical total construction, reasonable to posit the existence of a definite point of time for the transformation of a single atom?" One may not even ask what this assertion *means*. One can only ask whether such a proposition, within the framework of the chosen conceptual system—with a view to its ability to grasp theoretically what is empirically given—is reasonable or not.

Insofar, then, as a quantum-theoretician takes the position that the description by means of a ψ-function refers only to an ideal systematic totality but in no wise to the individual system, he may calmly assume a definite point of time for the transformation. But, if he represents the assumption that his description by way of the ψ-function is to be taken as the *complete* description of the individual system, then he must reject the postulation of a specific decay-time. He can justifiably point to the fact that a determination of the instant of disintegration is not possible on an isolated system, but would require disturbances of such a character that they must not be neglected in the critical examination of the situation. It would, for example, not be possible to conclude from the empirical statement that the transformation has already taken place, that this would have been the case if the disturbances of the system had not taken place.

As far as I know, it was E. Schrödinger who first called attention to a modification of this consideration, which shows an interpretation of this type to be impracticable. Rather than considering a system which comprises only a radioactive atom (and its process of transformation), one considers a system which includes also the means for ascertaining the radioactive transformation—for example, a Geiger-counter with automatic registration-mechanism. Let this latter include a registration-strip, moved by a clockwork, upon which a mark is made by tripping the counter. True, from the point of view of quantum mechanics this total system is very complex and its configuration space is of very high dimension. But there is in principle no objection to treating this entire system from the standpoint of quantum mechanics. Here too the theory determines the probability of each configuration of all its co-ordinates for every time instant. If one considers all configurations of the co-ordinates, for a time large compared with the average decay-time of the radioactive atom, there will be (at most) *one* such registration-mark on the paper strip. To each co-ordinate-configuration corresponds a definite position of the mark on the paper strip. But, inasmuch as the theory yields only the relative probability of the thinkable co-ordinate-configurations, it also offers only relative probabilities for the positions of the mark on the paperstrip, but no definite location for this mark.

In this consideration the location of the mark on the strip plays the role played in the original consideration by the time of the disintegration. The reason for the introduction of the system supplemented by the registration-mechanism lies in the following. The location of the mark on the registration-strip is a fact which belongs entirely within the sphere of macroscopic concepts, in contradistinction to the instant of disintegration of a single atom. If we attempt [to work with] the interpretation that the quantum-theoretical description is to be understood as a complete description of the individual system, we are forced to the interpretation that the location of the mark on the strip is nothing which belongs to the system *per se*, but that the existence of that location is essentially dependent upon the carrying out of an observation made on the registration-strip. Such an interpretation is certainly by no means absurd from a purely logical standpoint; yet there is hardly likely to be anyone who would be inclined to consider it seriously. For, in the macroscopic sphere it simply is considered certain that one must adhere to the program of a realistic description in space and time; whereas in the sphere of microscopic situations one is more readily inclined to give up, or at least to modify, this program.

This discussion was only to bring out the following. One arrives at very implausible theoretical conceptions, if one attempts to maintain the thesis that the statistical Quantum Theory is in principle capable of producing a complete description of an individual physical system. On the other hand, those difficulties of theoretical interpretation disappear, if one views the quantum-mechanical description as the description of ensembles of systems.

I reached this conclusion as the result of quite different types of considerations. I am convinced that everyone who will take the trouble to carry through such reflections conscientiously will find himself finally driven to this interpretation of quantum-theoretical description (the ψ-function is to be understood as the description not of a single system but of an ensemble of systems).

Roughly stated the conclusion is this: Within the framework of statistical Quantum Theory there is no such thing as a complete description of the individual system. More cautiously it might be put as follows: The attempt to conceive the quantum-theoretical description as the complete description of the individual systems leads to unnatural theoretical interpretations, which become immediately unnecessary if one accepts the interpretation that the description refers to ensembles of systems and not to individual systems. In that case the whole "egg-walking" performed in order to avoid the "physically real" becomes superfluous. There exists, however, a simple psychological reason for the fact that this most nearly obvious interpretation is being shunned. For if the statistical Quantum Theory does not pretend to describe the individual system (and its development in time) completely, it appears unavoidable to look elsewhere for a complete description of the individual system; in doing so it would be clear from the very beginning that the elements of such a description are not contained within the conceptual scheme of the statistical Quantum Theory. With this one would admit that, in principle, this scheme could not serve as the basis of theoretical physics. Assuming the success of efforts to accomplish a complete physical description, the statistical Quantum Theory would, within the framework of future physics, take an approximately analogous position to the statistical mechanics within the framework of classical mechanics. I am rather firmly convinced that the development of theoretical physics will be of this type; but the path will be lengthy and difficult.

I now imagine a quantum theoretician who may even admit that the quantum-theoretical description refers to ensembles of systems and not to individual systems, but who, nevertheless, clings to the idea that the type of description of the statistical Quantum Theory will, in its essential features, be retained in the future. He may argue as follows: True, I admit that the quantum-theoretical description is an incomplete description of the individual system. I even admit that a complete theoretical description is, in principle, thinkable. But I consider it proven that the search for such a complete description would be aimless. For the lawfulness of nature is thus constituted that the laws can be completely and suitably formulated within the framework of our incomplete description.

To this I can only reply as follows: Your point of view—taken as theoretical possibility—is incontestable. For me, however, the expectation that the adequate formulation of the universal laws involves the use of *all* conceptual elements which are necessary for a complete description, is more natural. It is furthermore not at all surprising that, by using an incomplete description, (in the main) only statistical statements can be obtained out of such description. If it should be possible to move forward to a complete description, it is likely that the laws would represent relations among all the conceptual elements of this description which, *per se,* have nothing to do with statistics.

8

— Werner Heisenberg —

(1901–1976)

As a German physicist known primarily for his formulation of the matrix mechanics as well as the Uncertainty Principle, Heisenberg was a key figure in the development of the Copenhagen Interpretation of Quantum Theory. After study-

ing under *Arnold Sommerfeld*, *Heisenberg became professor of physics at Leipzig University, and eventually director of the Max Planck Institutes in Berlin, Göttingen, and Munich.*

Heisenberg began work in the field of quantum physics early on, developing a nonrelativistic form of quantum mechanics at the age of twenty four, and receiving the Nobel Prize for physics in 1932 at the age of thirty three. His initial attempts to resolve some of the difficulties encountered in Quantum Theory involved taking as "real" only those properties that are in principle observable. This position was instrumental in his eventual formulation of the matrix mechanics, which turned out to be mathematically equivalent to the wave mechanics of Schrödinger. Heisenberg would use this theory to predict successfully (among other things) the observed frequencies and intensities of atomic and molecular spectral lines. Nevertheless, as his concern for the philosophical and conceptual problems inherent in the quantum situation grew more intense, Heisenberg would soon abandon the matrix mechanics as an insufficient explanation of quantum phenomena.

Heisenberg worked very closely with Niels Bohr in Copenhagen and was instrumental in the discussions that took place there. In 1927, Heisenberg proposed his Uncertainty Principle, which declares that it is impossible to determine the exact position and momentum of a particle simultaneously. This principle would be extended to a series of canonically conjugate pairs, including time and energy, such that the more precisely one determined the first, the less accurately one was able to measure the second. This Uncertainty Principle soon became an integral part of the Copenhagen Interpretation of Quantum Theory.

In this reading Heisenberg provides his own account of the history of Quantum Theory, one that is spiced with personal involvement in the various developments. Of particular note is Heisenberg's description of the reluctance of Max Planck to make public his discovery of the quantum of action. Heisenberg then goes on to describe the various moments in the development of Quantum Theory and the corresponding conundrums they posed. This selection, taken from Heisenberg's book Physics and Philosophy, *shows the author at his philosophic best.*

A BRIEF HISTORY OF QUANTUM THEORY

The origin of Quantum Theory is connected with a well-known phenomenon, which did not belong to the central parts of atomic physics. Any piece of matter when it is heated starts to glow, gets red hot and white hot at higher temperatures. The color does not depend much on the surface of the material, and for a black body it depends solely on the temperature. Therefore, the radiation emitted by such a black body at high temperatures is a suitable object for physical research; it is a simple phenomenon that should find a simple explanation in terms of the known laws for radiation and heat. The attempt made at the end of the nineteenth century by Lord Rayleigh and Jeans failed, however, and revealed serious difficulties. It would not be possible to describe these difficulties here in simple terms. It must be sufficient to state that the application of the known laws did not lead to sensible results. When Planck, in 1895, entered this line of research he tried to turn the problem from radiation to the radiating atom. This turning did not remove any of the difficulties inherent in the problem, but it simplified the interpretation of the empirical facts. It was just at this time, during the summer of 1900, that Curlbaum and Rubens in Berlin had made very accurate new measurements of the spectrum of heat radiation. When Planck heard of these results he tried to represent them by simple mathematical formulas which looked plausible from his research on the general connection between heat and radiation. One day Planck and Rubens met for tea in Planck's home and compared Rubens's latest results with a new formula suggested by Planck. The comparison showed a complete agreement. This was the discovery of Planck's law of heat radiation.

It was at the same time the beginning of intense theoretical work for Planck. What was the correct physical interpretation of the new formula? Since Planck could, from his earlier work, translate his formula easily into a statement about the radiating atom (the so-called oscillator), he must soon have found that his formula looked as if the oscillator could only contain discrete quanta of energy—a result that was so different from anything known in classical physics that he certainly must have refused to believe it in the beginning. But in a period of most intensive work during the summer of 1900 he finally convinced himself that there was no way of escaping from this conclusion. It was told by Planck's son that his father spoke to him about his new ideas on a long walk through the Grunewald, the wood in the suburbs of Berlin. On this walk he explained that he felt he had possibly made a discovery of the first rank, comparable perhaps only to the discoveries of Newton. So Planck must have realized at this time that his formula had touched the foundations of our description of nature, and that these foundations would one day start to move from their traditional present location toward a new and as yet unknown position of stability. Planck, who was conservative in his whole outlook, did not like this consequence at all, but he published his quantum hypothesis in December of 1900.

The idea that energy could be emitted or absorbed only in discrete energy quanta was so new that it could not be fitted into the traditional framework of physics. An attempt by Planck to reconcile his new hypothesis with the older laws of radiation failed in the essential points. It took five years until the next step could be made in the new direction.

This time it was the young Albert Einstein, a revolutionary genius among the physicists, who was not afraid to go further away from the old concepts. There were two problems in which he could make use of the new ideas. One was the so-called photoelectric effect, the emission of electrons from metals under the influence of light. The experiments, especially those of Lenard, had shown that the energy of the emitted electrons did not depend on the intensity of the light, but only on its color or, more precisely, on its frequency. This could not be understood on the basis of the traditional theory of radiation. Einstein could explain the observations by interpreting Planck's hypothesis as saying that light consists of quanta of energy traveling through space. The energy of one light quantum should, in agreement with Planck's assumptions, be equal to the frequency of the light multiplied by Planck's constant.

The other problem was the specific heat of solid bodies. The traditional theory led to values for the specific heat which fitted the observations at higher temperatures but disagreed with them at low ones. Again Einstein was able to show that one could understand this behavior by applying the quantum hypothesis to the elastic vibrations of the atoms in the solid body. These two results marked a very important advance, since they revealed the presence of Planck's quantum of action—as his constant is called among the physicists—in several phenomena, which had nothing immediately to do with heat radiation. They revealed at the same time the deeply revolutionary character of the new hypothesis, since the first of them led to a description of light completely different from the traditional wave picture. Light could either be interpreted as consisting of electromagnetic waves, according to Maxwell's theory, or as consisting of light quanta, energy packets traveling through space with high velocity. But could it be both? Einstein knew, of course, that the well-known phenomena of diffraction and interference can be explained only on the basis of the wave picture. He was not able to dispute the complete contradiction between this wave picture and the idea of the light quanta; nor did he even attempt to remove the inconsistency of this interpretation. He simply took the contradiction as something which would probably be understood only much later.

In the meantime the experiments of Becquerel, Curie and Rutherford had led to some clarification concerning the structure of the atom. In 1911 Rutherford's observations on the interaction of α-rays penetrating through matter resulted in his famous atomic model. The atom is pictured as consisting of a nucleus, which is positively charged and contains nearly the total mass of the atom, and electrons, which circle around the nucleus like the planets circle around the sun. The chemical bond between atoms of different elements is explained as an interaction between the outer electrons of the neighboring atoms; it has not directly to do with the atomic nucleus. The nucleus determines the chemical behavior of the atom through its charge which in turn fixes the number of electrons in the neutral atom. Initially this model of the atom could not explain the most characteristic feature of the atom, its enormous stability. No planetary system following the laws of Newton's mechanics would ever go back to its original configuration after a collision with another such system. But an atom of the element carbon, for instance, will still remain a carbon atom after any collision or interaction in chemical binding.

The explanation for this unusual stability was given by Bohr in 1913, through the application of Planck's quantum hypothesis. If the atom can change its energy only by discrete energy quanta, this must mean that the atom can exist only in discrete stationary states, the lowest of which is the normal state of the atom. Therefore, after any kind of interaction the atom will finally always fall back into its normal state.

By this application of Quantum Theory to the atomic model, Bohr could not only explain the stability of the atom but also, in some simple cases, give a theoretical interpretation of the line spectra emitted by the atoms after the excitation through electric discharge or heat. His theory rested upon a combination of classical mechanics for the motion of the electrons with quantum conditions, which were imposed upon the classical motions for defining the discrete stationary states of the system. A consistent mathematical formulation for those conditions was later given by Sommerfeld. Bohr was well aware of the fact that the quantum conditions spoil in some way the consistency of Newtonian mechanics. In the simple case of the hydrogen atom one could calculate from Bohr's theory the frequencies of the light emitted by the atom, and the agreement with the observations was perfect. Yet these frequencies were different from the orbital frequencies and their harmonics of the electrons circling around the nucleus, and this fact showed at once that the theory was still full of contradictions. But it contained an essential part of the truth. It did explain qualitatively the chemical behavior of the atoms and their line spectra; the existence of the discrete stationary states was verified by the experiments of Franck and Hertz, Stern and Gerlach.

Bohr's theory had opened up a new line of research. The great amount of experimental material collected by spectroscopy through several decades was now available for information about the strange quantum laws governing the motions of the electrons in the atom. The many experiments of chemistry could be used for the same purpose. It was from this time on that the physicists learned to ask the right questions; and asking the right question is frequently more than halfway to the solution of the problem.

What were these questions? Practically all of them had to do with the strange apparent contradictions between the results of different experiments. How could it be that the same radiation that produces interference patterns, and therefore must consist of waves, also produces the photoelectric effect, and therefore must consist of moving particles? How could it be that the frequency of the orbital motion of the electron in the atom does not show up in the frequency of the emitted radiation? Does this mean that there is no orbital motion? But if the idea of orbital motion should be incorrect, what happens to the electrons inside the atom? One can see the electrons move through a cloud chamber, and sometimes they are knocked out of an atom; why should they not also move within the atom? It is true that they might be at rest in the normal state of the atom, the state of lowest energy. But there are many states of higher energy, where the electronic shell has an angular momentum. There the electrons cannot possible be at rest. One could add a number of similar examples. Again and again one found that the attempt to describe atomic events in the traditional terms of physics led to contradiction.

Gradually, during the early twenties, the physicists became accustomed to these difficulties, they acquired a certain vague knowledge about where trouble would occur, and they learned to avoid contradictions. They knew which description of an atomic event would be the correct one for the special experiment under discussion. This was not sufficient to form a consistent general picture of what happens in a quantum process, but it changed the minds of the physicists in such a way that they somehow got into the spirit of Quantum Theory. Therefore, even some time before one had a consistent formulation of Quantum Theory one knew more or less what would be the result of any experiment.

One frequently discussed what one called ideal experiments. Such experiments were designed to answer a very critical question irrespective of whether or not they could actually be carried out. Of course it was important that it should be possible in principle to carry out the experiment, but the technique might be extremely complicated. These ideal experiments could be very useful in clarifying certain problems. If there was no agreement among the physicists about the result of such an ideal experiment, it was frequently possible to find a similar but simpler experiment that could be carried out, so that the experimental answer contributed essentially to the clarification of Quantum Theory.

The strangest experience of those years was that the paradoxes of Quantum Theory did not disappear during this process of clarification; on the contrary, they became even more marked and more exciting. There was, for instance, the experiment of Compton on the scattering of X-rays. From earlier experiments on the interference of scattered light there could be no doubt that scattering takes place essentially in the following way: The incident light wave makes an electron in the beam vibrate in the frequency of the wave; the oscillating electron then emits a spherical wave with the same frequency and thereby produces the scattered light. However, Compton found in 1923 that the frequency of scattered X-rays was different from the frequency of the incident X-ray. This change of frequency could be formally understood by assuming that scattering is to be described as collision of a light quantum with an electron. The energy of the light quantum is changed during the collision; and since the frequency times Planck's constant should be the energy of the light quantum, the frequency also should be changed. But what happens in this interpretation of the light wave? The two experiments—one on the interference of scattered light and the other on the change of frequency of the scattered light—seemed to contradict each other without any possibility of compromise.

By this time many physicists were convinced that these apparent contradictions belonged to the intrinsic structure of atomic physics. Therefore, in 1924 de Broglie in France tried to extend the dualism between wave description and particle description to the elementary particles of matter, primarily to the electrons. He showed that a certain matter wave could "correspond" to a moving electron, just as a light wave corresponds to a moving light quantum. It was not clear at the time what the word "correspond" meant in this connection. But de Broglie suggested that the quantum condition in Bohr's theory should be interpreted as a statement about the matter waves. A wave circling around a nucleus can for geometrical reasons only be a stationary wave; and the perimeter of the orbit must be an integer multiple of the wave length. In this way de Broglie's idea connected the quantum condition, which always had been a foreign element in the mechanics of the electrons, with the dualism between waves and particles.

In Bohr's theory the discrepancy between the calculated orbital frequency of the electrons and the frequency of the emitted radiation had to be interpreted as a limitation to the concept of the electronic orbit. This concept had been somewhat doubtful from the beginning. For the higher orbits, however, the electrons should move at a large distance from the nucleus just as they do when one sees them moving through a cloud chamber. There one should speak about electronic orbits. It was therefore very satisfactory that for these higher orbits the frequencies of the emitted radiation approach the orbital frequency and its higher harmonics. Also Bohr had already suggested in his early papers that the intensities of the emitted spectral lines approach the intensities of the corresponding harmonics. This principle of correspondence had proved very useful for the approximative calculation of the intensities of spectral lines. In this way one had the impression that Bohr's theory gave a qualitative but not a quantitative description of what happens inside the atom; that some new feature of the behavior of matter was qualitatively expressed by the quantum conditions, which in turn were connected with the dualism between waves and particles.

The precise mathematical formulation of Quantum Theory finally emerged from two different developments. The one started from Bohr's principle of correspondence. One had to give up the concept of the electronic orbit but still had to maintain it in the limit of high quantum numbers, i.e., for the large orbits. In this latter case the emitted radiation, by means of its frequencies and intensities, gives a picture of the electronic orbit; it represents what the mathematicians call a Fourier expansion of the orbit. The idea suggested itself that one should write down the mechanical laws not as equations for the positions and velocities of the electrons but as equations for the frequencies and amplitudes of their Fourier expansion. Starting from such equations and changing them very little one could hope to come to relations for those quantities which correspond to the frequencies and intensities of the emitted radiation, even for the small orbits and the ground state of the atom. This plan could actually be carried out; in the summer of 1925 it led to a mathematical formalism called matrix mechanics or, more generally, quantum mechanics. The equations of motion in Newtonian mechanics were replaced by similar equations between matrices; it was a strange experience to find that many of the old results of Newtonian mechanics, like conservation of energy, etc., could be derived also in the new scheme. Later the investigations of Born, Jordan and Dirac showed that the matrices representing position and momentum of the electron do not commute. This latter fact demonstrated clearly the essential difference between quantum mechanics and classical mechanics.

The other development followed de Broglie's idea of matter waves. Schrödinger tried to set up a wave equation for de Broglie's stationary waves around the nucleus. Early in 1926 he succeeded in deriving the energy values of the stationary states of the hydrogen atom as "Eigenvalues" of his wave equation and could give a more general prescription for transforming a given set of classical equations of motion into a corresponding wave equation in a space of many dimensions. Later he was able to prove that his formalism of Wave Mechanics was mathematically equivalent to the earlier formalism of quantum mechanics.

Thus one finally had a consistent mathematical formalism, which could be defined in two equivalent ways starting either from relations between matrices or from wave equations. This formalism gave the correct energy values for the hydrogen atom; it took less than one year to show that it was also successful for the helium atom and the more complicated problems of the heavier atoms. But in what sense did the new formalism describe the atom? The paradoxes of the dualism between wave picture and particle picture were not solved; they were hidden somehow in the mathematical scheme.

A first and very interesting step toward a real understanding of Quantum Theory was taken by Bohr, Kramers, and Slater in 1924. These authors tried to solve the apparent contradiction between the wave picture and the particle picture by the concept of the probability wave. The electromagnetic waves were interpreted not as "real" waves but as probability waves, the intensity of which determines in every point the probability for the absorption (or induced emission) of a light quantum by an atom at this point. This idea led to the conclusion that the laws of conservation of energy and momentum need not be true for the single event, that they are only statistical laws and are true only in the statistical average. This conclusion was not correct, however, and the connections between the wave aspect and the particle aspect of radiation were still more complicated.

But the paper of Bohr, Kramers and Slater revealed one essential feature of the correct interpretation of Quantum Theory. This concept of the probability wave was something entirely new in theoretical physics since Newton. Probability in mathematics or in statistical mechanics means a statement about our degree of knowledge of the actual situation. In throwing dice we do not know the fine details of the motion of our hands which determine the fall of the dice and therefore we say that the probability for throwing a special number is just one in six. The probability wave of Bohr, Kramers, and Slater, however, meant more than that; it meant a tendency for something. It was a quantitative version of the old concept of "potentia" in Aristotelian philosophy. It introduced something standing in the middle between the idea of an event and the actual event, a strange kind of physical reality just in the middle between possibility and reality.

Later when the mathematical framework of Quantum Theory was fixed, Born took up this idea of the probability wave and gave a clear definition of the mathematical quantity in the formalism, which was to be interpreted as the probability wave. It was not a three-dimensional wave like elastic or radio waves, but a wave in the many-dimensional configuration space, and therefore a rather abstract mathematical quantity.

Even at this time, in the summer of 1926, it was not clear in every case how the mathematical formalism should be used to describe a given experimental situation. One knew how to describe the stationary states of an atom, but one did not know how to describe a much simpler event—as for instance an electron moving through a cloud chamber.

When Schrödinger in that summer had shown that his formalism of Wave Mechanics was mathematically equivalent to quantum mechanics he tried for some time to abandon the idea of quanta and "quantum jumps" altogether and to replace the electrons in the atoms simply by his three-dimensional matter waves. He was inspired to this attempt by his result, that the energy levels of the hydrogen atom in his theory seemed to be simply the eigen frequencies of the stationary matter waves. Therefore, he thought it was a mistake to call them energies; they were just frequencies. But in the discussions which took place in the autumn of 1926 in Copenhagen between Bohr and Schrödinger and the Copenhagen group of physicists it soon became apparent that such an interpretation would not even be sufficient to explain Planck's formula of heat radiation.

During the months following these discussions an intensive study of all questions concerning the interpretation of Quantum Theory in Copenhagen finally led to a complete and, as many physicists believe, satisfactory clarification of the situation. But it was not a solution which one could easily accept. I remember discussions with Bohr which went through many hours till very late at night and ended almost in despair; and when at the end of the discussion I went alone for a walk in the neighboring park I repeated to myself again and again the question: Can nature possibly be as absurd as it seemed to us in these atomic experiments?

The final solution was approached in two different ways. The one was a turning around of the question. Instead of asking: How can one in the known mathematical scheme express a given experimental situation? The other question was put: Is it true, perhaps, that only such experimental situations can arise in nature as can be expressed in the mathematical formalism? The assumption that this was actually true led to limitations in the use of those concepts that had been the basis of classical physics since Newton. One could speak of the position and of the velocity of an electron as in Newtonian mechanics and one could observe and measure these quantities. But one could not fix both quantities simultaneously with an arbitrarily high accuracy. Actually the product of these two inaccuracies turned out to be not less than Planck's constant divided by the mass of the particle. Similar relations could be formulated for other experimental situations. They are usually called relations of uncertainty or principle of indeterminacy. One had learned that the old concepts fit nature only inaccurately.

The other way of approach was Bohr's concept of complementarity. Schrödinger had described the atom as a system not of a nucleus and electrons but of a nucleus and matter waves. This picture of the matter waves certainly also contained an element of truth. Bohr considered the two pictures—particle picture and wave picture—as two complementary descriptions of the same reality. Any of these descriptions can be only partially true, there must be limitations to the use of the particle concept as well as of the wave concept, else one could not avoid contradictions. If one takes into account those limitations which can be expressed by the uncertainty relations, the contradictions disappear.

In this way since the spring of 1927 one has had a consistent interpretation of Quantum Theory, which is frequently called the "Copenhagen Interpretation." This interpretation received its crucial test in the autumn of 1927 at the Solvay conference in Brussels. Those experiments which had always led to the worst paradoxes were again and again discussed in all details, especially by Einstein.

New ideal experiments were invented to trace any possible inconsistency of the theory, but the theory was shown to be consistent and seemed to fit the experiments as far as one could see. . . .

It should be emphasized at this point that it has taken more than a quarter of a century to get from the first idea of the existence of energy quanta to a real understanding of the quantum theoretical laws. This indicates the great change that had to take place in the fundamental concepts concerning reality before one could understand the new situation.

9
— Werner Heisenberg —

This essay by Heisenberg continues to reveal his concern for and preoccupation with the philosophical implications of Quantum Theory and provides a retelling of Young's "double slit experiment," which is considered by most physicists to be the impetus behind wave-particle dualism as it comes to bear on the "true" description of light. At the end of the essay, Heisenberg stands steadfastly behind the more philosophical interpretation of Quantum Theory as espoused by Niels Bohr and company—the so-called "Copenhagen interpretation." This reading also comes from Physics and Philosophy.

THE COPENHAGEN INTERPRETATION

The Copenhagen Interpretation of Quantum Theory starts from a paradox. Any experiment in physics, whether it refers to the phenomena of daily life or to atomic events, is to be described in the terms of classical physics. The concepts of classical physics form the language by which we describe the arrangement of our experiments and state the results. We cannot and should not replace these concepts by any others. Still the application of these concepts is limited by the relations of uncertainty. We must keep in mind this limited range of applicability of the classical concepts while using them, but we cannot and should not try to improve them.

For a better understanding of this paradox it is useful to compare the procedure for the theoretical interpretation of an experiment in classical physics and in Quantum Theory. In Newton's mechanics, for instance, we may start by measuring the position and the velocity of the planet whose motion we are going to study. The result of the observation is translated into mathematics by deriving numbers for the co-ordinates and the momenta of the planet from the observation. Then the equations of motion are used to derive from these values of the co-ordinates and momenta at a given time the values of these co-ordinates or any other properties of the system at a later time, and in this way the astronomer can predict the properties of the system at a later time. He can, for instance, predict the exact time for an eclipse of the moon.

In Quantum Theory the procedure is slightly different. We could for instance be interested in the motion of an electron through a cloud chamber and could determine by some kind of observation the initial position and velocity of the electron. But this determination will not be accurate; it will at least contain the inaccuracies following from the uncertainty relations and will probably contain still larger errors due to the difficulty of the experiment. It is the first of these inaccuracies which allows us to translate the result of the observation into the mathematical scheme of Quantum Theory. A probability function is written down which represents the experimental situation at the time of the measurement, including even the possible errors of the measurement.

This probability function represents a mixture of two things, partly a fact and partly our knowledge of a fact. It represents a fact in so far as it assigns at the initial time the probability unity (i.e., complete certainty) to the initial situation: the electron moving with the observed velocity at the observed position; "observed" means observed within the accuracy of the experiment. It represents our knowledge in so far as another observer could perhaps know the position of the electron more accurately. The error in the experiment does—at least to some extent—not represent a property of the electron but a deficiency in our knowledge of the electron. Also this deficiency of knowledge is expressed in the probability function.

In classical physics one should in a careful investigation also consider the error of the observation. As a result one would get a probability distribution for the initial values of the co-ordinates and velocities and therefore something very similar to the probability function in quantum mechanics. Only the necessary uncertainty due to the uncertainty relations is lacking in classical physics.

When the probability function in Quantum Theory has been determined at the initial time from the observation, one can from the laws of Quantum Theory calculate the probability function at any later time and can thereby determine the probability for a measurement giving a specified value of the measured quantity. We can, for instance, predict the probability for finding the electron at a later time at a given point in the cloud chamber. It should be emphasized, however, that the probability function does not in itself represent a course of events in the course of time. It represents a tendency for events and our knowledge of events. The probability function can be connected with reality only if one essential condition is fulfilled: if a new measurement is made to determine a certain property of the system. Only then does the probability function allow us to calculate the probable result of the new measurement. The result of the measurement again will be stated in terms of classical physics.

Therefore, the theoretical interpretation of an experiment requires three distinct steps: (1) the translation of the initial experimental situation into a probability function; (2) the following up of this function in the course of time; (3)

the statement of a new measurement to be made of the system, the result of which can then be calculated from the probability function. For the first step the fulfillment of the uncertainty relations is a necessary condition. The second step cannot be described in terms of the classical concepts; there is no description of what happens to the system between the initial observation and the next measurement. It is only in the third step that we change over again from the "possible" to the "actual."

Let us illustrate these three steps in a simple ideal experiment. It has been said that the atom consists of a nucleus and electrons moving around the nucleus; it has also been stated that the concept of an electronic orbit is doubtful. One could argue that it should at least in principle be possible to observe the electron in its orbit. One should simply look at the atom through a microscope of a very high resolving power, then one would see the electron moving in its orbit. Such a high resolving power could to be sure not be obtained by a microscope using ordinary light, since the inaccuracy of the measurement of the position can never be smaller than the wave length of the light. But a microscope using γ-rays with a wave length smaller than the size of the atom would do. Such a microscope has not yet been constructed but that should not prevent us from discussing the ideal experiment.

Is the first step, the translation of the result of the observation into a probability function, possible? It is possible only if the uncertainty relation is fulfilled after the observation. The position of the electron will be known with an accuracy given by the wave length of the γ-ray. The electron may have been practically at rest before the observation. But in the act of observation at least one light quantum of the γ-ray must have passed the microscope and must first have been deflected by the electron. Therefore, the electron has been pushed by the light quantum, it has changed its momentum and its velocity, and one can show that the uncertainty of this change is just big enough to guarantee the validity of the uncertainty relations. Therefore, there is no difficulty with the first step.

At the same time one can easily see that there is no way of observing the orbit of the electron around the nucleus. The second step shows a wave pocket moving not around the nucleus but away from the atom, because the first light quantum will have knocked the electron out from the atom. The momentum of light quantum of the γ-ray is much bigger than the original momentum of the electron if the wave length of the γ-ray is much smaller than the size of the atom. Therefore, the first light quantum is sufficient to knock the electron out of the atom and one can never observe more than one point in the orbit of the electron; therefore, there is no orbit in the ordinary sense. The next observation—the third step—will show the electron on its path from the atom. Quite generally there is no way of describing what happens between two consecutive observations. It is of course tempting to say that the electron must have been somewhere between the two observations and that therefore the electron must have described some kind of path or orbit even if it may be impossible to know which path. This would be a reasonable argument in classical physics. But in Quantum Theory it would be a misuse of the language which, as we will see later, cannot be justified. We can leave it open for the moment, whether this warning is a statement about the way in which we should talk about atomic events or a statement about the events themselves, whether it refers to epistemology or to ontology. In any case we have to be very cautious about the wording of any statement concerning the behavior of atomic particles.

Actually we need not speak of particles at all. For many experiments it is more convenient to speak of matter waves; for instance, of stationary matter waves around the atomic nucleus. Such a description would directly contradict the other description if one does not pay attention to the limitations given by the uncertainty relations. Through the limitations the contradiction is avoided. The use of "matter waves" is convenient, for example, when dealing with the radiation emitted by the atom. By means of its frequencies and intensities the radiation gives information about the oscillating charge distribution in the atom, and there the wave picture comes much nearer to the truth than the particle picture. Therefore, Bohr advocated the use of both pictures, which he called "complementary" to each other. The two pictures are of course mutually exclusive, because a certain thing cannot at the same time be a particle (i.e., substance confined to a very small volume) and a wave (i.e., a field spread out over a large space), but the two complement each other. By playing with both pictures, by going from the one picture to the other and back again, we finally get the right impression of the strange kind of reality behind our atomic experiments. Bohr uses the concept of "complementarity" at several places in the interpretation of Quantum Theory. The knowledge of the position of a particle is complementary to the knowledge of its velocity or momentum. If we know the one with high accuracy we cannot know the other with high accuracy; still we must know both for determining the behavior of the system. The space-time description of the atomic events is complementary to their deterministic description. The probability function obeys an equation of motion as the co-ordinates did in Newtonian mechanics; its change in the course of time is completely determined by the quantum mechanical equation, but it does not allow a description in space and time. The observation, on the other hand, enforces the description in space and time but breaks the determined continuity of the probability function by changing our knowledge of the system.

Generally the dualism between two different descriptions of the same reality is no longer a difficulty since we know from the mathematical formulation of the theory that contradictions cannot arise. The dualism between the two complementary pictures—waves and particles—is also clearly brought out in the flexibility of the mathematical scheme. The formalism is normally written to resemble Newtonian mechanics, with equations of motion for the co-ordinates and the momenta of the particles. But by a simple transformation it can be rewritten to resemble a wave equation for an ordinary three-dimensional matter wave. Therefore, this possibility of playing with different complementary pictures has its analogy in the different transformations of the mathematical scheme; it does not lead to any difficulties in the Copenhagen interpretation of Quantum Theory.

A real difficulty in the understanding of this interpretation arises, however, when one asks the famous question: But what happens "really" in an atomic event? It has been said before that the mechanism and the results of an observation can always be stated in terms of the classical concepts. But what one deduces from an observation is a probability function, a mathematical expression that combines statements about possibilities or tendencies with statements about our knowledge of facts. So we cannot completely objectify the result of an observation, we cannot describe what "happens" between this observation and the next. This looks as if we had introduced an element of subjectivism into the theory, as if we meant to say: what happens depends on our way of observing it or on the fact that we observe it. Before discussing this problem of subjectivism it is necessary to explain quite clearly why one would get into hopeless difficulties if one tried to describe what happens between two consecutive observations.

For this purpose it is convenient to discuss the following ideal experiment: We assume that a small source of monochromatic light radiates toward a black screen with two small holes in it. The diameter of the holes may be not much bigger than the wave length of the light, but their distance will be very much bigger. At some distance behind the screen a photographic plate registers the incident light. If one describes this experiment in terms of the wave picture, one says that the primary wave penetrates through the two holes; there will be secondary spherical waves starting from the holes that interfere with one another, and the interference will produce a pattern of varying intensity on the photographic plate.

The blackening of the photographic plate is a quantum process, a chemical reaction produced by single light quanta. Therefore, it must also be possible to describe the experiment in terms of light quanta. If it would be permissible to say what happens to the single light quantum between its emission from the light source and its absorption in the photographic plate, one could argue as follows: The single light quantum can come through the first hole or through the second one. If it goes through the first hole and is scattered there, its probability for being absorbed at a certain point of the photographic plate cannot depend upon whether the second hole is closed or open. The probability distribution on the plate will be the same as if only the first hole was open. If the experiment is repeated many times and one takes together all cases in which the light quantum has gone through the first hole, the blackening of the plate due to these cases will correspond to this probability distribution. If one considers only those light quanta that go through the second hole, the blackening should correspond to a probability distribution derived from the assumption that only the second hole is open. The total blackening, therefore, should just be the sum of the blackenings in the two cases; in other words, there should be no interference pattern. But we know this is not correct, and the experiment will show the interference pattern. Therefore, the statement that any light quantum must have gone *either* through the first *or* through the second hole is problematic and leads to contradictions. This example shows clearly that the concept of the probability function does not allow a description of what happens between two observations. Any attempt to find such a description would lead to contradictions; this must mean that the term "happens" is restricted to the observation.

Now, this is a very strange result, since it seems to indicate that the observation plays a decisive role in the event and that the reality varies, depending upon whether we observe it or not. To make this point clearer we have to analyze the process of observation more closely.

To begin with, it is important to remember that in natural science we are not interested in the universe as a whole, including ourselves, but we direct our attention to some part of the universe and make that the object of our studies. In atomic physics this part is usually a very small object, an atomic particle or a group of such particles, sometimes much larger—the size does not matter; but it is important that a large part of the universe, including ourselves, does *not* belong to the object.

Now, the theoretical interpretation of an experiment starts with the two steps that have been discussed. In the first step we have to describe the arrangement of the experiment, eventually combined with a first observation, in terms of classical physics and translate this description into a probability function. This probability function follows the laws of Quantum Theory, and its change in the course of time, which is continuous, can be calculated from the initial conditions; this is the second step. The probability function combines objective and subjective elements. It contains statements about possibilities or better tendencies ("potentia" in Aristotelian philosophy), and these statements are completely objective, they do not depend on any observer; and it contains statements about our knowledge of the system, which of course are subjective in so far as they may be different for different observers. In ideal cases the subjective element in the probability function may be practically negligible as compared with the objective one. The physicists then speak of a "pure case."

When we now come to the next observation, the result of which should be predicted from the theory, it is very important to realize that our object has to be in contact with the other part of the world, namely, the experimental arrangement, the measuring rod, etc., before or at least at the moment of observation. This means that the equation of motion for the probability function does now contain the influence of the interaction with the measuring device. This influence introduces a new element of uncertainty, since the measuring device is necessarily described in the terms of classical physics; such a description contains all the uncertainties concerning the microscopic structure of the device which we know from thermodynamics, and since the device is connected with the rest of the world, it contains in fact the uncertainties of the microscopic structure of the whole world. These uncertainties may be called objective in so far as they are simply a consequence of the description in the terms of classical physics and do not depend on any observer. They may be called subjective in so far as they refer to our incomplete knowledge of the world.

After this interaction has taken place, the probability function contains the objective element of tendency and the subjective element of incomplete knowledge, even if it has been a "pure case" before. It is for this reason that the result of the observation cannot generally be predicted with certainty; what can be predicted is the probability of a certain result of the observation, and this statement about the probability can be checked by repeating the experiment many times. The probability function does—unlike the common procedure in Newtonian mechanics—not describe a certain event but, at least during the process of observation, a whole ensemble of possible events.

The observation itself changes the probability function discontinuously; it selects of all possible events the actual one that has taken place. Since through the observation our knowledge of the system has changed discontinuously, its mathematical representation also has undergone the discontinuous change and we speak of a "quantum jump." When the old adage "Natura non facit saltus" is used as a basis for criticism of Quantum Theory, we can reply that certainly our knowledge can change suddenly and that this fact justifies the use of the term "quantum jump."

Therefore, the transition from the "possible" to the "actual" takes place during the act of observation. If we want to describe what happens in an atomic event, we have to realize that the word "happens" can apply only to the observation, not to the state of affairs between two observations. It applies to the physical, not the psychical act of observation, and we may say that the transition from the "possible" to the "actual" takes place as soon as the interaction of the object with the measuring device, and thereby with the rest of the world, has come into play; it is not connected with the act of registration of the result by the mind of the observer. The discontinuous change in the probability function, however, takes place with the act of registration, because it is the discontinuous change of our knowledge in the instant of registration that has its image in the discontinuous change of the probability function.

To what extent, then, have we finally come to an objective description of the world, especially of the atomic world? In classical physics science started from the belief—or should one say from the illusion?—that we could describe the world or at least parts of the world without any reference to ourselves. This is actually possible to a large extent. We know that the city of London exists whether we see it or not. It may be said that classical physics is just that idealization in which we can speak about parts of the world without any reference to ourselves. Its success has led to the general ideal of an objective description of the world. Objectivity has become the first criterion for the value of any scientific result. Does the Copenhagen interpretation of Quantum Theory still comply with this ideal? One may perhaps say that Quantum Theory corresponds to this ideal as far as possible. Certainly Quantum Theory does not contain genuine subjective features, it does not introduce the mind of the physicist as a part of the atomic event. But it starts from the division of the world into the "object" and the rest of the world, and from the fact that at least for the rest of the world we use the classical concepts in our description. This division is arbitrary and historically a direct consequence of our scientific method; the use of the classical concepts is finally a consequence of the general human way of thinking. But this is already a reference to ourselves and in so far our description is not completely objective.

It has been stated in the beginning that the Copenhagen interpretation of Quantum Theory starts with a paradox. It starts from the fact that we describe our experiments in the terms of classical physics and at the same time from the knowledge that these concepts do not fit nature accurately. The tension between these two starting points is the root of the statistical character of Quantum Theory. Therefore, it has sometimes been suggested that one should depart from the classical concepts altogether and that a radical change in the concepts used for describing the experiments might possibly lead back to a nonstatical, completely objective description of nature.

This suggestion, however, rests upon a misunderstanding. The concepts of classical physics are just a refinement of the concepts of daily life and are an essential part of the language which forms the basis of all natural science. Our actual situation in science is such that we *do* use the classical concepts for the description of the experiments, and it was the problem of Quantum Theory to find theoretical interpretation of the experiments on this basis. There is no use in discussing what could be done if we were other beings than we are. At this point we have to realize, as von Weizsäcker has put it, that "Nature is earlier than man, but man is earlier than natural science." The first part of the sentence justifies classical physics, with its ideal of complete objectivity. The second part tells us why we cannot escape the paradox of Quantum Theory, namely, the necessity of using the classical concepts.

We have to add some comments on the actual procedure in the quantum-theoretical interpretation of atomic events. It has been said that we always start with a division of the world into an object, which we are going to study, and the rest of the world, and that this division is to some extent arbitrary. It should indeed not make any difference in the final result if we, e.g., add some part of the measuring device or the whole device to the object and apply the laws of Quantum Theory to this more complicated object. It can be shown that such an alteration of the theoretical treatment would not alter the predictions concerning a given experiment. This follows mathematically from the fact that the laws of Quantum Theory are for the phenomena in which Planck's constant can be considered as a very small quantity, approximately identical with the classical laws. But it would be a mistake to believe that this application of the quantum-theoretical laws to the measuring device could help to avoid the fundamental paradox of Quantum Theory.

The measuring device deserves this name only if it is in close contact with the rest of the world, if there is an interaction between the device and the observer. Therefore, the uncertainty with respect to the microscopic behavior of the world will enter into the quantum-theoretical system here just as well as in the first interpretation. If the measuring device would be isolated from the rest of the world, it would be neither a measuring device nor could it be described in the terms of classical physics at all.

With regard to this situation Bohr has emphasized that it is more realistic to state that the division into the object and the rest of the world is not arbitrary. Our actual situation in research work in atomic physics is usually this: we wish to understand a certain phenomenon, we wish to recognize how this phenomenon follows from the general laws of nature. Therefore, that part of matter or radiation which takes part in the phenomenon is the natural "object" in the theoretical treatment and should be separated in this respect from the tools used to study the phenomenon. This again emphasizes a subjective element in the description of atomic events, since the measuring device has been constructed by the observer, and we have to remember that what we observe is not nature in itself but nature exposed to our method of questioning. Our scientific work in physics consists in asking questions about nature in the language that we possess and trying to get an answer from experiment by the means that are at our disposal. In this way Quantum Theory reminds us, as Bohr has put it, of the old wisdom that when searching for harmony in life one must never forget that in the drama of existence we are ourselves both players and spectators. It is understandable that in our scientific relation to nature our own activity becomes very important when we have to deal with parts of nature into which we can penetrate only by using the most elaborate tools.

10
— Erwin Schrödinger —
(1887–1961)

Erwin Schrödinger, an Austrian-born physicist, was educated at Vienna University and held a series of academic appointments there and elsewhere before becoming professor of physics at the Institute for Advanced Studies in Dublin.

Influenced by de Broglie's postulation of the matter waves, Schrödinger noted that a partial differential equation (wave equation) would describe the motion of a particle. Because this equation was difficult to apply in practice, he then put William Hamilton's method for describing particle motion into wave form. This equation was much easier to visualize and to apply to real situations, and became known as Schrödinger's equation. It was the basis of his wave mechanics and proved to be very successful in predicting the energy levels of an electron in a hydrogen atom without the ad hoc assumptions of the Bohr model. Schrödinger received the Nobel Prize for physics in 1933 (along with Dirac) for this achievement. Eventually, Schrödinger's wave mechanics was shown to be mathematically equivalent to Heisenberg's matrix mechanics by Dirac and von Neumann.

This reading is taken from Schrödinger's book Science, Theory and Man. *Here Schrödinger describes the basic principle of wave mechanics and its foundation in optical theory.*

THE FUNDAMENTAL IDEA OF WAVE MECHANICS

When a ray of light passes through an optical instrument, such as a telescope or a photographic lens, it undergoes a change of direction as it strikes each refractive or reflective surface. We can describe the path of the light ray once we know the two simple laws which govern the change of direction. One of these is the law of refraction, which was discovered by Snell about three hundred years ago; and the other is the law of reflection, which was know to Archimedes nearly two thousand years before. Figure 1 gives a simple example of a ray, A—B passing through two lenses and undergoing a change of direction at each of the four surfaces in accordance with Snell's law.

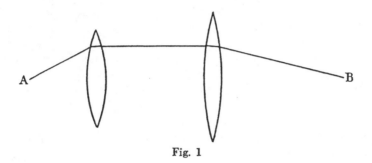

Fig. 1

From a much more general point of view, Fermat summed up the whole career of a light ray. In passing through media of varying optical densities light is propagated at correspondingly varying speeds, and the path which it follows is such as would have to be chosen by the light if it had the purpose of arriving *within the shortest possible time* at the destination which it actually reaches. (Here it may be remarked, in parenthesis, that any two points along the path of the light ray can be chosen as the points of departure and arrival respectively.) Any deviation from the path which the ray has actually chosen would mean a delay. This is Fermat's famous *Principle of Least Time*. In one admirably concise statement it defines the whole career of a ray of light, including also the more general case where the nature of the medium does not change suddenly but alters gradually from point to point. The atmosphere surrounding our earth is an example of this. When a ray of light, coming from outside, enters the earth's atmosphere, the ray travels more slowly as it penetrates into deeper and increasingly denser layers. And although the difference in the speed of propagation is extremely small, yet under these circumstances Fermat's Principle demands that the ray of light must bend earthwards (see Fig. 2), because by doing so it travels for a somewhat longer time in the higher "speedier" layers and comes sooner to its destination than if it were to choose the straight and shorter way (the dotted line in Fig. 2, the small quadrangle WW W^{1}W^1 to be ignored for the present). Most people will have noticed how the sun no longer presents the shape of a circular disk when it is low on the horizon, but is somewhat flattened, its vertical

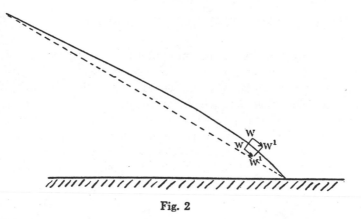

Fig. 2

diameter appearing shortened. That phenomenon is caused by the bending of the light rays as they traverse the earth's atmosphere.

According to the wave theory of light, what we call light rays have, correctly speaking, only a fictitious meaning. They are not the physical tracks of any particles of light, but a purely mathematical construction. The mathematician calls them "orthogonal trajectories" of the wave-fronts, that is, lines which at every point run at right angles to the wave-surface. Hence they point in the direction in which the light is propagated and, as it were, guide the light's propagation. (See Fig. 3, which represents the simplest case of concentric spherical wave-fronts and the corresponding rectilinear rays, while Fig. 4 illustrates the case of bent rays.) It seems strange that a general principle of such great

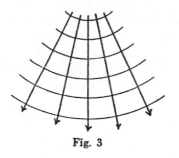

Fig. 3

importance as that of Fermat should be stated directly in reference to these mathematical lines, which are only a mental construction, and not in reference to the wave-fronts themselves. One might therefore be inclined to take it merely for a mathematical curiosity. But that would be a serious mistake. For only from the viewpoint of the wave theory does this principle become directly and immediately intelligible and cease to be a miracle. What we called *bending* of the light ray presents itself to the wave theory as a *turning* of the wave-front, and is much more readily understood. For that is just what we must expect in consequence of the fact that neighboring portions of the wave-front advance at various speeds; the turning is effected in the same way as with a company of soldiers marching in line, who are ordered to "right wheel." Here the soldiers in each rank take steps of varying lengths, the man on the right wing taking the shortest steps and the man on the left taking the longest. In the case of atmospheric refraction (Fig. 2) consider a small portion WW of the wave-surface. This portion must necessarily perform a "right wheel" towards W^1W,[1] because its left part is in the somewhat higher and rarer air and therefore is moving forward faster than the right, which is in the deeper layer. Now in examining the case more closely it is found that the statement made in Fermat's Principle is virtually identical with the trivial and obvious assertion that, because the velocity of light varies from point to point, the wave-front must turn, as in the instance I have referred to. I cannot prove that here; but I shall try to show that it is quite reasonable.

Let us revert to the row of soldiers marching in line. To prevent the front rank losing its perfect alignment, let us suppose that a long pole is placed abreast of the men and that each man holds it firmly with his hand against his chest. No word of command as to direction is given, but simply the order that each man must march or run as fast as he can. If the condition of the ground slowly changes from place to place, then either the left or the right section of the line advances more quickly than the other and this inevitably produces quite spontaneously a wheeling of the whole line to the right or left respectively. After a time it will be noticed that the line of advance, when looked upon as a whole, is not straight, but shows a definite curvature. Now this curved route is precisely the one along which the soldiers reach any place on their way *in the shortest*

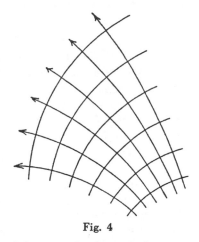

Fig. 4

possible time, taking into account the nature of the ground. Although this may seem remarkable there is actually nothing strange about it, for after all, by hypothesis, each soldier has done his best to travel as quickly as possible. And it may be further noticed that the bending will always have taken place in the direction towards which the condition of the ground underfoot is less favorable; so that finally it will appear as if the marchers had purposely avoided unfavorable conditions by making a detour around those regions where they would have found their forward pace slackened.

Thus Fermat's Principle directly appears as the *trivial quintessence* of the wave theory. Hence it was a very remarkable event when Hamilton one day made the theoretical discovery that the orbit of a mass point moving in a field of force (for instance, of a stone thrown in the gravitational field of the earth or of some planet in its course around the sun) is governed by a very similar general principle, which thenceforth bore the name of the discoverer and made him famous. Although Hamilton's principle does not precisely consist in the statement that the mass point chooses the quickest way, yet it states something *so similar*— that is to say, it is *so closely* analogous to the principle of minimum light time—that one is faced with a puzzle. It seemed as if Nature had effected exactly the same thing twice, but in two very different ways—once, in the case of light, through a fairly transparent wave-mechanism, and on the other occasion, in the case of mass points, by methods which were utterly mysterious, unless one was prepared to believe in some underlying undulatory character in the second case also. But at first sight this idea seemed impossible. For the laws of mechanics had at that time only been established and confirmed experimentally on a large scale for bodies of visible and (in the case of the planets) even huge dimensions which played the role of "mass points," so that something like an "undulatory nature" here appeared to be inconceivable.

The smallest and ultimate constructive elements in the constitution of matter, which we now call "mass points" in a much more particular sense, were at that time purely hypothetical. It was not until the discovery of radio-activity that the process of steadily refining our methods of measurement inaugurated a more detailed investigation of these corpuscles or particles; the development was crowned by C.T.R. Wilson's highly ingenious method, which succeeded in taking snap-shots of the track of a single particle and measuring it very accurately by means of stereometric photographs. As far as the measurements go they confirm, in the case of corpuscles, the validity of the same mechanical laws that hold on a large scale, as with planets, etc. Moreover, it was found that neither the molecules nor the atoms are to be considered as the ultimate building stones of matter, but that the atom itself is an extremely complicated composite system. Definite ideas were formed of the way in which atoms are composed of corpuscles, leading to models that closely resembled the celestial planetary system. And it was natural that in the theoretical construction of these tiny systems the attempt was at first made to use the same laws of motion as had been so successfully proved to hold good on a large scale. In other words, we endeavored to conceive the "inner" life of the atoms in terms of Hamiltonian mechanics, which, as I have said, have their culmination in the Hamiltonian principle. Meanwhile the very close analogy between the latter and Fermat's optical principle had been almost entirely forgotten. Or if any thought were given to this at all, the analogy was looked upon as merely a curious feature of the mathematical theory of the subject.

Now it is very difficult, without going closely into details, to give a correct notion of the success or failure encountered in the attempt to explain the structure of matter by this picture of the atom which was based on classical mechanics. On the one hand the Hamiltonian principle directly proved itself to be the truest and most reliable guide; so much so as to be considered absolutely indispensable. On the other hand, in order to account for certain facts, one had to tolerate the "rude intrusion" (*groben Eingriff*) of quite new and incomprehensible postulates, which were called quantum conditions and quantum postulates. These were gross dissonances in the symphony of classical mechanics—and yet they were curiously chiming in with it, as if they were being played on the same instrument. In mathematical language, the situation may be stated thus: The Hamiltonian principle demands only that a certain integral must be a minimum, without laying down the numerical value of the minimum in this demand; the new postulates require that the numerical value of the minimum must be a whole multiple of a universal constant, which is Planck's Quantum of Action. But this, only in parenthesis. The situation was rather hopeless. If the old mechanics had failed entirely, that would have been tolerable, for thus the ground would have been cleared for a new theory. But as it was, we were faced with the difficult problem of saving its *soul*, whose breath could be palpably detected in this microcosm, and at the same time persuading it, so to speak, not to consider the quantum conditions "rude intruders" but something arising out of the inner nature of the situation itself.

The way out of the difficulty was actually (though unexpectedly) found in the possibility I have already mentioned, namely, that in the Hamiltonian Principle we might also assume the manifestation of a "wave-mechanism," which we supposed to lie at the basis of events in point mechanics, just as we have been long accustomed to acknowledge it in the phenomena of light and in the governing principle enunciated by Fermat. By this, of course, the individual "path" of a mass point absolutely loses its inherent physical significance and becomes something fictitious, just as the individual light ray. Yet the "soul" of the theory, the minimum principle, not only remains inviolate but we could even never reveal its true and simple meaning, as was stated above, *without*, introducing the wave theory. The new theory is in reality no *new* theory, but is a thorough organic expansion and development, one might almost say merely a re-statement of the old theory in more subtle terms.

But how could this new and more "subtle" interpretation lead to results that are appreciably different? When applied to the atom, how could it solve any difficulty which the old interpretation could not cope with? How can this new standpoint make that "rude intruder" not merely tolerable but even a welcome guest and part of the household, as it were?

These questions, too, can best be elucidated by reference to the analogy with optics. Although I have asserted, and with good reason, that Fermat's principle is the quintessence of the wave theory of light, yet that principle is not such as to render superfluous a more detailed study of wave processes. The optical phenomena of *diffraction* and *interference* can be understood only when we follow up the particulars of the wave process; because these phenomena depend not merely upon where the wave finally arrives but also on whether at a given moment it arrives there as a wave-crest or a wave-trough. To the older and cruder methods of investigation interference phenomena appeared as only small details and escaped observation. But as soon as they were observed and properly accounted for by means of the undulatory theory, quite a number of experimental devices could be easily arranged in which the undulatory character of light was prominently displayed, not only in the finer details but also in the general character of the experiment.

To explain this I shall bring forward two examples: the first is that of an optical instrument, such as a telescope or a microscope. With such an instrument we aim at obtaining a sharp image. This means that we endeavor to focus all the rays emitted from an object point and re-unite them at what is called the image point (see Fig. 1a). Formerly it was thought that the difficulties which stood in the way were only those of geometrical optics, which are actually very considerable. Later it turned out that even in the best constructed instruments

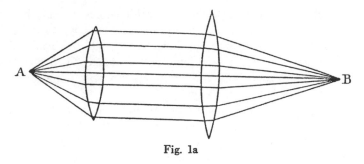

Fig. 1a

lack of precise focusing was considerably greater than might have been expected if in reality each ray, independently of its neighboring ray, followed Fermat's principle exactly. The light which is emitted from a luminous point and received by an instrument does not focus at an exact point after it has passed the instru-

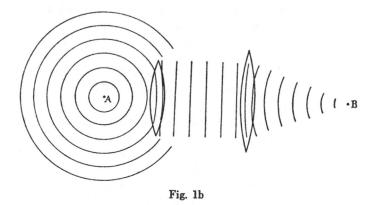

Fig. 1b

ment. Instead of this, it covers a small circular area, which is called the diffraction image and which is mostly circular only because the diaphragms and the circumference of the lenses are usually circular. For diffraction results from the fact that the instrument cannot possibly receive the whole of the spherical waves which are emitted from a luminous point. The borders of the lenses, and sometimes the diaphragms, cut off a part of the wave surface (Fig. 1b) and—if I may use a somewhat crude expression—the torn edges of the wound prevent an exact focus at a point and bring about the indistinctness or blurring of the image. This blurring is closely connected with the *wave-length* of the light and is absolutely unavoidable, owing to this deep seated theoretical connection. This phenomenon, originally scarcely noticed, now completely governs and inescapably limits the efficiency of the modern microscope, all the other causes of a lack of distinctness in the image having been successfully overcome. With respect to details, which are not much more coarse-grained than the wave-length of light, the optical image can only reach a distant similarity to the original, and none at all whenever the structural details in the object are *finer* than the wave-length.

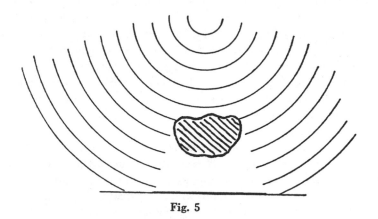

Fig. 5

The second example is of a simpler nature. Let us take a tiny source of light, just a point only. If we place an opaque body between it and a screen we find a shadow thrown on the screen. To construct the shadow theoretically we should follow each ray of light emitted from the point and should ascertain whether the opaque body prevents it from reaching the screen. The *rim* of the shadow is formed by those light rays which just graze and pass by the outline of the opaque body. But it can be shown by experiment that even where the light source is made as minute as possible, and the outline of the opaque body as sharp as possible, the outer rim of the shadow cast by the opaque body on the screen is not really sharp. The cause of this is again the same as in the former example. The wave-front is split, as it were (Fig. 5), by the outline of the opaque body; and the traces of this lesion blur the rim of the shadow. This would be inexplicable if the individual light rays were independent in themselves and traveled independently with no reference to one another.

This phenomenon, which is also called *diffraction*, is generally speaking not very noticeable where larger bodies are concerned. But if the opaque body which throws the shadow be very small, at least in one dimension, then the diffraction has two effects, first, nothing like a true shadow is produced and, secondly— which is far more striking—the tiny body seems to be glowing with its own light and emitting rays in all directions (predominantly, however, at very narrow angles with the incoming rays). Everybody is familiar with the so-called "motes" that appear in the track of a sunbeam entering a dark room. In the same way the filigree of tiny strands and cobwebs that appear around the brow of a hill behind which the sun is hidden, or even the hair of a person standing against the sun, sometimes glows marvelously with diffracted light. The visibility of smoke and fog is due to the same phenomenon. In all these cases the light does not really issue from the opaque body itself but from its immediate surroundings, that is to say, from the area in which the body produces a considerable perturbation of the incident wave-fronts. It is interesting, and for what follows very important, to note that the area of perturbation is always and in every direction at least as large as one or a few wave-lengths, no matter how small the opaque body may be. Here again, therefore, we see the close relation between wave-length and the phenomenon of diffraction. Perhaps this can be more palpably illustrated by reference to another wave process, namely, that of sound. Here, on account of the much longer wave-length, which extends into centimeters and meters, the shadow loses all distinctness and the diffraction predominates to a degree that is of practical importance. We can distinctly *hear* a call from behind a high wall or around the corner of a solid building, although we cannot *see* the person who calls.

Let us now return from optics to mechanics and try to develop the analogy fully. The optical parallel of the *old* mechanics is the method of dealing with isolated rays of light, which are supposed not to influence one another. The new Wave Mechanics has its parallel in the undulatory theory of light. The advantage of changing from the old concept to the new must obviously consist in clearer insight into diffraction phenomena, or rather into something that is strictly analogous to the diffraction of light, although ordinarily even less significant; for otherwise the old mechanics could not have been accepted as satisfactory for so long a time. But it is not difficult to conjecture the conditions in which the neglected phenomenon must become very prominent, entirely dominate the mechanical process and present problems that are insoluble under the old concept. This occurs inevitably *whenever the entire mechanical system is comparable in its extension with the wave-lengths* of "material waves," which play the same role in mechanical processes as light waves do in options.

That is the reason why, in the tiny system of the atom, the old concept is bound to fail. In mechanical phenomena on a large scale it will retain its validity as an excellent approximation, but it must be replaced by the new concept if we wish to deal with the fine interplay which takes place within regions of the order of magnitude of only one or a few wave-lengths. It was amazing to see all the strange additional postulates, which I have mentioned, arising quite automatically from the new undulatory concept, whereas they had to be artificially grafted onto the old one in order to make it fit in with the internal processes of the atom and yield a tolerable explanation of its actually observed manifestations.

In this connection it is, of course, of outstanding importance that the diameter of the atom and the wave-length of these hypothetical "material" waves should be very nearly of the same order of magnitude. And you will undoubtedly ask whether we are to consider it as purely an accident that in the progressive analysis of the structure of matter we should just here encounter the wavelength order of magnitude, or whether this can be explained. Is there any further evidence of the equality in question? Since the material waves are an entirely new requisite of this theory, which had not been hitherto discerned elsewhere, one might suspect that it is merely a question of suitable *assumption* as to their wave-length, an assumption forced upon us in order to support the preceding arguments.

Well, the coincidence between the two orders of magnitude is by no means a mere accident, and there is no necessity to make any particular assumption in this regard; the coincidence follows naturally from the theory, on account of the following remarkable circumstances. Let us begin by stating that Rutherford's and Chadwick's experiments on the dispersion of Alpha rays have firmly established the fact that the heavy *nucleus* of the atom is very much smaller than the atom, which justifies us in treating it as a point-like center of attraction in all the argument which follows. Instead of the *electron* we introduce hypothetical waves, the wave-length of which is left an open question as yet, because we do not know anything about it. It is true that this introduces into our calculations a symbol, say *a*, which represents a number as yet undefined. But in such calculations we are accustomed to that sort of thing and it does not hinder us from inferring that the nucleus of the atom will inevitably produce a sort of diffraction phenomenon of these waves, just as a minute mote does with light waves. Precisely as with light waves, here too the extension of the perturbed area surrounding the nucleus turns out to bear a close relation to the wave-length and to be of the same order of magnitude. Remember that the latter had to be left an open question! But now comes the most important step: *we identify the perturbed area, the diffraction halo, with the atom; the atom being thus regarded as really nothing more than the diffraction phenomenon arising from an electron wave that has been intercepted by the nucleus of the atom.* Thus it is no longer an accident that the size of the atom is of the same order of magnitude as the wave-length. It is in the nature of the case itself. Of course numerically we know neither the one nor the other; because in our calculation there always remains this *one* undefined constant which we have called *a*. It can, however, be determined in two ways, which control one another reciprocally. Either we can choose for *a* that value which will quantitatively account for the observable effects produced by the atom, especially for the emitted spectral lines, which can be measured with extreme accuracy; or, in the second place, the value of *a* can be adapted in order to give to the diffraction halo the right size, which from other evidence is to be expected for the atom. These two ways of defining *a* (of which the second is, of course, much less definite, because the phrase "size of the atom" is somewhat indefinite) are *in perfect accord with one another.* Thirdly, and finally, it may be remarked that the constant which has remained indeterminate has not really the physical dimension of Length, but of Action, that is, energy multiplied by time. It is, then, very tempting to assign to it the numerical value of Planck's universal Quantum of Action, which is known with fair accuracy from the laws of heat radiation. The result is that with all desirable exactitude, *we now fall back upon the first (the most exact) method of determining* a.

Thus, from the quantitative point of view, the theory answers its purpose with a minimum of new assumptions. It involves a single available constant, to which we only have to assign a numerical value that is already quite familiar to us in the earlier Quantum Theory, in order, first, to give the proper magnitude to the diffraction halos and therewith render possible their identification with the atoms; and, secondly, to calculate with quantitative exactitude all the observable effects produced by the atoms, their radiation of light, the energy required for ionization, etc., etc.

I have tried to explain to you in the simplest possible manner the fundamental concept on which this wave theory of matter is based. Let me confess that, in order to avoid bringing the subject before you in an abstruse form at the very outset, I have embellished it somewhat. Not indeed as regards the thoroughness with which conclusions properly deduced from the theory have been corroborated by experiment, but rather as regards the conceptual simplicity and absence of difficulty in the chain of reasoning which lead to these conclusions. In saying this I do not refer to the mathematical difficulties, which eventually are always trivial, but rather to the conceptual difficulties. Naturally it does not call for a great mental effort to pass from the idea of a path to a system of wave-fronts perpendicular to the path (see Fig. 6). But the wave-surfaces, even when we

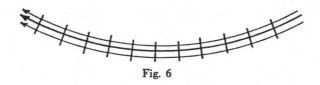

Fig. 6

restrict them to small elements of surface, still involve at least a slender *bundle* of possible paths, to all of which they stand in the same relation. According to the traditional idea, in each concrete case one of these paths is singled out as the one "really traveled," in contradistinction to all the other "merely possible" paths. According to the new concept the case is quite different. We are confronted with the profound logical antithesis between

Either this or that (Particle Mechanics)
(aut—aut) and

This as well as that (Wave Mechanics)
(et—et)

Now this would not be so perplexing if it were really a question of abandoning the old concept and *substituting* the new one for it. But unfortunately that is not the state of affairs. From the standpoint of Wave Mechanics the innumerable multitude of possible particle paths would be only fictitious and no single one would have the special prerogative of being that actually traveled in the individual case. But, as I have already remarked, we have in some cases actually observed such individual tracks of a particle. The wave theory cannot meet this case, except in a very unsatisfactory way. We find it extraordinarily difficult to regard the track, whose trace we actually *see*, only as a slender bundle of equally possible (*Gleichberechtigten*) tracks between which the wave-fronts form a lateral connection. And yet these lateral connections are necessary to the understanding of diffraction and interference phenomena, which the very same particles produce before our eyes with equal obviousness—that is to say, produce experimentally on a large scale and not only in those concepts of the interior of the atom discussed previously. It is true that we can deal with every concrete individual case without the two contrasted aspects leading to different expectations as to the result of any given experiment. But with the old and cherished and apparently indispensable concepts, such as "really" and "merely possible," we cannot advance. We can never say what really *is* or what really *happens*, but only what is *observable*, in each concrete case. Shall we content ourselves with this as a permanent feature? In principle, yes. It is by no means a new demand to claim that, in principle, the ultimate aim of exact science must be restricted to the description of what is really observable. The question is only whether we must henceforth forgo connecting the description, as we did hitherto, with a definite hypothesis as to the real structure of the Universe. To-day there is a widespread tendency to insist on this renunciation. But I think that this is taking the matter somewhat too lightly.

I would describe the present state of our knowledge as follows: The light ray, or track of the particle, corresponds to a *longitudinal* continuity of the propagating process (that is to say, *in the* direction of the spreading); the wave-front, on the other hand, to a *transverse* one, that is to say, perpendicular to the direction of spreading. *Both* continuities are undoubtedly real. The one has been proved by photographing the particle tracks, and the other by interference experiments. As yet we have not been able to bring the two together into a uniform scheme. It is only in extreme cases that the transverse—the spherical—continuity or the longitudinal—the ray-continuity shows itself so predominantly that we *believe* we can avail ourselves either of the wave scheme or of the particle scheme alone.

11
— Erwin Schrödinger —

In this reading, Schrödinger argues against the notion of discontinuity, known as "quantum jumps," and shows how what on first glance may be interpreted as evidence of the energy parcel or packet view (requiring the postulation of discontinuous jumps) could easily be reinterpreted and understood from the point of view of resonance, thus eliminating any lack of continuity. He also points out how his position is parasitic upon de Broglie's conception of matter-waves. While Schrödinger is best known as a nemesis of Quantum Theory, this alternative approach remains an important moment in the development of Quantum Theory and had to be reckoned with by all those who opposed it. The reading comes from an article entitled "Are There Quantum Jumps?" that appeared in the British Journal for the Philosophy of Science in 1952.

ARE THERE QUANTUM JUMPS?

The Cultural Background

Physical science, which aims not only at devising fascinating new experiments, but at obtaining a rational understanding of the results of observations, incurs at present, so I believe, the grave danger of getting severed from its historical background. The innovations of thought in the last 50 years, great and momentous and unavoidable as they were, are usually overrated compared with those of the preceding century; and the disproportionate foreshortening by time-perspective, of previous achievements on which all our enlightenment in modern times depends, reaches a disconcerting degree according as earlier and earlier centuries are considered. Along with this disregard for historical linkage there is a tendency to forget that all science is bound up with human culture in general, and that scientific findings, even those which at the moment appear the most advanced and esoteric and difficult to grasp, are meaningless outside their cultural context. A theoretical science, unaware that those of its constructs considered relevant and momentous are destined eventually to be framed in concepts and words that have a grip on the educated community and become part and parcel of the general world picture—a theoretical science, I say, where this is forgotten, and where the initiated continue musing to each other in terms that are, at best, understood by a small group of close fellow travelers, will necessarily be cut off from the rest of cultural mankind; in the long run it is bound to atrophy and ossify, however virulently esoteric chat may continue within its joyfully isolated groups of experts. . . .

The disregard for historical connectedness, nay the pride of embarking on new ways of thought, of production and of action, the keen endeavor of shaking off, as it were, the indebtedness to our predecessors, are no doubt a general trend of our time. In the fine arts we notice strong currents quite obviously informed by this vein; we witness its results in modern painting, sculpture, architecture, music and poetry. There are many who look upon this as a new buoyant rise, while others regard it as a flaring up that inaugurates decay. It is not here the place to dwell on this question, and my personal views on it might interest nobody. But I may say that whenever this trend enters science, it ought to be opposed. There obviously is a certain danger of its intruding into science in general, which is not an isolated enterprise of the human spirit, but grows on the same historic soil as the others and participates in the mood of the age. There is, however, so I believe, no other nearly so blatant example of this happening as the theories of physical science in our time. I believe that we are here facing a development which is the precise counterpart of that in the fine arts alluded to above. . . .

What has all this to do with quantum jumps? I have been trying to produce a mood that makes one wonder what parts of contemporary science will still be of interest to others than historians 2,000 years hence. There have been ingenious constructs of the human mind that gave an exceedingly accurate description of observed facts and have yet lost all interest except to historians. I am thinking of the theory of epicycles. I confess to the heretical view that their modern counterpart in physical theory are the quantum jumps. Or rather these correspond to the *circles* which the sun, the moon and the stars were thought to describe around the earth in 24 hours after earlier and better knowledge had been condemned. I am reminded of *epicycles* of various orders when I am told of the hierarchy of *virtual* quantum transitions. But let these rude remarks not deter you. We shall now come to grips with the subject proper.

The Discontinuous States as Proper Modes

Max Planck's essential step in 1900, amounted, as we say now, to laying the foundation of Quantum Theory; it was his discovery, by abstract thought, of a *discontinuity* where it was least expected, namely in the exchange of energy between an elementary material system (atom or molecule) and the radiation of light and heat. He was at first very reluctant to draw the much more incisive conclusion that each atom or molecule had only to choose between a *discrete* set of "states"; that it could normally only harbor certain discrete amounts of energy, sharply defined and characteristic of its nature; that it would normally find itself on one of these "energy levels" (as the modern expression runs)— except when it changes over more or less abruptly from one to another, radiating its surplus energy to the surrounding, or absorbing the required amount from there, as the case may be. Planck was even more hesitant to adopt the view that radiation itself be divided up into portions or light-quanta or "photons," to use the present terminology. In all this his hesitance had good reasons. Yet only a few years later (1905) Einstein advanced the hypothesis of light-quanta, clinching it with irresistible arguments; and in 1913 Niels Bohr, by taking the discrete states of the atoms seriously and extending Planck's assumptions in two directions with great ingenuity, but irrefutable consistency, could explain quantitatively some of the atomic line spectra, which are all patently *discrete*, and which had in their entirety formed a great conundrum up to then: Bohr's theory turned them into the ultimate and irrevocable direct evidence, that the discrete states are a genuine and real fact. Bohr's theory held the ground for about a dozen of years, scoring a grand series of so marvelous and genuine successes, that we may well claim excuses for having shut our eyes to its one great deficiency: while describing minutely the so-called "stationary" states which the atom had normally, i.e. in the comparatively uninteresting periods when *nothing happens*, the theory was silent about the periods of transition or "quantum jumps" (as one then began to call them). Since intermediary states had to remain disallowed, one could not but regard the transition as instantaneous; but on the other hand, the radiating of a coherent wave train of 3 or 4 feet length, as it can be observed in an interferometer, would use up just about the average interval between two transitions, leaving the atom no time to "be" in those stationary states, the only ones of which the theory gave a description.

This difficulty was overcome by quantum mechanics, more especially by Wave Mechanics, which furnished a new description of the *states*; this was precisely what was still missing in the earliest version of the new theory which had preceded Wave Mechanics by about one year. The previously admitted discontinuity was not abandoned, but it shifted from the *states* to something else, which is most easily grasped by the simile of a vibrating string or drumhead or metal plate, or of a bell that is tolling. If such a body is struck, it is set vibrating, that is to say it is slightly deformed and then runs in rapid succession through a continuous series of slight deformations again and again. There is, of course, an infinite variety of ways of striking a given body, say a bell, by a hard or soft, sharp or blunt instrument, at different points or at several points at a time. This produces an infinite variety of initial deformations and accordingly a truly infinite variety of shapes of the ensuing vibration: the rapid "succession of cinema pictures," so we might call it, which describes the vibration following on a particular initial deformation is infinitely manifold. But in every case, however complicated the actual motion is, it can be mathematically analysed as being the *superposition* of a discrete series of comparatively simple "proper vibrations," each of which goes on with a quite definite frequency. This discrete series of frequencies depends on the shape and on the material of the body, its density and elastic properties. It can be computed from the theory of elasticity, from which the existence and the discreteness of proper modes and proper frequencies, and the fact that any possible vibration of that body can be analysed into a superposition of them, are very easily deduced quite generally, i.e. for an elastic body of any shape whatsoever.

The achievement of wave mechanics was, that it found a general model picture in which the "stationary" states of Bohr's theory take the role of proper vibrations, and their discrete "energy levels" the role of the proper frequencies of these proper vibrations; and all this follows from the new theory, once it is accepted, as simply and neatly as in the theory of elastic bodies, which we mentioned as a simile. Moreover, the radiated frequencies, observed in the line spectra, are in the new model, equal to the *differences* of the proper frequencies; and this is easily understood, when two of them are acting simultaneously, on simple assumptions about the nature of the vibrating "something."

The Alleged Energy Balance—A Resonance Phenomenon

But to me the following point has always seemed the most relevant, and it is the one I wish to stress here, because it has been almost obliterated—if words mean something, and if certain words now in general use are taken to mean what they say. The principle of superposition not only bridges the gaps between the 'stationary' states, and allows, nay compels us, to admit intermediate states without removing the discreteness of the "energy levels" (because they have become proper frequencies); but it completely *does away with the prerogative of the stationary states*. The epithet stationary has become obsolete. Nobody who would get acquainted with Wave Mechanics without knowing its predecessor (the Planck-Einstein-Bohr-theory) would be inclined to think that a wave-mechanical system has a predilection for being affected by only one of its proper modes at a time. Yet this is implied by the continued use of the words "energy levels," "transitions," "transition probabilities."

The perseverance in this way of thinking is understandable, because the great and genuine successes of the idea of energy parcels has made it an ingrained habit to regard the product of Planck's constant h and a frequency as a bundle of energy, lost by one system and gained by another. How else should one understand the exact dove-tailing in the great "double-entry" book-keeping in nature? I maintain that it can in all cases be understood as a resonance phenomenon. One ought at least to try, and look upon atomic frequencies just as frequencies and drop the idea of energy-parcels. I submit that the word "energy" is at present used with two entirely different meanings, macroscopic and microscopic. Macroscopic energy is a "quantity-concept" (Quantitätsgrösse). Microscopic energy meaning (hv) is a "quality-concept" or "intensity-concept" (Intensitätsgrösse); it is quite proper to speak of high-grade and low-grade energy according to the value of the frequency v. True, the macroscopic energy is, strangely enough, obtained by a certain weighted summation over the frequencies, and in this relation the constant h is operative. But this does not necessarily entail that in every single case of microscopic interaction a whole portion hv of *macroscopic* energy is exchanged. I believe one is allowed to regard microscopic interaction as a continuous phenomenon without losing either the precious results of Planck and Einstein on the equilibrium of (macroscopic) energy between radiation and matter, or any other understanding of phenomena that the parcel-theory affords.

The one thing which one has to accept and which is the inalienable consequence of the wave-equation as it is used in every problem, under the most various forms, is this: that the interaction between two microscopic physical systems is controlled by a peculiar law of resonance. This law requires that the *difference* of two proper frequencies of the one system be equal to the difference of two proper frequencies of the other:

$$v_1 - v'_1 = v_2' - v_2 \qquad \qquad \dots (1)$$

The interaction is appropriately described as a gradual change of the amplitudes of the four proper vibrations in question. People have kept to the habit of multiplying this equation by h and saying it means, that the first system (index 1) has dropped from the energy level hv_1 to the level hv_1' the balance being transferred to the second system, enabling it to rise from hv_2 to hv_2'. This interpretation is obsolete. There is nothing to recommend it, and it bars the understanding of what is actually going on. It obstinately refuses to take stock of the principle of superposition, which enables us to envisage simultaneous gradual changes of any and all amplitudes without surrendering the essential discontinuity, if any, namely that of the frequencies. To be accurate we must add, that the condition of resonance, equation (1), may include three or more interacting systems. It may for example read

$$v_1 - v_1' = v_2' - v_2 + v_3' - v_3 \qquad \ldots (2)$$

Moreover we may adopt the view that the two or more interacting systems are regarded as *one* system. One is then inclined to write equations (1) and (2), respectively, as follows:

$$v_1 + v_2 = v_1' + v_2' \qquad \ldots \ldots (1')$$

$$v_1 + v_2 + v_3 = v_1' + v_2' + v_3', \qquad \ldots (2')$$

and to state the resonance condition thus: the interaction is restricted to constituent vibrations of the *same* frequency. This is a familiar state of affairs, of old. Unfamiliar is the tacit admission that frequencies are *additive*, when two or more systems are considered as forming *one* system. It is an inevitable consequence of Wave Mechanics. Is it so very repugnant to common sense? If I smoke 25 cigarettes per day, and my wife smokes 10, and my daughter 12—is not the family consumption 47 per day—on the average?

A Typical Experiment

Jokes aside, I wish to consider some typical experiments that ostensibly force the energy parcel view upon us, and I wish to show that this is an illusion. A beam of cathode rays of uniform velocity, which can be gradually increased, is passed through sodium vapour. Behind the vessel containing the vapour the beam passes an electric field which deflects it and tells us the velocity of the particles after the passage. At the same time a spectrometer inspects the light, if any, emitted by the vapour. For small initial velocity nothing happens: no light, no change of velocity in the cathode beam. But when the initial velocity is increased beyond a sharply defined limit, two things happen. The vapour begins to glow, radiating the frequency of the first line of the "principal series"; and the beam of cathode rays emerging from the vapour is split into two by the deflecting electric field, one indicating the initial velocity unchanged, and another slow one has "lost an amount of energy" equal to the frequency of the said spectral line multiplied by Planck's constant h. If the velocity is further increased the story repeats itself when the incident cathode ray energy increases beyond the "energy level" that is responsible for the second line (or rather the "level-difference" in question); this line appears and a third beam of cathode rays with correspondingly reduced speed occurs; and so on. This was, and still is, regarded as blatant evidence of the energy parcel view.

But it is just as easily understood from the resonance point of view. A cathode ray of particles with uniform velocity is a monochromatic beam of de Broglie waves. Only when its frequency (v_1) surpasses the frequency difference ($v_2' - v_2$) between the lowest (v_2) and the second (v_2') proper frequencies of the sodium atom is there a de Broglie frequency $v_1' > 0$ that fulfils the resonance demand, equation (1). Then the vibration v_1' appears in the de Broglie wave and v_2' among the atoms which begin to glow with frequency $v_2' - v_2$, since Maxwell's "electromagnetic vacuum" is prepared for resonance with anything. The splitting of the cathode ray beam in the deviating electric field, after passing the vapour, is accounted for by de Broglie's wave equation. An electric field has for de Broglie waves an "index of refraction" that *depends* on their frequency ("dispersion") and has a gradient in the direction of the field (which thus acts as an "inhomogeneous medium"). Any further events that might happen, for instance a transfer of some of the "energy quanta" $h(v_2' - v_2)$ from the sodium atoms to other gas molecules by "impacts of the second kind," are just as easily understood as resonance phenomena, provided only one keeps to the wave picture throughout and for all particles involved.

Many similar cases of apparent transfer of energy-parcels can be reduced to resonance—for instance photochemical action. The pattern is always the same: you may either take equations like (1) or (2) as they stand (resonance), or multiply them by h and think they express an energy balance of every single micro-transition. In the preceding example one point is of particular interest. One is able by an external agent (the electric field) to *separate in space* the two or more frequencies which have arisen in the cathode ray by the interaction; for they behave differently towards this agent and the different behaviour is completely understood from de Broglie's wave equation; one thus obtains two or more beams of homogenous frequency (or velocity). It is extremely valuable that there are simple cases of this kind in which the separation into two "phases" has nothing enigmatic; it is an immediate consequence of the principles laid down in L. de Broglie's earliest work on material waves. I say, this is fortunate; for there is a vast domain of phenomena in which the separation in space either takes place in the natural conditions of observation, or can easily be brought about by simple appliances; but it is not as easily explained on first principles. This might dishearten one in accepting the view of gradually changing amplitudes, that I put forward here; for the separation into different phases that produces itself before our eyes seems to confirm the belief that a discontinuous abrupt and *complete* transition occurs in every single microscopic interaction.

12
— P . A . M . D i r a c —
(1902–1984)

P.A.M. Dirac was a British physicist and professor of mathematics at Cambridge University, and a central figure in the development of quantum mechanics. In 1928 Dirac predicted the existence of the positron, a positively charged particle of the same mass as the negatively charged electron. He was awarded the Nobel Prize for physics in 1933 (jointly with Schrödinger).

While Dirac's scientific work contributed significantly to the field of quantum mechanics, his penchant for the philosophical places him in the company of Bohr and Heisenberg in terms of importance for a full understanding of the development of Quantum Theory. In this essay, Dirac provides his own account of having to grapple with the conceptual difficulties of Quantum Theory. He evinces his philosophic spirit most poignantly when he notes that "it is more important to have beauty in one's equations than to have them fit experiment." In the end, Dirac emphasizes the evolutionary nature of modern physics and muses about just where it might be headed.

The reading comes from an article entitled "The Evolution of the Physicist's Picture of Nature," which appeared in Scientific American *in 1963.*

THE CONCEPTUAL DIFFICULTIES OF QUANTUM THEORY

In this article I should like to discuss the development of general physical theory: how it developed in the past and how one may expect it to develop in the future. One can look on this continual development as a process of evolution, a process that has been going on for several centuries.

The first main step in this process of evolution was brought about by Newton. Before Newton, people looked on the world as being essentially two-dimensional—the two dimensions in which one can walk about—and the up-and-down dimension seemed to be something essentially different. Newton showed how one can look on the up-and-down direction as being symmetrical with the other two directions, by bringing in gravitational forces and showing how they take their place in physical theory. One can say that Newton enables us to pass from a picture with two-dimensional symmetry to a picture with three-dimensional symmetry.

Einstein made another step in the same direction, showing how one can pass from a picture with three-dimensional symmetry to a picture with four-dimensional symmetry. Einstein brought in time and showed how it plays a role that is in many ways symmetrical with the three space dimensions. However, this symmetry is not quite perfect. With Einstein's picture one is led to think of the world from a four-dimensional point of view, but the four dimensions are not completely symmetrical. There are some directions in the four-dimensional picture that are different from others: directions that are called null directions, along which a ray of light can move; hence the four-dimensional picture is not completely symmetrical. Still, there is a great deal of symmetry among the four dimensions. The only lack of symmetry, so far as concerns the equations of physics, is in the appearance of a minus sign in the equations with respect to the time dimension as compared with the three space dimensions.

We have, then, the development from the three-dimensional picture of the world to the four-dimensional picture. The reader will probably not be happy with this situation, because the world still appears three-dimensional to his consciousness. How can one bring this appearance into the four-dimensional picture that Einstein requires the physicist to have?

What appears to our consciousness is really a three-dimensional section of the four-dimensional picture. We must take a three-dimensional section to give us what appears to our consciousness at one time; at a later time we shall have a different three-dimensional section. The task of the physicist consists largely of relating events in one of these sections to events in another section referring to a later time. Thus the picture with four-dimensional symmetry does not give us the whole situation. This becomes particularly important when one takes into account the developments that have been brought about by Quantum Theory. Quantum Theory has taught us that we have to take the process of observation into account, and observations usually require us to bring in the three-dimensional sections of the four-dimensional picture of the universe.

The special Theory of Relativity, which Einstein introduced, requires us to put all the laws of physics into a form that displays four-dimensional symmetry. But when we use these laws to get results about observations, we have to bring in something additional to the four-dimensional symmetry, namely the three-dimensional sections that describe our consciousness of the universe at a certain time.

Einstein made another most important contribution to the development of our physical picture: he put forward the General Theory of Relativity, which requires us to suppose that the space of physics is curved. Before this physicists had always worked with a flat space, the three-dimensional flat space of Newton which was then extended to the four-dimensional flat space of special relativity. General relativity made a really important contribution to the evolution of our physical picture by requiring us to go over to curved space. The general requirements of this theory mean that all the laws of physics can be formulated in curved four-dimensional space, and that they show symmetry among the four dimensions. But again, when we want to bring in observations, as we must if we look at things from the point of view of Quantum Theory, we have to refer to a section of this four-dimensional space. With the four-dimensional space curved, any section that we make in it also has to be curved, because in general we cannot give a meaning to a flat section in a curved space. This leads us to a picture in which we have to take curved three-dimensional sections in the curved four-dimensional space and discuss observations in these sections.

During the past few years people have been trying to apply quantum ideas to gravitation as well as to the other phenomena of physics, and this has led to a rather unexpected development, namely that when one looks at gravitational theory from the point of view of the sections, one finds that there are some degrees of freedom that drop out of the theory. The gravitational field is a tensor field with 10 components. One finds that six of the components are adequate for describing everything of physical importance and the other four can be dropped out of the equations. One cannot, however, pick out the six important components from the complete set of 10 in any way that does not destroy the four-dimensional symmetry. Thus if one insists on preserving four-dimensional symmetry in the equations, one cannot adapt the theory of gravitation to a discussion of measurements in the way Quantum Theory requires without being forced to a more complicated description than is needed by the physical situation. This result has led me to doubt how fundamental the four-dimensional requirement in physics is. A few decades ago it seemed quite certain that one had to express the whole of physics in four-dimensional form. But now it seems that four-dimensional symmetry is not of such overriding importance, since the description of nature sometimes gets simplified when one departs from it.

Now I should like to proceed to the developments that have been brought about by Quantum Theory. Quantum Theory is the discussion of very small things, and it has formed the main subject of physics for the past 60 years. During this period physicists have been amassing quite a lot of experimental information and developing a theory to correspond to it, and this combination of theory and experiment has led to important developments in the physicist's picture of the world.

The quantum first made its appearance when Planck discovered the need to suppose that the energy of electromagnetic waves can exist only in multiples of a certain unit, depending on the frequency of the waves, in order to explain the law of black-body radiation. Then Einstein discovered the same unit of energy occurring in the photoelectric effect. In this early work on Quantum Theory one simply had to accept the unit of energy without being able to incorporate it into a physical picture.

The first new picture that appeared was Bohr's picture of the atom. It was a picture in which we had electrons moving about in certain well-defined orbits and occasionally making a jump from one orbit to another. We could not picture how the jump took place. We just had to accept it as a kind of discontinuity. Bohr's picture of the atom worked only for special examples, essentially when there was only one electron that was of importance for the problem under consideration. Thus the picture was an incomplete and primitive one.

The big advance in the Quantum Theory came in 1925, with the discovery of quantum mechanics. This advance was brought about independently by two men, Heisenberg first and Schrödinger soon afterward, working from different points of view. Heisenberg worked keeping close to the experimental evidence about spectra that was being amassed at that time, and he found out how the experimental information could be fitted into a scheme that is now known as matrix mechanics. All the experimental data of spectroscopy fitted beautifully into the scheme of matrix mechanics, and this led to quite a different picture of the atomic world. Schrödinger worked from a more mathematical point of view, trying to find a beautiful theory for describing atomic events, and was helped by de Broglie's ideas of waves associated with particles. He was able to extend de Broglie's ideas and to get a very beautiful equation, known as Schrödinger's wave equation, for describing atomic processes. Schrödinger got this equation by pure thought, looking for some beautiful generalization of de Broglie's ideas, and not by keeping close to the experimental development of the subject in the way Heisenberg did.

I might tell you the story I heard from Schrödinger of how, when he first got the idea for this equation, he immediately applied it to the behavior of the electron in the hydrogen atom, and then he got results that did not agree with experiment. The disagreement arose because at that time it was not known that the electron has a spin. That, of course, was a great disappointment to Schrödinger, and it caused him to abandon the work for some months. The he noticed that if he applied the theory in a more approximate way, not taking into account the refinements required by relativity, to this rough approximation his work was in agreement with observation. He published his first paper with only this rough approximation, and in that way Schrödinger's wave equation was presented to the world. Afterward, of course, when people found out how to take into account correctly the spin of the electron, the discrepancy between the results of applying Schrödinger's relativistic equation and the experiments was completely cleared up.

I think there is a moral to this story, namely that it is more important to have beauty in one's equations than to have them fit experiment. If Schrödinger had been more confident of his work, he could have published it some months earlier, and he could have published a more accurate equation. That equation is now known as the Klein-Gordon equation, although it was really discovered by Schrödinger, and in fact was discovered by Schrödinger before he discovered his nonrelativistic treatment of the hydrogen atom. It seems that if one is working from the point of view of getting beauty in one's equations, and if one has really a sound insight, one is on a sure line of progress. If there is not complete agreement between the results of one's work and experiment, one should not allow oneself to be too discouraged, because the discrepancy may well be due to minor features that are not properly taken into account and that will get cleared up with further developments of the theory.

That is how quantum mechanics was discovered. It led to a drastic change in the physicist's picture of the world, perhaps the biggest that has yet taken place. This change comes from our having to give up the deterministic picture we had always taken for granted. We are led to a theory that does not predict with certainty what is going to happen in the future but gives us information only about the probability of occurrence of various events. This giving up of determinacy has been a very controversial subject, and some people do not like it at all. Einstein in particular never liked it. Although Einstein was one of the great contributors to the development of quantum mechanics, he still was always rather hostile to the form that quantum mechanics evolved into during his lifetime and that it still retains.

The hostility some people have to the giving up of the deterministic picture can be centered on a much discussed paper by Einstein, Podolsky and Rosen dealing with the difficulty one has in forming a consistent picture that still gives results according to the rules of quantum mechanics. The rules of quantum mechanics are quite definite. People know how to calculate results and how to compare the results of their calculations with experiment. Everyone is agreed on the formalism. It works so well that nobody can afford to disagree with it. But still the picture that we are to set up behind this formalism is a subject of controversy.

I should like to suggest that one not worry too much about this controversy. I feel very strongly that the stage physics has reached at the present day is not the final stage. It is just one stage in the evolution of our picture of nature, and we should expect this process of evolution to continue in the future, as biological evolution continues into the future. The present stage of physical theory is merely a steppingstone toward the better stages we shall have in the future. One can be quite sure that there will be better stages simply because of the difficulties that occur in the physics of today.

I should now like to dwell a bit on the difficulties in the physics of the present day. The reader who is not an expert in the subject might get the idea that because of all these difficulties physical theory is in pretty poor shape and that the Quantum Theory is not much good. I should like to correct this impression by saying that Quantum Theory is an extremely good theory. It gives wonderful agreement with observation over a wide range of phenomena. There is no doubt that it is a good theory, and the only reason physicists talk so much about the difficulties in it is that it is precisely the difficulties that are interesting. The successes of the theory are all taken for granted. One does not get anywhere simply by going over the successes again and again, whereas by talking over the difficulties people can hope to make some progress.

The difficulties in Quantum Theory are of two kinds. I might call them Class One difficulties and Class Two difficulties. Class One difficulties are the difficulties I have already mentioned: How can one form a consistent picture behind the rules for the present Quantum Theory? These Class One difficulties do not really worry the physicist. If the physicist knows how to calculate results and compare them with experiment, he is quite happy if the results agree with his experiments, and that is all he needs. It is only the philosopher, wanting to have a satisfying description of nature, who is bothered by Class One difficulties.

There are, in addition to the Class One difficulties, the Class Two difficulties, which stem from the fact that the present laws of Quantum Theory are not always adequate to give any results. If one pushes the laws to extreme conditions—to phenomena involving very high energies or very small distances—one sometimes gets results that are ambiguous or not really sensible at all. Then it is clear that one has reached the limits of application of the theory and that some further development is needed. The Class Two difficulties are important even for the physicist, because they put a limitation on how far he can use the rules of Quantum Theory to get results comparable with experiment.

I should like to say a little more about the Class One difficulties. I feel that one should not be bothered with them too much, because they are difficulties that refer to the present stage in the development of our physical picture and are almost certain to change with future development. There is one strong reason, I think, why one can be quite confident that these difficulties will change. There are some fundamental constants in nature: the charge on the electron (designated e), Planck's constant divided by 2π (designated $\hbar$) and the velocity of light (c). From these fundamental constants one can construct a number that has no dimensions: the number $\hbar c/e^2$. That number is found by experiment to have the value 137, or something very close to 137. Now, there is no known reason why it should have this value rather than some other number. Various people have put forward ideas about it, but there is no accepted theory. Still, one can be fairly sure that someday physicists will solve the problem and explain why the number has this value. There will be a physics in the future that works when $\hbar c/e^2$ has the value 137 and that will not work when it has any other value.

The physics of the future, of course, cannot have the three quantities $\hbar$, e and c all as fundamental quantities. Only two of them can be fundamental, and the third must be derived from those two. It is almost certain that c will be one of the two fundamental ones. The velocity of light, c, is so important in the four-dimensional picture, and it plays such a fundamental role in the special Theory of Relativity, correlating our units of space and time, that it has to be fundamental. Then we are faced with the fact that of the two quantities $\hbar$ and e, one will be fundamental and one will be derived. If $\hbar$ is fundamental, e will have to be explained in some way in terms of the square root of $\hbar$, and it seems most unlikely that any fundamental theory can give e in terms of a square root, since square roots do not occur in basic equations. It is much more likely that e will be the fundamental quantity and that $\hbar$ will be explained in terms of e^2. Then there will be no square root in the basic equations. I think one is on safe ground if one makes the guess that in the physical picture we shall have at some future stage e and c will be fundamental quantities and $\hbar$ will be derived.

If $\hbar$ is a derived quantity instead of a fundamental one, our whole set of ideas about uncertainty will be altered: $\hbar$ is the fundamental quantity that occurs in the Heisenberg uncertainty relation connecting the amount of uncertainty in a position and in a momentum. This uncertainty relation cannot play a fundamental role in a theory in which $\hbar$ itself is not a fundamental quantity. I think one can make a safe guess that uncertainty relations in their present form will not survive in the physics of the future.

Of course there will not be a return to the determinism of classical physical theory. Evolution does not go backward. It will have to go forward. There will have to be some new development that is quite unexpected, that we cannot make a guess about, which will take us still further from classical ideas but which will alter completely the discussion of uncertainty relations. And when this new development occurs, people will find it all rather futile to have had so much of a discussion on the role of observation in the theory, because they will have then a much better point of view from which to look at things. So I shall say that if we can find a way to describe the uncertainty relations and the indeterminacy of present quantum mechanics that is satisfying to our philosophical ideas, we can count ourselves lucky. But if we cannot find such a way, it is nothing to be really disturbed about. We simply have to take into account that we are at a transitional stage and that perhaps it is quite impossible to get a satisfactory picture for this stage.

I have disposed of the Class One difficulties by saying that they are really not so important, that if one can make progress with them one can count oneself lucky, and that if one cannot it is nothing to be genuinely disturbed about. The Class Two difficulties are the really serious ones. They arise primarily from the fact that when we apply our Quantum Theory to fields in the way we have to if we are to make it agree with special relativity, interpreting it in terms of the three-dimensional sections I have mentioned, we have equations that at first look all right. But when one tries to solve them, one finds that they do not have any solutions. At this point we ought to say that we do not have a theory. But physicists are very ingenious about it, and they have found a way to make progress in spite of this obstacle. They find that when they try to solve the equations, the trouble is that certain quantities that ought to be finite are actually infinite. One gets integrals that diverge instead of converging to something definite. Physicists have found that there is a way to handle these infinities according to certain rules, which makes it possible to get definite results. This method is known as the renormalization method.

I shall merely explain the idea in words. We start out with a theory involving equations. In these equations there occur certain parameters: the charge of the electron, e, the mass of the electron, m, and things of a similar nature. One then finds that these quantities, which appear in the original equations, are not equal to the measured values of the charge and the mass of the electron. The measured values differ from these by certain correcting terms—Δe, Δm and so on—so that the total charge is $e + \Delta e$ and the total mass $m + \Delta m$. These changes in charge and mass are brought about through the interaction of our elementary particle with other things. Then one says that $e + \Delta e$ and $m + \Delta m$, being the observed things, are the important things. The original e and m are just mathematical parameters; they are unobservable and therefore just tools one can discard when one has got far enough to bring in the things that one can compare with observation. This would be a quite correct way to proceed if Δe and Δm were small (or even if they were not so small but finite) corrections. According to the actual theory, however, Δe and Δm are infinitely great. In spite of that fact one can still use the formalism and get results in terms of $e + \Delta e$ and $m + \Delta m$, which one can interpret by saying that the original e and m have to be minus infinity of a suitable amount to compensate for the Δe and Δm that are infinitely great. One can use the theory to get results that can be compared with experiment, in particular for electrodynamics. The surprising thing is that in the case of electrodynamics one gets results that are in extremely good agreement with experiment. The agreement applies to many significant figures—the kind of accuracy that previously one had only in astronomy. It is because of this good agreement that physicists do attach some value to the renormalization theory, in spite of its illogical character.

It seems to be quite impossible to put this theory on a mathematically sound basis. At one time physical theory was all built on mathematics that was inherently sound. I do not say that physicists always use sound mathematics; they often use unsound steps in their calculations. But previously when they did so it was simply because of, one might say, laziness. They wanted to get results as quickly as possible without doing unnecessary work. It was always possible for the pure mathematician to come along and make the theory sound by bringing in further steps, and perhaps by introducing quite a lot of cumbersome notation and other things that are desirable from a mathematical point of view in order to get everything expressed rigorously but do not contribute to the physical ideas. The earlier mathematics could always be made sound in that way, but in the renormalization theory we have a theory that has defied all the attempts of the mathematician to make it sound. I am inclined to suspect that the renormalization theory is something that will not survive in the future, and that the remarkable agreement between its results and experiment should be looked on as a fluke.

This is perhaps not altogether surprising, because there have been similar flukes in the past. In fact, Bohr's electron-orbit theory was found to give very good agreement with observation as long as one confined oneself to one-electron problems. I think people will now say that this agreement was a fluke, because the basic ideas of Bohr's orbit theory have been superseded by something radically different. I believe the successes of the renormalization theory will be on the same footing as the successes of the Bohr orbit theory applied to one-electron problems.

The renormalization theory has removed some of these Class Two difficulties, if one can accept the illogical character of discarding infinities, but it does not remove all of them. There are a good many problems left over concerning particles other than those that come into electrodynamics: the new particles—mesons of various kinds and neutrinos. There the theory is still in a primitive stage. It is fairly certain that there will have to be drastic changes in our fundamental ideas before these problems can be solved.

One of the problems is the one I have already mentioned about accounting for the number 137. Other problems are how to introduce the fundamental length to physics in some natural way, how to explain the ratios of the masses of the elementary particles and how to explain their other properties. I believe separate ideas will be needed to solve these distinct problems and that they will be solved one at a time through successive stages in the future evolution of physics. At this point I find myself in disagreement with most physicists. They are inclined to think one master idea will be discovered that will solve all these problems together. I think it is asking too much to hope that anyone will be able to solve all these problems together. One should separate them one from another as much as possible and try to tackle them separately. And I believe the future development of physics will consist of solving them one at a time, and that after any one of them has been solved, there will still be a great mystery about how to attack further ones.

I might perhaps discuss some ideas I have had about how one can possibly attack some of these problems. None of these ideas has been worked out very far, and I do not have much hope for any one of them. But I think they are worth mentioning briefly.

One of these ideas is to introduce something corresponding to the luminiferous ether, which was so popular among the physicists of the 19th century. I said earlier that physics does not evolve backward. When I talk about reintroducing the ether, I do not mean to go back to the picture of the ether that one had in the 19th century, but I do mean to introduce a new picture of the ether that will conform to our present ideas of Quantum Theory. The objection to the old idea of the ether was that if you suppose it to be a fluid filling up the whole of space, in any place it has a definite velocity, which destroys the four-dimensional symmetry required by Einstein's special principle of relativity. Einstein's special relativity killed this idea of the ether.

But with our present Quantum Theory we no longer have to attach a define velocity to any given physical thing, because the velocity is subject to uncertainty relations. The smaller the mass of the thing we are interested in, the more important are the uncertainty relations. Now, the ether will certainly have very little mass, so that uncertainty relations for it will be extremely important. The velocity of the ether at some particular place should therefore not be pictured as definite, because it will be subject to uncertainty relations and so may be anything over a wide range of values. In that way one can get over the difficulties of reconciling the existence of an ether with the special Theory of Relativity.

There is one important change this will make in our picture of a vacuum. We would like to think of a vacuum as a region in which we have complete symmetry between the four dimensions of space-time as required by special relativity. If there is ether subject to uncertainty relations, it will not be possible to have this symmetry accurately. We can suppose that the velocity of the ether is equally likely to be anything within a wide range of values that would give the symmetry only approximately. We cannot in any precise way proceed to the limit of allowing all values for the velocity between plus and minus the velocity of light, which we would have to do in order to make the symmetry accurate. Thus the vacuum becomes a state that is unattainable. I do not think that this is a physical objection to the theory. It would mean that the vacuum is a state we can approach very closely. There is no limit as to how closely we can approach it, but we can never attain it. I believe that would be quite satisfactory to the experimental physicist. It would, however, mean a departure from the notion of the vacuum that we have in the Quantum Theory, where we start off with the vacuum state having exactly the symmetry required by special relativity.

That is one idea for the development of physics in the future that would change our picture of the vacuum, but change it in a way that is not unacceptable to the experimental physicist. It has proved difficult to continue with the theory, because one would need to set up mathematically the uncertainty relations for the ether and so far some satisfactory theory along these lines has not been discovered. If it could be developed satisfactorily, it would give rise to a new kind of field in physical theory, which might help in explaining some of the elementary particles.

Another possible picture I should like to mention concerns the question of why all the electric charges that are observed in nature should be multiples of one elementary unit, e. Why does one not have a continuous distribution of charge occurring in nature? The picture I propose goes back to the idea of Faraday lines of force and involves a development of this idea. The Faraday lines of force are a way of picturing electric fields. If we have an electric field in any region of space, then according to Faraday we can draw a set of lines that have the direction of the electric field. The closeness of the lines to one another gives a measure of the strength of the field—they are close where the field is strong and less close where the field is weak. The Faraday lines of force give us a good picture of the electric field in classical theory.

When we go over to Quantum Theory, we bring a kind of discreteness into our basic picture. We can suppose that the continuous distribution of Faraday lines of force that we have in the classical picture is replaced by just a few discrete lines of force with no lines of force between them.

Now, the lines of force in the Faraday picture end where there are charges. Therefore with these quantized Faraday lines of force it would be reasonable to suppose the charge associated with each line, which has to lie at the end if the line of force has an end, is always the same (apart from its sign), and is always just the electronic charge, $-e$ or $+e$. This leads us to a picture of discrete Faraday lines of force, each associated with a charge, $-e$ or $+e$. There is a direction attached to each line, so that the ends of a line that has two ends are not the same, and there is a charge $+e$ at one end and a charge $-e$ at the other. We may have lines of force extending to infinity, of course, and then there is no charge.

If we suppose that these discrete Faraday lines of force are something basic in physics and lie at the bottom of our picture of the electromagnetic field, we shall have an explanation of why charges always occur in multiples of e. This happens because if we have any particle with some lines of force ending on it, the number of these lines must be a whole number. In that way we get a picture that is qualitatively quite reasonable.

We suppose these lines of force can move about. Some of them, forming closed loops or simply extending from minus infinity to infinity, will correspond to electromagnetic waves. Others will have ends, and the ends of these lines will be the charges. We may have a line of force sometimes breaking. When that happens, we have two ends appearing and there must be charges at the two ends. This process—the breaking of a line of force—would be the picture for the creation of an electron (e^-) and a positron (e^+). It would be quite a reasonable picture, and if one could develop it, would provide a theory in which e appears as a basic quantity. I have not yet found any reasonable system of equations of motion for these lines of force, and so I just put forward the idea as a possible physical picture we might have in the future.

There is one very attractive feature in this picture. It will quite alter the discussion of renormalization. The renormalization we have in our present quantum electrodynamics comes from starting off with what people call a bare electron—an electron without a charge on it. At a certain stage in the theory one brings in the charge and puts it on the electron, thereby making the electron interact with the electromagnetic field. This brings a perturbation into the equations and causes a change in the mass of the electron, the Δm, which is to be added to the previous mass of the electron. The procedure is rather roundabout because it starts off with the unphysical concept of the bare electron. Probably in the improved physical picture we shall have in the future the bare electron will not exist at all.

Now, that state of affairs is just what we have with the discrete lines of force. We can picture the lines of force as strings, and then the election in the picture is the end of a string. The string itself is the Coulomb force around the electron. A bare electron means an electron without the Coulomb force around it. That is inconceivable with this picture, just as it is inconceivable to think of the end of a piece of string without thinking of the string itself. This, I think, is the kind of way in which we should try to develop our physical picture—to bring in ideas that make inconceivable the things we do not want to have. Again we have a picture that looks reasonable, but I have not found the proper equations for developing it.

I might mention a third picture with which I have been dealing lately. It involves departing from the picture of the electron as a point and thinking of it as a kind of sphere with a finite size. Of course, it is really quite an old idea to picture the electron as a sphere, but previously one had the difficulty of discussing a sphere that is subject to acceleration and to irregular motion. It will get distorted, and how is one to deal with the distortions? I propose that one should allow the electron to have, in general, an arbitrary shape and size. There will be some shapes and sizes in which it has less energy than in others, and it will tend to assume a spherical shape with a certain size in which the electron has the least energy.

This picture of the extended electron has been stimulated by the discovery of the mu meson, or muon, one of the new particles of physics. The muon has the surprising property of being almost identical with the electron except in one particular, namely, its mass is some 200 times greater than the mass of the electron. Apart from this disparity in mass the muon is remarkably similar to the electron, having, to an extremely high degree of accuracy, the same spin and the same magnetic moment in proportion to its mass as the electron does. This leads to the suggestion that the muon should be looked on as an excited electron. If the electron is a point, picturing how it can be excited becomes quite awkward. But if the electron is the most stable state for an object of finite size, the muon might just be the next most stable state in which the object undergoes a kind of oscillation. That is an idea I have been working on recently. There are difficulties in the development of this idea, in particular the difficulty of bringing in the correct spin.

I have mentioned three possible ways in which one might think of developing our physical picture. No doubt there will be others that other people will think of. One hopes that sooner or later someone will find an idea that really fits and leads to a big development. I am rather pessimistic about it and am inclined to think none of them will be good enough. The future evolution of basic physics—that is to say, a development that will really solve one of the fundamental problems, such as bringing in the fundamental length or calculating the ratio of the masses—may require some much more drastic change in our physical picture. This would mean that in our present attempts to think of a new physical picture we are setting our imaginations to work in terms of inadequate physical concepts. If that is really the case, how can we hope to make progress in the future?

There is one other line along which one can still proceed by theoretical means. It seems to be one of the fundamental features of nature that fundamental physical laws are described in terms of a mathematical theory of great beauty and power, needing quite a high standard of mathematics for one to understand it. You may wonder: Why is nature constructed along these lines? One can only answer that our present knowledge seems to show that nature is so constructed. We simply have to accept it. One could perhaps describe the situation by saying that God is a mathematician of a very high order, and He used very advanced mathematics in constructing the universe. Our feeble attempts at mathematics enable us to understand a bit of the universe, and as we proceed to develop higher and higher mathematics we can hope to understand the universe better.

This view provides us with another way in which we can hope to make advances in our theories. Just by studying mathematics we can hope to make a guess at the kind of mathematics that will come into the physics of the future. A good many people are working on the mathematical basis of Quantum Theory, trying to understand the theory better and to make it more powerful and more beautiful. If someone can hit on the right lines along which to make this development, it may lead to a future advance in which people will first discover the equations and then, after examining them, gradually learn how to apply them. To some extent that corresponds with the line of development that occurred with Schrödinger's discovery of his wave equation. Schrödinger discovered the equation simply by looking for an equation with mathematical beauty. When the equation was first discovered, people saw that it fitted in certain ways, but the general principles according to which one should apply it were worked out only some two or three years later. It may well be that the next advance in physics will come about along these lines: people first discovering the equations and then needing a few years of development in order to find the physical ideas behind the equations. My own belief is that this is a more likely line of progress than trying to guess at physical pictures.

Of course, it may be that even this line of progress will fail, and then the only line left is the experimental one. Experimental physicists are continuing their work quite independently of theory, collecting a vast storehouse of information. Sooner or later there will be a new Heisenberg who will be able to pick out the important features of this information and see how to use them in a way similar to that in which Heisenberg used the experimental knowledge of spectra to build his matrix mechanics. It is inevitable that physics will develop ultimately along these lines, but we may have to wait quite a long time if people do not get bright ideas for developing the theoretical side.

13
— John Archibald Wheeler —
(1911–)

John Wheeler, an American-born physicist, helped to develop the Institute of Theo-retical Physics in Austin, Texas, into one of the premier centers for quantum reality research. As a young man, Wheeler went to Copenhagen to study quantum physics with Niels Bohr. While there he published (along with his mentor) the first paper that successfully explained nuclear fission in terms of quantum physics.

In addition to his unique—and sometimes controversial—interpretation of quantum physics, Wheeler is best known for coining the term "black hole." He was awarded the Wolf Prize in Physics for 1996–97 for leading the development of black-hole physics.

In this reading, Wheeler describes a variation of the classic two-slit experiment that reveals the dual nature of quantum phenomena. Wheeler finds evidence for his unique interpretation of quantum physics in terms of a "participatory uni-verse" or "observer-created reality," where an act of observation is required for any phenomenon to be considered real. Comparing the quantum situation to a version of the game "Twenty Questions," Wheeler affirms time and again the claim that "no elementary phenomenon is a real phenomenon until it is an observed phenomenon." The reading comes from an article entitled "The 'Past' and the 'Delayed Choice' Double-Slit Experiment," which appeared in 1978.

OBSERVER-CREATED REALITY

Partway down the optic axis of the traditional double-slit experiment stands the central element, the doubly-slit screen. Can one choose whether the photon (or electron) *shall have* come through both of the slits, or only one of them, after it has *already* transversed this screen? That is the new question raised and analyzed here.

Known since the days of Young is the possibility to use the receptor at the end of the apparatus to record well defined interference fringes. How can they be formed unless the electromagnetic energy has come through both slits? In later times Einstein noted that in principle one can determine the lateral kick given to the receptor by each arriving quantum. How can this kick be understood unless the energy came through only a single slit?

Einstein's further reasoning as reported by Bohr is familiar. Record both the kicks and the fringes. Conclude from the kicks that each quantum of energy comes through a single slit alone; from the fringes, that it nevertheless also comes through both slits. But this conclusion is self-contradictory. Therefore Quantum Theory destroys itself by internal inconsistency.

Bohr's reply has become by now a central lesson of Quantum Physics. One can record the fringes or the kicks but not both. The arrangement for the recording of the one automatically rules out the recording of the other. The quantum has momentum p, de Broglie wave length $\lambda = h/p$, and reduced wave length $\lambdabar = \hbar/p$. To record for it well defined interference fringes one must fix the location of the receptor within a latitude

$$\Delta y < (fringe\ spacing)/2\pi = (L/2s)\lambdabar.$$

To tell from which slit the quantum of energy arrives one must register the transverse kick it gives to the receptor within a latitude small enough to distinguish clearly between a momentum $p = \hbar/x$ coming from below, at the inclination S/L, and a momentum coming from above at a like inclination; thus,

$$\Delta p_y < (S/L)(\hbar/\lambdabar).$$

However, for the receptor simultaneously to serve both functions would be incompatible with what the principle of indeterminacy has to say about receptor dynamics in the Y-direction,

$$\Delta y\ \Delta p_y > \hbar/2.$$

Not being able to observe simultaneously the two complementary features of the radiation, it is natural to focus on the one and forego examination of the other. Either one will insert the pin through the hole shown in Fig. 1. It will couple the receptor to the rest of the device. It will give the receptor a well defined location. Then one will be able to check on the predicted pattern of interference fringes. Or one will remove the pin. Then one can measure the through-the-slot component of momentum of the receptor before and after the impact of the quantum. Then one will say that one knows through which slit the energy came.

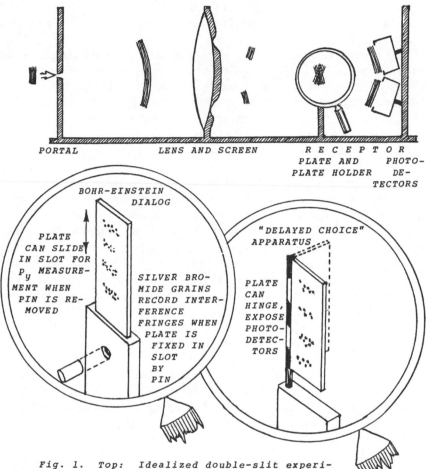

PORTAL   *LENS AND SCREEN*

RECEPTOR
PLATE AND   PHOTO-
PLATE HOLDER   DE-
TECTORS

BOHR-EINSTEIN
DIALOG

PLATE
CAN SLIDE
IN SLOT FOR
P_y MEASURE-
MENT WHEN
PIN IS RE-
MOVED

SILVER BRO-
MIDE GRAINS
RECORD INTER-
FERENCE
FRINGES WHEN
PLATE IS
FIXED IN
SLOT
BY
PIN

"DELAYED CHOICE"
APPARATUS

PLATE
CAN
HINGE,
EXPOSE
PHOTO-
DETEC-
TORS

Fig. 1. Top: *Idealized double-slit experi-*
ment. *Distance of each slit from optic axis, S; from*
photographic plate, L. For simplicity, details of the plate
and plate holder are omitted from the circle encompassed by
the magnifying glass and are presented below, magnified and
in perspective. Lower left: The version of the Bohr-Einstein
dialog. The plate catches every photon. It registers pre-
cisely the y-coordinate of impact or the y-component of im-
pulse delivered--but does not and cannot do both. Omit the
photodetectors. Lower right: The present "delayed choice"
varsion. Include the photodetectors. One or other of them
is sure to catch the quantum of energy when the plate is
swung aside. Whether to expose the plate or expose the photo-
detectors, whether thus to infer that a single quantum of
energy shall have *gone through both slits in the screen or*
only one, is subject to the free choice of the observer after
the energy has already *traversed the screen.*

Pin in or pin out: when may the choice be made? Must it be made before the quantum of energy passes through the doubly slit screen? Or may it be made after? That is the central question in this paper as that question first seems to impose itself. However, a closer look shows that the measurement of transverse momentum kick, in principle conceivable, is practically almost out of the question. Therefore it is appropriate to alter the idealized experiment before taking up the question of "before" *versus* "after." What is the difficulty and what is the change? . . .

The origin of the difficulty is not far to seek. The measuring equipment—the photographic plate—has itself been made the subject of measurement. One might as well put a Geiger counter or Schrödinger's cat or Wigner's friend into a box and try to measure the lateral momentum communicated to that enormity! An irreversible act of amplification, yes; but make it part of the system under study? No. Principle does not call for it and practicality forbids it.

The difficulty is overcome by a simple change:

1. Give up measuring the y-component of the momentum of the photographic plate.
2. Hold its y-coordinate fixed.
3. By means of a hinge parallel to the y-axis arrange that this high narrow plate can be swung out of the way of the incident light—at the last minute option of the observer—quicker than the flight of light from screen to plate. (Switch from "operative" to "open" position.)
4. Sufficiently far beyond the region of the plate, the beams from upper and lower slits cease to overlap and become well separated. There place photodetectors. Let each have an opening such that it records with essentially 100 percent probability a quantum of energy arriving in its own beam, and with essentially zero probability a quantum arriving in the other beam.

Now the choice is clear; and the objective, too. We today cannot argue, and Einstein in his later years would not even have wanted to argue, his erstwhile case of logical inconsistency against Quantum Theory: the photon goes through both slits, as evidenced in interference fringes, and yet simultaneously through only one, as evidenced in lateral momentum kick. Choose we know we must between the two complementary features open to study; and choose we do, by putting the plate athwart the light or turning it out of the line of fire. In the one case the quantum will transform a grain of silver bromide and contribute to the record of a two-slit interference fringe. In the other case one of the two counters will go off and signal in which beam—and therefore from which slit—the photon has arrived.

In our arrangement the photographic plate registers only the point of impact of a photon. In the earlier idealized experiment it could additionally (Einstein) or alternatively (Bohr) record the transverse momentum delivered by the impact. We have assigned the two distinct kinds of measurement to two distinct kinds of register. We have demoted the plate from a privileged status. That demotion is irrelevant to any question now at issue. Equally irrelevant is the different distance—and time of flight—from entry portal to plate, or photodetector, according as the one or other register is exposed. But the essential new point is the timing of the *choice*—between observing a two-slit effect and a one-slit one—until after the single quantum of energy in question has *already* passed through the screen.

Let the reasoning be passed in review that leads to this at first sight strange inversion of the normal order of time. Then let the general lesson of this apparent time inversion be drawn: "No phenomenon is a phenomenon until it is an observed phenomenon." In other words, it is not a paradox that we choose what *shall* have happened after "it has *already* happened." It has not really happened, it is not a phenomenon, until it is an observed phenomenon. . . .

"Delayed Choice" as an Additive Option in Other Idealized Experiments

To be forced to choose between complementary modes of observation is familiar, but it is unfamiliar to make this choice after the relevant interaction has already come to an end. Moreover, one can assert this "voice in what shall have happened, after it appears already to have happened" in illustrations of complementarity other than the double slit, by suitable modification of the idealized apparatus.

In the gamma ray microscope as described by Bohr and Heisenberg, a lens of angular opening ε, receiving and bringing to a focus light of the reduced wave length x, tells the position of the electron that scatters the light into the lens within a latitude of the order of

$$\Delta x \sim \lambda\varepsilon. \tag{39}$$

When the lens is thus used to fix position, the quantum of energy scattered into the lens gives the electron a lateral kick, the amount of which is subject to an uncertainty of the order

$$\Delta p \sim (photon\ momentum)(angular\ opening\ of\ the\ lens) \sim (\hbar/\lambda)\varepsilon. \tag{40}$$

This magnitude is coordinated with the Δx of (39) by the usual indeterminacy relation. However, the uncertainty in the lateral kick can be reduced to a very small fraction of (40) by placing a sufficiently great collection of sufficiently small photodetectors at a little distance above the lens. Whichever one of them goes off signals the direction of the scattered photon and thus the momentum imparted to the electron. When the lens is operated in this mode it ceases to serve as a lens in the true sense of the word. Lost then is the possibility to know the position of the electron within anything like the narrowness of limits implied by the Δx of (39). All this is the standard and well known lesson of complementarity.

We now add the feature of delayed choice. Only after the quantum of electro-magnetic energy has already passed through the lens do we decide which lattice work of photodetectors to swing into action. One lattice is located in the focal plane of the lens. Let it be the lattice that is swung into action. Then one of the counters in this lattice goes off. This irreversible act of amplification tells us where the electron was when it scattered the radiation, within the latitude Δx of (39). In drawing this conclusion about resolving power we accept the fact that the radiation made use of the entire aperture of the lens.

Or let the other lattice of photodetectors be swung into action. They are located in a plane some small fraction of the way up from lens to focal plane. One of these devices thereupon registers an event. This indelible record tells us within a certain range, much smaller than the Δp of (40), what lateral kick was given to the electron in the Compton process. The reasoning is simple. The kick to the electron is deduced from the direction of the scattered photon. The direction of this photon is revealed by the coordinates of the photocounter that registered, because it responds to a quantum of energy only when that energy goes through a highly restricted portion of the aperture of the lens.

Shall we say that "the whole aperture transmitted the energy" or that "only a very small fraction of the aperture transmitted the energy?" We can freely decide one way or the other, according as we activate one set of photocounters or the other, after the lens has *already* finished transmitting the energy. That is the unfamiliar feature that "delayed choice" brings to the idealized gamma-ray microscope.

The split-beam experiment provides another example, Bohr tells us, "to which Einstein very early called attention and often has reverted. If a semi-reflecting mirror is placed in the way of a photon, leaving two possibilities for its direction of propagation, the photon may either be recorded on one, and only one, of two photographic plates situated at great distances in the two directions in question, or else we may, by replacing the plates by mirrors, observe effects exhibiting an interference between the two reflected wave-trains. In any attempt of a pictorial representation of the behaviour of the photon we would, thus, meet with the difficulty: to be obliged to say, on the one hand, that the photon always chooses *one* of the two ways and, on the other hand, that it behaves as if it has passed *both* ways." . . .

Lesson of Delayed-Choice Experiments

The double slit experiment, like the other six idealized experiments (microscope, split beam, tilt-teeth, radiation pattern, one photon polarization and polarization of paired photons), imposes a choice between complementary modes of observation. In each experiment we have found a way to delay that choice of type of phenomenon to be looked for up to the very final stage of development of the phenomenon, *whichever* type we then fix upon. That delay makes no difference in the experimental predictions. On this score everything we find was foreshadowed in that solitary and pregnant sentence of Bohr, " . . . it . . . can make no difference, as regards observable effects obtainable by a definite experimental arrangement, whether our plans for constructing or handling the instruments are fixed beforehand or whether we prefer to postpone the completion of our planning until a later moment when the particle is already on its way from one instrument to another."

Not one of the seven delayed choice experiments has yet been done. There can hardly be one that the student of physics would not like to see done. In none is any justification whatsoever evident for doubting the obvious predictions.

We search here, not for new experiments or new predictions, but for new insight. Experiments dramatize and predictions spell out the quantum's consequences; but what is its central idea? A pedant of Copernican times could have calculated planetary positions from the equations of Copernicus as well as Copernicus himself; but what would we think of him if his eyes were closed to the main point, that the "Earth goes around the Sun"?

In the absence of an equally simple statement of *its* central idea, Quantum Theory appears to many as strange, unwelcome, and forced on physics as it were from outside against its will. In contrast, if the essential point could be grasped in a single phrase, we can well believe that the quantum would seem so natural that we would recognize at once that the universe could not even have come into being without it.

Special relativity's findings shower out like fireworks from a single compact package, "The laws of physics are the same in every inertial reference system." No leap of the imagination to a comparably compact and explodable formulation of Quantum Theory being forthcoming, experience recommends step-by-step progress. Of such steps none in recent times moved our understanding forward more than the Einstein-Bohr dialog. Out of that dialog no concept emerged of greater fruitfulness than "phenomenon": " . . . [In my discussions with Einstein] I advocated the application of the word *phenomenon* exclusively to refer to the observations obtained under specific circumstances, including an account of the whole experimental arrangement." No other point does the present analysis of idealized delayed-choice experiments have but to investigate what "phenomenon" means as applied to the "past."

After the quantum of energy has *already* gone through the doubly slit screen, a last-instant free choice on our part—we have found—gives at will a double-slit-interference record or a one-slit-beam count. Does this result mean that present choice influences past dynamics, in contravention of every formulation of causality? Or does it mean, calculate pedantically and don't ask questions? Neither; the lesson presents itself rather as this, that the past has no existence except as it is recorded in the present. It has no sense to speak of what the quantum of electromagnetic energy was doing except as it is observed. More generally, we would seem forced to say that no phenomenon is a phenomenon until—by observation, or some proper combination of theory and observation—it is an observed phenomenon. The universe does not "exist, out there," independent of all acts of observation. Instead, it is in some strange sense a participatory universe.

That present choice of mode of observation in the double-slit experiment should influence what we say about the "past" of the photon; that the "past" is undefined and undefinable without the observation, may be illustrated by a little story.

The "game of twenty questions" will be recalled. One of the party is sent out of the room. The others agree on a word. The one fated to be questioner returns and starts his questions. "Is it a living object?" "No." "Does it belong to the mineral kingdom?" "Yes." So the questions go from respondent to respondent around the room until at length the word emerges: victory if in twenty tries or less; otherwise, defeat.

Well does one participant recall the evening when he, fourth to be sent out, returned to find a smile on everyone's face, sign of a joke or a plot. He innocently started his questions. But each question he put took longer in the answering—strange, when the answer itself was only a simple "yes" or "no." At length, feeling hot on the trail, he asked, "Is the word 'cloud'?" "Yes," came the reply and everyone burst out laughing. They explained that when he had gone out, they had agreed not to agree in advance on any word at all. Everyone could respond "yes" or "no" as he pleased to whatever question was put to him. But however he answered, he had to have a word in mind compatible with his own reply—and with all the replies that went before. No wonder it took time to answer!

It is natural to compare the game in its two versions with physics in its two formulations, classical and quantum. First, the puzzled participant thought the word already existed "out there" as physics once thought that the position and momentum of the electron existed "out there," independent of any act of observation. Second, the information about the word was brought into being step by step through the questions that the interrogator raised, as the information about the electron is brought into being step by step by the experiments that the observer chooses to make. Third, if the participant had chosen to ask different questions he would have ended up with a different word, as the experimenter would have ended up with a different story for the doings of the electron if he had done different measurements or the same measurements in a different sequence. Fourth, whatever power the interrogator had in influencing the outcome for the word was partial only. A major part of the decision lay in the hands of the other participants. Similarly, the experimenter has some substantial influence on what will happen to the electron by the choice of experiments he will do on it; but he is well aware that there is much impredictability about what any given one of his measurements will disclose. Fifth, there was a "rule of the game" that required of every participator that his choice of yes or no should be compatible with *some* word. Similarly, there is a consistency about the observations made in physics. One person can tell another in plain language what he finds and the second person can verify the observation. Interesting though this comparison is between the world of physics and the world of the game, there is an important point of difference. The game has a finite number of participants and terminates after a finite number of steps. In contrast, the making of observations is a continuing process. Moreover, it is extraordinarily difficult to state sharply and clearly (1) where the community of observer-participators begins and where it ends, and (2) what the degree of amplification must be to define an observation: "The amplification of atomic effects, which makes it possible to base the account on measurable quantities and which gives the phenomenon a peculiar closed character, only emphasizes the irreversibility characteristic of the very concept of observation."

It is not necessary to understand every point about the quantum principle in order to understand something about it. Of all the points that stand forth from comparing the world of quantum observations with the game of twenty questions, none is more central than this: As in the game no word was the word until that word had been promoted to reality by the choice of questions asked and answers given, so no phenomenon is a phenomenon until it is an observed phenomenon.

THE COMPLETENESS DEBATE

//

14
Albert Einstein,
— Boris Podolsky, —
and Nathan Rosen

In what is considered the most serious of Einstein's objections to the Copenhagen Interpretation of Quantum Theory, he and his colleagues propose here a thought experiment that attempts to demonstrate the theory's incompleteness. The authors argue that because Quantum Theory makes only statistical predictions, it cannot help but leave out certain "elements of reality" that a more adequate theory of the world must include. This claim is best expressed in the basic criterion for reality set forth in the article: "If, without in any way disturbing a system, we can predict with certainty (i.e., with probability equal to unity) the value of a physical quantity, then there exists an element of physical reality corresponding to this physical quantity"—a criterion that the authors claim Quantum Theory violates. In the end, Einstein, Podolsky, and Rosen maintain the belief that a more complete theory of reality remains to be discovered.

This paper, originally entitled "Can Quantum Mechanical Description of Physical Reality Be Considered Complete?" appeared in the Physical Review in 1935. Although the paper is quite technical it is included here in its entirety because of its historical significance and because any abridgment would have compromised its coherence. The paper gave rise to the famous Completeness Debate, which is chronicled in the readings that follow.

THE EPR PARADOX

In a complete theory there is an element corresponding to each element of reality. A sufficient condition for the reality of a physical quantity is the possibility of predicting it with certainty, without disturbing the system. In quantum mechanics in the case of two physical quantities described by non-commuting operators, the knowledge of one precludes the knowledge of the other. Then either (1) the description of reality given by the wave function in quantum mechanics is not complete or (2) these two quantities cannot have simultaneous reality. Consideration of the problem of making predictions concerning a system on the basis of measurements made on another system that had previously interacted with it leads to the result that if (1) is false then (2) is also false. One is thus led to conclude that the description of reality as given by a wave function is not complete.

1

Any serious consideration of a physical theory must take into account the distinction between the objective reality, which is independent of any theory, and the physical concepts with which the theory operates. These concepts are intended to correspond with the objective reality, and by means of these concepts we picture this reality to ourselves.

In attempting to judge the success of a physical theory, we may ask ourselves two questions: (1) "Is the theory correct?" and (2) "Is the description given by the theory complete?" It is only in the case in which positive answers may be given to both of these questions, that the concepts of the theory may be said to be satisfactory. The correctness of the theory is judged by the degree of agreement between the conclusions of the theory and human experience. This experience, which alone enables us to make inferences about reality, in physics takes the form of experiment and measurement. It is the second question that we wish to consider here, as applied to quantum mechanics.

Whatever the meaning assigned to the term *complete*, the following requirement for a complete theory seems to be a necessary one: *every element of the physical reality must have a counterpart in the physical theory.* We shall call this the condition of completeness. The second question is thus easily answered, as soon as we are able to decide what are the elements of the physical reality.

The elements of the physical reality cannot be determined by *a priori* philosophical considerations, but must be found by an appeal to results of experiments and measurements. A comprehensive definition of reality is, however, unnecessary for our purpose. We shall be satisfied with the following criterion, which we regard reasonable. *If, without in any way disturbing a system, we can predict with certainty (i.e. with probability equal to unity) the value of a physical quantity, then there exists an element of physical reality corresponding to this physical quantity.* It seems to us that this criterion, while far from exhausting all possible ways of recognizing a physical reality, at least provides us with one such way, whenever the conditions set down in it occur. Regarded not as a necessary, but merely as a sufficient, condition of reality, this criterion is in agreement with classical as well as quantum-mechanical ideas of reality.

To illustrate the ideas involved let us consider the quantum-mechanical description of the behavior of a particle having a single degree of freedom. The fundamental concept of the theory is the concept of *state*, which is supposed to be completely characterized by the wave function ψ, which is a function of the variables chosen to describe the particle's behavior. Corresponding to each physically observable quantity A there is an operator, which may be designated by the same letter.

If ψ is an eigenfunction of the operator A, that is, if

$$\psi' \equiv A\psi = a\psi, \tag{1}$$

where a is a number, then the physical quantity A has with certainty the value a whenever the particle is in the state given by ψ. In accordance with our criterion of reality, for a particle in the state given by ψ for which Eq. (1) holds, there is an element of physical reality corresponding to the physical quantity A. Let, for example,

$$\psi = e^{(2\pi i/h)p_o x},\tag{2}$$

where h is Planck's constant, p_o is some constant number, and x the independent variable. Since the operator corresponding to the momentum of the particle is

$$P = (h/2\pi\ i)\ \partial/\partial x,\tag{3}$$

we obtain

$$\psi' = p\psi = (h/2\pi\ i)\ \partial\ \psi/\partial x = p_o\ \psi.\tag{4}$$

Thus, in the state given by Eq. (2), the momentum has certainly the value p_o. It thus has meaning to say that the momentum of the particle in the state given by Eq. (2) is real.

On the other hand if Eq. (1) does not hold, we can no longer speak of the physical quantity A having a particular value. This is the case, for example, with the coordinate of the particle. The operator corresponding to it, say q, is the operator of multiplication by the independent variable. Thus,

$$q\psi = x\psi \neq a\psi\tag{5}$$

In accordance with quantum mechanics we can only say that the relative probability that a measurement of the coordinate will give a result lying between a and b is

$$P(a,b) = \int_a^b \overline{\psi}\psi dx = \int_a^b dx = b - a.\tag{6}$$

Since this probability is independent of a, but depends only upon the difference $b - a$, we see that all values of the coordinate are equally probable.

A definite value of the coordinate, for a particle in the state given by Eq. (2), is thus not predictable, but may be obtained only by a direct measurement. Such a measurement however disturbs the particle and thus alters its state. After the coordinate is determined, the particle will no longer be in the state given by Eq. (2). The usual conclusion from this in quantum mechanics is that *when the momentum of a particle is known, its coordinate has no physical reality.*

More generally, it is shown in quantum mechanics that, if the operators corresponding to two physical quantities, say A and B, do not commute, that is, if $AB \neq BA$, then the precise knowledge of one of them precludes such a knowledge of the other. Furthermore, any attempt to determine the latter experimentally will alter the state of the system in such a way as to destroy the knowledge of the first.

From this follows that either *(1) the quantum-mechanical description of reality given by the wave function is not complete* or *(2) when the operators corresponding to two physical quantities do not commute the two quantities cannot have simultaneous reality.* For if both of them had simultaneous reality—and thus definite values—these values would enter into the complete description, according to the condition of completeness. If then the wave function provided such a complete description of reality, it would contain these values; these would then be predictable. This not being the case, we are left with the alternatives stated.

In quantum mechanics it is usually assumed that the wave function *does* contain a complete description of the physical reality of the system in the state to which it corresponds. At first sight this assumption is entirely reasonable, for the information obtainable from a wave function seems to correspond exactly to what can be measured without altering the state of the system. We shall show, however, that this assumption, together with the criterion of reality given above, leads to a contradiction.

2

For this purpose let us suppose that we have two systems, I and II, which we permit to interact from the time $t = 0$ to $t = T$, after which time we suppose that there is no longer any interaction between the two parts. We suppose further that the states of the two systems before $t = 0$ were known. We can then calculate with the help of Schrödinger's equation the state of the combined system I + II at any subsequent time; in particular, for any $t > T$. Let us designate the corresponding wave function by Ψ. We cannot, however, calculate the state in which either one of the two systems is left after the interaction. This, according to quantum mechanics, can be done only with the help of further measurements, by a process known as the *reduction of the wave packet.* Let us consider the essentials of this process.

Let $a_1, a_2, a_3, \ldots$ be the eigenvalues of some physical quantity A pertaining to system I and $u_1(x_1), u_2(x_1), u_3(x_1), \ldots$ the corresponding eigenfunctions, where x_1 stands for the variables used to describe the first system. Then Ψ, considered as a function of x_1, can be expressed as

$$\Psi(x_1, x_2) = \sum_{n=1}^{\infty} \psi_n(x_2)\, u_n(x_1), \tag{7}$$

where x_2 stands for the variables used to describe the second system. Here $\psi_n(x_2)$ are to be regarded merely as the coefficients of the expansion of Ψ into a series of orthogonal functions $u_n(x_1)$. Suppose now that the quantity A is measured and it is found that it has the value a_k. It is then concluded that after the measurement the first system is left in the state given by the wave function $u_k(x_1)$. and that the second system is left in the state given by the wave function $\psi_k(x_2)$. This is the process of reduction of the wave packet; the wave packet given by the infinite series (7) is reduced to a single term $\psi_k(x_2)u_k(x_1)$.

The set of functions $u_n(x_1)$ is determined by the choice of the physical quantity A. If, instead of this, we had chosen another quantity, say B, having the eigenvalues b_1, b_2, b_3 . . . and eigenfunctions $v_1\,(x_1)$, $v_2\,(x_1)$ $v_3\,(x_1)$ . . . we should have obtained, instead of Eq. (7), the expansion

$$\Psi(x_1, x_2) = \sum_{s=1}^{\infty} \varphi_s(x_2)v_s(x_1), \qquad (8)$$

where φ_s's are the new coefficients. If now the quantity B is measured and is found to have the value b we conclude that after the measurement the first system is left in the state given by $v_r\,(x_1)$ and the second system is left in the state given by $\varphi\,(x_2)$.

We see therefore that, as a consequence of two different measurements performed upon the first system, the second system may be left in states with two different wave functions. On the other hand, since at the time of measurement the two systems no longer interact, no real change can take place in the second system in consequence of anything that may be done to the first system. This is, of course, merely a statement of what is meant by the absence of an interaction between the two systems. Thus, *it is possible to assign two different wave functions* (in our example ψ_k and φ_r) *to the same reality* (the second system after the interaction with the first).

Now, it may happen that the two wave functions, ψ_k *and* φ_r, are eigenfunctions of two non-commuting operators corresponding to some physical quantities P and Q, respectively. That this may actually be the case can best be shown by an example. Let us suppose that the two systems are two particles, and that

$$\Psi(x_1, x_2) = \int_{-\infty}^{\infty} e^{2\pi i/(h)(x_1-x_2+x_0)p}dp, \qquad (9)$$

where x_o is some constant. Let A be the momentum of the first particle; then, as we have seen in Eq. (4), its eigenfunctions will be

$$u_p(x_1) = e^{(2\pi i/h)px_1} \qquad (10)$$

corresponding to the eigenvalue p. Since we have here the case of a continuous spectrum, Eq. (7) will now be written

$$\Psi(x_1, x_2) = \int_{-\infty}^{\infty} \psi_p(x_2)u_p(x_1)dp, \qquad (11)$$

where

$$\psi_p(x_2) = e^{-(2\pi i/h)(x_2-x_0)p}. \qquad (12)$$

This ψ_p however is the eigenfunction of the operator

$$P = (h/2\pi i)\partial/\partial x_2, \qquad (13)$$

corresponding to the eigenvalue $-p$ of the momentum of the second particle. On the other hand, if B is the coordinate of the first particle, it has for eigenfunctions

$$v_x(x_1) = \delta(x_1 - x), \tag{14}$$

corresponding to the eigenvalue x, where $\delta(x_1 - x)$ is the well-known Dirac delta-function. Eq. (8) in this case becomes

$$\Psi(x_1, x_2) = \int_{-\infty}^{\infty} \varphi_x(x_2) v_x(x_1) dx, \tag{15}$$

where

$$\varphi_x(x_2) = \int_{-\infty}^{\infty} e^{(2\pi i/h)(x-x_2+x_0)p} dp = h\delta(x - x_2 + x_0). \tag{16}$$

This φ_x, however, is the eigenfunction of the operator

$$Q = x_2 \tag{17}$$

corresponding to the eigenvalue $x + x_0$ of the coordinate of the second particle. Since

$$PQ - QP = h/2\pi i, \tag{18}$$

we have shown that it is in general possible for ψ_k and φ_r to be eigenfunctions of two noncommuting operators, corresponding to physical quantities.

Returning now to the general case contemplated in Eqs. (7) and (8), we assume that ψ_x and φ_r are indeed eigenfunctions of some noncommuting operators P and Q, corresponding to the eigenvalues p_k and q_r, respectively. Thus, by measuring either A or B we are in a position to predict with certainty, and without in any way disturbing the second system, either the value of the quantity P (that is p_k) or the value of the quantity Q (that is q_r). In accordance with our criterion of reality, in the first case we must consider the quantity P as being an element of reality, in the second case the quantity Q is an element of reality. But, as we have seen, both wave functions ψ_k and φ_r belong to the same reality.

Previously we proved that either (1) the quantum-mechanical description of reality given by the wave function is not complete or (2) when the operators corresponding to two physical quantities do not commute the two quantities cannot have simultaneous reality. Starting then with the assumption that the wave function does give a complete description of the physical reality, we arrived at the conclusion that two physical quantities, with noncommuting operators, can have simultaneous reality. Thus the negation of (1) leads to the negation of the only other alternative (2). We are thus forced to conclude that the quantum-mechanical description of physical reality given by wave functions is not complete.

One could object to this conclusion on the grounds that our criterion of reality is not sufficiently restrictive. Indeed, one would not arrive at our conclusion if one insisted that two or more physical quantities can be regarded as simultaneous elements of reality *only when they can be simultaneously measured or predicted*. On this point of view, since either one or the other, but not both simultaneously, of the quantities P and Q can be predicted, they are not simultaneously real. This makes the reality of P and Q depend upon the process of measurement carried out on the first system, which does not disturb the second system in any way. No reasonable definition of reality could be expected to permit this.

While we have thus shown that the wave function does not provide a complete description of the physical reality, we left open the question of whether or not such a description exists. We believe, however, that such a theory is possible.

15
— Albert Einstein —

In 1936, a year after the EPR paper was published, Einstein wrote an essay entitled "Physics and Reality." He devoted part of that essay to Quantum Theory, again explaining his claim that the theory is incomplete. The reasons given for incompleteness are, first, the statistical (and therefore in principle incomplete) nature of its laws, and, second, its incompatibility with the Theory of Relativity. At times, in his critique of both Quantum Theory and its adherents, Einstein resorts to very strong language, an indication that the matter had become of great personal consequence to all involved. In sum, Einstein's position regarding Quantum Theory is perhaps best expressed in his assertion that "it is so very contrary to my scientific instinct that I cannot forego the search for a more complete conception." To his dying day, Einstein maintained the hope that his search would be fulfilled. The essay "Physics and Reality" originally appeared in the Journal of the Franklin Institute *and has been reprinted in several collections of Einstein's essays.*

THE ARGUMENT FOR INCOMPLETENESS

The theoretical physicists of our generation are expecting the erection of a new theoretical basis for physics which would make use of fundamental concepts greatly different from those of the field theory considered up to now. The reason is that it has been found necessary to use—for the mathematical representations of the so-called quantum phenomena—new sorts of methods of consideration.

While the failure of classical mechanics, as revealed by the theory of relativity, is connected with the finite speed of light (its avoidance of being ∞), it was discovered at the beginning of our century that there were other kinds of inconsistencies between deductions from mechanics and experimental facts, which inconsistencies are connected with the finite magnitude (the avoidance of being zero) of Planck's constant h. In particular, while molecular mechanics requires that both, heat content and (monochromatic) radiation density, of solid bodies should decrease in *proportion* to the decreasing absolute temperature, experience has shown that they decrease much more rapidly than the absolute temperature. For a theoretical explanation of this behavior it was necessary to assume that the energy of a mechanical system cannot assume any sort of value, but only certain discrete values whose mathematical expressions were always dependent upon Planck's constant h. Moreover, this conception was essential for the theory of the atom (Bohr's theory). For the transitions of these states into one another— with or without emission or absorption of radiation—no causal laws could be given, but only statistical ones; and, a similar conclusion holds for the radioactive decomposition of atoms, which decomposition was carefully investigated about the same time. For more than two decades physicists tried vainly to find a uniform interpretation of this "quantum character" of systems and phenomena. Such an attempt was successful about ten years ago, through the agency of two entirely different theoretical methods of attack. We owe one of these to Heisenberg and Dirac, and the other to de Broglie and Schrödinger. The mathematical equivalence of the two methods was soon recognized by Schrödinger. I shall try here to sketch the line of thought of de Broglie and Schrödinger, which lies closer to the physicist's method of thinking, and shall accompany the description with certain general considerations.

The question is first: How can one assign a discrete succession of energy value H_σ to a system specified in the sense of classical mechanics (the energy function is a given function of the coordinates q_r and the corresponding momenta p_r)? Planck's constant h relates the frequency H_σ/h to the energy values H_σ. It is therefore sufficient to give to the system a succession of discrete *frequency* values. This reminds us of the fact that in acoustics, a series of discrete frequency values is coordinated to a linear partial differential equation (if boundary values are given) namely the sinusoidal periodic solutions. In corresponding manner, Schrödinger set himself the task of coordinating a partial differential equation for a scalar function ψ to the given energy function ε (q_r, p_r), where the q_r and the time t are independent variables. In this he succeeded (for a complex function ψ) in such a manner that the theoretical values of the energy H_σ, as required by the statistical theory, actually resulted in a satisfactory manner from the periodic solution of the equation.

To be sure, it did not happen to be possible to associate a definite movement, in the sense of mechanics of material points, with a definite solution ψ (q_r, t) of the Schrödinger equation. This means that the ψ function does not determine, at any rate *exactly*, the story of the q_r as functions of the time t. According to Born, however, an interpretation of the physical meaning of the ψ functions was shown to be possible in the following manner: $\psi \bar{\psi}$ (the square of the absolute value of the complex function ψ) is the probability density at the point under consideration in the configuration-space of the q_r, at the time t. It is therefore possible to characterize the content of the Schrödinger equation in a manner, easy to be understood, but not quite accurate, as follows: it determines how the probability density of a statistical ensemble of systems varies in the configuration-space with the time. Briefly: the Schrödinger equation determines the alteration of the function ψ of the q_r with the time.

It must be mentioned that the result of this theory contains—as limiting values—the result of the particle mechanics if the wave-length encountered during the solution of the Schrödinger problem is everywhere so small that the potential energy varies by a practically infinitely small amount for a change of one wavelength in the configuration-space which, although large (in every dimension) in relation to the wave length, is small in relation to the practical dimensions of the configuration-space. Under these conditions it is possible to choose a function of ψ for an initial time t_0 in such a manner that it vanishes outside of the region G_0, and behaves, according to the Schrödinger equation, in such a manner that it retains this property—approximately at least—also for a latter time, but with the region G_0 having passed at that time t into another region G. In this manner one can, with a certain degree of approximation, speak of the motion of the region G as a whole, and one can approximate this motion by the motion of a point in the configuration-space. This motion then coincides with the motion which is required by the equations of classical mechanics.

Experiments on interference made with particle rays have given a brilliant proof that the wave character of phenomena of motion as assumed by the theory does, really, correspond to the facts. In addition to this, the theory succeeded, easily, in demonstrating the statistical laws of the transition of a system from one quantum condition to another under the action of external forces, which, from the standpoint of classical mechanics, appears as a miracle. The external forces were here represented by small additions of the potential energy as functions of the time. Now, while in classical mechanics, such additions can produce only correspondingly small alterations of the system, in the quantum mechanics they produce alterations of any magnitude however large, but with correspondingly small probability, a consequence in perfect harmony with experience. Even an understanding of the laws of radioactive decomposition, at least in their broad lines, was provided by the theory.

Probably never before has a theory been evolved which has given a key to the interpretation and calculation of such a heterogeneous group of phenomena of experience as has the Quantum Theory. In spite of this, however, I believe that the theory is apt to beguile us into error in our search for a uniform basis for physics, because, in my belief, it is an *incomplete* representation of real things, although it is the only one which can be built out of the fundamental concepts of force and material points (quantum corrections to classical mechanics). The incompleteness of the representation is the outcome of the statistical nature (incompleteness) of the laws. I will now justify this opinion.

I ask first: How far does the ψ function describe a real condition of a mechanical system? Let us assume the ψ_r to be the periodic solutions (put in the order of increasing energy values) of the Schrödinger equation. I shall leave open, for the time being, the question as to how far the individual ψ_r are *complete* descriptions of physical conditions. A system is first in the condition ψ_1 of the lowest energy ε_1. Then during a finite time a small disturbing force acts upon the system. At a later instant one obtains then from the Schrödinger equation a ψ function of the form

$$\psi = \Sigma c_r \psi_r$$

where the c_r are (complex) constants. If the ψ_r are "normalized," then $|c_1|$ is nearly equal to 1, $|c_2|$ etc. is small compared with 1. One may now ask: Does ψ describe a real condition of the system? If the answer is yes, then we can hardly do otherwise than ascribe to this condition with a definite energy ε, and, in particular, such an energy as exceeds ε_1 by a small amount (in any case $\varepsilon_1 < \varepsilon < \varepsilon_2$). Such an assumption is, however, at variance with the experiments on electron impact such as have been made by J. Franck and G. Hertz, if, in addition to this, one accepts Millikan's demonstration of the discrete nature of electricity. As a matter of fact, these experiments lead to the conclusion that energy values of a state lying between the quantum values do not exist. From this it follows that our function ψ does not in any way describe a homogeneous condition of the body, but represents rather a statistical description in which the c_r represent probabilities of the individual energy values. It seems to be clear, therefore, that the Born statistical interpretation of the Quantum Theory is the only possible one. The ψ function does not in any way describe a condition which could be that of a single system; it relates rather to many systems, to "an ensemble of systems" in the sense of statistical mechanics. If, except for certain special cases, the ψ function furnishes only *statistical* data concerning measurable magnitudes, the reason lies not only in the fact that *the operation of measuring* introduces unknown elements, which can be grasped only statistically, but because of the very fact that the ψ function does not, in any sense, describe the condition of one *single* system. The Schrödinger equation determines the time variations which are experienced by the ensemble of systems which may exist with or without external action on the single system.

Such an interpretation eliminates also the paradox recently demonstrated by myself and two collaborators, and which relates to the following problem.

Consider a mechanical system constituted of two partial systems A and B which have interaction with each other only during limited time. Let the ψ function before their interaction be given. Then the Schrödinger equation will furnish the ψ function after the interaction has taken place. Let us now determine the physical condition of the partial system A as completely as possible by measurements. Then the quantum mechanics allows us to determine the ψ function of the partial system B from the measurements made, and from the ψ function of the total system. This determination, however, gives a result which depends upon *which* of the determining magnitudes specifying the condition of A has been measured (for instance coordinates *or* momenta). Since there can be only *one* physical condition of B after the interaction and which can reasonably not be considered as dependent on the particular measurement we perform on the system A separated from B it may be concluded that the ψ function is not unambiguously coordinated with the physical condition. This coordination of several ψ functions with the same physical condition of system B shows again that the ψ function cannot be interpreted as a (complete) description of a physical condition of a unit system. Here also the coordination of the ψ function to an ensemble of systems eliminates every difficulty.

The fact that quantum mechanics affords, in such a simple manner, statements concerning (apparently) discontinuous transitions from one total condition to another without actually giving a representation of the specific process, this fact is connected with another, namely the fact that the theory, in reality, does not operate with the single system, but with a totality of systems. Their coefficients c_r of our first example are really altered very little under the action of the external force. With this interpretation of quantum mechanics one can understand why this theory can easily account for the fact that weak disturbing forces are able to produce alterations of any magnitude in the physical condition of a system. Such disturbing forces produce, indeed, only correspondingly small alterations of the *statistical density* in the ensemble of systems, and hence only infinitely weak alterations of the ψ functions, the mathematical description of which offers far less difficulty than would be involved in the mathematical representation of finite alterations experience by part of the single systems. What happens to the single system remains, it is true, entirely unclarified by this mode of consideration; this enigmatic happening is entirely eliminated from the representation by the statistical manner of consideration.

But now I ask: Is there really any physicist who believes that we shall never get any inside view of these important alterations in the single systems, in their structure and their causal connections, and this regardless of the fact that these single happenings have been brought so close to us, thanks to the marvelous inventions of the Wilson chamber and the Geiger counter? To believe this is logically possible without contradiction; but, it is so very contrary to my scientific instinct that I cannot forego the search for a more complete conception.

To these considerations we should add those of another kind which also voice their plea against the idea that the methods introduced by quantum mechanics are likely to give a useful basis for the whole of physics. In the Schrödinger equation, absolute time, and also the potential energy, play a decisive role, while these two concepts have been recognized by the theory of relativity as inadmissible in principle. If one wishes to escape from this difficulty he must found the theory upon field and field laws instead of upon forces of interaction. This leads us to transpose the statistical methods of quantum mechanics to fields, that is to systems of infinitely many degrees of freedom. Although the attempts so far made are restricted to linear equations, which, as we know from the results of the general theory of relativity, are insufficient, the complications met up to now by the very ingenious attempts are already terrifying. They certainly will risk sky high if one wishes to obey the requirements of the general theory of relativity, the justification of which in principle nobody doubts.

To be sure, it has been pointed out that the introduction of a space-time continuum may be considered as contrary to nature in view of the molecular structure of everything which happens on a small scale. It is maintained that perhaps the success of the Heisenberg method points to a purely algebraical method of description of nature, that is to the elimination of continuous functions from physics. Then, however, we must also give up, by principle, the space-time continuum. It is not unimaginable that human ingenuity will some day find methods which will make it possible to proceed along such a path. At the present time, however, such a program looks like an attempt to breathe in empty space.

There is no doubt that quantum mechanics has seized hold of a beautiful element of truth, and that it will be a test stone for any future theoretical basis, in that it must be deducible as a limiting case from that basis, just as electrostatics is deducible from the Maxwell equations of the electromagnetic field or as thermodynamics is deducible from classical mechanics. However, I do not believe that quantum mechanics will be the *starting point* in the search for this basis, just as, vice versa, one could not go from thermodynamics (esp. statistical mechanics) to the foundations of mechanics.

In view of this situation, it seems to be entirely justifiable seriously to consider the question as to whether the basis of field physics cannot by *any* means be put into harmony with the facts of the Quantum Theory. Is this not the only basis which, consistently with today's possibility of mathematical expression, can be adapted to the requirements of the general theory of relativity? The belief, prevailing among the physicists of today, that such an attempt would be hopeless, may have its root in the unjustifiable idea that such a theory should lead, as a first approximation, to the equations of classical mechanics for the motion of corpuscles, or at least to total differential equations. As a matter of fact up to now we have never succeeded in representing corpuscles theoretically by fields free of singularities, and we can, a priori, say nothing about the behavior of such entities. *One thing*, however, is certain: if a field theory results in a representation of corpuscles free of singularities, then the behavior of these corpuscles with time is determined solely by the differential equations of the field.

16
— Niels Bohr —

In this reading, written in direct response to the earlier article by Einstein, Podolsky, and Rosen, Bohr defends Quantum Theory against the charge of incompleteness. In maintaining the mathematical completeness of Quantum Theory, he criticizes the EPR paper for using an inappropriate analogy between quantum mechanics and classical mechanics and for using an ambiguous criterion for physical reality. According to Bohr, it was this ambiguity that led to EPR's failure to recognize the influence of experimental conditions upon future predictions—as stipulated by the measurement problem. The recognition and influence of such conditions, Bohr argues, is an inherent feature of the quantum situation and comes to bear on any and all descriptions of physical reality at the quantum level. Therefore, according to Bohr, the charge of incompleteness is both misleading and inappropriate.

This reading comes from an article entitled "Can Quantum Mechanical Description of Physical Reality Be Considered Complete?" It appeared in the issue of Physical Review *that followed that presented the original EPR paper.*

RESPONSE TO EPR

In a recent article under the above title A. Einstein, B. Podolsky and N. Rosen have presented arguments which lead them to answer the question at issue in the negative. The trend of their argumentation, however, does not seem to me adequately to meet the actual situation with which we are faced in atomic physics. I shall therefore be glad to use this opportunity to explain in somewhat greater detail a general viewpoint, conveniently termed "complementarity," which I have indicated on various previous occasions, and from which quantum mechanics within its scope would appear as a completely rational description of physical phenomena, such as we meet in atomic processes.

The extent to which an unambiguous meaning can be attributed to such an expression as "physical reality" cannot of course be deduced from *a priori* philosophical conceptions, but—as the authors the article cited themselves emphasize—must be founded on a direct appeal to experiments and measurements. For this purpose they propose a "criterion of reality" formulated as follows: "If, without in any way disturbing a system, we can predict with certainty the value of a physical quantity, then there exists an element of physical reality corresponding to this physical quantity." By means of an interesting example, to which we shall return below, they next proceed to show that in quantum mechanics, just as in classical mechanics, it is possible under suitable conditions to predict the value of any given variable pertaining to the description of a mechanical system from measurements performed entirely on other systems which previously have been in interaction with the system under investigation. According to their criterion the authors therefore want to ascribe an element of reality to each of the quantities represented by such variables. Since, moreover, it is a well-known feature of the present formalism of quantum mechanics that it is never possible, in the description of the state of a mechanical system, to attach definite values to both of two canonically conjugate variables, they consequently deem this formalism to be incomplete, and express the belief that a more satisfactory theory can be developed.

Such an argumentation, however, would hardly seem suited to affect the soundness of quantum-mechanical description, which is based on a coherent mathematical formalism covering automatically any procedure of measurement like that indicated. The apparent contradiction in fact discloses only an essential inadequacy of the customary viewpoint of natural philosophy for a rational account of physical phenomena of the type with which we are concerned in quantum mechanics. Indeed the *finite interaction between object and measuring agencies* conditioned by the very existence of the quantum of action entails—because of the impossibility of controlling the reaction of the object on the measuring instruments if these are to serve their purpose—the necessity of a final renunciation of the classical ideal of causality and a radical revision of our attitude towards the problem of physical reality. In fact, as we shall see, a criterion of reality like that proposed by the named authors contains—however cautious its formulation may appear—an essential ambiguity when it is applied to the actual problems with which we are here concerned. In order to make the argument to this end as clear as possible, I shall first consider in some detail a few simple examples of measuring arrangements.

Let us begin with the simple case of a particle passing through a slit in a diaphragm, which may form part of some more or less complicated experimental arrangement. Even if the momentum of this particle is completely known before it impinges on the diaphragm, the diffraction by the slit of the plane wave giving the symbolic representation of its state will imply an uncertainty in the momentum of the particle, after it has passed the diaphragm, which is the greater the narrower the slit. Now the width of the slit, at any rate if it is still large compared with the wave-length, may be taken as the uncertainty Δq of the position of the particle relative to the diaphragm, in a direction perpendicular to the slit. Moreover, it is simply seen from de Broglie's relation between momentum and wavelength that the uncertainty Δp of the momentum of the particle in this direction is correlated to Δq by means of Heisenberg's general principle

$$\Delta p \Delta q \sim h,$$

which in the quantum-mechanical formalism is a direct consequence of the commutation relation for any pair of conjugate variables. Obviously the uncertainty Δp is inseparably connected with the possibility of an exchange of momentum between the particle and the diaphragm; and the question of principal interest for our discussion is now to what extent the momentum thus exchanged can be taken into account in the description of the phenomenon to be studied by the experimental arrangement concerned, of which the passing of the particle through the slit may be considered as the initial stage.

Let us first assume that, corresponding to usual experiments on the remarkable phenomena of electron diffraction, the diaphragm, like the other parts of the apparatus,—say a second diaphragm with several slits parallel to the first and a photographic plate,—is rigidly fixed to a support which defines the space frame of reference. Then the momentum exchanged between the particle and the diaphragm will, together with the reaction of the particle on the other bodies, pass into this common support, and we have thus voluntarily cut ourselves off from any possibility of taking these reactions separately into account in predictions regarding the final result of the experiment,—say the position of the spot produced by the particle on the photographic plate. The impossibility of a closer analysis of the reactions between the particle and the measuring instrument is indeed no peculiarity of the experimental procedure described, but is rather an essential property of any arrangement suited to the study of the phenomena of the type concerned, where we have to do with a feature of *individuality* completely foreign to classical physics. In fact, any possibility of taking into account the momentum exchanged between the particle and the separate parts of the apparatus would at once permit us to draw conclusions regarding the "course" of such phenomena,—say through what particular slit of the second diaphragm the particle passes on its way to the photographic plate—which would be quite incompatible with the fact that the probability of the particle reaching a given element of area on this plate is determined not by the presence of any particular slit, but by the positions of all the slits of the second diaphragm within reach of the associated wave diffracted from the slit of the first diaphragm.

By another experimental arrangement, where the first diaphragm is not rigidly connected with the other parts of the apparatus, it would at least in principle be possible to measure its momentum with any desired accuracy before and after the passage of the particle, and thus to predict the momentum of the latter after it has passed through the slit. In fact, such measurements of momentum require only an unambiguous application of the classical law of conservation of momentum, applied for instance to a collision process between the diaphragm and some test body, the momentum of which is suitably controlled before and after the collision. It is true that such a control will essentially depend on an examination of the space-time course of some process to which the ideas of classical mechanics can be applied; if, however, all spatial dimensions and time intervals are taken sufficiently large, this involves clearly no limitation as regards the accurate control of the momentum of the test bodies, but only a renunciation as regards the accuracy of the control of their space-time coordination. This last circumstance is in fact quite analogous to the renunciation of the control of the momentum of the fixed diaphragm in the experimental arrangement discussed above, and depends in the last resort on the claim of a purely classical account of the measuring apparatus, which implies the necessity of allowing a latitude corresponding to the quantum-mechanical uncertainty relations in our description of their behavior.

The principal difference between the two experimental arrangements under consideration is, however, that in the arrangement suited for the control of the momentum of the first diaphragm, this body can no longer be used as a measuring instrument for the same purpose as in the previous case, but must, as regards its position relative to the rest of the apparatus, be treated, like the particle traversing the slit, as an object of investigation, in the sense that the quantum-mechanical uncertainty relations regarding its position and momentum must be taken explicitly into account. In fact, even if we knew the position of the diaphragm relative to the space frame before the first measurement of its momentum, and even though its position after the last measurement can be accurately fixed, we lose, on account of the uncontrollable displacement of the diaphragm during each collision process with the test bodies, the knowledge of its position when the particle passed through the slit. The whole arrangement is therefore obviously unsuited to study the same kind of phenomena as in the previous case. In particular it may be shown that, if the momentum of the diaphragm is measured with an accuracy sufficient for allowing definite conclusions regarding the passage of the particle through some selected slit of the second diaphragm, then even the minimum uncertainty of the position of the first diaphragm compatible with such a knowledge will imply the total wiping out of any interference effect—regarding the zones of permitted impact of the particle on the photographic plate—to which the presence of more than one slit in the second diaphragm would give rise in case the positions of all apparatus are fixed relative to each other.

In an arrangement suited for measurements of the momentum of the first diaphragm, it is further clear that even if we have measured this momentum before the passage of the particle through the slit, we are after this passage still left with a *free choice* whether we wish to know the momentum of the particle or its initial position relative to the rest of the apparatus. In the first eventuality we need only to make a second determination of the momentum of the diaphragm, leaving unknown forever its exact position when the particle passed. In the second eventuality we need only to determine its position relative to the space frame with the inevitable loss of the knowledge of the momentum exchanged between the diaphragm and the particle. If the diaphragm is sufficiently massive in comparison with the particle, we may even arrange the procedure of measurements in such a way that the diaphragm after the first determination of its momentum will remain at rest in some unknown position relative to the other parts of the apparatus, and the subsequent fixation of this position may therefore simply consist in establishing a rigid connection between the diaphragm and the common support.

My main purpose in repeating these simple, and in substance well-known considerations, is to emphasize that in the phenomena concerned we are not dealing with an incomplete description characterized by the arbitrary picking out of different elements of physical reality at the cost of sacrificing other such elements, but with a rational discrimination between essentially different experimental arrangements and procedures which are suited either for an unambiguous use of the idea of space location, or for a legitimate application of the conservation theorem of momentum. Any remaining appearance of arbitrariness concerns merely our freedom of handling the measuring instruments, characteristic of the very idea of experiment. In fact, the renunciation in each experimental arrangement of the one or the other of two aspects of the description of physical phenomena,—the combination of which characterizes the method of classical physics, and which therefore in this sense may be considered as *complementary* to one another,—depends essentially on the impossibility, in the field of Quantum Theory, of accurately controlling the reaction of the object on the measuring instruments, i.e., the transfer of momentum in case of position measurements, and the displacement in case of momentum measurements. Just in this last respect any comparison between quantum mechanics and ordinary statistical mechanics,—however useful it may be for the formal presentation of the theory,—is essentially irrelevant. Indeed we have in each experimental arrangement suited for the study of proper quantum phenomena not merely to do with an ignorance of the value of certain physical quantities, but with the impossibility of defining these quantities in an unambiguous way.

The last remarks apply equally well to the special problem treated by Einstein, Podolsky and Rosen, which has been referred to above, and which does not actually involve any greater intricacies than the simple examples discussed above. The particular quantum-mechanical state of two free particles, for which they give an explicit mathematical expression, may be reproduced, at least in principle, by a simple experimental arrangement, comprising a rigid diaphragm with two parallel slits, which are very narrow compared with their separation, and through each of which one particle with given initial momentum passes independently of the other. If the momentum of this diaphragm is measured accurately before as well as after the passing of the particles, we shall in fact know the sum of the components perpendicular to the slits of the momenta of the two escaping particles, as well as the difference of their initial positional coordinates in the same direction; while of course the conjugate quantities, i.e., the difference of the components of their momenta, and the sum of their positional coordinates, are entirely unknown. In this arrangement, it is therefore clear that a subsequent single measurement either of the position or of the momentum of one of the particles will automatically determine the position or momentum, respectively, of the other particle with any desired accuracy; at least if the wave-length corresponding to the free motion of each particle is sufficiently short compared with the width of the slits. As pointed out by the named authors, we are therefore faced at this stage with a completely free choice whether we want to determine the one or the other of the latter quantities by a process which does not directly interfere with the particle concerned.

Like the above simple case of the choice between the experimental procedures suited for the prediction of the position or the momentum of a single particle which has passed through a slit in a diaphragm, we are, in the "freedom of choice" offered by the last arrangement, just concerned with a *discrimination between different experimental procedures which allow of the unambiguous use of complementary classical concepts.* In fact to measure the position of one of the particles can mean nothing else than to establish a correlation between its behavior and some instrument rigidly fixed to the support which defines the space frame of reference. Under the experimental conditions described such a measurement will therefore also provide us with the knowledge of the location, otherwise completely unknown, of the diaphragm with respect to this space frame when the particles passed through the slits. Indeed, only in this way we obtain a basis for conclusions about the initial position of the other particle relative to the rest of the apparatus. By allowing an essentially uncontrollable momentum to pass from the first particle into the mentioned support, however, we have by this procedure cut ourselves off from any future possibility of applying the law of conservation of momentum to the system consisting of the diaphragm and the two particles and therefore have lost our only basis for an unambiguous application of the idea of momentum in predictions regarding the behavior of the second particle. Conversely, if we choose to measure the momentum of one of the particles, we lose through the uncontrollable displacement inevitable in such a measurement any possibility of deducing from the behavior of this particle the position of the diaphragm relative to the rest of the apparatus, and have thus no basis whatever for predictions regarding the location of the other particle.

From our point of view we now see that the wording of the above-mentioned criterion of physical reality proposed by Einstein, Podolsky and Rosen contains an ambiguity as regards the meaning of the expression "without in any way disturbing a system." Of course there is in a case like that just considered no question of a mechanical disturbance of the system under investigation during the last critical stage of the measuring procedure. But even at this stage there is essentially the question of *an influence on the very conditions which define the possible types of predictions regarding the future behavior of the system.* Since these conditions constitute an inherent element of the description of any phenomenon to which the term "physical reality" can be properly attached, we see that the argumentation of the mentioned authors does not justify their conclusion that quantum-mechanical description is essentially incomplete. On the contrary this description, as appears from the preceding discussion, may be characterized as a rational utilization of all possibilities of unambiguous interpretation of measurement, compatible with the finite and uncontrollable interaction between the objects and the measuring instruments in the field of Quantum Theory. In fact, it is only the mutual exclusion of any two experimental procedures, permitting the unambiguous definition of complementary physical quantities, which provides room for new physical laws, the coexistence of which might at first sight appear irreconcilable with the basic principles of science. It is just this entirely new situation as regards the description of physical phenomena, that the notion of *complementarity* aims at characterizing.

The experimental arrangements hitherto discussed present a special simplicity on account of the secondary role which the idea of time plays in the description of the phenomena in question. It is true that we have freely made use of such words as "before" and "after" implying time-relationships; but in each case allowance must be made for a certain inaccuracy, which is of no importance, however, so long as the time intervals concerned are sufficiently large compared with the proper periods entering in the closer analysis of the phenomenon under investigation. As soon as we attempt a more accurate time description of quantum phenomena, we meet with well-known new paradoxes, for the elucidation of which further features of the interaction between the objects and the measuring instruments must be taken into account. In fact, in such phenomena we have no longer to do with experimental arrangements consisting of apparatus essentially at rest relative to one another, but with arrangements containing moving parts,—like shutters before the slits of the diaphragms,—controlled by mechanisms serving as clocks. Besides the transfer of momentum, discussed above, between the object and the bodies defining the space frame, we shall therefore, in such arrangements, have to consider an eventual exchange of energy between the object and these clock-like mechanisms.

The decisive point as regards time measurements in Quantum Theory is now completely analogous to the argument concerning measurements of positions outlined above. Just as the transfer of momentum to the separate parts of the apparatus,—the knowledge of the relative positions of which is required for the description of the phenomenon—has been seen to be entirely uncontrollable, so the exchange of energy between the object and the various bodies, whose relative motion must be known for the intended use of the apparatus, will defy any closer analysis. Indeed it is *excluded in principle to control the energy which goes into the clocks without interfering essentially with their use as time indicators*. This use in fact entirely relies on the assumed possibility of accounting for the functioning of each clock as well as for its eventual comparison with other clocks on the basis of the methods of classical physics. In this account we must therefore obviously allow for a latitude in the energy balance, corresponding to the quantum-mechanical uncertainty relation for the conjugate time and energy variables. Just as in the question discussed above of the mutually exclusive character of any unambiguous use in Quantum Theory of the concepts of position and momentum, it is in the last resort this circumstance which entails the complementary relationship between any detailed time account of atomic phenomena on the one hand and the unclassical features of intrinsic stability of atoms, disclosed by the study of energy transfers in atomic reactions on the other hand.

This necessity of discriminating in each experimental arrangement between those parts of the physical system considered which are to be treated as measuring instruments and those which constitute the objects under investigation may indeed be said to form a *principal distinction between classical and quantum-mechanical description of physical phenomena*. It is true that the place within each measuring procedure where this discrimination is made is in both cases largely a matter of convenience. While, however, in classical physics the distinction between object and measuring agencies does not entail any difference in the character of the description of the phenomena concerned, its fundamental importance in Quantum Theory, as we have seen, has its root in the indispensable use of classical concepts in the interpretation of all proper measurements, even though the classical theories do not suffice in accounting for the new types of regularities with which we are concerned in atomic physics. In accordance with this situation there can be no question of any unambiguous interpretation of the symbols of quantum mechanics other than that embodied in the well-known rules which allow to predict the results to be obtained by a given experimental arrangement described in a totally classical way, and which have found their general expression through the transformation theorems, already referred to. By securing its proper correspondence with the classical theory, these theorems exclude in particular any imaginable inconsistency in the quantum-mechanical description, connected with a change of the place where the discrimination is made between object and measuring agencies. In fact it is an obvious consequence of the above argumentation that in each experimental arrangement and measuring procedure we have only a free choice of this place within a region where the quantum-mechanical description of the process concerned is effectively equivalent with the classical description.

Before concluding I should still like to emphasize the bearing of the great lesson derived from general relativity theory upon the question of physical reality in the field of Quantum Theory. In fact, notwithstanding all characteristic differences, the situations we are concerned with in these generalizations of classical theory present striking analogies which have often been noted. Especially, the singular position of measuring instruments in the account of quantum phenomena, just discussed, appears closely analogous to the well-known necessity in relativity theory of upholding an ordinary description of all measuring processes, including a sharp distinction between space and time coordinates, although the very essence of this theory is the establishment of new physical laws, in the comprehension of which we must renounce the customary separation of space and time ideas. The dependence on the reference system, in relativity theory, of all readings of scales and clocks may even be compared with the essentially uncontrollable exchange of momentum or energy between the objects of measurements and all instruments defining the space-time system of reference, which in Quantum Theory confronts us with the situation characterized by the notion of complementarity. In fact this new feature of natural philosophy means a radical revision of our attitude as regards physical reality, which may be paralleled with the fundamental modification of all ideas regarding the absolute character of physical phenomena, brought about by the general Theory of Relativity.

17
— D a v i d   B o h m —
(1917–1992)

David Bohm was an American-born physicist who studied at The University of California, Berkeley, under J. Robert Oppenheimer and who learned Quantum Theory in Copenhagen from Niels Bohr. He taught at Princeton University before leaving the United States and settling in England in the late 1950s.

Bohm's primary contribution to Quantum Theory lies in his attempt to construct an ordinary reality interpretation in terms of "hidden variables." Bohm considered this interpretation of Quantum Theory to be a significant advance in the development of the theory, one that might appease the many critics who, having aligned themselves with Einstein, still maintained that Quantum Theory is incomplete. This claim as to the possible existence of "hidden variables" would serve as a springboard for future analysis and experimentation by people like Bell and Aspect, who would be instrumental in finally closing the book on the completeness debate.

In this reading Bohm describes the "hidden variables" hypothesis which is at the heart of his ordinary reality interpretation. It originally appeared as two separate articles, published together in the January 1952 edition of Physical Review *under the title "A Suggested Interpretation of the Quantum Theory in Terms of Hidden Variables, I and II."*

THE HIDDEN VARIABLES HYPOTHESIS

I

The usual interpretation of the Quantum Theory is based on an assumption having very far-reaching implications, viz., that the physical state of an individual system is completely specified by a wave function that determines only the probabilities of actual results that can be obtained in a statistical ensemble of similar experiments. This assumption has been the object of severe criticisms, notably on the part of Einstein, who has always believed that, even at the quantum level, there must exist precisely definable elements or dynamical variables determining (as in classical physics) the actual behavior of each individual system, and not merely its probable behavior. Since these elements or variables are not now included in the quantum theory and have not yet been detected experimentally, Einstein has always regarded the present form of the Quantum Theory as incomplete, although he admits its internal consistency.

Most physicists have felt that objections such as those raised by Einstein are not relevant, first, because the present form of the Quantum Theory with its usual probability interpretation is in excellent agreement with an extremely wide range of experiments, at least in the domain of distances larger than 10^{-18} cm, and, secondly, because no consistent alternative interpretations have as yet been suggested. The purpose of this paper (and of a subsequent paper hereafter denoted by II) is, however, to suggest just such an alternative interpretation. In contrast to the usual interpretation, this alternative interpretation permits us to conceive of each individual system as being in a precisely definable state, whose changes with time are determined by definite laws, analogous to (but not identical with) the classical equations of motion. Quantum-mechanical probabilities are regarded (like their counterparts in classical statistical mechanics) as only a practical necessity and not as a manifestation of an inherent lack of complete determination in the properties of matter at the quantum level. As long as the present general form of Schrödinger's equation is retained, the physical results obtained with our suggested alternative interpretation are precisely the same as those obtained with the usual interpretation. We shall see, however, that our alternative interpretation permits modifications of the mathematical formulation which could not even be described in terms of the usual interpretation. Moreover, the modifications can quite easily be formulated in such a way that their effects are insignificant in the atomic domain, where the present Quantum Theory is in such good agreement with experiment, but of crucial importance in the domain of dimensions of the order of 10^{-18}cm, where, as we have seen, the present theory is totally inadequate. It is thus entirely possible that some of the modifications describable in terms of our suggested alternative interpretation, but not in terms of the usual interpretation, may be needed for a more thorough understanding of phenomena associated with very small distances. We shall not, however, actually develop such modifications in any detail in these papers. . . .

The Usual Physical Interpretation of Quantum Theory

The usual physical interpretation of the Quantum Theory centers around the Uncertainty Principle. Now, the Uncertainty Principle can be derived in two different ways. First, we may start with the assumption already criticized by Einstein, namely, that a wave function that determines only probabilities of actual experimental results nevertheless provides the most complete possible specification of the so-called "quantum state" of an individual system. With the aid of this assumption and with the aid of the de Broglie relation, $p = hk$, where k is the wave number associated with a particular fourier component of the wave function, the Uncertainty Principle is readily deduced. From this derivation, we are led to interpret the Uncertainty Principle as an inherent and irreducible limitation on the precision with which it is correct for us even to conceive of momentum and position as simultaneously defined quantities. For if, as is done in the usual interpretation of the Quantum Theory, the wave intensity is assumed to determine only the probability of a given position, and if the kth Fourier component of the wave function is assumed to determine only the probability of a corresponding momentum, $p = hk$, then it becomes a contradiction in terms to ask for a state in which momentum and position are simultaneously and precisely defined.

A second possible derivation of the Uncertainty Principle is based on a theoretical analysis of the processes with the aid of which physically significant quantities such as momentum and position can be measured. In such an analysis, one finds that because the measuring apparatus interacts with the observed system by means of indivisible quanta, there will always be an irreducible disturbance of some observed property of the system. If the precise effects of this disturbance could be predicted or controlled, then one could correct for these effects, and thus one could still in principle obtain simultaneous measurements of momentum and position, having unlimited precision. But if one could do this, then the Uncertainty Principle would be violated. The Uncertainty Principle is, as we have seen, however, a necessary consequence of the assumption that the wave function and its probability interpretation provide the most complete possible specification of the state of an individual system. In order to avoid the possibility of a contradiction with this assumption, Bohr and others have suggested an additional assumption, namely, that the process of transfer of a single quantum from observed system to measuring apparatus is inherently unpredictable, uncontrollable, and not subject to a detailed rational analysis or description. With the aid of this assumption, one can show that the same uncertainty principle that is deduced from the wave function and its probability interpretation is also obtained as an inherent and unavoidable limitation on the precision of all possible measurements. Thus, one is able to obtain a set of assumptions which permit a self-consistent formulation of the usual interpretation of the Quantum Theory.

The above point of view has been given its most consistent and systematic expression by Bohr, in terms of the "principle of complementarity." In formulating this principle, Bohr suggests that at the atomic level we must renounce our hitherto successful practice of conceiving of an individual system as a unified and precisely definable whole, all of whose aspects are, in a manner of speaking, simultaneously and unambiguously accessible to our conceptual gaze. Such a system of concepts, which is sometimes called a "model," need not be restricted to pictures, but may also include, for example, mathematical concepts, as long as these are supposed to be in a precise (i.e., one-to-one) correspondence with the objects that are being described. The principle of complementarity requires us, however, to renounce even mathematical models. Thus, in Bohr's point of view, the wave function is in no sense a conceptual model of an individual system, since it is not in a precise (one-to-one) correspondence with the behavior of this system, but only in a statistical correspondence.

In place of a precisely defined conceptual model, the principle of complementarity states that we are restricted to complementarity pairs of inherently imprecisely defined concepts, such as position and momentum, particle and wave, etc. The maximum degree of precision of definition of either member of such a pair is reciprocally related to that of the opposite member. This need for an inherent lack of complete precision can be understood in two ways. First, it can be regarded as a consequence of the fact that the experimental apparatus needed for a precise measurement of one member of a complementary pair of variables must always be such as to preclude the possibility of a simultaneous and precise measurement of the other member. Secondly, the assumption that an individual system is completely specified by the wave function and its probability interpretation implies a corresponding unavoidable lack of precision in the very conceptual structure, with the aid of which we can think about and describe the behavior of the system.

It is only at the classical level that we can correctly neglect the inherent lack of precision in all of our conceptual models; for here, the incomplete determination of physical properties implied by the Uncertainty Principle produces effects that are too small to be of practical significance. Our ability to describe classical systems in terms of precisely definable models is, however, an integral part of the usual interpretation of the theory. For without such models, we would have no way to describe, or even to think of, the result of an observation, which is of course always finally carried out at a classical level of accuracy. If the relationships of a given set of classically describable phenomena depend significantly on the essentially quantum-mechanical properties of matter, however, then the principle of complementarity states that no single model is possible which could provide a precise and rational analysis of the connections between these phenomena. In such a case, we are not supposed, for example, to attempt to describe in detail how future phenomena arise out of past phenomena. Instead, we should simply accept without further analysis the fact that future phenomena do in fact somehow manage to be produced, in a way that is, however, necessarily beyond the possibility of a detailed description. The only aim of a mathematical theory is then to predict the statistical relations, if any, connecting these phenomena.

Criticism of the Usual Interpretation of the Quantum Theory

The usual interpretation of the Quantum Theory can be criticized on many grounds. In this paper, however, we shall stress only the fact that it requires us to give up the possibility of even conceiving precisely what might determine the behavior of an individual system at the quantum level, without providing adequate proof that such a renunciation is necessary. The usual interpretation is admittedly consistent; but the mere demonstration of such consistency does not exclude the possibility of other equally consistent interpretations, which would involve additional elements or parameters permitting a detailed causal and continuous description of all processes, and not requiring us to forego the possibility of conceiving the quantum level in precise terms. From the point of view of the usual interpretation, these additional elements or parameters could be called "hidden" variables. As a matter of fact, whenever we have previously had recourse to statistical theories, we have always ultimately found that the laws governing the individual members of a statistical ensemble could be expressed in terms of just such hidden variables. For example, from the point of view of macroscopic physics, the coordinates and momenta of individual atoms are hidden variables, which in a large scale system manifest themselves only as statistical averages. Perhaps then, our present quantum-mechanical averages are similarly a manifestation of hidden variables, which have not, however, yet been detected directly.

Now it may be asked why these hidden variables should have so long remained undetected. To answer this question, it is helpful to consider as an analogy the early forms of the atomic theory, in which the existence of atoms was postulated in order to explain certain large-scale effects, such as the laws of chemical combination, the gas laws, etc. On the other hand, these same effects could also be described directly in terms of existing macrophysical concepts (such as pressure, volume, temperature, mass, etc.); and a correct description in these terms did not require any reference to atoms. Ultimately, however, effects were found which contradicted the predictions obtained by extrapolating certain purely macrophysical theories to the domain of the very small, and which could be understood correctly in terms of the assumption that matter is composed of atoms. Similarly, we suggest that if there are hidden variables underlying the present Quantum Theory, it is quite likely that in the atomic domain, they will lead to effects that can also be described adequately in the terms of the usual quantum-mechanical concepts; while in a domain associated with much smaller dimensions, such as the level associated with the "fundamental length" of the order of 10^{-18}cm, the hidden variables may lead to completely new effects not consistent with the extrapolation of the present quantum theory down to this level.

If, as is certainly entirely possible, these hidden variables are actually needed for a correct description at small distances, we could easily be kept on the wrong track for a long time by restricting ourselves to the usual interpretation of the Quantum Theory, which excludes such hidden variables as a matter of principle. It is therefore very important for us to investigate our reasons for supposing that the usual physical interpretation is likely to be the correct one. To this end, we shall begin by repeating the two mutually consistent assumptions on which the usual interpretation is based (see Sec. 2):

1. The wave function with its probability interpretation determines the most complete possible specification of the state of an individual system.
2. The process of transfer of a single quantum from observed system to measuring apparatus is inherently unpredictable, uncontrollable, and unanalyzable.

Let us now inquire into the question of whether there are any experiments that could conceivably provide a test for these assumptions. It is often stated in connection with this problem that the mathematical apparatus of the Quantum Theory and its physical interpretation form a consistent whole and that this combined system of mathematical apparatus and physical interpretation is tested adequately by the extremely wide range of experiments that are in agreement with predictions obtained by using this system. If assumptions (1) and (2) implied a unique mathematical formulation, then such a conclusion would be valid, because experimental predictions could then be found which, if contradicted, would clearly indicate that these assumptions were wrong. Although assumptions (1) and (2) do limit the possible forms of the mathematical theory, they do not limit these forms sufficiently to make possible a unique set of predictions that could in principle permit such an experimental test. Thus, one can contemplate practically arbitrary changes in the Hamiltonian operator, including, for example, the postulation of an unlimited range of new kinds of meson fields each having almost any conceivable rest mass, charge, spin, magnetic moment, etc. And if such postulates should prove to be inadequate, it is conceivable that we may have to introduce nonlocal operators, nonlinear fields, S-matrices, etc. This means that when the theory is found to be inadequate (as now happens, for example, at distances of the order of 10^{-18}cm), it is always possible, and, in fact, usually quite natural, to assume that the theory can be made to agree with experiment by some as yet unknown change in the mathematical formulation alone, not requiring any fundamental changes in the physical interpretation. This means that as long as we accept the final physical interpretation of the Quantum Theory, we cannot be led by any conceivable experiment to give up this interpretation, even if it should happen to be wrong. The usual physical interpretation therefore presents us with a considerable danger of falling into a trap, consisting of a self-closing chain of circular hypotheses, which are in principle unverifiable if true. The only way of avoiding the possibility of such a trap is to study the consequences of postulates that contradict assumptions (1) and (2) at the outset. Thus, we could, for example, postulate that the precise outcome of each individual measurement process is in principle determined by some at present "hidden" elements or variables; and we could then try to find experiments that depended in a unique and reproducible way on the assumed state of these hidden elements or variables. If such predictions are verified, we should then obtain experimental evidence favoring the hypothesis that hidden variables exist. If they are not verified, however, the correctness of the usual interpretation of the quantum theory is not necessarily proved, since it may be necessary instead to alter the specific character of the theory that is supposed to describe the behavior of the assumed hidden variables.

We conclude then that a choice of the present interpretation of the quantum theory involves a real physical limitation on the kinds of theories that we wish to take into consideration. For the arguments given here, however, it would seem that there are no secure experimental or theoretical grounds on which we can base such a choice because this choice follows from hypotheses that cannot conceivably be subjected to an experimental test and because we now have an alternative interpretation.

II

In this paper, we shall show how the theory of measurements is to be understood from the point of view of a physical interpretation of the Quantum Theory in terms of "hidden" variables, developed in a previous paper. We find that in principle, these "hidden" variables determine the precise results of each individual measurement process. In practice, however, in measurements that we now know how to carry out, the observing apparatus disturbs the observed system in an unpredictable and uncontrollable way, so that the Uncertainty Principle is obtained as a practical limitation on the possible precision of measurements. This limitation is not, however, inherent in the conceptual structure of our interpretation. We shall see, for example, that simultaneous measurements of position and momentum having unlimited precision would in principle be possible if, as suggested in the previous paper, the mathematical formulation of the quantum theory needs to be modified at very short distances in certain ways that are consistent with our interpretation but not with the usual interpretation.

We give a simple explanation of the origin of quantum-mechanical correlations of distant objects in the hypothetical experiment of Einstein, Podolsky, and Rosen, which was suggested by these authors as a criticism of the usual interpretation.

Finally, we show that von Neumann's proof that Quantum Theory is not consistent with hidden variables does not apply to our interpretation, because the hidden variables contemplated here depend both on the state of the measuring apparatus and the observed system and therefore go beyond certain of von Neumann's assumptions. . . .

Quantum Theory of Measurements

We shall now show how the Quantum Theory of measurements is to be expressed in terms of our suggested interpretation of the Quantum Theory.

In general, a measurement of any variable must always be carried out by means of an interaction of the system of interest with a suitable piece of measuring apparatus. The apparatus must be so constructed that any given state of the system of interest will lead to a certain range of states of the apparatus. Thus, the interaction introduces correlations between the state of the observed system and the state of the apparatus. The range of indefiniteness in this correlation may be called the uncertainty, or the error, in the measurement

Let us now consider an observation designed to measure an arbitrary (hermitian) "observable" Q, associated with an electron. Let x represent the position of the electron, y that of the significant apparatus coordinate (or coordinates if there are more than one). Now, one can show that it is enough to consider an impulsive measurement, i.e., a measurement utilizing a very strong interaction between apparatus and system under observation, which lasts for so short a time that the changes of the apparatus and the system under observation that would have taken place in the absence of interaction can be neglected. Thus, at least while the interaction is taking place, we can neglect the parts of the Hamiltonian associated with the apparatus alone and with the observed system alone, and we need retain only the part of the Hamiltonian, H_1 representing the interaction. Moreover, if the Hamilton operator is chosen to be a function only of quantities that commute with Q, the interaction process will produce no uncontrollable changes in the observable, Q, but only in observables that do not commute with Q. In order that the apparatus and the system under observation shall be coupled, however, it is necessary that H_1 shall also depend on operators involving y. . . .

On the Possibility of Measurements of Unlimited Precision

We have seen that the so-called "observables" do not measure any very readily interpretable properties of a system. For example, the momentum "observable" has in general no simple relation to the actual particle momentum. It may therefore be fruitful to consider how we might try to measure properties which, according to our interpretation, are (along with the ψ-field) the physically significant properties of an electron, namely, the actual particle position and momentum. In connection with this problem, we shall show that if, as suggested in Paper I, we give up three mutually consistent special assumptions leading to the same results as those of the usual interpretation of the quantum theory, then in our interpretation, the particle position and momentum can in principle be measured simultaneously with unlimited precision.

Now, for our purposes, it will be adequate to show that precise predictions of the future behavior of a system are in principle possible. In our interpretation, a sufficient condition for precise predictions is as we have seen that we shall be able to prepare a system in a state in which the ψ-field and the initial particle position and momentum are precisely known. We have shown that it is possible, by measuring the "observable," Q, with the aid of methods that are now available, to prepare a state in which the ψ-field is effectively transformed into a known form, $\psi_q(x)$; but we cannot in general predict or control the precise position and momentum of the particle. If we could now measure the position and momentum of the particle without altering the ψ-field, then precise predictions would be possible. However, the results of I (above) prove that as long as the three special assumptions indicated above are valid, we cannot measure the particle position more accurately without effectively transforming the ψ-function into an incompletely predictable and controllable packet that is much more localized than $\psi_q(x)$. Thus, efforts to obtain more precise definition of the state of the system will be defeated. But it is clear that the difficulty originates in the circumstance that the potential energy of interaction between electron and apparatus, $V(x,y)$, plays two roles. For it not only introduces a direct interaction between the two particles, proportional in strength to $V(x,y)$ itself, but introduces an indirect interaction between these particles, because this potential also appears in the equation governing the ψ-field. This indirect interaction may involve rapid and violent fluctuations, even when $V(x,y)$ is small. Thus, we are led to lose control of the effects of this interaction, because no matter how small $V(x,y)$ is, very large and chaotically complicated disturbances in the particle motion may occur.

If, however, we give up the three special assumptions mentioned previously, then it is not inherent in our conceptual structure that every interaction between particles must inevitably also produce large and uncontrollable changes in the ψ-field. Thus, in Paper I, we give an example in which we postulate a force acting on a particle that is not necessarily accompanied by a corresponding change in the ψ-field. The equation in Paper I is concerned only with a one-particle system, but similar assumptions can be made for systems of two or more particles. In the absence of any specific theory, our interpretation permits an infinite number of kinds of such modifications, which can be chosen to be important at small distances but negligible in the atomic domain. For the sake of illustration, suppose that it should turn out that in certain processes connected with very small distances, the force acting on the apparatus variable is

$$F_y = ax,$$

where a is a constant. Now if "a" is made large enough so that the interaction is impulsive, we can neglect all changes in y that are brought about by the forces that would have been present in the absence of this interaction. Moreover, for the sake of illustration of the principles involved, we are permitted to make the assumption, consistent with our interpretation, that the force on the electron is zero. The equation of motion of y is then

$$\bar{y} = ax/m,$$

The solution is

$$y - y_o = (axt^2/2m) + \dot{y}_o t,$$

where $\dot{y}_o$ is the initial velocity of the apparatus variable and y_o its initial position. Now, if the product, at^2, is large enough, then $y - y_o$ can be made much larger than the uncertainty in y arising from the uncertainty of y_0 and the uncertainty of $\dot{y}_o$. Thus, $y - y_o$ will be determined primarily by the particle position, x. In this way, it is conceivable that we could obtain a measurement of x that does not significantly change x, $\dot{x}$, or the ψ-function. The particle momentum can then be obtained from the relation, $p = \nabla S(x)$, where $S/\hbar$ is the phase of the ψ-function. Thus, precise predictions would in principle be possible. . . .

The Hypothetical Experiment of Einstein, Podolsky, and Rosen

The hypothetical experiment of Einstein, Podolsky, and Rosen is based on the fact that if we have two particles, the sum of their momenta, $p = p_1 + p_2$, commutes with the difference of their positions, $\xi = x_1 - x_2$. We can therefore define a wave function in which p is zero, while ξ has a given value, a. Such a wave function is

$$\psi = \delta(x_1 - x_2 - a). \tag{17}$$

In the usual interpretation of the quantum theory, $p_1 - p_2$ and $x_1 + x_2$ are completely undetermined in a system having the above wave function.

The whole experiment centers on the fact that an observer has a choice of measuring either the momentum or the position of any one of the two particles. Whichever of these quantities he measures, he will be able to infer a definite value of the corresponding variable in the other particle, because of the fact that the above wave function implies correlations between variables belonging to each particle. Thus, if he obtains a position x_1 for the first particle, he can infer a position of $x_2 = a - x_1$ for the second particle; but he loses all possibility of making any inferences about the momenta of either particle. On the other hand, if he measures the momentum of the first particle and obtains a value of p_1, he can infer a value of $p_2 = -p_1$ for the momentum of the second particle; but he loses all possibility of making any inferences about the position of either particle. Now, Einstein, Podolsky, and Rosen believe that this result is itself probably correct, but they do not believe that quantum theory as usually interpreted can give a complete description of how these correlations are propagated. Thus, if these were classical particles, we could easily understand the propagation of correlations because each particle would then simply move with a velocity opposite to that of the other. But in the usual interpretation of Quantum Theory, there is no similar conceptual model showing in detail how the second particle, which is not in any way supposed to interact with the first particle, is nevertheless able to obtain either an uncontrollable disturbance of its position or an uncontrollable disturbance of its momentum depending on what kind of measurement the observer decided to carry out on the first particle. Bohr's point of view is, however, that no such model should be sought and that we should merely accept the fact that these correlations somehow manage to appear. We must note, of course, that the quantum-mechanical description of these processes will always be consistent, even though it gives us no precisely definable means of describing and analyzing the relationships between the classically describable phenomena appearing in various pieces of measuring apparatus.

In our suggested new interpretation of the Quantum Theory, however, we can describe this experiment in terms of a single precisely definable conceptual model, for we now describe the system in terms of a combination of a six-dimensional wave field and a precisely definable trajectory in a six-dimensional space (see Paper I). If the wave function is initially equal to Eq. (17), then since the phase vanishes, the particles are both at rest. Their possible positions are, however, described by an ensemble, in which $x_1 - x_2 = a$. Now if we measure the position of the first particle, we introduce uncontrollable fluctuations in the wave function for the entire system, which, through the "quantum-mechanical" forces, bring about corresponding uncontrollable fluctuations in the momentum of each particle. Similarly, if we measure the momentum of the first particle, uncontrollable fluctuations in the wave function for the system bring about, through the "quantum-mechanical" forces, corresponding uncontrollable changes in the position of each particle. Thus, the "quantum-mechanical" forces may be said to transmit uncontrollable disturbances instantaneously from one particle to another through the medium of the ψ–field.

What does this transmission of forces at an infinite rate mean? In nonrelativistic theory, it certainly causes no difficulties. In a relativistic theory, however, the problem is more complicated. We first note that as long as the three special assumptions mentioned in Sec. 2 are valid, our interpretation can give rise to no inconsistencies with relativity, because it leads to precisely the same predictions for all physical processes as are obtained from the usual interpretation (which is known to be consistent with relativity). The reason why no contradictions with relativity arise in our interpretation despite the instantaneous transmission of momentum between particles is that no signal can be carried in this way. For such a transmission of momentum could constitute a signal only if there were some practical means of determining precisely what the second particle would have done if the first particle had not been observed, and as we have seen, this information cannot be obtained as long as the present form of the Quantum Theory is valid. To obtain such information, we require conditions (such as might perhaps exist in connection with distances of the order of 10^{-13}cm) under which the usual form of the Quantum Theory breaks down, so that the positions and momenta of the particles can be determined simultaneously and precisely. If such conditions should exist, then there are two ways in which contradictions might be avoided. First, the more general physical laws appropriate to the new domains may be such that they do not permit the transmission of controllable aspects of interparticle forces faster than light. In this way, Lorentz covariance could be preserved. Secondly, it is possible that the application of the usual criteria of Lorentz covariance may not be appropriate when the usual interpretation of Quantum Theory breaks down. Even in connection with gravitational theory, general relativity indicates that the limitation of speeds to the velocity of light does not necessarily hold universally. If we adopt the spirit of general relativity, which is to seek to make the properties of space dependent on the properties of the matter that moves in this space, then it is quite conceivable that the metric, and therefore the limiting velocity, may depend on the ψ-field as well as on the gravitational tensor $g^{u,p}$. In the classical limit, the dependence on the ψ-field could be neglected, and we would get the usual form of covariance. In any case, it can hardly be said that we have a solid experimental basis for requiring the same form of covariance at very short distances that we require at ordinary distances.

To sum up, we may assert that wherever the present form of the Quantum Theory is correct, our interpretation cannot lead to inconsistencies with relativity. In the domains where the present theory breaks down, there are several possible ways in which our interpretation could continue to treat the problem of covariance consistently. The attempt to maintain a consistent treatment of covariance in this problem might perhaps serve as an important heuristic principle in the search for new physical laws. . . .

Summary and Conclusions

The usual interpretation of the Quantum Theory implies that we must renounce the possibility of describing an individual system in terms of a single precisely defined conceptual model. We have, however, proposed an alternative interpretation which does not imply such a renunciation, but which instead leads us to regard a quantum-mechanical system as a synthesis of a precisely definable particle and a precisely definable ψ-field which exerts a force on this particle. An experimental choice between these two interpretations cannot be made in a domain in which the present mathematical formulation of the Quantum Theory is a good approximation; but such a choice is conceivable in domains, such as those associated with dimensions of the order of 10^{-18}cm, where the extrapolation of the present theory seems to break down and where our suggested new interpretation can lead to completely different kinds of predictions.

At present, our suggested new interpretation provides a consistent alternative to the usual assumption that no objective and precisely definable description of reality is possible at the quantum level of accuracy. For, in our description, the problem of objective reality at the quantum level is at least in principle not fundamentally different from that at the classical level, although new problems of measurement of the properties of an individual system appear, which can be solved only with the aid of an improvement in the theory, such as the possible modifications in the nuclear domain suggested above. In this connection, we wish to point out that what we can measure depends not only on the type of apparatus that is available, but also on the existing theory, which determines the kind of inference that can be used to connect the directly observable state of the apparatus with the state of the system of interest. In other words, our epistemology is determined to a large extent by the existing theory. It is therefore not wise to specify the possible forms of future theories in terms of purely epistemological limitations deduced from existing theories.

18
— John Stewart Bell —
(1928–1990)

John Stewart Bell was born in Belfast, and was for many years a theoretical physicist at CERN, the particle accelerator center in Geneva. In the 1960s Bell formulated a theorem that was to have a major impact on quantum mechanics and the Completeness Debate. It states that any hidden variable theory that satisfies the condition of locality cannot possibly reproduce all the statistical predictions of quantum mechanics. If true, this would prove once and for all not only

*the tenability of quantum physics, but more important that any reality underlying
the quantum facts must be nonlocal, thereby refuting the EPR paradox.*

*In this reading Bell criticizes the hidden variables hypothesis and suggests the
possibility of experimental tests that might resolve the controversy. The reading
is taken from an article entitled "Introduction to the Hidden Variable Question,"
published in 1971 in the* Proceedings of the International School of Physics,
"Enrico Fermi," course IL; Foundations of Quantum Mechanics, *and later
appeared in his collection* Speakable and Unspeakable in Quantum Mechanics.

COMMENT ON THE HIDDEN VARIABLES HYPOTHESIS

Theoretical physicists live in a classical world, looking out into a quantum-
mechanical world. The latter we describe only subjectively, in terms of procedures and results in our classical domain. This subjective description is effected
by means of quantum-mechanical state functions ψ, which characterize the classical conditioning of quantum-mechanical systems and permit predictions about
subsequent events at the classical level. The classical world of course is described
quite directly—"as it is." We could specify for example the actual positions Λ_1,
$\Lambda_2, \ldots$ of material bodies, such as the switches defining experimental conditions
and the pointers, or print, defining experimental results. Thus in contemporary
theory the most complete description of the state of the world as a whole, or of
any part of it extending into our classical domain, is of the form

$$(\Lambda_1, \Lambda_2, \ldots, \psi)$$

with both classical variables and one or more quantum-mechanical wave functions.

Now nobody knows just where the boundary between the classical and quantum domain is situated. Most feel that experimental switch settings and pointer
readings are on this side. But some would think the boundary nearer, others
would think it farther, and many would prefer *not* to think about it. In fact, the
matter is of very little importance in practice. This is because of the immense
difference in scale between things for which quantum-mechanical description is
numerically essential and those ordinarily perceptible by human beings. Nevertheless, the movability of the boundary is of only approximate validity; demonstrations of it depend on neglecting numbers which are small, but not zero,
which might tend to zero for infinitely large systems, but are only very small for
real finite systems. A theory founded in this way on arguments of manifestly
approximate character, however good the approximation, is surely of provisional
nature. It seems legitimate to speculate on how the theory might evolve. But of
course no one is obliged to join in such speculation.

A possibility is that we find exactly where the boundary lies. More plausible
to me is that we will find that there is no boundary. It is hard for me to envisage

intelligible discourse about a world with no classical part—no base of given events, be they only mental events in a single consciousness, to be correlated. On the other hand, it is easy to imagine that the classical domain could be extended to cover the whole. The wave functions would prove to be a provisional or incomplete description of the quantum-mechanical part, of which an objective account would become possible. It is this possibility, of a homogeneous account of the world, which is for me the chief motivation of the study of the so-called "hidden variable" possibility.

A second motivation is connected with the statistical character of quantum-mechanical predictions. Once the incompleteness of the wave-function description is suspected, it can be conjectured that the seemingly random statistical fluctuations are determined by the extra "hidden" variables—"hidden" because at this stage we can only conjecture their existence and certainly cannot control them. Analogously, the description of Brownian motion for example might first have been developed in a purely statistical way, the statistics becoming intelligible later with the hypothesis of the molecular constitution of fluids, this hypothesis then pointing to previously unimagined experimental possibilities, the exploitation of which made the hypothesis entirely convincing. For me the possibility of determinism is less compelling than the possibility of having one world instead of two. But, by requiring it, the programme becomes much better defined and more easy to come to grips with.

A third motivation is in the peculiar character of some quantum-mechanical predictions, which seem almost to cry out for a hidden variable interpretation. This is the famous argument of Einstein, Podolsky and Rosen. Consider the example, advanced by Bohm, of a pair of spin -½ particles formed somehow in the singlet spin state and then moving freely in opposite directions. Measurements can be made, say by Stern-Gerlach magnets, on selected components of the spins σ_1 and σ_2. If measurement of $\sigma_1 \cdot a$, where a is some unit vector, yields the value +1, then, according to quantum mechanics, measurement $\sigma_2 \cdot a$ must yield the value −1, and *vice versa*. Thus we can know in advance the result of measuring any component of σ_2 by previously, and possibly at a very distant place, measuring the corresponding component of σ_1. This strongly suggests that the outcomes of such measurements, along arbitrary directions, are actually determined in advance, by variables over which we have no control, but which are sufficiently revealed by the first measurement so that we can anticipate the result of the second. There need then be no temptation to regard the performance of one measurement as a causal influence on the result on the second, distant, measurement. The description of the situation could be manifestly "local." This idea seems at least to merit investigation.

We will find, in fact, that no local deterministic hidden-variable theory can reproduce all the experimental predictions of quantum mechanics. This opens the possibility of bringing the question into the experimental domain, by trying to approximate as well as possible the idealized situations in which local hidden variables and quantum mechanics cannot agree. However, before coming to this, we must clear the ground by some remarks on various mathematical investigations that have been made on the possibility of hidden variables in quantum mechanics without any reference to locality.

The Absence of Dispersion-Free States in Various Formalisms Derived from Quantum Mechanics

Consider first the usual Heisenberg Uncertainty Principle. It says that for quantum-mechanical states the predictions for measurements for at least one of a pair of conjugate variables must be statistically uncertain. Thus no quantum-mechanical state can be "dispersion-free" for every observable. It follows that if a hidden-variable account is possible, in which the results of all observations are fully determined, each quantum-mechanical state must correspond to an ensemble of states each with different values of the hidden variables. Only these component states will be dispersion-free. So one way to formulate the hidden-variable problem is a search for a formalism permitting such dispersion-free states.

An early, and very celebrated, example of such an investigation was that of von Neumann. He observed that in quantum mechanics an observable whose operator is a linear combination of operators for other observables

$$A = \beta B + \gamma C$$

has for expectation value the corresponding linear combination of expectation values:

$$<A> = \beta <B> + \gamma <C>.$$

He considered more general schemes in which this particular feature was preserved. Now for the hypothetical dispersion-free states there is no distinction between expectation values and eigenvalues—for each such state must yield with certainty a particular one of the possible results for any measurement. But eigenvalues are not additive. Consider for example components of spin for a particle of spin ½. The operator for the component along the direction half-way between the x and y axes is

$$(\sigma_x + \sigma_y)/\sqrt{2},$$

whose eigenvalues ± 1 are certainly not the corresponding linear combinations

$$(\pm 1 \pm 1)/\sqrt{2}$$

of eigenvalues of σ_x and σ_y. Thus the requirement of additive expectation values excludes the possibility of dispersion-free states. Von Neumann concluded that a hidden-variable interpretation is not possible for quantum mechanics: "it is therefore not, as is often assumed, a question of reinterpretation of quantum mechanics—the present system of quantum mechanics would have to be objectively false in order that another description of the elementary process than the statistical one be possible."

It seems therefore that von Neumann considered the additivity (2) more as an obvious axiom than as a possible postulate. But consider what it means in terms of the actual physical situation. Measurements of the three quantities

$$\sigma_x, \ \sigma_y, \ (\sigma_x + \sigma_y)/\sqrt{2},$$

require three different orientations of the Stern-Gerlach magnet, and cannot be performed simultaneously. It is just this which makes intelligible the non-additivity of the eigenvalues—the values observed in specific instances. It is by no means a question of simply measuring different components of a pre-existing vector, but rather of observing different products of different physical procedures. That the statistical averages should then turn out to be additive is really a quite remarkable feature of quantum-mechanical states, which could not be guessed *a priori*. It is by no means a "law of thought" and there is no *a priori* reason to exclude the possibility of states for which it is false. It can be objected that although the additivity of expectation values is not a law of thought, it *is* after all experimentally true. Yes, but what we are now investigating is precisely the hypothesis that the states presented to us by nature are in fact mixtures of component states which we cannot (for the present) prepare individually. The component states need only have such properties that ensembles of them have the statistical properties of observed states.

It has subsequently been shown that in various other mathematical schemes, derived from quantum mechanics, dispersion-free states are not possible. The persistence in these schemes of a kind of Uncertainty Principle is of course useful and interesting to the people working with those schemes. However, the importance of these results, for the question that we are concerned with, is easily exaggerated. The postulates often have great intrinsic appeal to those approaching quantum mechanics in an abstract way. Translated into assumptions about the behavior of actual physical equipment, they are again seen to be of a far from trivial or inevitable nature.

On the other hand, if no restrictions whatever are imposed on the hidden variables, or on the dispersion-free states, it is trivially clear that such schemes can be found to account for any experimental results whatever. Ad hoc schemes of this kind are devised every day when experimental physicists, to optimize the design of their equipment, simulate the expected results by deterministic computer programmes drawing on a table of random numbers. Such schemes, from our present point of view, are not very interesting. Certainly what Einstein wanted was a comprehensive account of physical processes evolving continuously and locally in ordinary space and time. . . .

What happens to the hidden variables during and after the measurement is a delicate matter. Note only that a prerequisite for a specification of what happens to the hidden variables would be a specification of what happens to the wave function. But it is just at this point that the notoriously vague "reduction of the wave packet" intervenes, at some ill-defined time, and we come up against the ambiguities of the usual theory, which for the moment we aim only to reinterpret rather than to replace. It would indeed be very interesting to go beyond this point. But we will not make the attempt here, for we will find a very striking difficulty at the level to which the scheme has been developed already. Before coming to this, a number of instructive features of the scheme are worth indicating.

One such feature is this. We have here a picture in which although the wave has two components, the particle has only position λ. The particle does not "spin," although the experimental phenomena associated with spin are reproduced. Thus the picture resulting from a hidden-variable account of quantum mechanics need not very much resemble the traditional classical picture that the researcher may, secretly, have been keeping in mind. The electron need not turn out to be a small spinning yellow sphere.

A second way in which the scheme is instructive is in the explicit picture of the very essential role of apparatus. The result of a "spin measurement," for example, depends in a very complicated way on the initial position λ of the particle and on the strength and geometry of the magnetic field. Thus the result of the measurement does not actually tell us about some property previously possessed by the system, but about something which has come into being in the combination of system and apparatus. Of course, the vital role of the complete physical set-up we learned long ago, especially from Bohr. When it is forgotten, it is more easy to expect that the results of the observations should satisfy some simple algebraic relations and to feel that these relations should be preserved even by the hypothetical dispersion-free states of which quantum-mechanical states may be composed. The model illustrates how the algebraic relations, valid for the statistical ensembles, which are the quantum-mechanical states may be built up in a rather complicated way. Thus the contemplation of this simple model could have a liberalizing effect on mathematical investigators.

Finally, this simple scheme is also instructive in the following way. Even if the infamous boundary, between classical and quantum worlds, should not go away, but rather become better defined as the theory evolves, it seems to me that some classical variables will remain essential (they may describe "macroscopic" objects, or they may be finally restricted to apply only to my sense data). Moreover, it seems to me that the present "Quantum Theory of measurement" in which the quantum and classical levels interact only fitfully during highly idealized "measurements" should be replaced by an interaction of a continuous, if variable, character.

A Difficulty

The difficulty is this. Looking at (the mathematics) one sees that the behaviour of a given variable λ_1 is determined not only by the conditions in the immediate neighbourhood (in ordinary three-space) but also by what is happening at the other positions $\lambda_2, \lambda_3, \ldots$. That is to say, that although the system of equations is "local" in an obvious sense in the $3n$-dimensional space, it is not at all local in ordinary three-space. As applied to the Einstein-Podolsky-Rosen situation, we find that this scheme provides an explicit causal mechanism by which operations on one of the two measuring devices can influence the response of the distant device. This is quite the reverse of the resolution hoped for by EPR, who envisaged that the first device could serve only to reveal the character of the information already stored in space, and propagating in an undisturbed way towards the other equipment.

The question arises: can we not find another hidden-variable scheme with the desired local character? . . . It can be shown that this is not possible. . . .

Thus, the quantum-mechanical result cannot be reproduced by a hidden-variable theory which is local in the way described.

This result opens up the possibility of bringing the questions that we have been considering into the experimental area. Of course, the situation envisaged above is highly idealized. It is supposed that the system is initially in a known spin state, that the particles are known to proceed toward the instruments, and to be measured there with complete efficiency. The question then is whether the inevitable departures from this ideal situation can be kept sufficiently small in practice that the quantum-mechanical prediction still violated the inequality.

In this connection other systems, for example the two-photon system of the two-kaon system, may be more promising than that of two-spin ½ particles. A very serious study of the photon case will be reported to this meeting by Shimony. The experiment described by him, and now under way, is not sufficiently close to the ideal to be conclusive for a quite determined advocate of hidden variables. However, for most a confirmation of quantum-mechanical predictions, which is only to be expected given the general success of quantum mechanics, would be severe discouragement.

19
— John Stewart Bell —

In this reading Bell analyzes the EPR thought experiment of David Bohm and offers a variety of interpretations that are presented in an amusing tone accessible to the lay reader. At each pass, Bell reaffirms the untenability of both the EPR charge of incompleteness and the "hidden variables" theory of Bohm, thereby reconfirming both the completeness of Quantum Theory as well as the finality of the assertions of Bell's theorem.

The reading is taken from an article entitled "Bertlmann's Socks and the Nature of Reality," originally published in 1981 in the Journal de Physique. *It later appeared in Bell's collection of essays* Speakable and Unspeakable in Quantum Mechanics.

A CONCEPTUAL ANALYSIS OF THE EPR THOUGHT EXPERIMENT OF DAVID BOHM

The philosopher in the street, who has not suffered a course in quantum mechanics, is quite unimpressed by Einstein-Podolsky-Rosen correlations. He can point to many examples of similar correlations in everyday life. The case of Bertlmann's socks is often cited. Dr. Bertlmann likes to wear two socks of different colours. Which colour he will have on a given foot on a given day is quite unpredictable. But when you see that the first sock is pink you can be already sure that the second sock will not be pink. Observation of the first, and experience of Bertlmann, gives immediate information about the second. There is no accounting for tastes, but apart from that there is no mystery here. And is not the EPR business just the same?

Consider for example the particular EPR gedanken experiment of Bohm (Fig. 2). Two suitable particles, suitably prepared (in the "singlet spin state"), are directed from a common source towards two widely separated magnets followed by detecting screens. Each time the experiment is performed each of the two particles is deflected either up or down at the corresponding magnet. Whether either particle separately goes up or down on a given occasion is quite unpredictable. But when one particle goes up the other always goes down and vice-versa. After a little experience it is enough to look at one side to know also about the other.

Fig. 1.

Les chaussettes
de M. Bertlmann
et la nature
de la réalité

Fondation Hugot
juin 17 1980

pink

not
pink

So what? Do we not simply infer that the particles have properties of some kind, detected somehow by the magnets, chosen á la Bertlmann by the source—differently for the two particles? Is it possible to see this simple business as obscure and mysterious? We must try.

To this end it is useful to know how physicists tend to think intuitively of particles with "spin," for it is with such particles that we are concerned. In a crude classical picture it is envisaged that some internal motion gives the particle an angular momentum about some axis, and at the same time generates a magnetization along that axis. The particle is then like a little spinning magnet with north and south poles lying on the axis of rotation. When a magnetic field is applied to a magnet the north pole is pulled one way and the south pole is pulled the other way. If the field is uniform the net force on the magnet is zero. But in a non-uniform field one pole is pulled more than the other and the magnet as a whole is pulled in the corresponding direction. The experiment in

Fig. 2. Einstein–Podolsky–Rosen–Bohm gedanken experiment with two spin ½ particles and two Stern–Gerlach magnets.

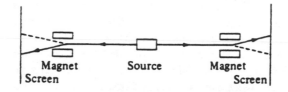

Fig. 3. Forces on magnet in non-uniform magnetic field. The field points towards the top of the page and increases in strength in that direction.

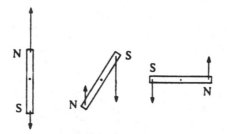

question involves such non-uniform fields—set up by so-called "Stern-Gerlach" magnets. Suppose that the magnetic field points up, and that the strength of the field increases in the upward direction. Then a particle with south-north axis pointing up would be pulled up (Fig. 3). One with axis pointing down would be pulled down. One with axis perpendicular to the field would pass through the field without deflection. And one oriented at an intermediate angle would be deflected to an intermediate degree. (All this is for a particle of zero electric charge; when a charged particle moves in a magnetic field there is an additional force which complicates the situation.)

A particle of given species is supposed to have a given magnetization. But because of the variable angle between particle axis and field there would still be a range of deflections possible in a given Stern-Gerlach magnet. It could be expected then that a succession of particles would make a pattern something like Fig. 4 on a detecting screen. But what is observed in the simplest case is more like Fig. 5, with two distinct groups of deflections (i.e., up or down) rather than a more or less continuous band. (This simplest case, with just two groups of deflections, is that of so-called "spin-½" particles; for "spin-j" particles there are $(2j + 1)$ groups).

Fig. 4. Naive classical expectation for pattern on detecting screen behind Stern–Gerlach magnet.

Fig. 5. Quantum mechanical pattern on screen, with vertical Stern–Gerlach magnet.

The pattern of Fig. 5 is very hard to understand in naïve classical terms. It might be supposed for example that the magnetic field first pulls the little magnets into alignment with itself, like compass needles. But even if this were dynamically sound it would account for only one group of deflections. To account for the second group would require "compass-needles" pointing in the wrong direction. And anyway it is not dynamically sound. The internal angular momentum, by gyroscopic action, should stabilize the angle between particle axis and magnetic field. Well then, could it not be that the source for some reason delivers particles with axes pointing just one way or the other and not in between? But this is easily tested by turning the Stern-Gerlach magnet. What we get (Fig. 6) is just the same split pattern as before, but turned around with the Stern-Gerlach magnet. To blame the absence of intermediate deflections on the source we would have to imagine that it anticipated somehow the orientation of the Stern-Gerlach magnet.

Phenomena of this kind made physicists despair of finding any consistent space-time picture of what goes on on the atomic and subatomic scale. Making a virtue of necessity, and influenced by positivistic and instrumentalist philosophies, many came to hold not only that it is difficult to find a coherent picture but that it is wrong to look for one—if not actually immoral then certainly unprofessional. Going further still, some asserted that atomic and subatomic particles do not *have* any definite properties in advance of observation. There is nothing, that is to say, in the particles approaching the magnet, to distinguish those subsequently deflected up from those subsequently deflected down. Indeed even the particles are not really there.

For example, "Bohr once declared when asked whether the quantum mechanical algorithm could be considered as somehow mirroring an underlying quantum reality: 'There is no quantum world. There is only an abstract quantum mechanical description. It is wrong to think that the task of physics is to find out how Nature *is*. Physics concerns what we can say about Nature.'"

And for Heisenberg " . . . in the experiments about atomic events we have to do with things and facts, with phenomena that are just as real as any phenomena in daily life. But the atoms or the elementary particles are not as real; they form a world of potentialities or possibilities rather than one of things or facts."

And "Jordan declared, with emphasis, that observations not only *disturb* what has to be measured, they *produce* it. In a measurement of position, for example, as performed with the gamma ray microscope, 'the electron is forced to a decision. We compel it *to assume a definite position*; previously it was, in general, neither here nor there, it had not yet made its decision for a definite position. . . . If by another experiment the *velocity* of the electron is being measured, this means: the electron is compelled to decide itself for some exactly defined value of the velocity. . . . we ourselves produce the results of measurement.'"

Fig. 6. Quantum mechanical pattern with rotated Stern–Gerlach magnet.

It is in the context of ideas like these that one must envisage the discussion of the Einstein-Podolsky-Rosen correlations. Then it is a little less unintelligible that the EPR paper caused such a fuss, and that the dust has not settled even now. It is as if we had come to deny the reality of Bertlmann's socks, or at least of their colours, when not looked at. And as if a child has asked: How come they always choose different colours when they *are* looked at? How does the second sock know what the first has done?

Paradox indeed! But for the others, not for EPR. EPR did not use the word "paradox." They were with the man in the street in this business. For them these correlations simply showed that the quantum theorists had been hasty in dismissing the reality of the microscopic world. In particular Jordan had been wrong in supposing that nothing was real or fixed in that world before observation. For after observing only one particle the result of subsequently observing the other (possibly at a very remote place) is immediately predictable. Could it be that the first observation somehow fixes what was unfixed, or makes real what was unreal, not only for the near particle but also for the remote one? For EPR that would be an unthinkable "spooky action at a distance." To avoid such action at a distance they have to attribute, to the space-time regions in question, *real* properties in advance of observation, correlated properties, which *predetermine* the outcomes of these particular observations. Since these real properties, fixed in advance of observation, are not contained in quantum formalism, that formalism for EPR is *incomplete*. It may be correct, as far as it goes, but the usual quantum formalism cannot be the whole story.

It is important to note that to the limited degree to which *determinism* plays a role in the EPR argument, it is not assumed but *inferred*. What is held sacred is the principle of "local causality"—or "no action at a distance." Of course, mere *correlation* between distant events does not by itself imply action at a distance, but only correlation between the signals reaching the two places. These signals, in the idealized example of Bohm, must be sufficient to *determine* whether the particles go up or down. For any residual undeterminism could only spoil the perfect correlation. . . .

Illustration

Let us illustrate the *possibility* of what Einstein had in mind in the context of the particular quantum mechanical predictions already cited for the EPRB gedanken experiment. These predictions make it hard to believe in the completeness of quantum formalism. But of course outside that formalism they make no difficulty whatever for the notion of local causality. To show this explicitly we exhibit a trivial *ad hoc* space-time picture of what might go on. It is a modification of the naive classical picture already described. Certainly something must be modified in that, to reproduce the quantum phenomena. Previously, we implicitly assumed for the net force in the direction of the field gradient (which we always take to be in the same direction as the field) a form

$$F \cos \theta \tag{1}$$

where θ is the angle between magnetic field (and field gradient) and particle axis. We change this to

$$F \cos \theta / |\cos \theta|. \tag{2}$$

Whereas previously the force varied over a continuous range with θ, it takes now just two values, $\pm F$, the sign being determined by whether the magnetic axis of the particle points more nearly in the direction of the field or in the opposite direction. No attempt is made to explain this change in the force law. It is just an *ad hoc* attempt to account for the observations. And of course it accounts immediately for the appearance of just two groups of particles, deflected either in the direction of the magnetic field or in the opposite direction. To account then for the Einstein-Podolsky-Rosen Bohm correlations we have only to assume that the two particles emitted by the source have oppositely directed magnetic axes. Then if the magnetic axis of one particle is more nearly along (than against) one Stern-Gerlach field, the magnetic axes of the other particle will be more nearly against (than along) a parallel Stern-Gerlach field. So when one particle is deflected up, the other is deflected down, and vice versa. There is nothing whatever problematic or mind-boggling about these correlations, with parallel Stern-Gerlach analyzers, from the Einsteinian point of view.

So far so good. But now go a little further than before, and consider *non*-parallel Stern-Gerlach magnets. Let the first be rotated away from some standard position, about the particle line of flight, by an angle a. Let the second be rotated likewise by an angle b. Then if the magnetic axis of either particle separately is randomly oriented, but if the axes of the particles of a given pair are always oppositely oriented, a short calculation gives for the probabilities of the various possible results, in the *ad hoc* model,

$$\left. \begin{array}{l} P(\text{up, up}) = P(\text{down, down}) = \dfrac{|a-b|}{2\pi} \\[3mm] P(\text{up, down}) = P(\text{down, up}) = \dfrac{1}{2} - \dfrac{|a-b|}{2\pi} \end{array} \right\} \tag{3}$$

where "up" and "down" are defined with respect to the magnetic fields of the two magnets. However, a quantum mechanical calculation gives

$$\left. \begin{array}{l} P(\text{up, up}) = P(\text{down, down}) = \dfrac{1}{2}\left(\sin\dfrac{a-b}{2}\right)^2 \\[3mm] P(\text{up, down}) = P(\text{down, up}) = \dfrac{1}{2} - \dfrac{1}{2}\left(\sin\dfrac{a-b}{2}\right)^2 \end{array} \right\} \tag{4}$$

Thus the *ad hoc* model does what is required of it (i.e., reproduces quantum mechanical results) only at $(a-b) = 0$, $(a-b) = \pi/2$ and $(a-b) = \pi$, but not at intermediate angles.

Of course this trivial model was just the first one we thought of, and it worked up to a point. Could we not be a little more clever, and devise a model which reproduces the quantum formulae completely? No. It cannot be done, so long as action at a distance is excluded. This point was realized only subsequently. Nether EPR nor their contemporary opponents were aware of it. . . .

Let us summarize once again the logic that leads to the impasse. The EPRB correlations are such that the result of the experiment on one side immediately foretells that on the other, whenever the analyzers happen to be parallel. If we do not accept the intervention on one side as a causal influence on the other, we seem obliged to admit that the results on both sides are determined in advance anyway, independently of the intervention on the other side, by signals from the source and by the local magnet setting. But this has implications for non-parallel settings which conflict with those of quantum mechanics. So we *cannot* dismiss intervention on one side as a causal influence on the other.

It would be wrong to say "Bohr wins again"; the argument was not known to the opponents of Einstein, Podolsky and Rosen. But certainly Einstein could no longer write so easily, speaking of local causality " . . . I still cannot find any fact anywhere which would make it appear likely that that requirement will have to be abandoned." . . .

Envoi

By way of conclusion I will comment on four possible positions that might be taken on this business—without pretending that they are the only possibilities.

First, and those of us who are inspired by Einstein would like this best, quantum mechanics may be *wrong* in sufficiently critical situations. Perhaps Nature is not so queer as quantum mechanics. But the experimental situation is not very encouraging from this point of view. It is true that practical experiments fall far short of the ideal, because of counter inefficiencies, or analyzer inefficiencies or geometrical imperfections, and so on. It is only with added assumptions, or conventional allowance for inefficiencies and extrapolation from the real to the ideal, that one can say the inequality is violated. Although there is an escape route there, it is hard for me to believe that quantum mechanics works so nicely for inefficient practical set-ups and is yet going to fail badly when sufficient refinements are made. Of more importance, in my opinion, is the complete absence of the vital *time* factor in existing experiments. The analyzers are not rotated during the flight of the particles. Even if one is obliged to admit some long range influence, it need not travel faster than light—and so would be much less indigestible. For me, then, it is of capital importance that Aspect is engaged in an experiment in which the time factor is introduced.

Secondly, it may be that it is not permissible to regard the experimental settings a and b in the analyzers as independent variables, as we did. We supposed them in particular to be independent of the supplementary variables λ, in that a and b could be changed without changing the probability distribution $\rho(\lambda)$. Now even if we have arranged that a and b are generated by apparently random radioactive devices, housed in separate boxes and thickly shielded, or by Swiss national lottery machines, or by elaborate computer programmes, or by apparently free willed experimental physicists, or by some combination of all of these, we cannot be *sure* that a and b are not significantly influenced by the same factors λ that influence A and B. But this way of arranging quantum mechanical correlations would be even more mind boggling than one in which causal chains go faster than light. Apparently separate parts of the world would be deeply and conspiratorially entangled, and our apparent free will would be entangled with them.

Thirdly, it may be that we have to admit that causal influences *do* go faster than light. The role of Lorentz invariance in the completed theory would then be very problematic. An "aether" would be the cheapest solution. But the unobservability of this aether would be disturbing. So would the impossibility of "messages" faster than light, which follows from ordinary relativistic quantum mechanics in so far as it is unambiguous and adequate for procedures we can actually perform. The exact elucidation of concepts like "message" and "we," would be a formidable challenge.

Fourthly and finally, it may be that Bohr's intuition was right—in that there is no reality below some "classical" "macroscopic" level. Then fundamental physical theory would remain fundamentally vague, until concepts like "macroscopic" could be made sharper than they are today.

20
— **Abner Shimony** —
(1928–)

Abner Shimony is an American philosopher and physicist who is currently Pro-
fessor Emeritus at Boston University. He holds Ph.D.s from both Yale and
Princeton Universities and has taught at MIT, the University of Paris XI, the
University of Geneva, and Mount Holyoke College. His most significant contribu-
tion to the development of Quantum Theory came from his collaboration with
John Clauser and others, when they were able to derive a version of Bell's in-
equality that is testable with low-efficiency detectors. This experiment, known as
the "CHSH inequality" (after John F. Clauser, Michael A. Horne, Shimony, and
Richard A. Holt), was a crucial step toward not only wider acceptance of the
definitive nature of Quantum Theory, but perhaps more important, the validity
of Bell's theorem. Shimony's most recent work is the two-volume The Search for
a Naturalistic World View.

In the final reading of Part IV, Shimony reflects upon Bohr's philosophy of
quantum mechanics and more specifically his principle of complementarity. After
presenting his account of the essential features of the epistemological debate be-
tween Bohr and Einstein, Shimony then goes on to explain why he aligns himself
more on the side of Einstein—this despite the preponderance of scientific evi-
dence that would seem to corroborate the position of Bohr and company. Then,
after laying out the foundations for his own epistemological view, Shimony
readily admits that there are still many questions in Quantum Theory left to be
answered. Most significant, he points to the important role of philosophy in any
significant attempt at answering those questions.

The reading is taken from an article entitled "Conceptual Foundations of
Quantum Mechanics," which appeared in The New Physics, *edited by Paul*
Davies.

PHILOSOPHICAL REFLECTIONS
ON THE COMPLETENESS DEBATE

The Philosophy of Bohr

More than any other person, Niels Bohr formulated a defense of the intelligibility
of quantum mechanics within a few years after its discovery and laid to rest the
philosophical scruples of most of a generation of physicists. His principle of
complementarity provided a point of view within which the wave-particle duality and the Heisenberg uncertainty relations could be understood, and his an-

swer to the argument of EPR was generally accepted by the community of physicists as a vindication of the completeness of quantum mechanical descriptions.

Bohr maintained, as a general principle of interpretation in physics, that theoretical concepts, including assertions of the reality of entities or of their properties, cannot be used unambiguously without careful reference to the experimental arrangement in which the concepts are applied. This principle introduces no relativity to the individual observer, because the irreversible registration that occurs at the conclusion of a measurement is accessible to many observers, and the information obtained can be communicated without ambiguity. What quantum mechanics adds to this general principle is the indivisible quantum of action, which implies that the interaction between the object and the apparatus cannot be made negligible or entirely compensated for. Consequently, properties of an object which are measured by incompatible experimental arrangements must not be regarded as simultaneously real, as was possible in classical physics. The experimental arrangements exhibit a relationship of complementarity to each other, and the resulting limitation upon the simultaneous realisability of properties of a specified object is the principle of complementarity. Furthermore, the phenomena which result from the interaction of an object and an apparatus cannot be subdivided in such a way that the contributions of the two physical systems can be discerned, because "any attempt at a well-defined subdivision would demand a change in the experimental arrangement incompatible with the definition of the phenomena under investigation." Bohr's conception of complementarity thus has two distinguishable elements. One, the indivisibility of the quantum, is purely physical. The other, concerning the condition of unambiguous use of concepts, is more generally philosophical, making it possible for Bohr to extend his principle of complementarity to other domains, especially biology and psychology, in which an appropriate surrogate for the indivisibility of the quantum can be found.

Bohr believed that the apparent anomalies of quantum mechanics are consequences of misinterpretation, resulting from the use of theoretical concepts without detailed attention to the experimental arrangement and from neglect of the fact that the arrangement must be described in the language of classical physics. The use of expressions like "creation of physical attributes of objects by observation" are manifestations of such misinterpretation, and we conjecture that he would object equally strenuously against the phrase "actualisation of potentialities," which has been used repeatedly in this chapter.

Can the anomalies of quantum mechanics be resolved or, more accurately, prevented from arising, in the way that Bohr suggests? To do justice to this question would require a longer excursion into philosophy and a more detailed examination of Bohr's subtle and condensed assertions than space permits. All that can be said briefly is that an answer depends crucially upon the choice between two radically different conceptions of legitimate demands upon a theory of knowledge. It is essential to Bohr's theory of knowledge that the ordinary human elements in experience be accepted without challenge or revision; he wrote "we must not forget that, in spite of their limitation, we can by no means dispense with those forms of perception which colour our whole language and in terms of which all experience must ultimately be expressed." In ordinary experience there are objects with definite macroscopic properties which are perceivable by all properly endowed observers. It is illegitimate to demand a complete explanation of how these definite perceptions are permitted by the basic laws of nature, because the definiteness is the presupposition of the whole scientific enterprise of systematising experience. The enterprise requires that there are objects with definite macroscopic properties which are perceivable by all properly endowed observers. Bohr wrote: "Without entering into metaphysical speculations, I may perhaps add that an analysis of the very concept of explanation would, naturally, begin and end with a renunciation as to explaining our own conscious activity." The theme of renunciation and submission to the unavoidable limitations of the human condition is recurrent in Bohr's writings, and places him, more perhaps than he realised himself, in a philosophical tradition of renunciation of excessive claims to human knowledge, including Hume and Kant. The latter systematically maintained that human beings have no knowledge of "things in themselves," but only of the objects of experience.

There is, however, another philosophical tradition, according to which the presuppositions of human knowledge are open to full rational investigation. A coherent philosophy, according to this tradition, has not one but two starting points. One starting point is what is given in ordinary human experience, on the basis of which inferences are made about the constitution of the world beyond us. The other starting point consists of the fundamental principles of the constitution of the world, among which are the principles of physics. According to this philosophical tradition, a coherent philosophical system shows how the two starting points are compatible and connected. In particular, the cognitive power of the knowing subject is a legitimate object of investigation, and there is no "renunciation as to explaining our own conscious activity." Of the very diverse philosophers belonging to this tradition (including Aristotle, Locke, Leibniz, and perhaps Einstein), not many would claim that a coherent philosophical system has been achieved, but all would regard coherence in the sense defined as a philosophical desideratum, and they discern no limitation of human faculties which in principle precludes its achievement.

The problem of the actualisation of potentialities serves as an illustration of challenge to the achievement of philosophical coherence. Suppose, as Bohr insists, that definite results of observations employing macroscopic apparatus provide the indispensable data to which physical theories refer. Then, according to the second tradition, the principles governing the interactions of microscopic objects and macroscopic apparatus must be such as to guarantee the definiteness of the physicists' observations. We cannot be content with Bohr's assertion that the definiteness of these observations is the starting point of scientific investigation, but should be able to explain in terms of natural principles how this definiteness occurs. It would follow, then, that if the unlimited validity of quantum mechanics throughout the physical world prevents the explanation of the definiteness of measurement results, some modification or limitation of quantum mechanics is needed.

The debate for nearly three decades between Bohr and Einstein concerning human knowledge and the foundations of quantum mechanics can be understood as a conflict between the two philosophical traditions which have just been cited. It should be obvious that the bulk of the present paper was written with sympathy for the second tradition, which aims at a "coherent" philosophy in which the two distinct starting points are connected. The problem of the actualisation of potentialities has turned out to be a serious obstacle to this philosophical programme. There remain avenues to be explored, and it is surely premature to say that the programme as a whole is a failure. Nevertheless, the prospects of the programme are less favourable than they were a generation ago, and Bohr's philosophical views on limitations of explanation and the inexplicability of the starting point of human knowledge have been favoured by the outcomes of analysis and experimentation. It should be emphasised, however, that even if in the long run Bohr's general theses continue to be supported by scientific developments, a major task will remain of expanding his condensed dicta into a detailed account of human knowledge.

Conclusions

Intensive research on the foundations of quantum mechanics has yielded quite firm answers to some questions, but has left others undecided. The experimental disconfirmation of local hidden variables theories is unlikely to be reversed by further experimentation. Consequently, it does not seem feasible to interpret quantum mechanical indefiniteness, chance, probability, entanglement and nonlocality merely as features of the observer's knowledge of a physical system. Rather, they seem to be objective features of the systems themselves. Thus the conceptual innovations of quantum mechanics are likely to remain a permanent part of the physical world view. To be sure, the family of nonlocal hidden variable theories has not been refuted, but the prospects for such theories are not at present promising, and, even if one turns out to be successful, it would still have the consequence of recognising nonlocality as a feature of the physical world.

There is, however, no consensus among students of the foundations of quantum mechanics concerning a solution to the problem of the actualisation of potentialities, or, to its special case, the measurement problem. Possibly a solution will be obtained by yet another radical modification of physics, such as a change of the dynamical law of quantum mechanics. Another possibility is a further radical modification of the conception of physical reality: for instance, the elimination of the dichotomy between potentiality and actuality proposed by Everett, or the attribution to consciousness of the power to actualise potentialities. By contrast with these radical proposals, Bohr's treatment of the problem of measurement may seem conservative, but that judgment would be inaccurate, for Bohr himself recognised that he was proposing radical limitations on the scope of human knowledge.

Whatever the outcome of our present uncertainties may be, it is sure to be philosophically significant. Those who have deplored the rift between science and philosophy which began to develop in the eighteenth century may take comfort in the mutual relevance of these disciplines exhibited in the foundations of quantum mechanics. They will find here a vindication of the old sense of "Natural Philosophy."

SELECTED BIBLIOGRAPHY

Albert, D. *Quantum Mechanics and Experience*. Cambridge: Harvard University Press, 1992.

Baggott, J. E. *The Meaning of Quantum Theory*. Oxford: Oxford University Press, 1992.

Bohm, D. *Quantum Theory*. Englewood Cliffs, N.J.: Prentice-Hall, 1951.

Cushing, J. T. *Quantum Mechanics: Historical Contingency and the Copenhagen Hegemony*. Chicago: University of Chicago Press, 1994.

Dirac, P.A.M. *The Principles of Quantum Mechanics*. 4th ed. Oxford: Clarendon Press, 1958.

Feynman, R. *The Character of Physical Law*. Cambridge: MIT Press, 1965.

Jammer, M. *The Conceptual Development of Quantum Mechanics*. 2nd ed. New York: Tomash Publishing Co., 1989.

Kragh, H. *Quantum Generations: A History of Physics in the Twentieth Century*. Princeton: Princeton University Press, 1999.

Kuhn, T. S. *Black-Body Theory of Quantum Discontinuity: 1894–1912*. Oxford: Claredon, 1978.

Omnes, R. *The Interpretation of Quantum Mechanics*. Princeton: Princeton University Press, 1994.

Peebles, P. J. *Quantum Mechanics*. Princeton: Princeton University Press, 1992.

Rae, A. E. *Quantum Mechanics*. 2nd ed. Bristol: Adam Hilger, 1986.

Whitaker, A. *Einstein, Bohr, and the Quantum Dilemma*. Cambridge: Cambridge University Press, 1995.

PART V

BIG BANG COSMOLOGICAL THEORY

//

INTRODUCTION

/////////////////////////////////////

Cosmology is the study of the origin, evolution, and structure of the universe, or at least that part of it with which we have some empirical contact. The currently accepted cosmological theory, called Big Bang Theory, suggests that the universe as it now exists is the result of some evolutionary process that began sometime between eight and twenty billion years ago. The beginning of that process is said to have been a kind of explosion in which the matter, energy, space, and time of the universe were created. The universe has been expanding and developing ever since.

The Problem of Gravity and Olbers's Paradox

Although Big Bang Theory is a product of the twentieth century, serious cosmological speculation, at least in the Western tradition, began in ancient Greece, where philosophers, with no experience of something coming into existence from nothing, assumed that the universe was eternal. They held different ideas concerning the overall structure of the universe, but one view, which was described in detail by Aristotle, became dominant. This view considered the universe to be a large but finite sphere, with the earth, also a sphere, geometrically at its center. On the earth were the living species, whose flourishing seemed to be the goal of the universe. The elements water, air and fire (a gas lighter than air) stayed in the vicinity of the earth to support life. This sublunar region was enclosed by a hollow transparent sphere on which the moon was attached. Additional spheres carried the sun and the five known planets. The entire universe was enclosed by the outermost or celestial sphere on which were attached all the stars. Each of the spheres turned in great circles about a motionless earth. Although these spheres were unobservable, Aristotle's cosmology was accepted over any moving earth theory primarily because of the lack of empirical evidence for stellar parallax and the lack of any experiential evidence for the motion of the earth.

When Copernicus reintroduced the moving earth theory in the sixteenth century, he retained the celestial sphere to enclose a finite universe despite the fact that it was no longer needed to carry the stars around the earth. Tycho Brahe's observation of the supernova of 1572 and his tracing of the orbits of comets, along with Galileo's telescopic observations, demonstrated clearly that the hollow spheres of Aristotle's universe did not exist and that stars were independent bodies at various distances from the earth. Once the celestial sphere was "broken," the universe was seen as indefinitely large. Newton's concept of absolute space, a key element in his science of mechanics, required that the universe be infinitely large because space must be "always and everywhere the same."

The fact that the stars were now seen as independent bodies at various distances from the earth and from one another presented Newton with a very difficult problem. According to his Law of Universal Gravitation the stars should be collapsing on one another. Yet the universe appeared to be static and the stars appeared to be fixed. To "solve" this problem Newton assumed that there were an infinite number of stars spread fairly evenly throughout the infinite space of the universe. Thus the net gravitational force on any one star would be zero and the stars would retain their fixed status. Unfortunately there were both theoretical and observational difficulties with this explanation. Stars appear to be of different sizes and they do not appear to be evenly distributed. Thus at least local collapse should be occurring. Further, the physical implications of positing the existence of an infinite number of stars remained quite mysterious. It was not until the discovery of the expansion of the universe that a satisfactory explanation of the gravitational problem was achieved.

Newton's conception of an infinitely large universe with an infinite number of stars also presented another problem. This problem, first noticed by Kepler in 1610 and discussed by Edmund Halley later in the seventeenth century, was clearly and succinctly stated by an astronomer named Heinrich Olbers in 1823 and ever since has been called Olbers's Paradox. Olbers, following Newton, assumed that the universe was composed of an infinite number of stars of similar luminosity uniformly distributed through space. From an observer's vantage point there should not be a single point in the sky that is not occupied by the surface of a star. The night sky should be bright, perhaps as bright as the daytime sky. Yet the night sky is, of course, quite dark. It was not until the universe was determined to have a finite age that a satisfactory explanation of Olbers's Paradox was achieved.[1]

The Discovery of the Large-Scale Structure of the Universe

The movement away from Newtonian cosmology began with the work of the great German-English astronomer William Herschel in the eighteenth century. Using his engineering skills Herschel built a twenty-foot reflecting telescope

---

1. The finite age of the universe combined with the finite velocity of light would eliminate Olbers's Paradox even if the universe were infinite in size.

with a twenty-inch aperture. This extraordinary instrument enabled Herschel to observe numerous fuzzy patches of light in the heavens, which he called nebulae. Although he was unable to measure the distances to these nebulae, he considered them all to be island universes—star systems similar to, but independent of, the Milky Way, which was the system in which our sun and the solar system of planets were located. By the end of his career Herschel had catalogued five thousand of these nebulae, but he never realized that only some of them were island universes while others were within the Milky Way. Herschel's work led the philosopher Immanuel Kant in 1755 to speculate that the universe contained an infinite number of island universes or galaxies similar in structure to the Milky Way. Unfortunately, astronomers at the time were unable to measure distances to the nebulae, nor were they able to show that the nebulae were indeed star systems. Thus astronomers were unable to either confirm or disconfirm the Herschel-Kant theory that the universe is composed of large star systems.

The great breakthrough occurred in 1912 when the astronomer Henrietta Leavitt of the Harvard Observatory made an extraordinary discovery concerning variable stars in the Magellanic Clouds. These stars were known to vary in brightness over very specific periods of time, ranging from several days to one month. By observing a cluster of these stars essentially the same distance from earth, Leavitt discovered that their cycles depended upon their luminosity: the brighter the star, the longer the period of variation. In 1917 Harlow Shapley was able to measure the actual distances to these stars, which he identified as Cepheid variables, and thus was able to determine their intrinsic luminosity. Once this was achieved Shapley had a method of measuring distances to other Cepheids. From their period of variation he could determine their intrinsic luminosity, and by comparing their intrinsic luminosity to their observed luminosity he could calculate their distance.

In the early 1920s Shapley used the Cepheids to map out the Milky Way and develop a rough estimate of its size. Because he failed to consider the absorption of light from the Cepheids by interstellar dust, Shapley's estimate of 300,000 light years for the diameter of the Milky Way was much too large. Today we consider its diameter to be between 90,000 and 100,000 light years. Although wrong by a factor of three, Shapley's estimate proved to be very valuable in the determination of the large-scale structure of the universe. In 1923 the astronomer Edwin Hubble used the telescope on Mount Wilson to resolve the Andromeda nebula into individual stars, thus proving that at least some of the nebula was star systems. Since some of the stars in the Andromeda nebula were Cepheids, Hubble was able to use the period-luminosity-distance relationship to calculate that the nebula was 900,000 light years away. Obviously this nebula was a galaxy unto itself. Further investigations showed other nebulae to be independent galaxies as well. Herschel's insight and Kant's speculation had turned out to be quite prophetic. The universe was indeed filled with innumerable galaxies at great distances from one another.

The Red Shift and the Expansion of the Universe

While Leavitt, Shapley, and Hubble were determining the large-scale structure and dimensions of the universe, another astronomer, Vesto Slipher, was making an equally important discovery. Slipher had been hired by Percival Lowell to come to the Lowell Observatory in Flagstaff, Arizona, to determine whether spiral nebulae were actually developing planetary systems in development. This he hoped to accomplish by analyzing the spectra of these nebulae for the purpose of detecting their spinning motion.

The theoretical foundation for Slipher's task was something called the Doppler effect as applied to light. It had been determined that sound waves coming from a source moving toward an observer would be compressed, causing an increase in frequency and a higher pitch, while sound waves coming from a source moving away from an observer would be elongated, decreasing the frequency and lowering the pitch. Such a phenomenon is commonly observed. By analogy it was thought that wavelengths of light from a source moving toward an observer would be compressed, or shifted toward the blue end of the visible spectrum, while wavelengths of light from a source moving away from an observer would be elongated, or shifted toward the red end of the visible spectrum. If spiral nebulae were indeed spinning, light from one edge should be blue-shifted and light from the other edge red-shifted. When Slipher made his measurements for the Andromeda nebula in 1912 he discovered that light from the entire nebula was blue-shifted, indicating that the nebula as a whole was moving toward us. Over the next ten years, however, Slipher's analyses of the spectra of forty one nebulae produced a very definite pattern. Light from thirty six of them exhibited a red shift. Most of the nebulae were moving away from us. Further, since the amount of red shift in the spectra indicated how fast the nebulae were moving away from us, it followed that there was a variety of recessional velocities.

In 1929 Hubble decided to compare Slipher's measurements of recessional velocities with his own measurements of nebular distances. What he discovered was that the farther from us the nebulae, now known to be galaxies, the greater the red shift of their light and thus the greater their recessional velocity. This velocity-distance relationship came to be known as Hubble's Law. Hubble's work further revealed that the recession of the galaxies was the same in all directions. The universe of galaxies was indeed an expanding universe. If this expansion were traced backward in time, the galaxies would seem to collapse to a single point at some definite moment in the past. The concept of a great cosmic explosion, taking place at that moment, and starting the expansion process, was obviously implied. The era of Big Bang Cosmology had begun.

Theoretical Development After Relativity

Theoretical developments in cosmology during the first thirty years of the twentieth century proceeded independently of the experimental developments just outlined. While developing his theories of relativity Einstein seemed to be un-aware of the discoveries of Henrietta Leavitt and Vesto Slipher. However, the relativity theories have obvious cosmological ramifications. Special Relativity suggests that space and time are not rigid and immutable but rather are com-bined in a dynamic interrelationship. General Relativity is a theory of gravity in which the structure of space and the passage of time depend upon the local distribution of matter. Changes in this distribution would produce changes in the structure of space and the passage of time. Because of this it was impossible to construct a mathematical model of a static universe in the context of Relativity Theory.

Einstein was convinced as of 1915, however, that the stars are fixed and that the universe is static. If this were so, the problem of gravity that bedeviled Newton would be an even greater problem for him. His solution was to add a cosmological constant to his equations. This constant amounted to a repulsive force that operated at large distances and just balanced the force of gravity, preserving a static model of the universe.

As acceptance of the Theory of Relativity grew, scientists began to use it for the development of cosmological theory. In 1917 the Dutch astronomer Willem de Sitter constructed a cosmological theory that was an alternative to Einstein's. De Sitter thought the mass density of the universe was so low as to be irrelevant. Keeping the cosmological constant he found a static solution to the equations of relativity for a universe that contained no matter. In itself that static solution was not very fruitful. However, in 1923 Arthur Eddington and Herman Weyl decided to see what would happen if they introduced some particles of matter into the de Sitter universe. They discovered that these particles should fly apart from one another. The idea of an expanding universe had become part of cosmo-logical theory.

In a separate but even more important development Alexander Friedmann, a Russian mathematician, discovered a series of solutions to Einstein's equations during the years 1922 to 1924. Friedmann rejected the idea of a static universe and thus was able to work out solutions that were free of the cosmological constant. He found that in those solutions the universe was expanding and that the process of expansion depended upon the mass density of the universe. If the mass density of the universe was sufficiently high, then gravity would cause the expansion to slow down, stop, and be followed by a period of contrac-tion. Such a universe would be *closed*, and finite in space and time. If the mass density of the universe was sufficiently low, gravity would be too weak to halt the expansion. This would result in an *open* universe that is infinite in space and time. If the mass density happened to be at the boundary between an open and a closed universe, the universe would theoretically expand forever but the expansion rate would approach zero. Such a universe would be called a *flat* universe.

The first theoretical cosmologist to work with the knowledge that the universe was actually expanding was Georges Lemaître. At MIT in 1924 and 1925 he was privy to the work of Hubble and Shapley. In 1927 Lemaître wrote a paper describing the evolution of an expanding universe. The paper included information about the velocity-distance relationship actually before Hubble himself had released it. Lemaître found that he had to include Einstein's cosmological constant to make the theory fit the observational data. It turned out that the cosmological constant was necessary because he was using an erroneous value for the Hubble constant. This constant is a measure of the rate of expansion of the universe and appears in Hubble's Law relating the distance of a galaxy to its recessional velocity. Hubble's original calculation of the value of the constant was too large by a factor of ten.

Lemaître's paper had suggested that the expansion of the universe could not have been going on forever and thus there must have been a beginning to the process. Although accepting the reality of expansion, Eddington found the idea of a beginning scientifically repugnant and so stated in an article in *Nature* in 1931. Lemaître responded immediately. Using the rapidly developing field of quantum physics as support, he suggested that the idea of a beginning to the universe was not only not repugnant but consistent with the state of science at that time. Two years later, in a lecture at the Mt. Wilson Observatory, Lemaître mused that the expansion of the universe had begun with a great fiery explosion in the distant past, and the Big Bang hypothesis was born. By 1935 the hypothesis had developed into a detailed and mature theory. In that theory the universe was understood to have begun with the radioactive disintegration (explosion) of a primeval atom of great concentration and extremely small size. Lemaître refined his theory for the next fifteen years and a popular account of his final thoughts on the subject was published in 1950.

Although the father of Big Bang cosmology, Lemaître was not the one who coined the phrase. The phrase "Big Bang" was first used in the 1940s by the astronomer Fred Hoyle as a sign of derision. Not only did Hoyle despise the idea that the universe had some kind of beginning, but he along with Herman Bondi and Thomas Gold, developed an alternative hypothesis which came to be known as the Steady State Theory, which considered the universe to be infinitely large and infinitely old. As the expansion proceeded, new matter was created in the intervening space. By 1950 the Steady State Theory could explain the known phenomena as well as Big Bang Theory. Some empirical evidence was needed to distinguish them.

Confirmation of Big Bang Cosmology

An advocate of Big Bang Theory, George Gamow, offered a suggestion as to what kind of evidence might allow for a choice between the two theories. Gamow was a former student of Alexander Friedmann and so was well schooled in theoretical cosmology. But he was also a student of atomic and nuclear physics. Gamow felt that if the Big Bang cosmology of Lemaître was correct, then all the elements that now appear in the universe must have been produced in the first moments of that cosmic explosion. As a consequence of the formation of these elements there should be the remnants of some primordial radiation present throughout the universe. The detection of this radiation would be evidence in favor of the Big Bang.

It turned out that Gamow was wrong in thinking that all the elements were formed in the Big Bang. In fact, further developments demonstrated that helium could have been so synthesized but not the heavier elements. When it was discovered that helium constitutes one-third the mass of the universe, second only in abundance to hydrogen, Big Bang theorists felt that it must have been synthesized in the Big Bang. Thus there should be some primordial radiation available to detect, although not exactly of the type that Gamow had originally suggested.

Through the 1950s and into the 1960s further refinements were made concerning the nature and location of this primordial radiation. For example, it was felt that the radiation would be black- body radiation possessing a perfect black-body spectrum. Further, its peak wavelength should correspond to a temperature of 3° above absolute zero. Despite the fact that they had a very precise prediction to test, astronomers made no immediate effort to do so. The discovery of the background radiation was finally made in 1964 through research with communications satellites. Physicists Arno Penzias and Robert Wilson at the Bell Telephone Laboratories in New Jersey, using a relatively unsophisticated radio antenna, kept detecting background noise that they could not eliminate. Further investigation showed that this background noise fitted the prediction of the Big Bang cosmologists precisely.

As Steven Weinberg was detailed in his famous book *The First Three Minutes*, this background noise quickly came to be understood as the radiation echo of the Big Bang. The Big Bang Theory thus secured its victory over the Steady State Theory. It was now clear that the universe had some kind of beginning at a definite moment in the past. Even Hoyle became a supporter and admitted that the Steady State Theory was wrong. But there were still problems in the Big Bang model that prevented its final acceptance.

The first of these problems was called the *horizon problem*. The horizon is the radius of the visible universe and is equal to the age of the universe times the speed of light. If the expansion of the universe had been uniform from its beginning and we retrace its history, then we come to a time 10^{-35} seconds after the Big Bang when the visible universe was 3 mm in radius. But at this time light and any other electromagnetic signals would have traveled only 10^{-26} mm. If we assume relativity theory to be accurate, then there would be regions of the universe that would be totally independent of one another. The problem would be simply this: how did all these separate regions come to have precisely the same temperature and expansion rate when there was insufficient time for the transfer of heat and energy that would have been necessary to produce such homogeneity?

The second of these problems, called the problem of *magnetic monopoles*, arose because of the incorporation of the newly developed grand unified theories into cosmology. Grand unified theories are theories that unify three of the four known forces, namely, the electromagnetic force with the weak and strong nuclear interaction forces (only gravity is omitted). When these theories were combined with the standard big bang model they predicted that a large number of magnetic monopoles would have been produced in the early universe. A magnetic monopole would be a magnet with just one pole, either a north or a south. The problem is that no such monopoles have ever been detected.

The *flatness problem* is related to something called the omega number. This number is the ratio of the actual mass density of the universe to the critical mass density. The critical mass density would be the boundary mass density between an open and a closed universe. If the omega number is greater than one, the universe would be closed and the galaxies now receding from one another would eventually be recalled. If the omega number is less than one, the universe would be open and the expansion would continue forever. In the original Big Bang model the omega number would have to be exactly one or else the universe would not have turned out the way ours did. Big Bang Theory offered no explanation for this precise value but simply assumed it to be the case as an arbitrary postulate. This problem, as of 1980, was further compounded by the fact that the measured mass density of the universe produced an omega number of much less than one.

Inflationary Theory

The solutions to these problems became possible when an MIT physicist, Alan Guth, developed Inflationary Theory and attached to it the suggestion that the universe contains huge quantities of dark matter, which cannot be observed with optical telescopes. Inflationary theory suggests that the universe, when it was only 10^{-35} seconds old, underwent something like a phase change that released tremendous amounts of energy. This energy fueled an incredibly rapid expansion that caused the universe to move from the size of a proton to the size of a basketball (an increase by a factor of 10^{50}) between the times of 10^{-35} seconds and 10^{-32} seconds. In other words, the size of the universe increased by an incredible amount in an infinitesimal amount of time. At the end of this inflationary era the rate of expansion slowed down to approximately the pace we observe today.

Inflationary Theory goes a long way toward solving the problems that had arisen in the Big Bang model. Before the inflationary era the minute parts of the universe were in sufficiently close proximity to communicate with (influence) one another. After inflation the parts were too distant from one another to communicate, but the influence had already taken place. Further, very rapid inflation would have the tendency to smooth out any irregularities, allowing the universe to appear homogeneous and isotropic in all directions. So the horizon problem is eliminated. Inflation also explains our inability to observe the magnetic monopoles whose existence is predicted by the grand unified theories. The inflationary era would have so diluted the magnetic monopole density that there might be only one such monopole in the entire visible portion of the universe. Obviously the chance of detecting such monopoles would be very slight.

Inflation also solves the flatness problem by showing that any observable part of the universe would appear to be flat after such a rapid expansion. Further, Inflationary Theory clearly predicts that the omega number is one, such that the average mass density and the critical mass density are equal. Although the average mass density currently observed and measured is less than the critical mass density by a factor of ten, the theory shows that they could not be so close today if they had not been equal to one another in the early stages of the universe. That is, the omega number must have been one in the early stages of the universe, and if it had started out as one it would be at one today.

According to Inflationary Theory, the missing "dark" matter that was needed to bring the measured omega number to one played a very important role in the early universe. Much of the energy that produced the rapid expansion would have been turned into matter as predicted by the Theory of Relativity. Some of that matter would not be in the form of protons and electrons, and in fact would be in much greater abundance than the protons and electrons or the hydrogen and helium that they formed. This undefined dark matter would, through its gravitational attraction, cause the clumping of hydrogen and helium that was necessary for the formation of stars and galaxies over the next few billion years.

The Final Piece of the Puzzle

If Inflationary Theory is correct there should be very small irregularities in the cosmic background radiation produced by the clumping of hydrogen and helium. As of 1989 measurements of the background radiation made from the surface of the earth indicated a radiation distribution too smooth to allow for the clumping necessary for the production of galaxies. If the irregularities were actually there, however, they could only be detected by instruments free of the interference of earthbound noise. So in late 1989 researchers launched a satellite designed to detect the irregularities. The satellite was called the Cosmic Background Explorer, or simply COBE.

Essentially the fate of the entire Big Bang cosmology depended upon the results of the COBE experiment. If the irregularities did not exist, then Inflationary Theory could not explain the existence of stars and galaxies and thus would be called into serious question. Without Inflationary Theory, Big Bang Cosmology could not solve its other problems. One can only imagine the excitement among cosmologists when, in April 1992, it became clear that COBE had detected precisely the irregularities predicted by the theory. Astronomer George Smoot, the leader of the COBE project, announced to the world that they had confirmed Big Bang cosmology: "It was like looking at the face of God." The British physicist Stephen Hawking called the COBE results the discovery of the century, if not of all time.

Perhaps the science writer John Gribbin captured the magnitude of the moment best when he began the prologue to his book *In the Beginning* with the following words:

> Astronomers breathed a huge collective sigh of relief in the spring of 1992, when the most important prediction they had ever made was proved correct at the eleventh hour. The dramatic discovery of ripples in the structure of the universe dating back almost 15 thousand million years has set the seal on twentieth-century science's greatest achievement—the Big Bang Theory—which explains the origin of the universe and everything in it, including ourselves. Slotting in this missing piece of the cosmic jigsaw puzzle confirms that the universe really was born out of a tiny, hot fireball all that time ago, and has been expanding ever since.[2]

Big Bang Cosmological Theory is now a permanent fixture in science. Its role in the history of science for the indefinite future is guaranteed.

1

— Henrietta Leavitt —

(1868–1921)

American astronomer Henrietta Leavitt joined the Harvard Observatory in 1895. From 1904 to 1908, she made a thorough study of variable stars in the Small and Large Magellanic Clouds. In 1912, she was able to confirm that there is a simple relationship between the period of variability and the intrinsic luminosity of stars known as Cepheid variables. The relationship was later used to measure stellar distances and became a key element in the mapping of the large-scale structure of the universe. In 1915, Harlow Shapley used the relationship to develop the first picture of the size and structure of our galaxy. Later, Edwin Hubble used the relationship to prove that some nebulae were actually distant galaxies.

In this reading Leavitt describes her initial discovery of variable stars in the Magellanic Clouds and relates their characteristics. It appeared in the Annals of the Harvard College Observatory *in 1908.*

---

2. John Gribbin, *In the Beginning* (Boston: Little, Brown, 1993), ix.

VARIABLES IN THE MAGELLANIC CLOUDS

In the spring of 1904, a comparison of two photographers of the Small Magellanic Cloud, taken with the 24-inch Bruce Telescope, led to the discovery of a number of faint variable stars. As the region appeared to be interesting, other plates were examined, and although the quality of most of these was below the usual high standard of excellence of the later plates, 57 new variables were found, and announced in Circular 79. In order to furnish material for determining their periods, a series of sixteen plates, having exposures of from two to four hours, was taken with the Bruce Telescope the following autumn. When they arrived at Cambridge, in January, 1905, a comparison of one of them with an early plate led immediately to the discovery of an extraordinary number of new variable stars. It was found, also, that plates, taken within two or three days of each other, could be compared with equally interesting results, showing that the periods of many of the variables are short. The number thus discovered, up to the present time, is 969. Adding to these 23 previously known, the total number of variables in this region is 992. The Large Magellanic Cloud has also been examined on 18 photographs taken with the 24-inch Bruce Telescope, and 808 new variables have been found, of which 152 were announced in Circular 82. As much time will be required for the discussion of these variables, the provisional catalogues given below have been prepared.

The labor of determining the precise right ascensions and declinations of nearly eighteen hundred variables and several hundred comparison stars would be very great, and as many of the objects are faint, the resulting positions could not readily be used in locating them. Accordingly, their rectangular coordinates have been employed. A reticule was prepared by making a photographic enlargement of a glass plate ruled accurately in squares, a millimetre on a side. The resulting plate measured 14×17 inches, the size of the Bruce places, and was covered with squares measuring a centimetre on a side. Great care was taken to have the scale uniform in all parts of this plate, which was designed to furnish a standard reticule, not only for the Magellanic Clouds, but for any other region in which it may be desirable to measure a large number of objects. A glass positive was then made from a photograph of each of the Magellanic Clouds, and from this a negative on glass was printed, upon which a print from the plate containing the reticule was superposed. The resulting photograph in each case was a duplicate of the original negative, with the addition of a reticule whose lines are one centimetre apart, a distance corresponding, on these plates, to ten minutes of arc. For measuring objects on the plates, a scale was made in the manner described in Volume 26, 238, but having each division equal to a third of a millimetre. This was attached to a positive eye-piece having a focal length of an inch and a half, and the measures were made and reduced as described in Volume 26, 238. One division of the scale equals $20''$, and estimates were made to tenths. This amount, however, is reduced one half when the differences of the measures from opposite sides of the squares are taken, and the measurements, therefore, were made to single seconds. The measured positions have not been reduced to standard coordinates, but a number of catalogue stars were measures, and furnish the means of making this reduction if desired. Practically, however, the labor involved seems unnecessary for the present purpose, as reproductions of the plates measured accompany this article, and the approximate positions of objects in the catalogues can readily be found on them.

The variables appear to fall into three or four distinct groups. The majority of the light curves have a striking resemblance, in form, to those of cluster variables. As a rule, they are faint during the greater part of the time, the maxima being very brief, while the increase of light usually does not occupy more than from one-sixth to one-tenth of the entire period. It is worthy of notice that the brighter variables have the longer periods. It is also noticeable that those having the longest periods appear to be as regular in their variations as those which pass through their changes in a day or two. This is especially striking in the case of No. 821, which has a period of 127 days, as 89 observations with 45 returns of maximum give an average deviation from the light curve of only six hundredths of a magnitude. Six of the sixteen variables are brighter at maximum than the fourteenth magnitude, and have periods longer than eight days. The number which have been measured up to the present time is 59, and of these the brighter stars were first selected for discussion, as the material for them was more abundant. A few of the fainter variables, selected at random, were then studied but no attempt has yet been made to determine periods for the remainder. While, therefore, the light curves thus far obtained have characteristics to which the majority of the variables will probably be found to conform, no inference can be drawn with regard to the prevalence of any particular type, until many more of the periods have been determined.

The distribution of the variables in the two regions differs greatly though not more than the regions themselves differ. The Small Cloud is very compact in its formation, the preceding and southern edges being sharply defined, while the limits on the north and following sides are less definite. The distribution of variables closely follows this configuration, as has already been shown in Circular 96, Table I. The formation of the Large Magellanic Cloud is more open. Its most characteristic feature is a stream of faint stars extending between the group of clusters near N.G.C. 1850, and the remarkable nebula, N.G.C. 2070. The majority of the variables, especially of those fainter at maximum than the fourteenth magnitude, are found in this region. There are numerous other collections of stars, in connection with which groups of variables are found. Comparatively few variables have been discovered, however, in regions where large numbers of stars are bright as the thirteenth magnitude are collected.

2
— Henrietta Leavitt —

In this reading, Leavitt describes her discovery of the remarkable relationship between the period of variability and the intrinsic luminosity of these variable stars. As mentioned, this relationship eventually gave astronomers the ability to measure great distances. The reading comes from a Harvard College Observatory Circular *published in 1912.*

THE VARIABILITY-LUMINOSITY RELATIONSHIP

A Catalogue of 1,777 variable stars in the two Magellanic clouds is given in H. A. 60, No. 4. The measurement and discussion of these objects present problems of unusual difficulty, on account of the large area covered by the two regions, the extremely crowded distribution of the stars contained in them, the faintness of the variables, and the shortness of their periods. As many of them never become brighter than the fifteenth magnitude, while very few exceed the thirteenth magnitude at maximum, long exposures are necessary, and the number of available photographs is small. The determination of absolute magnitudes for widely separated sequences of comparison stars of this degree of faintness may not be satisfactorily completed for some time to come. With the adoption of an absolute scale of magnitudes for stars in the North Polar Sequence, however, the way is open for such a determination.

Fifty-nine of the variables in the Small Magellanic Cloud were measured in 1904, using a provisional scale of magnitudes, and the periods of seventeen of them were published in H. A. 60, No. 4, Table VI. They resemble the variables found in globular clusters, diminishing slowly in brightness, remaining near minimum for the greater part of the time, and increasing very rapidly to a brief maximum. Table I gives all the periods which have been determined thus far, 25 in number, arranged in the order of their length. The first five columns contain the Harvard Number, the brightness at maximum and at minimum as read from the light curve, the epoch expressed in days following J. D. 2,410,000, and the length of the period expressed in days. The Harvard Numbers in the first column are placed in Italics, when the period has not been published hitherto. A remarkable relation between the brightness of these variables and the length of their periods will be noticed. In H. A. 60, No. 4, attention was called to the fact that the brighter variables have the longer periods, but at that time it was felt that the number was too small to warrant the drawing of general conclusions. The periods of 8 additional variables which have been determined since that time, however, conform to the same law.

TABLE I.

PERIODS OF VARIABLE STARS IN THE SMALL MAGELLANIC CLOUD.

H.	Max.	Min.	Epoch.	Period.	Res. M.	Res. m.	H.	Max.	Min.	Epoch.	Period.	Res. M.	Res. m.
			d.	d.						d.	d.		
1505	14.8	16.1	0.02	1.25336	−0.6	−0.5	1400	14.1	14.8	4.0	6.650	+0.2	−0.3
1436	14.8	16.4	0.02	1.6637	−0.3	+0.1	1355	14.0	14.8	4.8	7.483	+0.2	−0.2
1446	14.8	16.4	1.38	1.7620	−0.3	+0.1	1374	13.9	15.2	6.0	8.397	+0.2	−0.3
1506	15.1	16.3	1.08	1.87502	+0.1	+0.1	818	13.6	14.7	4.0	10.336	0.0	0.0
1413	14.7	15.6	0.35	2.17352	−0.2	−0.5	1610	13.4	14.6	11.0	11.645	0.0	0.0
1460	14.4	15.7	0.00	2.913	−0.3	−0.1	1365	13.8	14.8	9.6	12.417	+0.4	+0.2
1422	14.7	15.9	0.6	3.501	+0.2	+0.2	1351	13.4	14.4	4.0	13.08	+0.1	−0.1
842	14.6	16.1	2.61	4.2897	+0.3	+0.6	827	13.4	14.3	11.6	13.47	+0.1	−0.2
1425	14.3	15.3	2.8	4.547	0.0	−0.1	822	13.0	14.6	13.0	16.75	−0.1	+0.3
1742	14.3	15.5	0.95	4.9866	+0.1	+C.2	823	12.2	14.1	2.9	31.94	−0.3	+0.4
1646	14.4	15.4	4.30	5.311	+0.3	+0.1	824	11.4	12.8	4.	65.8	−0.4	−0.2
1649	14.3	15.2	5.05	5.323	+0.2	−0.1	821	11.2	12.1	97.	127.0	−0.1	−0.4
1492	13.8	14.8	0.6	6.2926	−0.2	−0.4							

The relation is shown graphically in Figure 1, in which the abscissas are equal to the periods, expressed in days, and the ordinates are equal to the corresponding magnitudes at maxima and at minima. The two resulting curves, one for maxima and one for minima, are surprisingly smooth, and of remarkable form. In Figure 2, the abscissas are equal to the logarithms of the periods, and the ordinates to the corresponding magnitudes, as in Figure 1. A straight line can readily be drawn among each of the two series of points corresponding to maxima and minima, thus showing that there is a simple relation between the brightness of the variables and their periods. The logarithm of the period increases by about 0.48 for each increase of one magnitude in brightness. The residuals of the maximum and minimum of each star from the lines in Figure 2 are given in the sixth and seventh columns of Table I. It is possible that the deviations from a straight line may become smaller when an absolute scale of magnitudes is used, and they may even indicate the corrections that need to be applied to the provisional scale. It should be noticed that the average range, for bright and faint variables alike, is about 1.2 magnitudes. Since the variables are probably at nearly the same distance from the Earth, their periods are apparently associated with their actual emission of light, as determined by their mass, density, and surface brightness.

The faintness of the variables in the Magellanic Clouds seems to preclude the study of their spectra, with our present facilities. A number of brighter variables have similar light curves, as UY Cygni, and should repay careful study. The class of spectrum ought to be determined for as many such objects as possible. It is to be hoped, also, that the parallaxes of some variables of this type may be measured. Two fundamental questions upon which light may be thrown by such inquiries are whether there are definite limits to the mass of variable stars of the cluster type, and if the spectra of such variables having long periods differ from those of variables whose periods are short.

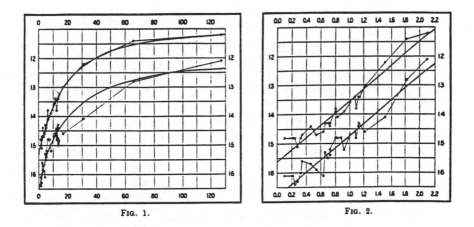

FIG. 1.   FIG. 2.

The facts known with regard to these 25 variables suggest many other questions with regard to distribution, relations to star clusters and nebulae, differences in the forms of the light curves, and the extreme range of the length of the periods. It is hoped that a systematic study of the light changes of all the variables, nearly two thousand in number, in the two Magellanic Clouds may soon be undertaken at this Observatory.

3
— Vesto Slipher —
(1875–1969)

Vesto Slipher, an American astronomer, was the first to discover that light from galaxies was shifted either to the blue or red end of the visible spectrum. Slipher had been hired by the Lowell Observatory in Arizona to measure the rotational velocities of galaxies when he made his extraordinary discovery. Ultimately Slipher found that light from the vast majority of galaxies exhibited a red shift, suggesting that these galaxies are moving away from us and that the universe is expanding.

In this reading Slipher describes the discovery of the blue shift of light from the Andromeda nebula. He concludes that the nebula is approaching the solar system with a velocity of 300 kilometers per second. The reading comes from the Bulletin of the Lowell Observatory of 1912.

THE RADIAL VELOCITY OF THE ANDROMEDA NEBULA

Keeler, by his splendid researches on the nebulae, showed, among other things that the nebulae are generally spiral in form, and that such nebulae exist in far vaster numbers than had been supposed. These facts seem to suggest that the spiral nebula is one of the important products of the forces of nature. The spectra of these objects, it was recognized, should convey valuable information, and they have been studied, photographically, first by Huggins and Scheiner, and recently more extensively by Fath and Wolf; but no attempt has to my knowledge been made to determine their radial velocity, although the value of such observations has doubtless occurred to many investigators.

The one obstacle in the way of the success of this undertaking is the faintness of these nebulae. The extreme feebleness of their dispersed light is difficult to realize by one not experienced in such observing, and it no doubt appears strange that the magnificent Andromeda Spiral, which under a transparent sky is so evident to the naked eye, should be so faint spectrographically. The contest is with the low intrinsic brightness of the nebular surface, a condition which no choice of telescope can relieve. However, the proper choice of parts in the spectrograph will make the best of this difficulty. The collimator must of course fit the telescope, but the dispersion-piece and the camera may and should be carefully selected for their special fitness for the work. While the speed of the camera is all-important in recording the spectrum, the detail in the spectrum depends upon the dispersion, for obviously a line, no matter how dark it may be, must have a certain magnitude or else it cannot be recorded by the granular surface of the photographic plate. Hence the light must be concentrated by a camera of very short focus and the dimension of the spectral line be increased by using a high angular dispersion and a wider slit, as one in this way attains a higher resolving power in the photographed spectrum.

Although I had made spectrograms of the Andromeda Nebula a few years ago, using the short camera, it was not until last summer that I thought to employ the higher dispersion and the wider slit. The early attempts recorded well the continuous spectrum crossed by a few Fraunhofer groups, and were particularly encouraging as regards the exposure time required. The first of the recent plates was exposed for 6 hours and 50 minutes, on September 17, 1912, using a very dense 64 degree prism, the instrument having already been tried out on some globular star clusters. When making this exposure the brightness of the nebula on the slit-plate compared with that of the clusters indicated that one night's exposure should suffice for the single-prism, and suggested that, by extending the exposure through several nights, one could employ the battery of three dense flint prisms whose dispersion would make it possible to observe the velocity of the nebula. The success of the plate bore out this suggestion. Indeed, upon subsequent examination of this plate it was seen that the nebular lines were perceptibly displaced with reference to the comparison lines. The next plate secured showed the same displacement. Still other single-prism plates were obtained during the autumn and early winter, but the observing program with the 24-inch telescope did not allow an opportunity to carry out the original plan to make the longer exposure spectrogram with the prism-train.

These spectrograms were measured with the Hartmann spectrocomparator, using a magnification of fifteen diameters. A similar plate of Saturn was employed as a standard. The observations were as follows:

1912, September 17,	Velocity, 284 km.
November 15–16,	" 296
December 3–4,	" 308
December 29–30-31,	" 301

Mean velocity: 300 km.

Tests for determining the degree of accuracy of such observations have not been completed, but in rounding off to 300 kilometers in taking the mean one is doubtless well within the accuracy of the observations. The measures extended over the region of spectrum from F to H.

The conditions were purposely varied in making the observations. This was done although it was early noted that the shift at the violet end of the spectrum was fully twice that at the blue end, which should be the case if it were due to velocity.

The magnitude of this velocity, which is the greatest hitherto observed, raises the question whether the velocity-like displacement might not be due to some other cause, but I believe we have at the present no other interpretation for it. Hence we may conclude that the Andromeda Nebula is approaching the solar system with a velocity of about 300 kilometers per second.

This result suggests that the nebula, in its swift flight through space, might have encountered a dark "star," thus giving rise to the peculiar nova that appeared near the nucleus of the nebula in 1885.

That the velocity of the first spiral observed should be so high intimates that the spirals as a class have higher velocities than do the stars and that it might not be fruitless to observe some of the more promising spirals for proper motion. Thus extension of the work to other objects promises results of fundamental importance, but the faintness of the spectra makes the work heavy and the accumulation of results slow.

4
— Vesto Slipher —

In this reading Slipher describes the discovery of a very strong red shift in the spectral lines of Nebula 1068 and concludes that it is receding from the solar system at a velocity of 1,100 kilometers per second. He also announces red shifts in the spectral lines of the vast majority of nebulae studies. This reading also comes from the Bulletin of the Lowell Observatory *of 1912.*

THE DISCOVERY OF THE RED SHIFT FOR NEBULAE

The nebula numbered 1068 in Dreyer's *New General Catalogue* of nebulae and clusters (Messier 77), was first photographed by Roberts and he reproduced a plate of it, made November 26, 1892, in Plate 10 of his first volume of *Celestial Photographs*. Lord Rosse and Lassell regarded the nebula as a spiral and the photographs show this to be its form. A good photograph of this object by Keeler is reproduced in Plate 7, Volume VIII, *Publications of the Lick Observatory*.

Fath was the first to observe the spectrum of this nebula. He made two plates of it in August and September, 1908, and found a composite spectrum, some dark lines and the chief emission lines typical of the gaseous nebulae.

I first photographed the spectrum at Flagstaff on November 6, 1913, with an exposure of six and one-half hours. Because of the interesting nature of this plate a second one was made on the nights of November 22 and 23, 1913, with about double the exposure of the first plate, but the sky was poor both nights and the negative is not much stronger than the first plate. For both these plates the slit stood E and W across the nebula.

These plates confirmed Fath's analysis of the spectrum and showed that the dark lines were Fraunhofer lines. In addition they revealed in both series of lines a very large shift toward the red. Measurement of the displacement implied that the nebula was receding from the *Sun* with a velocity of about 1100 km. per second. A brief account of these observations was given—along with similar results for other nebulae—at the Evanston meeting (1914) of the American Astronomical Society.

Since then the spectrum has been observed at Mount Wilson by Pease, who secured a somewhat less high value for the velocity based upon measures of the two chief emission lines. Later Moore at the Lick Observatory from the three nebular emission lines got a radial velocity intermediate between the Flagstaff and Mount Wilson values.

Recently I have again observed the spectrum of the nebula under more favorable conditions and secured, with a narrower slit, a plate on the nights of November 6, 7, and 8. This plate revealed a further interesting feature of the spectrum—namely, that the emission lines typical of gaseous nebulae are not images of the slit of the spectrograph but are instead small disks. This could be inferred from the better of the 1913 plates, but it seemed desirable to secure still another plate under the higher dispersion of two 64°.5 prisms. An exposure was made from November 12 to 16, 1917, with a duration of thirty-five hours. Both of the 1917 plates were made with the slit of the spectrograph turned upon the major axis of the brighter central part of the nebula. In these and other spectrographic observations, I have had for some months assistance from Mr. G. H. Hamilton.

In addition to confirming the earlier plates in regard to the composite nature of the spectrum, the extremely high radial velocity and the peculiar disk-like images of the emission lines, the high scale spectrogram brings out what had not been recognized before—that the hydrogen lines extend farther into the fainter parts of the nebula than do the nebulium lines N_1 and N_2, and that these lines are strongly inclined. (They are perhaps also somewhat concave toward the violet.) On it the dark lines are faintly traceable into the fainter parts of the nebula and these, too, apparently incline with the long emission lines.

The nebula is elongated in a south-preceding north-following direction and the inclination of the lines implies that the southwestern portion of the nebula is receding and the northeastern portion approaching, relatively, in consequence of rotation about an axis through the shorter diameter of the nebula.

The lines are inclined through an angle of about 5°, an amount corresponding to the extreme speed of rotation of about 300km. at 1′ from the nucleus. Further observations are necessary to give accurately the velocity of rotation. It seems that it is even higher for this nebula than for the Virgo Nebula N.G.C. 4594 which has the highest rotational speed of any object hitherto observed.

These recent spectrograms have been measured for radial velocity. The results for the four spectrograms—two of 1913 and two of 1917—are as follows:

Plate 1913, November 6,	Velocity +1060 km.
Plate 1913, November 22, 23,	Velocity +1150
Plate 1917, November 6, 7, 8,	Velocity +1080
Plate 1917, November 12, 16 (two prisms)	Velocity +1130 s
	Velocity +1145 c
	Velocity +1135 c

Mean Velocity: +1120 km.

The measures were made upon the emission lines N_1, N_2, H_β, H_γ, and upon the dark lines and bands chiefly in the H_γ region which apparently are the same as those of the solar spectrum used as standard in the comparator measurements.

The last plate was thrice measured, once with the screw measuring engine and twice with the spectrocomparator, and the three results are given separately. The single prism plates were measured with the comparator. Because of the higher accuracy of the last plate the mean of the six measures was taken, which is practically what would have resulted from weighting the plates. The values from the several plates are in good agreement. But the velocities are somewhat larger than those got at Mount Wilson (765 km.) and at Lick (910 km.). However, there seems otherwise to be no evidence that my velocities of nebulae are too large, for they have been verified at Mount Wilson and at Lick for the Great Andromeda Nebula and at the former for N.G.C. 4594, with results slightly larger than mine. In the present case the Mount Wilson and Lick observers were doubtless at some disadvantage since they were limited to measures on two and three lines, respectively, which were at the end of the spectrum of lowest dispersion.

I wish to add one further comment on the accuracy of these nebular velocity observations lest the divergence in results for this nebula be taken too seriously. The mean of the six plates—four Lowell, one Mount Wilson and one Lick—is about 1020 km. and the most divergent of the individual values differs from the mean by only 25% of the quantity observed. Now it sometimes happens that different observers disagree on the radial velocity of a bright star by as great a percentage of the star's velocity. Thus, owing to the great magnitude of the velocity of the spiral nebulae, the percentage accuracy with which these brighter ones are observable is indeed comparable with the percentage accuracy with which stars are sometimes observed. When observations in a new field are accurate within a small fraction of the quantity observed one is encouraged to believe the results will be of fundamental value.

The disk-form images of the chief nebular emissions which is particularly striking in N_1 and N_2 is peculiar to this nebula. We had become accustomed to finding the gaseous emissions of nebulae to be images of the spectroscope slit. Although I found here a few years ago that the emission lines in the spectrum of the Crab Nebula were apparently widely resolved into two components with suggestive resemblance to the Stark electric field resolution of lines of hydrogen and helium, the disk character of the lines in N.G.C. 1068 furnishes the first striking case of its kind.

Perhaps pressure increasing towards the center or nucleus of the nebula might be a sufficient cause for the great broadening of the lines, but if so it is singular that well developed cases have not been found in the planetary nebulae of strong central condensation. The case hardly lends itself to ordinary radial velocity interpretation, as is also the case of the Crab Nebula mentioned above. The lines do not present uniformly intense disks and it is possible that a similar cause operates in both these nebulae. At present it is not possible to decide to what extent the dark lines are broadened.

Returning to the matter of rotation implied in the inclined lines of the spectrum of this nebula, it will be of interest to consider the direction of rotation relative to the spiral arms. As I have previously pointed out it seems to be possible to infer by indirect means which side of a spiral nebula is the nearer us. Spiral nebulae presenting an edge view are commonly crossed longitudinally by a dark band, which obviously belongs to the nebula and has its origin in absorbing or occulting matter on our edge of the nebula. Nebulae inclined a little to the line of sight will still show this dark lane, but it will be somewhat shifted to the side: and still more inclination of the nebular plane to the line of view causes the dark lane finally to disappear as such. It may then resolve into rifts, and its presence still be evident in these and the lesser extension of the nebulosity on its side of the nebula. This dissymmetry is a common feature of the inclined spiral nebulae, and is often so striking as to leave little doubt that it is due to the same cause as the dark lane of edge view spirals and is therefore an index to the orientation of the nebulae.

Examination of a good negative of the nebula N.G.C. 1068 made by Mr. Lampland with the 40-inch Lowell reflector led to the conclusion that the south-eastern side is the nearer us. Applying this orientation of the nebula to the inclination of its spectrum showed that the inner part of the nebula (all that is recorded on the spectrograms) is turning into the arms of the spiral like a winding spring. The Great Andromeda Nebula and a few other spirals, all that I have found to show inclined lines, show rotation in the same direction relative to the spiral arms. (Unfortunately the rotation of N.G.C. 4594 is not interpretable since it is viewed so nearly edge on as to leave invisible its arms.) This agreement in direction of rotation is confirmatory evidence of the interpretation of the orientation of the spiral nebulae, for dynamics lead us to expect that all spiral nebulae rotate in the same one direction relative to their arms.

5
— Harlow Shapley —
(1885–1972)

Harlow Shapley was an American astronomer who perfected the Cepheid variables method for measuring stellar distances. Using this method in 1915 Shapley was able to provide the first reasonably accurate picture of the Milky Way galaxy. In 1921 Shapley became the director of the Hale Observatory at Harvard, a position he held until 1952.

In this reading Shapley explains how he built upon the discovery of variable stars in the Magellanic Clouds by Henrietta Leavitt to develop a revolutionary new method for measuring great astronomical distances. Shapley also identifies most of the variables in the Magellanic Clouds as Cepheid variables. With this new astronomical tool Shapley was able to give a rough estimate of the size of the Milky Way and to measure the distances to nearby galaxies. The reading comes from Shapley's book Galaxies.

THE MEASUREMENT OF GREAT DISTANCES

The two Clouds of Magellan, as remarked in the preceding chapter, are satisfactorily located in space for the effective study of the structure of galaxies, even though inconveniently far south for an expedient exploitation by astronomers. Their distances of less than a hundred thousand light-years give us access to all of their giant and supergiant stars; their considerable angular separations from the star clouds of the Milky Way keep them clear not only of most of the absorbing dust and gas in low galactic latitudes, but also of the confusingly rich foreground of stars near the Milky Way.

During the past generation we have had high profit from our studies of these nearby galaxies, for they have turned out to be veritable treasure chests of sidereal knowledge, and astronomical tool houses of great merit. We shall see that the hypotheses, deductions, and techniques that arise from studies of the "magellanic" stars can be used to explore the mysteries of our own galactic system, and also to interpret outward to the more distant galaxies. It seems inevitable that additional discoveries will reward the future investigations of these two external systems that can be studied objectively and in detail because of their nearness and externality.

The usefulness of the Magellanic Clouds in the larger problems of cosmogony can be illustrated by citing, without stopping now to explain the meaning of the items or their significance, a partial list of the contributions to knowledge of stars and galaxies that have already come from studies of the Clouds, or are on the way.

1. The period-luminosity relation.
2. The general luminosity curve—that is, the relative number of stars in successive intervals of brightness.
3. The internal motions of irregular galaxies.
4. A comparative study of the sizes, luminosities, and types of open star clusters.
5. The incidence of Cepheid variation—that is, the number of Cepheid variables compared with the numbers of other types of giant stars of the same mass and brightness.
6. The dependence, for Cepheid variables, of the characteristics of the light-curve on length of period.
7. The "true" frequencies of the periods of Cepheid variables.
8. The dependence of a Cepheid's period on location in the galaxy.
9. The total absolute magnitudes of globular star clusters.
10. The "star haze" surrounding galaxies.

Nearly all of these subjects can be studied more successfully in the Magellanic Clouds than elsewhere. And many can better be read about in the technical reports than here. Some involve the problems of stellar evolution; others, galactic dimensions and structure. Several of the items have their solutions in the near future, rather than in the past; and although many are important in cosmogony, only a few can be considered fully in this chapter.

The Abundance of Cepheid Variables

The outstanding phenomenon associated with the Magellanic Clouds is undoubtedly the relatively great number of giant variable stars, of which the majority are of the Cepheid class. They are easily available for detailed investigation since they stand out conspicuously among the brighter stars.

In each of the Clouds there are more classical Cepheid variables than are as yet known in our much larger galaxy. The survey in the Clouds approaches completeness, the survey in the galactic system is fragmentary and seriously hindered by the interstellar dust along the Milky Way where classical Cepheids are concentrated. Probably fewer than half of the Cepheid variables of our Milky Way system have been detected.

Of the variable stars in the Magellanic Clouds that have been worked up, about 80 percent are "classical" Cepheids. In the neighborhood of the sun, there are only a few of these pulsating stars; among them are Polaris and Delta Cephei, the latter being the star that gives a name to the class. In the solar neighborhood, as elsewhere in the galactic system, variables of other types are considerably more numerous than cepheids, for instance, here are hundreds of eclipsing binaries; whereas only a few score are known in the Magellanic Clouds. Also we have found in our Galaxy more than 3000 "cluster" (RR Lyrae) variables, which are variables with periods less than a day, but only a few have been identified with certainty in the Magellanic Clouds. In the galactic system there are a great many long-period variables—the kind of stars that are carefully watched by the organized variable—star observers—but such stars are not yet abundant in the records of the Clouds.

Does this richness in the Magellanic Clouds of classical cepheid variables, with periods between 1 and 50 days, indicate that the population differs fundamentally in such irregular galaxies from that in the Milky Way spiral? Not necessarily so. The relative scarcity of cluster-type cepheids, long-period variables and eclipsing stars in the present records of the Clouds is best accounted for by the relatively low candlepower of variable stars of those types. Even at maximum such variables are not quite bright enough to get numerously into our eighteenth-magnitude pictures of the Magellanic Clouds. Until recently we have photographed almost exclusively the giants that are two hundred times or more brighter than the sun. The larger reflectors are beginning to explore among the fainter stars and possibly will soon reveal many cluster variables at the nineteenth magnitude, and eventually get down to stars of the sun's brightness.

The Period-Luminosity Relation and the Light Curves of Cepheids

Some years after Miss Leavitt had discovered and published 1777 variable stars in the two Clouds, she presented the results of a study of the periods of some of the variables. For the investigation she had selected the brightest of the variables as well as a few fainter ones. At once there appeared the interesting fact that, when the average brightness of a given variable is high, the period, which is the time-interval separating successive maxima of brightness, is long compared with the interval for fainter stars. The fainter the variable, the shorter the period.

The graph of her results for 25 variables is reproduced in Figure 32. It is of historic significance. Miss Leavitt and Professor Pickering recognized at once that if the periods of variation depend on the brightness, they must also be associated with other physical characteristics of the stars, such as mass and density and size. But apparently they did not foresee that this relation between brightness and period for cepheids in the small cloud would be the preliminary blueprint of one of astronomy's most potent tools for measuring the universe; nor did they, in fact, identify these variables of the Magellanic Cloud with the already well-known Cepheid variables of the solar neighborhood. They merely had found a curiosity among the variables of the Small Magellanic Cloud.

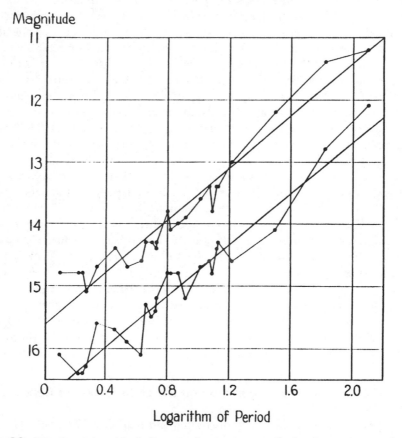

Fig. 32. Miss Leavitt's original diagram showing, separately for the maxima and the minima of 25 variable stars in the Small Cloud, the relation between photographic magnitude (vertical coordinate) and the logarithm of the period (horizontal coordinate).

Soon after Miss Leavitt's announcement of the period-magnitude relation for this small fraction of the variables that she had discovered in the Small Magellanic Cloud, Ejnar Hertzspring and others pointed out that the Cepheid variable stars of the Milky Way are giants—a fact that was readily deduced from their small cross motions and from spectral peculiarities. If the galactic cepheids and the Magellanic variables are closely comparable in luminosities, these fifteenth- and sixteenth-magnitude objects in the Clouds must also be giants, and in order to appear so faint, they must be very remote, and so also must be the Clouds.

Shapley and others pursued the inquiry and supplemented Miss Leavitt's work by studies of the variable stars that Bailey and others had detected in the globular star clusters. The many variables of the globular clusters are mostly cepheids of the cluster type with periods less than a day. But also in clusters are a few longer-period cepheids, and it was eventually possible to bring together all the data necessary for a practical but tentative period-luminosity curve. The new investigation appeared to connect the typical or "classical" cepheids with the cluster variables. Shapley then derived a zero point from trigonometric measurers of the distance of the nearby cepheids and thereby changed the Leavitt relation from period and apparent magnitude to period and absolute luminosity, thus making distance determinable from light measurers only, as will be shown below.

Using the apparently similar cepheid variables found in globular star clusters as guides to the zero point, Shapley derived the first period-luminosity relation for cepheid variables in 1917 (Fig. 33). He was able to establish a true and absolute luminosity for cepheids by taking advantage of the fact that what appeared to be normal cepheids existed in globular star clusters together with the cluster-type RR Lyrae variables. The latter are fairly common in our Galaxy and there are enough near the sun that the distances to the RR Lyrae variables could be established by statistical consideration based on their motions as seen over the sky. Therefore, it was well known that the RR Lyrae variables all have absolute magnitudes of approximately zero and thus Shapley was able to establish magnitudes for the cepheid variables in the globular clusters and, by comparison, also in Magellanic Clouds.

More recent studies of the true luminosity of RR Lyrae stars have confirmed rather well these early results utilized by Shapley more than 50 years ago. The accurate absolute luminosities obtained for globular star clusters for which we can measure distance by comparison of their main-sequence stars with main-sequence stars in nearby clusters, such as the Hyades, have led to the conclusion that in the mean, RR Lyrae variables have an absolute magnitude of approximately +0.5 with a spread from cluster to cluster of 0.2 or 0.3 magnitude. However, it is an entirely different story with regard to the longer period cepheids found in the globular clusters. In the 1950s it was established that there are two kinds of cepheid variables, one belonging to the Population I, young, spiral-arm component of the Galaxy and the other belonging to the Population II component, which includes the halo and the globular star clusters. The Population II cepheids were found on the average to be 1.5 magnitudes fainter for a

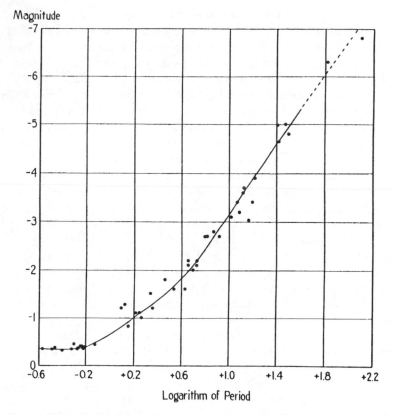

Fig. 33. The early Harvard period-luminosity relation, based on the 25 Small Cloud variables and cepheids from globular clusters in the local galaxy. The ordinate gives visual magnitudes on the absolute scale.

given period than the Population I cepheids and therefore the period-luminosity relation for the two have a very different zero point. This meant that the cepheid variables in the Magellanic Clouds, which were clearly Population I cepheids, were very much more luminous than had originally been thought and therefore the distance to the Clouds must be twice as great as had been computed in 1917. It may be well at this point to show how one uses the period-luminosity curve of Figure 34 to measure the distance of the classical cepheid variables in our Milky Way, or the distance to some remote external galaxy, like the Andromeda Nebula. The procedure is very simple, once the period-luminosity relation is set up and accurately calibrated. First must come the discovery of the variable star, and then, through the making of a hundred or so observations of the brightness at scattered times, comes the verification from the shape of the mean light curve, that the variable belongs to the cepheid class. On a correct magnitude scale we next determine the amplitude (range) of variation, and the value of the magnitude half-way between maximum and minimum. This *median apparent magnitude, m* which is now almost always determined photographically or photoelectrically, constitutes one half of the needed observational material. The other half, namely the period P, is also determined from the observations of magnitudes.

With the period and its logarithm known, the relative absolute luminosity, M is then derived directly. For example the simple formula

$$\dot{M} = -1.78 - 1.74 \log P$$

is satisfactory for getting the relative absolute magnitudes of all cepheids with periods between 1.2 and 40 days.

When we have thus derived the relative absolute magnitude from the period, we compute the distance *d* from the equally simple relation

$$\log d = 0.2 \, (\dot{m} - \dot{M} - \delta m) + 1,$$

where the distance is expressed in parsecs (1 parsec = 3.26 light years, or about 19 trillion miles), and δm is the correction one must make to the observed median magnitude because of the scattering and absorption of starlight by the dust and gas of interstellar space. . . .

6
— Willem de Sitter —
(1872–1934)

Dutch cosmologist and mathematician Willem de Sitter was the first to propose an expanding universe solution to the equations of General Relativity. Einstein himself had proposed a static universe solution to his equations. Initially, matter was not part of de Sitter's universe, but eventually it was successfully introduced into what came to be known as the Einstein-de Sitter universe.

This reading is a chapter of a book entitled Kosmos, published in 1932. In the chapter de Sitter describes the cosmological models that arose subsequent to the General Theory of Relativity, including his "empty universe" model. He acknowledges that General Relativity could support either an expanding or contracting universe and states that we do not have sufficient evidence to make a judgment. He even considers explanations of the red shift without expansion.

RELATIVITY AND COSMOLOGY

The theory of relativity may be considered as the logical completion of Newton's theory of gravitation, the direct continuation of the line of thought which dominates the development of the science of mechanics, from Archimedes—who may be considered as the first relativist—through Galileo to Newton. Newton's theory had celebrated its greatest triumphs in the eighteenth and nineteenth centuries; one after another all the irregularities in the motions of the planets and the moon had been explained by the mutual gravitational action of these bodies. In the beginning of the nineteenth century Laplace's monumental work completed the application of the theory on the motions of the planets.

The final triumph came in 1846 by the discovery of Neptune, verifying the prediction by Adams and Leverrier, based on the theory of gravitation.

Gradually Newton's law of gravitation had become a model on which physical laws were framed, and all physical phenomena were reduced to laws which were formulated as attractions or repulsions inversely proportional to some power of the distance, such as, e.g., Laplace's theory of capillarity, which was even published as a chapter in his "Mécanique céleste." Gradually, however, during the second half of the nineteenth century, the uncomfortable feeling of dislike of the action at a distance, which had been so strong in Huygens and other contemporaries of Newton, but had subsided during the eighteenth century, began to emerge again, and gained strength rapidly.

This was favored by the purely mathematical transformation (which can be compared in a sense with that from the Ptolemaic to the Copernican system), replacing Newton's finite equations by the differential equations, the potential becoming the primary concept, instead of the force, which is only the gradient of the potential. These ideas, of course, arose first in the theory of electricity and magnetism—or perhaps one should say in the brain of Faraday. In electromagnetism also the law of the inverse square had been supreme, but, as a consequence of the work of Faraday and Maxwell, it was superseded by the field. And the same change took place in the theory of gravitation. By and by the material particles, electrically charged bodies and magnets—which are the things that we actually observe—come to be looked upon only as "singularities" in the field. So far this transformation from the force to the potential, from the action at a distance to the field, is only a purely mathematical operation. Whether we talk of a "particle of matter" or of a "singularity in the gravitational field" is only a question of a name. But this giving of names is not so innocent as it looks. It has opened the gate for the entrance of hypotheses. Very soon the field is materialized, and is called aether. From the mathematical point of view, of course, "aether" is still just another word for "field," or, perhaps better, for "space"—the absolute space of Newton—in which there may or may not be a field. From the point of view of physical theory (and it is especially in the theory of electromagnetism that this evolution took place), however, the "aether of space," as it used to be called about forty years ago, is not simply space, it is something substantial, it is the carrier of the field, and mechanical models, consisting of racks and pinions and cogwheels, are devised to explain how it does the carrying. These mechanical models have, of course, been given up long ago: they were too crude. But hypotheses have kept cropping up on all sides: electrons, atomic nuclei, protons, wave-packets, etc. At first the imagining of mechanical models went on. Fifteen, or even ten, years ago, although an atom was no longer, as the name implies, just a piece of matter that could not be cut into smaller pieces, atoms, electrons, and protons were still thought of as mechanical structures, models of the atom were imagined, having the mechanical properties of ordinary matter. The inconsistency of first explaining matter by atoms and then explaining atoms by matter was only slowly realised, and it is only comparatively recently that we have come to see that there is nothing paradoxical in the fact that an atom or an electron, which are not matter, may have properties different from those of matter, and must be allowed to do things that a material particle could not do.

However, whilst in all other domains of physics hypotheses have been found successful in accounting for the observed facts, and replacing the formal laws, the case of gravitation stands apart. Gravitation has been insusceptible to this general infection. By using this word I do not mean to suggest that the luxuriant growth of hypotheses in physics is a contagious disease,—it is not a disease, but a natural development,—but it is certainly contagious. Gravitation, however, seems to be immune to it. In the course of history a great number of hypotheses have been proposed in order to "explain" gravitation, but not one of these has ever had the least chance, they have all been failures. Why is that? How does it come about that we have been able to find satisfactory hypotheses to explain electricity and magnetism, light and heat, in short all other physical phenomena, but have been unsuccessful in the case of gravitation? The explanation must be sought in the peculiar position that gravitation occupies amongst the laws of nature. In the case of other physical phenomena there is something to get hold of, there are circumstances on which the action depends. Gravitation is entirely independent of everything that influences other natural phenomena. It is not subject to absorption or refraction, no velocity of propagation has been observed. You can do whatever you please with a body, you can electrify or magnetise it, you can heat it, melt or evaporate it, decompose it chemically, its behavior with respect to gravitation is not affected. Gravitation acts on all bodies in the same way, everywhere and always we find it in the same rigorous and simple form, which frustrates all our attempt to penetrate into its internal mechanism. Gravitation is, in its generality and rigour, entirely similar to inertia, which has never been considered to require a particular hypothesis for its explanation, as any ordinary special physical law or phenomenon. Inertia has from the beginning been admitted as one of the fundamental facts of nature, which have to be accepted without explanation, like the axioms of geometry.

But gravitation is not only similar to inertia in its generality, it is also measured by the same number, called the mass. The inertial mass is what Newton calls the "quantity of Matter": it is a measure for the resistance offered by a body to a force trying to alter its state of motion. It might be called the "passive mass." The gravitational mass, on the other had, is a measure of the force exerted by the body in attracting other bodies. We might call it the "active" mass. The equality of active and passive, or gravitational and inertial, mass was in Newton's system a most remarkable accidental co-incidence, something like a miracle.

Newton himself decidedly felt it as such, and made experiments to verify it, by swinging a pendulum with a hollow bob which could be filled with different materials. The force acting on the pendulum is proportional to its gravitational mass, the inertia to its inertial mass: the period of its swing thus depends on the ratio between these two masses. The fact that the period is always the same therefore proves that the gravitational and inertial masses are equal. Gradually, during the eighteenth century, physicists and philosophers had become so accustomed to Newton's law of gravitation, and to the equality of gravitational and inertial mass, that the miraculousness of it was forgotten and only an acute mind like Bessel's perceived the necessity of repeating those experiments. By the experiments of Bessel about 1830 and of Eötvös in 1909 the equality of gravitational and inertial mass has become one of the best ascertained empirical facts in physics.

In Einstein's general theory of relativity the identity of these two coefficients, the gravitational and the inertial mass, is no longer a miracle, but a necessity, because gravitation and inertia are identical.

There is another side to the theory of relativity. We have pointed out in the beginning how the development of science is in the direction to make it less subjective, to separate more and more in the observed facts that which belongs to the reality behind the phenomena, the absolute, from the subjective element, which is introduced by the observer, the relative. Einstein's theory is a great step in that direction. We can say that the theory of relativity is intended to remove entirely the relative and exhibit the pure absolute.

The physical world has three space dimensions and one time dimension; the position of a material particle at a certain time t is defined by three space coordinates, x, y, z. In Newton's system of mechanics this is unhesitatingly accepted as a property of the outside world: there is an absolute space and an absolute time. In Einstein's theory time and space are interwoven, and the way in which they are interwoven depends on the observer. Instead of three plus one we have four dimensions.

Is the fact that we observe the outside world as a four-dimensional continuum a property of this outside world, or is it a consequence of the particular nature of our consciousness, does it belong to the absolute or to the relative? I do not think the answer to that question can yet be given. For the present we may accept it as an empirically ascertained fact.

The sequence of different positions of the same particle at different times forms a one-dimensional continuum in the four-dimensional space-time, which is called the *world-line* of the particle. All that physical experiments or observations can teach us refers to intersections of world-lines of different material particles, light-pulsations, etc., and how the course of the world-line is between these points of intersection is entirely irrelevant and outside the domain of physics. The system of intersecting world-lines can thus be twisted about at will, so long as no points of intersection are destroyed or created, and their order is not changed. It follows that the equations expressing the physical laws must be invariant for arbitrary transformations.

This is the mathematical formulation of the theory of relativity. The metric properties of the four-dimensional continuum are described, as is shown in treatises on differential geometry, by a certain number (ten, in fact) of quantities, denoted by $g_{\alpha\beta}$, and commonly called "potentials." The physical status of matter and energy, on the other had, is described by ten other quantities, denoted by $T_{\alpha\beta}$, the set of which is called the "material tensor." This special tensor has been selected because it has the property which is mathematically expressed by saying that its divergence vanishes, which means that it represents something permanent. The fundamental fact of mechanics is the law of inertia, which can be expressed in its most simple form by saying that it requires the fundamental laws of nature to be differential equations of the second order. Thus the problem was to find a differential equation of the second order giving a relation between the metric tensor $g_{\alpha\beta}$ and the material tensor $T_{\alpha\beta}$. This is a purely mathematical problem, which can be solved without any reference to the physical meaning of the symbols. The simplest possible equation (or rather set of ten equations, because there are ten g's) of that kind that can be found was adopted by Einstein as the fundamental equation of his theory. It defines the space-time continuum, or the "field." The world-lines of material particles and light quanta are the geodesics in the four-dimensional continuum defined by the solutions $g_{\alpha\beta}$ of these field-equations. The equations of the geodesic thus are equivalent to the equations of motion of mechanics. When we come to solve the field-equations and substitute the solutions in the equations of motion, we find that in the first approximation, i.e. for small material velocities (small as compared with the velocity of light), these equations of motion are the same as those resulting from Newton's theory of gravitation. The distinction between gravitation and inertia has disappeared; the gravitational action between two bodies follows from the same equations, and is the same thing, as the inertia of one body. A body, when not subjected to an extraneous force (i.e. a force other than gravitation), describes a geodesic in the continuum, just as it described a geodesic, or straight line, in the absolute space of Newton under the influence of inertia alone.

The field-equations and the equations of the geodesic together contain the whole science of mechanics, including gravitation.

In the first approximation, as has been said just now, the new theory gives the same results as Newton's theory of gravitation. The enormous wealth of experimental verification of Newton's law, which has been accumulated during about two and a half centuries, is therefore at the same time an equally strong verification of the new theory. In the second approximation there are small differences, which have been confirmed by observations, so far as they are large enough for such a confirmation to be possible. Thus especially the anomalous motion of the perihelion of Mercury, which had baffled all attempts at explanation for over half a century is now entirely accounted for. Further the theory of relativity has predicted some new phenomena, such as the deflection of rays of light that pass near the sun, which has actually been observed on several occasions during eclipses; and the redshift of spectral lines originating in a strong gravitational field, which is also confirmed by observations, e.g. in the spectrum of the sun, and also in the spectrum of the companion of Sirius, which, being a so-called white dwarf, i.e., a small star with very high density and consequently a strong gravitational field, gives a considerable redshift. We cannot stop to explain these phenomena in detail. It must suffice just to mention them.

Two points should be specially emphasized in connection with the general theory of relativity.

First that it is a purely *physical* theory, invented to explain empirical physical facts, especially the identity of gravitational and inertial mass, and to coordinate and harmonise different chapters of physical theory, and simplify the enuncia- tion of the fundamental laws. There is nothing metaphysical about its origin. It has, of course, largely attracted the attention of philosophers, and has, on the whole, had a very wholesome influence of metaphysical theories. But that is not what it set out to do, that is only a by-product.

Second that it is a pure generalization, or abstraction, like Newton's system of mechanics and law of gravitation. It contains *no hypothesis*, as contrasted with other modern physical theories, electron theory, quantum theory, etc., which are full of hypotheses. It is, as already been said, to be considered as the logical sequence and completion of Newton's Principia.

A special feature of the development of physics in the nineteenth century has been the arising of general principles beside the special laws, such as the princi- ples of conservation of mass and of energy, the principle of least action, and the like. These differ from the special laws, not only by being more general, but they aspire, so to say, to a higher status than the laws. Their claim is that they express fundamental facts of nature, general rules, to which all special laws have to conform. And they accordingly exclude a priori all attempts at "explanation" by hypotheses or mechanical models. It is characteristic of the theory of relativity that it enables us to include all these principles of conservation in one single equation.

We have a direct knowledge only of that part of the universe of which we can make observations. I have already called this "our neighborhood." Even within the confines of this province our knowledge decreases very rapidly as we get away from our own particular position in space and time. It is only within the solar system that our empirical knowledge of the quantities determining the state of the universe, the potentials $g_{\alpha\beta}$, extends to the second order of smallness (and that only for g_{44}, and not for the others), the first order corresponding to about one unit in the eighth decimal place. How the $g_{\alpha\beta}$ outside our neighbor- hood are, we do not know, and how they are at infinity, of either space of time, we shall never know, otherwise it would not be infinity. That is what Archi- medes meant when he said that the universe could not be infinite. The universe that we know cannot be infinite, because we ourselves are finite. Infinity is not a physical, but a mathematical concept, introduced to make our equations more symmetrical and elegant. From the physical point of view everything that is outside our neighborhood is pure extrapolation, philosophical or aesthetical pre- dilections—or prejudices. It is true that some of these prejudices are so deeply rooted that we can hardly avoid believing them to be above any possible suspicion of doubt, but this conviction is not founded on any physical basis. One of the convictions, on which extrapolation is naturally based, is that the particular part of the universe in which we happen to be is in no way exceptional or privileged, in other words that the universe, when considered on a large enough scale, is isotropic and homogeneous. It should, however, be remembered that there have been epochs in the evolution of mankind when this was by no means thought self-evident, and the contrary conviction was rather generally held.

During the last years the limits of our "neighborhood" have been enormously extended by the observations of extragalactic nebulae, made chiefly at the Mount Wilson Observatory. These wonderful observations have enabled us to make fairly reliable estimates of the distances of these objects and to say something about their distribution in space. It appears that they are distributed approximately evenly over "our neighborhood." They also are all of roughly the same size, so that we can make an estimate of the density of matter in space. Further, the observations have disclosed the remarkable fact that in their spectra there is a displacement of the lines towards the red corresponding to a receding velocity increasing with the distance, and, so far as the determinations of the distances are reliable, proportional with it. If the velocity is proportional to the distance, then not only the distance of any nebula from us is increasing, but *all* mutual distances between any two of them are increasing at the same rate. Our own galactic system is only one of a great many, and observations made from any of the others would show exactly the same thing: all systems are receding, not from any particular centre, but *from each other*: the whole system of galactic systems is *expanding*.

It is perhaps somewhat difficult to imagine the expansion of three-dimensional space. A two-dimensional analogy may help to make it clear. Let the universe have only two dimensions, and let it be the surface of an india rubber ball. It is only the *surface* that is the universe, not the ball itself. Observations can only be made, distances can only be measured, along the surface, and evidently no point of the surface is different from any other point. Let there be specks of dust fixed to the surface to represent the different galactic systems. If the ball is inflated, the universe expands, and these specks of dust will recede from each other, their mutual distances, measured along the surface, will increase in the same rate as the radius of the ball. An observer in any one of the specks will see all the others receding from himself, but it does not follow that he is the centre of the universe. The universe (which is the surface of the ball, not the ball itself) has no centre.

It is, of course, not essential that we have chosen for our illustration a rubber *ball*. We might just as well have taken any other surface; it is not even necessary that it should be a closed surface. Even a plane sheet of rubber might do just as well, if only the stretching to which it must be subjected to illustrate the expanding universe is the same in all directions.

These then are the two observational facts about our neighborhood, which have to be accounted for by theory: there is a finite density of matter, and there is expansion, i.e. the mutual distances are increasing, and therefore the density is decreasing. Of course we can only be certain of these facts so far as our observations reach, i.e., for our "neighborhood," but, in agreement with our principle of extrapolation, we extend these statements to the whole of the universe.

We have thus to find a universe—i.e., a set of potentials $g_{\alpha\beta}$ satisfying the field-equations of the general theory of relativity—that has both a finite density of matter and an expansion. And, since we only consider the universe on a very large scale, and make abstraction of all details and local irregularities, our universe must be homogeneous and isotropic. It follows at once from this condition of homogeneity and isotropy that the three-dimensional space of it must be what mathematicians call a space of constant curvature. Even so mathematics offers us a free choice between different kinds of space. The curvature may be positive, negative, or zero. It is not possible to picture, or imagine, the different kinds of three-dimensional space. We think that we have a mental picture of euclidian, or flat, space, i.e., space of which the curvature is everywhere zero, but I am not sure that this is not a self-deception, caused by the fact that the geometry of this special space has been taught in the schools for the last two thousand or more years. It is certain that for physical phenomena on the scale which our sense organs are able to perceive, i.e., neither too small nor too large, the euclidian space is a very close approximation to the true physical space, but for the electron, and for the universe, the approximation breaks down. To help us to understand three-dimensional spaces, two-dimensional analogies may be very useful (though also sometimes misleading). We can imagine different kinds of two-dimensional space, since we are able to place ourselves outside them. A two-dimensional space of zero curvature is a plane, say a sheet of paper. The two-dimensional space of positive curvature is a convex surface, such as the shell of an egg. It is bent away from the plane towards the same side in all directions. The curvature of the egg, however, is not constant: it is strongest at the small end. The surface of constant positive curvature is the sphere, say our india rubber ball of a moment ago. The two-dimensional space of negative curvature is a surface that is convex in some directions and concave in others, such as the surface of a saddle or the middle part of an hour glass. Of these two-dimensional surfaces we can form a mental picture because we can view them from outside, living, as we do, in three-dimensional space. But for a being, who would be unable to leave the surface on which he was living, that would be impossible. He could only decide of which kind his surface was by studying the properties of geometrical figures drawn on it. For the geometrical figures have different properties on the different surfaces. On the sheet of paper the sum of the three angles of a triangle is equal to two right angles, on the egg, or the sphere, it is larger, on the saddle it is smaller. On the flat paper—and on the saddle-shaped surface—we can proceed indefinitely in the same direction, on the egg or the sphere, if we continue to move in the same direction we ultimately come back to our starting point. The spaces of zero and negative curvature are infinite, that of positive curvature is finite. Thus the inhabitant of the two-dimensional surface could determine its curvature if he were able to study very large triangles or very long straight lines. If the curvature were so minute that the sum of the angles of the largest triangle that he could measure would still differ from two right angles by an amount too small to be appreciable with the means at his disposal, then he would be unable to determine the curvature, unless he had some means of communicating with somebody living in the third dimension. Now our case with reference to three-dimensional space is exactly similar. We have no intuitive knowledge of the kind of space we live in. So we must find out which kind it is by studying the triangles and other geometrical figures in it. As we are concerned with *physical* space, the triangles that we must investigate are those formed by the tracks of material particles and rays of light, and naturally, in order to be able to distinguish different kinds of space, we must study very large triangles and rays of light coming from very great distances. Thus the decision must necessarily depend on astronomical observations.

Even the most refined astronomical observations, however, fail to show any trace of curvature. The triangles that we can measure are not large enough, and, I fear, never will be large enough, to detect the curvature. Fortunately, however, we are, in a way, able to communicate with the fourth dimension. The theory of relativity has given us an insight into the structure of the real universe: it does not consist of a three-dimensional space *and* a one-dimensional time, existing independently of each other, as in Newton's system of mechanics, but is a *four*-dimensional structure. The study of the way in which the three space-dimensions are interwoven with the time-dimension affords a kind of outside point of view of the three-dimensional space, and it is conceivable that from this outside point of view we might be able to perceive the curvature of the three-dimensional world.

At one time it was thought that this was so, and that we could actually prove that the curvature must be positive. However, recent mathematical investigation has shown that this was a mistake. We shall never be able to say anything about the curvature without introducing certain hypotheses. These rather vague statements will become clearer when we penetrate more deeply into the nature of the cross-connection between space and time in the four-dimensional universe. It is, of course, difficult to explain these without the use of mathematical formulae, but, by following the historical line of development, I shall endeavor to lead up to an understanding of the present position without using too technical language.

Let us begin by considering the finite density of matter in the universe. The average density is very small. Matter is actually distributed very unevenly, it is conglomerated into stars and galactic systems. The average density is the density that we should get if all these great systems could be evaporated into atoms of hydrogen, or protons, and these distributed evenly over the whole of space. There would then probably not be more than three or four protons in every cubic foot. That is a very small density indeed: it is about a million million times less than that of the most perfect vacuum that we can produce in our physical laboratories. The universe thus consists mostly of emptiness, and it appears natural to consider a universe without any matter at all, an empty universe, as a good approximation to begin with for our grand scale model. The galactic systems are details which can be put in afterwards. But we may also take as our first approximation a universe containing the same amount of matter as the actual one, but equally distributed, i.e. having a finite average density of three or four protons per cubic foot. The local deviations from the average, caused by the conglomeration of matter into stars and stellar systems, are then disregarded in the grand scale model, and are only taken into account when we come to study details.

Now fifteen years ago, in the beginning of 1917, two solutions of the field-equations for a homogeneous isotropic universe had been found, which I shall provisionally call the solutions "A" and "B." It should be mentioned that at that time only *static* solutions were looked for. It was thought that the universe must be a stable structure, which would retain its large scale properties unchanged for all time, or at least change them so slowly that the change could be disregarded. In one of these solutions (B) the average density was zero, it was empty; the other one (A) had a finite density. Both, of course, were, as was well appreciated, only approximations to the actual universe. In B, to get the real universe, we should have to put in a few galactic systems, in A we should have to condense the evenly distributed matter into galactic systems. The universe A is really and essentially static, there can be no systematic motions in it. It has an average density, but no expansion. It is therefore called the *empty universe*. Thus we had two approximations: the static universe with matter and without expansion, and the empty one without matter and with expansion. The actual universe, as we have seen, has both matter and expansion, and can, therefore, be neither A nor B. In 1917 this dilemma had not yet become urgent, and was hardly realized. The actual value of the density was still entirely unknown, and the expansion had not yet been discovered.

Now in both the solutions A and B the curvature is positive, in both three-dimensional space is finite: the universe has a definite size, we can speak of its radius, and in the case A, of its total mass. In the case A, the static universe, there is a definite relation between the curvature and the density, in fact the density is proportional to the curvature, the factor of proportionality being a pure number ($1/4\pi$, if appropriate units are used). Thus, if we wish to have a finite density in a static universe, we must have a finite positive curvature.

At this point we must say a few words about the famous *lambda*. The field-equations, in their most general form, contain a term multiplied by a constant, which is denoted by the Greek letter λ (lambda), and which is sometimes called the "cosmical constant." This is a name without any meaning, which was only conferred upon it because it was thought appropriate that it should have a name, and because it appeared to have something to do with the constitution of the universe; but it must not be inferred that, since we have given it a name, we know what it means. We have, in fact, not the slightest inkling of what its real significance is. It is put in the equations in order to give them the greatest possible degree of mathematical generality, but, so far as its mathematical function is concerned, it is entirely undetermined: it may be positive or negative, it might also be zero. Purely mathematical symbols have no meaning by themselves; it is the privilege of pure mathematicians, to quote Bertrand Russell, not to know what they are talking about. They—the symbols—only get a meaning by the interpretation that is put on the equations when they are applied to the solution of physical problems. It is the physicist, and not the mathematician, who must know what he is talking about. At first, in Einstein's paper of November 1915, in which the theory reached its final form, the term with λ was simply omitted, in other words λ was supposed to have the special value *zero*. That was the simplest way of avoiding the responsibility of attaching a label to it, and, of course, an entirely legitimate way: it is very bad physics to introduce more arbitrary constants than are needed for the representation of the phenomena. But fifteen months later, in February 1917, it was found that a static solution with a positive curvature—the solution A—was not possible without the λ. In fact the curvature is proportional to λ (in solution A, λ is equal to the curvature; in B, when treated as a static solution, it is three times the curvature). Thus at the time when we had only the two static solutions A and B, and thought that these were the only possible ones, here was a plausible physical interpretation of the meaning of λ: it was the curvature of the world, and the square root of its reciprocal, the radius of curvature, could be conceived as providing a natural unit of length. It gave the electron something to measure itself by, so that it might know how large it ought to be, as Sir Arthur Eddington has expressed it. Recently Eddington has replaced this interpretation by another one of the same nature, involving the gravitational mass and the electric charge of the electron. But recent developments have made it very difficult to maintain these interpretations, as we shall presently see.

We must now take up the thread of the narrative where we left it a little while ago. We were in the position of having two possible solutions: the static universe with matter but without expansion, and the empty universe with expansion but without matter.

Now the observed rate of expansion is large: the universe doubles its size in about fifteen hundred million years, which is a short time, astronomically speaking. In the "static" universe expansion is impossible; the "empty" universe does expand. Therefore we may be tempted to consider the empty universe as the most likely approximation; and we can proceed to compute the radius of curvature of the universe, supposing it to be of the empty type, from the observed rate of expansion. It comes out as about two thousand million light-years.

The universe, however, is not empty, but contains matter. The point is how much matter. Is the density anywhere near that corresponding to the static universe, or is it so small that we can consider the empty universe as a good approximation? We have seen that the universe is some million million times as empty as our most perfect vacuum. But this is not the correct way to measure the emptiness of the universe. We must use as a standard of comparison, not our terrestrial experience, but the theoretical density of the static universe. It is easy to compute the density of a static universe of a radius of two thousand million light years, and it comes out only very little larger than the observed density. The actual universe is thus very far from empty, it is, on the contrary, nearly full.

We thus come to the conclusion, which was already foreshadowed above, that the actual universe is neither the static nor the empty one. It differs so much from both of these that neither can be used as an appropriate grand scale model. We must thus look for other solutions of the general field-equations. On account of the expansion our solution must necessarily be a non-static one, and it must have a finite density. There is only one possible static solution possessing a finite density, viz. our old friend A, but of non-static solutions with finite density there exists a great variety. I will now depart from the strictly historical narrative and enumerate these different possible solutions, not in the order in which they have been discovered, but in the sequence of a natural classification.

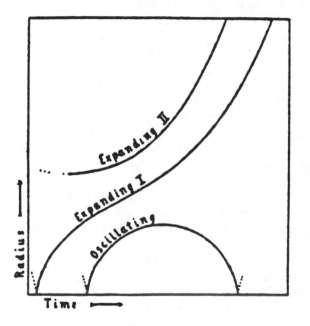

FIGURE I

The three families of non-static universes

In the solutions A and B the curvature of three-dimensional space was necessarily positive, and the mysterious "cosmical constant" λ was also positive. In the non-static solutions this is not so. At first this was not realized. We had become so accustomed to think of λ as an essentially positive quantity, and of a finite world with positive curvature, that the idea of investigating the possibility of solutions with negative or zero values of λ and of the curvature simply did not arise. But when this oversight was corrected, it appeared at once that in the non-static case both λ and the curvature need not be positive, but can be negative or zero quite as well. I will therefore use the value of λ and the sign of the curvature as the principles of classification. The instantaneous state of the universe is characterized by a certain quantity occurring in the equations, which is denoted by the letter R, and which, if there is a curvature, can be interpreted as the radius of curvature, or the "radius" for short. The way in which the universe expands is determined by the variation of this R with the time. There are three types, or families, of non-static universes, which I will call the oscillating universes, and the expanding universes of the first and of the second kind. They are represented graphically in Figure 1. The horizontal coordinate is the time, the vertical coordinate is the "radius." Of each type only one example is given in the figure, but it should be realized that each of these is a representative of a family, comprising an infinite number of members differing in size and shape.

In the oscillating universes the "radius" R increases from zero to a certain maximum size, which is different for each member of the family, and then decreases again to zero. The period of oscillation has a certain finite (and rather short) value, different for each member of the family. In the expanding family of the first kind the radius is continually increasing from a certain initial time, when it was zero, to become infinitely large after an infinite time. In the expanding series of the second type the radius has at the initial time a certain minimum value, different for the different members of the family, and increases to become infinite after an infinite time.

If λ is negative, only oscillating universes are possible, whatever the curvature may be, positive, negative or zero. The only choice is between the different members of the oscillating family, and to each special value of λ corresponds one member of the family.

If λ is zero, and the curvature positive, we have still an oscillating universe; if the curvature is zero or negative, the universe is one of the first family of expanding universes.

If λ is positive, then for positive curvatures the three families are possible; for negative and zero curvatures only the first family of expanding universes exists. The different possibilities are given in the following small table.

	Curvature		
λ	*Negative*	*Zero*	*Positive*
Negative	Oscillating	Oscillating	Oscillating
Zero	Expanding I	Expanding I	Oscillating
Positive	Expanding I	Expanding I	Oscillating Expanding I Expanding II

We do not know to which of the three possible families our own universe belongs, and there is nothing in our observational data to guide us in making the choice. And even if we have decided on the family, we have still the freedom to select any particular member of it. This is not because the data are not accurate enough, but because they are deficient in number. The observations give us *two* data, viz. the rate of expansion and the average density, and there are *three* unknowns: the value of λ, the sign of the curvature, and the scale of the figure, i.e. the units of R and of the time. The problem is indeterminate. If we make an hypothesis regarding either λ or the curvature, then we can find the other from the observed data, or rather we would be able to do so, if the data were sufficiently accurate. We might, e.g., decide a priori that the curvature should be zero, i.e. that the three dimensional space should be Euclidean. If we make that hypothesis, we have a sufficient number of data to enable us to determine the value of λ. We might, on the other hand, wish to get rid of the λ, on the ground that it ought never to have been there, being unfortunately introduced into the equations in a past stage of the development of the theory, when we were mistakenly trying to find a static solution. In other words we might make the hypothesis that the true value of λ is zero. In that case the data of observation will allow us, if they are sufficiently accurate, to determine the curvature.

As a matter of fact neither the average density nor the rate of expansion are at the present time known with sufficient accuracy to make an actual *determination* possible, even if an hypothesis of this kind is adopted. All we can say is that, if the curvature is small (as we know it must be, because it is imperceptible by ordinary geometric methods in our neighborhood), then λ must be small, and *if* the curvature is *very* small, then λ must be very small. On the other hand, *if* λ is very small, or zero, then the curvature must be very small, and may even be zero, for aught we can say at present.

The interpretation of the expanding universe, the making of a mental picture, or a model, of it, which was, or appeared to be, easy when we knew that the universe was finite, is not such a simple matter now that we do not even know whether the curvature is positive, zero or negative, whether the universe is finite or infinite. It sounds rather strange to talk of an infinite universe still expanding. If we were certain that the curvature was negative, we might still, as in the case of positive curvature, replace the phrase "the universe expands" by the equivalent one "the curvature of the universe decreases." But if the curvature is zero, and remains zero throughout, what sort of meaning are we to attach to the "expansion"? The real meaning is, of course, that the mutual distances between the galactic systems, measured in so-called natural measure, increase proportionally to a certain quantity R appearing in the equations, and varying with the time. The interpretation of R as the "radius of curvature" of the universe, though still possible if the universe has a curvature, evidently does not go down to the fundamental meaning of it. The manner in which time and space are bound up with each other in the four-dimensional continuum is variable. It is difficult to express this variability of the cross-connections between space and time in simple language, and different interpretations of it are possible, corresponding to different mathematical transformations of the fundamental line-element, e.g. a different choice of the variable which we interpret as "time." Perhaps the best way we can express it is by saying that the solution of the field-equations of the theory of relativity shows that there is in the universe a tendency to change its scale, which at the present time results in an expansion, but may perhaps at other times become, or have been, a shrinking. This is true of the grand scale model of the universe. If we put in the details, the singularities of the field, viz. the galactic systems and the stars, we find that there is also a tendency, called gravitation, to decrease the mutual distances of these "singularities." At short distances, within the confines of a galactic system, this second tendency is by far the strongest, and the galactic systems retain their size independent of the expansion or contraction of the universe; at large distances, such as those separating one galactic system from the next, the first preponderates.

The interpretation of λ as providing a natural unit of length also, of course, becomes difficult. The radius of curvature, or even the more vaguely defined quantity R, can no longer provide the electron with the means of knowing how large it ought to be, for R changes and the electron does not change its size. The electron could still measure itself in some mysterious way against the square root of the reciprocal of λ—or of *minus* λ, if λ happened to be negative,—but if λ were zero that also would not help it.

Also other interpretations, involving the total mass of, or the number of protons, in the universe, become untenable, if the universe may, for aught we know, be just as well infinite and have no total mass.

The theory of the expanding universe is at the present moment much less definite than we supposed it to be a few months ago, but that does not affect its significance—it only brings out more clearly what this significance is, and what it is not.

Some consequences of the theory merit a special mention, however briefly.

A question which has long troubled astronomers and physicists is what becomes of the energy that is continually being poured out into space by the sun and the stars. To this question a complete answer is given by the new theory. It is used up, diluted, or degraded, by the expansion of the universe. Just as a man running to catch a bus or tram-car gets out of breath and spends his energy, or a projectile thrown after a moving train hits it with less force than it would hit a stationary object, so the light travelling through the expanding universe and, so to say, trying to reach a particular star, or stellar system, which is continually receding with great velocity, is losing energy in trying to catch up with it. It is this degradation of the light, technically known as the redshift of the spectral lines, by which we become aware of the receding velocities of the extra-galactic nebulae. It can be shown that the decrease of the total amount of radiant energy in the universe by this degradation exceeds the increase by the radiation of the stars. It would not be correct, however, to conclude that the expansion is caused *by* the energy thus lost by the radiation, any more than it would be correct to say that the tram-car is propelled by the energy expended by the man who runs after it.

There is one very serious difficulty presented by the theory of the expanding universe, which we shall have to face with careful deliberation.

In all solutions there is a certain minimum value of the "radius" R, either zero, or in the expanding family of the second kind a finite value, which the universe had at a definite time in the past. There appears to be a definite "beginning of time," a few thousand million years back in history, as there is a definite "absolute zero" of temperature, corresponding to *minus* 273 degrees on the ordinary scale. What is the meaning of this?

The temptation is strong to identify the epoch of the beginning of the expansion with the "beginning of the world," whatever that may mean. Now astronomically speaking this beginning of the expansion took place only yesterday, not much longer ago than the formation of the oldest rocks on the earth. According to all our modern views the evolution of a star, of a double star, or a star cluster, requires intervals of time which are enormously longer. The stars and the stellar systems must be some thousands of times older than the universe!

What must be our attitude with regard to this paradox? It would appear that, if two theories are in contradiction, we must give up either the one or the other. The conflict apparently is between the modern theories of stellar evolution and the dynamical theories of the evolution of double stars and star clusters on the one hand, and the general theory of relativity on the other hand. If this were the real contest, there could be no doubt about the issue: the theory of relativity would come out of the trial victorious, and the theories of evolution would have to be revised. This seems to be Sir Arthur Eddington's standpoint, as he writes: "we must accept this alarmingly rapid dispersal of the nebulae with its important consequences in limiting the time available for evolution." I am afraid, however, that very few astronomers, not to speak of geophysicists, will be prepared to accept this drastic reduction of the time scale.

It is possible to relegate the epoch of the starting of the expansion to minus infinity, e.g. by using instead of the ordinary time the logarithm of the time elapsed since the beginning. But this is only a mathematical trick. We call zero minus infinity, but that only means that we allow the universe an infinite time to get well started on its course of expansion, but it does not make the time during which anything really happens any longer.

Mention should be made of a suggestion that has sometimes been advanced, viz. that the observed shift of the spectral lines towards the red might not indicate a receding motion of the spiral nebulae, but might be accounted for in some different way. In fact, all that the observations tell us is that light coming from great distances—and which therefore has been a long time under way—is redder when it arrives than when it left its source. *Light is reddened by age*, it loses energy as it gets older, travelling through space. Or expressed mathematically: the wavelength of light is proportional to a certain quantity R, which increases with the time. By the general equations of the theory of relativity the naturally measured distances in a homogeneous and isotropic world are then necessarily proportional to the same quantity R, unless some extraneous cause for the increase of wavelength, or the loss of energy, is present. By extraneous I mean foreign to the theory of relativity and the conception of the nature of light consistent with that theory. Moreover this hypothetical cause should have no other observable consequences, especially it should produce loss of energy without any concomitant dispersion, which would blur the images and make the faint nebulae unobservable. It would require an hypothesis *ad hoc*, and a very carefully framed one too, so as not to overshoot the mark. No such hypothesis, deserving serious consideration, has yet been forthcoming.

It appears to me that there is no way out of the dilemma. It is an unavoidable consequence of the equations that the time taken by the radius to increase from anywhere near its minimum to its present value is of the same order of magnitude as the present radius itself, if we adopt corresponding units of space and time, e.g. years for time and light-years for space. The scale is determined by the observed rate of expansion. I am afraid all we can do is to accept the paradox and try to accommodate ourselves to it, as we have done to so many paradoxes lately in modern physical theories. We shall have to get accustomed to the idea that the change of the quantity R, commonly called the "radius of the universe," and the evolutionary changes of stars and stellar systems are two different processes, going on side by side without any apparent connection between them. After all the "universe" is an hypothesis, like the atom, and must be allowed the freedom to have properties and to do things which would be contradictory and impossible for a finite material structure. What we observe are the stars and nebulae constituting "our neighborhood." All that goes beyond that, in time or in space, or both, is pure extrapolation. The conclusions derived about the expanding universe depend on the assumed homogeneity and isotropy, i.e. on the hypothesis that the observed finite material density and expansion of our neighborhood are not local phenomena, but properties of the "universe." It is not inconceivable that this hypothesis may at some future stage of the development of science have to be given up, or modified, or at least differently interpreted.

Our conception of the structure of the universe bears all the marks of a transitory structure. Our theories are decidedly in a state of continuous, and just now very rapid, evolution. It is not possible to predict how long our present views and interpretations will remain unaltered and how soon they will have to be replaced by perhaps very different ones, based on new observational data and new critical insight in their connection with other data.

Meanwhile the simple workers in science go on quietly, each working at his own particular problem, undisturbed by the many strange and contradictory things that are happening around them and in their own house. And it is on this quiet and unostentatious work that the great advances of science are based. Especially in astronomy two characteristics are common to all data on which the solution of the great problems depends. The first is the extreme minuteness of the quantities to be measured. The determinations of the distances, of the proper motions and radial velocities, which form the materials out of which the great structures of the theories of the galactic system and the universe have been built, all require very accurate measurement of extremely small quantities. It is always a struggle for the last decimal place, and the great triumphs of science are gained when, by new methods or new instruments, the last decimal is made into the penultimate. Thus, as we have seen, new epochs were inaugurated in the beginning of the seventeenth century by the invention of the telescope, and in the last third of the nineteenth by the discovery of photography and spectroscopy.

The other characteristic is that astronomy always requires a very large number of data. As a consequence of the fact that direct experiment is impossible in astronomy, we must have observations of very many stars, and extended to ever fainter objects, in order to enable us to draw reliable conclusions.

These two characteristics of the data that the astronomer requires to build his science on make two things more necessary in astronomy than in any other science: patience and organized cooperation. Patience because many of the phenomena develop so slowly that a long time is necessary for them to become measurable, cooperation because the material is too large and too various to be mastered by one man, or even by one institute. And cooperation not only between different workers and institutions all over the world, but also cooperation with predecessors and successors for the solution of problems that require, by their very nature, more than one man's lifetime. The astronomer—each working at his own task, whether performing long calculations, making theories and hypotheses, or patiently collecting observations in daily routine—is always conscious of belonging to a community, whose members, separated in space and time, nevertheless feel joined by a very real tie, almost of kinship. He does not work for himself alone, he is not guided exclusively, and not even in the first place, by his own insight or preferences, his work is always coordinated with that of others as a part of an organized whole. He knows that, whatever his special work may be it is always a link in a chain, which derives its value from the fact that there is another link to the left and one to the right of it. It is the chain that is important, not the separate links.

It cannot be doubted that at the present moment science stands high in the popular esteem, much higher than it did even a short time ago. But I am afraid that often in the popular mind a sufficiently sharp distinction is not drawn between science and the applications of science to practical ends. Science is praised and esteemed because to it are due the wonderful technical developments that we have witnessed in the last century, the railway and the steamship, the motor car and the airplane, telephone and radio, matches, bathrooms, and so on. All this of course is true, and very wonderful and most important. The general standard of life has been enormously raised; every citizen now enjoys and takes for granted, daily comforts and commodities that would have been undreamt-of luxuries even for the most fastidious millionaire a hundred years ago. If we ask, however, why this increase in our comfort and general standard of life is good the only answer can be that it gives us more leisure to devote ourselves to those things that give real value to life, the pursuit of science and art. It is as Poincaré said: "Je ne dis pas: la Science est utile, parce qu'elle nous apprend à construire des machines; je dis: les machines sont utiles, parce qu'en temps pour faire de las science." And, of course, as we have repeatedly pointed out, also the improved technique provides science with more powerful tools, and thereby stimulates its development. It is a significant fact, proved by history, that all the great technical advances have been based on scientific discoveries which at the time appeared to be utterly useless, and were made by men who studied science for its own sake, without giving the slightest thought to the possibility of application. They are the reward accruing to mankind from the disinterestedness of its greatest representatives.

A moment ago I coupled science and art in one sentence. They are the highest manifestations of the human mind, science of the intellectual side of it, art of the emotional side. There is however much misunderstanding. It happens now and then that those in whom natural gifts or circumstances have developed an exclusively artistic or emotional view of life are filled with pity for the miserable, unimaginative scientists, crawling in the dust of matter-of-fact measurements and calculations, instead of soaring in the pure ether of mystic contemplation and understanding. On the other hand it also happens now and then, but perhaps not so often, that persons with a specially scientific outlook and education look down on the poor unpractical and illogical artist. Both these judgements, of course, are very much mistaken. Science and art approach the great problems of the understanding of nature each in its own way, but both require, and use, the full attributes of the human mind. Imagination is as indispensable for the physicist or astronomer as for the poet; logic is as necessary for the architect or the musician as for the mathematician.

The great men of science, as well as the great artists, are filled with a spirit of reverence, with a consciousness of the presence of mystery and sublimity in the simplest and smallest as well as the greatest of things and phenomena, and with faith in the order and unity of all things. Only the way in which they try to comprehend this order and penetrate to its deeper meaning is different.

It is a rather common misapprehension that science, by analysing and dissecting nature, by subjecting it to the rigorous rule of mathematical formulae and numerical expression, would lose the sense of its beauty and sublimity. On the contrary, even the purely aesthetical appreciation, say of a landscape, or of a thunderstorm, is in my opinion, helped rather than impeded by the knowledge, so far as it goes, that the scientific beholder has of the inner structure and the connection of the phenomena. And the measurement and reduction to numbers, "Pointer readings," as Sir Arthur Eddington says, is not the ultimate aim of science, but its means to an end. By the use of mathematics, that most nearly perfect and most immaterial tool of the human mind, we try to transcend as much as possible the limitations imposed by our finiteness and materiality, and to penetrate ever nearer to the understanding of the mysterious unity of the Kosmos.

7

— Edwin Hubble —

(1889–1953)

Edwin Hubble was an American astronomer and cosmologist credited with dis-
covering that the universe is expanding and with the development of Hubble's law
which relates recessional velocity and distance. Using the hundred-inch reflecting
telescope at Mt. Wilson, Hubble was also able to measure the size and age of
the universe.

In this reading Hubble describes the evidence that the nebulae are island
universes at great distances from us. The use of Cepheid Variables was the main
but not the only tool for measuring those distances. Hubble shows his familiarity
with early nebular hypotheses of Wright, Kant, Derham, Maupertuis, and of
course William Herschel. The reading is the first chapter of Hubble's book The
Realm of the Nebula.

THE STRUCTURE OF THE UNIVERSE

The exploration of space has penetrated only recently into the realm of the
nebulae. The advance into regions hitherto unknown has been made during
the last dozen years with the aid of great telescopes. The observable region
of the universe is now defined and a preliminary reconnaissance has been
completed. The chapters which follow are reports on various phases of the
reconnaissance.

The earth we inhabit is a member of the solar system—a minor satellite of
the sun. The sun is a star among the many millions which form the stellar
system. The stellar system is a swarm of stars isolated in space. It drifts through
the universe as a swarm of bees drifts through the summer air. From our position somewhere within the system, we look out through the swarm of stars, past
the borders, into the universe beyond.

The universe is empty, for the most part, but here and there, separated by
immense intervals, we find other stellar systems, comparable with our own.
They are so remote that, except in the nearest systems, we do not see the individual stars of which they are composed. These huge stellar systems appear as
dim patches of light. Long ago they were named "nebulae" or "clouds"—mysterious bodies where nature was a favorite subject for speculation.

But now, thanks to great telescopes, we know something of their nature,
something of their real size and brightness, and their mere appearance indicates
the general order of their distances. They are scattered through space as far as
telescopes can penetrate. We see a few that appear large and bright. These are

the nearer nebulae. Then we find them smaller and fainter, in constantly increasing numbers, and we know that we are reaching out into space, farther and ever farther, until, with the faintest nebulae that can be detected with the greatest telescope, we arrive at the frontiers of the known universe.

This last horizon defines the observable region of space. It is a vast sphere, perhaps a thousand million light-years in diameter. Throughout the sphere are scattered a hundred million nebulae-stellar systems—in various stages of their evolutionary history. The nebulae are distributed singly, in group, and occasionally in great clusters, but when large volumes of space are compared, the tendency to cluster averages out. To the very limits of the telescope, the large scale distribution of nebulae is approximately uniform.

One other general characteristic of the observable region has been found. Light which reaches us from the nebulae is reddened in proportion to the distance it has traveled. This phenomenon is known as the velocity-distance relation, for it is often interpreted, in theory, as evidence that the nebulae are all rushing away from our stellar system, with velocities that increase directly with distances.

Receding Horizons

This sketch roughly indicates the current conception of the realm of the nebulae. It is the culmination of a line of research that began long ago. The history of astronomy is a history of receding horizons. Knowledge has spread in successive waves, each wave representing the exploitation of some new clue to the interpretation of observational data.

The exploration of space presents three such phases. At first the explorations were confined to the realm of the planets, then they spread through the realm of the stars, and finally they penetrated into the realm of the nebulae.

The successive phases were separated by long intervals of time. Although the distance of the moon was well known to the Greeks, the order of the distance of the sun and the scale of planetary distances was not established until the latter part of the seventeenth century. Distances of stars were first determined almost exactly a century ago, and distances of nebulae, in our own generation. The distances were the essential data. Until they were found, no progress was possible.

The early explorations halted at the edge of the solar system, facing a great void that stretched away to the nearer stars. The stars were unknown quantities. They might be little bodies, relatively near, or they might be gigantic bodies, vastly remote. Only when the gap was bridged, only when the distances of a small, sample collection of stars had been actually measured, was the nature determined of the inhabitants of the realm beyond the solar system. Then the explorations, operating from an established base among the now familiar stars, swept rapidly through the whole of the stellar system.

Again there was a halt, in the face of an even greater void, but again, when instruments and technique had sufficiently developed, the gap was bridged by the determination of the distances of a few of the nearer nebulae. Once more, with the nature of the inhabitants known, the explorations swept even more rapidly through the realm of the nebulae and halted only at the limits of the greatest telescope.

The Theory of Island Universes

This is the story of the explorations. They were made with measuring rods, and they enlarged the body of factual knowledge. They were always preceded by speculations. Speculations once ranged through the entire field, but they have been pushed steadily back by the explorations until now they lay undisputed claim only to the territory beyond the telescopes, to the dark unexplored regions of the universe at large.

The speculations took many forms and most of them have long since been forgotten. The few that survived the test of the measuring rod were based on the principle of the uniformity of nature—the assumption that any large sample of the universe is much like any other. The principle was applied to stars long before distances were determined. Since the stars were too far away for the measuring instruments, they must necessarily be very bright. The brightest object known was the sun. Therefore, the stars were assumed to be like the sun, and distances could be estimated from their apparent faintness. In this way, the conception of a stellar system, isolated in space, was formulated as early as 1750; the author was Thomas Wright (1711–1786), an English instrument maker and private tutor.

But Wright's speculations went beyond the Milky Way. A single stellar system, isolated in the universe, did not satisfy his philosophical mind. He imagined other, similar systems and, as visible evidence of their existence, referred to the mysterious clouds called "nebulae."

Five years later, Immanuel Kant (1724–1804) developed Wright's conception in a form that endured, essentially unchanged, for the following century and a half. Some of Kant's remarks concerning the theory furnish an excellent example of reasonable speculation based on the principle of uniformity. A rather free translation runs as follows:

> I come now to another part of my system, and because it suggests a lofty idea of the plan of creation, it appears to me as the most seductive. The sequence of ideas that led us to it is very simple and natural. They are as follows: let us imagine a system of stars gathered together in a common plane, like those of the Milky Way, but situated so far away from us that even with the telescope we cannot distinguish the stars composing it; let us assume that its distance, compared to that separating us from the stars of the Milky Way, is in the same proportion as the distance of the Milky Way is to the distance from the earth to the sun; such a stellar world will appear to the observer, who contemplates it at so enormous a distance, only as a little spot feebly illumined and subtending a very small angle; its

shape will be circular, if its plane is perpendicular to the line of sight, elliptical, if it is seen obliquely. The faintness of its light, its form, and its appreciable diameter will obviously distinguish such a phenomenon from the isolated stars around it.

We do not need to seek far in the observations of astronomers to meet with such phenomena. They have been seen by various observers, who have wondered at their strange appearance, have speculated about them, and have suggested sometimes the most amazing explanations, sometimes theories which were more rational, but which had no more foundation than the former. We refer to the nebulae, or, more precisely, to a particular kind of celestial body which M. de Maupertius describes as follows:

"These are small luminous patches, only slightly more brilliant than the dark background of the sky; they have this in common, that their shapes are more or less open ellipses; and their light is far more feeble than that of any other objects to be perceived in the heavens."

Kant then mentions and rejects the views of Derham that the patches are openings in the firmament, through which the fiery Empyrean is seen, and of Maupertius that the nebulae are enormous single bodies, flattened by rapid rotation. Kant then continues:

It is much more natural and reasonable to assume that a nebula is not a unique and solitary sun, but a system of numerous suns, which appear crowded, because of their distance, into a space so limited that their light, which would be imperceptible were each of them isolated, suffices, owing to their enormous numbers, to give a pale and uniform luster. Their analogy with our own system of stars; their form, which is precisely what it should be according to our theory; the faintness of their light, which denotes an infinite distance; all are in admirable accord and lead us to consider these elliptical spots as systems of the same order as our own-in a word, to be Milky Ways similar to the one whose constitution we have explained. And if these hypotheses, in which analogy and observation consistently lend mutual support, have the same merit as formal demonstrations, we must consider the existence of such systems as demonstrated. . . .

We see that scattered through space out to infinite distances, there exist similar systems of stars [nebulous stars, nebulae], and that creation, in the whole extent of its infinite grandeur, is everywhere organized into systems whose members are in relation with one another. . . . A vast field lies open to discoveries, and observation alone will give the key.

The theory, which came to be called the theory of island universes, found a permanent place in the body of philosophical speculation. The astronomers themselves took little part in the discussions: they studied the nebulae. Toward the end of the nineteenth century, however, the accumulation of observational data brought into prominence the problem of the status of the nebulae and, with it, the theory of island universes as a possible solution.

The Nature of the Nebulae

(a) The Formulation of the Problem

A few nebulae had been known to the naked-eye observers and, with the development of telescopes, the numbers grew, slowly at first, then more and more rapidly. At the time Sir William Herschel (1738–1822), the first outstanding leader in nebular research, began his surveys, the most extensive published lists were those by Messier, the last of which (1784) contained 103 of the most conspicuous nebulae and clusters. These objects are still known by the Messier numbers—for example, the great spiral in Andromeda is M31. Sir William Herschel catalogued 2,500 objects, and his son, Sir John (1792–1871), transporting the telescopes to the southern hemisphere (near Capetown is south Africa) added many more. Positions of about 20,000 nebulae are now available, and perhaps ten times that number have been identified on photographic plates. The mere size of catalogues has long since ceased to be important. Now the desirable data are the numbers of nebulae brighter than successive limits of apparent faintness, in sample areas widely distributed over the sky.

Galileo, with his first telescopes, resolved a typical "cloud"—Præsepe—into a cluster of stars. With larger telescopes and continued study, many of the more conspicuous nebulae met the same fate. Sir William Herschel concluded that all nebulae could be resolved into star-clusters, if only sufficient telescopic power were available. In his later days, however, he revised his position and admitted the existence, in certain cases, of a luminous "fluid" which was inherently unresolvable. Ingenious attempts were made to explain away these exceptional cases until Sir William Huggins (1824–1910), equipped with a spectrograph, fully demonstrated in 1864 that some of the nebulae were masses of luminous gas.

Huggins's results clearly indicated that nebulae were not all members of a single, homogeneous group and that some kind of classification would be necessary before they could be reduced to order. The nebulae actually resolved into stars—the star-clusters—were weeded out of the lists to form a separate department of research. They were recognized as component parts of the galactic system, and thus had no bearing on the theory of island universes.

Among unresolved nebulae, two entirely different types were eventually differentiated. One type consisted of the relatively few nebulae definitely known to be unresolvable-clouds of dust and gas mingled among, and intimately associated with, the stars in the galactic system. They were usually found within the belt of the Milky Way and were obviously, like the star-clusters, members of the galactic system. For this reason, they have since been called "galactic" nebulae. They are further subdivided into two groups, "planetary" nebulae and "diffuse" nebulae, frequently shortened to "planetaries" and "nebulosities."

The other type consisted of the great numbers of small, symmetrical objects found everywhere in the sky except in the Milky Way. A spiral structure was found in most, although not in all, of the conspicuous objects. They had many features in common and appeared to form a single family. They were given various names but, to anticipate, they are now know as "extragalactic" nebulae and will be called simply "nebulae."

The status of the nebulae, as the group is now defined, was undetermined because the distances were wholly unknown. They were definitely beyond the limits of direct measurement, and the scanty, indirect evidence bearing on the problem could be interpreted in various ways. The nebulae might be relatively nearby objects and hence members of the stellar system, or they might be very remote and hence inhabitants of outer space. At this point, the development of nebular research came into immediate contact with the philosophical theory of island universes. The theory represented, in principle, one of the alternative solutions of the problem of nebular distances. The question of distances was frequently put in the form: Are nebulae island universes?

(b) The Solution of the Problem

The situation developed during the years between 1885 and 1914; from the appearance of the bright nova in the spiral M31, which stimulated a new interest in the question of distances, to the publication of Slipher's first extensive list of radial velocities of the nebulae, which furnished data of a new kind and encouraged serious attempts to find a solution of the problem.

The solution came ten years later, largely with the help of a great telescope, the 100-inch reflector, that had been completed in the interim. Several of the most conspicuous nebulae were found to be far beyond the limits of the galactic system—they were independent, stellar systems in extragalactic space. Further investigations demonstrated that the other, fainter nebulae were similar systems at greater distances, and the theory of island universes was confirmed.

The 100-inch reflector partially resolved a few of the nearest, neighboring nebulae into swarms of stars. Among these stars various types were recognized which were well known among the brighter stars in the galactic system. The intrinsic luminosities (candle powers) were known, accurately in some cases, approximately in others. Therefore, the apparent faintness of the stars in the nebulae indicated the distances of the nebulae.

The most reliable results were furnished by Cepheid variables, but other types of stars furnished estimates of orders of distance, which were consistent with the Cepheids. Even the brightest stars, whose intrinsic luminosities appear to be nearly constant in certain types of nebulae, have been used as statistical criteria to estimate mean distances for groups of systems.

The Inhabitants of Space

The nebulae whose distances were known from the stars involved, furnished a sample collection from which new criteria, derived from the nebulae and not from their contents, were formulated. It is now known that the nebulae are all of the same order of intrinsic luminosity. Some are brighter than others, but at least half of them are within the narrow range from one half to twice the mean value, which is 85 million times the luminosity of the sun. Thus, for statistical purposes, the apparent faintness of nebulae indicates their distances.

With the nature of the nebulae known and the scale of nebular distances established, the investigations proceeded along two lines. In the first place the general features of the individual nebulae were studied; in the second, the characteristics of the observable region as a whole were investigated.

The detailed classification of nebular forms has led to an ordered sequence ranging from globular nebulae, through flattening, ellipsoidal figures, to a series of unwinding spirals. The fundamental pattern of rotational symmetry changes smoothly through the sequence in a manner that suggests increasing speed of rotation. Many features are found which vary systematically along the sequence, and the early impression that the nebulae were members of a single family appears to be confirmed. The luminosities remain fairly constant through the sequence (mean value, 8.5×10^7 suns, as previously mentioned), but the diameters steadily increase from about 1,800 light-years for the globular nebulae to about 10,000 light-years for the most open spirals. The masses are uncertain, the estimates ranging from 2×10^9 to 2×10^{11} times the mass of the sun.

The Realm of the Nebulae

(a) The Distribution of Nebulae

Investigations of the observable region as a whole have led to two results of major importance. One is the homogeneity of the region—the uniformity of the large-scale distribution of nebulae. The other is the velocity-distance relation.

The small-scale distribution of nebulae is very irregular. Nebulae are found singly, in pairs, in groups of various sizes, and in clusters. The galactic system is the chief component of a triple nebula in which the Magellanic Clouds are the other members. The triple system, together with a few additional nebulae, forms a typical small group that is isolated in the general field of nebulae. The members of this local group furnished the first distances, and the Cepheid criterion of distance is still confined to the group.

When large regions of the sky, or large volumes of space, are compared, the irregularities average out and the large-scale distribution is sensibly uniform. The distribution over the sky is derived by comparing the numbers of nebulae brighter than a specified limit of apparent faintness, in sample areas scattered at regular intervals.

The true distribution is confused by local obscuration. No nebulae are seen within the Milky Way, and very few along the borders. Moreover, the apparent distribution thins out, slightly but systematically, from the poles to the borders of the Milky Way. The explanation is found in the great clouds of dust and gas which are scattered throughout the stellar system, largely in the galactic plane. These clouds hide the more distant stars and nebulae. Moreover, the sun is embedded in a tenuous medium which behaves like a uniform layer extending more or less indefinitely along the galactic plane. Light from nebulae near the galactic poles is reduced about one-fourth by the obscuring layer, but in the lower latitudes, where the light-paths through the medium are longer, the absorption is correspondingly greater. It is only when these various effects of galactic obscuration are evaluated and removed, that the nebular distribution over the sky is revealed as uniform, or isotropic (the same in all directions).

The distribution in depth is found by comparing the numbers of nebulae brighter than successive limits of apparent faintness, that is to say, the numbers within successive limits of distance. The comparison is effectively between numbers of nebulae and the volumes of space which they occupy. Since the numbers increase directly with the volumes (certainly as far as the surveys have been carried, probably as far as telescopes will reach), the distribution of the nebulae must be uniform. In this problem, also, certain corrections must be applied to the apparent distribution in order to derive the true distribution. These corrections are indicated by the velocity-distance relation, and their observed values contribute to the interpretation of that strange phenomenon.

Thus the observable region is not only isotropic but homogeneous as well—it is much the same everywhere and in all directions. The nebulae are scattered at average intervals of the order of two million light-years or perhaps two hundred times the mean diameters. The pattern might be represented by tennis balls fifty feet apart.

The order of the mean density of matter in space can also be roughly estimated if the (unknown) material between the nebulae is ignored. If the nebular material were spread evenly through the observable region, the smoothed-out density would be of the general order of 10^{-29} or 10^{-28} grams per cubic centimeter-about one grain of sand per volume of space equal to the size of the earth.

The size of the observable region is a matter of definition. The dwarf nebulae can be detected only to moderate distances, while giants can be recorded far out in space. There is no way of distinguishing the two classes, and thus the limits of the telescope are most conveniently defined by average nebulae. The faintest nebulae that have been identified with the 100-inch reflector are at an average distance of the order of 500 million light-years, and to this limit about 100 million nebulae would be observable except for the effects of galactic obscuration. Near the galactic pole, where the obscuration is least, the longest exposures record as many nebulae as stars.

(b) The Velocity-Distance Relation

The foregoing sketch of the observable region has been based almost entirely upon results derived from direct photographs. The region is homogeneous and the general order of the mean density is known. The next—and last—property to be discussed, the velocity-distance relation, emerged from the study of spectrograms.

When a ray of light passes through a glass prism (or other suitable device) the various colors of which the light is composed are spread out in an ordered sequence called a spectrum. The rainbow is, of course, a familiar example. The sequence never varies. The spectrum may be long or short, depending on the apparatus employed, but the order of the colors remains unchanged. Position in the spectrum is measured roughly by colors, and more precisely by wave-lengths, for each color represents light of a particular wave-length. From the short waves of the violet, they steadily lengthen to the long waves of the red.

The spectrum of a light source shows the particular colors or wave-lengths which are radiated, together with their relative abundance (or intensity), and thus gives information concerning the nature and the physical condition of the light source. An incandescent solid radiates all colors, and the spectrum is *continuous* from violet to red (and beyond in either direction). An incandescent gas radiates only a few isolated colors and the pattern, called an *emission* spectrum, is characteristic for any particular gas.

A third type, called an *absorption* spectrum and of special interest for astronomical research, is produced when an incandescent solid (or equivalent source), giving a continuous spectrum, is surrounded by a cooler gas. The gas absorbs from the continuous spectrum just those colors which the gas would radiate if it were itself incandescent. The result is a spectrum with a continuous background interrupted by dark spaces called absorption lines. The pattern of dark absorption lines indicates the particular gas or gases that are responsible for the absorption.

The sun and the stars give absorption spectra and many of the known elements have been identified in their atmospheres. Hydrogen, iron, and calcium produce very strong lines in the solar spectrum, the most conspicuous being a pair of calcium lines in the violet, known as H and K.

The nebulae in general show absorption spectra similar to the solar spectrum, as would be expected for systems of stars among which the solar type predominated. The spectra are necessarily short—the light is too faint to be spread over long spectra—but the H and K lines of calcium are readily identified and, in addition, the G-band of iron and a few hydrogen lines can generally be distinguished.

Nebular spectra are peculiar in that the lines are not in the usual positions found in nearby light sources. They are displaced toward the red of their normal position, as indicated by suitable comparison spectra. The displacements, called red-shifts, increase, on the average, with the apparent faintness of the nebula that is observed. Since apparent faintness measures distance, it follows that red-shifts increase with distance. Detailed investigation shows that the relation is linear.

Small microscopic shifts, either to the red or to the violet, have long been known in the spectra of astronomical bodies other than nebulae. These displacements are confidently interpreted as the results of motion in the line of sight—radial velocities of recession (red-shifts) or of approach (violet-shifts). The same interpretation is frequently applied to the red-shifts in nebular spectra and has led to the term "velocity-distance" relation for the observed relation between red-shifts and apparent faintness. On this assumption, the nebulae are supposed to be rushing away from our region of space, with velocities that increase directly with distance.

Although no other plausible explanation of red-shifts has been found, the interpretation as velocity-shifts may be considered as a theory still to be tested by actual observations. Critical tests can probably be made with existing instruments. Rapidly receding light sources should appear fainter than stationary sources at the same distances, and near the limits of telescopes the "apparent" velocities are so great that the effects should be appreciable.

The Observable Region as a Sample of the Universe

A completely satisfactory interpretation of red-shifts is a question of great importance, for the velocity-distance relation is a property of the observable region as a whole. The only other property that is known is the uniform distribution of nebulae. Now the observable region is our sample of the universe. If the sample is fair, its observed characteristics will determine the physical nature of the universe as a whole.

And the sample may be fair. As long as explorations were confined to the stellar system, the possibility did not exist. The system was known to be isolated. Beyond lay a region, unknown, but necessarily different from the star-strewn space within the system. We now observe that region—a vast sphere, through which comparable stellar systems are uniformly distributed. There is no evidence of a thinning-out, no trace of a physical boundary. There is not the slightest suggestion of a supersystem of nebulae isolated in a larger world. Thus, for purposes of speculation, we may apply the principle of uniformity, and suppose that any other equal portion of the universe, selected at random, is much the same as the observable region. We may assume that the realm of the nebulae is the universe and that the observable region is a fair sample.

The conclusion, in a sense, summarizes the results of empirical investigations and offers a promising point of departure for the realm of speculation. That realm, dominated by cosmological theory, will not be entered in the present summary. The discussions will be largely restricted to the empirical data—to reports of the actual explorations—and their immediate interpretations.

Yet observation and theory are woven together, and it is futile to attempt their complete separation. Observations always involve theory. Pure theory may be found in mathematics but seldom in science. Mathematics, it has been said, deals with possible worlds—logically consistent systems. Science attempts to discover the actual world we inhabit. So in cosmology, theory presents an infinite array of possible universes, and observation is eliminating them, class by class, until now the different types among which our particular universe must be included have become increasingly comprehensible.

The reconnaissance of the observable region has contributed very materially to this process of elimination. It has described a large sample of the universe, and the sample may be fair. To this extent the study of the structure of the universe may be said to have entered the field of empirical investigation.

8
— Edwin Hubble —

In this reading, Chapter 5 from The Realm of the Nebula, *Hubble develops his velocity-distance relationship, which declares that the farther a nebula is from us, the faster it is receding. It is, in Hubble's own words, "the observational basis for an expanding universe."*

THE VELOCITY-DISTANCE RELATION

Early Spectrograms of Nebulae

Spectra of nebulae were first investigated visually in 1864 by Sir William Huggins (1824–1910). Those of white nebulae, as extragalactic systems were then called, were apparently continuous, but so faint that no details could be determined with certainty. Prolonged study of the brightest, M31, led to the surmise that both absorption and emission lines or bands were present, and a very faint photograph, achieved in 1888, seemed to confirm the tentative conclusion. No report of the photograph had been published in 1899 when Scheiner settled the question with readable spectrograms of M31. They showed a solar type spectrum with no emission. He concluded that the spiral was probably a stellar system and thus revived the waning interest in the controversy over island universes. Fath and Wolf extended the investigations to other nebulae with similar results, and eventually the prevalence of solar types among spectra of the brighter spirals, was generally recognized.

The First Radial Velocity

The radial velocity of a nebula was measured for the first time in 1912 by V. M. Slipher at the Lowell Observatory. Although the general character of the spectra had been established, the more difficult problem of determining the precise positions of the absorption lines had not been solved. The difficulties arose from the dim surface-brightness of the nebular images. Unlike the stars, whose light is concentrated into practically point-images by all telescopes, the nebulae form relatively large images and the areas increase with the focal length of the telescope employed. Larger telescopes, if the focal ratios are constant, merely spread more light over larger images, leaving the surface-brightness unchanged.

The difficulties are met in direct photography by shortening the focus for a given aperture and thus concentrating the light into smaller images. When the images are photographed through a prism, however, this modification of the telescope offers no advantage. The explanation is simple, but as it involves properties of optical instruments it need not be presented in detail. For large, uniform surfaces, all telescopes are about equally efficient. No advantage can be gained except in the camera behind the prism, which actually photographs the spectra. The rule breaks down in the case of small surfaces, and for the concentrated, semistellar images of the fainter nebulae the larger telescopes are increasingly efficient. Nevertheless, the most important single factor in the photography of spectra of faint light sources is the speed of the camera.

Slipher exploited this principle and adapted a very fast, short-focus camera to a small-dispersion spectrograph attached to the 24-inch refractor at the Lowell Observatory. With this equipment he was able to record the spectrum of M31 with good definition and on a scale which, although small, was sufficient to show that the absorption lines were not quite in their customary positions. The displacements were toward the violet end of the spectrum, indicating that the radial component of motion was toward the earth. Precise measures revealed that the velocity of approach was about 190 miles (300 kilometers) per second. Four spectrograms secured in the autumn of 1912 gave consistent velocities, and the results could be published with complete confidence in their reliability.

Slipher's List of Radial Velocities

The determination of the velocity of M31 has been discussed at some length on the general principle that the first steps in a new field are the most difficult and the most significant. Once the barrier is forced, further development is comparatively simple. But the accumulation of nebular velocities was a slow process and became increasingly laborious after the brightest objects had been observed. Slipher carried on the work almost alone. In 1914 he presented a list of thirteen velocities, and by 1925 the number of his contributions had grown to forty-one. A few of the velocities had been redetermined at other observatories, sufficient to establish the validity of the data beyond any reasonable doubt, but only four new velocities had been added to Slipher's list. In 1925, a total of forty-five nebular velocities was available for discussion.

Although the first velocity was negative, indicting motion toward the observer, positive velocities, indicating motion away from the observer, were found in increasing numbers and soon they completely dominated the list. Moreover, after the most conspicuous nebulae had been observed, the numerical values of the new velocities were found to be surprisingly large. The complete list ranged from −190 miles/sec. to +1, 125 and averaged about +375. The velocities were of an entirely different order from those of any other known type of astronomical body. They were so large that the nebulae were probably beyond the control of the gravitational field of the stellar system. The nebulae, it appeared, were independent bodies and this conclusion was consistent with the theory of island universes.

Interpretation of the Data

Solar Motion with Respect to the Nebulae

Actually no other theory was seriously considered in attempting to interpret the data. The stellar system, carrying the sun along with it was supposed to be moving rapidly through the realm of the nebulae, which themselves were rushing about with comparable speeds in random directions. Each observed velocity was thus a combination of (a) the "peculiar motion" of the nebula, as the individual motion is called, and (b) the reflection of the solar motion. If sufficient nebulae were observed, their random peculiar motions would tend to cancel out, leaving only the reflection of the solar motion to emerge from the totality of the data.

The principle was a familiar one and had worked very well within the stellar system for the determination of the motion of the sun with respect to the stars. It was first applied to the nebulae by Truman in 1916, when only a dozen nebular velocities were known. Others also solved the equations, including Slipher, when, in 1917, he had twenty-five velocities at his disposal. The numerical results were all rather similar—a solar motion, interpreted as effectively the motion of the stellar system, of the general order of 420 miles/sec., in the general direction of the constellation Capricornus.

It was expected that, when the solar motion was removed, the residual, peculiar motions of the nebulae would be much smaller than the observed velocities and, furthermore, that they would be distributed at random—that velocities of approach would be as numerous as velocities of recession. Actually, the residual motions were still large and predominantly positive. The unsymmetrical distribution indicated the presence of some systematic effect in addition to the motion of the sun. It was for this reason that Wirtz, in 1918, introduced a seemingly arbitrary K-term—a constant velocity to be subtracted from all observed velocities before the search for the solar motion was undertaken.

The conception of a K-term was not new. It had been used, for instance, in determining the solar motion with respect to the B-stars. In that case it amounted to about four kilometers per second and was supposed to represent some effect of atmospheric pressure, gravitational field, or other condition peculiar to the blue giants. In the case of the nebulae, however, a term of fantastic dimensions—of the order of a hundred times four kilometers—would be required in order to effect seriously the distribution of residuals. The introduction was a logical step but it requires some boldness to make such a venture.

Wirtz's formulation of the problem included the K-term together with the solar motion as unknowns to be determined from the observational data. He knew of only fifteen velocities at the time of his first solution, but three years later (1921), he repeated the investigation using twenty-nine velocities. The new values were of the same general order as the earlier results. The K-term amounted to about 500 miles/sec. The solar motion was again about 440 miles/sec., but was now in the general direction of the north pole of the heavens.

More important, however, the scatter of the residuals or, in other words, of the peculiar velocities of the individual nebulae, was more or less random. The evidence of systematic effects had almost vanished. The problem was not completely solved—the residuals were not wholly satisfactory—but the improvement was so marked that the K-term was accepted as a characteristic feature of nebular velocities. All subsequent discussions of the problem included the K-term as a matter of course.

The K-Term as a Function of Distance

When Wirtz first introduced the K-term, he merely stated that it was necessary because of the preponderance of positive signs and the large numerical values of the velocities. He was fully aware of the consequence that, if the displacements of the spectral lines were literally interpreted as actual velocity-shifts, the K-term must represent a systematic recession of all nebulae from the vicinity of the galactic system. He did not fully commit himself to this interpretation, but left the question open and used the term as though it were an arbitrary device for "saving the phenomena." The explanation might be found later.

It seemed possible, however, that current theory had already indicated the significance of the K-term. Einstein, in 1915, had formulated his cosmological equation which expressed the relation between the contents of space and the geometry of space, as derived from the theory of general relativity. On the assumption that the universe is static (does not vary systematically with time), he had found a solution to the equation and had, therefore, described a particular kind of universe. De Sitter, in 1916–17, using the same equation, had found another solution. It was later shown that no other solutions were possible on the particular assumptions. The two possible universes were carefully studied in order to see which of them more closely corresponded to the universe we actually inhabit. One outstanding difference between the two was the fact that de Sitter's solution predicted positive displacements (red-shifts) in the spectra of distant light sources, which, on the average, should increase with distance from the observer. De Sitter knew of only three velocities at the time and could not make an extensive comparison between theory and observation. Nevertheless, it was clear, as he stated, that the large, positive velocities of the two fainter nebulae (NGC 1068 and 4594) as contrasted with the negative velocity of M31, the brightest of all the spirals, were consistent with the prediction.

The de Sitter universe is no longer considered as a representation of the actual universe, but at the time it served the important purpose of directing attention to the possibility of a variable K-term. The numerical rate of increase of red-shifts with distance was not predicted by the theory; the rate might be large or small, conspicuous or imperceptible, and the question could be determined only by observations. But among the necessary data were distances of nebulae, and distances were then unknown. This fact, together with perhaps a natural inertia in the face of revolutionary ideas couched in the unfamiliar language of general relativity, discouraged immediate investigation. It was not until later, when Eddington and others had, as it were, "popularized" the new ideas, that the problem was seriously considered.

If velocities increased with distance, the large constant K-term might represent the velocity corresponding to the mean distance of the particular group of nebulae that had been observed. This possibility was generally recognized, although no one seems to have made the statement specifically. The problem was formulated as follows: Is the K-term constant for all nebulae or does it vary with the distance?

Absolute distances of nebulae were very uncertain. The only available criteria of relative distances were apparent diameters and apparent luminosities. Neither were reliable because the ranges in intrinsic size and brightness were wholly unknown, and the ranges were believed to be considerable. In the triple system of M31 and its two companions, for instance, the diameters ranged from sixty to one, and the luminosities from a hundred to one. There was no evidence that even these ranges applied to nebulae at large. Nevertheless, in a general way, the smaller, fainter nebulae were doubtless at greater mean distances than the larger, brighter objects. The criteria might be useful provided the range in distance covered by the velocities was large compared to the scatter introduced by the criteria.

Wirtz, the leader in the field, made the first attempt, in 1924, to express the K-term as a function of distance using apparent diameters and velocities of forty-two nebulae. A plausible correlation appeared in the expected direction—velocities tended to increase as diameters diminished. The results, however, were suggestive rather than definitive. No only were they subject to uncertainties arising from the unknown scatter in real diameters, but they included the effects of an apparent correlation between velocity and concentration. The highly concentrated globular nebulae, as a class, exhibited the largest mean velocity, and the large, faint irregular nebulae and open spirals exhibited the smallest mean velocity. In between these limits, the velocities increased with the concentration.

The correlation was generally known, and had inspired unsuccessful attempts to account for the K-term as Einstein-shifts produced by strong, gravitational fields—analogous to the red-shift in the solar spectrum which had served as one of the crucial tests of general relativity. Eventually it was realized that the correlation was a simple effect of selection. The concentrated objects, because of their great surface-brightness, were given preference in the laborious task of photographing nebular spectra. So, although these objects are relatively rare, there was a natural tendency to select them for the investigations of faint nebulae. They represented, on the average, the faintest and most distant of the nebulae observed and for this reason they exhibited the largest mean velocity. But the explanation came much later. At the time, it was believed that the progression in diameters might signify a progression either in concentration or in distance or in both; therefore the correlation between diameters and velocities was ambiguous.

Furthermore, Wirtz had not used the simple diameters, but their logarithms. The choice was a convenient one, but it led him to express his results as a linear relation between velocities and log diameters or, as he considered, log distances. Such a relation differed in principle from the relation predicted by de Sitter. Hence, in view of the possibility of an alternative explanation as a concentration effect, there was a tendency among astronomers to defer judgment until more information should become available.

Wirtz presented some arguments suggesting that his correlation could not be entirely due to variations either in real diameters or surface-brightness, and Dose, shortly afterwards, showed that a similar, although less-pronounced, correlation existed between velocities and simple diameters. Nevertheless, the later investigations of Lundmark and of Strömberg failed to establish any definite relation between velocity and distance. Lundmark in 1924, using the same nebulae as Wirtz and employing diameters and luminosities in combination as criteria of distances, concluded somewhat optimistically that "there may be a relation between the two quantities (velocity and distance), although not a very definite one." Strömberg, in 1925, using luminosities alone as criteria of distances, made an especially clear-cut analysis of the data and found "no sufficient reason to believe that there exists any dependence of radial motion upon distance from the sun." This statement, of course, referred to the situation as indicated by the information then at hand. It represented the observer's point of view—that whatever the ultimate truth of the matter might be, the data did not establish a relation. Further discussion would probably contribute little; the important desiderata were additional data and more precise criteria of distances. Strömberg did bring out rather clearly, however, that although the K-term did not seem to vary systematically with distance, it probably varied from nebula to nebula, being small for M31 and the Magellanic Clouds, but large for NGC 584 (for which the largest velocity, +1,125 miles/sec., had been measured).

Shortly afterwards, Lundmark made a final attempt to uncover a variable K-term. He used the same data as before but replaced the constant K-term in the equations by a power series,

$$K = k + lr + mr^2$$

where r was the distance in terms of the undetermined distance of M31 as a unit. The results were disappointing. The constant, k, in the series, was found to be 320 miles/sec., somewhat smaller than, but still of the same general order as, the former values of K. The coefficient, l, was small and uncertain, about +6 miles/sec., indicating a slight distance-effect (about 8 per cent of the current value), but the coefficient m, very small and still more uncertain, was negative, −0.047. Lundmark considered that m, although its precise value was uncertain, expressed a real phenomenon which obviously set an upper limit to the velocities of recession that nebulae could ever attain (aside from their peculiar motions). He concluded that "one would scarcely expect to find any radial velocities larger than +3000 km/sec. among the spirals."

The Velocity-Distance Relation

Here the matter rested until 1929. Slipher had turned to other problems and only two or three new velocities had been determined. But new criteria of distances had been developed which were much more reliable than those derived from apparent size and brightness. The new criteria were furnished by stars involved in nebulae and not by the nebulae themselves. Nebulae were now recognized as independent stellar systems scattered through extragalactic space. In a few of the very nearest neighbors, swarms of stars could be photographed and among them various types, well known in the galactic system, could be identified. The apparent faintness of these stars furnished reliable distances of the nebulae in which they were involved.

Less precise distances were indicated by the apparent faintness of the very brightest stars in the nebulae. This criterion could be applied as far as the nearest of the great clusters, the Virgo cluster, at a distance of the order of six or seven million light-years. The scatter in the new criterion was reasonably small compared to the range in distance covered by the velocities. The new development led inevitably to a reinvestigation of the K-term as a function of distance.

Although velocities of forty-six objects were available in 1929, the new criteria gave distances for only eighteen of the isolated nebulae and for the Virgo cluster. Nevertheless, the uncertainties in the distances were so small compared to the range over which they were distributed, that the velocity-distance relation (Fig. 9) emerged from the data in essentially its present form.

The motion of the sun with respect to the nebulae was found to be about 175 miles/sec. in the general direction of the bright star Vega. This result is not very different from the solar motion due to galactic rotation—the orbital motion of the sun around the galactic center. The agreement clearly indicates that the motion of the galactic system among the nebulae must be small. The data are not yet sufficient to determine this motion with any precision.

The K-term was closely represented as a linear function of distance. Velocities, on the average, increased at the rate of roughly 100 miles/sec., per million light-years of distance, over the observed range of about 6.5 million light-years. When the distance effects as well as the solar motion were eliminated from the observed velocities, the residuals, which represented the peculiar motion of the nebulae, averaged about 100 miles/sec. Moreover, the velocities of approach were about as numerous as the velocities of recession. The observed velocities were thus reduced to order and the distribution of residuals was satisfactory.

The velocity-distance relation, once established, could evidently be used as a criterion of distance for all nebulae whose velocities were known. The first application of the new criterion was made to the nebulae in Slipher's list in which no stars could be detected. Observed velocities, divided by the K-term, indicated distances whose only uncertainties were those introduced by peculiar motions. Distances and apparent faintness, taken together, indicated intrinsic luminosities. Intrinsic luminosities derived in this manner were strictly comparable with

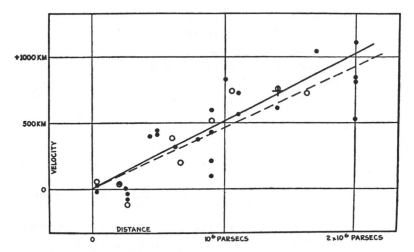

Fig. 9. *The Formulation of the Velocity-Distance Relation.*

The radial velocities (in km/sec.), corrected for solar motion, are plotted against distances (in parsecs) estimated from involved stars and, in the case of the Virgo cluster (represented by the four most distant nebulæ), from the mean luminosity of all nebulæ in the cluster. The black disks and full line represent a solution for the solar motion using the nebulæ individually; the circles and dashed line, a solution combining the nebulæ into groups.

those of the nebulae in which stars were observed; both the mean luminosities, and the ranges through which they were scattered, agreed within the uncertainties of the determinations. The nebulae whose velocities had been measured appeared to form a homogeneous group, and the nebulae in which stars could be observed, were a fair sample of the group. The consistency of these results was additional evidence of the validity of the velocity-distance relation.

Significance of the Velocity-Distance Relation

As a mere criterion of distance the relation is a valuable aid to nebular research. The only serious errors are those introduced by the peculiar motions. These average between 100 and 150 miles/sec., and are presumably independent of distance. As the K-term increases with distance, the peculiar motion, remaining constant, is an ever-smaller fraction of the K-term. Thus the percentage-accuracy of the determinations increases as the distances increase—a welcome contrast to methods which involve photometric measures.

The velocity-distance relation is not merely a powerful aid to research, it is also a general characteristic of our sample of the universe—one of the very few that are known. Until lately, the explorations of space had been confined to relatively short distances and small volumes—in a cosmic sense, to comparatively microscopic phenomena. Now, in the realm of the nebulae, large-scale, macroscopic phenomena of matter and radiation could be examined. Expectations ran high. There was a feeling that almost anything might happen and, in fact, the velocity-distance relation did emerge as the mists receded. This was of the first importance for, if it could be fully interpreted, the relation would probably contribute an essential clue to the problem of the structure of the universe.

Observations show that details in nebular spectra are displaced toward the red from their normal positions, and that the red-shifts increase with apparent faintness of the nebulae. Apparent faintness is confidently interpreted in terms of distance. Therefore, the observational result can be restated—red-shifts increase with distance.

Interpretations of the red-shifts themselves do not inspire such complete confidence. Red-shifts may be expressed as fractions, $d\lambda/\lambda$, where $d\lambda$ is the displacement of a spectral line whose normal wave-length is λ. The displacements, $d\lambda$, vary systematically through any particular spectrum, but the variation is such that the fraction, $d\lambda/\lambda$, remains constant. Thus $d\lambda/\lambda$ specifies the shift for any nebula, and it is the fraction which increases linearly with distances of the nebulae. From this point, the term red-shift will be employed for the fraction $d\lambda/\lambda$.

Moreover, the displacements, $d\lambda$, are always positive (toward the red) and so the wave-length of a displaced line, $\lambda + d\lambda$, is always greater than the normal wave-length, λ. Wave-lengths are increased by the factor $(\lambda + d\lambda)/\lambda$, or the equivalent $1 + d\lambda/\lambda$. Now there is a fundamental relation in physics which states that the energy of any light quantum, multiplied by the wave-length of the quantum, is constant. Thus

$$\text{Energy X wave-length} = \text{constant.}$$

Obviously, since the product remains constant, red-shifts, by increasing wave-lengths, must reduce the energy in the quanta. Any plausible interpretation of red-shifts must account for the loss of energy. The loss must occur either in the nebulae themselves or in the immensely long paths over which the light travels on its journey to the observer.

Thorough investigation of the problem has led to the following conclusions. Several ways are known in which red-shifts might be produced. Of them all, only one will produce large shifts without introducing other effects which should be conspicuous, but which are not observed. This explanation interprets red-shifts as Doppler effects, that is to say, as velocity-shifts indicating actual motion of recession. It may be stated with some confidence that red-shifts are velocity-shifts or else they represent some hitherto unrecognized principle in physics.

The interpretation as velocity-shifts is generally adopted by theoretical investigators, and the velocity-distance relation is considered as the observational basis for theories of an expanding universe. Such theories are widely current. They represent solutions of the cosmological equation, which follow from the assumption of a nonstatic universe. They supersede the earlier solutions made upon the assumption of a static universe, which are now regarded as special cases in the general theory.

Nebular red-shifts, however, are on a very large scale, quite new in our experience, and empirical confirmation of their provisional interpretation as familiar velocity-shifts, is highly desirable. Critical tests are possible, at least in principle, since rapidly receding nebulae should appear fainter than stationary nebulae at the same distances. The effects of recession are inconspicuous until the velocities reach appreciable fractions of the velocity of light. This condition is fulfilled, and hence the effects should be measurable, near the limits of the 100-inch reflector.

The problem will be discussed more fully in the concluding chapter. The necessary investigations are beset with difficulties and uncertainties, and conclusions from data now available are rather dubious. They are mentioned here in order to emphasize the fact that the interpretation of red-shifts is at least partially within the range of empirical investigation. For this reason the attitude of the observer is somewhat different from that of the theoretical investigator. Because the telescopic resources are not yet exhausted, judgment may be suspended until it is known from observations whether or not red-shifts do actually represent motion.

Meanwhile, red-shifts may be expressed on a scale of velocities as a matter of convenience. They behave as velocity-shifts behave and they are very simply represented on the same familiar scale, regardless of the ultimate interpretation. The term "apparent velocity" may be used in carefully considered statements, and the adjective always implied where it is omitted in general usage.

9

— Arthur Eddington —

In this reading, the first chapter of his book The Expanding Universe, *Eddington explores the observational and theoretical evidence for such a universe.*

THE EXPANDING UNIVERSE

I

The first hint of an "expanding universe" is contained in a paper published in November 1917 by Prof. W. de Sitter. Einstein's general theory of relativity had been published two years before, but it had not yet attained notoriety; it was not until the eclipse expeditions of 1919 obtained confirmation of its prediction of the bending of light that public interest was aroused. Meanwhile many investigators had been examining the various consequences of the new theory. Prominent among them was de Sitter who was interested especially in the astronomical consequences. In the course of a highly technical discussion he found that the relativity theory led to an expectation *that the most remote celestial objects would be moving away from us,* or at least they would deceive the observer into thinking that they were moving away.

De Sitter was perhaps a tipster rather than a prophet. He would not promise anything definitely; but suggested that we ought to keep a look out for the recession as a rather likely phenomenon. Theory was at the crossroads, and desired guidance from observation as to which of two possible courses should be pursued. If astronomers were to find a general motion of recession of the most distant objects visible, it would be a strong indication that the road rather fancied by de Sitter was the one to follow. If not, the inference was more doubtful; it might mean that the other road should be followed, or it might only mean that our astronomical survey had not yet been extended to sufficient distance.

Subsequent researches in the field opened up by de Sitter's pioneer investigation have developed and modified his theory. A new point of view has been discovered which renders the results less paradoxical than they appeared originally. We are still led to expect a recession of remote objects, though the recession now predicted is not the original de Sitter effect, which has turned out to be of minor importance. It varies with the distance according to a different law. Moreover, it is a genuine receding motion of remote objects, whereas the phenomenon predicted by de Sitter might be regarded as an imitation recession, and generally was so regarded.

We shall put aside theory for the present, and consider first what astronomical observation tells us. Practically all that I have to relate has been discovered since de Sitter's forecast, much of it within the last four years. These observational results are in some ways so disturbing that there is a natural hesitation in accepting them at their face value. But they have not come upon us like a bolt from the blue, since theorists for the last fifteen years have been half expecting that a study of the most remote objects of the universe might yield a rather sensational development.

The spiral nebulae are the most remote objects known. Rough measurements of their distances have been made, and we place them from 1 million to 150 million light years away; they doubtless extend far beyond the latter distance, but at present it is the limit of our survey. The name "nebula" is applied to different classes of astronomical objects which have nothing in common except a cloudy appearance. There are *gaseous nebulae,* shown by their spectrum to be extremely rarefied gas, either attached to and controlled by a single star or spreading irregularly through a region containing many stars; these are not particularly remote. The *spiral nebulae* on the other hand are extra-galactic objects; that is to say, they lie beyond the limits of the Milky Way aggregation of stars which is the system to which our sun belongs, and are separated from it by wide gulfs of empty space. When we have taken together the sun and all the naked-eye stars and many hundreds of millions of telescopic stars, we have not reached the end of things; we have explored only one island—one oasis in the desert of space. Other islands lie beyond. It is possible with the naked eye to make out a hazy patch of light in the constellation Andromeda which is one of the other islands. A telescope shows many more—an archipelago of island galaxies stretching away one behind another until our sight fails. It is these island galaxies which appear to us as spiral nebulae.

Each island system is believed to be an aggregation of thousands of millions of stars with a general resemblance to our own Milky Way system. As in our own system there may be along with the stars great tracts of nebulosity, sometimes luminous, sometimes dark and obscuring. Many of the nearest systems are seen to have a beautiful double-spiral form; and it is believed that the coils of the Milky Way would give the same spiral appearance to our own system if it were viewed from outside. The term "spiral nebula" is, however, to be regarded as a name rather than a description, for it is generally applied to all external galaxies whether they show traces of spiral structure or not.

The island systems are exceedingly numerous. From sample counts it is estimated that more than a million of them are within reach of our present telescopes. If the theory treated in this book is to be trusted, the total number of them must be of the order 100,000,000,000.

In order to fix in our minds the vastness of the system that we shall have to consider, I will give you a "celestial multiplication table." We start with a star as the unit most familiar to us, a globe comparable to the sun. Then—

> A hundred thousand million stars make one Galaxy;
> A hundred thousand million Galaxies made one Universe.

These figures may not be very trustworthy, but I think they give a correct impression.

The lesson of humility has so often been brought home to us in astronomy that we almost automatically adopt the view that our own galaxy is not specially distinguished—not more important in the scheme of nature than the millions of other island galaxies. But astronomical observation scarcely seems to bear this out. According to the present measurements the spiral nebulae, though bearing a general resemblance to our Milky Way system, are distinctly smaller. It has been said that if the spiral nebulae are islands, our own galaxy is a continent. I suppose that my humility has become a middle-class pride, for I rather dislike the imputation that we belong to the aristocracy of the universe. The earth is a middle-class planet, not a giant like Jupiter, nor yet one of the smaller vermin like the minor planets. The sun is a middling sort of star, not a giant like Capella but well above the lowest classes. So it seems wrong that we should happen to belong to an altogether exceptional galaxy. Frankly I do not believe it; it would be too much of a coincidence. I think that this relation of the Milky Way to the other galaxies is a subject on which more light will be thrown by further observational research, and that ultimately we shall find that there are many galaxies of a size equal to and surpassing our own. Meanwhile the question does not much affect the present discussion. If we are in a privileged position, we shall not presume upon it.

I promised to leave aside theory for the present, but I must revert to it for a moment to try to focus our conception of this super-system of galaxies. It is a vista not only of space but of time. A faint cluster of nebulae in Gemini, which at present marks the limit of our soundings of space, takes us back 150 million years into the past—to the time when the light now reaching us started on its journey across the gulf of space. Thus we can scarcely isolate the thought of vast extension from the thought of time and change; and the problem of form and organization becomes merged in the problem of origin and development. We must, I suppose, imagine the island galaxies to have been formed by gradual condensation of primordial matter. Perhaps in the first stage only the rudiments of matter existed—protons and electrons traversing the void—and the evolution of the elements has progressed simultaneously with the evolution of worlds. Slight condensations occurring here and there by accident would by their gravitating power draw more particles to themselves. Some would quickly disperse again, but some would become firmly established—

> Champions fierce,
> Strive here for mastery, and to battle bring
> Their embryon atoms. . . . To whom these most adhere,
> He rules a moment: Chaos umpire sits,
> And by decision more embroils the fray
> By which he reigns: next him, high arbiter,
> Chance governs all.
> —*Paradise Lost*, Book II

By such conflict the matter of the universe would slowly be collected into islands, leaving comparatively empty spaces from which it had been drained away. We think that one of these original islands has become our Milky Way system, having subdivided again and again into millions of stars. The other islands similarly developed into galaxies, which we see today shining as spiral nebulae. It is to these prime units of subdivision of the material universe that our discussion here will relate.

II

If a spiral nebula is not too faint it is possible to measure its radial velocity in the line of sight by measuring the shift of the lines in its spectrum. A valuable early series of such determinations was made by Prof. V. M. Slipher at the Lowell Observatory.

More recently the distances of some of the spiral nebulae have been determined by a fairly trustworthy method. In the nearest spirals it is possible to make out some of the individual stars; but only the most luminous stars, some hundreds or thousands of times brighter than the sun, can be seen at so great a distance. Fortunately among the very brightest of the stars there is a particularly useful class called the Cepheid variables. They vary periodically in brightness owing to an actual pulsation or physical change of the star, the period being anything from a few hours to a few weeks. It has been ascertained from observational study that Cepheids which have the same period are nearly alike in their other properties—luminosity, radius, spectral type, etc. The period is thus a badge, easily recognizable at a distance, which labels the star as having a particular luminosity. For example, if the star is seen to have a period of 10 days, we immediately recognize it as a star of luminosity 950 times greater than the sun. Knowing then its real brightness we put the question, How far off must it be situated so as to be reduced to the faint point of light which we see? The answer gives the distance of the star and of the galaxy in which it lies. This method uses the Cepheid variables as standard candles. If you see a standard candle anywhere and note how bright it appears to you, you can calculate how far off it is; in the same way an astronomer observes his "standard candle" in the midst of a nebula, notes its apparent brightness or magnitude, and deduces the distance of the nebula.

Dr. E. P. Hubble at Mount Wilson Observatory was able to discover Cepheid variables in two or three of the nearest spiral nebulae, and so obtained the first real measurement of their distances. Unfortunately this method is not available for the more distant galaxies, and he has had to use more indirect devices for extending the survey. I think that, apart from those distances actually determined by the Cepheid method, we must regard the distances assigned to the spiral nebulae as rather risky estimates; but there is reason to believe that they cannot be entirely misleading, and we shall provisionally accept them here.

When the collected data as to radial velocities and distances are examined a very interesting feature is revealed. The velocities are large, generally very much larger than ordinary stellar velocities. The more distant nebulae have the bigger velocities; the results seem to agree very well with a linear law of increase, the velocity being simply proportional to the distance. The most striking feature is that the galaxies are almost unanimously running away from us.

Let us consider especially the last result and state the observational evidence in more detail. The light of the spiral nebulae, being compounded of the light of a great variety of stars, does not give a good spectrum for measurement. For this reason and because of its faintness the deduced velocities are inaccurate as judged by ordinary standards; but, except for the nearest nebulae, the velocities are themselves so enormous that the error of measurement is comparatively unimportant. Taking the results as published, the present position is that line-of-sight velocities of about 90 galaxies have been measured, and of these only five are moving towards us. At first sight it may seem wrong to pass over the minority as insignificant. But the five exceptions are confined to the very nearest of the nebulae, and their approaching velocities are not large. Since the phenomenon is one which depends on distance (the effect increasing with distance), it is natural that we should have to go out to a fair distance before we find it strong enough to prevail over all other effects (including observational error) so as to display itself uniformly. The five approaching velocities are at least partly attributable to the use of an inappropriate standard of reference. Line-of-sight velocities as published are relative to the sun; but it would be more satisfactory to discuss the velocities relative to our Milky Way system as a whole. It has been found that the sun is pursuing an orbit round the center of the Milky Way system and has an orbital speed from 200 to 300 kilometers per second. When we correct for this so as to obtain the velocities referred to our galaxy as a whole, the approaching velocities are reduced or disappear. I think it will turn out ultimately that, after all corrections are applied, these nearest nebulae have small receding velocities; for the existence of even one genuine exception would be difficult to explain.

In saying that the speeds of the nebulae are large, the velocities of ordinary stars are our standard of comparison. For stars in our neighborhood the individual speed averages 10 to 50 km. per sec. If the speed exceeds 100 km. per sec. the star is described as a "runaway." (We do not here include above-mentioned orbital motion about the center of the galaxy which is shared by all stars in the neighborhood of the sun.) Slipher's first determination of the radial velocities of 40 nebulae included a dozen with velocities from 800–1800 km. per sec. The survey has since been extended to fainter and more distant nebulae by M. L. Humason at Mount Wilson Observatory, and much higher velocities have been found. The speed record is continually being broken. The present holder of the trophy is a nebula forming one of a faint cluster in the constellation Gemini, which is receding with a velocity of 25,000 km. per sec. (15,000 miles per second). This is about the speed of an Alpha particle. Its distance is estimated at 150,000,000 light-years. Doubtless a faster and more distant nebula will have been announced by the time these words are in print.

The simple proportionality of speed to distance was first found by Hubble in 1929. This law is also predicted by relativity theory. According to the original investigation of de Sitter a velocity proportional to the square of the distance would have been expected; but the theory had become better understood since then, and it was already known (though perhaps only to a few) that simple proportionality to the distance was the correct theoretical result.

According to Hubble's most recent determination, the speed of recession amounts to 550 km. per sec. per megaparsec. That is to say, a nebula at 1 megaparsec distance should have a speed 550 km. per sec.; at 10 megaparsecs distance, 5500 km. per sec.; and so on. It has been claimed that this determination is accurate to 20 percent, but I do not think many astronomers take so optimistic a view. The uncertainty lies almost entirely in the scale of nebular distances; there are weak links in the long chain of connection between these vast distances and our terrestrial standard meter. Corrections which have been suggested mostly tend to increase the result; and perhaps the fairest statement is that the velocity of recession is probably between 500 and 1000 km. per sec. per mp.

Specimens of the spectra from which these radial velocities are obtained are shown in Plate II. In the lower four photographs the spectra of the nebulae are the torpedo-shaped black patches; they have terrestrial comparison spectra above and below, which are used to place them in correct vertical alignment. Practically the only recognizable features in the nebular spectra are the H and K lines—two interruptions in the tail of the torpedo where it is fading away. It will be seen that these interruptions move to the right, i.e. to the red end of the spectrum, as we go down the plate. It is this displacement which is measured and gives the receding velocities stated at the foot of the plate.

III

We can exclude the spiral nebulae which are more or less hesitating as to whether they shall leave us by drawing a sphere of rather more than a million light-years radius round our galaxy. *In the region beyond, more than 80 have been observed to be moving outwards, and not one has been found coming in to take their place.*

The inference is that in the course of time all the spiral nebulae will withdraw to a greater distance, evacuating the part of space that we now survey. Ultimately they will be out of reach of our telescopes unless telescopic power is increased to correspond. I find that the observer of nebulae will have to double the aperture of his telescope every 1300 million years merely to keep up with their recession. If we have been thinking that the human race has still billions of years before it in which to find out all that can be found out about the universe, we must count the problem of the spiral nebulae as one of urgency. Let us make haste to study them before they disappear into the distance!

The unanimity with which the galaxies are running away looks almost as though they had a pointed aversion to us. We wonder why we should be shunned as though our system were a plague spot in the universe. But that is too hasty an inference, and there is really no reason to think that the animus is especially directed against our galaxy. If this lecture-room were to expand to twice its present size, the seats all separating from each other in proportion, you would notice that everyone had moved away from you. Your neighbor who was 2 feet away is now 4 feet away; the man over yonder who was 40 feet away is now 80 feet away. It is not *you* they are avoiding; everyone is having the same experience. In a general dispersal or expansion every individual observes every other individual to be moving away from him. The law of a general uniform expansion is that each individual recedes from you at a rate proportional to his distance from you—precisely the law which we observe in the receding motions of the spiral nebulae.

We shall therefore no longer regard the phenomenon as a movement away from our galaxy. It is a general scattering apart, having no particular center of dispersal.

I do not wish to insist on these observational facts dogmatically. It is granted that there is a possibility of error and misinterpretation. The survey is just beginning, and things may appear in a different light as it proceeds. But if you ask what is the picture of the universe now in the minds of those who have been engaged in practical exploration of its large-scale features—men not likely to be moved overmuch by ideas of bending of space or the gauge-invariance of the Riemann-Christoffel tensor—I have given you their answer. Their picture is the picture of an *expanding universe*. The super-system of the galaxies is dispersing as a puff of smoke disperses. Sometimes I wonder whether there may not be a greater scale of existence of things, in which it is no more than a puff of smoke.

For the present I make no reference to any "expansion of space." I am speaking of nothing more recondite than the expansion or dispersal of a material system. Except for the large scale of the phenomenon the expansion of the universe is as commonplace as the expansion of a gas. But nevertheless it gives very serious food for thought. It is perhaps in keeping with the universal change we see around us that time should set a term even to the greatest system of all; but what is startling is the rate at which it is found to be melting away. We do not look for immutability, but we had certainly expected to find a permanence greater than that of terrestrial conditions. But it would almost seem that the earth alters less rapidly than the heavens. The galaxies separate to double their original distances in 1,300 million years. That is only of the order of geological time; it is approximately the age assigned to the older rocks in the earth's crust. This is a rude awakening from our dream of leisured evolution through billions of years.

Such a conclusion is not to be accepted lightly; and those who have cast about for some other interpretation of what seems to have been observed have displayed no more than a proper caution. If the apparent recession of the spiral nebulae is treated as an isolated discovery it is too slender a thread on which to hang far-reaching conclusions; we can only state the bare results of observation, contemplate without much conviction the amazing possibility they suggest, and await further information on the subject.

If that is not my own attitude, it is because the motion of the remote nebulae does not present itself to me as an isolated discovery. Following de Sitter, I have for fifteen years been awaiting these observational results to see how far they would fall into line with and help to develop the physical theory, which though at first merely suggestive has become much more cogent in the intervening years. After Prof. Weyl's famous extension of the relativity theory I became convinced that the scale of structure of atoms and electrons is determined by the same physical agent that was concerned in de Sitter's prediction. So that hope of progress of a really fundamental kind in our understanding of electrons, protons and quanta is bound up with this investigation of the motions of remote galaxies. Therefore when Dr. Hubble hands over a key which he has picked up in intergalactic space, I am not among those who are turning it over and over unable to decide from the look of it whether it is good metal or base metal. The question for me is, Will it unlock the door?

If the observed radial velocities are accepted as genuine, there is no evading the conclusion that the nebulae are rapidly dispersing. The velocities are direct evidence of a hustle which (according to the usual ideas of the rate of evolutionary change) is out of keeping with the character of our staid old universe. Thus the only way of avoiding a great upset of ideas would be to explain away these radial velocities as spurious. What is actually observed is a shifting of the spectrum of the nebula towards the red. Such a shift is commonly caused by the Doppler effect of a receding velocity, in the same way that the pitch of a receding whistle is lowered; but other causes are imaginable. The reddening signifies lower frequency of the light-waves and (in accordance with quantum theory) lower energy; so that if for any cause a light-quantum loses some of its energy in travelling to reach us, the reddening is accounted for without assuming any velocity of the source. For example, the light coming to us from an atom on the sun uses up some of its energy in escaping from the sun's gravitational attraction, and consequently becomes slightly reddened as compared with the light of a terrestrial atom which does not suffer this loss; this is the well-known red shift predicted by Einstein.

In one respect this hypothesis of the loss of energy of nebular light is attractive. If the loss occurs during the passage of the light from the nebula to the observer, we should expect it to be proportional to the distance; thus the red-shift, misinterpreted as a velocity, should be proportional to the distance—which is the law actually found. But on the other hand there is nothing in the existing theory of light (wave theory or quantum theory) which justifies the assumption of such a loss. We cannot without undue dogmatism exclude the possibility of modifications of the existing theory. Light is a queer thing—queerer than we imagined twenty years ago—but I should be surprised if it is as queer as all that.

A theory put forward by Dr. Zwicky, that light, by its gravitational effects, parts with its energy to the material particles thinly strewn in intergalactic space which it passes on its way, at one time attracted attention. But the numerical accordance alleged to support his theory turned out to be fallacious, and the suggestion seems definitely untenable.

I think then we have no excuse for doubting the genuineness of the observed velocities—except in so far as they share the general uncertainty that surrounds all our attempts to probe into the secrets of nature.

IV

Now let us turn to theory.

A scientist commonly professes to base his beliefs on observations, not theories. Theories, it is said, are useful in suggesting new ideas and new lines of investigation for the experimenter; but "hard facts" are the only proper ground for conclusion. I have never come across anyone who carries this profession into practice—certainly not the hard-headed experimentalist, who is the more swayed by his theories because he is less accustomed to scrutinize them. Observation is not sufficient. We do not believe our eyes unless we are first convinced that what they appear to tell us is credible.

It is better to admit frankly that theory has, and is entitled to have, an important share in determining belief. For the reader resolved to eschew theory and admit only definite observational facts, *all* astronomical books are banned. *There are no purely observational facts about the heavenly bodies.* Astronomical measurements are, without exception, measurements of phenomena occurring in a terrestrial observatory or station; it is only by theory that they are translated into knowledge of a universe outside.

When an observer reports that he has discovered a new star in a certain position, he is probably unaware that he is going beyond the simple facts of observation. But he does not intend his announcement to be taken as a description of certain phenomena that have occurred in his observatory; he means that he has located a celestial body in a definite direction in interstellar space. He looks on the location as an observational fact—on a surer footing therefore than theoretical inferences such as have been deduced from Einstein's theory. We must break it to him that his supposed "fact," far from being purely observational, is actually an inference based on Einstein's theory unless, indeed, he has based it on some earlier theory which is even more divorced from observational facts. The observer has given a theoretical interpretation to his measurements by assuming for theoretical reasons that light travels through interstellar space approximately in a straight line. Perhaps he will reply that, in assuming the rectilinear propagation of light, he is not concerned with any theory but is using a fact established by direct experiment. That begs the question how far an experiment under terrestrial conditions can be extrapolated to apply to interstellar space. Surely a reasoned theory is preferable to blind extrapolation. But indeed the observer is utterly mistaken in supposing that the straightness of rays of light assumed in astronomy has been verified by terrestrial experiment. If the rays in interstellar space were no straighter than they are on the earth, the direction in which a star is seen would be no guide to its actual position. Light would in fact curl round and come back again before traversing the distance to the nearest star.

Our warrant for concluding that the celestial body is nearly in the direction in which it is seen, is Einstein's theory, which determines the deviation of light from a straight line. Coupled with other theoretical deductions as to the density of matter in interstellar space, it allows us to conclude that the deviation in this case is inappreciable. So if we are willing to use both fact and theory as a basis for belief, we can accept the observer's announcement; but it is not a "hard fact of observation." Although it is a minor point, we may also insist that the theory concerned is Einstein's theory. There was an earlier theory according to which light in empty space travels in straight lines in all circumstances; but since this has been found experimentally to be untrue, it can scarcely be the basis of our observer's conclusion. Perhaps, however, the observer is one of those who do not credit the eclipse observations of the deflection of light, or who deem them insufficient ground for quitting the old theory. If so, he illustrates my dictum that with the hard-headed experimentalist the basis of belief is often theory rather than observation.

My point is that in astronomy it is not a question of whether we are to rely on observation or on theory. The so-called facts are in any case theoretical interpretations of the observations. The only question is, Shall we for this interpretation use the fullest resources of modern theory? For my own part I can see no more reason for preferring the theories of fifty years ago than for preferring the observational data of fifty years ago.

In turning now to the more theoretical side of the problem of the expanding universe, I do not think that we should feel that we are stepping from solid ground into insecurity. Perhaps we are a little safer, for we no longer depend on the interpretation of one type of observation; and our theory comes from the welding together of different lines of physical research. I do not, however, promise security. An explorer is jealous of his reputation for proper caution, but he can never aspire to the quintessence of caution displayed by the man who entrenches himself at home.

V

In 1915 Einstein had by his general theory of relativity brought a large section of the domain of physics into good order. The theory covered *field-physics*, which includes the treatment of matter, electricity, radiation, energy, etc., on the ordinary macroscopic scale perceptible to our senses, but not the phenomena arising from the minute subdivision into atoms, electrons, quanta. For the study of microscopic structure another great theory was being developed—the quantum theory. At that time it lagged far behind, and even now it has not reached the clearness and logical perfection of the relativity theory. It is recognized that the two theories will meet, and that they must ultimately coalesce into one comprehensive theory. The first bridge between them was made by Prof. P.A.M. Dirac in 1928 by his relativity wave-equation of an electron. I hope to show in the last chapter that the recession of the spiral nebulae leads us to the borderland territory between the two theories, where a number of interesting problems await solution. At present, however, we are concerned only with its relation to the theory of relativity.

The central result of Einstein's theory was his law of gravitation, generally expressed in the form $G_{\mu\nu} = 0$, which has the merit of brevity if not of lucidity. We naturally hear most about those rate phenomena in which Einstein's law gives results appreciably different from Newton's law; but it is to be remembered that for ordinary practical purposes the two laws come to the same thing. So if we take $G_{\mu\nu} = 0$ to be the law governing the motions of the spiral nebulae, that is as much as to say they exert the ordinary Newtonian attraction on one another varying as the inverse square of their distance apart. The law throws no light on why the nebulae are running away from us and from one another. The tendency would rather be for the whole system to fall together—though this tendency to collapse might be counteracted as it is in the solar system, for example.

A year or so later Einstein made a slight amendment to his law to meet certain difficulties that he encountered in his theory. There was just one place where the theory did not seem to work properly, and that was—infinity. I think Einstein showed his greatness in the simple and drastic way in which he disposed of difficulties at infinity. He abolished infinity. He slightly altered his equations so as to make space at great distances bend round until it closed up. So that, if in Einstein's space you keep going right on in one direction, you do not get to infinity; you find yourself back at your starting-point again. Since there was no longer any infinity, there could be no difficulties at infinity. Q.E.D.

However, at present we are not concerned with this new kind of space. I only mention it here because I want to speak of the alteration that Einstein made in his law of gravitation. The amended law is written $G_{\mu\nu} = \lambda g_{\mu\nu}$, and contains a natural constant λ called the *cosmical constant*. The term $\lambda g_{\mu\nu}$ is called the *cosmical term*. The constant is so small that in ordinary applications to the solar system, etc., we set it equal to zero, and so revert to the original law $G_{\mu\nu} = 0$. But however small λ may be, the amended law presents the phenomenon of gravitation to us in a new light, and has greatly helped to an understanding of its real significance; moreover, we have no reason to think that λ is not so small as to be entirely beyond observation. The nature of the alteration can be stated as follows: the original law stated that a certain geometrical characteristic ($G_{\mu\nu}$) of empty space is always zero, the revised law states that it is always in a constant ratio to another geometrical characteristic ($g_{\mu\nu}$). We may say that the first form of the law utterly dissociates the two characteristics by making one of them zero and therefore independent of the other; the second form intimately connects them. Geometers can invent spaces which have not either of these properties; but actual space, surveyed by physical measurement, is not of so unlimited a nature.

We have already said that the original term in the law gives rise to what is practically the Newtonian attraction between material objects. It is found similarly that the added term ($\lambda g_{\mu\nu}$) gives rise to a repulsion directly proportional to the distance. Distance from what? Distance from *anywhere*; in particular, distance from the observer. It is a dispersive force like that which I imagined as scattering apart the audience in the lecture-room. Each thinks it is directed away from him. We may say that the repulsion has no center, or that every point is a center of repulsion.

Thus in straightening out his law of gravitation to satisfy certain ideal conditions, Einstein almost inadvertently added a repulsive scattering force to the Newtonian attraction of bodies. We call this force the *cosmical repulsion*, for it depends on and is proportional to the cosmical constant. It is utterly imperceptible within the solar system or between the sun and neighboring stars. But since it increases proportionately to the distance we have only to go far enough to find it appreciable, then strong, and ultimately overwhelming. In practical observation the farthest we have yet gone is 150 million light-years. Well within that distance we find that celestial objects are scattering apart as if under a dispersive force. Provisionally we conclude that here cosmical repulsion has become dominant and is responsible for the dispersion.

We have no *direct* evidence of an outward acceleration of the nebulae, since it is only the velocities that we observe. But it is reasonable to suppose that the nebulae, individually as well as collectively, follow the rule—the greater the distance the faster the recession. If so, the velocity increases as the nebula recedes, so that there is an outward acceleration. Thus from the observed motions we can work backwards and calculate the repulsive force, and so determine observationally the cosmical constant λ.

Much turns on whether Einstein was really justified in making the change in his law of gravitation which introduced this cosmical repulsion. His original reason was not very convincing, and for some years the cosmical term was looked on as a fancy addition rather than as an integral part of the theory. Einstein has been as severe a critic of his own suggestion as anyone, and he has not invariably adhered to it. But the cosmical constant has now a secure position owing to a great advance made by Prof. Weyl, in whose theory it plays an essential part. Not only does it unify the gravitational and electromagnetic fields, but it renders the theory of gravitation and its relation to space-time measurement so much more illuminating, and indeed self-evident, that return to the earlier view is unthinkable. I would as soon think of reverting to Newtonian theory as of dropping the cosmical constant.

VI

Let us now review the position. According to relativity theory the complete field of force contains besides the ordinary Newtonian attraction a repulsive (scattering) force varying directly as the distance. It is well known that Einstein's law differs slightly from Newton's, giving for example an extra effect which has been detected in the orbit of the fast-moving planet Mercury; the cosmical repulsion is another point of difference between them, detectable only in the motions of remote objects. From a theoretical standpoint I think there is no more doubt about the cosmical repulsion than about the force which perturbs Mercury; but it does not admit of so decisive an observational test. As regards Mercury the theoretical prediction was quantitative; but relativity theory does not indicate any particular magnitude for the cosmical repulsion. A merely qualitative test is never very conclusive.

However, so far as it goes, the test is satisfactory. We do find observationally a dispersion of the system of the galaxies such as would be caused by the predicted repulsion. The motions are extremely large and the effect stands out clearly above all minor irregularities. The theory thus clears its first hurdle with some *éclat*; whether it will win the race is another question. Although the test is not quantitative it is more far-reaching than is sometimes supposed. There are only two ways of accounting for large receding velocities of the nebulae: (1) they have been produced by an outward directed force as we here suppose, or (2) as large or larger velocities have existed from the beginning of the present order of things. Several rival explanations of the recession of the nebulae, which do not accept it as evidence of repulsive force, have been put forward. These necessarily adopt the second alternative, and postulate that the large velocities have existed from the beginning. This might be true; but it can scarcely be called an *explanation* of the large velocities.

Our best hope of further progress is to discover some additional test for the theory—if possible, a stringent quantitative test. We want to predict the actual magnitude of the cosmical repulsion, and see if the observed motions of the nebulae confirm the predicted value. Relativity theory alone cannot do this, but when relativity is combined with wave-mechanics the quantitative prediction seems possible. This development is explained in Chapter IV.

Thus far we have been treating a fairly straightforward subject. Apart from the vast magnitudes involved there is nothing that particularly taxes the imagination. . . .

If I introduce a different kind of outlook it is because I am going on to treat regions of the universe beyond those that we have hitherto considered. Primarily the present chapter deals with the region actually explored, up to 150 million light-years' distance. If the galaxies come to an end there, no more need be said. . . . But there is no sign that the system of the galaxies is coming to an end, and presumably it extends considerably beyond 150 million light-years. It might extend to, say, five times that distance without any important new feature; but if we have to go much beyond that, there is trouble in store. The appropriate speed of recession would be beginning to approach the velocity of light—a point which evidently requires looking into. We have a force of cosmical repulsion, increasing with the distance, which is already rather powerful; if we go on to a vastly greater distance something must give way at last—only Einstein has taken the precaution of closing up the universe to prevent us from going too far.

The object of the ensuing development is to deal with questions which arise as to the possible extension of the system of the galaxies beyond the region at present explored. We shall consider extrapolation in time as well as in space, and discuss the history of evolution of the system.

What is the object of making these risky extrapolations to regions of space and time remote from our practical experience? It might be a sufficient answer to say that we are *explorers*. But there is another and more urgent reason. The man who for the first time sees an aeroplane passing overhead doubtless wonders how it goes. I do not think he can be accused of eccentricity if he also wonders *how it stops*. It is true that he sees no signs of its stopping; he is mentally extrapolating the flight beyond the range that is visible. He cannot be sure of his extrapolation; outside his range of vision there may be conditions, of which he is unaware, which will stop the flight in a manner different from his conjecture. But he will have much more confidence in his conclusions as to the mechanism of the aeroplane if they will explain the flight from start to stop without

postulating some unknown intervention. At first sight it seems a reasonable programme for science to tidy up the region of space and time of which we have
some experience and not to theorize about what lies beyond; but the danger of
such a limitation is that the tidying up may consist in taking the difficulties and
inexplicabilities and dumping them over the border instead of really straightening them out.

We have seen that there is a force of cosmical repulsion growing larger as the
distance from us increases. At the greatest distance yet explored it is still increasing. The foregoing theory explains how it goes. But we have still a desire to
understand how it stops.

10
— Georges Lemaître —
(1894–1966)

*Georges Lemaître was a Belgian astronomer and cosmologist who introduced the
Big Bang theory of the universe. Although he did not use the "Big Bang," Lemaître
found a solution to the equations of Relativity Theory that resulted in an expanding universe. Tracing the expansion back in time, Lemaître speculated that the
entire process began with the radioactive decay of a primordial atom.*

*In this reading Lemaître offers his "Big Bang" cosmological theory, named for
the bang that results from the radioactive disintegration of the "Primeval Atom."
Lemaître suggests a method for determining the moment in the past when the
disintegration occurred based upon the discovery of evidence of maximum energy
concentration. The passage is from his book* The Primeval Atom: An Essay on
Cosmology.

THE BEGINNING OF BIG BANG COSMOLOGY

Introduction

The Primeval Atom hypothesis is a cosmogonic hypothesis which pictures the
present universe as the result of the radioactive disintegration of an atom.

I was led to formulate this hypothesis, some fifteen years ago, from thermodynamic considerations while trying to interpret the law of degradation of energy
in the frame of quantum theory. Since then, the discovery of the universality
of radioactivity shown by artificially provoked disintegrations, as well as the

establishment of the corpuscular nature of cosmic rays, manifested by the force which the Earth's magnetic field exercises on these rays, made more plausible an hypothesis which assigned a radioactive origin to these rays, as well as to all existing matter.

Therefore, I think that the moment has come to present the theory in deductive form. I shall first show how easily it avoids several major objections which would tend to disqualify it from the start. Then I shall strive to deduce its results far enough to account, not only for cosmic rays, but also for the present structure of the universe, formed of stars and gaseous clouds, organized into spiral or elliptical nebulae, sometimes grouped in large clusters of several thousand nebulae which, more often, are composed of isolated nebulae, receding from one another according to the mechanism known by the name of the expanding universe.

For the exposition of my subject, it is indispensable that I recall several elementary geometric conceptions, such as that of the closed space of Riemann, which led to that of space with a variable radius, as well as certain aspects of the theory of relativity, particularly the introduction of the cosmological constant and of the cosmic repulsion which is the result of it.

Closed Space

All particle space is open space. It is comprised in the interior of a surface, its boundary, beyond which there is an exterior region. Our habit of thought about such open regions impels us to think that this is necessarily so, however large the regions being considered may be. It is to Riemann that we are indebted for having demonstrated that total space can be closed. To explain this concept of closed space, the most simple method is to make a small-scale model of it in an open space. Let us imagine, in such a space, a sphere in the interior of which we are going to represent the whole of closed space. On the rim surface of the sphere, each point of closed space will be supposed to be represented twice, by two points, A and A′, which for example, will be two antipodal points, that is two extremities of the same diameter. If we join these two points A and A′ by a line located in the interior of the sphere, this line must be considered as a closed line, since the two extremities A and A′ are two distinct representations of the same, single point. The situation is altogether analogous to that which occurs with the Mercator projection, where the points on the 180th meridian are represented twice, at the eastern and western edges of the map. One can thus circulate indefinitely in this space without ever having to leave it.

It is important to notice that the points represented by the outer surface of the sphere, in the interior of which we have represented all space, are not distinguished by any properties of the other points of space, any more than is the 180th meridian for the geographic map. In order to account for that, let us imagine that we displaced the sphere in such a manner that point A is superposed on B, and the antipodal point A′ on B′. We shall then suppose that the entire segment AB and the entire segment A'B' are two representations of a similar segment in closed space. Thus we shall have a portion of space which has already been represented in the interior of the initial sphere which is now represented a second time at the exterior of this sphere. Let us disregard the interior representation as useless; a complete representation of the space in the interior of the new sphere will remain. In this representation, the closed contours will be soldered into a point which is twice represented, namely, by the points B and B', mentioned above, instead of being welded, as they were formerly, to point A and A'. Therefore, these latter are not distinguished by an essential property.

Let us notice that when we modify the exterior sphere, it can happen that a closed contour which intersects the first sphere no longer intersects the second, or, more generally, that a contour no longer intersects the finite sphere at the same number of points. Nevertheless, it is evident that the number of points of intersection can only vary by an even number. Therefore, there are two kinds of closed contours which cannot be continuously distorted within one another. Those of the first kind can be reduced to a point. They do not intersect the outer sphere or they intersect it at an even number of points. The others cannot be reduced to one point; we call them the *odd contours* since they intersect the sphere at an odd number of points.

If, in a closed space, we leave a surface which we can suppose to be horizontal, in going toward the top we can, by going along an odd contour, return to our point of departure from the opposite direction without having deviated to the right or left, backward or forward, without having traversed the horizontal plane passing through the point of departure.

Elliptical Space

That is the essential of the topology of closed space. It is possible to complete these topological ideas by introducing, as is done in a geographical map, scales which vary from one point to another and from one direction to another. That can be done in such a manner that all the points of space and all the directions in it may be perfectly equivalent. Thus, Riemann's homogeneous space, or elliptical space, is obtained. The straight line is an odd contour of minimum length. Any two points divide it into two segments, the sum of which has a length which is the same for all straight lines and which is called the tour of space.

All elliptical spaces are similar to one another. They can be described by comparison with one among them. The one in which the tour of the straight line is equal to $\pi = 3.1416$ is chosen as the standard elliptical space. In every elliptical space, the distances between two points are equal to the corresponding distances in standard space, multiplied by the number R which is called the radius of elliptical space under consideration. The distances in standard space, called space of unit radius, are termed angular distances. Therefore, the true distances, or linear distances, are the product of the radius of space times the angular distances.

Space of Variable Radius

When the radius of space varies with time, space of variable radius is obtained. One can imagine that material points are distributed evenly in it, and that spatio-temporal observations are made on these points. The angular distance of the various observers remains invariant, therefore the linear distances vary proportionally to the radius of space. All the points in space are perfectly equivalent. A displacement can bring any point into the center of the representation. The measurements made by the observers are thus also equivalent, each one of them makes the same map of the universe.

If the radius increases with time, each observer sees all points which surround him receding from him, and that occurs at velocities which become greater as they recede further. It is this which has been observed for the extra-galactic nebulae that surround us. The constant ratio between distance and velocity has been determined by Hubble and Humason. It is equal to $T_H = 2 \times 10^9$ Years.

If one makes a graph, plotting as abscissa the values of time and as ordinate the value of radius, one obtains a curve, the sub-tangent of which at the point representing the present instant in precisely equal to T_H.

The Primeval Atom

These are the geometric concepts that are indispensable to us. We are now going to imagine that the entire universe existed in the form of an atomic nucleus which filled elliptical space of convenient radius in a uniform manner.

Anticipating that which is to follow, we shall admit that, when the universe had a density of 10^{-27} gram per cubic centimeter, the radius of space was about a billion light-years, that is, 10^{27} centimeters. Thus the mass of the universe is 10^{54} grams. If the universe formerly had a density equal to that of water, its radius was then reduced to 10^{18} centimeters, say, one light-year. In it, each proton occupied a sphere of one angstrom, say, 10^{-8} centimeter. In an atomic nucleus, the protons are contiguous and their radius is 10^{-13}, thus about 100,000 times smaller. Therefore, the radius of the corresponding universe is 10^{13} centimeters, that is to say, an astronomical unit.

Naturally, too much importance must not be attached to this description of the primeval atom, a description which will have to be modified, perhaps, when our knowledge of atomic nuclei is more perfect.

Cosmogonic theories propose to seek out initial conditions which are ideally simple, from which the present world, in all its complexity, might have resulted, through the natural interplay of known forces. It seems difficult to conceive of conditions which are simpler than those which obtained when all matter was unified in an atomic nucleus. The future of atomic theories will perhaps tell us, some day, how far the atomic nucleus must be considered as a system in which associated particles still retain some individuality of their own. The fact that particles can issue from a nucleus, during radioactive transformations, certainly does not prove that these particles pre-existed as such. Photons issue from an atom of which they were not constituent parts, electrons appear there, where they were not previously, and the theoreticians deny them an individual existence in the nucleus. Still more protons or alpha particles exist there, without doubt. When they issue forth, their existence becomes more independent, nevertheless, and their degrees of freedom more numerous. Also, their existence, in the course of radioactive transformations, is a typical example of the degradation of energy, with an increase in the number of independent quanta or increase in entropy.

That entropy increases with the number of quanta is evident in the case of electromagnetic radiation in thermodynamic equilibrium. In fact, in black body radiation, the entropy and the total number of photons are both proportional to the third power of the temperature. Therefore, when one mixes radiations of different temperatures and one allows a new statistical equilibrium to be established, the total number of photons has increased. The degradation of energy is manifested as a pulverization of energy. The total quantity of energy is maintained, but it is distributed in an ever larger number of quanta, it becomes broken into fragments which are ever more numerous.

If, therefore, by means of thought, one wishes to attempt to retrace the course of time, one must search in the past for energy concentrated in a lesser number of quanta. The initial condition must be a state of maximum concentration. It was in trying to formulate this condition that the idea of the primeval atom was germinated. Who knows if the evolution of theories of the nucleus will not, some day, permit the consideration of the primeval atom as a single quantum?

Formation of Clouds

We picture the primeval atom as filling space which has a very small radius (astronomically speaking). Therefore, there is no place for superficial electrons, the primeval atom being nearly an *isotope of a neutron*. This atom is conceived as having existed for an instant only, in fact, it was unstable and, as soon as it came into being, it was broken into pieces which were again broken, in their turn; among these pieces electrons, protons, alpha particles, etc., rushed out. An increase in volume resulted, the disintegration of the atom was thus accompanied by a rapid increase in the radius of space which the fragments of the primeval atom filled, always uniformly. When these pieces became too small, they ceased to break up; certain ones, like uranium, are slowly disintegrating now, with an average life of four billion years, leaving us a meager sample of the universal disintegration of the past.

In this first phase of the expansion of space, starting asymptotically with a radius practically zero, we have particles of enormous velocities (as a result of recoil at the time of the emission of rays) which are immersed in radiation, the total energy of which is, without doubt, a notable fraction of the mass energy of the atoms.

The effect of the rapid expansion of space is the attenuation of this radiation and also the diminution of the relative velocities of the atoms. This latter point requires some explanation. Let us imagine that an atom has, along the radius of the sphere in which we are representing closed space, a radial velocity which is greater than the velocity normal to the region in which it is found. Then this atom will depart faster from the center than the ideal material particle which has normal velocity. Thus the atom will reach, progressively, regions where its velocity is less abnormal, and its proper velocity, that is, its excess over normal velocity, will diminish. Calculation shows that proper velocity varies in this way in inverse ratio to the radius of space. We must therefore look for a notable attenuation of the relative velocities of atoms in the first period of expansion. From time to time, at least, it will happen that, as a result of favorable chances, the collisions between atoms will become sufficiently moderate so as not to give rise to atomic transformations or emissions of radiation, but that these collisions will be elastic collisions, controlled by superficial electrons, so considered in the theory of gases. Thus we shall obtain, at least locally, a beginning of statistical equilibrium, that is, the formation of gaseous clouds. These gaseous clouds will still have considerable velocities, in relation to one another, and they will be mixed with radiations that are themselves attenuated by expansion.

It is these radiations which will endure until our time in the form of cosmic rays, while the gaseous clouds will have given place to stars and to nebulae by a process which remains to be explained.

Cosmic Repulsion

For that explanation, we must say a few words about the theory of relativity. When Einstein established his theory of gravitation, or generalized theory, he admitted, under the name of the principle of equivalence, that the ideas of special relativity were approximately valid in a sufficiently small domain. In the special theory, the differential element of space-time measurements had for its square a quadratic form with four coordinates, the coefficients of which had special constant values. In the generalization, this element will still be the square root of a quadratic form, but the coefficients, designated collectively by the name of *metric tensors*, will vary from place to place. The geometry of space-time is then the general geometry of Riemann at three plus one dimensions. The spaces with variable radii are a particular case in this general geometry, since the theory of spatial homogeneity or of the equivalence of observers is introduced here.

It can be that this geometry differs only apparently from that of special relativity. This is what happens when the quadratic form can be transformed, by a simple change of coordinates, into a form having constant coefficients. Then one says with Riemann that the corresponding variety (that is, space-time) is flat or Euclidean. For that, it is necessary that certain expressions, expressed by components of a tensor with four indices called Riemann's tensor, vanish completely at all points. When it is not so, the tensor of Riemann expresses the departure from flatness. Reimann's tensor is calculated by the average of second derivatives of the metric tensor. Starting with Riemann's tensor with four indices, it is easy to obtain a tensor which has only two indices like the metric tensor; it is called the contracted Riemannian tensor. One can also obtain a scalar, the totally contracted Riemannian tensor.

In special relativity, a free point describes a straight line with uniform motion, that is the principle of inertia. One can also say that, in an equivalent manner, it describes a geodesic of space-time. In the generalization, it is again presumed that a free point describes a geodesic. These geodesics are no longer representable by a uniform, rectilinear motion, they now represent a motion of a point under the action of the forces of gravitation. Since the field of gravitation is caused by the presence of matter, it is necessary that there be a relation between the density of the distribution of matter and Riemann's tensor which expresses the departure from flatness. The density is, in itself, considered as the principal component of a tensor with two indices called the *material tensor*; thus one obtains as a possible expression of the material tensor $T_{\mu\nu}$ as a function of the metric tensor $g_{\mu\nu}$ and of the two tensors of Riemann, contracted to $R_{\mu\nu}$ and totally contracted to R,

$$T_{\mu\nu} = aR_{\mu\nu} + bRg_{\mu\nu} + cg_{\mu\nu}$$

where a, b, and c are three constants.

But this is not all; certain identities must exist between the components of the material tensor and its derivatives. These identities can be interpreted, for a convenient choice of coordinates, a choice which corresponds, moreover, to the practical conditions of observations, as expressing the principles of conservation, that of energy and that of momentum. In order that such identities may be satisfied, it is no longer possible to choose arbitrarily the values of the three constants. b must be taken as equal to $-a/2$. Theory cannot predict either their magnitude or their sign. It is only observation which can determine them.

The constant a is linked to the constant of gravitation. In fact, when theory is applied to conditions which are met in the applications (in particular, the fact that astronomical velocities are small in comparison to the speed of light) and when one profits from these conditions by introducing coordinates which facilitate comparison with experiment, one finds that the geodesics differ from rectilinear motion by an acceleration which can be interpreted as an attraction in inverse ratio to the square of the distances, and which is exercised by the masses represented by the material tensor. This is simply the principal effect foreseen by the theory; this theory predicts small departures which, in favorable cases, have been confirmed by observation.

A good agreement with planetary observations is obtained by leaving out the term in c. That does not prove that this term may not have experimental consequence. In fact, in the conditions which were employed to obtain Newton's law as an approximation of the theory, the term in c would furnish a force varying, not in the inverse square ratio of the distance, but proportionally to this distance. This force could therefore have a marked action at very great distances although, for the distances of the planets, its action would be negligible. Also, the relation c/a, designated customarily by the letter *lambda*, is called the cosmological constant. When λ is positive, the additional force proportional to the distance is called *cosmic repulsion*.

The theory of relativity has thus unified the theory of Newton. In Newton's theory, there were two principles posed independently of one another: universal attraction and the conservation of mass. In the theory of relativity, these principles take a slightly modified form, while being practically identical to those of Newton in the case where these have been confronted with the facts. But universal attraction is now a result of the conservation of mass. The size of the force, the constant of gravitation, is determined experimentally.

The theory again indicates that the constancy of mass has, as a result, besides the Newtonian force of gravitation, a repulsion proportional to the distance of which the size and even the sign can only be determined by observation and by observation requiring great distances.

Cosmic repulsion is not a special hypothesis, introduced to avoid the difficulties which are presented in the study of the universe. If Einstein has reintroduced it in his work on cosmology, it is because he remembered having arbitrarily dropped it when he had established the equations of gravitation. To suppress it amounts to determining it arbitrarily by giving it a particular value: zero.

The Universe of Friedmann

The theory of relativity allows us to complete our description of space with a variable radius by introducing here some dynamic considerations. As before, we shall represent it as being in the interior of a sphere, the center of which is a point which we can choose arbitrarily. This sphere is not the boundary of the system, it is the edge of the map or of the diagram which we have made of it. It is the place at which the two opposite, half-straight lines are soldered into a closed straight line. Cosmic repulsion is manifested as a force proportional to the distance to the center of the diagram. As for the gravitational attraction, it is known that in the case of distribution involving spherical symmetry around a point, and that is certainly the case here, the regions farther away from the center than the point being considered have no influence upon its motion; as for the interior points, they act as though they were concentrated at the center. By virtue of the homogeneity of the distribution of matter, the density is constant, the force of attraction which results is thus proportional to the distance, just as is cosmic repulsion.

Therefore, a certain density exists, which we shall call the density of equilibrium or the *cosmic density*, for which the two forces will be in equilibrium.

These elementary considerations permit recognition, in a certain measure, of the result which calculation gives and which is contained in Friedmann's equation:

$$\left(\frac{dR}{dt}\right)^2 = -1 + \frac{2M}{R} + \frac{R^2}{T^2}$$

The last term represents cosmic repulsion (it is double the function of the forces of this repulsion). T is a constant depending on the value of the cosmological constant and being able to replace this. The next-to-last term is double the potential of attraction due to the interior mass. The radius of space R is the distance from the origin of a point of angular distance $\sigma = 1$. If one multiplied the equation by σ^2, one would have the corresponding equation for a point at any distance.

That which is remarkable in Friedmann's equation is the first term -1. The elementary considerations which we have just advanced would allow us to assign it a value which is more or less constant; it is the constant of energy in the motion which takes place under the action of two forces. The complete theory determines this constant and thus links the geometric properties to the dynamic properties.

Einstein's Equilibrium

Since, by virtue of equations, the radius R remains constant, the state of the universe in equilibrium, or Einstein's universe, is reached. The conditions of the universe in equilibrium are easily deduced from Friedmann's equation:

$$R_E = \frac{T}{\sqrt{3}} \; ; \; \rho_E = \frac{3}{4\pi} \frac{1}{T^2} \; ; \; M = \frac{T}{\sqrt{3}}$$

In these formulas, the distances are calculated in light-time, which amounts to taking the velocity of light c as equal to unity, but, in addition, the unit of mass is chosen in such a way that the constant of gravitation may also be equal to unity. It is easy to pass on to the numerical values in C.G.S. by re-establishing in the formulas the constants c and G in such a manner as to satisfy the equations of dimension. In particular, if one takes T as being equal to 2×10^9 years, as we shall suppose in a moment, one finds that the density ρ_E is equal to 10^{-27} gram per cubic centimeter.

These considerations can be extended to a region in which distribution is no longer homogeneous and where even the spherical symmetry is no longer verified, provided that the region under consideration be of small dimension. In fact, it is known that, in a small region, Newtonian mechanics is always a good approximation. Naturally, it is necessary, in applying Newtonian mechanics, to take account of cosmic repulsion but, aside from this easy modification, π is perfectly legitimate to utilize the intuition acquired by the practice of classic mechanics and its application to systems which are more or less complicated. Among other things, it can be noted that the equilibrium of which we have just spoken is unstable and that the equilibrium can even be disturbed in one sense, in one place, and in the opposite sense in another region.

Perhaps it is necessary to mention here that Friedmann's equation is only rigorously exact if the mass M remains constant. While one takes account of the radiation which circulates in space and also of the characteristic velocities of the particles which cross one another in the manner of molecules in a gas and, as in a gas, give rise to pressure, it is necessary to consider the work of this pressure during the expansion of space, in the evaluation of the mass or the energy. But it is apparent that such an effect is generally negligible, as detailed researches elsewhere have shown.

The Significance of Clusters of Nebulae

We are now in a position to take up again the description which we had begun of the expansion of space, following the disintegration of the primeval atom. We had shown how, in a first period of rapid expansion, gaseous clouds must have been formed, animated by great, characteristic velocities. We are now going to suppose that the mass M is slightly larger than $\dfrac{T}{\sqrt{3}}$.

The second member of Friedmann's equation will thus be able to become smaller, but it will not be able to vanish. Thus, we may distinguish three phases in the expansion of space. The first rapid expansion will be followed by a period of deceleration, during the course of which attraction and repulsion will virtually bring themselves into equilibrium. Finally, repulsion will definitely prevail over attraction, and the universe will enter into the third phase, that of the resumption of expansion under the dominant action of cosmic repulsion.

Let us consider the phase of slow expansion in more detail. The gaseous clouds are undoubtedly not distributed in a perfectly uniform manner. Let us consider in a region sufficiently small, and that only from the point of view of classic mechanics, the conflict between the forces of repulsion and attraction which almost produces equilibrium. We easily see that as a result of local fluctuations of density, there will be regions where attraction will finally prevail over repulsion, in spite of the fact that we have supposed that, for the universe in its entirety, it is the contrary which takes place. These regions in which attraction has prevailed will thus fall back upon themselves, while the universe will be entering upon a period of renewed expansion. We shall obtain a universe formed of regions of condensations which are separated from one another. Will not these regions of condensations be elliptical or spiral nebulae? We shall come back to this question in a moment.

Let us note that, although it is of rare occurrence, it will be possible for large regions where the density or the speed of expansion differ slightly from the average to hesitate between expansion and contraction, and remain in equilibrium, while the universe has resumed expansion. Could these regions not be identified with the clusters of nebulae, which are made up of several hundred nebulae located at relative distances from one another, which are a dozen times smaller than those of isolated nebulae? According to this interpretation, these clusters are made up of nebulae which are retarded in the phase of equilibrium; they represent a sample of the distribution of matter, as it existed everywhere, when the radius of space was a dozen times smaller than it is at present when the universe was passing through equilibrium.

The Findings of de Sitter

This interpretation gives the explanation for a remarkable coincidence upon which de Sitter insisted strongly, in the past. Calculating the radius of the universe in the hypothesis which bears his name, that is, ignoring the presence of matter and introducing into the formulas the value T_H given by the observation of expansion, he obtained a result which scarcely differs from that which is obtained, in Einstein's totally different hypothesis of the universe, by introducing into the formulas the observed value of the density of matter. The explanation of this coincidence is, according to our interpretation of the clusters of nebulae, that, for a value of the radius which is a dozen times the radius of equilibrium, the last term in Friedmann's formula greatly prevails over the others. The constant T which figures in it is therefore practically equal to the observed value T_H: but since, in addition, the clusters are a fragment of Einstein's universe, it is legitimate to use the relationship existing between the density and the constant T for them. For $T = T_H$ one finds, as we have seen, that the density in the clusters must be 10^{-27} gram per cubic centimeter, which is the value given by observation. This observation is based on counts of nebulae and on the estimate of their mass indicated by their spectroscopic velocity of rotation.

In addition to this argument of a quantitative variety, the proposed interpretation also takes account of important facts of a qualitative order. It explains why the clusters do not show any marked central condensations and have vague forms, with irregular extensions, all things which it would be difficult to explain if they formed dynamic structures controlled by dominant forces, as is manifestly the case for the starclusters or the elliptical and spiral nebulae. It also takes into account a manifest fact which is the existence of large fluctuations of density in the distribution of the nebulae, even outside the clusters. This must be so, in fact, if the universe has just passed through a state of unstable equilibrium, a whole gamut of transition between the properly-termed clusters which are still in equilibrium, while passing through regions where the expansion, without being arrested, has nevertheless been retarded, in such a manner that these regions have a density which is greater than the average.

This interpretation permits the value of the radius at the moment of equilibrium to be determined at a billion light-years, and thus 10^{10} light-years for the present value of the radius. Since American telescopes prospect the universe as far as half a billion light-years, one sees that this observed region already constitutes a sample of a size which is not at all negligible compared to entire space; hence, it is legitimate to hope that the values of the coefficient of expansion T_H and of the density, obtained for this restricted domain, are representative of the whole.

The only indeterminate which exists is that which is relative to the degree of approximation with which the situation of equilibrium has been approached. It is on this value which the estimate of the duration of expansion depends. Perhaps it will be possible to estimate this value by means of statistical considerations regarding the relative frequency of the clusters, compared to the isolated nebulae.

The Proper Motion of Nebulae

Now we must come back to the question of the formation of nebulae from the regions of condensation. We have seen that the characteristic velocities, or the relative velocities of gaseous clouds, which cross one another in the same place, must have been very large. Since certain of them, because of a density which is a little too large, form a nucleus of condensation, they will be able to retain the clouds which have about the same velocity as this nucleus. The proper velocity of the cloud so formed will hence be determined by the velocity of the nucleus of condensation. The nebulae formed by such a mechanism must have large relative velocities. In fact, that is what is observed in the clusters of nebulae. In the one which has been best studied, that of *Virgo*, the dispersion of the velocities about the mean velocity is 650 kilometers per second. The proper velocity must have been the proper velocity of all the nebulae at the moment of passage through equilibrium. For isolated nebulae, this velocity has been reduced to about one-twelfth, as a result of expansion, by the same mechanism which we have explained with reference to the formation of gaseous clouds.

The Formation of Stars

The density of the clouds is, on the average, the density of equilibrium 10^{-27}. For this density of distribution, a mass such as the Sun would occupy a sphere of one hundred light-years in radius. These clouds have no tendency to contract. In order that a contraction due to gravitation can be initiated, their density must be notably increased. This is what can occur if two clouds happen to collide with great velocities. Then the collision will be an inelastic collision, giving rise to ionization and emission of radiation. The two clouds will flatten one another out, while remaining in contact, the density will be easily doubled and condensation will be definitely initiated. It is clear that a solar system or a simple or multiple star may arise from such a condensation, through known mechanisms. That which characterizes the mechanism to which we are led is the greatness of the dimensions of the gaseous clouds, the condensation of which will form a star. The circumstance takes account of the magnitude of the angular momentum, which is conserved during the condensation and whose value could only be nil or negligible if the initial circumstances were adjusted in a wholly improbable manner. The least initial rotation must give rise to an energetic rotation in a concentrated system, a rotation incompatible with the presence of a single body but assuming either multiple stars turning around one another or, simply, one star with one or several large planets turning in the same direction.

The Distribution of Densities in Nebulae

Here is the manner in which we can picture for ourselves the evolution of the regions of condensation. The clouds begin by falling toward the center, and by describing a motion of oscillation following a diameter from one part and another of the center. In the course of these oscillations, they will encounter one another with velocities of several hundreds of kilometers per second and will give rise to stars. At the same time, the loss of energy due to these inelastic collisions will modify the distribution of the clouds and stars already formed in such a manner that the system will be further condensed. It seems likely that this phenomenon could be submitted to mathematical analysis. Certain hypotheses will naturally have to be introduced, in such a way as to simplify the model, so as to render the calculation possible and also so as artificially to eliminate secondary phenomena. There is scarcely any doubt that there is a way of thus obtaining the law of final distribution of the stars formed by the mechanism described above. Since the distribution of brilliance is known for the elliptical nebulae and from that one can deduce the densities in these nebulae, one sees that such a calculation is susceptible of leading to a decisive verification of the theory.

One of the complications to which I alluded, a moment ago, is the eventual presence of a considerable angular momentum. In excluding it, we have restricted the theory to condensations respecting spherical symmetry, that is, nebulae which are spherical or slightly elliptical. It is easy to see what modification will bring about the presence of considerable angular momentum. It is evident that one will obtain, in addition to a central region analogous to the elliptical nebulae, a flat system analogous to the ring of Saturn or the planetary systems, in other words, something resembling the spiral nebulae. In this theory, the spiral or elliptical character of the nebula is a matter of chance; it depends on the fortuitous value of the angular momentum in the region of condensation. It can no longer be a question of the evolution of one type into another. Moreover, the same thing obtains for stars where the type of the star is determined by the accidental value of its mass, that is, of the sum of the masses of the clouds whose encounter produced the star.

Distribution of the Supergiant Stars

If the spirals have this origin, it must follow that the stars are formed by an encounter of clouds in two very distinct processes. In the first place, and especially in the central region, the clouds encounter one another in their radial movement, and this is the phenomenon which we have invoked for the elliptical nebulae. Kapteyn's preferential motion may be an indication of it. But besides this relatively rapid process, there must be a slower process of star formation, beginning with the clouds which escaped from the central region as a result of their angular momentum. These will encounter one another in a to-and-fro motion, from one side to another of the plane of the spiral. The existence of these two processes, with different ages, is perhaps the explanation of the fact that supergiant stars are not found in the elliptical nebulae or in the nucleus of spirals, but that one observes them only in the exterior region of the spirals. In fact, it is known that the stars radiate energy which comes from the transformation of their hydrogen into helium. The supergiant stars radiate so much energy that they could only maintain this output during a hundred million years. It should be understood, thus, that, for the oldest stars, the supergiants may be extinct for lack of fuel, whereas they still shine where they have been recently formed.

The Uniform Abundance of the Elements

But it is doubtless not worthwhile to allow ourselves to be prematurely led to the attempted pursuit of the theory in such detail, but rather to restrict ourselves, for the moment, to the more general consequences of the hypothesis of the primeval atom. We have seen that the theory takes account of the formations of stars in the nebulae. It also explains a very remarkable circumstance which could be demonstrated by the analysis of stellar spectra. It concerns the quantitative composition of matter, or the relative abundance of the various chemical elements, which is the same in the Sun, in the stars, on the Earth and in the meteorites. This fact is a necessary consequence of the hypothesis of the primeval atom. Products of the disintegration of an atom are naturally found in very definite proportions, determined by the laws of radioactive transformations.

Cosmic Rays

Finally, we said in the beginning that the radiations produced during the disintegrations, during the first period of expansion, could explain cosmic rays. These rays are endowed with an energy of several billion electron-volts. We know no other phenomenon currently taking place which may be capable of such effects. That which these rays resemble most is the radiation produced during present radioactive disintegrations, but the individual energies brought into play are enormously greater. All that agrees with rays of superradioactive origin. But it is not only by their quality that these rays are remarkable, it is also by their total quantity. In fact, it is easy, from their observed density which is given in ergs per centimeter, to deduce their density of energy by dividing by c, then their density in grams per cubic centimeter by dividing by c^2. Thus one finds 10^{-34} grams per cubic centimeter, about one ten-thousandth the present density of the matter existing in the form of stars. It seems impossible to explain such an energy which represents one part in ten thousand of all existing energy, if these rays had not been produced by a process which brought into play all existing matter. In fact, this energy, at the moment of its formation, must have been at least ten times greater, since a part of it was able to be absorbed and the remainder has been reduced as a result of the expansion of space. The total intensity observed for cosmic rays is therefore just about that which might be expected.

Conclusion

The purpose of any cosmogonic theory is to seek out ideally simple conditions which could have initiated the world and from which, by the play of recognized physical forces, that world, in all its complexity, may have resulted.

I believe that I have shown that the hypothesis of the primeval atom satisfies the rules of the game. It does not appeal to any force which is not already known. It accounts for the actual world in all its complexity. By a single hypothesis it explains stars arranged in galaxies within an expanding universe as well as those local exceptions, the clusters of nebulae. Finally, it accounts for that mighty phenomenon, the ultrapenetrating rays. They are truly cosmic, they testify to the primeval activity of the cosmos. In their course through wonderfully empty space, during billions of years, they have brought us evidence of the superradioactive age, indeed they are a sort of fossil rays which tell us what happened when the stars first appeared.

I shall certainly not pretend that this hypothesis of the primeval atom is yet proved, and I would be very happy if it has not appeared to you to be either absurd or unlikely. When the consequences which result from it, especially that which concerns the law of the distribution of densities in the nebulae, are available in sufficient detail, it will doubtless be possible to declare oneself definitely for or against.

11

— Arno Penzias —

(1933–)

and Robert Wilson

(1936–)

Arno Penzias, an astrophysicist, and Robert Wilson, a physicist, co-discovered the cosmic microwave background radiation. While employed at the Bell Telephone Laboratories in Holmdel, New Jersey, in 1964, they had occasion to scan the sky with a radio telescope that had been designed for satellite communication. They found more radio noise than they expected and could not account for it by any known terrestrial source.

In the following letter to the Astrophysical Journal, *Penzias and Wilson announce their discovery of this background radiation. The letter appeared in November 1965.*

THE DISCOVERY OF BACKGROUND RADIATION

Measurements of the effective zenith noise temperature of the 20 foot horn-reflector antenna (Crawford, Hogg, and Hunt 1961) at the Crawford Hill Laboratory, Holmdel, New Jersey, at 4080 Mc/s have yielded a value about 3.5°K higher than expected. This excess temperature is, within the limits of our observations, isotropic, unpolarized, and free from seasonal variations (July, 1964–April, 1965). A possible explanation for the observed excess noise temperature is the one given by Dicke, Peebles, Roll, and Wilkinson (1965) in a companion letter in this issue.

The total antenna temperature measured at the zenith is 6.7°K of which 2.3°K is due to atmospheric absorption. The calculated contribution due to ohmic losses in the antenna and back-lobe response is 0.9°K.

The radiometer used in this investigation has been described elsewhere (Penzias and Wilson 1965). It employs a traveling-wave maser, a low-loss (0.027-db) comparison switch, and a liquid helium-cooled reference termination (Penzias 1965). Measurements were made by switching manually between the antenna input and the reference termination. The antenna, reference termination, and radiometer were well matched so that a round-trip return loss of more than 55db existed throughout the measurement; thus errors in the measurement of the effective temperature due to impedance mismatch can be neglected. The estimated error in the measured value of the total antenna temperature is 0.3°K and comes largely from uncertainty in the absolute calibration of the reference termination.

The contribution to the antenna temperature due to atmospheric absorption was obtained by recording the variation in antenna temperature with elevation angle and employing the secant law. The result, 2.3° ± 0.3°K, is in good agreement with published values (Hogg 1959; DeGrasse, Hogg, Ohm, and Scovil 1959; Ohm 1961).

The contribution to the antenna temperature from ohmic losses is computed to be 0.8° ± 0.4°K. In this calculation we have divided the antenna into three parts: (1) two non-uniform tapers approximately 1 m in total length which transform between the 2⅛ inch round output waveguide and the 6-inch-square antenna throat opening; (2) a double-choke rotary joint located between these two tapers; (3) the antenna itself. Care was taken to clean and align joints between these parts so that they would not significantly increase the loss in the structure. Appropriate tests were made for leakage and loss in the rotary joint with negative results.

The possibility of losses in the antenna horn due to imperfections in its seams was eliminated by means of a taping test. Taping all the seams in the section near the throat and most of the others with aluminum tape caused no observable change in antenna temperature.

The backlobe response to ground radiation is taken to be less than 0.1°K for two reasons: (1) Measurements of the response of the antenna to a small transmitter located on the ground in its vicinity indicate that the average back-lobe level is more than 30db below isotropic response. The horn-reflector antenna was pointed to the zenith for these measurements, and complete rotations in azimuth were made with the transmitter in each of ten locations using horizontal and vertical transmitted polarization from each position. (2) Measurements on smaller horn-reflector antennas at these laboratories, using pulsed measuring sets on flat antenna ranges, have consistently shown a back-lobe level of 30db below isotropic response. Our larger antenna would be expected to have an even lower back-lobe level.

From a combination of the above, we compute the remaining unaccounted-for antenna temperature to be 3.5° = 1.0°K at 4080 Mc/s. In connection with this result it should be noted that DeGrasse et al. (1959) and Ohm (1961) give total system temperatures at 5650 Mc/s and 23900 Mc/s, respectively. From these it is possible to infer upper limits to the background temperatures at these frequencies. These limits are, in both cases, of the same general magnitude as our value.

We are grateful to R. H. Dicke and his associates for fruitful discussions of their results prior to publication. We also wish to acknowledge with thanks the useful comments and advice of A. B. Crawford, D. C. Hogg, and E. A. Ohm in connection with the problems associated with this measurement.

Note Added in Proof

The highest frequency at which the background temperature of the sky had been measured previously was 404 Mc/s (Pauliny-Toth and Shakeshaft 1962) where a minimum temperature of 16°K was observed. Combining this value with our result, we find that the average spectrum of the background radiation over this frequency range can be no steeper that $\lambda^{0.7}$. This clearly eliminates the possibility that the radiation we observe is due to radio sources of types known to exist, since in this event, the spectrum would have to be very much steeper.

12
R. H. Dicke
(1916–1997)

— P.J.E. Peebles, P. G. Roll, —
(1935–)   (1933–)

and D. T. Wilkinson
(1935–2002)

Robert Dicke was an American physicist who was appointed professor of physics at Princeton University in 1957. He became the Albert Einstein Professor of Science there in 1975. Dicke's most important work was in the area of gravitation where, among other things, he was able to demonstrate with extraordinary accuracy the equivalence of gravitational mass and inertial mass. In 1961 he collaborated with Carl Brans to develop a new theory of gravity that is an alternative to Einstein's General Theory of Relativity.

In 1964 Dicke speculated to some colleagues at Princeton that there should be some radiation background permeating the universe that was produced when the expansion began. He assigned Peebles the task of determining the theoretical consequences of this initial phase of the expansion and asked Roll and Wilkinson to devise a method for detecting the background radiation. They were in the midst of their work when they heard of the discovery of Penzias and Wilson.

This reading comes from a letter to the same 1965 edition of the Astrophysical Journal in which the Penzias and Wilson letter appears. In the letter Dicke and his colleagues describe their preparations for the detection of the background radiation and their explanation of the Penzias and Wilson discovery.

AN EXPLANATION OF THE PENZIAS
AND WILSON DISCOVERY

One of the basic problems of cosmology is the singularity characteristic of the familiar cosmological solutions of Einstein's field equations. Also puzzling is the presence of matter in excess over antimatter in the universe, for baryons and leptons are thought to be conserved. Thus, in the framework of conventional theory we cannot understand the origin of matter or of the universe. We can distinguish three main attempts to deal with these problems.

1. The assumption of continuous creation (Bondi and Gold 1948; Hoyle 1948), which avoids the singularity by postulating a universe expanding for all time and a continuous but slow creation of new matter in the universe.
2. The assumption (Wheeler 1964) that the creation of new matter is intimately related to the existence of the singularity, and that the resolution of both paradoxes may be found in a proper quantum mechanical treatment of Einstein's field equations.
3. The assumption that the singularity results from a mathematical over-idealization, the requirement of strict isotropy or uniformity, and that it would not occur in the real world (Wheeler 1958; Lifshitz and Khalatnikov 1963).

If this third premise is accepted tentatively as a working hypothesis, it carries with it a possible resolution of the second paradox, for the matter we see about us now may represent the same baryon content of the previous expansion of a closed universe, oscillating for all time. This relieves us of the necessity of understanding the origin of matter at any finite time in the past. In this picture it is essential to suppose that at the time of maximum collapse the temperature of the universe would exceed 10^{10}°K, in order that, the ashes of the previous cycle would have been reprocessed back to the hydrogen required for the stars in the next cycle.

Even without this hypothesis it is of interest to inquire about the temperature of the universe in these earlier times. From this broader viewpoint we need not limit the discussion to closed oscillating models. Even if the universe had a singular origin it might have been extremely hot in the early stages.

Could the universe have been filled with black-body radiation from this possible high-temperature state? If so, it is important to notice that as the universe expands the cosmological redshift would serve to adiabatically cool the radiation, while preserving the thermal character. The radiation temperature would vary inversely as the expansion parameter (radius) of the universe.

The presence of thermal radiation remaining from the fireball is to be expected if we can trace the expansion of the universe back to a time when the temperature was of the order of 10^{10}°K ($\sim m_e c^2$). In this state, we would expect to find that the electron abundance had increased very substantially, due to thermal electron-pair production, to a density characteristic of the temperature only. One readily verifies that, whatever the previous history of the universe, the photon absorption length would have been short with this high electron density, and the radiation content of the universe would have promptly adjusted to a thermal equilibrium distribution due to pair-creation and annihilation processes. This adjustment requires a short time interval compared with the characteristic expansion time of the universe, whether the cosmology is general relativity or more rapidly evolving Brans-Dicke theory (Brans and Dicke 1961).

The above equilibrium argument may be applied also to the neutrino abundance. In the epoch where $T > 10^{10} \,^\circ K$, the very high thermal electron and photon abundance would be sufficient to assure an equilibrium thermal abundance of electron-type neutrinos, assuming the presence of neutrino-antineutrino pair-production processes. This means that a strictly thermal neutrino and antineutrino distribution, in thermal equilibrium with the radiation, would have issued from the highly contracted phase. Conceivably, even gravitational radiation could be in thermal equilibrium.

Without some knowledge of the density of matter in the primordial fireball we cannot predict the present radiation temperature. However, a rough upper limit is provided by the observation that black-body radiation at a temperature of 40° K provides an energy density of 2×10^{-29} gm cm^3, very roughly the maximum total energy density compatible with the observed Hubble constant and acceleration parameter. Evidently, it would be of considerable interest to attempt to detect this primeval thermal radiation directly.

Two of us (P.G.R. and D.T.W.) have constructed a radiometer and receiving horn capable of an absolute measure of thermal radiation at a wavelength of 3 cm. The choice of wavelength was dictated by two considerations, that at much shorter wavelengths atmospheric absorption would be troublesome, while at longer wavelengths galactic and extragalactic emission would be appreciable. Extrapolating from the observed background radiation at longer wavelengths ($\sim$100 cm) according to the power-law spectra characteristic of synchrotron radiation or bremsstrahlung, we can conclude that the total background at 3 cm due to the Galaxy and the extragalactic sources should not exceed 5×10^{-30} K when averaged over all directions. Radiation from stars at 3 cm is $< 10^{-90}$ K. The contribution to the background due to the atmosphere is expected to be approximately 3.5° K, and this can be accurately measured by tipping the antenna (Dicke, Beringer, Kyhl, and Vane 1946).

While we have not yet obtained results with our instrument, we recently learned that Penzias and Wilson (1965) of the Bell Telephone Laboratories have observed background radiation at 7.3-cm wavelength. In attempting to eliminate (or account for) every contribution to the noise seen at the output of their receiver, they ended with a residual of $3.5^\circ \pm 1^\circ$ K. Apparently this could only be due to radiation of unknown origin entering the antenna.

It is evident that more measurements are needed to determine a spectrum, and we expect to continue our work at 3 cm. We also expect to go to a wavelength of 1 cm. We understand that measurements at wavelengths greater than 7 cm may be filled in by Penzias and Wilson. . . .

Conclusions

While all the data are not yet in hand we propose to present here the possible conclusions to be drawn if we tentatively assume that the measurements of Penzias and Wilson (1965) do indicate black-body radiation at 3.5°K. We also assume that the universe can be considered to be isotropic and uniform, and that the present energy density in gravitational radiation is a small part of the whole. Wheeler (1958) has remarked that gravitational radiation could be important.

For purpose of obtaining definite numerical results we take the present Hubble redshift age to be 10^{10} years.

Assuming the validity of Einstein's field equations, the above discussion and numerical values impose severe restrictions on the cosmological problem. The possible conclusions are conveniently discussed under two headings, the assumption of a universe with either an open or a closed space.

Open universe.—From the present observations we cannot exlude the possibility that the total density of matter in the universe is substantially below the minimum value 2×10^{-29} gm cm^3 required for a closed universe. Assuming general relativity is valid, we have concluded from the discussion of the connection between helium production and the present radiation temperature that the present density of material in the universe must be $\leq 3 \times 10^{-32}$ gm cm^3, a factor of 600 smaller than the limit for a closed universe. The thermal-radiation energy density is even smaller, and from the above arguments we expect the same to be true of neutrinos.

Apparently, with the assumption of general relativity and a primordial temperature consistent with the present 3.5°K, we are forced to adopt an open space, with very low density. This rules out the possibility of an oscillating universe. Furthermore, as Einstein (1950) remarked, this result is distinctly non-Machian, in the sense that, with such a low mass density, we cannot reasonably assume that the local inertial properties of space are determined by the presence of matter, rather than by some absolute property of space.

Closed universe.—This could be the type of oscillating universe visualized in the introductory remarks, or it could be a universe expanding from a singular state. In the framework of the present discussion the required mass density in excess of 2×10^{-29} gm cm^3 could not be due to thermal radiation, or to neutrinos, and it must be presumed that it is due to ordinary matter, perhaps intergalactic gas uniformly distributed or else in large clouds (small protogalaxies) that have not yet generated stars.

With this large matter content, the limit placed on the radiation temperature by the low helium content of the solar system is very severe. The present black-body temperature would be expected to exceed 30°K (Peebles 1965). One way that we have found reasonably capable of decreasing this lower bound to 3.5°K is to introduce a zero-mass scalar field into the cosmology. It is convenient to do this without invalidating the Einstein field equation, and the form of the theory for which the scalar interaction appears as an ordinary matter interaction (Dicke 1962) has been employed. The cosmological equation (Brans and Dicke 1961) was originally integrated for a cold universe only, but a recent investigation of the solutions for a hot universe indicates that with the scalar field the universe would have expanded through the temperature range $T \sim 10^{9°}$ K so fast that essentially no helium would have been formed. The reason for this is that the static part of the scalar field contributes a pressure just equal to the scalar-field energy density. By contrast, the pressure due to incoherent electro-magnetic radiation or to relativistic particles is one third of the energy density. Thus, if we traced back to a highly contracted universe, we would find that the scalar-field energy density exceeded all other contributions, and that this fast increasing scalar-field energy caused the universe to expand through the highly contracted phase much more rapidly than would be the case if the scalar field vanished. The essential element is that the pressure approaches the energy density, rather than one third of the energy density. Any other interaction which would cause this such as the model given by Zeldovich (1962), would also prevent appreciable helium production in the highly contracted universe.

Returning to the problem stated in the first paragraph, we conclude that it is possible to save baryon conservation in a reasonable way if the universe is closed and oscillating. To avoid a catastrophic helium production, either the present matter density should be $<3 \times 10^{-32}$gm/cm^3, or there should exist some form of energy content with very high pressure, such as the zero-mass scalar, capable of speeding the universe through the period of helium formation. To have a closed space, an energy density of 2×10^{-29} gm/cm^3 is needed. Without a zero-mass scalar, or some other "hard" interaction, the energy could not be in the form of ordinary matter and may be presumed to be gravitational radiation (Wheeler 1958).

One other possibility for closing the universe, with matter providing the energy content of the universe, is the assumption that the universe contains a net electron-type neutrino abundance (in excess of antineutrinos) greatly larger than the nucleon abundance. In this case, if the neutrino abundance were so great that these neutrinos are degenerate, the degeneracy would have forced a negligible equilibrium neutron abundance in the early, highly contracted universe, thus removing the possibility of nuclear reactions leading to helium formation. However, the required ratio of lepton to baryon number must be $>10^9$.

We deeply appreciate the helpfulness of Drs. Penzias and Wilson of the Bell Telephone Laboratories, Crawford Hill, Holmdel, New Jersey, in discussing with us the result of their measurements and in showing us their receiving system. We are also grateful for several helpful suggestions of Professor J. A. Wheeler.

13
— Steven Weinberg —
(1933–)

American physicist Steven Weinberg received the Nobel Prize in Physics in 1979 for developing a theory uniting electromagnetic and weak nuclear interaction forces. His award winning book, The First Three Minutes *(1977), is considered the classic account of the Big Bang.*

In this reading, taken from Chapter 3 of The First Three Minutes, *Weinberg discusses the background radiation that has come to be considered the empirical evidence for the Big Bang. He regards this radiation echo the most important cosmological discovery since the discovery of the red shift.*

THE COSMIC MICROWAVE RADIATION BACKGROUND

Now we come to a different kind of astronomy, to a story that could not have been told a decade ago. We will be dealing not with observations of light emitted in the last few hundred million years from galaxies more or less like our own, but with observations of a diffuse background of radio static left over from near the beginning of the universe. The setting also changes, to the roofs of university physics buildings, to balloons or rockets flying above the earth's atmosphere, and to the fields of northern New Jersey.

In 1964 the Bell Telephone Laboratories was in possession of an unusual radio antenna on Crawford Hill at Holmdel, New Jersey. The antenna had been built for communication via the *Echo* satellite, but its characteristics—a 20-foot horn reflector with ultra low noise—made it a promising instrument for radio astronomy. A pair of radio astronomers, Arno A. Penzias and Robert W. Wilson, set out to use the antenna to measure the intensity of the radio waves emitted from our galaxy at high galactic latitudes, i.e., out of the plane of the Milky Way.

This kind of measurement is very difficult. The radio waves from our galaxy, as from most astronomical sources, are best described as a sort of *noise*, much like the "static" one hears on a radio set during a thunderstorm. This radio noise is not easily distinguished from the inevitable electrical noise that is produced by the random motions of electrons within the radio antenna structure and the amplifier circuits, or from the radio noise picked up by the antenna from the earth's atmosphere. The problem is not so serious when one is studying a relatively "small" source of radio noise, like a star or a distant galaxy. In this case one can switch the antenna beam back and forth between the source and the neighboring empty sky; any spurious noise coming from the antenna structure, amplifier circuits, or the earth's atmosphere will be about the same whether the antenna is pointed at the source or the nearby sky, so it would cancel out when the two are compared. However, Penzias and Wilson were intending to measure the radio noise coming from our own galaxy—in effect, from the sky itself. It was therefore crucially important to identify any electrical noise that might be produced within their receiving system.

Previous tests of this system had in fact revealed a little more noise than could be accounted for, but it seemed likely that this discrepancy was due to a slight excess of electrical noise in the amplifier circuits. In order to eliminate such problems, Penzias and Wilson made use of a device known as a "cold load"—the power coming from the antenna was compared with the power produced by an artificial source cooled with liquid helium, about four degrees above absolute zero. The electrical noise in the amplifier circuits would be the same in both cases, and would therefore cancel out in the comparison, allowing a direct measurement of the power coming from the antenna. The antenna power measured in this way would consist only of contributions from the antenna structure, from the earth's atmosphere, and from any astronomical sources of radio waves.

Penzias and Wilson expected that very little electrical noise would be produced within the antenna structure. However, in order to check this assumption, they started their observations at a relatively short wavelength of 7.35 centimeters, where the radio noise from our galaxy should have been negligible. Some radio noise could naturally be expected at this wavelength from our earth's atmosphere, but this would have a characteristic dependence on direction: it would be proportional to the thickness of atmosphere along the direction in which the antenna was pointed—less toward the zenith, more toward the horizon. It was expected that, after subtraction of an atmospheric term with this characteristic dependence on direction, there would be essentially no antenna power left over, and this would confirm that the electrical noise produced within the antenna structure was indeed negligible. They would then be able to go on to study the galaxy itself at a longer wavelength, around 21 centimeters, where the galactic radio noise was expected to be appreciable.

(Incidentally, radio waves with wavelengths like 7.35 centimeters or 21 centimeters, and up to 1 meter, are known as "microwave radiation." This is because these wavelengths are shorter than those of the VHF band used by radar at the beginning of World War II.)

To their surprise, Penzias and Wilson found in the spring of 1964 that they were receiving a sizable amount of microwave noise at 7.35 centimeters that was independent of direction. They also found that this "static" did not vary with the time of day or, as the year went on, with the season. It did not seem that it could be coming from our galaxy; if it were, then the great galaxy M31 in Andromeda, which is in most respects similar to our own, would presumably also be radiating strongly at 7.35 centimeters, and this microwave noise would already have been observed. Above all, the lack of any variation of the observed microwave noise with direction indicated very strongly that these radio waves, if real, were not coming from the Milky Way, but from a much larger volume of the universe.

Clearly, it was necessary to reconsider whether the antenna itself might be producing more electrical noise than expected. In particular, it was known that a pair of pigeons had been roosting in the antenna throat. The pigeons were caught; mailed to the Bell Laboratories Whippany site; released; found back in the antenna at Holmdel a few days later; caught again; and finally discouraged by more decisive means. However, in the course of their tenancy, the pigeons had coated the antenna throat with what Penzias delicately calls "a white dielectric material," and this material might at room temperature be a source of electrical noise. In early 1965 it became possible to dismantle the antenna throat and clean out the mess, but this, and all other efforts, produced only a very small decrease in the observed noise level. The mystery remained: Where was this microwave noise coming from?

The one piece of numerical data that was available to Penzias and Wilson was the intensity of the radio noise they had observed. In describing this intensity they used a language that is common among radio engineers, but which turned out in this case to have unexpected relevance. Any sort of body at any temperature above absolute zero will always emit radio noise, produced by the thermal motions of electrons within the body. Inside a box with opaque walls, the intensity of the radio noise at any given wavelength depends only on the temperature of the walls—the higher the temperature, the more intense the static. Thus, it is possible to describe the intensity of radio noise observed at a given wavelength in terms of an "equivalent temperature"—the temperature of the walls of a box within which the radio noise would have the observed intensity. Of course, a radio telescope is not a thermometer; it measures the strength of radio waves by recording the tiny electric currents that the waves induce in the structure of the antenna. When a radio astronomer says that he observes radio noise with such and such an equivalent temperature, he means only that this is the temperature of the opaque box into which the antenna would have to be placed to produce the observed radio noise intensity. Whether or not the antenna is in such a box is of course another question.

(To forestall objections from experts, I should mention that radio engineers often describe the intensity of radio noise in terms of so-called antenna temperature, which is slightly different from the "equivalent temperature" described above. For the wavelengths and intensities observed by Penzias and Wilson, the two definitions are virtually identical.)

Penzias and Wilson found that the equivalent temperature of the radio noise they were receiving was about 3.5 degrees centigrade above absolute zero (or more accurately, between 2.5 and 4.5 degrees above absolute zero). Temperatures measured on the centigrade scale, but referred to absolute zero rather than the melting point of ice, are reported in "degrees Kelvin." Thus, the radio noise observed by Penzias and Wilson could be described as having an "equivalent temperature" of 3.5 degrees Kelvin, or 3.5° K for short. This was much greater than expected, but still very low in absolute terms, so it is not surprising that Penzias and Wilson brooded over their result for a while before publishing it. It certainly was not immediately clear that this was the most important cosmological advance since the discovery of red shifts.

The meaning of the mysterious microwave noise soon began to be clarified through the operation of the "invisible college" of astrophysicists. Penzias happened to telephone a fellow radio astronomer, Bernard Burke of M.I.T., about other matters. Burke had just heard from yet another colleague, Ken Turner of the Carnegie Institution, of a talk that Turner had in turn heard at Johns Hopkins, given by a young theorist from Princeton, P.J.E. Peebles. In this talk Peebles argued that there ought to be a background of radio noise left over from the early universe, with a present equivalent temperature of roughly 10° K.

Burke already knew that Penzias was measuring radio noise temperatures with the Bell Laboratories horn antenna, so he took the occasion of the telephone conversation to ask how the measurements were going. Penzias said that the measurements were going fine, but that there was something about the results he didn't understand. Burke suggested to Penzias that the physicists at Princeton might have some interesting ideas on what it was that his antenna was receiving.

In his talk, and in a preprint written in March 1965, Peebles had considered the radiation that might have been present in the early universe. "Radiation" is of course a general term, encompassing electromagnetic waves of all wavelengths—not only radio waves, but infrared light, visible light, ultraviolet light, X-rays, and the very short-wavelength radiation called gamma rays. There are no sharp distinctions; with changing wavelength one kind of radiation blends gradually into another. Peebles noted that if there had not been an intense background of radiation present during the first few minutes of the universe, nuclear reactions would have proceeded so rapidly that a large fraction of the hydrogen present would have been "cooked" into heavier elements, in contradiction with the fact that about three-quarters of the present universe is hydrogen. This rapid nuclear cooking could have been prevented only if the universe was filled with radiation having an enormous equivalent temperature at very short wavelengths, which could blast nuclei apart as fast as they could be formed.

We are going to see that this radiation would have survived the subsequent expansion of the universe, but that its equivalent temperature would continue to fall as the universe expanded, in inverse proportion to the size of the universe. (As we shall see, this is essentially an effect of the red shift.) It follows that the present universe should also be filled with radiation, but with an equivalent temperature vastly less than it was in the first few minutes. Peebles estimated that, in order for the radiation background to have kept the production of helium and heavier elements in the first few minutes within known bounds, it would have to have been so intense that its present temperature would be at least 10 degrees Kelvin.

The figure of 10° K was somewhat of an overestimate, and this calculation was soon supplanted by more elaborate and accurate calculations by Peebles and others. Peebles's preprint was in fact never published in its original form. However, the conclusion was substantially correct: from the observed abundance of hydrogen we can infer that the universe must in the first few minutes have been filled with an enormous amount of radiation which could prevent the formation of too much of the heavier elements; the expansion of the universe since then would have lowered its equivalent temperature to a few degrees Kelvin, so that it would appear now as a background of radio noise, coming equally from all directions. This immediately appeared as the natural explanation of the discovery of Penzias and Wilson. Thus, in a sense the antenna at Holmdel *is* in a box—the box is the whole universe. However, the equivalent temperature recorded by the antenna is not the temperature of the present universe, but rather the temperature that the universe had long ago, reduced in proportion to the enormous expansion that the universe has undergone since then.

Peebles's work was only the latest in a long series of similar cosmological speculations. In fact, in the late 1940s a "big bang" theory of nucleosynthesis had been developed by George Gamow and his collaborators, Ralph Alpher and Robert Herman, and was used in 1948 by Alpher and Herman to predict a radiation background with a present temperature of about 5° K. Similar calculations were carried out in 1964 by Ya. B. Zeldovich in Russia and independently by Fred Hoyle and R. J. Tayler in England. This earlier work was not at first known to the groups at Bell Laboratories and Princeton, and it did not have an effect on the actual discovery of the radiation background. . . .

Peebles's 1965 calculation had been instigated by the ideas of a senior experimental physicist at Princeton, Robert H. Dicke. (Among other things, Dicke had invented some of the key microwave techniques used by radio astronomers.) Sometime in 1964 Dicke had begun to wonder whether there might not be some observable radiation left over from a hot dense early stage of cosmic history. Dicke's speculations were based on an "oscillating" theory of the universe. He apparently did not have a definite expectation of the temperature of this radiation, but he did appreciate the essential point, that there was something worth looking for. Dicke suggested to P. G. Roll and D. T. Wilkinson that they mount a search for a microwave radiation background, and they began to set up a small low-noise antenna on the roof of the Palmer Physical Laboratory at Princeton. (It is not necessary to use a large radio telescope for this purpose because the radiation comes from all directions, so that nothing is gained by having a more tightly focused antenna beam.)

Before Dicke, Roll, and Wilkinson could complete their measurements, Dicke received a call from Penzias, who had just heard from Burke of Peebles's work. They decided to publish a pair of companion letters in the *Astrophysical Journal*, in which Penzias and Wilson would announce their observations, and Dicke, Peebles, Roll, and Wilkinson would explain the cosmological interpretation. Penzias and Wilson, still very cautious, gave their paper the modest title "A Measurement of Excess Antenna Temperature at 4,080 Mc/s." (The frequency to which the antenna was tuned was 4,080 Mc/s, or 4,080 million cycles per second, corresponding to the wavelength of 7.35 centimeters.) They announced simply that "Measurements of the effective zenith noise temperature . . . have yielded a value of about 3.5° K higher than expected," and they avoided all mention of cosmology, except to note that "A possible explanation for the observed excess noise temperature is the one given by Dicke, Peebles, Roll, and Wilkinson in a companion letter in this issue."

14
— Alan Guth —
(1947–)
and Paul Steinhardt
(1952–)

Alan Guth, an American astrophysicist, is the author of Inflationary Theory, which addressed the original Big Bang Theory's inability to account for certain features of the universe. Guth suggested that the expansion of the universe had not been uniform but had a very rapid phase early on in the process.

In this reading Guth and his collaborator Paul Steinhardt spell out the Inflationary Hypothesis and its implications. This article, which originally appeared in May of 1984 in Scientific American *(vol. 250.5, 16–28), clearly explains how the Inflationary Hypothesis solves the problems of the original Big Bang Theory; it is reprinted here by permission of* Scientific American.

THE INFLATIONARY UNIVERSE

Introduction

In the past few years certain flaws in the standard big-bang theory of cosmology have led to the development of a new model of the very early history of the universe. The model, known as the inflationary universe, agrees precisely with the generally accepted description of the observed universe for all times after the first 10^{-30} s. For this first fraction of a second, however, the story is dramatically different. According to the inflationary model, the universe had a brief period of extraordinarily rapid inflation, or expansion during which its diameter increased by a factor perhaps 10^{50} times larger than had been thought. In the course of this stupendous growth spurt all the matter and energy in the universe could have been created from virtually nothing. The inflationary process also has important implications for the present universe. If the new model is correct, The observed universe is only a very small fraction of the entire universe.

The inflationary model has many features in common with the standard big-bang model. In both models the universe began between ten and fifteen billion years ago as a primeval fireball of extreme density and temperature, and it has been expanding and cooling ever since. This picture has been successful in explaining many aspects of the observed universe including the red-shifting of the light from distant galaxies, the cosmic microwave background radiation and the primordial abundances of the lightest elements. All these predictions have to do only with events that presumably took place after the first second, when the two models coincide.

Until about five years ago there were few serious attempts to describe the universe during its first second. The temperature in this period is thought to have been higher than ten billion degrees Kelvin, and little was known about the properties of matter under such conditions. Relying on recent developments in the physics of elementary particles, however, cosmologists are now attempting to understand the history of the universe back to 10^{-45} second after its beginning. (At even earlier times the energy density would have been so great that Einstein's general theory of relativity would have to be replaced by a quantum theory of gravity, which so far does not exist.) When the standard big-bang model is extended to these earlier times, various problems arise. First, it becomes clear that the model requires a number of very stringent, unexplained assumptions about the initial conditions of the universe. In addition most of the new theories of elementary particles imply that the standard model would lead to a tremendous overproduction of the exotic particles called magnetic monopoles (each of which corresponds to an isolated north or south magnetic pole).

The inflationary universe was invented to overcome these problems. The equations that describe the period of inflation have a very attractive feature: from almost any initial conditions the universe evolves to precisely the state that had to be assumed as the initial one in the standard model. Moreover, the predicted density of magnetic monopoles becomes small enough to be consistent with observations. In the context of the recent developments in elementary particle theory, the inflationary model seems to be a natural solution to many of the problems of the standard big-bang picture.

The standard model is based on several assumptions. First, it is assumed that the fundamental laws of physics do not change with time and that the effects of gravitation are correctly described by Einstein's theory of general relativity. It is also assumed that the early universe was filled with an almost perfectly uniform, expanding, intensely hot gas of elementary particles in thermal equilibrium. The gas filled all of space, and the gas and space expanded together at the same rate. When they are averaged over large regions, the densities of matter and energy have remained nearly uniform from place to place as the universe has evolved. It is further assumed that any changes in the state of the matter and radiation have been so smooth that they have had a negligible effect on the thermodynamic history of the universe. The violation of the last assumption is a key to the inflationary universe model.

The big-bang model leads to three important, experimentally testable predictions. First, the model predicts that as the universe expands, galaxies recede from one another with a velocity proportional to the distance between them. In the 1920s Edwin P. Hubble inferred just such an expansion law from his study of the red-shifts of distant galaxies. Second, the big-bang model predicts that there should be a background of microwave radiation bathing the universe as a remnant of the intense heat of its origin. The universe became transparent to this radiation several hundred thousand years after the big bang. Ever since then the matter has been clumping into stars, galaxies, and the like, but the radiation has simply continued to expand and red-shift, and in effect to cool. In 1964

Arno A. Penzias and Robert W. Wilson of the Bell Telephone Laboratories discovered a background of microwave radiation, received uniformly from all directions with an effective temperature of about three degrees K. Third, the model leads to successful predictions for the formation of light atomic nuclei from protons and neutrons during the first minutes after the big bang. Successful predictions can be obtained in this way for the abundance of helium 4, deuterium, helium 3 and lithium 7. (Heavier nuclei are thought to have been produced much later in the interior of stars.)

Unlike the successes of the big-bang model, all of which pertain to events a second or more after the big bang, the problems all concern times when the universe was much less than a second old. One set of problems has to do with the special conditions the model requires as the universe emerged from the big bang.

The first problem is the difficulty of explaining the large-scale uniformity of the observed universe. The large-scale uniformity is most evident in the microwave background radiation, which is known to be uniform in temperature to about one part in 10,000. In the standard model the universe evolves much too quickly to allow this uniformity to be achieved by the usual processes whereby a system approaches thermal equilibrium. The reason is that no information or physical process can propagate faster than a light signal. At any given time there is a maximum distance, known as the horizon distance, that a light signal could have traveled since the beginning of the universe. In the standard model the sources of the microwave background radiation observed from opposite directions in the sky were separated from each other by more than 90 times the horizon distance when the radiation was emitted. Since the regions could not have communicated, it is difficult to see how they could have evolved conditions so nearly identical.

The puzzle of explaining why the universe appears to be uniform over distances that are large compared with the horizon distance is known as the horizon problem. It is not a genuine inconsistency of the standard model, if the uniformity is assumed in the initial conditions, then the universe will evolve uniformly. The problem is that one of the most salient features of the observed universe—its large-scale uniformity—cannot be explained by the standard model; it must be assumed as an initial condition.

Even with the assumption of large-scale uniformity, the standard big-bang model requires yet another assumption to explain the nonuniformity observed on smaller scales. To account for the clumping of matter into galaxies, clusters of galaxies, superclusters of clusters, and so on, a spectrum of primordial inhomogeneities must be assumed as part of the initial conditions. The fact that the spectrum of inhomogeneities has no explanation is a drawback in itself, but the problem becomes even more pronounced when the model is extended back to 10^{-45} second after the big bang. The incipient clumps of matter develop rapidly with time as a result of their gravitational self-attraction, and so a model that begins at a very early time must begin with very small inhomogeneities. To begin at 10^{-45} second the matter must start in a peculiar state of extraordinary but not quite perfect uniformity. A normal gas in thermal equilibrium would be far too inhomogeneous, owing to the random motion of particles. This peculiarity of the initial state of matter required by the standard model is called the smoothness problem.

Another subtle problem of the standard model concerns the energy density of the universe. According to general relativity, the space of the universe can in principle be curved, and the nature of the curvature depends on the energy density. If the energy density exceeds a certain critical value, which depends on the expansion rate, the universe is said to be closed; space curves back on itself to form a finite volume with no boundary. (A familiar analogy is the surface of a sphere, which is finite in area and has no boundary.) If the energy density is less than the critical density, the universe is open: space curves but does not turn back on itself, and the volume is infinite. If the energy density is just equal to the critical density, the universe is flat: space is described by the familiar Euclidean geometry (again with infinite volume).

The ratio of the energy density of the universe to the critical density is a quantity cosmologists designate by the capital Greek letter Ω. The value $\Omega = 1$ (corresponding to a flat universe) represents a state of unstable equilibrium. If Ω was ever exactly equal to 1, it would remain exactly equal to 1 forever. If Ω differed slightly from 1 an instant after the big bang, however, the deviation from 1 would grow rapidly with time. Given this instability, it is surprising that Ω is measured today as being between .1 and 2. (Cosmologists are still not sure whether the universe is open, closed, or flat.) In order for Ω to be in this rather narrow range today, its value a second after the big bang had to equal 1 to within one part in 10^{15}. The standard model offers no explanation of why Ω began so close to 1 but merely assumes the fact as an initial condition. This shortcoming of the standard model, called the flatness problem, was first pointed out in 1979 by Robert H. Dicke and P.J.E. Peebles of Princeton University.

The successes and drawbacks of the big-bang model we have considered so far involve cosmology, astrophysics and nuclear physics. As the big-bang model is traced backward in time, however, one reaches an epoch for which these branches of physics are no longer adequate. In this epoch all matter is decomposed into its elementary particle constituents. In an attempt to understand this epoch cosmologists have made use of recent progress in the theory of elementary particles. Indeed, one of the important developments of the past decade has been the fusing of interests in particle physics, astrophysics and cosmology. The result for the big-bang model appears to be at least one more success and at least one more failure.

Perhaps the most important development in the theory of elementary particles over the past decade has been the notion of grand unified theories, the prototype of which was proposed in 1974 by Howard M. Georgi and Sheldon Lee Glashow of Harvard University. The theories are difficult to verify experimentally because their most distinctive predictions apply to energies far higher than those that can be reached with particle accelerators. Nevertheless, the theories have some experimental support, and they unify the understanding of elementary-particle interactions so elegantly that many physicists find them extremely attractive.

The basic idea of a grand unified theory is that what were perceived to be three independent forces—the strong, the weak and the electromagnetic—are actually parts of a single unified force. In the theory a symmetry relates one force to another. Since experimentally the forces are very different, in strength and character, the theory is constructed so that the symmetry is spontaneously broken in the present universe.

A spontaneously broken symmetry is one that is present in the underlying theory describing a system but is hidden in the equilibrium state of the system. For example, a liquid described by physical laws that are rotationally symmetric is itself rotationally symmetric: the distribution of molecules looks the same no matter how the liquid is turned. When the liquid freezes into a crystal, however, the atoms arrange themselves along crystallographic axes and the rotational symmetry is broken. One would expect that if the temperature of a system in a broken-symmetry state were raised, it could undergo a kind of phase transition to a state in which the symmetry is restored, just as a crystal can melt into a liquid. Grand unified theories predict such a transition at a critical temperature of roughly 10^{27} degrees.

One novel property of the grand unified theories has to do with the particles called baryons, a class whose most important members are the proton and the neutron. In all physical processes observed up to now the number of baryons minus the number of antibaryons does not change; in the language of particle physics the total baryon number of the system is said to be conserved. A consequence of such a conservation law is that the proton must be absolutely stable; because it is the lightest baryon, it cannot decay into another particle without changing the total baryon number. Experimentally the lifetime of the proton is now known to exceed 10^{31} years.

Grand unified theories, imply that baryon number is not exactly conserved. At low temperature, in the broken-symmetry phase, the conservation law is an excellent approximation, and the observed limit on the proton lifetime is consistent with at least many versions of grand unified theories. At high temperatures, however, processes that change the baryon number of a system of particles are expected to be quite common.

One direct result of combining the big-bang model with grand unified theories is the successful prediction of the asymmetry of matter and antimatter in the universe. It is thought that all the stars, galaxies and dust observed in the universe are in the form of matter rather than antimatter; their nuclear particles are baryons rather than antibaryons. It follows that the total baryon number of the observed universe is about 10^{78}. Before the advent of grand unified theories, when baryon number was thought to be conserved, this net baryon number had to be postulated as yet another initial condition of the universe. When grand unified theories and the big-bang picture are combined, however, the observed excess of matter over antimatter can be produced naturally by elementary-particle interactions at temperatures just below the critical temperature of the phase transition. Calculations in the grand unified theories depend on too many arbitrary parameters for a quantitative prediction, but the observed matter-antimatter asymmetry can be produced with a reasonable choice of values for the parameters.

A serious problem that results from combining grand unified theories with the big-bang picture is that a large number of defects are generally formed during the transition from the symmetric phase to the broken-symmetry phase. The defects are created when regions of symmetric phase undergo a transition to different broken-symmetry states. In an analogous situation, when a liquid crystallizes, different regions may begin to crystallize with different orientations of the crystallographic axes. The domains of different crystal orientation grow and coalesce, and it is energetically favorable for them to smooth the misalignment along their boundaries. The smoothing is often imperfect, however, and localized defects remain.

In the grand unified theories there are serious cosmological problems associated with pointlike defects, which correspond to magnetic monopoles, and surfacelike defects called domain walls. Both are expected to be extremely stable and extremely massive. (The monopole can be shown to be about 10^{16} times as heavy as the proton.) A domain of correlated broken-symmetry phase can be shown to be about 10^{16} times as heavy as the proton.) A domain of correlated broken-symmetry phase cannot be much larger than the horizon distance at that time, and so the minimum number of defects created during the transition can be estimated. The result is that there would be so many defects after the transition that their mass would dominate the energy density of the universe and thereby speed up its subsequent evolution. The microwave background radiation would reach its present temperature of three degrees K. only 30,000 years after the big bang instead of 10 billion years, and all the successful predictions of the big-bang model would be lost. Thus any successful union of grand unified theories and the big-bang picture must incorporate some mechanism to drastically suppress the production of magnetic monopoles and domain walls.

The inflationary universe model appears to provide a satisfactory solution to these problems. Before the model can be described, however, we must first explain a few more of the details of symmetry breaking and phase transitions in grand unified theories.

All modern particle theories, including the grand unified theories, are examples of quantum field theories. The best-known field theory is the one that describes electromagnetism. According to the classical (nonquantum) theory of electromagnetism developed by James Clerk Maxwell in the 1860s, electric and magnetic fields have a well-defined value at every point in space, and their variation with time is described by a definite set of equations. Maxwell's theory was modified early in the twentieth century in order to achieve consistency with the quantum theory. In the classical theory it is possible to increase the energy of an electromagnetic field by any amount, but in the quantum theory the increases in energy can come only in discrete lumps, the quanta, which in this

Type of Universe	Ratio of Energy Density to Critical Density (Ω)	Spatial Geometry	Volume	Temporal Evolution
Closed	>1	Positive Curvature (Spherical)	Finite	Expands and Recollapses
Open	<1	Negative Curvature (Hyperbolic)	Infinite	Expands Forever
Flat	1	Zero Curvature (Euclidean)	Infinite	Expands Forever, But Expansion Rate Approaches Zero

Three types of universe, classified as closed, open and flat, can arise from the standard big-bang model (under the usual assumption that the equations of general relativity are not modified by the addition of a cosmological term). The distinction between the different geometries depends on the quantity designated Ω, the ratio of the energy density of the universe to some critical density, whose value depends in turn on the rate of expansion of the universe. The value of Ω today is known to lie between .1 and 2, which implies that its value a second after the big bang was equal to 1 to within one part in 10^{15}. The failure of the standard big-bang model to explain why Ω began so close to 1 is called the "flatness problem."

case are called photons. The photons have both wavelike and particlelike properties, but in the lexicon of modern physics they are usually called particles. In general the formulation of a quantum field theory begins with a classical theory of fields, and it becomes a theory of particles when the rules of the quantum theory are applied.

As we have already mentioned, an essential ingredient of grand unified theories is the phenomenon of spontaneous symmetry breaking. The detailed mechanism of spontaneous symmetry breaking in grand unified theories is simpler in many ways than the analogous mechanism in crystals. In grand unified theory spontaneous symmetry breaking is accomplished by including in the formulation of the theory a special set of fields known as Higgs fields (after Peter W. Higgs of the University of Edinburgh). The symmetry is unbroken when all the Higgs fields have a value of zero, but it is spontaneously broken whenever at least one of the Higgs fields acquires a nonzero value. Furthermore, it is possible to formulate the theory in such a way that the Higgs field has a nonzero value in the state of lowest possible energy density, which in this context is known as the true vacuum. At temperatures greater than about 10^{27} degrees thermal fluctuations drive the equilibrium value of the Higgs field to zero, resulting in a transition to the symmetric phase.

We have now assembled enough background information to describe the inflationary model of the universe, beginning with the form in which it was first proposed by one of us (Guth) in 1980. Any cosmological model must begin with some assumptions about the initial conditions, but for the inflationary model the initial conditions can be rather arbitrary. One assumes, however, that the early universe included at least some regions of gas that were hot compared with the critical temperature of the phase transition and that were also expanding. In such a hot region the Higgs field would have a value of zero. As the expansion caused the temperature to fall it would become thermodynamically favourable for the Higgs field to acquire a nonzero value, bringing the system to its broken-symmetry phase.

For some values of the unknown parameters of the grand unified theories this phase transition would occur very slowly compared with the cooling rate. As a result the system could cool to well below 10^{27} degrees with the value of the Higgs field remaining at zero. This phenomenon, known as supercooling, is quite common in condensed-matter physics; water, for example, can be super-cooled to more than 20 degrees below its freezing point, and glasses are formed by rapidly supercooling a liquid to a temperature well below its freezing point.

As the region of gas continued to supercool, it would approach a peculiar state of matter known as a false vacuum. This state of matter has never been observed, but it has properties that are unambiguously predicted by quantum field theory. The temperature, and hence the thermal component of the energy density, would rapidly decrease and the energy density of the state would be concentrated entirely in the Higgs field. A zero value for the Higgs field implies a large energy density for the false vacuum. In the classical form of the theory such a state would be absolutely stable, even though it would not be the state of lowest energy density. States with a lower energy density would be separated from the false vacuum by an intervening energy barrier, and there would be no energy available to take the Higgs field over the barrier.

In the quantum version of the model the false vacuum is not absolutely stable. Under the rules of the quantum theory all the fields would be continually fluc-tuating. As was first described by Sidney R. Coleman of Harvard, a quantum fluctuation would occasionally cause the Higgs field in a small region of space to "tunnel" through the energy barrier, nucleating a "bubble" of the broken-symmetry phase. The bubble would then start to grow at a speed that would rapidly approach the speed of light, converting the false vacuum into the broken-symmetry phase. The rate at which bubbles form depends sensitively on the unknown parameters of the grand unified theory; in the inflationary model it is assumed that the rate would be extremely low.

The most peculiar property of the false vacuum is probably its pressure, which is both large and negative. To understand why, consider again the process by which a bubble of true vacuum would grow into a region of false vacuum. The growth is favored energetically because the true vacuum has a lower energy density than the false vacuum. The growth also indicates, however, that the pressure of the true vacuum must be higher than the pressure of the false vac-uum, forcing the bubble wall to grow outward. Because the pressure of the true vacuum is zero, the pressure of the false vacuum must be negative. A more detailed argument shows that the pressure of the false vacuum is equal to the negative value of its energy density (when the two quantities are measured in the same units).

The negative pressure would not result in mechanical forces within the false vacuum, because mechanical forces arise only from differences in pressure. Nevertheless, there would be gravitational effects. Under ordinary circumstances the expansion of the region of gas would be slowed by the mutual gravitational attraction of the matter within it. In Newtonian physics this attraction is proportional to the mass density, which in relativistic theories is equal to the energy density divided by the square of the speed of light. According to general relativity, the pressure also contributes to the attraction; to be specific, the gravitational force is proportional to the energy density plus three times the pressure. For the false vacuum the contribution made by the pressure would overwhelm the energy-density contribution and would have the opposite sign. Hence the bizarre notion of negative pressure leads to the even more bizarre effect of gravitational force that is effectively repulsive. As a result the expansion of the region would be accelerated and the region would grow exponentially, doubling in diameter during each interval of about 10^{-34} second.

This period of accelerated expansion is called the inflationary era, and it is the key element of the inflationary model of the universe. According to the model, the inflationary era continued for 10^{-32} second or longer, and during this period the diameter of the universe increased by a factor of 10^{50} or more. It is assumed that after this colossal expansion the transition to the broken-symmetry phase finally took place. The energy density of the false vacuum was then released, resulting in a tremendous amount of particle production. The region was reheated to a temperature of almost 10^{27} degrees. (In the language of thermodynamics the energy released is called latent heat; it is analogous to the energy released when water freezes.) From this point on the region would continue to expand and cool at the rate described by the standard big-bang model. A volume the size of the observable universe would lie well within such a region.

The horizon problem is avoided in a straightforward way. In the inflationary model the observed universe evolves from a region that is much smaller in diameter (by a factor of 10^{50} or more) than the corresponding region in the standard model. Before inflation begins the region is much smaller than the horizon distance, and it has time to homogenize and reach thermal equilibrium. This small homogeneous region is then inflated to become large enough to encompass the observed universe. Thus the sources of the microwave background radiation arriving today from all directions in the sky were once in close contact; they had time to reach a common temperature before the inflationary era began.

The flatness problem is also evaded in a simple and natural way. The equations describing the evolution of the universe during the inflationary era are different from those for the standard model, and it turns out that the ratio Ω is driven rapidly toward 1, no matter what value it had before inflation. This behaviour is most easily understood by recalling that a value of $\Omega = 1$ corresponds to a space that is geometrically flat. The rapid expansion causes the space to become flatter just as the surface of a balloon becomes flatter when it is inflated. The mechanism driving Ω toward 1 is so effective that one is led to an almost rigorous prediction: The value of Ω today should be very accurately equal to 1. Many astronomers (although not all) think a value of 1 is consistent with current observations, but a more reliable determination of Ω would provide a crucial test of the inflationary model.

In the form in which the inflationary model was originally proposed it had a crucial flaw: under the circumstances described, the phase transition itself would create inhomogeneities much more extreme than those observed today. As we have already described, the phase transition would take place by the random nucleation of bubbles of the new phase. It can be shown that the bubbles would always remain in finite clusters disconnected from one another, and that each cluster would remain dominated by a single largest bubble. Almost all the energy in the cluster would be initially concentrated in the surface of the largest bubble, and there is no apparent mechanism to redistribute energy uniformly. Such a configuration bears no resemblance to the observed universe.

For almost two years after the invention of the inflationary universe model it remained a tantalizing but clearly imperfect solution to a number of important cosmological problems. Near the end of 1981, however, a new approach was developed by A. D. Linde of the P. N. Lebedev Physical Institute in Moscow and independently by Andreas Albrecht and one of us (Steinhardt) of the University of Pennslyvania. This approach, known as the new inflationary universe, avoids all the problems of the original model while maintaining all its successes.

The key to the new approach is to consider a special form of the energy density function that describes the Higgs field, . . . there is no energy barrier separating the false vacuum from the true vacuum; instead the false vacuum lies at the top of a rather flat plateau. In the context of grand unified theories such an energy-density function is achieved by a special choice of parameters. As we shall explain below, this energy-density function leads to a special type of phase transition that is sometimes called a slow-rollover transition.

The scenario begins just as it does in the original inflationary model. Again one must assume the early universe had regions that were hotter than about 10^{27} degrees and were also expanding. In these regions thermal fluctuations would drive the equilibrium value of the Higgs fields to zero and the symmetry would be unbroken. As the temperature fell it would become thermodynamically favorable for the system to undergo a phase transition in which at least one of the Higgs fields acquired a nonzero value, resulting in a broken-symmetry phase. As in the previous case, however, the rate of this phase transition would be extremely low compared with the rate of cooling. The system would supercool to a negligible temperature with the Higgs field remaining at zero, and the resulting state would again be considered a false vacuum.

The important difference in the new approach is the way in which the phase transition would take place. Quantum fluctuations or small residual thermal fluctuations would cause the Higgs field to deviate from zero. In the absence of an energy barrier the value of the Higgs field would begin to increase steadily;

the rate of increase would be much like that of a ball rolling down a hill of the same shape as the curve of the energy-density function, under the influence of a frictional drag force. Since the energy density curve is almost flat near the point where the Higgs field vanishes, the early stage of the evolution would be very slow. As long as the Higgs field remained close to zero, the energy density would be almost the same as it is in the false vacuum. As in the original scenario, the region would undergo accelerated expansion, doubling in diameter every 10^{34} second or so. Now, however, the expansion would cease to accelerate when the value of the Higgs field reached the steeper part of the curve. By computing the time required for the Higgs field to evolve, the amount of inflation can be determined. An expansion factor of 10^{50} or more is quite plausible, but the actual factor depends on the details of the particle theory one adopts.

So far the description of the phase transition has been slightly oversimplified. There are actually many different broken-symmetry states, just as there are many possible orientations for the axes of a crystal. There are a number of Higgs fields, and the various broken-symmetry states are distinguished by the combination of Higgs fields that acquire nonzero values. Since the fluctuations that drive the Higgs fields from zero are random, different regions of the primordial universe would be driven toward different broken-symmetry states, each region forming a domain with an initial radius of roughly the horizon distance. At the start of the phase transition the horizon distance would be about 10^{-24} centimeter. Once the domain formed, with the Higgs fields deviating slightly from zero in a definite combination, it would evolve toward one of the stable broken-symmetry states and would inflate by a factor of 10^{50} or more. The size of the domain after inflation would then be greater than 10^{26} centimeters. The entire observable universe, which at that time would be only about 10 centimeters across, would be able to fit deep inside a single domain.

In the course of this enormous inflation any density of particles that might have been present initially would be diluted to virtually zero. The energy content of the region would then consist entirely of the energy stored in the Higgs field. How could this energy be released? Once the Higgs field evolved away from the flat part of the energy-density curve, it would start to oscillate rapidly about the true-vacuum value. Drawing on the relation between particles and fields implied by quantum field theory, this situation can also be described as a state with a high density of Higgs particles. The Higgs particles would be unstable, however: they would rapidly decay to lighter particles, which would interact with one another and possibly undergo subsequent decays. The system would quickly become a hot gas of elementary particles in thermal equilibrium, just as was assumed in the initial conditions for the standard model. The reheating temperature is calculable and is typically a factor of between two and 10 below the critical temperature of the phase transition. From this point on, the scenario coincides with that of the standard big-bang model, and so all the successes of the standard model are retained.

Note that the crucial flaw of the original inflationary model is deftly avoided. Roughly speaking, the isolated bubbles that were discussed in the original model are replaced here by the domains. The domains of the slow-rollover transition would be surrounded by other domains rather than by false vacuum, and they would tend not to be spherical. The term "bubble" is therefore avoided. The key difference is that in the new inflationary model each domain inflates in the course of its formation, producing a vast essentially homogeneous region within which the observable universe can fit.

Since the reheating temperature is near the critical temperature of the grand unified-theory phase transition, the matter-antimatter asymmetry could be produced by particle interactions just after the phase transition. The production mechanism is the same as the one predicted by grand unified theories for the standard big-bang model. In contrast to the standard model, however, the inflationary model does not allow the possibility of assuming the observed net baryon number of the universe as an initial condition; the subsequent inflation would dilute any initial baryon number density to an imperceptible level. Thus the viability of the inflationary model depends crucially on the viability of particle theories, such as the grand unified theories, in which baryon number is not conserved.

One can now grasp the solutions to the cosmological problems discussed above. The horizon and flatness problems are resolved by the same mechanisms as in the original inflationary universe model. In the new inflationary scenario the problem of monopoles and domain walls can also be solved. Such defects would form along the boundaries separating domains, but the domains would have been inflated to such an enormous size that the defects would lie far beyond any observable distance. (A few defects might be generated by thermal effects after the transition, but they are expected to be negligible in number.)

Thus with a few simple ideas the improved inflationary model of the universe leads to a successful resolution of several major problems that plague the standard big-bang picture: the horizon, flatness, magnetic monopole and domain-wall problems. Unfortunately the necessary slow-rollover transition requires the fine tuning of parameters; calculations yield reasonable predictions only if the parameters are assigned values in a narrow range. Most theorists (including both of us) regard such fine tuning as implausible. The consequences of the scenario are so successful, however, that we are encouraged to go on in the hope we may discover realistic versions of grand unified theories in which such a slow-rollover transition occurs without fine tuning.

The successes already discussed offer persuasive evidence in favour of the new inflationary model. Moreover, it was recently discovered that the model may also resolve an additional cosmological problem not even considered at the time the model was developed: the smoothness problem. The generation of density inhomogeneities in the new inflationary universe was addressed in the summer of 1982 at the Nuffield Workshop on the Very Early Universe by a number of theorists, including James M. Bardeen of the University of Washington, Stephen W. Hawking of the University of Cambridge, So-Young Pi of Boston University, Michael S. Turner of the University of Chicago, A. A. Starobinsky of the L. D. Landau Institute of Theoretical Physics in Moscow, and the two of us. It was found that the new inflationary model, unlike any previous cosmological model, leads to a definite prediction for the spectrum of inhomogeneities. Basically the process of inflation first smoothes out any primordial inhomogeneities that might have been present in the initial conditions. Then in the course of the phase transition inhomogeneities are generated by the quantum fluctuations of the Higgs field in a way that is completely determined by the underlying physics. The inhomogeneities are created on a very small scale of length, where quantum phenomena are important, and they are then enlarged to an astronomical scale by the process of inflation.

The predicted shape for the spectrum of inhomogeneities is essentially scale-invariant: that is, the magnitude of the inhomogeneities is approximately equal on all length scales of astrophysical significance. This prediction is comparatively insensitive to the details of the underlying grand unified theory. It turns out that a spectrum of precisely this shape was proposed in the early 1970s as a phenomenological model for galaxy formation by Edward R. Harrison at the University of Massachusetts at Amherst and Yakov B. Zeldovich of the Institute of Physical Problems in Moscow, working independently. The details of galaxy formation are complex and are still not well understood, but many cosmologists think a scale-invariant spectrum of inhomogeneities is precisely what is needed to explain how the present structure of galaxies and galactic clusters evolved. (See "The Large-Scale Structure of the Universe," by Joseph Silk, Alexander S. Szalay and Yakov B. Zeldovich; *Scientific American*, October 1983.)

The new inflationary model also predicts the magnitude of the density inhomogeneities, but the prediction is quite sensitive to the details of the underlying particle theory. Unfortunately the magnitude that results from the simplest grand unified theory is far too large to be consistent with the observed uniformity of the cosmic microwave background. This inconsistency represents a problem, but it is not yet known whether the simplest grand unified theory is the correct one. In particular the simplest grand unified theory predicts a lifetime for the proton that appears to be lower than present experimental limits. On the other hand, one can construct more complicated grand unified theories that result in density inhomogeneities of the desired magnitude. Many investigators imagine that with the development of the correct particle theory the new inflationary model will add the resolution of the smoothness problem to its list of successes.

One promising line of research involves a class of quantum field theories with a new kind of symmetry called supersymmetry. Supersymmetry relates the properties of particles with integer angular momentum to those of particles with half-integer angular momentum; it thereby highly constrains the form of the theory. Many theorists think supersymmetry might be necessary to construct a consistent quantum theory of gravity, and to eventually unify gravity with the strong, the weak, and the electromagnetic forces. A tantalizing property of models incorporating supersymmetry is that many of them give slow-rollover phase transitions without any fine tuning of parameters. The search is on to find a supersymmetry model that is realistic as far as particle physics is concerned and that also gives rise to inflation and to the correct magnitude for the density inhomogeneities.

In short, the inflationary model of the universe is an economical theory that accounts for many features of the observable universe lacking an explanation in the standard big-bang model. The beauty of the inflationary model is that the evolution of the universe becomes almost independent of the details of the initial conditions, about which little if anything is known. It follows, however, that if the inflationary model is correct, it will be difficult for anyone to ever discover observable consequences of the conditions existing before the inflationary phase transition. Similarly, the vast distance scales created by inflation would make it essentially impossible to observe the structure of the universe as a whole. Nevertheless, one can still discuss these issues, and a number of remarkable scenarios seem possible.

The simplest possibility for the very early universe is that it actually began with a big bang, expanded rather uniformly until it cooled to the critical temperature of the phase transition and then proceeded according to the inflationary scenario. Extrapolating the big-bang model back to zero time brings the universe to a cosmological singularity, a condition of infinite temperature and density in which the known laws of physics do not apply. The instant of creation remains unexplained. A second possibility is that the universe began (again without explanation) in a random, chaotic state. The matter and temperature distributions would be nonuniform, with some parts expanding and other parts contracting. In this scenario certain small regions that were hot and expanding would undergo inflation, evolving into huge regions easily capable of encompassing the observable universe. Outside these regions there would remain chaos, gradually creeping into the regions that had inflated. . . .

From a historical point of view probably the most revolutionary aspect of the inflationary model is the notion that all the matter and energy in the observable universe may have emerged from almost nothing. This claim stands in marked contrast to centuries of scientific tradition in which it was believed that something cannot come from nothing. The tradition, dating back at least as far as the Greek philosopher Parmenides in the fifth century B.C., has manifested itself in modern times in the formulation of a number of conservation laws, which state that certain physical quantities cannot be changed by any physical process. A decade or so ago the list of quantities thought to be conserved included energy, linear momentum, angular momentum, electric charge and baryon number.

Since the observed universe apparently has a huge baryon number and a huge energy, the idea of creation from nothing has seemed totally untenable to all but a few theorists. (The other conservation laws mentioned above present no such problems: the total electric charge and the angular momentum of the observed universe have values consistent with zero, whereas the total linear momentum depends on the velocity of the observer and so cannot be defined in absolute terms.) With the advent of grand unified theories, however, it now appears quite plausible that baryon number is not conserved. Hence only the conservation of energy needs further consideration.

The total energy of any system can be divided into a gravitational part and a nongravitational part. The gravitational part (that is, the energy of the gravitation field itself) is negligible under laboratory conditions, but cosmologically it can be quite important. The nongravitational part is not by itself conserved; in the standard big-bang model it decreases drastically as the early universe expands, and the rate of energy loss is proportional to the pressure of the hot gas. During the era of inflation, on the other hand, the region of interest is filled with a false vacuum that has a large negative pressure. In this case the nongravitational energy increases drastically. Essentially all the nongravitational energy of the universe is created as the false vacuum undergoes its accelerated expansion. This energy is released when the phase transition takes place, and it eventually evolves to become stars, planets, human beings and so forth. Accordingly, the inflationary model offers what is apparently the first plausible scientific explanation for the creation of essentially all the matter and energy in the observable universe.

Under these circumstances the gravitational part of the energy is somewhat ill-defined, but crudely speaking one can say that the gravitational energy is negative, and that it precisely cancels the nongravitational energy. The total energy is then zero and is consistent with the evolution of the universe from nothing.

If grand unified theories are correct in their prediction that baryon number is not conserved, there is no known conservation law that prevents the observed universe from evolving out of nothing. The inflationary model of the universe provides a possible mechanism by which the observed universe could have evolved from an infinitesimal region. It is then tempting to go one step further and speculate that the entire universe evolved from literally nothing.

15

— George Smoot —

(1945–)

and Keay Davidson

(1953–)

George Smoot, an American astronomer, led a team of investigators in a project designed to test Inflationary Theory. In 1992 the team discovered ripples in the cosmic background microwave radiation, as Inflationary Theory had predicted. Keay Davidson collaborated with Smoot in the writing of Wrinkles in Time, *a book describing the project.*

In this reading from Wrinkles in Time, *Smoot describes in detail the COBE satellite experiments and their results, and gives a fascinating look at the process and personalities involved in a scientific project. Smoot also gives the reader his philosophical reflections on the meaning of the results of the experiment.*

WRINKLES IN TIME

The search for the wrinkles had been a long endeavor, stretching back more than fifteen years; and the previous two years of the inexorable search for systematic errors inevitably precluded a Eureka-like experience. It was more like excavating a buried city, beginning by uncovering a first hint of interesting structure as the first layer of sediment is brushed aside. As further layers are removed, the form of the city's edifices begins to take shape, but slowly. Only when the final layers have been removed do we see the overall configuration of all the buildings. That evening I had the feeling that we were poised to sweep away the final obscuring layers on our quest for wrinkles.

All along I had been making simulations of how the wrinkles should look if the inflationary big bang theory was correct. There would be wrinkles of all sizes, from large to small, but with the same average area of sky occupied by each size class and the same average variation in amplitude—this is known as scale invariance. Inflation theory predicted such a size distribution, the products of quantum fluctuations at the instant of creation. This kind of size distribution of wrinkles would lead to the formation of structures of different sizes that we see in the universe today.

I ran the new program on these simulations as well as on our data. The results from the simulations were distinctive. And when I compared them with the results of subtracting the quadrupole from the map of the background radiation, it was clear that they matched the scale invariance predicted by inflation theory. The comparison was beautiful in an esoteric-modern-art way. Was I happy? Yes. Ecstatic, certainly. But not really surprised. By this time, my mind was so immersed in the data that the answer was obvious, if not consciously so. My immediate reaction was to shout it out—it looked so beautiful. A kaleidoscope of ideas was sorting itself out in my mind, making a perfect pattern. The big bang was correct; inflation theory worked; the pattern of wrinkles was about right for structure formation by cold dark matter; and the size distribution would yield the major structures of today's universe, under gravitational collapse through 15 billion years.

The pattern of wrinkles I saw on the map was primordial—I knew it in my bones. Some of the structures represented by the wrinkles were so large that they could only have been generated at the birth of the universe, not later on. I was staring at primordial wrinkles in time, the imprint of creation and the seeds of the modern universe. . . .

I planned to move toward the announcement in steps, because it was logical to do so and it added drama to the event. I was having some fun after not talking for so long. First I reminded the audience of the COBE result from two years earlier, that the cosmic background radiation matched the characteristics of blackbody radiation and therefore lent credence to big bang theory. I held up the placard labeled $L = 0$, which showed the blackbody spectrum, and then I passed it to John Mather. Next, I said that the differential microwave radiometer had precisely measured the dipole, which we had discovered and I had announced exactly fifteen years earlier. I held up the next placard, labeled $L = 1$, which showed the dipole: This demonstrated that the instrument was working properly. Then I passed it to Jon Aymon. Then, pausing just slightly for effect, I held up the next, $L = 2$, and announced, "We have a quadrupole." The sense of relief in the auditorium was palpable, as everyone knew the import of what I was saying. I passed it down to Chuck Bennett, pointing out that he would say more about the quadrupole in his talk.

Even those in the audience who had guessed that our big announcement was going to be the discovery of the quadrupole did not anticipate that we had gone even further. I held up the $L = 3$, $L = 4$, and $L = 5$ placards simultaneously, calling on the team members to hold up more placards, with ever-increasing L numbers, all the way up to $L = 20$. That was clearly a shock to most people, as we revealed that not only had we found the quadrupole, which would have been momentous, but that we had also detected a spectrum of wrinkles of all sizes.

For me, the moment marked the culmination of an eighteen-year search, and for cosmology a major milestone on the long journey to understanding the nature of the universe. Very simply, the discovery of the wrinkles salvaged big bang theory at a time when detractors were attacking in increasing numbers. The result indicated that gravity could indeed have shaped today's universe from the tiny quantum fluctuations that occurred in the first fraction of a second after creation. When Stephen Hawking later commented that we had made "the most important discovery of the century, if not of all time," he may have been exaggerating but it was still momentous. Before the discovery, our understanding of the origin and history of the universe rested on four major observations: first, the darkness of the night sky; second, the composition of the elements, with the great preponderance of hydrogen and helium over the heavier elements; third, the expansion of the universe; fourth, the existence of the cosmic background radiation, the afterglow of a fiery creation.

The discovery of the wrinkles that were present in the fabric of time at three hundred thousand years after creation becomes the fifth pillar in this intellectual edifice and gives us a way of understanding how structures of all sizes, from galaxies to superclusters, could have formed as the universe evolved during the past 15 billion years.

The evolution of the universe is effectively the change in distribution of matter through time—moving from a virtual homogeneity in the early universe to a very lumpy universe today, with matter condensed as galaxies, clusters, superclusters, and even larger structures. We can view that evolution as a series of phase transitions, in which matter passes from one state to another under the influence of decreasing temperature (or energy). We are all familiar with the way that steam, on cooling, condenses: This is a phase transition from a gaseous to a liquid state. Reduce the temperature further, and eventually the water freezes, making a phase transition from the liquid to the solid state. In the same way, matter has gone through a series of phase transitions since the first instant of the big bang.

At a ten-millionth of a trillionth of a trillionth of a trillionth (10^{-42}) of a second after the big bang—the earliest moment about which we can sensibly talk, and then only with some suspension of disbelief—all the universe we can observe today was the tiniest fraction of the size of a proton. Space and time had only just begun. (Remember, the universe did not expand *into* existing space after the big bang; its expansion created space-time as it went.) The temperature at this point was a hundred million trillion trillion (10^{32}) degrees, and the three forces of nature—electromagnetism and the strong and weak nuclear forces—were fused as one. Matter was undifferentiated from energy, and particles did not yet exist.

By a ten-billionth of a trillionth of a trillionth of a second (10^{-34} second) in-flation had expanded the universe (at an accelerating rate) a million trillion trillion (10^{30}) times, and the temperature had fallen to below a billion billion billion (10^{27}) degrees. The strong nuclear force had separated, and matter under-went its first phase transition, existing now as quarks (the building blocks of protons and neutrons), electrons, and other fundamental particles.

The next phase transition occurred at a ten-thousandth of a second, when quarks began to bind together to form protons and neutrons (and antiprotons and antineutrons). Annihilations of particles of matter and antimatter began, eventually leaving a slight residue of matter. All the forces of nature were now separate.

The temperature had fallen sufficiently after about a minute to allow protons and neutrons to stick together when they collided, forming the nuclei of hydro-gen and helium, the stuff of stars. This soup of matter and radiation, which initially was the density of water, continued expanding and cooling for another three hundred thousand years, but it was too energetic for electrons to stick to the hydrogen and helium nuclei to form atoms. The energetic photons existed in a frenzy of interactions with the particles in the soup. The photons could travel only a very short distance between interactions. The universe was essen-tially opaque.

When the temperature fell to about 3,000 degrees, at three hundred thousand years, a crucial further phase transition occurred. The photons were no longer energetic enough to dislodge electrons from around hydrogen and helium nuclei and so atoms of hydrogen and helium formed and stayed together. The photons no longer interacted with the electrons and were free to escape and travel great distances. With this decoupling of matter and radiation, the universe became transparent, and radiation streamed in all directions—to course through time as the cosmic background radiation we experience still. The radiation released at that instant gives us a snapshot of the distribution of matter within the universe at three hundred thousand years of age. Had all matter been distributed evenly, the fabric of space would have been smooth, and the interaction of photons with particles would have been homogeneous, resulting in a completely uniform cosmic background radiation. Our discovery of the wrinkles reveals that matter was not uniformly distributed, that it was already structured, thus forming the seeds out of which today's complex universe has grown.

Those regions of the universe with a higher concentration of matter exerted more gravitational attraction and therefore curved space positively; less dense areas had less gravitational attraction, resulting in less curvature of space. When radiation and matter decoupled three hundred thousand years later, the suddenly released flux of cosmic background photons bore the imprint of these distortions of space, showing the wrinkles we see in our maps: Radiation traveling from the denser areas looks cooler than the average background; that from less dense, warmer.

Matter in the universe, as we saw earlier, is of two kinds—dark matter and visible matter—and their role in gravitational formation of structure is different. Dark matter, which by its nature is unaffected by radiation but responsive to gravity, would have started forming structure much earlier than visible matter, which is buffeted by the energetic flux of photons. Molded by the contours of space that originated as quantum fluctuations in the inflationary universe, dark matter could have begun to aggregate, under the influence of gravity, as early as ten thousand years after the big bang. At three hundred thousand years, the decoupling of matter and radiation freed ordinary visible matter to be attracted to the structures formed by dark matter. As the visible matter aggregates, stars and galaxies form. A nice image here is the way cobwebs, often unseen in ordinary light, become strikingly visible when dew that settles on their strands during the night is lit by the morning sun. The gossamer network of galaxies we see in the night sky is the shimmering dew on a cosmic cobweb, as visible matter outlines the shape of structures of invisible dark matter, to which it has been drawn by gravitational attraction.

Because of the limitations of the differential microwave radiometer aboard COBE, the resolution of our maps is relatively poor. The smallest objects we can see as wrinkles are enormous, leading to structures as large as or larger than the Great Wall, a vast, concentrated sheet of galaxies stretching many hundreds of millions of light-years across. When we can achieve greater resolution, I expect us to be able to discern structures the size of galaxies. Despite current limitations, however, the message of our results—the message that engendered so much relief among cosmologists that April day—was clear. Fred Hoyle once claimed that big bang theory failed because it could not account for the early formation of galaxies. The COBE results prove him wrong. The existence of the wrinkles in time as we see them tell us that big bang theory, incorporating the effect of gravity, can explain not only the early formation of galaxies but also the aggregation within 15 billion years of the massive structures we know to be present in today's universe. This is a triumph for theory and observation.

We did not anticipate the extent of the reaction to our announcement, both immediately and in the subsequent weeks. I knew the occasion would be important for cosmologists. When Gary Hinshaw passed me on the way to the podium to give the talk after mine he muttered, "How am I supposed to follow *that*!" Later, when astrophysicist Sun Rhie rose to give her paper on topological defects, she opened by saying, "After the COBE announcement, I don't known if I should bother to give this talk." Just as our discovery supported gravitational accretion from primordial fluctuations and inflation, it undermined competing theories, particularly the theory of topological defects. This theory predicted a very different pattern of size and number of wrinkles from what we observed, and so was falsified. David Spergel, coauthor with Neil Turok of the textures version of topological defect theory, declared after the session, "We're dead." Paul Steinhardt, a leading inflation theorist said, "I'm glad I was here for this historic day. . . . I wish I had brought my students."

This was gratifying, of course, particularly when I was finally able to see the presentations by the MIT and South Pole teams. Although Lyman Page described the MIT data as contradicting COBE's, I could see from his final viewgraph that wasn't the case. They claimed there were no wrinkles, but it was clear to me that the correlation curve was consistent with COBE's data. A wave of relief passed over me as I realized there would be no controversy over this. And when Todd Gaier gave the South Pole results with his supervisor, Phil Lubin, looking on, they came close to disagreeing with us, but there were many possible explanations for the discrepancies. We were home free—for that day, at any rate.

I had a few moments to grab a sandwich, think about the press conference scheduled at noon, and search for the viewgraph that we had made up to explain our results to the press. It was delivered at the last moment fresh from the drawing board.

Then we walked down to the room where the press conference was to be held. There was a huge crowd there; in the back of the room a row of television cameras and a bank of lights, making me blink with more than just surprise. As Ned Wright, Chuck Bennett, Al Kogut, and I got up on the podium, I started to realize the immense public interest in our new results. I was still composing myself to express now, both to the press and the television simultaneously, the meaning of our results in words and metaphors that would be readily understandable.

I was first to speak and tried to present the high-level facts and significance of what we had done: "We had observed the oldest and largest structures ever seen in the early universe," I began. "These were the primordial seeds of modern-day structures such as galaxies, clusters of galaxies, and so on. Not only that, but they represented huge ripples in the fabric of space-time left from the creation period." Many other questions were raised, but primarily they centered on two issues: How big were the structures? and, What was the significance of our results?

I answered that superlatives failed to convey a real sense of these structures, which stretched across such vast distances that they dwarfed the Great Wall, itself hundreds of millions of light-years in extent. Some were so large that light had not yet traveled across them in the age of the universe. How fundamental? I gave many comparisons, but the one picked up most by the press was: "If you're religious, it's like seeing God." The big bang is a cultural icon, a scientific explanation of the creation. Out of forty minutes of questions from newspaper and television journalists, with repeated requests to rephrase what the COBE results meant to laypeople, that single line was the most quoted and remembered.

In cosmology there is a confluence of physics, metaphysics, and philosophy—when inquiry approaches the ultimate question of our existence, the lines between them inevitably become blurred. Einstein, who was devoted to a rational explanation of the world, once said: "I want to know how God created the world. I want to know his thoughts." He meant it metaphorically, as a measure of the profundity of his quest. My own much-quoted remark was cast in the same mold.

Metaphorically meant or not, my remarks, and the comments of other COBE team members and other cosmologists, appeared in newspapers around the world, affirming the deep public interest in understanding the origin of the universe and our place in it. Ned Wright spoke after me, explaining in more detail what our results meant in terms of theories of structure formation and the need for dark matter. Chuck Bennett gave a brief statement about the results and interpretation and answered questions. Finally, Al Kogut reviewed the efforts we had taken to process the data carefully and ensure that we had removed all instrumental effects. Al explained the great difficulty of the experiment. After Al finished, the conference was opened to general questions. The press conference went on and on until finally it was called to a halt after a couple of hours—whereupon it broke up into small knots of reporters interviewing us, other COBE team members, and people like Alan Guth and Paul Steinhardt, authors of inflation theory.

In a while Philip Schewe and the other American Physical Society press people led us back to the conference press office so that we could take calls and interviews from reporters not present. They set up six of us around the room to do radio, phone, TV, newspapers, and magazine interviews, and it went on with no letup through the afternoon into the evening. The only brief event I remember was around 4:30 or 5:00 P.M. I finished an interview and called across the noisy room to Philip Schewe, "Do you have press-piranha feedings like this very often?" Philip answered, "The only time I have ever seen it like this was for the cold fusion announcement." Suddenly, for a moment, the whole room was quiet. Everyone seemed to be thinking, What if this is wrong? What if we made a mistake? Then it was back to ringing phones and interviews.

Various COBE team members and other cosmologists were on TV, radio talk shows, and in newspapers for several days. The publicity and tremendous public interest provided a unique opportunity to discuss science with a very large audience and to promote the power of human endeavor in pursuing the mysteries of nature. I recognized from the experience people's profound hunger for the metaphors of creation, even if the science itself is intellectually taxing.

A powerful conviction for me, and one that I believe encourages confidence that one day we will understand the very essence of creation, is the idea that as we converge on the moment of creation, the constituents and laws of the universe become ever simpler. A useful analogy here is life itself, or, more simply, a single human being. Each of us is a vastly complex entity, assembled from many different tissues and capable of countless behaviors and thoughts. Trace that person back through his or her life, back beyond birth and finally to the moment of fertilization of a single ovum by a single sperm. The individual becomes ever simpler, ultimately encapsulated as information encoded in DNA in a set of chromosomes. The development that gradually transforms a DNA code into a mature individual is an unfolding, a complexification, as the information in the DNA is translated and manifested through many stages of life. So, I believe, it is with the universe. We can see how very complex the universe is now, and we are part of that complexity.

Cosmology—through the marriage of astrophysics and particle physics—is showing us that this complexity flowed from a deep simplicity as matter metamorphosed through a series of phase transitions. Travel back in time through those phase transitions, and we see an ever-greater simplicity and symmetry, with the fusion of the fundamental forces of nature and the transformation of particles to ever-more fundamental components. Go back further and we reach a point when the universe was nearly an infinitely tiny, infinitely dense concentration of energy, a fragment of primordial space-time. This increasing simplicity and symmetry of the universe as we near the point of creation gives me hope that we can understand the universe using the powers of reason and philosophy. The universe would then be comprehensible, as Einstein has yearned.

Go back further still, beyond the moment of creation—what then? What was there before the big bang? What was there before time began? Facing this, the ultimate question, challenges our faith in the power of science to find explanations of nature. The existence of a singularity—in this case the given, unique state from which the universe emerged—is anathema to science because it is beyond explanation. There can be no answer to *why* such a state existed. Is this, then, where scientific explanation breaks down and God takes over, the artificer of that singularity, that initial simplicity? The astrophysicist Robert Jastrow, in his book *God and the Astronomers*, described such a prospect as the scientist's nightmare: "He has scaled the mountains of ignorance; he is about to conquer the highest peak; as he pulls himself over the final rock, he is greeted by a band of theologians who have been sitting there for centuries."

Cosmologists have long struggled to avoid this bad dream by seeking explanations of the universe that avoid the necessity of a beginning. Einstein, remember, refused to believe the implication of his own equations—that the universe is expanding and therefore must have had a beginning—and invented the cosmological constant to avoid it. Only when Einstein saw Hubble's observations of an expanding universe could he bring himself to believe his equations. For many proponents of the steady state theory, one of its attractions was its provision that the universe had no beginning and no end, and therefore required no explanation of what existed before time = 0. It was known as the perfect cosmological principle.

A decade ago Stephen Hawking and Jim Hartle tried to resolve the challenge differently, by arguing the singularity out of existence. Flowing from an attempt at a theory of quantum gravity, they agreed that time is finite, but without a beginning. This is not as bizarre as it sounds, if you think of the surface of a sphere. The surface is finite, but it has no beginning or end—you can trace your finger over it continuously, perhaps finishing up where you began. Suppose the universe is a sphere of space-time. Travel around the surface, and again you may finish up where you started both in space and time. This, of course, requires time travel, in violation of Mach's principle. But the world of quantum mechanics, with its uncertainty principle, is an alien place in which otherworldly things can happen. It is so foreign a place that it may even be beyond human understanding, children as we are of a world of classical Newtonian mechanics.

We simply do not know yet whether there was a beginning of the universe, and so the origin of space-time remains in terra incognita. No question is more fundamental or more magical, whether cast in scientific or theological terms. My conviction—perhaps I should say my faith—is that science will continue to move ever closer to the moment of creation, facilitated by the ever-greater simplicity we find there. Some physicists argue that matter is ultimately reducible to pointlike objects with certain intrinsic properties. Others argue that fundamental particles are extraordinarily tiny strings that vibrate to produce their properties. Either way, in combination with certain concepts such as inflation, it is possible to envisage creation of the universe from almost nothing—not nothing, but practically nothing. Almost creation *ex nihilo* but not quite. That would be a great intellectual achievement, but it may still leave us with a limit to how far scientific inquiry can go, finishing with a description of the singularity, but not an explanation of it.

To an engineer, the difference between nothing and practically nothing might be close enough. To a scientist and certainly to a philosopher, such a difference, however miniscule, would be everything. We might find ourselves experiencing Jastrow's bad dream, facing a final question: Why? "Why" questions are not amenable to scientific inquiry and will always reside within philosophy and theology, which may provide solace if not material explication.

But what if the universe we see were the only one possible, the product of a singular initial state shaped by singular laws of nature? By now it is clear that the minutest variation in the value of a series of fundamental properties of the universe would have resulted in no universe at all, or at least a very alien universe. For instance, if the strong nuclear force had been slightly weaker, the universe would have been composed of hydrogen only; slightly stronger, and all the hydrogen would have been converted to helium. Slight variation in the excess of protons over antiprotons—one billion and one to one billion—might have produced a universe with no baryonic matter or cataclysmic plentitude of it. Had the expansion rate of the universe one second after the big bang been smaller by one part in a hundred thousand trillion, the universe would have recollapsed long ago. An expansion more rapid by one part in a million would have excluded the formation of stars and planets.

The list of cosmic coincidences required for our existence in this universe is long, moving Stephen Hawking to remark that "the odds against a universe like ours emerging out of something like the big bang are enormous." Princeton physicist Freeman Dyson went further, and said: "The more I examine the universe and the details of its architecture, the more evidence I find that the universe in some sense must have known we were coming." This concatenation of coincidences required for our presence in this universe has been termed the anthropic principle. In fact, it is merely a statement of the obvious: Had things been different, we would not exist. It may be that many different universes are possible, and many may exist in parallel with our own. Inflation theory can be interpreted in this way, with our universe budding off a larger fabric of space-time—like one strawberry in a patch of many strawberries. My speculation, however, is that because things become simpler as we near the moment of creation, there was only a limited range of possibilities; indeed, perhaps only one, with everything so perfect that it could have been no other way.

In this case, what could we say about the ultimate question? That God had no choice in how the universe would be, and therefore need not exist? Or that God was very smart, and got it just right? In any case science would still be left contemplating the question: Why these conditions and not others? Or perhaps the comprehensibility of the universe in these terms is sufficient explanation. The truth and treasure of the universe is its own existence, and our quest for that truth and treasure will be eternal, like the universe itself.

Our discovery of the wrinkles in the fabric of time is part of that eternal quest and marks an important step forward in this golden age of cosmology. Suddenly, pieces of a larger puzzle begin to fall together: Inflation looks stronger, and dark matter more real. Our faith in the big bang is revitalized: To the dark night sky, the composition of the elements, the evidence of an expanding universe, and the afterglow of creation is added a means by which the structures of today's universe could have formed. The creativity of the universe is its most potent force, forming through time the matter and structures of stars and galaxies, and, ultimately, us. The wrinkles are the core of that creativity, assembling structure from homogeneity.

The quest will continue, with the dual goals of discovering dark matter and understanding the origin of space-time. No one knows where the answers will come from, but if recent history is any guide, a combination of observational science and particle physics will be important. More satellite ventures are planned, and the superconducting supercollider (SSC), one of the boldest scientific endeavors of all time, is under construction in Texas. Without doubt it is expensive—perhaps as much as ten billion dollars. But the history of this science has been driven by the development of instrumentation. There may be no other way of understanding the first instants of creation than by re-entering big bangs or sorts in a massive machine like the fifty-four-mile circumference supercollider. As a culture, we have to decide whether the intellectual quest toward the ultimate question is worthwhile. The more we know about the history of the universe, the more we know about ourselves and the questions we are driven to ask.

In 1977, Steven Weinberg published *The First Three Minutes*, one of the finest popular books on cosmology ever written, and justly still in print. His book was based on a course he taught about gravitation and cosmology at MIT while I was a graduate student there. His class influenced my decision to enter cosmological research. Toward the end of his book, Weinberg muses on the questions we ask ourselves, particularly the conviction that, somehow, humans are not a mere cosmic accident, the chance outcome of a concatenation of physical processes in a universe that dwarfs us on every scale. He expresses his view on the matter this way: "It is very hard to realize that [this beautiful Earth] is all just a tiny part of an overwhelmingly hostile universe. It is even harder to realize that this present universe has evolved from an unspeakably unfamiliar early condition, and faces a future extinction of endless cold or intolerable heat. The more the universe seems comprehensible, the more it also seems pointless."

I must disagree with my old teacher. To me the universe seems quite the opposite of pointless. It seems that the more we learn, the more we see how it all fits together—how there is an underlying unity to the sea of matter and stars and galaxies that surround us. Likewise, as we study the universe as a whole, we realize that the "microcosm" and the "macrocosm" are, increasingly, the same subject. By unifying them, we are learning that nature is as it is not because it is the chance consequence of a random series of meaningless events; quite the opposite. More and more, the universe appears to be as it is because it *must* be that way; its evolution was written in its beginning—in its cosmic DNA, if you will. There is a clear order to the evolution of the universe, moving from simplicity and symmetry to greater complexity and structure. As time passes, simple components coalesce into more sophisticated building blocks spawning a richer, more diverse environment. Accidents and chance, if fact, are essential in developing the overall richness of the universe. In that sense (although not in the sense of quantum physics), Einstein had the right idea: God does not play dice with the universe. Though individual events happen as a matter of chance, there is an overall inevitability to the development of sophisticated complex systems. The development of beings capable of questioning and understanding the universe seems quite natural. I would be quite surprised if such intelligence has not arisen many places in our very large universe.

As I travel the world, I love to visit great art museums, to see classic sculptures, the works carved and painted and assembled by centuries of aesthetic visionaries. Cosmologists and artists have much in common: Both seek beauty, one in the sky and the other on canvas or in stone. When a cosmologist perceives how the laws and principles of the cosmos begin to fit together, how they are intertwined, how they display a symmetry that ancient mythologies reserved for their gods—indeed, how they imply that the universe *must* be expanding, *must* be flat, *must* be all that it is—then he or she perceives pure, unadulterated beauty.

The religious concept of creation flows from a sense of wonder at the existence of the universe and our place in it. The scientific concept of creation encompasses no less a sense of wonder: We are awed by the ultimate simplicity and power of the creativity of physical nature—and by its beauty on all scales.

16
— Stephen Hawking —
(1942–)

Stephen Hawking is a British theoretical physicist and the Lucasian Professor of Mathematics at Cambridge University. Hawking has made several significant contributions to Big Bang Cosmology, including the notion that the actual Big Bang was a space-time singularity, that is, a point of indefinitely high density and space-time curvature. Hawking has suggested that space-time and matter actually came into existence with the Big Bang.

Hawking has also done extensive work in the development of black hole theory and in the attempt to bring quantum mechanics into the field of cosmology.

In this reading Hawking combines Quantum Theory and Relativity Theory in the development of a "no boundary condition" cosmological theory. Hawking precedes his theory with a brief but insightful history of cosmology. The passage is taken from an article that appeared in The New Physics, *edited by Paul Davies.*

THE EDGE OF SPACETIME

The Big Bang Singularity

In the Friedmann model which recollapses eventually, space is finite but unbounded, like in the Einstein static model. In the other two Friedmann models, which expand forever, space is infinite. Time, on the other hand, has a boundary or edge in all three models. The expansion starts from a state of infinite density called the Big Bang singularity. The best way of visualising the Big Bang is to imagine a movie of the expanding universe being played backwards in time. A given spherical region of the universe then shrinks, more and more rapidly, until its radius reaches zero. At this juncture, all the matter and energy that was contained in that spherical volume of space will be compressed into a single point, or singularity. In this idealised model, the entire observable universe is considered to have started out compressed into such a point. Moreover, in the model which recollapses, the universe eventually returns to a singularity at the end—the so-called Big Crunch.

If one takes the singularities in these idealised models seriously, then some profound conclusions follow. Because of the infinite compression of matter and energy, the curvature of spacetime is infinite at the Friedmann singularities too. Under these circumstances the concepts of space and time cease to have any meaning. Moreover, because all present scientific theories are formulated on a spacetime background, all such theories will break down at these singularities. So, if there were events before the Big Bang, they would not enable one to predict the present state of the universe because predictability would break down at the Big Bang. Similarly, there is no way that one can determine what happened before the Big Bang from a knowledge of events after the Big Bang. This means that the existence or non-existence of events before the Big Bang is purely metaphysical; they have no consequences for the present state of the universe. One might as well apply the principle of economy, known as Occam's razor, to cut them out of the theory and say that time began at the Big Bang. Similarly, there is no way that we can predict or influence any events after the Big Crunch, so one might as well regard it as the end of time.

This beginning and possible end of time that are predicted by the Friedmann solutions are very different from earlier ideas. Prior to the Friedmann solutions, the beginning or end of time was something that had to be imposed from outside the universe; there was no necessity for a beginning or an end. In the Friedmann models, on the other hand, the beginning and end of time occur for dynamical reasons. One could still imagine the universe being created by an external agent in a state corresponding to some time after the Big Bang, but it would not have any meaning to say that it was created *before* the Big Bang. From the present rate of expansion of the universe we can estimate that the Big Bang should have occurred between ten- and twenty-thousand-million years ago.

Many people disliked the idea that time had a beginning or will have an end because it smacked of divine intervention. There were therefore a number of attempts to avoid this conclusion. One of these was the "steady state" model of the universe proposed in 1948 by Herman Bondi, Thomas Gold and Fred Hoyle. In this model it was suggested that, as the galaxies moved further away from each other, new galaxies were formed in between out of matter that was being "continually created." The universe would therefore look more or less the same at all time and the density would be roughly constant. This model had the great virtue that it made definite predictions that could be tested by observations. Unfortunately, observations of radio sources by Martin Ryle and his collaborators at Cambridge in the 1950s and early 1960s showed that the number of radio sources must have been greater in the past, contradicting the steady state model.

The final nail in the coffin of the steady state model came with the discovery of the microwave background radiation by Penzias and Wilson, which appears to bathe the entire cosmos uniformly. Measurements of the spectrum of this radiation reveal it to carry the unmistakable thumbprint of thermal equilibrium, i.e. it is a black body spectrum, corresponding to a temperature a little below 3 K (see figure 4.8). This thermal radiation has a very natural explanation in the Big Bang theory—it is a relic of the primeval heat which accompanied the Big Bang. In the steady state theory, however, there was no hot dense phase in the past from which this heat radiation could issue. Its presence had no natural explanation.

Another attempt to avoid a beginning of time was the suggestion that maybe the singularity was simply a consequence of the high degree of symmetry of the Friedmann solutions. This restricted the relative motion of any two galaxies to

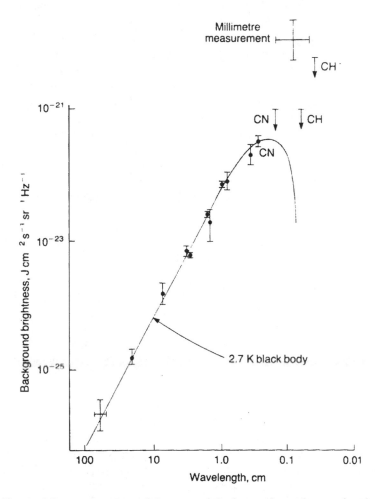

Figure 4.8. Measurements of the isotropic background at radio wavelengths. compared with a 2.7 K black body spectrum. The upper limits from interstellar molecules conflict with the millimetre measurement, which may refer to a discrete line superposed on the continuous background.

be along the line joining them. It would therefore not be surprising if they all collided with each other at some time. However, in the real universe, the galaxies would also have some random velocities perpendicular to the line joining them. These transverse velocities might be expected to cause the galaxies to miss each other and to allow the universe to pass from a contracting phase to an expanding one without the density ever becoming infinite. Indeed, in 1963 two Russian scientists, E. Lifshitz and I. Kalatnikov, claimed that this would happen in nearly every solution of the equations of general relativity. They based this claim on the fact that all the solutions with a singularity that they constructed had to satisfy some constraint or symmetry. They later realised, however, that there was a more general class of solutions with singularities which did not have to obey any constraint or symmetry.

This showed that singularities *could* occur in general solutions of general relativity but it did not answer the question whether they necessarily *would* occur. However, between 1965 and 1970 a number of theorems were proved which showed that any model of the universe which obeyed general relativity, satisfied one or two other reasonable assumptions and contained as much matter as we observe in the universe, must have a Big Bang singularity. The same theorems predict that there will be a singularity which will be an end of time if the whole universe recollapses. Even if the universe is expanding too fast to collapse in its entirety, we nevertheless expect some localised regions, such as massive burnt out stars, to collapse and form black holes. The theorems predict that the black holes will contain singularities which will be an end of time for anyone unfortunate or foolhardy enough to fall in.

Quantum Theory to the Rescue

Einstein's general theory of relativity is probably one of the two greatest intellectual achievements of the twentieth century. It is, however, incomplete because it is what is called a classical theory, that is it does not incorporate the uncertainty principle of the other great discovery of this century: quantum mechanics. The uncertainty principle states that certain pairs of quantities, such as the position and velocity of a particle, cannot be predicted simultaneously with an arbitrary high degree of accuracy. The more accurately one predicts the position of the particle, the less accurately one will be able to predict its velocity and vice versa. Quantum mechanics was developed in the early years of this century to describe the behaviour of very small systems such as atoms or individual elementary particles. In particular there was a problem with the structure of the atom which was supposed to consist of a number of electrons orbiting around the central nucleus, like the planets around the Sun. The previous classical theory predicted that each electron would radiate light waves because of its motion. The waves would carry away energy and so would cause the electrons to spiral inwards until they collided with the nucleus. However, such behaviour is not allowed by quantum mechanics because it would violate the uncertainty principle; if an electron were to sit on the nucleus, it would have both a definite position and a definite velocity. Instead, quantum mechanics predicts that the electron does not have a definite position but that the probability of finding it is spread out over some region around the nucleus with the probability density remaining finite even at the nucleus.

The prediction of classical theory of an infinite probability density of finding the electron at the nucleus is rather similar to the prediction of classical general relativity that there should be a Big Bang singularity of infinite density. Thus one might hope that if one was able to combine general relativity and quantum mechanics into a theory of quantum gravity one would find that the singularities of gravitational collapse or expansion were smeared out like in the case of the collapse of the atom.

The first indication that this might be so came with the discovery that black holes, formed by the collapse of localised regions such as stars, were not completely black if one took into account the uncertainty principle of quantum mechanics. Instead, a black hole would emit particles and radiation like a hot body with a temperature which was higher the smaller the mass of the black hole. The radiation would carry away energy and so would reduce the mass of the black hole. This in turn would increase the rate of emission. It seems that, eventually, the black hole will disappear completely in a tremendous burst of radiation. All the matter that collapsed to form the black hole and any astronauts who were unlucky enough to fall into the black hole would completely disappear, at least from our region of the universe. However, the energy that corresponded to their mass by Einstein's famous equation $E = mc^2$ would survive, to be emitted by the black hole in the form of radiation. Thus the astronaut's mass energy would be recycled to the universe. However, this would be rather a poor sort of immortality as the astronaut's subjective concept of time would almost certainly come to an end and the particles out of which he was composed would not in general be the same as the particles that were re-emitted by the black hole. Still, black hole evaporation did indicate that gravitational collapse might not lead to a complete end of time.

The Problem of Initial Conditions

The real problem with spacetime having an edge or boundary at a singularity is that the laws of science do not determine the initial state of the universe at the singularity but only how the universe evolves thereafter. This problem would remain even if there were no singularity and time continued back indefinitely; the laws of science would not fix what the state of the universe was in the infinite past. In order to pick out one particular state of the universe from among the set of all possible states that are allowed by the laws, one has to supplement the laws by boundary conditions which say what the state of the universe was at an initial singularity or in the infinite past. Many scientists are embarrassed at talking about the boundary conditions of the universe because they feel that it verges on metaphysics or religion. After all, they might say, the universe could have started off in a completely arbitrary state. That may be so, but in that case it could also have evolved in a completely arbitrary manner. Yet all the evidence that we have suggests it evolves in a well-determined way according to certain laws. It is therefore not unreasonable to suppose that there may also be simple laws that govern the boundary conditions and determine the state of the universe.

In the classical general theory of relativity, which does not incorporate the uncertainty principle, the initial state of the universe is a point of infinite density. It is very difficult to define what the boundary conditions of the universe should be at such a singularity. However, when quantum mechanics is taken into account, there is a fresh possibility, namely that the singularity might be smeared away. The question then arises as to what shape space and time may adopt instead of the point of infinite curvature.

In investigating this point, it is necessary to take into account a curious property that quantum mechanics can bestow upon spacetime. In the theory of relativity space and time are closely linked. In fact, physicists prefer to regard space and time together as forming a four-dimensional spacetime continuum, three dimensions of space plus one of time. In spite of this intimate association, there are still physical differences between space and time. One of these refers to the measurements of the four-dimensional distance or interval between two points in spacetime. If the points have a greater separation in time than they do in space (e.g. successive moments at the same spatial location), then the square of the four-dimensional separation is negative. By contrast, the four-dimensional separation of two points in spacetime for which the spatial separation exceeds the time separation (e.g. simultaneous events at different locations) has a positive square.

In the very early universe, when space was very compressed, the smearing effect of the uncertainty principle can change this basic distinction between space and time. It is possible for the square of the time separation to become positive under some circumstances. When this is the case, space and time lose their remaining distinction—we might say that time becomes fully spatialised— and it is then more accurate to talk, not of spacetime, but of a four-dimensional space. Calculations suggest that this state of affairs cannot be avoided when one considers the geometry of the universe during the first minute fraction of a second. The question then arises as to the geometry of the four-dimensional space which has to somehow smoothly join onto the more familiar spacetime once the quantum smearing effects subside.

One possibility is that this four-dimensional space curves around to form a closed surface, without any edge or boundary, in much the same way as the surface of a ball or the Einstein universe, but this time in *four* dimensions. In the case of the recontracting model universe, for which the three ordinary spatial dimensions are already closed into a hypersphere, this new proposal would imply that the whole of spacetime was finite and unbounded. This in turn would mean that the universe is completely self-contained and *did not require* boundary conditions. One would not have to specify the state in the infinite past and there would not be any singularities at which the laws of physics would break down. One could say that the boundary conditions of the universe are that it has no boundary.

It should be emphasised that this is simply a *proposal* for the boundary conditions of the universe. One cannot deduce them from some other principle but one can merely pick a reasonable set of boundary conditions, calculate what they predict for the present state of the universe and see if they agree with observations. The calculations are very difficult and have been carried out so far only in simple models with a high degree of symmetry. However, the results are very encouraging. They predict that the universe must have started out in a fairly smooth and uniform state. It would have undergone a period of what is called exponential or "inflationary" expansion during which its size would have increased by a very large factor but the density would have remained the same. The universe would then have become very hot and would have expanded to the state that we see it today, cooling as it expanded. It would be uniform and the same in every direction on very large scales but would contain local irregularities that would develop into stars and galaxies.

What happened at the beginning of the expansion of the universe? Did spacetime have an edge at the Big Bang? The answer is that, if the boundary conditions of the universe are that it has no boundary, time ceases to be well-defined in the very early universe just as the direction "north" ceases to be well-defined at the North Pole of the Earth. Asking what happens before the Big Bang is like asking for a point one mile north of the North Pole. The quantity that we measure as time had a beginning but that does not mean spacetime has an edge, just as the surface of the Earth does not have an edge at the North Pole, or at least, so I am told: I have not been there myself.

If spacetime is indeed finite but without boundary or edge, this would have important philosophical implications. It would mean that we could describe the universe by a mathematical model which was determined completely by the laws of science alone; they would not have to be supplemented by boundary conditions. We do not yet know the precise form of the laws; at the moment we have a number of partial laws which govern the behaviour of the universe under all but the most extreme conditions. However, it seems likely that these laws are all part of some unified theory that we have yet to discover. We are making progress and there is a reasonable chance that we will discover it by the end of the century. At first sight it might appear that this would enable us to predict everything in the universe. However, our powers of prediction would be severely limited, first by the uncertainty principle, which states that certain quantities cannot be exactly predicted but only their probability distribution, and, secondly, and even more importantly, by the complexity of the equations which makes them impossible to solve in any but very simple situations. Thus we would still be a long way from omniscience.

SELECTED BIBLIOGRAPHY

Barrow, John D. *The Origin of the Universe*. New York: Basic Books, 1994.

Bothun, G. *Modern Cosmological Observations and Problems*. London: Taylor and Francis, 1998.

Crowe, M. J. *Modern Theories of the Universe: From Herschel to Hubble*. New York: Dover, 1994.

Gribbin, J. *In Search of the Big Bang*. New York: Bantam, 1986.

Guth, A. *The Inflationary Universe*. Reading, Mass.: Addison-Wesley, 1997.

Harrison, E. *Cosmology*. Cambridge: Cambridge University Press, 1981.

Kolb, R. and Turner, M. *The Early Universe*. Reading, Mass.: Addison-Wesley, 1991.

Peebles, P.J.E. *The Large-Scale Structure of the Universe*. Princeton: Princeton University Press, 1980.

———. *Principles of Physical Cosmology*. Princeton: Princeton University Press, 1993.

Rowan-Robinson, M. *Cosmology*. 2nd ed. Oxford: Clarendon, 1981.

Sciama, D. W. *Modern Cosmology*. Cambridge: Cambridge University Press, 1971.

Silk, J. *The Big Bang*. New York: W. H. Freeman, 1980.

Epilogue

//

— Helge Kragh —
(1944–)

Helge Kragh is Professor of the History of Science at Aarhus University in Denmark. The following reading is the conclusion from his book Quantum Generations: A History of Physics in the Twentieth Century.

In the reading, Kragh describes the relationship between physics and the other sciences, defining the unique place of physics among the sciences of our times. He ends with a tempered but significant conclusion, that physical theory will not undergo radical revolutionary changes in the twenty-first century. Rather, he sees a large part of physical theory as firmly established. Although such theories may be subsumed under broader theories, he does not see them being overturned altogether. Perhaps the "tests of time" can guarantee a certain continuity in the development of physical theory.

PHYSICAL THEORY: PRESENT AND FUTURE

Physics and the Other Sciences

During the first half of the twentieth century, physics emerged as the number-one glamour science, a position it probably still has in spite of the troubles that it has faced more recently. The public came to see the great physicists as wizards with some kind of a direct connection to either God or nature. The most famous of the great thinkers who so fascinated the public was, of course, Einstein, but he was followed by other celebrated physicists, such as Bohr, Feynman, Gell-Mann and Hawking. The great and visionary theorists represented one side of the fascination of physics, the other and darker side being the power of physics as most dramatically symbolized by the mushroom clouds caused by nuclear bombs. Physics seemed to cover the entire spectrum, from deep quantum philosophy to technological devices such as radar and the laser. With respect to public fascination, no other science could compete with physics. A recent book undertakes to rank the world's one hundred most influential scientists in history, including psychologists and social scientists. Although the ranking should not be taken too seriously—how can one meaningfully compare Archimedes with Oppenheimer?—it is interesting to see how highly the author has ranked the physicists. The first three on the list (Newton, Einstein, and Bohr) are all physicists and among the twenty-five "most influential" scientists, twelve are physicists, eight of them belonging to the twentieth century.

Physics's dominant position among the sciences in the first three quarters of the century can be illustrated by the impact that physics exerted on the other classical sciences such as astronomy, chemistry, geology, and biology. The impact occurred mainly through three channels, the most direct of which was the migration of physicists to other scientific disciplines. In many cases, young physicists successfully migrated to, or did important work in, one of the other sciences. It is noteworthy that there have been very few cases of the reverse traffic. Another channel of impact has been the adoption of physical modes of thought in sciences that were traditionally foreign or even hostile to such attitudes to scientific work. No less important was the influence on the nonphysical sciences by the instruments and techniques provided by experimental physics. In the case of chemistry, in many ways a sister discipline to physics, this was not a new feature. It was well known at the time of Lavoisier and was an important part of the physical chemistry that emerged in the late 1880s. But in the twentieth century, chemistry came to rely even more closely on new experimental methods originating in physics, such as X-ray and electron diffraction, NMR spectroscopy, and mass spectrometry. On a more fundamental level, chemistry was even threatened with becoming a branch of physics, namely in the sense that some atomic and quantum physicists (including Born and Dirac) claimed that chemistry was merely applied quantum theory. Five years before the advent of quantum mechanics, Born wrote that "we have not penetrated far into the vast territory of chemistry," yet "we have travelled far enough to see before us in the distance the passes which must be traversed before physics can impose her laws upon her sister science." With the emergence in the late 1920s of quantum chemistry—a theory first developed by physicists, rather than chemists—Born's imperialistic hope (and the nightmare of many chemists) seemed to become a reality. However, it soon turned out that not even simple molecules could be reduced to quantum physics without empirical input from the chemist. All the same, theoretical chemistry was deeply affected by quantum mechanics (and other branches of physics) and the field can, to some extent, be regarded as "applied physics." . . .

In the case of geology, the impact was less direct, but nonetheless led to a drastic reorientation of this science, from its traditional status as natural history to a new "earth science" that was modeled on the standards of physics and made use of instruments and reasoning characteristic of physics. Part of the geological sciences, such as geophysics and seismology, were already "physicalized" in the early part of the century, especially under the influence of the German physicist Emil Wiechert. With the plate-tectonic revolution of the 1960s, the physics-inspired transformation was complete.

A somewhat similar story can be told about the impact of physics on biology, where the advent of molecular biology in the 1930s marked a further intrusion of physical and reductionist thought in the life sciences. This is hardly surprising, for several of the early leaders of molecular biology were trained as physicists, including Max Delbruck and Walter Elsasser, who had both made valuable contributions to physics before they left the field for biology (and, in Elsasser's case, the earth sciences). Francis Crick, of double-helix fame, graduated in physics in 1938 and turned to biology only after the end of the war, in part under the inspiration of Schrödinger's 1944 book *What is Life?* The elucidation of the structure of DNA in 1953, widely seen as the most important discovery of modern biology, was to a large extent the result of analysis of X-ray diffraction patterns made by Maurice Wilkins, another physicist-turned-biologist. The general trend of biology in this century, and molecular biology in particular, has been greatly inspired by physics and a reductionist thinking taken over from this science. In 1966, Crick wrote: "The ultimate aim of the modern movement in biology is in fact to explain *all* biology in terms of physics and chemistry. There is a very good reason for this. Since the revolution in physics in the mid-twenties, we have had a sound theoretical basis for chemistry and the relevant parts of physics. . . . And it is the realization that our knowledge on the atomic level is secure which has led to the great influx of physicists and chemists into biology." All in all, it would not be an exaggeration to claim that the overall pace and direction of the sciences in the twentieth century have been heavily influenced by the development of physics.

One of the most important results of this century's science is what appears to be the unlimited validity of the basic laws of physics. In the 1890s, it was still a matter of debate whether the second law of thermodynamics applied to living cells and, more generally, whether the laws of physics applied everywhere in nature and at any time. It is far from obvious that the laws have this wide range of validity, but many years of research seem to confirm that this is, in fact, the case. Not only do the laws apply to living organisms, but they also apply to the most distant parts of the universe, to the centers of stars, and to the supercompact state of the very early universe some ten billion years ago. All attempts to provide separate laws for separate strata of the world have failed. Physicists at the end of the twentieth century can claim with some confidence (not to be confused with certainty) that they know the fundamental laws and that these apply to all of nature. This does not mean that all of nature has been explained by physics, nor that the other sciences have been reduced to physics. But it does mean that there are no phenomena in nature whose explanation requires principles or laws that stand in contradiction to those accepted by the physicists. (Such phenomena may turn up, but so far they have not and we have no reason to assume that they will.) So, without suggesting any sort of simplistic reductionism, there is a sense in which physics can be said to be the most fundamental and general of all the sciences. This "imperialist" point of view is far from new, but it is only in this century that it has been substantiated and has become more than an article of faith and self-congratulation.

The Future of Physical Theory

As mentioned, the role played by physics in areas outside physics has changed completely during the twentieth century and turned the science into an integrated part of postindustrial society. This, and the effects it has had on the organization and performance of physics, is perhaps the biggest change that has occurred. When we look toward other aspects, it is fairly clear that the general picture has been on both continuity and discontinuity, both permanence and revolutionary changes. On the ontological level, the changes have been deep indeed, largely a result of the quantum revolution—according to Philip Anderson, "a dislocation which is yet to be mentally healed even for many physicists." Quantum mechanics has provided us with fundamental structures that have no resemblance at all to what can be perceived or measured directly. Our present beliefs in what the world ultimately consists of are a far cry from the beliefs of the 1890s, when it still made sense to think of matter as a collection of miniature blocks. The vacuum has turned out to be anything but "nothingness" and to be full of life, activity, and properties. This is a very important result of the new physics, but in itself it would not have shocked a physicist in 1900, who was accustomed to thinking of the vacuum as being filled with ether.

In some other respects, physics and physicists have not changed very much during the century. Thus, the basic rules of the game—the methodology of research—are much the same in the 1990s as they were in the 1890s. How to evaluate a claim, what counts as a good experiment, testing procedures, the function of mathematics in physical reasoning, and the use of thought experiments—these and other methodological topics have largely remained the same, although since the 1970s computer experiments have been added to the methods of modern physics. Had young Rutherford or Sommerfeld been catapulted into our world, they would have had great, but not insuperable, trouble in understanding many things about the theories and experiments of physics; they would easily have appreciated the methods of modern physics, so very close to those used in their own time. The same kind of continuity holds for the dreams and ultimate aspirations of physicists. Ideas of unification, mathematical beauty, and general principles as the basis of physics are not products of late twentieth-century physics. Although Planck or Mie would not have understood either the mathematics or the physics of GUT theories, they would have fully appreciated the general idea and aim of this class of modern theories.

I do not want to claim that there have been no changes in the methods of physics, only that methods and ideas widely different from those known in the nineteenth century have been relatively unimportant. As we noted, certain fields of high-energy physics (such as superstring theory and inflation cosmology) are so remote from experiment that they cannot be tested empirically. Mathematical consistency and aesthetic arguments therefore tend to become the means of demonstrating the "truth" of these theories. This is certainly an aberration from the commonly accepted methodology of science, and a potentially dangerous one at that. However, the situation should not be overdramatized. For one thing, this is only a tendency in a small corner of theoretical physics, and it does not affect the 99 percent of physics in which theory and experiment are in healthy contact. Moreover, it is not really a new problem. The ether vortex theory of the nineteenth century, the unified field theories of the early twentieth century, Eddington's fundamental theory of the 1930s, and most postwar theories of quantum gravity made use of standards that did not rely on experiment. Many years before the superstring theorists, there were physicists who argued for pure rationalism. For example, in a famous statement of 1933, Einstein suggested that "Nature is the realization of the simplest conceivable mathematical ideas . . . [and] we can discover, by means of purely mathematical constructions, those concepts and those lawful connections between them which furnish the key to the understanding of natural phenomena."

As the methods of doing physics have essentially remained the same, so have the ideals of what physics should be and how physicists should behave. Science has its uncodified culture norms—what the sociologist Robert Merton in 1942 called the scientific ethos or set of institutional imperatives. For example, scientists generally adhere to the idea of "universalism" (that the evaluation of scientific claims should be impersonal and objective), believe that secrecy should be avoided (a part of "communalism"), and accept "organized skepticism" as the proper attitude toward claims of new knowledge. These and other rules are occasionally violated, but they are nonetheless accepted as rules. The norms that contributed to end the N-ray affair in 1903 were largely the same norms that entered the scene when cold fusion was announced in 1989.

The great changes that have occurred in twentieth-century physics have built on existing knowledge and a healthy respect for tradition. There have been several attempts to base physics on an entirely new worldview (such as those proposed by Eddington and Milne in the 1930s), but they have all failed. It may seem strange that respect for traditions can produce revolutionary changes, but this is just what Thomas Kuhn described in 1962 under the label "normal science." On the other hand, the changes that sometimes follow paradigm-ruled or "normal" science are not revolutions in the strong sense that Kuhn suggested in 1962, namely, new paradigms incompatible with and totally different from the old ones. No such revolution has occurred in twentieth-century physics. After all, a theoretical physicist of the 1990s will have no trouble in understanding the spirit and details of Planck's work of 1900 in which the quantum discontinuity was introduced, nor will a modern experimentalist fail to appreciate J. J. Thompson's classical paper of 1897 in which the electron was announced. There is no insurmountable gap of communication, no deep incommensurability, between the physics of the 1990s and that of a century earlier.

The lesson to be extracted from the latest century of physics is that physical knowledge has greatly expanded and resulted in new and much-improved theories, but that these have been produced largely cumulatively and without a complete break with the past. It has always been important to be able to reproduce the successes of the old theories, and this sensible requirement guarantees a certain continuity in theoretical progress. The great discoveries and theories of our century have not, of course, left earlier knowledge intact, but neither have they turned it wholesale into non-knowledge. Most experimental facts continue to be facts even in the light of the new theories. The observation that Mercury's perihelion excess is 0.43″ per year was explained, not overthrown, by Einstein's theory of relativity, and any future theory of gravitation will have to accommodate the observational fact.

A large part of physics seems to be firmly stabilized. It becomes increasingly difficult to imagine that these parts, so thoroughly tested and so closely bound in a larger network of theories and experiments, will change drastically in the future. For several decades, it has been considered heretical, even ridiculous, to suggest that science develops "teleologically," that is, toward a certain state of knowledge that reflects the true structure of nature. It is correct, as the philosopher Nicholas Rescher pointed out, that "[s]ignificant scientific progress is generally a matter not of adding further facts—on the order of filling in of a crossword puzzle—but of changing the framework itself." Relativity, quantum mechanics, and the electro-weak theory are examples of such changes, frameworks that did not at first either rely on or suggest new experimental facts. Yet, not only does the view underrate the value of "adding further facts," but it also leaves open the question of whether or not there is a best possible framework that will leave a theory in a stable or "finished" state. We can smile at the naïveté of the fin-de-siècle physicists who believed that physics had essentially reached its final state, but their failure does not imply that no such final state exists. Because most physical theories proposed during history have turned out to be wrong, it does not follow that those accepted today are wrong as well and will be replaced by entirely new theories.

One might speculate that history might repeat itself and that tomorrow's physicists might find quite new phenomena in nature that would demand a major reframing of theoretical physics—a kind of analogy to the surprising discoveries of 1895–97. Is such a scenario plausible? It seems that although one can never preclude the possibility, it becomes still more unlikely that physicists have missed some big and important aspect of nature. The modern army of physicists and their arsenal of sophisticated high-precision instruments makes it much more difficult for such phenomena to remain hidden than in the case of radioactivity a century ago. It is many decades since a new discovery squarely contradicted fundamental theory. In 1986 the discovery of a "fifth force" was announced, a force of intermediate range that could not be accounted for without established theory. Had the discovery claim been accepted, it might have led to a major conceptual change in theoretical physics. But this was not what happened. After a few years of experiments and intense debate, it turned out that the fifth force did not pose a threat to the standard physics operating with four forces of nature. The fifth force does not exist.

Physics will undoubtedly continue to develop and make many interesting discoveries in the new century. But it is possible that the pattern of progress in physics will change, and that many of the most fundamental aspects will remain as they are now known. There will always be exciting work to do and discoveries to make, but it is far from certain that the development of physics in the twenty-first century will be as explosive as it has been in the twentieth century. Feynman . . . believed that "[t]he age in which we live is the age in which we are discovering the fundamental laws of nature, and that day will never come again." Whether Feynman's prophecy was correct or not can be shown only by developments in the next century. Perhaps a historian writing the history of physics in the twenty-first century will quote Feynman in order to demonstrate his wisdom; or perhaps he will quote him in order to show how utterly wrong he was.

Sources of the Readings

//

Part I: Heliocentric Theory

1. Aristotle, "On The Heavens." In *Collected Works of Aristotle*, the revised Oxford translation ed. Jonathan Barnes (Princeton: Princeton University Press, 1984), 268b10–269b16, 286a3–290a11, 293a15–296a20.
2. Aristarchus, "Aristarchus of Samos," in *The Library of Greek Thought* (London: J. M. Dent & Sons, 1932), 105–9.
3. Claudius Ptolemy. *Ptolemy's Almagest*, trans. and annotated by G. J. Toomer (Princeton: Princeton University Press, 1998), 35–45.
4. Nicholaus Copernicus, "Commentariolus," trans. Edward Rosen, in *Nicholas Copernicus: Minor Works*, ed. Pawel Czartoryski (London: Macmillan Press, Ltd., 1985), 81–83.
5. Nicholaus Copernicus, "On The Revolutions of the Heavenly Spheres," trans. Charles Glenn Wallis (Amherst, N.Y.: Prometheus Books, 1995), 3–30.
6. Tycho Brahe, *His Astronomical Coniectur of the New and Much Admired Star which Appeared in the Year 1572* (New York: Da Capo Press, 1969) 8–12, 16.
7. Tycho Brahe, *Tycho Brahe's Description of his Instruments and Scientific Work*, trans. and ed. Hans Raeder, Elis Stromgren, and Bengt Stromgren (Kobenhavn: Komission Ho5 Ejner Munksgaard, 1946), 116–17.
8. Johannes Kepler, *New Astronomy* trans. William H. Donahue (Cambridge: Cambridge University Press, 1992), 376–91.
9. Galileo Galilei, *Siderius Nuncius*, trans. Albert Van Helden (Chicago: University of Chicago Press, 1989), 35–66, 83–86. "Letters on Sunspots," in *Discoveries and Opinions of Galileo*, trans. Stillman Drake (New York: Anchor Doubleday, 1957), 93–94.
10. Johannes Kepler, *Epitome of Copernican Astronomy*, trans. Charles Glenn Wallis (New York: Prometheus Books, 1995), 12–13, 17–19, 19–22, 67–70.
11. Galileo Galilei, *Dialogue Concerning the Two Chief World Systems*, 2nd ed., trans. Stillman Drake (Berkeley: University of California Press, 1967), 114–20.
12. Isaac Newton, "The System of the World," from *Principia*, trans. Andrew Motte, rev. Florian Cajori (Berkeley: University of California Press, 1960), 549–59, 572–76, 406–15, 420–22, 596–97.
13. John Herschel, *Outlines of Astronomy* (New York: P. F. Collier & Son, 1902), 724–25, 728–29.

Part II: Electromagnetic Field Theory

1. William Gilbert, "On the Magnet," trans. E. Fleury Motteley, in William Francis Magie, *A Source Book in Physics* (New York: McGraw-Hill, 1935), 387–93.
2. Charles Coulomb, "Law of Electric Force," in William Francis Magie, *A Source Book in Physics* (New York: McGraw-Hill, 1935), 408–13.
3. Hans Christian Oersted, "Experiments on the Effect of a Current of Electricity on the Magnetic Needle," in *Selected Works of Hans Christian Ørsted*, trans. and ed. Karen Jelved, Andrew D. Jackson, and Ole Knudsen (Princeton: Princeton University Press, 1998), 417–20.
4. André Marie Ampére. "Ampere's Philosophy of Science," trans. O. M. Blunn, *Annales de Chimie et de Physique* Vol. 15 1820, in R.A.R. Tricker, *Early Electrodynamics* (London: Pergamon Press, 1965), 155–61.
5. Isaac Newton, "Opticks," Query 28–Query 29 in *Great Books of the Western World* (Chicago: Encyclopedia Britanica, 1952), 34: 525–30.
6. Christiaan Huygens, "Treatise on Light," trans. Silvanus P. Thompson, in *Great Books of the Western World* (Chicago: Encyclopedia Britannica, 1952), 34: 553–63,
7. Thomas Young, "Analogy Between Light and Sound," "Letter to Mr. Nicholson," "On the Theory of Light and Colours," "Production of Colours," "Phenomena of Polarization," in *Miscellaneous Works of the Late Thomas Young*, ed. George Peacock (John Murray, 1855) 1: 78–83; 131–33; 140–50; 166–69; 170–76, 412–17.
8. Dominique-Francois Arago and Augustin Fresnel, "Interference of Polarized Light," in William Francis Magie, *A Source Book in Physics* (New York: McGraw-Hill, 1935), 325–34.
9. Michael Faraday, "On the Induction of Electric Currents," in *Experimental Researches in Electricity* (New York: Dover, 1965), I: 1–16.
10. Michael Faraday, "The Physical Existence of Lines of Force," in *Experimental Researches in Electricity* (New York: Dover Publications, 1965), 3: 407–15, 438–43.
11. James Clerk Maxwell, "A Dynamical Theory of the Electromagnetic Field," in *The Scientific Papers of James Clerk Maxwell*, ed. W. D. Niven (Cambridge: Cambridge University Press, 1890), 1: 526–36, 563–64.
12. James Clerk Maxwell, "Electromagnetic Theory of Light," in *A Treatise on Electricity and Magnetism*, 3rd ed. (New York: Dover Publications, 1954), 431–37.
13. James Clerk Maxwell, "On Action at a Distance," in *The Scientific Papers of James Clerk Maxwell*, ed. W. D. Niven (Cambridge: Cambridge University Press, 1890), 2: 311–23.
14. Heinrich Hertz, "On the Finite Velocity of Propagation of Electromagnetic Action," in *Electric Waves*, trans. E. E. Jones (New York: Dover Publications, 1962), 107–11, 122–23, 124–31, 134–36.

Part III: The Theory of Relativity

1. James Clerk Maxwell, "Ether," in *The Scientific Papers of James Clerk Maxwell*, ed. W. D. Niven (Cambridge: Cambridge University Press, 1890), 2: 763–75.
2. Albert Michelson, *Light Waves and Their Uses* (Chicago: University of Chicago Press, 1903), 146–63.
3. George F. Fitzgerald, "The Ether and the Earth's Atmosphere," *Science* 13 (1889): 390.
4. H. A. Lorentz, "Michelson's Interference Experiment," in *The Principle of Relativity*, trans. W. Perrett and G. B. Jeffrey (New York: Dover Publications, 1952), 3–7.

5. Henri Poincaré, *The Value of Science*, trans. G. B. Holsted (New York: Dover Publications, 1958), 26–36, 110–11; *Science and Hypothesis* (New York: Dover Publications, 1952), 89–91.

6. Albert Einstein, "On the Electrodynamics of Moving Bodies," trans. Anna Beck, in *The Collected Papers of Albert Einstein*, Vol. 2 (Princeton: Princeton University Press, 1989), 140–45.

7. Herman Minkowski, "Space and Time," in *The Principle of Relativity*, trans. W. Perrett and G. B. Jeffrey (New York: Dover Publications 1952) 75–76, 79–81, 90–91.

8. Albert Einstein, "The Foundation of the General Theory of Relativity," trans. W. Perrett and G. B. Jeffrey, in *The Collected Papers of Albert Einstein* Vol. 6 (Princeton: Princeton University Press, 1997), 147–54.

9. Albert Einstein, "Relativity: The Special and General Theory" trans. Robert W. Lawson, in *The Collected Papers of Albert Einstein*, Vol. 6 (Princeton: Princeton University Press, 1997), 261–78, 314–22, 348–60.

10. Arthur Eddington, *Space, Time, and Gravitation* (New York: Cambridge University Press, 1920), 110–22.

11. Albert Einstein, "Ether and Relativity," in *Sidelights on Relativity*, trans. G. B. Jeffrey and W. Perrett (New York: Dover Publications, 1983), 3–24.

12. Albert Einstein, "The General Theory of Relativity," in *The Collected Papers of Albert Einstein*, Vol. 4 (Princeton: Princeton University Press, 1996), 260–63.

13. Albert Einstein. "$E = MC^2$—The Most Urgent Problem of Our Time," originally published in *Science Illustrated* I, 1946; "An Elementary Derivation of the Equivalence of Mass and Energy," originally published in *Technion Journal* V 1946. Both found in *Essays in Physics* (New York: Philosophical Library, 1950), 12–15, 70–73.

Part IV: The Quantum Theory

Historical and Conceptual Development

1. Max Planck, *A Survey of Physical Theory* (New York: Dover Publications, 1994) 46–47, 49–55.

2. Albert Einstein, "On a Heuristic Point of View about the Creation and Conversion of Light," in *The Collected Papers of Albert Einstein* Vol. 2 (Princeton: Princeton University Press, 1989), 86–87, 99–103.

3. Niels Bohr, "On the Constitution of Atoms and Molecules," *Philosophical Magazine* 26, no. 151 (July 1913): 1–3, 874.

4. Louis de Broglie, *Matter and Light*, (New York: Dover Publications, 1939) 165–69, 173–75, 179.

5. Niels Bohr, "The Quantum Postulate and the Recent Development of Atomic Theory," in *Nature* (Supplement), Vol. 121 (1928), 580–89.

6. Niels Bohr, "Discussions with Einstein on Epistemological Problems in Atomic Physics," in *Albert Einstein Philosopher-Scientist*, ed. Paul Arthur Schilpp (New York: Harper & Brothers, 1959), 1: 201–38.

7. Albert Einstein, "Einstein's Reply," in *Albert Einstein: Philosopher-Scientist*, ed. Paul Arthur Schilpp (New York: Harper & Brothers, 1959) 2: 666–72.

8. Werner Heisenberg, "The History of Quantum Theory," in *Physics and Philosophy* (New York: Harper & Row, 1962), 30–43.

9. Werner Heisenberg, "The Copenhagen Interpretation of Quantum Theory," in *Physics and Philosophy* (New York: Harper & Row, 1962), 44–58.

10. Erwin Schrödinger, *Science, Theory and Man* (New York: Dover Publications, 1957), 166–92.
11. Erwin Schrödinger, "Are There Quantum Jumps?" *British Journal of the Philosophy of Science* 3 (1952): 109–23.
12. P.A.M. Dirac, "The Evolution of the Physicists' Picture of Nature," *Scientific American* 208, 5 (May 1963): 45–53.
13. John Archibald Wheeler, "The 'Past' and the 'Delayed-Choice' Double-Slit Experiment," in *Mathematical Foundations of Quantum Theory*, ed. A. R. Marlow (New York: Academic Press, 1978), 9–14, 29–31, 39–43.

The Completeness Debate

14. Albert Einstein, Boris Podolsky, and Nathan Rosen, "Can Quantum Mechanical Description of Physical Reality Be Considered Complete?" *Physical Review* 47 (1935): 777–80.
15. Albert Einstein; "Physics and Reality," in *Essays in Physics* (New York: Philosophical Library, 1950), 41–49.
16. Niels Bohr, "Can Quantum Mechanical Description of Physical Reality Be Considered Complete?" (response to EPR), *Physical Review* 48 (1935): 696–702.
17. David Bohm, "A Suggested Interpretation of the Quantum Theory in Terms of Hidden Variables," *Physical Review* 85 (1952): 166–69, 180–87.
18. J. S. Bell, "Introduction to the Hidden-variable Question," *Foundations of Quantum Mechanics* (New York: Academic Press, 1971). Reprinted in J. S. Bell, *Speakable and Unspeakable in Quantum Mechanics* (New York: Cambridge University Press, 1987) 29–39.
19. J. S. Bell, "Bertlmann's Socks and the Nature of Reality," in Bell, *Speakable and Unspeakable in Quantum Mechanics* (New York: Cambridge University Press, 1987), 139–58.
20. Abner Shimony, "Conceptual Foundations of Quantum Mechanics," in *The New Physics*, ed. Paul Davies (New York: Cambridge University Press, 1989), 393–95.

Part V: The Big Bang Cosmological Theory

1. Henrietta Leavitt, *Annals of Harvard College Observatory* LX, no. IV (1908), 87–88, 107.
2. Henrietta Leavitt, *Harvard College Observatory*, Circular 173 (1912), 1–3.
3. Vesto Slipher, *Lowell Observatory Bulletin* No. 58, Vol. 2, No. 8 (1912), 56–57.
4. Vesto Slipher, *Lowell Observatory Bulletin* No. 80, Vol. 3, No. 5 (1913), 59–62.
5. Harlow Shapley, *Galaxies* (1943), revised by Paul W. Hodge (Cambridge: Harvard University Press, 1972), 57–64.
6. Willem de Sitter, *Kosmos* (Cambridge: Harvard University Press, 1932), 103–38.
7. Edwin Hubble, *The Realm of the Nebulae* (New Haven: Yale University Press, 1936), 20–35.
8. Edwin Hubble, *The Realm of the Nebulae* (New Haven: Yale University Press, 1936), 120–23.
9. Arthur Eddington, *The Expanding Universe* (Ann Arbor: University of Michigan Press, 1958), 1–28.
10. George Le Maitre, *The Primeval Atom: An Essay in Cosmogony*, trans. B. Korff and S. Korff (New York: D. Van Nostrand, 1950), 134–163.

11. Arno Penzias and Robert Wilson, "A Measurement of Excess Antenna Temperature at 4080 Mc/s," *Astrophysical Journal* 142 (Nov. 1965): 419–21.

12. R. H. Dicke, P.J.E. Peebles, P. G. Roll, D. T. Wilkinson, "Cosmic Blackbody Radiation" *Astrophysical Journal* 142 (Nov. 1965): 415–19.

13. Steven Weinberg, *The First Three Minutes* (New York: Basic Books, 1977), 44–52.

14. Alan Guth and Paul Steinhardt, "The Inflationary Universe," *Scientific American* 250: 5 (May 1984), 116–28.

15. George Smoot and Keay Davidson, *Wrinkles in Time* (New York: Avon Books, 1993), 277–79, 282–97.

16. Stephen Hawking, "The Edge of Space-Time," in *The New Physics*, ed. Paul Davies (New York: Cambridge University Press, 1992), 65–69.

Epilogue

Helge Kragh, *Quantum Generations* (Princeton: Princeton University Press, 2000), 444–51.

Index of Names

///

Adams, John Couch, 574
Aepinus, Franz, 135
Albrecht, Andreas, 661
Alpetragius, 60
Alpher, Ralph, 651
Ampère, André Marie, xxiv–xxv, 137, 140, 157, 210, 240, 241; "Exposé somaire des nouvelles expériences electro-magnétiques faites par plusieurs physiciens depuis le mois de mars 1821" (1822), 158; memoir presented to the Royal Academy of Sciences (1825), 157, 157–61 (text); "Observations in Electrodynamics," 158
Anaxagoras, 24
Anaximander, 51
Anaximenes, 24, 51
Anderson, Philip, 688
Andrade, Edward Neville de Costa, 293; *Leçons de Mécanique physique,* 298
Angström, Anders Jonas, 270
Apelles, 98–99
Aphrodiseus, Alexander, 150
Apollonius of Perga, 6, 7
Arago, Dominique, 139, 201, 210, 269; paper published in the *Annales de Chimie et de Physique* (1819), 202, 202–9 (text)
Archimedes, 455, 574, 578; *The Sand-Reckoner,* 27, 27–28 (text)
Aristarchus of Samos, xxxix, 5–6, 26–28, 105, 107; as the "Copernicus of antiquity," 5, 26; *On the Sizes and Distances of the Sun and Moon,* 27
Aristotle, xxii, 4–5, 7, 12, 13–14, 30, 56, 68, 101–3, 110, 126, 451, 542, 547; *De Animalibus,* 64; *Metaphysics,* 13; *Meteorology,* 13; "natura non facit saltus," 373, 376, 452; on natural states for bodies, 255; *On Coming-to-be and Passing-Away,* 13; *On the Heavens,* 14, 14–26 (text); *Physics,* 13; on the Unmoved Mover, 13–14

Aspect, Alain, xxx, xxxi, xxxvii, 369–70, 515
Averroes, 61

Bacon, Francis, xxxvi
Badovere, Jacques, 83
Bailey, Solon Irving, 571
Balmer, Johann Jakob, 387
Bardeen, James M., 663
al-Battani the Harranite, 61
Becquerel, Antoine-Henri, 440
Bell, John Stewart, xxx, xxxi, xxxvii, 369, 369n, 515, 526–27; Bell's inequalities, xxx, 369; "Bertlmann's Socks and the Nature of Reality" (1981), 532–33, 533–39 (text); "Introduction to the Hidden Variable Question" (1971), 527, 527–32 (text); and the locality assumption, 369; *Speakable and Unspeakable in Quantum Mechanics,* 533
Bellarmine, Robert, xxxi
Bennett, Chuck, 671, 672
Berkeley, George, xxvii, 435
Bernoulli, Daniel, 135
Bessel, Friedrich Wilhelm, 13, 128, 129, 576; discovery of stellar parallax (1838), xl, 13, 128–29
Biot, Jean-Baptiste, 159; Biot-Savart law, xxv
Bohm, David, xxix, xxx, 515, 528, 532, 533, 537; "A Suggested Interpretation of the Quantum Theory in Terms of Hidden Variables, I and II" (1952), 515, 515–26 (text); reconsideration of de Broglie's "hidden variables" hypothesis, xxxi, 368, 515
Bohr, Niels, xxviii, xxx, xxxvi, xli, 383–84, 389, 432, 433, 438, 440–41, 443, 444, 445, 446, 453, 472, 474, 479, 484, 485, 487, 488, 489, 490, 502, 515, 517, 524, 531, 535, 539, 540–44, 685, 686; "Can Quantum Mechanical Description of Physical Reality Be Considered Complete?" 507, 507–14 (text); Classical/Quantum theory of the atom

Bohr, Niels (*cont.*)
  (1913), xxxvi, xxxix, 361–62; Correspondence Principle, 363; director of the Institute for Theoretical Physics, 383; formulation of the Copenhagen Interpretation of Quantum Theory, 359, 366, 383; gold medal of the Royal Danish Academy of Science, 383; hidden variables debate with Einstein, xxviii, 407–32 (text), 490; and the Manhattan Project, 383; Nobel Prize in physics (1922), 383; "old" Quantum Theory (1913), 362, 364, 467; on complementarity and the new Quantum Theory, 392–406 (text); organizer of the Atoms for Peace Conference (1955), 383; as "philosopher," 383; the Principle of Complementarity (1927), xxix, 366, 383, 392, 407, 445, 449, 517–18, 540–42; on the quantum character of the atom, 384–87 (text); reaction to Heisenberg's Uncertainty Principle, 365–66; response to EPR paper, 368, 507
Boltzmann, Ludwig Eduard, 271, 408; Boltzmann's constant, 381
Bondi, Herman, xxxiii, 643, 678; development of Steady State Theory, 552
Born, Max, 364, 365, 391, 392, 401, 411, 432, 433, 443, 444, 504, 686
Boscovich, Ruggiero Giuseppe, 240
Boyle, Robert, 164, 173
Bradley, James, 276, 353
Brahe, Tycho, xxii, 10, 66–67, 70, 72, 73, 100–101, 107, 108, 118, 255; astronomical complex on Hveen, 66; *His Astronomical Coniectur of the New and Much Admired Star Which Appeared in the Year 1572*, 67, 67–69 (text); observation of the supernova of 1572, 10–11, 66, 69, 293–94, 548; *Progymnasmata,* 70; tracing of the orbits of comets, 11, 66, 548; *Tycho Brahe's Description of His Instruments and Scientific Work* (1598), 69, 69–70 (text)
Brans, Carl, and Brans-Dicke theory, 642, 644, 645
Brewster, David, 197, 198
Burke, Bernard, 649, 650

Cabeo, Nicolo, 134
Callippus, 39
Campbell, N. R., *Physics, the Elements* (1920), xxxvii
Capella, Martianus, 61, 106
Carnap, Rudolf, xxxvi
Cavendish, Henry, 135, 240, 244
Chadwick, James, 463

Cicero, 46
Clauser, John F., 369, 540; and the "CHSH inequality" experiment, 540
Coleman, Sidney R., 659
Compton, Arthur Holly, xxxvii, 362n, 389, 394, 399; experiment on the scattering of X-rays, 410–11, 413, 442
Copernicus, Nicholaus, xxii, 8, 9–10, 38–39, 72, 73, 77, 99, 102, 103–4, 105, 106, 107, 108, 110, 118, 170, 490, 547–48; *Commentariolus*, 10, 39, 39–42 (text); *De Revolutionibus Orbium Caelestium* (1543), 10, 39, 42–43, 44–66 (text); seven postulates for his new astronomy, 39, 40
Cornu, Marie-Alfred, 267
Cotes, Roger, 239
Cottingham, E. T., 334
Coulomb, Charles Augustin, xxiv–xxv, 135–36, 151, 159, 240, 244; invention of the torsion balance (1784), 151; Law of Electrostatic Force, xxxvii, 136, 151; memoir presented to the French Academy of Science (1785), 151, 151–53 (text)
Crawford, A. B., 641
Crick, Francis, 687
Crommelin, A.C.D., 334, 339
Curie, Pierre, 440
Cushing, James, xxxi

Davidson, C., 334
Davidson, Keay, 666; *Wrinkles in Time*, 666–67, 667–77 (text)
Davies, Paul, 677
Davisson, Clinton, xxxvii, 363
de Broglie, Louis, xxx, 363, 364, 387, 394, 395, 399, 401, 405, 406, 411, 442, 454, 465, 471, 475, 485, 502, 508, 516; hidden variables hypothesis, 368; *Matter and Light: The New Physics*, 387–88, 388–92 (text); Nobel Prize in physics (1929), 387
de Sitter, Willem, xxix, 316, 551, 573, 605, 611–12, 616, 618, 635–36; and the Einstein-de Sitter universe, 573; "empty universe" model, 573; *Kosmos* (1932), 573, 573–91 (text)
Delbruck, Max, 687
Democritus, 24, 51, 375
Derham, William, 592, 595
Descartes, René, xxxvi, 126, 158, 168, 169, 170, 171, 175, 179, 239, 265
Dicke, R. H., 264, 640, 641, 642, 645, 651, 655; and Brans-Dicke theory, 642, 644, 645; letter to the *Astrophysical Journal* (1965), 642, 642–46 (text)

Dirac, P.A.M., xxx, xxxvii, xli, 364, 391, 411, 422, 431, 443, 454, 472, 502, 621, 686; Nobel Prize in physics (with Schrödinger) (1933), 454, 472; prediction of the positron (1928), xxx, 472; "The Evolution of the Physicist's Picture of Nature" (1963), 472, 472–84 (text); *The Principles of Quantum Mechanics*, xxx–xxxi

Doppler, Christian Johann, 398; Doppler effect, 399

Dufay, Charles, 134

Duhem, Pierre, xxxvii

Dyson, Freeman, 675

Eddington, Arthur, 330–31, 551, 552, 583, 588, 591, 689; discovery of the relationship between mass and luminosity in stars, 330; expedition to photograph a solar eclipse (1919), xxxvii, 264, 331; *Space, Time and Gravitation* (1920), 331, 331–39 (text); *The Expanding Universe*, 611, 611–25 (text)

Ehrenfest, Paul, 401–2, 410, 415, 419, 426

Einstein, Albert, xxiii–xiv, xxvii–xxix, xxxvii, xxxviii, 145, 255, 299–300, 308, 331, 332, 335, 336, 337, 339, 340, 389, 390, 392, 393, 394, 406, 439–40, 467, 469, 472–73, 474, 485, 487, 489, 530, 540, 542, 543, 551, 573, 583, 619, 630, 645, 672, 673, 685, 686, 689; address at the University of Leyden (1920), 340, 340–46 (text); "An Elementary Derivation of the Relationship between Mass and Energy," 350, 350–55 (text); change in law of gravitation, 621–23; chief opponents of, 432; the cosmological constant (1915), 551, 552, 582–83, 605, 622–23, 632; "Die Kultur der Gegenwart, Ihre Entwicklung und ihre Ziele" (1925), 347, 347–50 (text); "E = MC²," 350, 350–52 (text); and the Einstein-de Sitter universe, 573; and the EPR paper, xxxi, 367–68, 428–30, 476, 495, 495–501 (text), 523–25, 528, 536–38; "Ether and Relativity" (1920), xxvii; *General Theory of Relativity* (1915), 299; on God's "playing dice," 407, 419, 676; hidden variables debate with Bohr, xxviii, 407–32, 490, 543; light quanta hypothesis, 380; Nobel Prize in physics (1921), 299; "On the Electrodynamics of Moving Bodies" (1905), 261; "On a Heuristic Point of View about the Creation and Conversion of Light" (1905), 381–82, 381–82 (text); opposition to Quantum Theory, 406, 445, 476, 487, 501, 515; paper announcing the Special Theory of Relativity (1905), 299,
300–304 (text); photon box experiment, 367; "Physics and Reality" (1936), 430, 501, 501–6 (text); and the photoelectric effect, 360–62; the Principle of Equivalence, xxxvii, xxxviii, 263; "Reply to Criticism," 432, 432–37 (text); "The Foundation of the General Theory of Relativity" (1916), 308, 309–13 (text); and the photoelectric effect, 360–62; *Relativity: The Special and General Theory* (1916), 314, 314–30 (text); theory of relativity, xxiv, xxvii–xxix, 255, 260–64, 299, 308, 313, 340, 347, 350, 473–74, 480, 576–78, 620, 621–22, 653, 680–81

Ekphantus, 46

Elsasser, Walter, 687

Empedocles, 23, 25, 51

Eötvös, Roland von, 263, 264, 332, 576

Euclid, *Optics*, 60

Eudoxus of Cnidus, 3–4, 5, 28, 39

Euler, Leonard, 180, 181, 182, 184, 189

Faraday, Michael, xxiv, xxv, xxvi, xxxvi, 141, 209, 226, 235, 240–41, 244, 250, 270, 272, 321, 346, 574; lack of mathematics in his work, 141, 209, 241; lines-of-force hypothesis, xxv, xxvi, 141–42, 209, 214, 241–43, 481; materials published in *Philosophical Magazine* (1852), 214, 214–23 (text); paper presented to the Royal Society (1831), 209, 209–14 (text); "Thoughts on Ray Vibrations," 231

Fermat, Pierre de, 455; the Principle of Least Time, 455, 457, 458, 459

Feynman, Richard, xxxvii, 685, 691; sums-over-paths approach to Quantum Theory, xxxii

Fine, A., xxviii

Fitzgerald, George F., 259–60, 285, 287; Lorentz-Fitzgerald contraction hypothesis, 260, 285; "The Ether and the Earth's Atmosphere" (1889), 285, 285 (text)

Fizeau, Armand-Hippolyte-Louis, 140, 270, 274, 277–78, 279, 341

Flamsteed, John, 116, 117–18, 129

Foucault, Jean-Bernard-Léon, 140, 201, 243

Fracastoro, Girolamo, 77

Franck, J., 409, 441, 504

Franklin, Benjamin, 134–35; conservation principle for electricity, 134; and the Leyden jar, 134–35, 135n; one-fluid theory, xxxix

Fraunhofer, Joseph von, 129

Fresnel, Augustin, xxv, 139–40, 199, 200–201, 201, 233, 244, 268, 270, 277, 278, 279, 286, 287, 345, 388; paper published in the *Annales de Chimie et Physique* (1819), 202, 202–9 (text)

Friedmann, Alexander, 551, 552, 632–34, 635, 677–78
Friedmann, Stuart, xxvii, 369

Gaier, Todd, 671
Galen, 149
Galilei, Galileo, xxii, xxiii, xxxi, xxxvi, 11–12, 106, 255, 309, 310, 323, 574, 596; development and use of the telescope, 11–12, 81, 548; *Dialogue Concerning the Two Chief World Systems* (1632), 12, 109, 109–14 (text); *Discourse Concerning Two New Sciences* (1638), 81; *Letter on Sunspots* (1613), 81, 82, 98–99 (text); *Starry Messenger* (1610), 81–82, 82–98 (text)
Galvani, Luigi, 136
Gamow, George, 434, 552–53, 651
Gauss, K. F., 136; Gauss co-ordinates, 326, 327, 328
Gell-Mann, Murray, 685; prediction of quarks, xxx
Georgi, Howard M., 655
Gerlach, Walther, and the Stern-Gerlach experiment (1922), 405, 410, 441, 528, 534–35, 537
Germer, Leotes, xxxvii, 363
Giese, Tiedeman, 44–45
Gilbert, William, 81, 133–34, 145; *De Magnete* (1600), 133, 145–46, 146–50 (text)
Gisin, Nicolas, 370
Glashow, Sheldon Lee, 655
Gödel, Kurt, xxix
Gold, Thomas, xxxiii, 643, 678; development of Steady State Theory, 552
Gray, Stephen, 134
Green, George, 233
Gribbin, John, 556
Grimaldi, Francesco Maria, xxxvii
Guth, Alan, 672; *Scientific American* article (1984), 652, 652–66 (text); and Inflationary Theory, xxxiii, xxxiv, 554, 652

Halley, Edmund, 118, 548
Hamilton, G. H., 565
Hamilton, William, 411, 454, 457, 458, 459
Harrison, Edward R., 664
Hartle, Jim, 674
Haüy, René-Just, 191
Hawking, Stephen W., 555, 663, 668, 674, 675, 677, 685; article published in *The New Physics*, 677, 677–83 (text)
Heath, Thomas, *Greek Astronomy*, 27
Heisenberg, Werner, xxix, xxx, xxxi, xxxviii, 383, 391, 395, 397–98, 400, 403, 404, 405, 411, 412, 422, 472, 474–75, 488, 502, 506, 508, 536; formulation of the Copenhagen Interpretation of Quantum Theory, 359, 366, 437; matrix mechanics, 364, 438, 454; Nobel Prize in physics (1932), 438; *Physics and Philosophy*, 438, 438–46 (text), 446, 446–54 (text); Uncertainty Principle (1927), 363, 365, 433, 438, 477, 529, 540
Helmholtz, Hermann von, 144, 373, 379
Hempel, Carl, xl
Henderson, Thomas, 129
Heracleitus, 51
Heraclides of Pontus, 5, 46
Herman, Robert, 651
Herschel, John, 128, 596; mapping of the southern sky (1834–38), 128; *Outlines of Astronomy* (1849), 128, 128–29 (text)
Herschel, William, 192, 548–49, 592, 596
Hertz, Heinrich, xxiii, xxvi, 144–45, 245, 340, 342, 409, 441, 504; *Electric Waves*, 144, 245, 245–51 (text)
Hertzspring, Ejnar, 571
Higgs, Peter W., 658; Higgs fields, 658–59, 661–62, 664
Hinshaw, Gary, 671
Hipparchus of Nicea, 6–7
Hogg, D. C., 641
Holt, Richard A., and the "CHSH inequality" experiment, 540
Hooke, Robert, xxxvii, 184, 185; *Micrographia,* 178
Horne, Michael A., and the "CHSH inequality" experiment, 540
Hoyle, Fred, xxxiii, 553, 643, 651, 670, 678; coinage of "Big Bang" phrase, 552; development of Steady State Theory, 552
Hubble, Edwin, 264, 549, 556, 592, 615, 616, 628, 653; Hubble's constant, xxxiii; Hubble's Law, 550, 552, 592; *The Realm of the Nebula,* 592, 592–601 (text), 602, 602–11 (text); and the velocity-distance relation, 602
Huggins, William, 244, 562, 596, 602
Humason, M. L., 616, 628
Hume, David, 542
Huygens, Christiaan, 118, 137–38, 163, 167–68, 180, 184, 190, 191, 233, 348–49, 574; *Treatise on Light* (1690), xxv, 162, 168, 168–79 (text); wave theory of light (1678), xxxvii, 167, 201, 265, 266

Jastrow, Robert, 673
Jeans, James Hopwood, 389, 438–39
Jenkin, F., 228
Jordan, Camille, 181, 195, 391, 411, 443, 536
Joule, James Prescott, 244

Kant, Immanuel, 542, 549, 592, 594–95
Kapteyn, Jacobus Cornelis, 638
Keeler, James Edward, 562, 564
Kelvin, Lord (William Thomson), 283
Kepler, Johannes, xxii, xxiii, 11, 67, 70–71, 99–100, 117, 118, 119, 158, 159, 548; *Astronomia Nova* (1600), xxiii, xxxviii, 70–71, 71–81 (text), 11; laws of planetary motion, 11, 13, 70; *Mysterium cosmographicum*, 79; support for Copernicus, 100; *The Epitome of Copernican Philosophy*, 99–100, 100–108 (text)
Khalatnikov, I., 643, 680
Kirchhoff, Gustav, 373
Kogut, Al, 671, 672
Kohlrausch, Friedrich Wilhelm Georg, 230
Kragh, Helge, 685; *Quantum Generations: A History of Physics in the Twentieth Century*, 685, 685–91 (text)
Kramers, Heinrich Anthony, 411–12, 444
Kuhn, Thomas, xli, 689

Lactantius, 47
Lakatos, Imre, xli
Laplace, Pierre, 359, 574
Lassell, William, 564
Laudan, Larry, xxxix, xli
Lavoisier, Antoine-Laurent, 686
Leavitt, Henrietta, 569–70; contribution to the *Annals of the Harvard College Observatory* (1908), 556, 556–59 (text); contribution to the *Harvard College Observatory Circular* (1912), 559, 559–61 (text); discovery of the variability-luminosity relationship, 549, 556, 567
Leibniz, Gottfried Wilhelm, xxvii, xxxviii, 350, 431, 542; Newton-Leibniz debate, xxvi–xxvii
Lemaître, Georges, 551–52, 625; as "father" of Big Bang Cosmology, 552, 625; *The Primeval Atom: An Essay on Cosmology*, 625, 625–39 (text)
Lenard, Philipp Eduard Anton, 382, 439
Leucippus, 51
Leverrier, Urbain-Jean-Joseph, 330, 574
Lifshitz, E., 643, 680
Linde, A. D., xxxiv, 661
Locke, John, 542
Lorentz, Hendrik A., xxvii, xli, 260, 280, 285, 286, 299, 307–8, 317, 340, 341, 342–43, 345, 349, 376, 388, 525, 539; Lorentz transformation equations, 261–62, 327; Lorentz-Fitzgerald contraction hypothesis, 260, 285; "The Ether and the Earth's Atmosphere,"
285, 285 (text); "Michelson's Inference Experiment," 286, 286–89 (text)
Lowell, Percival, 550
Lubin, Phil, 671

Mach, Ernst, xxvii, xxviii–xxix, 261, 309, 344–45, 674
Malebranche, Nicolas de, 184
Mandelbrot, Benoit, xli
Maupertius, Pierre-Louis, 592, 595
Maxwell, James Clerk, xxiii–xxiv, xxv, xxvi, xxvii, xxxvi, xxxviii, 133, 209, 224, 232, 235, 250, 251, 265, 280, 281, 286, 299, 300, 340, 341, 346, 354, 360, 376, 381, 440, 471, 574, 657; "A Dynamical Theory of the Electromagnetic Field" (1865), xxiv, 224, 224–32 (text); *A Treatise on Electricity and Magnetism* (1873), xxiv, 232, 232–35 (text); and black-body radiation, 360; Electromagnetic Field Theory (1865), xxxvi, 142–45, 258; *Encyclopedia Britannica* article, xxiv, 265, 265–73 (text); kinetic theory of gases, 224; Maxwell's equations, xxv, xxvi, 143, 224, 230 (text), 258, 506; "On Action at a Distance," 235, 235–45 (text); "On Physical Lines of Force" (1861), xxiv; rejection of Weber's theory, 142
Merton, Robert, 689
Messier, Charles, 596
Michelson, Albert, xxiii, xxiv, xxvi, xxvii, 273–74, 286–87; invention of the interferometer, 259, 273; *Light Waves and Their Uses* (1903), 274, 274–84 (text); Michelson-Morley experiment (1887), xxiii, xxvi, xxvii, xxxvi–xxxvii, 258–59, 274, 285, 286, 287, 288, 289, 307
Mie, Gustav, 688
Mill, John Stuart, 266
Millikan, Robert Andrews, xli, 504
Milne, Edward Arthur, 689
Milton, John, 614
Minkowski, Herman, xxvii, xxviii, 262, 304, 344; *Absolute World*, xxvii; "Space and Time" address (1908), 304, 304–8 (text); and the space-time continuum, 304
Moore, W., 564
Morley, Edward, xxiii, xxvi, xxvii, 274; Michelson-Morley experiment (1887), xxiii, xxvi, xxvii, xxxvi–xxxvii, 258–59, 274, 285, 286, 287, 288, 289, 307

Neumann, John von, xxxi, 224, 225, 268, 364, 454, 521, 529–30; "impossibility proof," 368

Newton, Isaac, xxii, xxiii, xxvi–xxvii, xxviii, xxxvi, xl, 114–15, 135, 137–38, 157–58, 159, 162, 180, 181, 182, 184, 189, 191, 194, 239, 266, 309, 329–30, 331, 344, 345, 348, 349, 388, 440, 446–47, 472, 574, 575–76, 686; absolute space concept, 256, 548, 574, 577; absolute time concept, 256; *De Motu,* xxxviii; Law of Universal Gravitation, xxxvii, 12, 237, 340–41, 548, 573, 574, 577; laws of motion, 12–13, 238, 256, 257, 305, 322; on natural states for material objects, 256; Newton-Leibniz debate, xxvi–xxvii; *Opticks* (1704), xxv, 115, 138, 162, 162–67 (text), 184, 193, 194, 331; particle theory of light, 167, 201, 359; passages cited by Young, 185 (text), 186 (text), 186–88 (text); *Principia* (1687), xxiii, xxxviii, 12, 115, 115–28 (text), 158, 239, 256, 578; science of mechanics, 115; table of colors, 237–38; thought experiment on real motion, 257; unwillingness to offer a causal explanation of gravity, 157
Newton-Smith, W. H., xxviii
Nicetas, 46
Nicholas of Cusa, 8
Nollet, Jean-Antoine, xxxix

Oersted, Hans Christian, xxv, xxvi, 136–37, 141, 153–54, 158, 240; pamphlet (1820), 153–54, 154–57 (text)
Ohm, E. A., 641
Olbers, Heinrich, 548; Olbers's Paradox, 548, 548n
Oppenheim, Paul, xl
Oppenheimer, J. Robert, 515
Oresme, Nicholas, 8; *Le Livre du ciel et du monde,* 8–9
Osiander, Andreas, xxii, 10, 39; introduction to *De Revolutionibus Orbium Caelestium* (1543), 42–43, 43–44 (text)
Ostriker, Jeremiah, xxxv

Page, Lyman, 671
Parmenides, 665
Pauli, Wolfgang, 391, 412, 432, 433; formulation of the Copenhagen Interpretation of Quantum Theory, 359, 366
Pease, Francis Goldheim, 564
Peebles, P.J.E., xxxv, 640, 642, 645, 649, 650, 651, 655; letter to the *Astrophysical Journal* (1965), 642, 642–46 (text)
Penzias, Arno, 553, 644, 645, 646, 647–51, 678–79; discovery of Cosmic Background Radiation (1964), xxxiii, 553, 640, 654; letter

to the *Astrophysical Journal* (1965), 640, 640–42 (text)
Perrin, Jean-Baptiste, xxxviii
Phidias, 28
Philolaus, 46
Picard, Jean, 172
Pickering, Edward Charles, 570
Pictet, Raoul-Pierre, *Essais de Physique* (1790), 192
Planck, Max, xxix, 373, 386, 388–89, 393, 408, 440, 467, 469, 474, 688, 689; and black-body radiation, xxxviii, 360, 389, 409, 439; and *Elementariquanta,* 360, 373, 407, 440, 458, 463; Nobel Prize in physics (1918), 373; "On the Law of Distribution of Energy in the Normal Spectrum" (1900), 373, 373–80 (text); Planck's constant (h), 360, 373, 381, 385, 386, 395, 408, 411, 440, 442, 445, 453, 469, 470, 477, 497, 502; president of the Kaiser Wilhelm Institute for the Advancement of Science, 373; reluctance to publicize his discovery of the quantum of action, 438, 439
Plato, xxi, 3, 7, 60; *Laws,* 48; *Meno,* xxxvi; *Timaeus,* 23, 60
Pliny, 148
Plücker, Julius, 218
Plutarch, 46, 49
Podolsky, Boris, and the EPR paper, xxxi, 367–68, 428–30, 476, 495, 495–501 (text), 523–25, 528, 536–38
Poincaré, Henri, 289–90, 389, 590; keynote address at the World's Fair, 290, 299 (text); "La Mesure du Temps," 289–90, 290–98 (text); *Science and Hypothesis* (1902), 290, 298 (text)
Poisson, Siméon-Denis, 240, 241
Popper, Karl, xxxvi, xxxix, xli
Pouillet, Claude-Servais-Mathias, 159
Poynting, J. H., xxiii; 143; Poynting's Vector, 143–44, 143n
Preston, S. Tolver, 272
Priestley, Joseph, 135, 192
Ptolemy, Claudius, xxii, 7–8, 9, 29, 40, 48–49, 56, 57, 60, 61, 72, 108, 110, 111; *Almagest,* 6, 7, 29, 29–38 (text); *Cosmography,* 51; introduction of the equant, 29; mathematical devices used by, 7–8; *Optics,* 29; *Planetary Hypothesis,* 29; reliance on Aristotle, 29
Putnam, Hilary, xxii

Rayleigh, Lord, 396, 438–39
Reichenbach, Hans, xxxvi, xl

Rescher, Nicholas, 690

Rheticus, Georg Joachim, *Narratio Prima* (1540), 10

Riemann, Georg Friedrich Bernhard, 626, 627, 630; Riemann's tensor, 631

Roberts, Isaac, 564

Robertson, Howard, xxvii, xxxv

Roll, P. G., 640, 642, 651; letter to the *Astrophysical Journal* (1965), 642, 642–46 (text)

Römer, Olaf, 117, 168, 171, 296–97

Rosen, Nathan, and the EPR paper, xxxi, 367–68, 428–30, 476, 495, 495–501 (text), 523–25, 528, 536–38

Rubin, Vera, xxxv

Rumford, Count (Benjamin Thomson), 182, 193

Russell, Bertrand, 583

Rutherford, Ernest, xxxvi, 362, 383, 384, 385, 386, 387, 463, 688; discovery of the atomic nucleus (1911), 408, 440

Rydberg, Johannes Robert, 387

Ryle, Martin, 678

Savart, Félix, 199; Biot-Savart law, xxv

Scaliger, Julius Caesar, 149

Scheiner, Christoph, 562, 602

Schelling, F. W., xxv–xxvi

Schewe, Philip, 672

Schonberg, Nicholas, 44

Schrödinger, Erwin, xxix, xxx, xxxvii, 391, 392, 401, 412, 435, 454, 474–75, 502–3; "Are There Quantum Jumps?" (1952), 465, 466–71 (text); the cat paradox, xxxi–xxxii, 487; as nemesis of Quantum Theory, 465; Nobel Prize in physics (with Dirac) (1933), 454; Schrödinger's equation, xxx, 411, 454, 475, 483, 498, 503, 504, 505, 516; *Science, Theory and Man,* 454, 455–65 (text); wave mechanics, xxx, 363–64, 364–65, 402–3, 404, 443, 444–45; *What is Life?* (1944), 687

Schwarzschild, Karl, 264

Shapley, Harlow, 556; director of the Hale Observatory, 567; *Galaxies,* 567, 568–73 (text); estimation of the diameter of the Milky Way, 549, 567; perfection of the Cepheid variables method for measuring stellar distances, 549, 556, 567

Shimony, Abner, 370, 532, 540; and the "CHSH inequality" experiment, 540; "Conceptual Foundations of Quantum Mechanics," 540, 540–44 (text)

Silk, Joseph, 664

Slipher, Vesto, 597, 602–3, 604, 608, 614, 616; contribution to the *Bulletin of the Lowell*

Observatory (1912), 561, 562–63 (text), 564, 564–67 (text); measurements of the Andromeda nebula (1912), 550; red shift/blue shift discovery, 550, 561, 564

Smart, J.J.C., xxii

Smoot, George, 555, 666; *Wrinkles in Time,* 666–67, 667–77 (text)

Snell, Willebrord, 455

So-Young Pi, 663

Sommerfeld, Arnold, 401, 410, 438, 441, 688

Sophocles, 64

Spergel, David, 671

Spinoza, Baruch, 431

Starobinsky, A. A., 663

Steinhardt, Paul, xxxiv, 652, 661, 671, 672; *Scientific American* article (1984), 652, 652–66 (text)

Stern, Otto, and the Stern-Gerlach experiment (1922), 405, 410, 441, 528, 534–35, 537

Stokes, George Gabriel, 269, 270

Sun Rhie, 671

Szalay, Alexander S., 664

Tayler, R. J., 651

Thales of Miletus, 23

Thompson, J. J., xli, 383, 384, 388, 689

Thomson, W., 226, 227, 236, 243, 272

Torricelli, Evangelista, 173–74

Trismegistus, 63–64

Turner, Ken, 649

Turner, Michael S., 663

Turok, Neil, 671

Virgil, 57

Volta, Alessandro, 136

Walker, Arthur, xxvii, xxxv

Weber, Wilhelm, xxvi, 142, 224, 225, 230

Weinberg, Steven, xxxiii, 553, 646, 676; Noble Prize in physics (1979), 646; *The First Three Minutes* (1977), 553, 646–47, 647–51 (text), 676

Weizsäcker, Carl Friedrich von, 453

Weyl, Herman, 346, 551, 618, 623

Wheatstone, Charles, 199

Wheeler, John Archibald, 366, 484, 643, 645, 646; and the term "black hole," 483; and the Institute of Theoretical Physics, 484; "The 'Past' and the 'Delayed Choice' Double-Slit Experiment" (1978), 484, 484–92 (text); Wolfe Prize in physics (1996–97), 484

Wiechert, Emil, 686–87

Wigner, Eugene, 487

Wilkins, Maurice, 687
Wilkinson, D. T., 640, 642, 651; letter to the *Astrophysical Journal* (1965), 642, 642–46 (text)
Wilson, C.T.R., 458
Wilson, Robert, 553, 644, 645, 646, 647–51, 678–79; discovery of Cosmic Background Radiation (1964), xxxiii, 553, 640, 654; letter to the *Astrophysical Journal* (1965), 640, 640–42 (text)
Wirtz, C., 604–5, 606–7
Witten, Edward, xxxviii
Wolf, Charles-Joseph-Etienne, 562, 602
Wright, Ned, 671, 672
Wright, Thomas, 592, 594

Xenophanes of Colophon, 23, 51

Young, Thomas, xxv, 138–39, 140, 179–80, 201, 202, 233, 244, 258, 360, 388, 485; and the deciphering of the Rosetta Stone inscription, 180; double slit experiment (1807), 139, 139n, 446; indebtedness to Newton, 180; writings on the wave theory of light (1801–23), 180, 180–201 (text); "Sound and Light" (1800), 138

Zeldovich, Yakov B., 646, 651, 664
Zweig, George, prediction of quarks (1963–64), xxx
Zwicky, Fritz xxxiv, 619

Index of Concepts

//

action-at-a-distance theories, xxv, 140; as framework for work done in electricity and magnetism, 136; Greek abhorrence of, 133

Big Bang Theory (BBT), xxxii–xxxvi, xl, 547, 552, 625–39; the Big Bang as a space-time singularity, 677–83; confirmation of Big Bang cosmology, 552–54; and the Cosmic Background Explorer (COBE) experiment, 555–56, 667–77; and the discovery of the large-scale structure of the universe, 548–49, 592–601, 611–25; and the flatness problem, 554; and the horizon problem, 553; and the magnetic monopoles problem, 553–54; and the problem of gravity, 547–48; and the red shift and the expansion of the universe, 550, 561, 564–67; Standard Model of, xxxii–xxxiii; theoretical development after relativity, 550–52; theoretical support for, xxxiii; and the velocity-distance relation, 602–11. *See also* Cosmic Background Radiation; Inflationary Theory
black holes, 264

Cosmic Background Radiation (CBR), xxxiii, xxxiv, 640–42, 642–46, 647–51

dark matter, xxxiv, xxxv–xxxvi, 555; candidates for, xxxv–xxxvi; evidence for, xxxv

Electromagnetic Theory, xxiii–xxvi; 133; early work on electrical current, 136–37, 154–57, 157–61; effluvium theory, 133–35; Faraday's lines-of-force concept, 141–42, 209–14, 214–23; and the Greek atomists, 133; Hertz's electromagnetic waves experiments, 245–51; Law of Electrostatic Force, 135–36, 151–53; Maxwell's electromagnetic field theory, 142–45, 224–32, 233–45; particle theory of light, 137–38, 162–67; pioneering work of Gilbert, 133–34, 146–50; prominence of in physics, 145; theoretical concerns, xxiii; wave theory of light, 138–40, 168–79, 180–201, 202–9

grand unified theories (GUT), xxxv, 553–54

Heliocentric Theory, xxii–xxiii, xl; Aristarchus's early version of, 27–28; the Aristotelian cosmology, 4–5, 14–26; the Copernican heliocentrism, 8–10, 39–42, 43–66, 100–108, 109–14; and the discovery of stellar parallax, 128–29; early theories of a moving earth, 5–6; the geocentric view of Eudoxus, 3–4; the geocentrism of Ptolemy, 6–8, 29–38; influence of Brahe on, 10–11, 67–70; influence of Galileo on, 11–12, 82–99, 109–14; influence of Kepler on, 11, 71–81, 100–108; influence of Newton on, 12–13, 115–28; and the invention of the telescope, 11–12; and the science of mechanics, 12–13, 115; summary of, 3

Inflationary Theory, xxxiv–xxxv, xl, 554–55, 652–66

Quantum Theory (QT), xxix–xxxii, xl, 359; and Bell's theorem, 369; and Bohr's theory of the atom, 362–63; brief history of, 438–46; and complementarity, 365–66, 380, 383, 392–406, 407; and the Completeness Debate, 495, 526–27; conceptual background of, 359–60; conceptual problems inherent in, 406–7, 472–84; Copenhagen Interpretation of, xxix, xxx, xxxi–xxxii, 359, 365–66, 383, 446–54; and the Correspon-

Quantum Theory (QT) (*cont.*)

dence Principle, 363; definitive experiments, 369–70; doubt regarding syntactic equivalence of formalizations of, xxx; and the EPR paper, 367–68, 496–501, 507–14, 527–32, 533–39; and the Heisenberg Uncertainty Principle, xxxi, 364–65, 438; and hidden variables theories, 368–69, 515–26; incompatibility with the Theory of Relativity, 501–6; and the "measurement problem," 383, 39, 507; "old" Quantum Theory (1913), 362, 380; and the photoelectric effect, 360–62; and Planck's *Elementariquanta*, 360, 373, 407, 440, 458, 463; the quantum character of the atom, 384–87; the Quantum Hypothesis, 373–80; statistical character of, 432, 496, 501; and wave and matrix mechanics, 363–64; wavering of opinion within, xxix–xxx

Realism vs. AntiRealism debate, xxii
reference frames, 257–58, 308

scientific discovery: analogy as a tool of discovery, xxxvii–xxxviii; problem of, xxxvi–xxxvii
scientific method, xxxvi
scientific theories, evaluation of, xxxviii–xlii
Steady State Theory (SST), xxxiii–xxxiv, 552, 553
Super String Theory, xxx, xxxvii; lack of guiding physical intuition, xxxviii

theoretical entities, existence of, xxi–xxii
Theory of Relativity, xxvi–xxix, xl, 225, 255, 260–62, 263–64, 551; and the apparent incompatibility with the law of the propagation of light, 316–17; confirmation of by the detection of black holes, 264; confirmation of by the detection of the lens effect of galaxies, 264; confirmation of by Dicke's experiments, 264; confirmation of by the Pound-Rebka experiment, 264; and cosmology, 573–91; and the displacement of spectral lines of light, 313; Einstein's later comments on, 347–40; and the equality of inertial and gravitational masses, 323–25; the equation of relativity, 262, 350–55; and ether, 340–46; exact formulation of, 327–28; and the gravitational field, 321–23; and the identification of mass with energy, 262; prediction of the magnitude of the bending of light rays, 264, 313; prediction of the precession of the planet Mercury, xxxvii, 263, 313; the Principle of General Covariance, 308; the Principle of Relativity, 314–15; relation to realism, xxix; and the relativity of simultaneity, 319–21; and the solution of the problem of gravitation, 328–30; and the space-time continuum, 262–63, 304–8, 325–27; and the theorem of the addition of velocities, 315–16; and time, 317–19; the two postulates of Special Relativity, 261, 300–304; and the velocity of light, 262

Permissions Acknowledgments

//

(by author, in the order of appearance in the text)

Aristotle

From "On the Heavens," in *The Collected Works of Aristotle*, vol. 1, edited by Jonathan Barnes, revised edition of the Oxford Translation; copyright © 1984 by Jowett Trustees. Reprinted by permission of Princeton University Press.

Aristarchus

From "Aristarchus of Samos," *Greek Astronomy*, by Sir Thomas Heath, originally published in 1932 by J. M. Dent & Sons, Ltd., as *The Library of Greek Thought*, vol. 10. Reprinted by permission of Dover Publications, Inc.

J. S. Bell

From "Introduction to the Hidden-variable Question," in *Foundations of Quantum Mechanics*; reprinted in J. S. Bell, *Speakable and Unspeakable in Quantum Mechanics*, copyright © Cambridge University Press 1987. Reprinted with permission of Cambridge University Press.

From "Bertlmann's Socks and the Nature of Reality," from *Journal de Physique*, 1981; reprinted in J. S. Bell, *Speakable and Unspeakable in Quantum Mechanics*, copyright © Cambridge University Press 1987. Reprinted by permission of Cambridge University Press.

David Bohm

From "A Suggested Interpretation of the Quantum Theory in Terms of Hidden Variables," from *Physical Review* 85 (1952), copyright 1952 by the American Physical Society. Reprinted by permission of the American Physical Society.

Niels Bohr

From "On the Constitution of Atoms and Molecules," in *Philosophical Magazine*, 1913. Reprinted by permission of Taylor & Francis, Ltd.: www.tandf.co.uk/journals.

From "The Quantum Postulate and the Recent Development of Atomic Theory," from *Nature*, supplement to vol. 121 (1928). Reprinted by permission granted by Macmillan Magazines Limited.

From "Discussions with Einstein on Epistemological Problems in Atomic Physics," *Albert Einstein, Philosopher-Scientist*, edited by Paul Arthur Schilpp, copyright 1949, 1951 by the Library of Living Philosophers, Inc. Reprinted by permission of Open Court Publishing Company, a division of Carus Publishing.

From "Can Quantum Mechanical Description of Physical Reality be Considered Complete?" from *Physical Review* 48, copyright 1935 by the American Physical Society. Reprinted by permission of the American Physical Society.

Tycho Brahe

From *Tycho Brahe's Description of His Instruments and Scientific Work*, translated and edited by Hans Raeder, Elis Stromgren, and Bengt Stromgren. Copyright © 1946 Munksgaard International Publishers, Ltd., Copenhagen, Denmark. Reprinted with permission of Munksgaard International Publishers Ltd.

Nicholaus Copernicus

From "Commentariolus," translated by Edward Rosen, in *Nicholas Copernicus: Minor Works*, edited by Pawel Czartoryski, copyright © 1985 by Edward Rosen. Reprinted with the permission of Palgrave Publishers Limited, Macmillan's Global Academic Publishing.

From *On The Revolutions of the Heavenly Spheres*, translated by Charles Glenn Wallis. Reprinted by permission of Prometheus Books.

Louis de Broglie

From *Matter and Light*, translated by W. H. Johnston, originally published in English by W. W. Norton & Company Inc., 1939. Reprinted by permission of Dover Publications, Inc.

Willem de Sitter

From "Relativity and Modern Theories of the Universe," in *Kosmos: A Course of Six Lectures On the Development of Our Insight into the Structure of the Universe,* by Willem de Sitter, copyright 1932 by the President and Fellows of Harvard College. Reprinted by permission of Harvard University Press.

R. H. Dicke, P.J.E. Peebles, P. G. Roll, and D. T. Wilkinson

From "Cosmic Blackbody Radiation," in the *Astrophysical Journal*, vol. 142, Nov. 1965. Reprinted by the kind permission of P.J.E. Peebles.

P.A.M. Dirac

From "The Evolution of the Physicists' Picture of Nature," found in *Scientific American*, vol. 208, 5 May 1963. Copyright © 1963 by Scientific American, Inc; all rights reserved. Reprinted with permission of the publisher.

Arthur Eddington

From *Space, Time, and Gravitation*, Cambridge University Press, 1929. Reprinted with the permission of Cambridge University Press.

From *The Expanding Universe*, Cambridge University Press, 1932. Reprinted with the permission of Cambridge University Press.

Albert Einstein

From "On the Electrodynamics of Moving Bodies," in *The Collected Papers of Albert Einstein*, vol. 2. Translated by Anna Beck, published 1989 by Princeton University Press. Copyright © 1989 by the Hebrew University of Jerusalem. Reprinted by permission of Princeton University Press.

From "The Foundation of the General Theory of Relativity," in *The Collected Papers of Albert Einstein*, vol. 6, translated by W. Perrett and G. B. Jeffrey, published in 1997 by Princeton University Press. Copyright © 1997 by the Hebrew University of Jerusalem. Reprinted by permission of Princeton University Press.

From "Relativity: The Special and General Theory," in *The Collected Papers of Albert Einstein*, vol. 6, translated by Robert W. Lawson, published in 1997 by Princeton University Press. Copyright © 1997 by the Hebrew University of Jerusalem. Reprinted by permission of Princeton University Press.

From "Ether and Relativity," translated by G. B. Jeffrey and W. Perrett, from *Sidelights on Relativity*, originally published by E. P. Dutton and Company, Publishers, in 1922. Reprinted with the permission of Dover Publications, Inc.

From "The General Theory of Relativity," translated by Anna Beck, in *The Collected Papers of Albert Einstein*, vol. 4, published in 1996 by Princeton University Press. Copyright © in 1996 by the Hebrew University of Jerusalem. Reprinted by permission of Princeton University Press.

From "E=MC²—The Most Urgent Problem of Our Time," first published in *Science Illustrated I*, 1946; reprinted in *Essays in Physics*, copyright © 1950 by the Philosophical Library, Inc. Permission granted by the Albert Einstein Archives, the Hebrew University of Jerusalem, Israel.

From "Elementary Derivation of the Equivalence of Mass and Energy," first published in *Technion Journal* V, 1946; reprinted in *Essays in Physics*. Permission granted by the Albert Einstein Archives, the Hebrew University of Jerusalem, Israel.

From "On a Heuristic Point of View About the Creation and Conversion of Light," from *The Collected Papers of Albert Einstein*, vol. 2, Princeton University Press, 1989. Copyright © 1989 by the Hebrew University of Jerusalem, Israel. Reprinted by permission of Princeton University Press.

"Einstein's Reply," reprinted by permission granted by the Albert Einstein Archives, the Hebrew University of Jerusalem, Israel.

From "Physics and Reality," from *Essays in Physics*; originally published in *Franklin Institute Journal* CCXX1 (1936), and reprinted by the Library of Living Philosophers, Inc. Permission kindly granted by the Albert Einstein Archives, the Hebrew University of Jerusalem, Israel.

Albert Einstein, Boris Podolsky, and Nathan Rosen

"Can Quantum Mechanical Description of Physical Reality be Considered Complete?" in *Physical Review* 47 (1935), copyright 1935 by the American Physical Society. Permission granted by the American Physical Society.

Michael Faraday

From "On the Induction of Electric Currents," in *Experimental Researches in Electricity*, vol. 1, originally published by Taylor and Francis in 1839 and 1855. Reprinted with the permission of Dover Publications, Inc.

From "The Physical Existence of Lines of Force," in *Experimental Researches in Electricity*, vol. 3, originally published by Taylor and Francis in 1839 and 1855. Reprinted by permission of Dover Publications, Inc.

Galileo Galilei

From *Siderius Nuncius*, translated by Albert Van Helden, copyright © 1989 by the University of Chicago Press. Reprinted by permission of the University of Chicago Press.

From "Letters on Sunspots," in *Discoveries and Opinions of Galileo*, translated by Stillman Drake, copyright © 1957 by Stillman Drake. Reprinted with permission of Doubleday, a division of Random House, Inc.

From *Dialogue Concerning the Two Chief World Systems*, 2d ed., translated by Stillman Drake, copyright © 1953, 1962, and 1967 by the Regents of the University of California. Reprinted with permission of the University of California Press.

Alan Guth and Paul Steinhardt

From "The Inflationary Universe," in the *Scientific American*, May 1984, copyright © 1984 by Scientific American, Inc. All Rights Reserved. Reprinted with permission of the publisher.

Stephen Hawking

From "The Edge of Space-Time," in *The New Physics*, edited by Paul Davies, copyright © Cambridge University Press. Reprinted with the permission of Cambridge University Press.

Werner Heisenberg

From *Physics and Philosophy*, copyright © 1958 by Werner Heisenberg, published by Harper & Row Publishers, 1962. Reprinted with permission of HarperCollins Publishers, Inc.

Heinrich Hertz

From "On the Finite Velocity of Propagation of Electromagnetic Action," in *Electric Waves*. Originally published by Macmillan and Company in 1893. Reprinted by permission of Dover Publications, Inc.

Edwin Hubble

From *The Realm of the Nebulae*, copyright 1936 by Yale University Press, 1936. Reprinted by permission of Yale University Press.

Christiaan Huygens

From "Treatise on Light," in *Great Books of the Western World*, vol. 34. William Benton, Publisher, copyright 1952, 1990 Encyclopaedia Britannica, Inc. Reprinted by permission of Encyclopaedia Britannica, Inc.

Johannes Kepler

From *New Astronomy*, translated by William H. Donahue. Copyright © 1992 by Cambridge University Press 1992. Reprinted with the kind permission of William H. Donahue and of Cambridge University Press.

From *Epitome of Copernican Astronomy & Harmonies of the World*, translated by Charles Glenn Wallis. Reprinted by permission of Prometheus Books.

Helge Kragh

From *Quantum Generations*, copyright © 2000 by Princeton University Press. Reprinted by permission of Princeton University Press.

Henrietta Leavitt

From the *Annals of Harvard College Observatory*, vol. LX, no. IV, 1908. Reprinted with the kind permission of the Library of the Harvard College Observatories.

From the *Harvard College Observatory*, Circular 173, 1912. Reprinted with the kind permission of the Library of the Harvard College Observatories.

George Lemaître

From *The Primeval Atom: An Essay in Cosmogony*, published by D. Van Nostrand Co., 1950. Reprinted by the kind permission of the family of Georges Lemaître.

H. A. Lorentz

From "Michelson's Interference Experiment," in *The Principle of Relativity*, translated by W. Perrett and G. B. Jeffrey. Originally published by Methuen and Company, Ltd., in 1923. Reprinted with the permission of Dover Publications, Inc.

James Clerk Maxwell

From "Electromagnetic Theory of Light," in *A Treatise on Electricity and Magnetism*, vol. 2, 3d ed., originally published by Constable and Company, Ltd., 1891. Reprinted by permission of Dover Publications, Inc.

From "On Action at a Distance," in *The Scientific Papers of James Clerk Maxwell*, edited by W. D. Niven, originally published by Cambridge University Press 1890. Reprinted with the permission of Cambridge University Press.

From "A Dynamical Theory of the Electromagnetic Field," in *The Scientific Papers of James Clerk Maxwell*, edited by W. D. Niven, originally published by Cambridge University Press 1890. Reprinted with the permission of Cambridge University Press.

"Ether," from *The Scientific Papers of James Clerk Maxwell*, edited by W. D. Niven, originally published by Cambridge University Press 1890. Reprinted with the permission of Cambridge University Press.

Herman Minkowski

From "Space and Time," in *The Principle of Relativity*, translated by W. Perrett and G. B. Jeffrey. Originally published by Methuen and Company, Ltd., in 1923. Reprinted with the permission of Dover Publications, Inc.

Isaac Newton

From "The System of the World," in *Newton's Mathematical Principles of Natural Philosophy and His System of the World*, vol. 2, translated 1729 by Andrew Motte, revised and supplied with an historical and explanatory appendixes by Florian Cajori, copyright 1934 (renewed 1962) by the Regents of the University of California. Reprinted with permission of the University of California Press.

From "Opticks," in *Great Books of the Western World*, vol. 34. William Benton, Publisher, copyright 1952, 1990 Encyclopaedia Britannica, Inc. Reprinted by permission of Encyclopaedia Britannica, Inc.

Hans Christian Oersted

From "Experiments on the Effect of a Current of Electricity on the Magnetic Needle," in *Selected Works of Hans Christian Ørsted*, translated and edited by Karen Jelved, Andrew D. Jackson, and Ole Knudsen, copyright © 1998 by Princeton University Press. Reprinted by permission of Princeton University Press.

Arno Penzias and Robert Wilson

From "A Measurement of Excess Antenna Temperature at 4080 Mc/s," in the *Astrophysical Journal*, vol. 142, Nov. 1965. Reprinted by the kind permission of Lucent Technologies.

Max Planck

From *A Survey of Physical Theory*, originally published by Methuen & Co., Ltd., as *A Survey of Physics* in 1925. Reprinted with permission of Dover Publications, Inc.

Henri Poincaré

From *The Value of Science*, translated by G. B. Holsted. Dover Publications, Inc. 1958 and from. Originally published by Walter Scott Publishing Company Ltd. 1905, reprinted in *Science and Hypothesis*, Dover Publications, Inc., 1952. Reprinted with permission of Dover Publications, Inc.

Claudius Ptolemy

From *Ptolemy's Almagest*, translated and annotated by G. J. Toomer, copyright © 1996 by Princeton University Press. Reprinted by permission of Princeton University Press

Erwin Schrödinger

From *Science, Theory, and Man*. Copyright © by Erwin Schrödinger 1935. Reprinted with permission of Dover Publications, Inc.

From "Are There Quantum Jumps?" in *British Journal for the Philosophy of Science* 3 (1952). Reprinted by permission of Oxford University Press.

Harlow Shapley

From "The Astronomical Toolhouse," in *Galaxies*, by Harlow Shapley, copyright 1943 by the President and Fellows of Harvard College; revised by Paul W. Hodge, copyright 1961, 1972. Reprinted by permission of Harvard University Press.

Abner Shimony

From "Conceptual Foundations of Quantum Mechanics," in *The New Physics*, edited by Paul Davies, copyright © Cambridge University Press, 1989. Reprinted with the permission of Cambridge University Press.

Vesto Slipher

From the *Lowell Observatory Bulletin*, 58, vol. I, no. 8, 1912. Reprinted with the kind permission of the Lowell Observatory.

From the *Lowell Observatory Bulletin*, 80, vol. III, no. 5, 1913. Reprinted with the kind permission of the Lowell Observatory.

George Smoot and Keay Davidson

From *Wrinkles in Time*, copyright © 1993 by George Smoot and Keay Davidson 1993. Reprinted in North America by permission of HarperCollins Publishers, Inc., and in the United Kingdom by permission of Time Warner Books UK.

Steven Weinberg

From *The First Three Minutes*, copyright © 1977, 1988 by Steven Weinberg. Reprinted in North America by permission of Basic Books, a member of Perseus Books, L.L.C., and in the United Kingdom by permission of HarperCollins Publishers Ltd.

John Archibald Wheeler

From "The 'Past' and the 'Delayed-Choice' Double-Slit Experiment," in *Mathematical Foundations of Quantum Theory*, edited by A. R. Marlow, copyright © Academic Press 1978. Reprinted by permission of the publisher.